A TALE OF TWO CRATONS:
THE SLAVE-KAAPVAAL WORKSHOP

A TALE OF TWO CRATONS:
THE SLAVE-KAAPVAAL WORKSHOP

EDITED BY:

ALAN G. JONES
GEOLOGICAL SURVEY OF CANADA,
OTTOWA, ONTARIO, CANADA

RICHARD W. CARLSON
DEPARTMENT OF TERRESTRIAL MAGNETISM,
CARNEGIE INSTITUTION OF WASHINGTON,
WASHINGTON, DC, USA

HERMAN GRÜTTER
MINERAL SERVICES CANADA INC.,
VANCOUVER, BC, CANADA

2004

ELSEVIER

Amsterdam – Boston – Heidelberg – London – New York – Oxford
Paris – San Diego – San Francisco – Singapore – Sydney – Tokyo

ELSEVIER B.V.
Sara Burgerhartstraat 25
P.O. Box 211, 1000 AE
Amsterdam, The Netherlands

ELSEVIER Inc.
525 B Street, Suite 1900
San Diego, CA 92101-4495
USA

ELSEVIER Ltd
The Boulevard, Langford Lane
Kidlington, Oxford OX5 1GB
UK

ELSEVIER Ltd
84 Theobalds Road
London WC1X 8RR
UK

First edition 2004

Library of Congress Cataloging in Publication Data
A catalog record is available from the Library of Congress.

British Library Cataloguing in Publication Data
A catalogue record is available from the British Library.

Reprinted from *LITHOS*, volume 71/2-4
ISBN: 0-444-51614-X

♾   The paper used in this publication meets the requirements of ANSI/NISO Z39.48-1992 (Permanence of Paper).
Printed in The Netherlands.

Contents

Contents

Preface

The Slave–Kaapvaal workshop: a tale of two cratons

When academic curiosity and commercial interest intersect, there can be a tremendous growth in knowledge. Such was and is the case with the Kaapvaal craton in South Africa, and such is the case now with the Slave craton in northern Canada, where the diamond industry is catalysing, promoting and funding scientific research activities aimed at understanding Archean lithospheric processes.

As a direct consequence of the diamond wealth discovered in South Africa in 1867, the Kaapvaal craton has been studied scientifically for over a hundred years (Bonney, 1899). Much is known about the craton's lithospheric mantle, in part because of the abundance of mantle xenoliths made available by mining activities in the numerous kimberlites in southern Africa. The regional coverage of this information has been greatly expanded recently as a consequence of the National Science Foundation's Continental Dynamics funded Kaapvaal Project.

In stark contrast, diamond interest in the Slave craton dates only from the early 1980s, when Chuck Fipke, Stu Blusson and Hugo Dummett followed glacially dispersed indicator mineral trains from west of the craton, and eventually discovered the kimberlite pipes in the Lac de Gras area (Fipke et al., 1995). Knowledge of the Slave's cratonic mantle petrologic, geochemical and geophysical structure was advanced through the 1990s as part of industrial and academic data acquisition, following on detailed studies of the ancient crustal section of the Slave conducted in the late 1980s.

Given the established nature of geoscientific knowledge of the Kaapvaal craton compared to the Slave craton, and given the exciting new interdisciplinary results coming from the Kaapvaal Project and from Slave craton studies, we thought it opportune to hold a workshop bringing together scientists from both cratons so that they could be compared and contrasted. The workshop was held in Merrickville,

Ontario, an hour's drive south of Ottawa, from 5–9 September 2001. There were almost 70 participants at the workshop, with 25 of them coming from outside North America, principally South Africa. The significant financial support we were able to raise for the workshop both facilitated attendance by many students and scientists and also paid for all the logistical costs.

A total of 52 presentations were given: 15 presentations on the Slave craton, 24 on the Kaapvaal craton and 13 comparing the two cratons with each other and with other cratons worldwide. Extended abstracts for these presentations can be found at www.ciw.edu/kaapvaal/abstracts. In addition, there were two evening presentations, one by Wouter Bleeker (included in this volume) discussing the Late Archean record of global cratons, and one by John Gurney giving humorous insight into diamond exploration of the Kaapvaal craton. Of those 54 papers, 24 are included in this volume presenting new results from cratons globally (5 papers), from southern Africa (12 papers) and from northern Canada (7 papers).

There are clearly major similarities and differences between these two Archean cratons. The crust of both was predominantly formed in the Mesoarchean, although the Slave craton does host older rocks than the Kaapvaal craton; the Acasta gneisses at 4.027 Ga (Stern and Bleeker, 1998) compared to the ca. 3.6 Ga Ancient Gneiss Complex of Swaziland (Kröner et al., 1993). Both contain crustal sections consisting of terranes of different ages welded together by Archean accretionary events. Both crustal sections are also underlain by lithospheric mantle sections consisting of peridotites that experienced extensive partial melt extraction between 2.9 and 3.2 Ga, but this is where the similarities between the cratons end.

One of the most striking differences between the Slave and Kaapvaal cratons is the apparent seismic homogeneity of the Kaapvaal craton's SCLM (James et al., 2001; Gao et al., 2002), whereas the Slave craton is seismically layered (Bostock, 1998; Snyder et al., this volume). The seismic layering in the centre of the craton correlates laterally and with depth with electrical layering (Jones et al., this volume) and geochemical layering (Griffin et al., 1999). These unique attributes of the central Slave SCLM apparently do not extend to the northern or southern Slave craton, despite being overlain by laterally continuous Archean crustal domains (Grütter et al., 1999). Fingerprinting the Slave's diverse crustal and mantle components through the multi-disciplinary studies represented at the workshop permits reconstruction of the Archean cratonic puzzle (Bleeker, this volume), and, by contrast, exposes the relative uniformity with which the Kaapvaal crust and mantle have been regarded for many years.

Detailed geophysical and geochemical studies of the Kaapvaal are currently unravelling the domainal structure of its Archean SCLM (e.g. Silver et al., 2001; Moser et al., 2001), the relationship with overlying Archean crustal domains (de Wit et al., 1992) and mounting evidence for post-Archean mantle reworking. While the influence of the Late Archean Ventersdorp magmatism remains partly enigmatic (Schmitz and Bowring, 2003), a pervasive metasomatic imprint in the northern and western Kaapvaal craton mantle is being correlated in space and time with intrusion of the Paleoproterozoic Bushveld Igneous Complex and western correlatives in Botswana (Shirey et al., this volume; Hoal, this volume). Metasomatic reworking during the Phanerozoic is considered so intense that the mantle sample derived from Kaapvaal Group-1 kimberlites may not represent primary Archean SCLM (Griffin et al., Bell et al., and Simon et al., this volume). The depleted high-Mg/Fe, orthopyroxene-rich peridotite characteristic of the Kaapvaal mantle is now also being regarded as relatively unique, and can no longer be thought of as being universally present in all Archean cratonic SCLMs (Francis, this volume). The preferential growth of diamond in Kaapvaal harzburgite (Gurney, 1984) appears valid also for harzburgite from the central Slave craton (Stachel et al., this volume).

Taken together, these differences suggest that SCLM formation was different for the two cratons, implying that the search for a single causative formation process is bound to fail. We are clearly making progress in our attempts to understand Archean tectonic processes for formation, deformation and destruction of cratonic lithosphere, and their influence on diamond genesis and preservation, but much work needs to be done.

Acknowledgements

We wish to acknowledge the financial support of a number of organizations that was crucial in making the workshop the success it was. The Gold sponsors were DeBeers Canada Exploration and the Continental Dynamics Program of the National Science Foundation. Silver sponsors were Lithoprobe, Kennecott Exploration, the Geological Survey of Canada, and Diamondex Resources. Ashton Mining of Canada was a Bronze sponsor.

This volume came together through many people working with a single objective. However, various factors, including illnesses, commitments and computer crashes for authors and reviewers alike all conspired to make the volume appear later than originally planned. Nevertheless, we thank all those involved in its production.

Finally, the workshop, and this volume, would not have happened without the tireless energy of Gwen Mason (Continental Geoscience Division, Geological Survey of Canada). Thanks Gwen!

References

Bonney, T.G., 1899. The parent rock of the diamond in South Africa. Geol. Mag., 309.

Bostock, M.G., 1998. Mantle stratigraphy and evolution of the Slave province. J. Geophys. Res. 103, 21183–21200.

de Wit, M.J., Roering, C., Hart, R.G., Armstrong, R.A., de Ronde, C.E.J., Green, R.W.E., Tredoux, M., Peberdy, E., Hart, R.A., 1992. Formation of an Archean continent. Nature 357, 553–562.

Fipke, C.E., Dummett, H.T., Moore, R.O., Carlson, J.A., Ashley R.M., Gurney, J.J., Kirkley, M.B., 1995. History of the discovery of diamondiferous kimberlites in the Northwest Territories, Canada. In: Sobolev, N.V. (Ed.), Sixth International Kimberlite Conference; Extended Abstracts. United Institute of Geology, Geophysics and Mineralogy, Siberian Branch of the Russian Academy of Sciences, Universitetsky pr. 3, Novosibirsk 90, 630090 Russia, pp. 158–160.

Gao, S.S., Silver, P.G., Liu, K.L., 2002. Mantle discontinuities beneath southern Africa. Geophys. Res. Lett. 29. (doi:10.1029/2001GL013834)

Griffin, W.L., Doyle, B.J., Ryan, C.G., Pearson, N.J., O'Reilly, S., Davies, R., Kivi, K., van Achterbergh, E., Natapov, L.M., 1999. Layered mantle lithosphere in the Lac de Gras area, Slave craton: composition, structure and origin. J. Petrol. 40, 705–727.

Grütter, H.S., Apter, D.B., Kong, J., 1999. Crust–mantle coupling: evidence from mantle-derived xenocrystic garnets. In: Gurney J.J., Gurney, J.L., Pascoe, M.D., Richardson, S.H. (Eds.), J.B. Dawson Volume, Proc. 7th Int. Kimb. Conf., Red Roof Design, Cape Town, pp. 307–313.

Gurney, J.J., 1984. A correlation between garnets and diamonds. In: Glover, J.E., Harris, P.G. (Eds.), Kimberlite Occurrence and Origins: A Basis for Conceptual Models in Exploration. Geology Department and University Extension, University of Western Australia, Publication, vol. 8, pp. 143–166.

James, D.E., Fouch, M.J., VanDecar, J.C., van der Lee, S, the Kaapvaal Seismic Group, 2001. Tectospheric structure beneath southern Africa. Geophys. Res. Lett. 28, 2485–2488.

Kröner, A., Hegner, E., Wendt, J.I., Byerly, G.R., 1993. The oldest part of the Barberton granitoid–greenstone terrain, South Africa: evidence for crust formation between 3.5 and 3.7 Ga. Precambrian Res. 78, 105–124.

Moser, D.E., Flowers, R.M., Hart, R.J., 2001. Birth of the Kaapvaal tectosphere 3.08 billion years ago. Science 291, 465–468.

Schmitz, M.D., Bowring, S.A., 2003. Ultrahigh-temperature metamorphism in the lower crust during Neoarchean Ventersdorp rifting and magmatism, Kaapvaal Craton, southern Africa. Bull. Geol. Soc. Am. 115, 533–548.

Silver, P.G., Gao, S.S., Liu, K.H., the Kaapvaal Seismic Group, 2001. Mantle deformation beneath southern Africa. Geophys. Res. Lett. 28, 2493–2496.

Stern, R.A., Bleeker, W., 1998. Age of the world's oldest rocks refined using Canada's SHRIMP: the Acasta Gneiss Complex, Northwest Territories, Canada. Geosci. Can. 25, 27–31.

Alan G. Jones*
Geological Survey of Canada, 615 Booth Street, Ottawa, Ontario, Canada K1A 0E9
E-mail address: ajones@nrcan.gc.ca

Richard W. Carlson
Department of Terrestrial Magnetism, Carnegie Institution of Washington, 5241 Broad Branch Road, NW, Washington, DC 20015, USA
E-mail address: carlson@dtm.ciw.edu

Herman Grütter
Mineral Services Canada Inc., #1300–409 Granville Street, Vancouver, BC, Canada V6C 1T2
E-mail address: herman.grutter@mineralservices.com

* Corresponding author. Fax: +1-613-943-9285.

Available online at www.sciencedirect.com

LITHOS

Lithos 71 (2003) 99–134

www.elsevier.com/locate/lithos

The late Archean record: a puzzle in ca. 35 pieces

Wouter Bleeker[*]

Continental Geoscience Division, Geological Survey of Canada, 601 Booth Street, Ottawa, Ontario, Canada K1A 0E8

Abstract

The global Archean record preserves ca. 35 large cratonic fragments and a less well defined number of smaller slivers. Most Archean cratons display rifted margins of Proterozoic age and therefore are mere fragments of supercratons, which are defined herein as large ancestral landmasses of Archean age with a stabilized core that on break-up spawned several independently drifting cratons. The tectonic evolution of individual Archean cratons, such as the Slave craton of North America or the Kaapvaal craton of southern Africa, should therefore always be considered in the context of their ancestral supercratons. This is particularly true for many of the smaller cratons, which are too limited in size to preserve the complete tectonic systems that led to their formation. These limitations not only apply to the crustal geology of Archean cratons but also to their underlying lithospheric mantle keels. If these keels are Archean in age, as their broad correlation with ancient surface rocks suggests, they also are rifted and drifted remains of larger keels that initially formed below ancestral supercratons. The study of Archean cratons and their lithospheric keels should thus be global in scope.

In the search for late Archean supercratons, a craton like the Slave, with three to four rifted margins, has a ca. 10% maximum probability of correlating with any of the ca. 35 remaining cratons around the globe, assuming wholesale recycling of Archean cratons has been limited since ca. 2.0 Ga. Alternatively, if the original number of independently drifting cratons was significantly larger, the probability of successful correlations is much less. Due to repeated cycles of break-up and plate tectonic dispersal since ca. 2.0 Ga, the probability of correlation is probably independent of present-day proximity, unless there is independent evidence that two neighbouring cratons were only separated by a narrow ocean. Nevertheless, most previously proposed craton correlations rely (erroneously?) on present-day proximity, implicitly extrapolating relatively recent paleogeography back to the Archean.

Considering the fundamental differences between some of the better known cratons such as the Slave, Superior, and Kaapvaal, it seems likely that these cratons originated from independent supercratons (Sclavia, Superia, Vaalbara) with distinct amalgamation and break-up histories. This is contrary to widely held opinion, based largely on an idealized view of the supercontinent cycle, that all cratons shared a common history in a single late Archean supercontinent.

In contrast to the Superior and Kaapvaal cratons, which are radically different from the Slave craton and each other, the Dharwar craton of peninsular India, the Zimbabwe craton of southern Africa, and the Wyoming craton of North America all show significant similarities to the Slave craton. It thus seems likely that at least some of these were nearest neighbours to the Slave craton in the ca. 2.6–2.2 Ga Sclavia supercraton.
© 2003 Elsevier B.V. All rights reserved.

Keywords: Slave craton; Supercontinent; Supercraton; Craton correlation; Archean

[*] Tel.: +1-613-995-7277; fax: +1-613-995-9273.
  E-mail address: WBleeker@NRCan.gc.ca (W. Bleeker).

1. Introduction

Despite rapid progress in recent years the study of Archean cratons commonly remains narrowly focussed, either by discipline or by geographic area. Although the reasons for such research biases are obvious this tends to impede a more complete understanding of the ancient cratons, their genesis, and the Archean Earth in general.

Archean cratons[1] are complex fragments of pre-2.5 Ga crust with, commonly, genetically linked upper mantle. Hence, their study requires a multidisciplinary approach at the lithospheric scale. Perhaps it is less obvious that this endeavor should be global in scope as most cratons show rifted or faulted margins and hence are merely fragments of larger ancestral landmasses (Williams et al., 1991). Individual cratons, now generally embedded as exotic bits in Proterozoic or younger collages (e.g., Laurentia, Hoffman, 1988, 1989), are in most cases too small to preserve the full complexity of the tectonic systems that led to their formation. Earth's major tectonic systems are typically developed at scales larger than 500 km or even 1000 km. Important examples are diffuse continental rift zones; complete subduction systems including an accretionary prism, arc, back-arc basin, and remnant arc; oceanic plateaus; collision zones with their associated sedimentary basins; and igneous provinces resulting from mantle plume impact or lithospheric delamination. The scale of these systems thus determines that our understanding of the tectonic settings in which individual cratons formed will remain underconstrained unless we are able to reconstruct the larger landmasses from which they originated. In other words, the diagnostic features that determined the tectonic evolution of one craton may no longer be preserved within its own realm, but could well be preserved within another

craton that has since rifted off and drifted around the globe. This limitation is equally true for understanding the crustal geology of Archean cratons as it is for understanding the architecture of their lithospheric mantle keels.

The necessity for a global approach is readily apparent in the case of the Slave craton located in the northwestern Canadian Shield (Fig. 1). Although one of the better exposed cratons in the world, its limited size (ca. 500 by 700 km), and the presence of three, if not four, rifted margins predetermine it to be merely a small fragment from a much larger crust–mantle system—the supercraton "Sclavia"[2,3] (Bleeker, 2001a) that is postulated to have existed from its amalgamation at ca. 2.6 Ga to its progressive break-up at ca. 2.2–2.0 Ga. Very similar considerations apply to the Kaapvaal craton of southern Africa, which is bordered by rifted and collisional margins of different ages. In the case of the Kaapvaal craton, a specific correlation has indeed been proposed: that the Kaapvaal craton was formerly connected to the Pilbara of Western Australia in a "Vaalbara" supercraton (e.g., Cheney, 1996). Presently, this is one of the leading contenders for an Archean craton correlation although details on the timing of a shared history, if any, remain a matter of debate (Cheney et al., 1988; Trendall et al., 1990; Cheney, 1996; Zegers et al., 1998; Wingate, 1998; Nelson et al., 1999; Strik et al., 2001; Byerly et al., 2002).

Hence, developing a better understanding of the Archean record not only involves a more detailed documentation of individual cratons, but also comparing and contrasting their geological records and, ultimately, testing potential correlations to reconstruct the larger Archean landmasses from which the present ensemble of cratons originated. The present volume on the Slave and Kaapvaal cratons is therefore timely as it focuses on two very different fragments of Archean lithosphere, both of which have seen intense research efforts in recent years, much of it multidisciplinary in scope (e.g., Bostock, 1997; Griffin et al., 1999; Cook et al., 1999; Bleeker and Davis, 1999a; Grütter et al., 1999; Carlson et al.,

---

[1] Craton: a segment of continental crust that has attained and maintained long-term stability, with tectonic reworking being confined to its margins. Although there is no strict age connotation in this definition, e.g., some segment of crust could have attained "cratonic" stability during the Proterozoic, the term is most commonly applied to stable segments of Archean crust. Long-term stability is thought to be a function, in part, of thicker lithosphere involving a relatively cool but compositionally buoyant keel of Fe-depleted upper mantle (i.e., tectosphere; Jordan, 1978, 1988).

[2] Supercraton: a large ancestral landmass of Archean age with a stabilized core that on break-up spawned several independently drifting cratons.

[3] Sclavia is derived from the Greek word for "slave": sclavos.

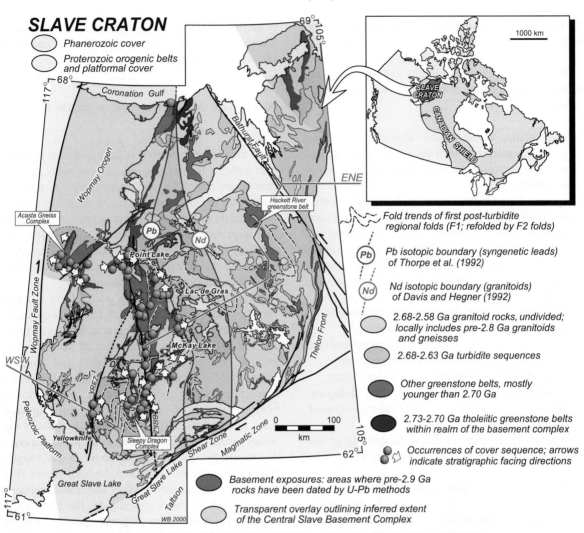

Fig. 1. Geology of the Slave craton. Transparent overlay outlines minimum extent of the Hadean to Mesoarchean basement complex underlying the central and western parts of the craton, the Central Slave Basement Complex of Bleeker et al. (1999a,b, 2000). Ancient basement may reappear in the northeastern part of the craton. WSW–ENE cross section is presented in Fig. 7. Pb isotopic boundary of Thorpe et al. (1992), Nd isotopic boundary of Davis and Hegner (1992).

2000; Kopylova and Russell, 2000; Irvine et al., 2001; Jones et al., 2001). In this contribution, I thus emphasize the need for a global approach in Archean studies and outline some of the basic logic that is relevant to craton correlations. I will introduce the Slave craton, particularly in terms of its first-order attributes that compare and contrast with those of other cratons. From this I conclude that the Slave craton is so different from another well-known craton in the Canadian Shield, the Superior, that these two cratons likely trace their ancestry to different super-cratons, Sclavia and Superia (Bleeker, 2001a). In searching for possible correlations with the Slave craton, I identify the Dharwar, Zimbabwe, and Wyoming cratons as the most likely candidates. Perhaps one or more of these "Slave-like" cratons may have been a "nearest neighbour" to the Slave in the late Archean supercraton Sclavia.

2. Craton correlation and supercratons: why bother?

As introduced above and briefly expanded on here, important reasons for a global approach to the Archean are:

- to compare, contrast, and ultimately correlate Archean cratons into their ancestral supercratons,
- and thus overcome the scale limitation of individual cratons;
- to simplify the highly fragmented Archean record;
- to address the fundamental question whether late Archean crustal evolution is best described by several transient, more or less independent, supercratons (as proposed in this study) or by a single late Archean supercontinent (Williams et al., 1991);
- and hence, in what form can the supercontinent cycle be projected back into the Archean?
- to provide major constraints on the reconstruction of any younger supercontinental aggregations, e.g., Nuna, Rodinia, or Gondwana (and vice versa);
- in an applied sense, to identify the dispersed "missing fragments" of mineral-rich crustal domains;
- and, finally, to improve our understanding of the genesis of diamondiferous lithospheric keels.

These reasons are clear. Even a single successful craton correlation will greatly simplify the highly fragmented Archean record and likely prompt other unforeseen correlations. At the same time, it also will put major constraints on the aggregation history of younger supercontinents such as Rodinia (McMenamin and McMenamin, 1990; Dalziel, 1991; Hoffman, 1991). Craton correlation also addresses important questions such as: where are the missing pieces or continuation of the richly endowed Abitibi greenstone belt of the Superior craton, a large Neoarchean crustal domain that is clearly truncated by younger boundaries on several sides? Or, where is the conjugate margin of a specific, fabulously rich, Paleoproterozoic break-up margin such as the Thompson Nickel Belt of the western Superior craton (Bleeker, 1990a,b)? And if many diamonds are Archean, as has been proposed (e.g., Richardson et al., 1984, 2001), and linked to formation of Archean lithospheric mantle keels, could

it be, for instance, that Slave diamonds and Indian diamonds (Dharwar craton) come from a once contiguous piece of mantle lithosphere?

3. The late Archean record: a puzzle in ca. 35 pieces

In the search for craton correlations, it is useful to start with the following question: how many Archean cratons are there? Or more specifically, in terms of reconstructing late Archean supercratons: how many independent fragments of Archean crust are preserved that experienced separate drift histories after the break-up of their ancestral late Archean supercratons but before their amalgamation in younger late Paleoproterozoic landmasses like Laurentia?

The Slave craton is clearly one of these independent fragments, as it appears to have broken out of its postulated ancestral supercraton Sclavia between 2.2 and 2.0 Ga based on the ages of marginal dyke swarms (LeCheminant et al., 1997) and marginal sedimentary sequences (Fig. 2). It probably drifted independently for ca. 200 million years prior to being amalgamated into the rapidly growing landmass of Laurentia (Hoffman, 1989) by 2.0–1.8 Ga, which forms the core to present-day North America. To answer the question posed above, today's highly fragmented state of the global Archean record (Fig. 3) is only a partial guide. Many additional fragments were created relatively recently by the break-up of the late Paleozoic supercontinent Pangaea. For instance, opening of the Atlantic Ocean separated the São Francisco craton in Brazil from Archean components in the Congo craton of central Africa (e.g., Hurley and Rand, 1969). Similarly, the Lewisian gneiss complex of western Scotland is merely an eastern fragment of the Nain craton, which underlies south–central Greenland (e.g., Park, 1995). Ancient gneisses in Labrador are a western fragment of the same craton left behind in North America with the aborted opening of the Labrador Sea. Hence, in any pre-Pangaea reconstructions these recently separated siblings should be combined and regarded as a single craton.

An accurate estimate of the number of independent cratons thus involves tracking the various bits and pieces of Archean crust back in time through previous

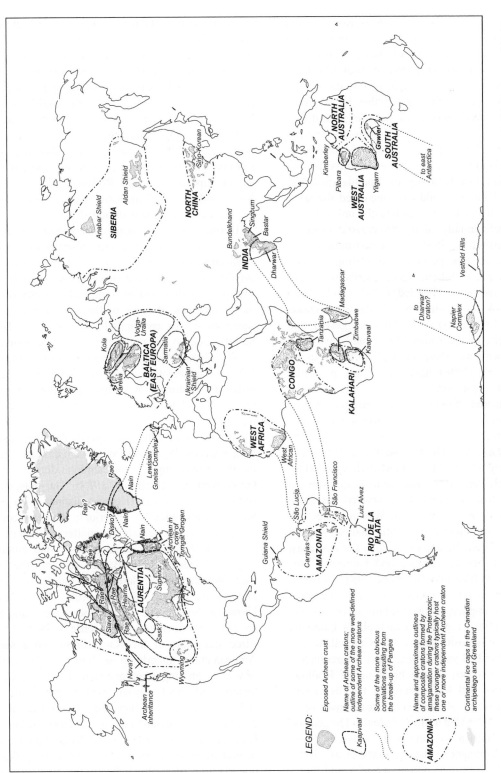

Fig. 3. Global distribution of exposed Archean crust (stippled pattern; from Geological Survey of Canada's Map of the World database). Map is annotated with (1) names of Archean cratons and shield areas (lower case, e.g., Slave), (2) obvious correlations resulting from the break-up of Pangaea, and (3) outlines of composite cratons that were amalgamated during the Proterozoic (upper case, e.g., LAURENTIA).

supercontinental aggregations such as Rodinia, and possibly Nuna (Hoffman, 1997), to evaluate when and where similar fragments of Archean crust may have joined up. However, proposed Rodinia configurations, guided by the principle of disparate continental blocks being stitched together by a global network of Grenvillian orogenic belts (e.g., Hoffman, 1991), tend to minimize contacts between Archean cratons, thus allowing only a limited number of craton correlations. This is counter-intuitive as one would predict that break-up of Rodinia, just like break-up of Pangea, would have increased the number of Archean fragments significantly, unless rifting occurred preferentially along Proterozoic orogenic belts. Nevertheless, Rodinia reconstructions did prompt at least one potential correlation: several authors have placed Siberia next to northern Laurentia at Rodinia time. This led Condie and Rosen (1994) to look for a counterpart to the Slave craton among the Siberian cratons. They suggested that Archean rocks of the Aldan Shield may have been contiguous with the Slave although further tests by Rainbird et al. (1998) have called this particular correlation into question.

Other potential craton correlations may result from the apparently long-lived mid-Proterozoic Laurentia–Baltica connection ("Nena" of Gower et al., 1990; see also Herz, 1969; Hurley and Rand, 1969). Although widely accepted (e.g., Gorbatschev and Bogdanova, 1993; Park, 1995; Karlstrom et al., 2001), it is intriguing that this connection survived unscathed through at least one, if not two, Wilson cycles: (1) break-up of mid-Proterozoic Nuna and subsequent formation of Grenville-age crust along the east coast of Greenland, between Baltica and Laurentia. This suggests that Baltica may be exotic relative to Laurentia, colliding only at Grenville time. To some extent this is implicit in the Rodinia reconstruction of Hoffman (1991, Fig. 1A), which places Laurentia and Baltica on opposite sides of the Grenvillian belt. (2) Several hundred million years later, Laurentia and Baltica are involved in the break-up of Rodinia, leading to 650–600 Ma opening of Iapetus Ocean. Eventually, closure of this wide ocean basin (e.g., Torsvik et al., 1996) led to the Caledonian Orogen and suturing of Baltica to Laurentia (Wilson, 1966). Although paleomagnetic data (Torsvik et al., 1996; Buchan et al., 2000) are permissive of a mid-Proterozoic Laurentia–Baltica fit, they also indicate complex relative motions between both continents during the life span of Iapetus Ocean (Torsvik et al., 1996), thus greatly reducing the probability that Baltica somehow came back to its pre-break-up position. This paradox strongly suggests that the mid-Proterozoic Laurentia–Baltica connection needs to be revisited.

From this brief discussion, it is clear that significant uncertainties remain in the reconstruction of Rodinia. Hence, the configuration of a possible precursor supercontinent to Rodinia, mid-Proterozoic Nuna, and its significance for Archean craton correlations therefore remain shrouded in mystery.

Despite these significant uncertainties, it appears that the present crustal record preserves ca. 35 large Archean fragments and a less well defined number of smaller, poorly preserved slivers (Figs. 3 and 4). With further age dating and an improved understanding of the architecture of some of the more complex Precambrian shields and buried platforms (e.g., Brazil, Siberia, Congo) this number may rise somewhat. On the other hand, a failure to recognize some of the post-1.7 Ga fragmentation events will tend to artificially inflate this estimate. A preliminary count of ca. 35 preserved, independent, Archean cratons is therefore reasonable.

At least five of these independent cratons (Slave, Rae, Superior, Wyoming, Nain) were amalgamated in the Proterozoic supercraton Laurentia during 2.0–1.8 Ga convergence (Hoffman, 1989). The independent status of three others (Hearne, distinct from Rae?; Sask, leading edge of Superior?; Archean core of Torngat Orogen) remains a matter of debate. As discussed above, this large post-1.8 Ga landmass may have extended into Baltica and into the larger East European craton. Here at least four large Archean cratons are present (Karelia, Kola, Volga–Uralia, Ukrainian Shield; Fig. 3). Whether Karelia and Kola are independent cratons in the sense defined above or rifted fragments of the Nain craton and a second Archean craton underlying much of central Greenland (Rae?) is an important question dependent on the history and exact configuration of the Laurentia–Baltica connection. If they indeed correlate, as suggested in the reconstruction by Gorbatschev and Bogdanova (1993), Karelia and Kola should not be treated as independent cratons. However, given the uncertainties described above, the present author

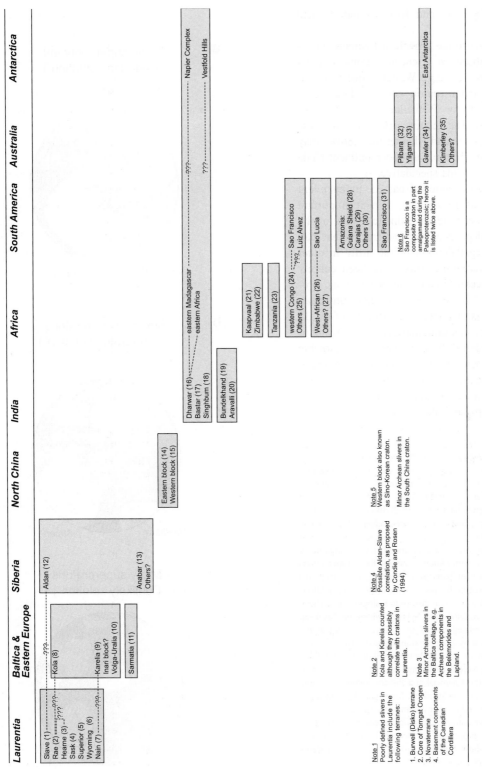

Fig. 4. A preliminary estimate of the number of independent Archean cratons preserved in the global Archean record. Names across the top are those of composite Proterozoic landmasses ("amalgamated cratons") or present-day continents. Names below these headings are Archean cratons or areas of significant Archean exposure. Shaded boxes represent "amalgamated cratons" with one or more Archean craton, some stretching across different continents due to break-up of Pangaea or Gondwana. Dashed lines further highlight possible correlations due to break-up of either Rodinia (e.g., Slave–Aldan correlation proposed by Condie and Rosen, 1994) or Gondwana. Although our knowledge of some complex shield areas remains insufficient, a preliminary count suggests that the Archean records represent a puzzle in ca. 35 significant pieces and a less well defined number of smaller slivers.

favours to treat Karelia and Kola as independent
cratons.

Processes similar to the growth of Laurentia amal-
gamated other subsets of Archean cratons in other
Proterozoic landmasses, e.g., Western Australia (Pil-
bara, Yilgarn; Gawler craton collided later; e.g.,
Myers et al., 1996) or Penisular India (Dharwar,
Bastar, Singbhum). The relative disposition of these
Proterozoic landmasses and whether they indeed may
have formed a Mesoproterozoic supercontinent Nuna
(Hoffman, 1997) remains speculative.

Accepting the estimate of ca. 35 independent
Archean cratons as reasonable, a next important
question is: how accurate an estimate is this of the
true number of independent Archean cratons that may
have existed during Paleoproterozoic drift? In other
words, how many cratons could have been entirely
erased from the record by erosion and tectonic processes in the last 2 billion years. As these processes
are irreversible it is clear that the estimate of ca. 35
cratons must represent a minimum. However, many of
the (preserved) cratons show a record of remarkable
relative stability, with the present erosion surface, at
the craton scale, being close to the enveloping surface
of folded Phanerozoic, Proterozoic, and in the case of
the Slave and Kaapvaal cratons, even Archean uncon-

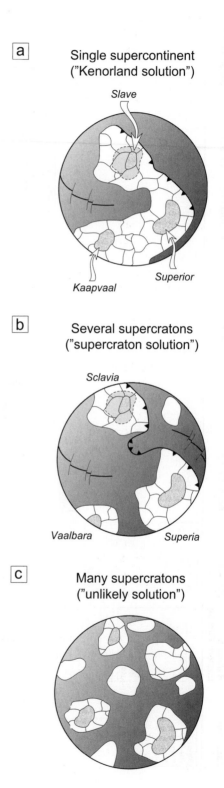

Fig. 5. Cartoons representing Earth in the latest Archean and across
the Archean–Proterozoic boundary. Three end member solutions
can be envisioned. (a) A single late Archean supercontinent,
Kenorland (Williams et al., 1991). Break-up of this postulated
supercontinent is thought to have spawned all rifted Archean cratons
preserved today. Three cratons are indicated schematically (Slave,
Superior, Kaapvaal). Each of these, with their respective nearest
neighbours, must have formed a supercraton. Due to significant
differences between the better-known cratons, these supercratons
must have occupied distant parts in a sprawling Kenorland. (b)
Alternatively, as preferred in this paper, some of these supercratons
may never have shared a common history in a single late Archean
supercontinent. A supercraton like Sclavia, which on break-up
spawned the Slave and related cratons (the "Slave clan" of cratons)
may have been underlain by a supercratonic lithospheric root (grey
outline), parts of which are now likely dispersed around the globe
with their host cratons. Note that Vaalbara may have rifted as early
as 2.77 Ga, spawning the Kaapvaal and Pilbara cratons, at a time
when Superia had not yet amalgamated. (c) A late Archean Earth
with many supercratons and smaller landmasses. This configuration
is unlikely for both geodynamic and geological reasons (see text). It
would predict a high proportion of "external" fragments, i.e.,
Archean cratons that show long-lived passive or active margins
across the Archean–Proterozoic boundary.

formity surfaces (e.g., Hoffman and Hall, 1993; Bleeker et al., 1999a,b; Tankard et al., 1982). A craton like the Slave thus appears to have been "bobbing" rather gently in the convective upper mantle since approximately 2.8 Ga, with effects of rifting, break-up, and collision events being restricted mostly to narrow margins since the Paleoproterozoic. Between ca. 2.6 and 2.2 Ga, it retained its stability as part of the late Archean supercraton Sclavia; after 1.8 Ga, it did so while being amalgamated within Laurentia. Maximum uplift of the Slave craton is constrained by high temperature–low pressure metamorphic facies across the craton, with andalusite being a characteristic index mineral over large areas. Intervening areas are characterized by either lower or higher grade. Over time, erosion has thus removed a maximum of 10–12 km from the surface of the craton and locally much less. Most of this surface erosion must have occurred between 2.6 and 2.2 Ga, during slow uplift accompanied by protracted cooling (Bethune et al., 1999). The size of the craton may have been further reduced by small crustal slivers being "filleted" off the edges during 2.0–1.8 Ga collisional events that amalgamated the Slave into Laurentia but the core of the craton remained stable. If similar stability and survivability is typical for Archean cratons in general, the remaining ensemble of ca. 35 cratons may be an accurate estimate of the number of independent Archean fragments that rifted and drifted around the globe at some time during the Paleoproterozoic.

Hence, if a craton like the Slave has three to four rifted margins bordered by Proterozoic orogenic belts, break-up of its supercraton Sclavia must have spawned circa three to four other cratons that were "nearest neighbours" to the Slave in the ancestral supercraton Sclavia (e.g., Fig. 5). This in turn suggests that 3–4 out of ca. 35 cratons or, in other words, approximately 1 in 10 cratons anywhere around the globe may represent a "missing piece" of the Slave. As at least half of the ca. 35 cratons are reasonably well characterized it should thus be possible, theoretically, to identify a significant number of likely correlations for well-known cratons like the Slave, Kaapvaal, Superior, and others. A reasonable approach then is to group known cratons into "clans" based on degree in similarity in their stratigraphic and structural histories, e.g., Kaapvaal and Pilbara, and then test these similarities in detail using a variety of

modern correlation tools, notably sequence stratigraphy (e.g., Cheney, 1996), precise U–Pb dating of stratigraphic marker horizons (e.g., Zegers et al., 1998; Nelson et al., 1999; Byerly et al., 2002), granitoid chronology, and dating and paleomagnetism of mafic dyke swarms (e.g., Heaman, 1997; Wingate, 1998, 1999, 2000).

4. End member solutions

Possible end member solutions to the problem of craton correlations are: (1) a single late Archean supercontinent; (2) a limited set of late Archean supercratons; and (3) many dispersed supercratons and craton-size landmasses.

The single supercontinent solution (e.g., Fig. 5a) is based in part on similar logic as discussed in this paper (i.e., most cratons have rifted margins and therefore must have come from a larger ancestral landmass) and on the general concept of a supercontinent cycle (e.g., Williams et al., 1991; Hoffman, 1992, 1997). Williams et al. (1991) introduced the name Kenorland[4] for this hypothetical late Archean supercontinent and proposed that all Archean cratons within 1.8 Ga Laurentia originated from diachronous early Proterozoic break-up of this landmass.

The second solution, that of a limited set of somewhat independent late Archean supercratons (Fig. 5b), is similar in nature but emphasizes the significant differences between many of the preserved cratons. Discriminating between these two solutions is difficult but could be achieved once several cratons have been correlated into ancestral supercratons. High-quality paleomagnetic data on precisely dated coeval rock formations ("key poles", Buchan et al., 2000), and other paleogeographic indicators, could then test relative proximity between these landmasses. Other possible tests are that a large supercontinent implies a Panthalassa-like ocean with abundant mature oceanic lithosphere. It is thus predicted that supercontinent aggregation should have been accompanied and followed by a eustatic fall in sea levels and

---

[4] Named after the Kenoran orogeny (Stockwell, 1961, 1964, 1972), the late Archean mountain building event that is thought to have progressively built the Superior Province.

widespread continental emergence. However, there are many uncertainties related to the overall issue of continental freeboard and, therefore, whether this test is sensitive enough to differentiate between a sprawling supercontinent and several somewhat independent supercratons is questionable. On the other hand, several somewhat independent supercratons would have a higher perimeter to surface ratio (e.g., Fig. 5a versus b). Hence, this would predict a higher proportion of cratons with long-lived (e.g., 2.6–2.0 Ga) passive or active margins.

The third solution of many dispersed supercratons and smaller landmasses (Fig. 5c) seems unlikely as it is in conflict with the observation of rifted margins and with general geodynamic understanding. Only sufficiently large landmasses (a single supercontinent or a small number of large supercratons) create the thermal insulation of the mantle that leads to mantle upwelling and continental rifting (e.g., Anderson, 1982; Gurnis, 1988; Davies, 1999).

Independent of configuration, a worst-case scenario would be that the original number of independent cratons was much higher than ca. 35 and that due to progressive fragmentation, erosion, and efficient Proterozoic recycling of Archean crust only a fairly unique set of ca. 35 nonmatching pieces has been preserved. This sobering possibility would imply that the amount of Archean continental crust at ca. 2.5 Ga was many times that what is preserved today. This in turn would be compatible with those end member models of continental growth which suggest that all of the net growth of continental crust happened early in Earth history (Fyfe, 1978; Armstrong, 1981, 1991). A further implication of this scenario would be that the preserved Archean cratons may comprise a nonrepresentative set of crustal (lithospheric) fragments that survived because of above-average stability. This "Darwinian solution" (i.e., survival of the fittest) would severely limit the prospects for successful craton correlations.

5. Craton "clans" based on degree in similarity

Of the five to eight Archean cratons that were amalgamated in Laurentia, several are sufficiently different that they are unlikely to have been nearest neighbours at any point in time prior to their amal-

gamation. This is particularly true for the two better-known Archean crustal fragments of Laurentia, the Slave and Superior cratons (Fig. 2). Although it is perhaps permissible that these two cratons were distant parts of a single large supercontinent at ca. 2.6 Ga, Kenorland (Williams et al., 1991; Hoffman, 1992, 1997; e.g., Fig. 5a), there is presently no compelling evidence to support such a notion. Perhaps it is more realistic to propose that the Slave and the Superior trace their ancestry to different ancestral supercratons, Sclavia and Superia, that were in existence across the Archean–Proterozoic boundary (Figs. 2 and 5b).

Similar arguments apply to the Pilbara and Yilgarn cratons of Western Australia, which show few if any correlatable features (e.g., Myers et al., 1996). Whereas the Pilbara craton may have been connected with the Kaapvaal craton of southern Africa (e.g., Zegers et al., 1998), the ancestry of the Yilgarn craton is unknown although it clearly is a substantial rifted fragment of a much larger ancestral supercraton. The latter is indicated by its rifted margins and by the fact that it is cross-cut, from edge to edge, by large, parallel trending, mafic dykes of the ca. 2.41–2.42 Ga Widgiemooltha swarm (Tyler, 1990; Nemchin and Pidgeon, 1998; Doehler and Heaman, 1998). One dyke of this swarm has been precisely dated at 2418 ± 3 Ma, an age that is identical to older Scourie dykes in the Lewisian gneiss complex of northwest Scotland (Nain craton; Heaman and Tarney, 1989). These identical ages and the scale of the Widgiemooltha dyke swarm raise the possibility that the Yilgarn and Nain cratons preserve different parts of a large late Archean supercraton (Fig. 6) that at 2418 Ma was intruded by a giant mafic dyke swarm (Doehler and Heaman, 1998; Nemchin and Pidgeon, 1998). This example not only illustrates the critical role that large mafic dyke swarms play in identifying and testing possible craton correlations, but also emphasizes that due to Paleoproterozoic and younger break-up and dispersal events the preserved Archean cratons have been so thoroughly "shuffled" that correlations among proximal and distal cratons are probably of equal likelihood. Hence, correlations largely based on current proximity, such as the first supercontinent "Ur" as proposed by Rogers (1996), the Kenorland configuration as proposed by Aspler and Chiarenzelli (1998), or the Laurentia–Baltica connection discussed earlier, should be viewed with suspicion. Rogers (1996) postulates his ca. 3 Ga Ur on the basis of the post-0.55 Ga paleogeography of east Gondwana (e.g., see Fitzsimons, 2000), whereas Aspler and Chiarenzelli (1998) apply the post-1.8 Ga paleogeography of Laurentia to the Archean. The random shuffling of the Archean record once again emphasizes the need for a global scope.

Based on the first-order differences between some of the best known cratons, I suggest there are at least three distinct "clans" of cratons, each exemplified by a type craton: a Slave-like clan, a Superior-like clan, and a Kaapvaal clan. Each clan likely traces its ancestry to a different supercraton (Table 1; Fig. 6). An important difference between these three clans is their relative age of cratonization. Granite–greenstone terrains of the Slave craton were cratonized after a terminal "granite bloom" that is dated at ca. 2600–2580 Ma across the craton (van Breemen et al., 1992; Davis and Bleeker, 1999; Davis et al., 2003). In contrast, much of the Superior craton (Card, 1990) experienced late stage granite plutonism between 2680 and 2640 Ma and hence was cratonized at least ca. 50 million years prior to the Slave craton. The Kaapvaal and Pilbara cratons were largely cratonized by 3.0 Ga (e.g., Moser et al., 2001). Other fundamental differences between these type cratons (or their clans) are the chronology of mafic dyke swarm events and the stratigraphy and chronology of Paleoproterozoic cover sequences. For instance, the Superior craton is intruded by the large Matachewan dyke swarm, which fans from a focal point along its southern margin. Dykes of this swarm have been dated at 2473 and 2446 Ma (Heaman, 1997). The significance of the two discrete ages is presently unknown but it seems likely that the composite swarm is associated with rifting along the southern margin of the Superior craton and deposition of the rift and passive margin sequence of the Huronian Supergroup (Young, 1973; Young et al., 2001). An important feature of the Huronian Supergroup is that it contains ca. 2.2–2.4 Ga glaciogenic deposits (e.g., Young et al., 2001). On the other hand, there is no indication of 2.47–2.44 Ma mafic dykes in the Slave craton, nor for marginal Proterozoic sequences of the right age to have recorded the 2.2–2.4 Ga glaciation events. Instead, the Slave craton appears to have broken out of its Sclavia supercraton starting at ca. 2.2 Ga (Fig. 2). In contrast with both the Superior

and Slave cratons, the Kaapvaal and Pilbara are overlain by extensive Paleoproterozoic cover sequences of the Transvaal and Hamersley basins with important carbonate and banded iron formation deposits (e.g., Cheney, 1996). Rifting of these two cratons, or rather of their ancestral supercraton Vaalbara (Fig. 6), may have occurred as early as 2775 Ma with the onset of voluminous mafic volcanism of the Fortescue Group (Wingate, 1998, 1999; Strik et al., 2001).

In addition to the three clans defined above there may be others. Perhaps the possible Yilgarn–Nain correlation based on 2418 Ma mafic dykes (Doehler and Heaman, 1998) suggests a fourth group. An interesting feature of at least parts of the Nain craton (Lewisian gneiss complex) is a significant ca. 2.5 Ga high-grade metamorphic and deformation event (Corfu et al., 1994; Friend and Kinny, 1995). A similar event is known from large parts of the Hearne craton in Laurentia, but appears absent from the Slave and Superior cratons. Other observations relevant to the Yilgarn craton are that the general chronology of ca. 2720–2630 Ma events is very similar to that of the southern Superior craton (e.g., Nelson, 1998). This view is reinforced by apparent similarities between the Widgiemooltha (Yilgarn) and Matachewan (Superior) dyke swarms, each of which cuts at nearly right angles through the belt-like arrangement of contrasting granite–greenstone domains in both cratons. However, the precise age dating of both swarms (2418 Ma for the Widgiemooltha swarm, versus 2446 and 2473 Ma for the Matachewan swarm) rules out a direct correlation.

In summary, first-order differences between some of the best-known cratons suggest an ancestry from different supercratons (Fig. 2). As an initial step towards identifying and testing such supercratons it is useful to group known cratons into different clans based on their degree of similarity (Table 1) using the chronology of latest Archean events, and Paleoproterozoic mafic dyke swarms and cover sequences. Among the well-characterized cratons, there appear at least three if not four distinct clans, with the Slave, Superior, Kaapvaal, and possibly Yilgarn, as type examples. Other clans may exist but an exhaustive survey is presently difficult due to the incomplete knowledge of all 35 cratons. It is also beyond the scope of this paper. Below, I will discuss the Slave

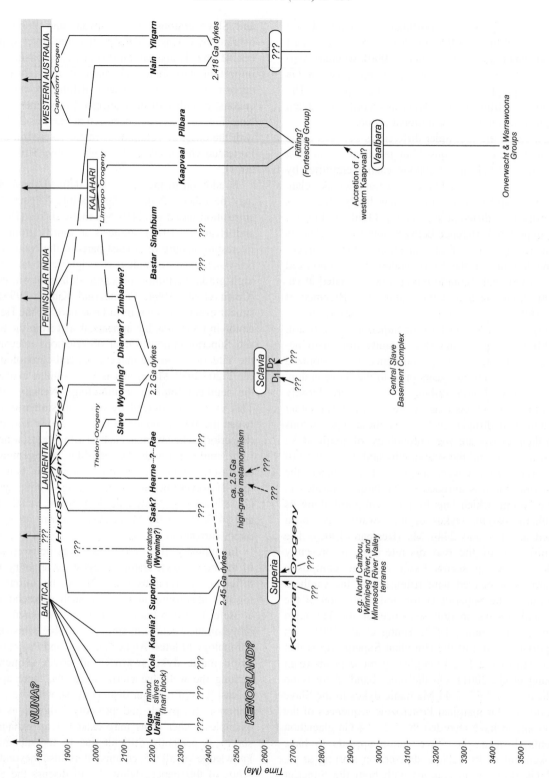

Table 1
Craton clans based on degree in similarity[a]

Craton clan and type craton	Slave	Superior	Kaapvaal	Yilgarn?[b]
Ancestral supercraton	Sclavia	Superia	Vaalbara	???
Approximate age of cratonization	2.58–2.50 Ga	2.68–2.63 Ga	ca. 3.0 Ga	2.68–2.63 Ga
Major dykes swarms and possible initiation of break-up	ca. 2.22 Ga	ca. 2.47–2.45 Ga	ca. 2.77 Ga	ca. 2.42–2.41 Ga
Other possible cratons in clan (i.e., possible "nearest neighbours" to type craton in ancestral supercraton)	Dharwar? Zimbabwe? Wyoming?[c] Others?	Karelia? Hearne? Wyoming?[c] Others? Yilgarn?[b]	Pilbara?	Nain?

[a] See text for explanation and references; question marks highlight significant uncertainties in all correlations.

[b] The Yilgarn clan is less distinct than the other three; e.g., age of cratonization of the Yilgarn is similar to that of the Superior and both clans may collapse into a single clan with more data.

[c] The Wyoming craton shows many similarities to the Slave craton and herein is considered a member of the Slave clan. Others have argued for proximity of the Wyoming craton to the Superior, based on lithostratigraphic comparisons of the undated Snowy Pass Supergroup with the Huronian Supergroup (e.g., Roscoe and Card, 1993).

clan in some detail before returning to broader correlation questions in the Discussion.

6. The Slave clan: possible progeny of a Sclavia supercraton

The Slave craton (Padgham and Fyson, 1992; Bleeker and Davis, 1999a) is characterized by a large Hadean to Mesoarchean (ca. 2.85 Ga) basement complex, which includes the oldest known, intact, terrestrial rocks, the Acasta gneisses (e.g., Bowring et al., 1989; Stern and Bleeker, 1998; Bowring and Williams, 1999). This basement is exposed in large antiformal basement culminations and is overlain by basalt-dominated Neoarchean greenstone belts and turbiditic sedimentary rocks (Figs. 7 and 8). Together with synvolcanic and several post-turbidite granitoid intrusive suites, these Neoarchean supracrustal rocks of the Yellowknife Supergroup dominate the surface

geology of the craton (Fig. 1). Although there are minor regional differences, the overall stratigraphy and granitoid geochronology is rather similar across the craton and can be characterized as follows (Fig. 8):

- Pre-2.86 Ga basement, overlain by a thin sequence of fuchsitic quartzites and banded iron formation (Central Slave Cover Group, Bleeker et al., 1999a,b; see Fig. 7b–e).
- 2730–2700 Ma, widespread basaltic volcanism across the basement complex (Kam Group, Fig. 7f), with minor komatiitic rocks, and locally intercalations of rhyolite, tuff, and reworked volcaniclastic rocks, particularly towards the top of the pile; precisely dated tuff layers at 2722 to 2700 Ma (e.g., Isachsen and Bowring, 1997).
- 2700–2660 Ma, transition to more intermediate, felsic, or bimodal volcanism (Banting Group, e.g., Henderson, 1985; Helmstaedt and Padgham, 1986); dacite–rhyolite complexes typically became

Fig. 6. Cladogram showing possible relationships between about half (17) of the ca. 35 independent Archean cratons. Archean supercratons, break-up of which spawned the independent Archean cratons, are shown at the bottom (Superia, Sclavia, Vaalbara, and possibly a fourth supercraton). Composite cratons that resulted from collisional amalgamation of the Archean cratons during the Paleoproterozoic are shown at the top (names in capital letters). In this cladogram, divergence represents break-up, whereas convergence represents amalgamation, typically associated with significant orogenies; question marks represent significant uncertainties or unknowns. Williams et al. (1991) argued that all the Archean cratons originated from break-up of a single late Archean supercontinent, Kenorland (possible life span of Kenorland represented by lower grey band). Alternatively, it is argued here that the cratons may have originated from break-up of several distinct supercratons, each supercraton spawning a clan of similar cratons. Supercraton Vaalbara may have rifted prior to amalgamation of Superia. The Sclavia supercraton was stabilized only after 2.6 Ga and remained intact for ca. 400 million years, after which it broke up and spawned some of the ca. 35 independently drifting Archean cratons. After ca. 1.9–1.8 Ga amalgamation, the resulting composite Paleoproterozoic cratons (e.g., Baltica, Laurentia, etc.) may have formed a mid-Proterozoic supercontinent, Nuna (Hoffman, 1992, 1997).

emergent; abundant volcaniclastic material (Fig. 7i); widespread tonalite–trondhjemite–granodiorite-type subvolcanic plutons.

- 2670–2650 Ma (locally as young as 2625 Ma), widespread turbidite sedimentation across the craton (e.g., Burwash Formation, Fig. 7j,k; Bleeker and Villeneuve, 1995; see also Isachsen and Bowring, 1994; Pehrsson and Villeneuve, 1999; and summary diagram in Davis and Bleeker, 1999, Fig. 6).
- 2645–2635 Ma, first major folds in Burwash Formation turbidites resulting in NE–SW trending F_1 fold belt (see Figs. 1 and 7k; e.g., Bleeker and Beaumont-Smith, 1995; Bleeker, 1996; Davis and Bleeker, 1999).
- 2630–2610 Ma, abundant tonalite–granodiorite ± diorite plutons, possibly diachronous across the craton from southeast to northwest (Defeat and Concession Suites; Davis and Bleeker, 1999).
- 2605–2590 Ma, more evolved granites, particularly two-mica granites in areas of down-folded turbiditic sedimentary rocks (e.g., Prosperous Suite; Henderson, 1985; Davis and Bleeker, 1999). Coeval craton-scale deformation resulting in north- to northwest-trending F_2 folds, which refold F_1 fold belt.
- 2590–2580 Ma, terminal "granite bloom" of evolved, commonly K-feldspar megacrystic granites across the craton (Fig. 7m); no apparent age trends (e.g., Morose Suite; Davis and Bleeker, 1999).
- ca. 2600–2580 Ma, late-kinematic conglomerate sequences (Fig. 7l; e.g., Corcoran et al., 1998), and large-scale strike-slip faulting.

Ca. 2.47–2.41 Ga mafic dyke swarms appear to be absent from the Slave craton. The Malley swarm, currently the oldest known Paleoproterozoic mafic dykes in the craton, has been dated at 2230 Ma (LeCheminant and van Breemen, 1994). Numerous other swarms have been dated between ca. 2.2 and 2.0 Ga (e.g., see Ernst and Buchan, 2001), suggesting break-up of the Sclavia supercraton during this time. Proterozoic marginal sequences such as the Coronation Supergroup are younger than 2.2 Ga and lack the ca. 2.4–2.2 Ga glaciogenic intervals of older cover sequences overlying other cratons such as the Superior. It should be pointed out, however, that the Paleoproterozoic rift-passive margin sequence overlying the eastern margin of the craton is inadequately known due to its involvement in the high-grade Thelon orogen.

Using these first-order characteristics of the Slave craton, I identify the Dharwar craton (southern India), Zimbabwe craton, and Wyoming craton as the most "Slave-like" and hence possibly as nearest neighbours to the Slave in the 2.6 Ga Sclavia supercraton. All three have extensive basement complexes going back to 3.3–3.5 Ga or older (e.g., Horstwood et al., 1999; Frost, 1993), which are overlain by lithostratigraphically remarkably similar Meso- to Neoarchean successions (Fig. 9): quartz pebble conglomerate and fuchsitic quartzites on heterogeneous basement including abundant ca. 2.9 Ga tonalites (e.g., for Zimbabwe: Bickle et al., 1975; Blenkinsop et al., 1993; Hunter et al., 1998). These basal quartzitic rocks are in turn overlain by thin banded iron formations or ferruginous cherts; ca. 2.73–2.70 Ga flood basalts; younger calc-alkaline volcanic rocks and turbidites; and late-kinematic conglomerates. In addition, these Slave-like cratons show a similar progression of late Archean granitoid suites, all culminating at 2.6 Ga or shortly thereafter. In the Zimbabwe craton this would be the widespread Chilimanzi Suite (e.g., Wilson et al., 1995; Jelsma et al., 1996), whereas in the Dharwar craton the late-stage Closepet granite suite would be the closest analog (Friend and Nutman, 1991), although where dated, in the high-grade migmatite and granulite terrain of the southern Dharwar craton, it appears to be younger.

In contrast, basement complexes, general stratigraphy, granitoid chronology, and the chronology of mafic dyke swarms of the Superior craton differ in detail and timing and the Superior is clearly not part of the same clan (e.g., Fig. 2). This raises an interesting point as it has been previously suggested that the Wyoming craton and Superior cratons were connected (Roscoe and Card, 1993). The potential Wyoming–Superior correlation has been largely based on the similarities in stratigraphy, including glaciogenic intervals, between the Snowy Pass Supergroup overlying the southern flank of the Wyoming craton and the Huronian Supergroup (Roscoe and Card, 1993). The Snowy Pass Supergroup remains poorly dated however. Furthermore, the postulated ca. 2.45 Ga mafic dykes (Heaman, 1997) remain to be identified in the Wyoming craton (Heaman, personal communication, 2002). The Archean geology of the Wyoming

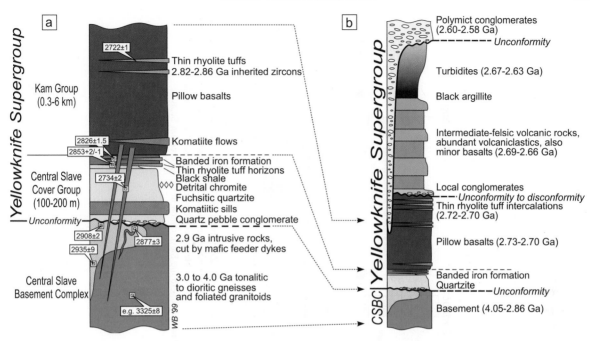

Fig. 8. Generalized stratigraphy of the Slave craton (precise zircon ages from Bleeker et al. (1999a,b), Isachsen and Bowring (1997), and Ketchum and Bleeker (unpublished). (a) Typical stratigraphy of the Central Slave Basement Complex and its overlying sedimentary and volcanic cover. (b) Entire stratigraphy; ca. 2.69–2.66 Ga volcanic rocks and overlying turbidites occur across the craton, their distribution overlapping isotopic boundaries (Bleeker, 2001b). Late-stage polymict conglomerates occur in small restricted basins.

craton (e.g., Houston et al., 1993), including highly evolved Pb isotopic compositions (Wooden and Mueller, 1988), and ancient detrital zircons up to ca. 4 Ga (Mueller et al., 1992, 1998), is more similar to that of the Slave (see also Frost, 1993; Frost et al., 1998, 2000). On the southern flank of the Wind River Range, Wyoming, Archean stratigraphy is exposed that matches that of the central Slave craton in both detail and known age constraints. Here, basement tonalite, cut by abundant mafic dykes, is overlain by fuchsitic quartzite, banded iron formation, and pillow basalts; ultramafic sills intrude near the basement contact (e.g., compare with Fig. 8); overlying units include ca. 2.65 Ga felsic volcanic rocks and a thick package of metaturbidites with abundant cordierite porphyroblasts. The latter are intruded by ca. 2.60–2.55 Ga late-stage granites, while lower in the section the hornblende–titanite bearing granodiorite of the ca. 2630 Ma Louis Lake batholith has an age and composition that is identical to that of Defeat Suite plutons in the Slave craton. Plutonism of this age is absent from the Superior craton.

Expanding on the Slave–Dharwar similarities, the Peninsular Gneiss complex of the Dharwar craton remains to be dated in detail but existing data leave little doubt its history goes back to ca. 3.5 Ga. In the Bababudan greenstone belt (Chadwick et al., 1985a,b; Chardon et al., 1998), a thick basaltic sequence overlies heterogeneous basement. Basalt extrusion was preceded by deposition of a thin, locally fuchsitic, quartz pebble conglomerate and, in some localities, a thin unit of ferruginous chert (Chadwick et al., 1985a). Rhyolite tuffs towards the top of the Bababudan basalt sequence are dated at ca. 2720 Ma (Trendall et al., 1997), an age that is within error of the first rhyolitic tuffs in the basalt sequence overlying the basement complex in the Slave craton. Younger stratigraphic units, including widespread turbidites in the northern part of the Dharwar craton[5] also match

[5] The western Dharwar appears to be gently tilted towards the north, presenting an oblique section from deep crustal granulites in the south to uppermost stratigraphic levels, including widespread turbidites, in the north.

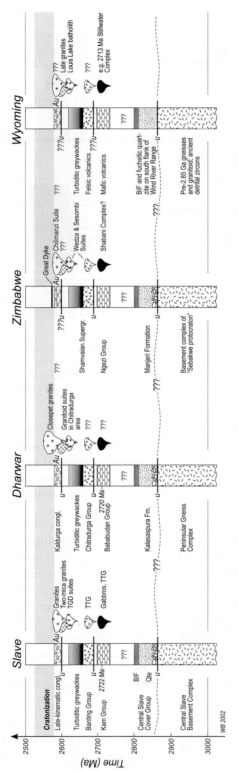

Fig. 9. Comparison of key stratigraphic elements and granitoid suites between cratons of the "Slave clan": Slave, Dharwar, Zimbabwe, and Wyoming. The column of the Slave craton is used as a template. Possible correlative features in the other cratons are identified. Features that have not (yet?) been identified are marked with question marks (e.g., ca. 2.60–2.58 Ga late-kinematic conglomerates in the Zimbabwe craton?). Although precise U–Pb dating is needed to test these possible correlations in detail, the overall similarities require an explanation. The explanation favoured here is that some or all of these cratons may have been nearest neighbours to the Slave craton in the supercraton Sclavia.

the broad stratigraphy of the Slave. And finally, the late-stage Kaldurga conglomerate of the Bababudan greenstone belt, with its abundant granitoid clasts, resembles the late-kinematic conglomerates at the top of the Slave craton stratigraphy (Fig. 8).

Clearly, the broad similarities identified above and illustrated in Fig. 9 require extensive further testing, particularly with high-precision U–Pb dating of key stratigraphic units, granitoid suites, dyke swarms, and Proterozoic cover sequences.

7. Discussion

Irrespective of whether some of the detailed potential correlations will stand the test of time, the broad similarities between certain cratons, but not others, demand an explanation. Although synchronized tectonic processes between distant localities ("parallel evolution") should be considered (e.g., superplume events; or global mantle overturn events), break-up of supercratons and dispersal of cratonic fragments around the globe seems a more likely solution, particularly where there are detailed matches involving several rock units across a broad time span: late Archean stratigraphy, the ages and nature of late Archean granitoid suites, and Paleoproterozoic dykes swarms and cover sequences. High-quality paleomagnetic data sets on coeval rock units across several cratons are needed to test for proximity of potentially correlating cratons.

The statistics of correlations are reasonable, unless the ca. 35 remaining cratonic fragments are just a small, unique, and possibly nonrepresentative sample of an originally much larger set of independent Archean crustal fragments in existence during Paleoproterozoic dispersal. This question is intimately related to models of crustal growth. If a Taylor and McLennan-type model of crustal growth is broadly correct (Taylor and McLennan, 1985), the ca. 35 cratonic fragments will be a reasonable representation of the total count of independent fragments. In that case, a craton such as the Slave, with three or four rifted margins, may have a ca. 10% probability of matching any other craton around the globe. Multiple break-up and dispersal events probably mean that distal and proximal correlations are equally likely. Hence, proposed correlations based on current prox-

imity should be viewed with suspicion, unless there are independent data to suggest that intervening ocean basins were restricted in size.

Based on distinct differences between some of the better known cratons, it seems likely that the ca. 35 preserved cratons trace their origin to several different late Archean supercratons, rather than a single supercontinent, although it is difficult to eliminate the possibility that some of these supercratons occupied distant positions in a sprawling Kenorland. However, several additional lines of evidence lend support for the interpretation of several independent supercratons.

Firstly, if the topology of Fig. 6 is broadly correct, it appears that Vaalbara may have broken up prior to amalgamation of a possible Kenorland, thus complicating the concept of a single late Archean supercontinent (Kenorland).

Secondly, and of a more fundamental nature, a hotter Archean Earth, with at least twice the heat production of today and with a more substantial fraction of its primordial heat budget retained, was probably characterized by more vigorous mantle convection and smaller, faster, plates (e.g., Pollack, 1997; see also Burke and Dewey, 1973; Burke et al., 1976; Hargraves, 1986). Such a regime, characterized by less organization and shorter length scales, is more likely to have favoured smaller, transient, continental aggregations in the form of several independent supercratons rather than a single large supercontinental aggregation. Later in Earth history, secular cooling and a resulting growth in average plate size would have increased the likelihood of supercontinental aggregations and mid-Proterozoic Nuna probably represents the first true supercontinent (Fig. 10). The same fundamental argument may also explain why, since the mid-Proterozoic, the time gap between successive supercontinental aggregations (e.g., Nuna to Rodinia, Rodinia to Gondwana/Pangaea) appears to have been decreasing. Presently, Earth is in the aftermath of the break-up of Pangaea and well on its way to a new supercontinental aggregation cored by Asia ("Amaesia"[6], Fig. 10). India has already collided with Eurasia; Africa, various Mediterranean micro-

---

[6] Contraction of the *Am*ericas, *A*frica, *E*urasia, *I*ndia and *A*ustralia. Changed from "Amasia" informally proposed by Hoffman (1997).

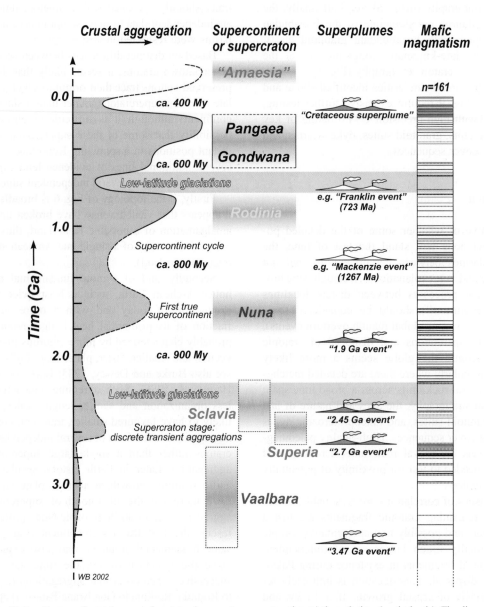

Fig. 10. Qualitative diagram of crustal aggregation states (supercratons, supercontinents) through time (vertical axis). The diagram also shows several other first-order events in Earth evolution that may relate, directly or indirectly, to the crustal aggregation cycle: superplume events (e.g., Condie, 2001), a compilation of mafic magmatic events (Ernst and Buchan, 2001), and the two Proterozoic time intervals during which low-latitude (global?) glaciations may have prevailed (e.g., Evans et al., 1997; Hoffman et al., 1998; Evans, 2000). Mid-Proterozoic Nuna was probably the first true supercontinent, whereas the late Archean may have been characterized by several discrete, transient, aggregations referred to here as supercratons: Vaalbara, Superia, Sclavia and possibly others. The diachronous break-up of these supercratons, in the Paleoproterozoic, spawned the present ensemble of ca. 35 Archean cratons, which now are variably incorporated into younger crustal assemblies. Since the assembly of Nuna, the time gaps between successive crustal aggregation maxima appear to have become shorter. Note the correlation of intervals of global glaciation with periods of continental break-up and dispersal, and with apparent minima in the frequency of mafic magmatic events in the continental record (legend for the latter: red line, well-established mantle plume event; black line, other mafic magmatic event; dashed line, poorly dated event; see Ernst and Buchan, 2001). Further note that inferred break-up of Nuna appears anomalous in this context, i.e., it is not followed by an interval of global glaciation.

plates, and Arabia are in the process of collision; and Australia, drifting rapidly northward, has just started to make contact with southeastern promontories of Eurasia. The exact trajectories of the Americas are more difficult to predict but, possibly, they will continue their westward drift to join the growing supercontinent 200–300 million years from now. Alternatively, the Atlantic basin may collapse due to aging of its oceanic lithosphere and rapid expansion of the Caribbean and Scotian arcs into the Atlantic realm. This would lead, perhaps on a similar time scale, to the Americas being welded to the Euro-African side of the future supercontinent. Hence, the time gap between maximum states of aggregation will have lasted 400–500 million years. This contrasts with a 500–600 million years time gap between the Rodinia and Gondwana/Pangaea maxima and an even longer time gap between the Nuna and Rodinia maxima. In summary, faster, smaller, and therefore more numerous Archean plates may have never amalgamated into a single aggregation, whereas large modern continental plates, after break-up of one supercontinent, quickly run out of room and thus rapidly re-aggregate.

Irrespective of which configuration will best represent the late Archean Earth, the grouping of cratons into clans based on broadly similar characteristics seems a useful step that will help guide formulation of specific tests. Based on some of the better known Archean cratons of Laurentia, Southern Africa and Western Australia, at least three, if not four distinct clans can be recognized, with type cratons being the Slave, Superior, Kaapvaal, and Yilgarn, although a possible Yilgarn–Superior connection deserves further investigation. Another interesting age match that has been noted in the literature is that of the 2449 ± 3 Ma Woongarra rhyolites and interlayered mafic volcanics of the Hamersley basin overlying the Pilbara craton (Barley et al., 1997) and coeval rift-related magmatism of the Huronian Supergroup in the southern Superior craton (e.g., the 2450+25/−10 Ma Copper Cliff rhyolite of the Sudbury area; Krogh et al., 1984). This apparent age match may reflect proximity between the Pilbara and Superior cratons at ca. 2.45 Ga. Heaman (1997) has suggested that 2.45 Ga magmatism may have been global in nature. Magmatism of this age appears absent in the Slave craton, however, and viewed from this craton it are the

striking similarities with the Dharwar, Zimbabwe, and Wyoming cratons that deserve further investigation.

An interesting point, particularly in the context of this volume, is the relationship between the Zimbabwe and Kaapvaal cratons. Several authors have proposed that these two cratons collided along the Limpopo orogen in the late Archean, some proposing a date as early as 2.68 Ga (e.g., de Wit et al., 1992 and references therein). This seems impossible as the Zimbabwe craton only became a craton after 2.60 Ga. Furthermore, major contrasts between Archean stratigraphy, structure, time of cratonization, chronology of mafic dyke swarms, and chronology and stratigraphy of Proterozoic cover sequences, appear to preclude a shared history prior to 2.0 Ga. The Transvaal Supergroup and Bushveld Complex have no expression in the Zimbabwe craton and, vice versa, the craton-wide ca. 2600 Ma Chilimanzi Suite and 2574 Ma Great Dyke do not cross the high-grade Limpopo orogenic belt into the Kaapvaal craton. From the perspective of a distant observer, these two cratons appear radically different and probably did not collide until ca. 2.0 Ga, the age of final high-grade metamorphism in the Limpopo belt (e.g., Holzer et al., 1998; see also Zhao et al., 2002, for a similar perspective).

I should end this discussion with a word of caution. Supercratons, like their dispersed fragments (the ca. 35 remaining cratons), were heterogeneous to variable degrees. Some of this heterogeneity, particularly where it manifests itself in a belt-like topology such as in the Superior craton, may have resulted from late Archean accretionary and collisional tectonics that contributed to the growth of the supercratons (but see discussion in Bleeker, 2002). Subsequent rifting and break-up in the Paleoproterozoic may have followed late Archean sutures between distinct terrains, thus spawning two or more distinct cratons that may share few elements in Archean stratigraphy but, nevertheless, were "nearest neighbours" in their ancestral late Archean supercraton (Fig. 11). In this end member scenario, successful correlation can only be based on shared remnants of pre-drift Paleoproterozoic cover sequences and mafic dyke swarms. This end member scenario may be rare though as most cratons show characteristic attributes that ignore terrain boundaries. In particular, the products of late stage granitoid plutonism (e.g., the 2590–2580 Ma

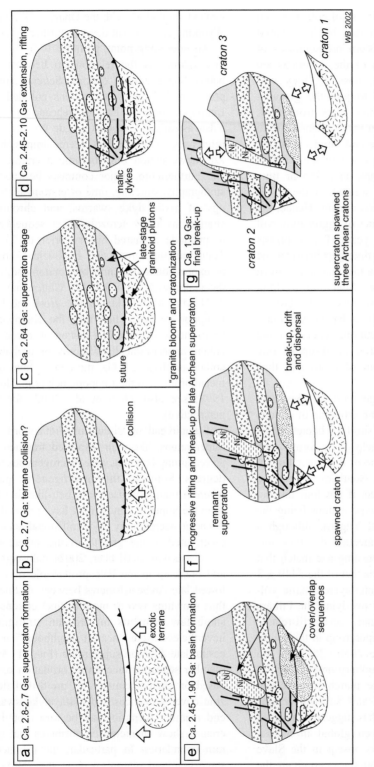

Fig. 11. Schematic illustration of the relationships between a heterogeneous, late Archean, supercraton and its progeny of three cratons (modeled here, in part, with the Superior craton in mind). (a–c) Progressive formation of a late Archean supercraton (e.g., Superia) involving the collision of a large, exotic, Mesoarchean terrain, and culminating in wide-spread, post-collisional, granitoid plutonism at ca. 2.64 Ga ("granite bloom"). (d,e) Progressive rifting, mafic dyke swarm intrusion, and deposition of Paleoproterozoic, rift-related, cover or overlap sequences (e.g., the Ospwagan Group of the western Superior craton margin, with important Ni deposits in ultramafic sills). (f) Rifting and dispersal of a first craton ("craton 1"), consisting largely of the Mesoarchean exotic terrain. (g) Rifting and dispersal of a second craton ("craton 2"). Because rifting and break-up of "craton 1" occurred along an old suture, its Archean stratigraphy is largely unique, whereas "craton 2" and "craton 3" are near-similar twins. In this end member scenario, "craton 1" only reveals its transient residence in the late Archean supercraton through the chronology of the late granitoid suite, and more definitively, through correlation of the dyke swarms and the sedimentary cover/overlap sequence.

"granite bloom" in the Slave, see Davis and Bleeker, 1999; the Chilimanzi Suite of the Zimbabwe craton) tend to occur across entire cratons and thus can be expected to be present also in rifted and drifted fragments of crust that otherwise are distinct in their Archean stratigraphy. Although this end member scenario should be kept in mind, it does not invalidate the cladistic approach of looking for common attributes among Archean cratons as an initial basis for correlating the ca. 35 remaining cratons into their ancestral supercratons.

8. Conclusions

The Archean record preserves ca. 35 cratons and a less well defined number of small crustal slivers. In general, these cratons are rifted fragments of larger ancestral landmasses and tectonic systems—supercratons—that must have existed in the late Archean and across the Archean–Proterozoic boundary.

Based on the marked differences between some of the better known cratons, some of these supercratons (e.g., Sclavia, Superia, Vaalbara) appear to have experienced distinct geological evolutions and therefore are unlikely to have been adjacent pieces of Archean crust. Whether these supercratons were entirely independent ("the supercraton solution") or part of a sprawling supercontinent Kenorland ("the Kenorland solution") remains to be established, although break-up of one supercraton, Vaalbara, appears to predate possible amalgamation of Kenorland. A strict interpretation of the Kenorland hypothesis (Williams et al., 1991; Hoffman, 1992, 1997) thus seems unlikely. Thermal arguments also favour several distinct, transient, late Archean supercratons rather than a single large supercontinent.

Diachronous 2.5–2.0 Ga break-up of these larger supercratons, perhaps preceded by earlier break-up of Vaalbara, spawned a minimum of ca. 35 independently drifting Archean cratons before their amalgamation in younger landmasses such as Laurentia and the potential Mesoproterozoic supercontinent Nuna (Hoffman, 1997). Nuna probably represents the first true supercontinent in Earth's history.

Our ability to correlate the remaining cratons into the original supercratons is strongly dependent on how many cratons were lost from the record. If the number of ca. 35 Archean cratons is a reasonable approximation of how many independent cratons existed during Paleoproterozoic dispersal, a typical craton with three to four rifted margins has ca. 10% maximum probability of correlating with any other craton in the world. This probability would be significantly reduced if the original number of independent cratons was much larger as would be implied by Armstrong-type crustal growth models. If such end member models are true, only the most stable Archean fragments may have survived, possibly representing a unique, nonmatching, and nonrepresentative set of Archean cratons (the "Darwinian solution"). Archean craton correlation problems are thus critically intertwined with crustal growth models and, ultimately, the success rate of such correlations may provide an important test between Taylor and McLennan-type and Armstrong-type models of crustal evolution.

Because of their rifted margins, the tectonic evolution of Archean cratons and that of their mantle keels should always be viewed in the context of the larger ancestral supercratons, even if we are not (yet) able to reconstruct these ancient landmasses in detail. Failure to take this into consideration leads to simplistic models for craton evolution. Due to their limited size, most cratons are too small to preserve their complete tectonic context and critical tectonic elements that shaped the evolution of one craton may now be preserved on a distant but correlating craton.

A first logical step towards craton correlations involves the grouping of cratons into clans based on their degree of similarity. This approach already led to the Vaalbara hypothesis (e.g., Cheney, 1996). Among the better known cratons, at least three obvious clans, and possibly more, can be established. Each clan may represent the progeny of a different supercraton. Similarities between the Slave, Dharwar, Wyoming, and Zimbabwe cratons suggest they may all be part of the Slave clan, which traces its origin to a postulated Sclavia supercraton. This supercraton stabilized in the very latest Archean (ca. 2.58 Ga or shortly thereafter), well after stabilization of another well-known craton, the Superior, and its inferred supercraton, Superia. Stability of Sclavia lasted until ca. 2.2 Ga, at which time it started to experience progressive break-up. By ca. 2.0 Ga the Slave craton had broken out of its ancestral supercraton and was drifting as an independent craton or microcontinent, just prior to its collision with the Rae craton and its eventual amalgamation into Laurentia.

A next critical step towards craton correlation involves establishing detailed chronostratigraphic profiles for all of the ca. 35 cratons (e.g., similar to Fig. 2), including precise ages for all first-order features that are essential to craton correlations: general stratigraphy, particularly the latest Archean events; chronology and character of late Archean granitoid suites; age and character of late Archean craton-scale faults; age of late-kinematic conglomerate basins; ages of all Paleoproterozoic mafic dyke swarms prior to dispersal; and sequence stratigraphy and chronology of Paleoproterozoic cover sequences. High-quality paleomagnetism on key rock units (e.g., precisely dated coeval mafic dyke swarms) is then required to test for proximity. Only then will we be able to view the Archean at the global scale required.

Acknowledgements

I thank my colleagues at the Geological Survey of Canada for stimulating discussions on the Archean record. In particular, I would like to thank my collaborators in the Slave craton, John Ketchum (Royal Ontario Museum), Bill Davis and Richard Stern (GSC), and Keith Sircombe (presently at Univeristy of Western Australia). Drs. Ganapathi Hegde and Vasudev guided the author in the Dharwar craton. The paper benefited from comments by Cees van Staal and Richard Ernst, and thorough reviews by Clark Isachsen and Paul Hoffman.

References

Anderson, D.L., 1982. Hotspots, polar wander, Mesozoic convection and the geoid. Nature 297 (5865), 391–393.

Armstrong, R.L., 1981. Radiogenic isotopes: the case for crustal recycling on a near-steady-state no-continental-growth Earth. Philosophical Transactions of the Royal Society of London, Series A: Mathematical and Physical Sciences 301 (1461), 443–472.

Armstrong, R.L., 1991. The persistent myth of crustal growth. Australian Journal of Earth Sciences 38 (5), 613–630.

Aspler, L.B., Chiarenzelli, J.R., 1998. Two Neoarchean supercontinents? Evidence from the Paleoproterozoic. Sedimentary Geology 120 (1–4), 75–104.

Barley, M.E., Pickard, A.L., Sylvester, P.J., 1997. Emplacement of a large igneous province as a possible cause of banded iron formation 2.45 billion years ago. Nature 385 (6611), 55–58.

Bethune, K.M., Villeneuve, M.E., Bleeker, W., 1999. Laser ^{40}Ar/^{39}Ar thermochronology of Archean rocks in Yellowknife Domain, southwestern Slave Province: insights into the cooling history of an Archean granite–greenstone terrane. Canadian Journal of Earth Sciences 36 (7), 1189–1206.

Bickle, M.J., Martin, A., Nisbet, E.G., 1975. Basaltic and peridotitic komatiites, stromatolites, and a basal unconformity in the Belingwe Greenstone Belt, Rhodesia. Earth and Planetary Science Letters 27, 155–162.

Bleeker, W., 1990a. Evolution of the Thompson Nickel Belt and its nickel deposits, Manitoba, Canada. PhD thesis, University of New Brunswick, Fredericton, New Brunswick. 400 pp.

Bleeker, W., 1990b. New structural-metamorphic constraints on Early Proterozoic oblique collision along the Thompson Nickel Belt, Manitoba, Canada. In: Lewry, J.F., Stauffer, M.R. (Eds.), The Early Proterozoic Trans-Hudson Orogen of North America. Geological Association of Canada Special Paper, vol. 37. Geological Association of Canada, pp. 57–73.

Bleeker, W., 1996. Thematic structural studies in the Slave Province, Northwest Territories: the Sleepy Dragon Complex. In: Current Research 1996-C. Geological Survey of Canada, pp. 37–48.

Bleeker, W., 1999. Structure, stratigraphy, and primary setting of the Kidd Creek volcanogenic massive sulphide deposits: a semi-quantitative reconstruction. In: Hannington, M.D., Barrie, C.T. (Eds.), The Giant Kidd Creek Volcanogenic Massive Sulfide Deposit, Western Abitibi Subprovince, Canada. Economic Geology Monograph, vol. 10. Economic Geology Publishing Company, Littleton, CO, pp. 71–121.

Bleeker, W., 2001a. Evolution of the Slave craton and the search for supercratons. Program with Abstracts, Geological Association of Canada, vol. 26, p. 14.

Bleeker, W., 2001b. The ca. 2680 Ma Raquette Lake Formation and correlative units across the Slave Province, Northwest Territories: evidence for a craton-scale overlap sequence. Geological Survey of Canada, Current Research 2001-C7. 26 pp.

Bleeker, W., 2002. Archean tectonics: a review, with illustrations from the Slave craton. In: Fowler, C.M.R., Ebinger, C.J., Hawkesworth, C.J. (Eds.), The Early Earth: Physical, Chemical and Biological Development. Special Publications. vol. 199, Geological Society, London, pp. 151–181.

Bleeker, W., Beaumont-Smith, C., 1995. Thematic structural studies in the Slave Province: preliminary results and implications for the Yellowknife Domain, Northwest Territories. In: Current Research 1995-C. Geological Survey of Canada, pp. 87–96.

Bleeker, W., Davis, W.J., 1999a. NATMAP slave province project. Canadian Journal of Earth Sciences 36 (7), 1033–1238.

Bleeker, W., Davis, W.J., 1999b. The 1991–1996 NATMAP slave province project: introduction. Canadian Journal of Earth Sciences 36 (7), 1033–1042.

Bleeker, W., Villeneuve, M., 1995. Structural studies along the Slave portion of the SNORCLE Transect. In: Cook, F., Erdmer, P. (Compilers), Slave-NORthern Cordillera Lithospheric Evolution (SNORCLE), Report of 1995 Transect Meeting, April 8–9, University of Calgary, LITHOPROBE Report No. 44, pp. 8–14.

Bleeker, W., Ketchum, J.W.F., Jackson, V.A., Villeneuve, M.E.,

1999a. The Central Slave Basement Complex: Part I. Its structural topology and autochthonous cover. Canadian Journal of Earth Sciences 36 (7), 1083–1109.

Bleeker, W., Ketchum, J.W.F., Davis, W.J., 1999b. The Central Slave Basement Complex: Part II. Age and tectonic significance of high-strain zones along the basement-cover contact. Canadian Journal of Earth Sciences 36 (7), 1111–1130.

Bleeker, W., Stern, R., Sircombe, K., 2000. Why the Slave Province, Northwest Territories, got a little bigger. Geological Survey of Canada, Current Research 2000-C2. 9 pp.

Blenkinsop, T.G., Fedo, C.M., Bickle, M.J., Eriksson, K.A., Martin, A., Nisbet, E.G., Wilson, J.F., 1993. Ensialic origin for the Ngezi Group, Belingwe greenstone belt, Zimbabwe. Geology 21 (12), 1135–1138.

Bostock, M.E., 1997. Anisotropic upper-mantle stratigraphy and architecture of the Slave Craton. Nature 390 (6658), 392–395.

Bowring, S.A., Williams, I.S., 1999. Priscoan (4.00–4.03 Ga) orthogneisses from northwestern Canada. Contributions to Mineralogy and Petrology 134 (1), 3–16.

Bowring, S.A., Williams, I.S., Compston, W., 1989. 3.96 Ga gneisses from the Slave Province, Northwest Territories, Canada. Geology 17 (11), 971–975 (with Suppl. Data 89-17).

Buchan, K.L., Mertanen, S., Park, R.G., Pesonen, L.J., Elming, S.A., Abrahamsen, N., Bylund, G., 2000. Comparing the drift of Laurentia and Baltica in the Proterozoic. The importance of key palaeomagnetic poles. Tectonophysics 319 (3), 167–198.

Burke, K., Dewey, J.F., 1973. An outline of Precambrian plate development. In: Tarling, D.H., Runcorn, S.K. (Eds.), Implications of Continental Drift to the Earth Sciences, vol. 2. Academic Press, New York, pp. 1035–1045.

Burke, K., Dewey, J.F., Kidd, W.S.F., 1976. Dominance of horizontal movements, arc and microcontinental collisions during the later permobile regime. In: Windley, B.F. (Ed.), The Early History of the Earth. John Wiley & Sons, New York, pp. 113–129.

Byerly, G.R., Lowe, D.R., Wooden, J.L., Xie, X., 2002. An Archean impact layer from the Pilbara and Kaapvaal cratons. Science 297 (5585), 1325–1327.

Card, K.D., 1990. A review of the Superior Province of the Canadian Shield, a product of Archean accretion. Precambrian Research 48 (1–2), 99–156.

Carlson, R.W., Boyd, F.R., Shirey, S.B., Janney, P.E., Grove, T.J., Bowring, S.A., Schmitz, M.D., Dann, J.C., Bell, D.R., Gurney, J.J., Richardson, S.H., Tredoux, M., Menzies, A.H., Pearson, D.G., Hart, R.J., Wilson, A.H., Moser, D., 2000. Continental growth, preservation, and modification in Southern Africa. GSA Today 10 (2), 1–7.

Chadwick, B., Ramakrishnan, M., Viswanatha, M.N., 1985a. Bababudan—a late Archaean intracratonic volcanosedimentary basin, Karnataka, southern India: Part I. Stratigraphy and basin development. Journal of the Geological Society of India 26, 769–801.

Chadwick, B., Ramakrishnan, M., Viswanatha, M.N., 1985b. Bababudan—a late Archaean intracratonic volcanosedimentary basin, Karnataka, southern India: Part II. Structure. Journal of the Geological Society of India 26, 802–821.

Chardon, D., Choukroune, P., Jayananda, M., 1998. Sinking of the Dharwar Basin (South India): implications for Archaean tectonics. Precambrian Research 91 (1–2), 15–39.

Cheney, E.S., 1996. Sequence stratigraphy and plate tectonic significance of the Transvaal succession of Southern Africa and its equivalent in Western Australia. Precambrian Research 79 (1–2), 3–24.

Cheney, E.S., Roering, C., Stettler, E., 1988. Vaalbara. Geological Society of South Africa Geocongress '88. Extended Abstracts, pp. 85–88.

Condie, K.C., 2001. Mantle plumes and their record in Earth history. Cambridge Univ. Press, Cambridge. 306 pp.

Condie, K.C., Rosen, O.M., 1994. Laurentia–Siberia connection revisited. Geology 22 (2), 168–170.

Cook, F.A., van der Velden, A.J., Hall, K.W., Roberts, B.J., 1999. Frozen subduction in Canada's Northwest Territories. Lithoprobe deep lithospheric reflection profiling of the western Canadian Shield. Tectonics 18 (1), 1–24.

Corcoran, P.L., Mueller, W.U., Chown, E.H., 1998. Climatic and tectonic influences on fan deltas and wave- to tide-controlled shoreface deposits. Evidence from the Archaean Keskarrah Formation, Slave Province, Canada. Sedimentary Geology 120 (1–4), 125–152.

Corfu, F., 1993. The evolution of the southern Abitibi greenstone belt in light of precise U–Pb geochronology. Economic Geology 88 (6), 1323–1340.

Corfu, F., Heaman, L.M., Rogers, G., 1994. Polymetamorphic evolution of the Lewisian Complex, NW Scotland, as recorded by U–Pb isotopic compositions of zircon, titanite and rutile. Contributions to Mineralogy and Petrology 117 (3), 215–228.

Dalziel, I.W.D., 1991. Pacific margins of Laurentia and East Antarctica–Australia as a conjugate rift pair. Evidence and implications for an Eocambrian supercontinent. Geology 19 (6), 598–601.

Davies, G.F., 1999. Dynamic Earth. Cambridge Univ. Press, Cambridge, United Kingdom. 458 pp.

Davis, W.J., Bleeker, W., 1999. Timing of plutonism, deformation, and metamorphism in the Yellowknife Domain, Slave Province, Canada. Canadian Journal of Earth Sciences 36 (7), 1169–1187.

Davis, W.J., Hegner, E., 1992. Neodymium isotopic evidence for the tectonic assembly of late Archean crust in the Slave Province, Northwest Canada. Contributions to Mineralogy and Petrology 111 (4), 493–504.

Davis, W.J., Jones, A.G., Bleeker, W., Grütter, H., 2003. Lithosphere development in the Slave craton: a linked crustal and mantle perspective. Lithos 71, 575–589 (this issue).

de Wit, M.J., Roering, C., Hart, R.J., Armstrong, R.A., de, R.C.E.J., Green, R.W.E., Tredoux, M., Peberdy, E., Hart, R.A., 1992. Formation of an Archaean continent. Nature 357 (6379), 553–562.

Doehler, J.S., Heaman, L.M., 1998. 2.41 Ga U–Pb baddeleyite ages for two gabbroic dykes from the Widgiemooltha Swarm, Western Australia. A Yilgarn–Lewisian connection? In: Geological Society of America 1998 Annual Meeting, Abstracts with Programs, Geological Society of America, vol. 30, pp. 291–292.

Ernst, R.E., Buchan, K.L., 2001. Large mafic magmatic events through time and links to mantle–plume heads. In: Ernst, R.E., Buchan, K.L. (Eds.), Mantle Plumes: Their Identification Through Time. Special Paper, vol. 352. Geological Society of America, Boulder, CO, pp. 483–575.

Evans, D.A., 2000. Stratigraphic, geochronological, and paleomagnetic constraints upon the Neoproterozoic climatic paradox. American Journal of Science 300, 347–433.

Evans, D.A., Beukes, N.J., Kirschvink, J.L., 1997. Low-latitude glaciation in the Paleoproterozoic era. Nature 386, 262–266.

Fitzsimons, I.C.W., 2000. Grenville-age basement provinces in East Antarctica. Evidence for three separate collisional orogens. Geology 28 (10), 879–882.

Friend, C.R.L., Kinny, P.D., 1995. New evidence for protolith ages of Lewisian granulites, Northwest Scotland. Geology 23 (11), 1027–1030.

Friend, C.R.L., Nutman, A.P., 1991. SHRIMP U–Pb geochronology of the Closepet Granite and Peninsular Gneiss, Karnataka, South India. Journal of the Geological Society of India 38 (4), 308–357.

Frost, C.D., 1993. Nd isotopic evidence for the antiquity of the Wyoming Province. Geology 21 (4), 351–354.

Frost, C.D., Frost, B.R., Chamberlain, K.R., Hulsebosch, T.P., 1998. The late Archean history of the Wyoming Province as recorded by granitic magmatism in the Wind River Range, Wyoming. Precambrian Research 89 (3–4), 145–173.

Frost, B.R., Chamberlain, K.R., Swapp, S., Frost, C.D., Hulsebosch, T.P., 2000. Late Archean structural and metamorphic history of the Wind River Range. Evidence for a long-lived active margin on the Archean Wyoming Craton. Geological Society of America Bulletin 112 (4), 564–578.

Fyfe, W.S., 1978. The evolution of the Earth's crust. Modern plate tectonics to ancient hot spot tectonics? Chemical Geology 23 (2), 89–114.

Gorbatschev, R., Bogdanova, S., 1993. Frontiers in the Baltic Shield. Precambrian Research 64 (1–4), 3–21.

Gower, C.F., Rivers, T., Ryan, B., 1990. Mid-Proterozoic Laurentia–Baltica: an overview of its geological evolution and a summary of the contributions made by this volume. In: Gower, C.F., Rivers, T., Ryan, B. (Eds.), Mid-Proterozoic Laurentia–Baltica. Geological Association of Canada Special Paper, vol. 38, pp. 1–22.

Griffin, W.L., Doyle, B.J., Ryan, C.G., Pearson, N.J., O'Reilly, S.Y., Davies, R., Kivi, K., van Achterberg, E., Natapov, L.M., 1999. Layered mantle lithosphere in the Lac de Gras area, Slave Craton. Composition, structure and origin. Journal of Petrology 40 (5), 705–727.

Grütter, H.S., Apter, D.B., Kong, J., 1999. Crust–mantle coupling: evidence from mantle-derived xenocrystic garnets. In: Gurney, J.J., Gurney, J.L., Pascoe, M.D, Richardson, S.H. (Eds.), The J.B. Dawson volume. Proceedings of the VIIth International Kimberlite Conference, vol. 1, pp. 307–313.

Gurnis, M., 1988. Large-scale mantle convection and the aggregation and dispersal of supercontinents. Nature 332 (6166), 695–699.

Hargraves, R.B., 1986. Faster spreading or greater ridge length in the Archean. Geology 14, 750–752.

Heaman, L.M., 1997. Global mafic magmatism at 2.45 Ga. Remnants of an ancient large igneous province? Geology 25 (4), 299–302.

Heaman, L.M., Tarney, J., 1989. U–Pb baddeleyite ages for the Scourie dyke swarm, Scotland. Evidence for two distinct intrusion events. Nature 340 (6236), 705–708.

Helmstaedt, H., Padgham, W.A., 1986. A new look at the stratigraphy of the Yellowknife Supergroup at Yellowknife, N.W.T. implications for the age of gold-bearing shear zones and Archean basin evolution. Canadian Journal of Earth Sciences 23 (4), 454–475.

Henderson, J.B., 1985. Geology of the Yellowknife–Hearne Lake area, District of Mackenzie: a segment across an Archean basin. Geological Survey of Canada Memoir, vol. 414. Geological Survey of Canada. 135 pp.

Herz, N., 1969. Anorthosite belts, continental drift, and the anorthosite event. Science 164 (3882), 944–947.

Hoffman, P.F., 1988. United plates of America, the birth of a craton. Early Proterozoic assembly and growth of Laurentia. Annual Review of Earth and Planetary Sciences 16, 543–603.

Hoffman, P.F., 1989. Precambrian geology and tectonic history of North America. In: Bally, A.W., Palmer, A.R. (Eds.), The Geology of North America: An Overview. Geological Society of America, Boulder, CO, pp. 447–512.

Hoffman, P.F., 1991. Did the breakout of Laurentia turn Gondwanaland inside-out? Science 252 (5011), 1409–1412.

Hoffman, P.F., 1992. Supercontinents. Encyclopedia of Earth System Science, vol. 4. Academic Press, London, pp. 323–328.

Hoffman, P.F., 1997. Tectonic genealogy of North America. In: van der Pluijm, B.A., Marshak, S. (Eds.), Earth Structure and Introduction to Structural Geology and Tectonics. McGraw Hill, New York, pp. 459–464.

Hoffman, P.F., Hall, L., 1993. Geology, Slave craton and environs, District of Mackenzie, Northwest Territories. Geological Survey of Canada, Open File 2559.

Hoffman, P.F., Kaufman, A.J., Halverson, G.P., Schrag, D.P., 1998. A neoproterozoic snowball earth. Science 281, 1342–1346.

Holzer, L., Frei, R., Barton Jr., J.M., Kramers, J.D., 1998. Unraveling the record of successive high grade events in the Central Zone of the Limpopo Belt using Pb single phase dating of metamorphic minerals. Precambrian Research 87, 87–115.

Horstwood, M.S.A., Nesbitt, R.W., Noble, S.R., Wilson, J.F., 1999. U–Pb zircon evidence for an extensive early Archean craton in Zimbabwe: a reassessment of the timing of craton formation, stabilization, and growth. Geology 27 (8), 707–710.

Houston, R.S., Erslev, E.A., Frost, C.D., Karlstrom, K.E., Page, N.J., Zientek, M.L., Reed Jr., J.C., Snyder, G.L., Worl, R.G., Bryant, B., Reynolds, M.W., Peterman, Z.E., 1993. The Wyoming province. In: Reed Jr., J.C., Bickford, M.E., Houston, R.S., Link, P.K., Rankin, D.W., Sims, P.K., Van Schmus, W.R. (Eds.), Precambrian: Conterminous U.S., The Geology of North America C-2. Geological Society of America, Boulder, CO, pp. 121–170.

Hunter, M.A., Bickle, M.J., Nisbet, E.G., Martin, A., Chapman, H.J., 1998. Continental extensional setting for the Archean Belingwe greenstone belt, Zimbabwe. Geology 26 (10), 883–886.

Hurley, P.M., Rand, J.R., 1969. Pre-drift continental nuclei. Science 164 (3885), 1229–1242.

Irvine, G.J., Pearson, D.G., Carlson, R.W., 2001. Lithospheric mantle evolution of the Kaapvaal Craton: a Re–Os isotope study of peridotite xenoliths from Lesotho kimberlites. Geophysical Research Letters 28 (13), 2505–2508.

Isachsen, C.E., Bowring, S.A., 1994. Evolution of the Slave Craton. Geology 22 (10), 917–920.

Isachsen, C.E., Bowring, S.A., 1997. The Bell Lake Group and Anton Complex. A basement-cover sequence beneath the Archean Yellowknife greenstone belt revealed and implicated in greenstone belt formation. Canadian Journal of Earth Sciences 34 (2), 169–189.

Jelsma, H.A., Vinyu, M.L., Valbracht, P.J., Davies, G.R., Wijbrans, J.R., Verdurmen, E.A.T., 1996. Constraints on Archaean crustal evolution of the Zimbabwe Craton. A U–Pb zircon, Sm–Nd and Pb–Pb whole-rock isotope study. Contributions to Mineralogy and Petrology 124 (1), 55–70.

Jones, A.G., Ferguson, I.J., Chave, A.D., Evans, R.L., McNeice, G.W., 2001. Electric lithosphere of the Slave Craton. Geology 29 (5), 423–426.

Jordan, T.H., 1978. Composition and development of the continental tectosphere. Nature 274 (5671), 544–548.

Jordan, T.H., 1988. Structure and formation of the continental tectosphere. Journal of Petrology, Special Lithosphere Issue, 11–37.

Karlstrom, K.E., Ahäll, K.I., Harlan, S.S., Williams, M.L., McLelland, J., Geissman, J.W., 2001. Long-lived (1.8–1.0 Ga) convergent orogen in southern Laurentia, its extensions to Australia and Baltica, and implications for refining Rodinia. Precambrian Research 111 (1–4), 5–30.

Kopylova, M.G., Russell, J.K., 2000. Chemical stratification of cratonic lithosphere. Constraints from the northern Slave Craton, Canada. Earth and Planetary Science Letters 181 (1–2), 71–87.

Krogh, T.E., Davis, D.W., Corfu, F., 1984. Precise U–Pb zircon and baddeleyite ages for the Sudbury area. In: Pye, E.G., Naldrett, A.J., Giblin, P.E. (Eds.), The Geology and Ore Deposits of the Sudbury Structure. Ontario Geological Survey Special, vol. 1, pp. 431–446.

LeCheminant, A.N., van Breemen, O., 1994. U–Pb ages of Proterozoic dyke swarms, Lac de Gras area, N.W.T.: evidence for progressive break-up of an Archean supercontinent. Program with Abstracts, Geological Association of Canada, vol. 19, p. A-62.

LeCheminant, A.N., Buchan, A.N., van Breemen, O., Heaman, L.M., 1997. Paleoproterozoic continental break-up and reassembly: evidence from 2.19 Ga diabase dyke swarms in the Slave and western Churchill provinces. Abstract Volume, Geological Association of Canada, vol. 22, p. A-86.

McMenamin, M.A.A., McMenamin, D.L.S., 1990. The Emergence of Animals: The Cambrian Breakthrough. Columbia University, New York. 217 pp.

Moser, D.E., Flowers, R.M., Hart, R.J., 2001. Birth of the Kaapvaal tectosphere 3.08 billion years ago. Science 291, 465–468.

Mueller, P.A., Wooden, J.L., Nutman, A.P., 1992. 3.96 Ga zircons from an Archean quartzite, Beartooth Mountains, Montana. Geology 20 (4), 327–330.

Mueller, P.A., Wooden, J.L., Nutman, A.P., Mogk, D.W., 1998. Early Archean crust in the northern Wyoming Province: evidence from U–Pb ages of detrital zircons. Precambrian Research 91 (3–4), 295–307.

Myers, J.S., Shaw, R.D., Tyler, I.M., 1996. Tectonic evolution of Proterozoic Australia. Tectonics 15 (6), 1431–1446.

Nelson, D.R., 1998. Granite–greenstone crust formation on the Archaean Earth. A consequence of two superimposed processes. Earth and Planetary Science Letters 158 (3–4), 109–119.

Nelson, D.R., Trendall, A.F., Altermann, W., 1999. Chronological correlations between the Pilbara and Kaapvaal cratons. Precambrian Research 97, 165–189.

Nemchin, A.A., Pidgeon, R.T., 1998. Precise conventional and SHRIMP baddeleyite U–Pb age for the Binneringie Dyke, near Narrogin, Western Australia. Australian Journal of Earth Sciences 45 (5), 673–675.

Padgham, W.A., Fyson, W.K., 1992. The Slave Province. A distinct Archean craton. Canadian Journal of Earth Sciences 29 (10), 2072–2086.

Park, R.G., 1995. Palaeoproterozoic Laurentia–Baltica relationships: a view from the Lewisian. In: Coward, M.P., Ries, A.C. (Eds.), Early Precambrian Processes. Geological Society Special Publications, vol. 95, pp. 211–224.

Pehrsson, S.J., Villeneuve, M.E., 1999. Deposition and imbrication of a 2670–2629 Ma supracrustal sequence in the Indin Lake area, southwestern Slave Province, Canada. Canadian Journal of Earth Sciences 36, 1149–1168.

Pollack, H.N., 1997. Thermal characteristics of the Archaean. In: de Wit, M.J., Ashwal, L.D. (Eds.), Greenstone Belts. Oxford Monographs on Geology and Geophysics, vol. 35. Oxford University Press, Oxford, pp. 223–232.

Rainbird, R.H., Stern, R.A., Khudoley, A.K., Kropachev, A.P., Heaman, L.M., Sukhorukov, V.I., 1998. U–Pb geochronology of Riphean sandstone and gabbro from Southeast Siberia and its bearing on the Laurentia–Siberia connection. Earth and Planetary Science Letters 164 (3–4), 409–420.

Richardson, S.H., Gurney, J.J., Erlank, A.J., Harris, J.W., 1984. Origin of diamonds in old enriched mantle. Nature 310 (5974), 198–202.

Richardson, S.H., Shirey, S.B., Harris, J.W., Carlson, R.W., 2001. Archean subduction recorded by Re–Os isotopes in eclogitic sulfide inclusions in Kimberley diamonds. Earth and Planetary Science Letters 191 (3–4), 257–266.

Rogers, J.J.W., 1996. A history of continents in the past three billion years. The Journal of Geology 104, 91–107.

Roscoe, S.M., Card, K.D., 1993. The reappearance of the Huronian in Wyoming. Rifting and drifting of ancient continents. Canadian Journal of Earth Sciences 30 (12), 2475–2480.

Sircombe, K.N., Bleeker, W., Stern, R.A., 2001. Detrital zircon geochronology and grain-size analysis of ~ 2800 Ma Mesoarchean proto-cratonic cover succession, Slave Province, Canada. Earth and Planetary Science Letters 189 (3–4), 207–220.

Stern, R.A., Bleeker, W., 1998. Age of the world's oldest rocks refined using Canada's SHRIMP. The Acasta gneiss complex, Northwest Territories, Canada. Geoscience Canada 25 (1), 27–31.

Stockwell, C.H., 1961. Structural provinces, orogenies, and time classification of rocks of the Canadian Precambrian Shield. Age Determinations by the Geological Survey of Canada. Geological Survey of Canada Paper, vol. 61–17, pp. 108–118.

Stockwell, C.H., 1964. Fourth report on structural provinces, orogenies, and time-classification of rocks of the Canadian Precambrian Shield. Age Determinations and Geological

Studies. Geological Survey of Canada Paper, vol. 64-17, pp. 1–21. Part II.

Stockwell, C.H., 1972. Revised Precambrian time scale for the Canadian Shield. Geological Survey of Canada Paper, vol. 72-52. 4 pp.

Strik, G., Blake, T.S., Langereis, C.G., 2001. The Fortescue and Ventersdorp Groups: a paleomagnetic comparison of two cratons. In: Cassidy, K.F., Dunphy, J.M., Van Kranen-donk, M.J. (Eds.), 4th International Archaean Symposium, 24–28 September 2001, Extended Abstracts, AGSO-Geoscience Australia, Record 2001/37, Perth, Western Australia, pp. 532–533.

Tankard, A.J., Jackson, M.P.A., Eriksson, K.A., Hobday, D.K., Hunter, D.R., Minter, W.E.L., 1982. Crustal Evolution of Southern Africa. 3.8 Billion Years of Earth history. Springer-Verlag, New York. 523 pp.

Taylor, S.R., McLennan, S.M., 1985. The continental crust: its composition and evolution. Blackwell, Oxford. 312 pp.

Thorpe, R.I., Cumming, G.L., Mortensen, J.K., 1992. A significant Pb isotope boundary in the Slave Province and its probable relation to ancient basement in the western Slave Province. In: Richardson, D.G., Irving, M. (Eds.), Project Summaries, Cana-da-Northwest Territories Mineral Development Subsidiary Agreement 1987–1991, Open-File Report. Geological Survey of Canada, pp. 179–184.

Torsvik, T.H., Smethurst, M.A., Meert, J.G., Van, D.V.R., McKer-row, W.S., Brasier, M.D., Sturt, B.A., Walderhaug, H.J., 1996. Continental break-up and collision in the Neoproterozoic and Palaeozoic. A tale of Baltica and Laurentia. Earth-Science Reviews 40 (3–4), 229–258.

Trendall, A.F., Compston, W., Williams, I.S., Armstrong, R.A., Arndt, N.T., McNaughton, N.J., Nelson, D.R., Barley, M.E., Beukes, N.J., de Laeter, J.R., Retief, E.A., Thorne, A.M., 1990. Precise zircon U–Pb chronological comparison of the volcano-sedimentary sequences of the Kaapvaal and Pilbara cratons between about 3.1 and 2.4 Ga. In: Glover, J.E., Ho, S.E. (Compilers), Third International Archaean Symposium, Perth, 1990, Extended Abstracts Volume, pp. 81–84.

Trendall, A.F., de, L.J.R., Nelson, D.R., Mukhopadhyay, D., 1997. A precise zircon U–Pb age for the base of the BIF of the Mulaingiri Formation, (Bababudan Group, Dharwar Super-group) of the Karnataka Craton. Journal of the Geological Society of India 50 (2), 161–170.

Tyler, I.M., 1990. Mafic dyke swarms. Geology and mineral resources of Western AustraliaMemoir. vol. 3, Geological Survey of Western Australia, Perth, Australia, pp. 191–194.

van Breemen, O., Davis, W.J., King, J.E., 1992. Temporal distribution of granitoid plutonic rocks in the Archean Slave Province, Northwest Canadian Shield. Canadian Journal of Earth Sciences 29 (10), 2186–2199.

Williams, H., Hoffman, P.F., Lewry, J.F., Monger, J.W.H., Rivers, T., 1991. Anatomy of North America: thematic geologic portrayals of the continent. Tectonophysics 187 (1–3), 117–134.

Wilson, J.T., 1966. Did the Atlantic close and then re-open? Nature 211 (5050), 676–681.

Wilson, J.F., Nesbitt, R.W., Fanning, C.M., 1995. Zircon geochronology of Archaean felsic sequences in the Zimbabwe craton; a revision of greenstone stratigraphy and a model for crustal growth. Geological Society Special Publications 95, 109–126.

Wingate, M.T.D., 1998. A palaeomagnetic test of the Kaapvaal–Pilbara (Vaalbara) connection at 2.78 Ga. South African Journal of Geology 101 (4), 257–274.

Wingate, M.T.D., 1999. Ion microprobe baddeleyite and zircon ages for Late Archaean mafic dykes of the Pilbara Craton, Western Australia. Australian Journal of Earth Sciences 46, 493–500.

Wingate, M.T.D., 2000. Ion microprobe U–Pb zircon and baddeleyite ages for the Great Dyke and its satellite dykes, Zimbabwe. South African Journal of Geology 103 (1), 74–80.

Wooden, J.L., Mueller, P.A., 1988. Pb, Sr, and Nd isotopic compositions of a suite of late Archean, igneous rocks, eastern Beartooth Mountains. Implications for crust–mantle evolution. Earth and Planetary Science Letters 87 (1–2), 59–72.

Young, G.M., 1973. Huronian stratigraphy and sedimentation. Special Paper, Geological Association of Canada, vol. 12. 271 pp.

Young, G.M., Long, D.G.F., Fedo, C.M., Nesbitt, H.W., 2001. Paleoproterozoic Huronian basin: product of a Wilson cycle punctuated by glaciations and a meteorite impact. Sedimentary Geology 141–142, 233–254.

Zegers, T.E., de Wit, M.J., Dann, J., White, S.H., 1998. Vaalbara, earth's oldest assembled continent? A combined structural, geochronological, and palaeomagnetic test. Terra Nova 10, 250–259.

Zhao, G., Cawood, P.A., Wilde, S.A., Sun, M., 2002. Review of global 2.1–1.8 Ga orogens: implications for a pre-Rodinia supercontinent. Earth Science Reviews 59, 125–162.

Available online at www.sciencedirect.com

SCIENCE  DIRECT°

ELSEVIER

Lithos 71 (2003) 135–152

LITHOS

www.elsevier.com/locate/lithos

Cratonic mantle roots, remnants of a more chondritic Archean mantle?

Don Francis[*]

Earth and Planetary Sciences, McGill University, 3450 University Street, Montreal, Quebec, Canada H3A 2A7

Abstract

The Earth's continents are cored by Archean cratons underlain by seismically fast mantle roots descending to depths of 200^+ km that appear to be both more refractory and colder than the surrounding asthenospheric mantle. Low-temperature mantle xenoliths from kimberlite pipes indicate that the shallow parts of these cratonic mantle roots are dominated by refractory harzburgites that are very old (3^+ Ga). A fundamental mass balance problem arises, however, when attempts are made to relate Archean high-Mg lavas to a refractory restite equivalent to the refractory lithospheric mantle roots beneath Archean cratons. The majority of high-Mg Archean magmas are too low in Al and high in Si to leave behind a refractory residue with the composition of the harzburgite xenoliths that constitute the Archean mantle roots beneath continental cratons, if a Pyrolitic primitive mantle source is assumed. The problem is particularly acute for 3^+ Ga Al-depleted komatiites and the Si-rich harzburgites of the Kaapvaal and Slave cratons, but remains for cratonic harzburgites that are not anomalously rich in orthopyroxene and many Al-undepleted komatiites. This problem would disappear if fertile Archean mantle was richer in Fe and Si, more similar in composition to chondritic meteorites than the present Pyrolitic upper mantle of the Earth. Accepting the possibility that the Earth's convecting upper mantle has become poorer in Fe and Si over geologic time not only provides a simpler way of relating Archean high-Mg lavas to the lithospheric mantle roots that underlie Archean cratons, but could lead to new models for the nature Archean magmatism and the lower mantle sources of modern hot-spot volcanism.
© 2003 Elsevier B.V. All rights reserved.

Keywords: Archean; Craton; Lithosphere; Mantle; Root; Chondrite

1. Introduction

Jordan (1988) was the first to recognize that the Earth's continents are cored by Archean cratons underlain by seismically fast mantle roots descending to depths of 200^+ km that are both colder than the surrounding asthenospheric mantle and very old. He proposed that their long-term stability might reflect their relatively refractory chemical composition, which

renders them buoyant with respect to the asthenospheric mantle. Comparisons of mantle xenolith suites in kimberlites from Archean cratonic terranes with mantle xenoliths from off-craton alkaline basalts in Proterozoic or younger terranes have confirmed that cratonic mantle roots have higher Mg numbers [Mg/(Mg + Fe)] and are low in Al, which controls the abundance of garnet, the densest major phase in the lithospheric mantle (Boyd, 1989). The refractory nature of the lithospheric mantle roots beneath Archean cratons, however, raises new problems in terms of their relationship to komatiitic magmas and the nature of the

* Fax: +1-514-398-4680.

E-mail address: donf@eps.mcgill.ca (D. Francis).

Earth's fertile mantle in the Archean. This paper uses a comparison of the compositional differences between the Proterozoic lithospheric mantle beneath the Canadian Cordillera and the adjacent Archean mantle roots beneath the Slave and Churchill Provinces of the North American craton, as well as other Archean cratons on one hand and the compositional arrays of Archean and Tertiary high-Mg lavas on the other, to argue that a serious mass balance problem exists in the Earth whose resolution may be hindered by current Pyrolitic models for the Earth's primitive mantle. High-Mg Archean magmas, such as komatiites and ferropicrites, are too low in Al and high in Si to leave behind a refractory residue with the composition of the harzburgite xenoliths that constitute the mantle roots beneath Archean cratons, if a Pyrolitic primitive mantle source is assumed. This paper demonstrates that the problem would be resolved if early Archean fertile mantle had a composition closer to that of chondritic meteorites, as opposed to the Pyrolitic composition of the present upper mantle. This would require that the

chemical composition of the Earth's upper mantle has changed since the Hadean. Forsaking conventional geodynamic wisdom that the composition of the Earth's mantle has been in a steady state since the Hadean, and accepting the possibility that the Earth's convecting upper mantle has become poorer in Fe and Si since then, not only provides a simple way of relating Archean high-Mg lavas to Archean cratonic mantle roots, but would have important implications for mantle processes both in the Archean and today.

2. Data

2.1. Cordilleran versus cratonic mantle xenoliths

Mantle xenoliths from 16 recent off-craton alkaline volcanic suites along the Canadian Cordillera (Fig. 1) range from depleted harzburgites, similar to those along mid-ocean ridges, to fertile lherzolites (Figs. 2 and 3) that approach the estimated composition of

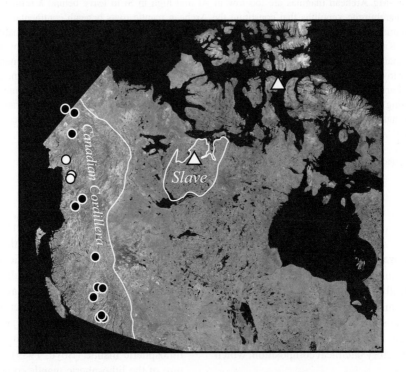

Fig. 1. Radar mosaic map of northwestern Canada showing the locations of mantle xenolith localities hosted by recent alkaline basalts along the Canadian Cordillera, the Jericho kimberlite pipe in the Slave Province, and the Nikos kimberlite pipe of Somerset Island. Symbols: bimodal Cordilleran xenolith suites—white circles, unimodal Cordilleran xenolith suites—black circles, Jericho pipe and Nikos pipes—white triangles.

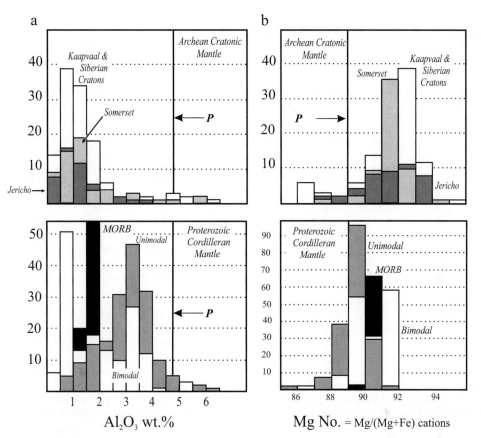

Fig. 2. Histograms of (a) Al_2O_3 content and (b) whole-rock Mg number (Mg/Mg + Fe) of kimberlite-hosted cratonic mantle xenolith suites from the Canadian shield, and other cratons, Proterozoic Canadian Cordilleran mantle xenolith suites hosted in recent alkaline basalts, and MORB peridotites dredged from the MARK region of the mid-Atlantic Ridge. Fields: MARK region peridotites—black, Kaapvaal and Siberian craton xenolith suites—white, Somerset Island xenolith suite—light gray, Jericho xenolith suite—dark gray, Cordilleran bimodal xenolith suites—white, Cordilleran unimodal xenolith suites—gray, Primitive mantle—P (data sources: Boyd, 1987; Boyd et al., 1997; Carswell et al., 1979; Casey, 1997; Chesley et al., 1999; Cox et al., 1987; Danchin, 1979; Francis, 1987; Kopylova and Russell, 2000; Lee and Rudnick, 1999; Nixon and Boyd, 1973; Schmidberger and Francis, 1999; Shi et al., 1998; Peslier et al., 2002).

primitive mantle (McDonough and Sun, 1995). Most of these mantle xenolith suites exhibit unimodal populations with a prominent mode corresponding to relatively fertile spinel lherzolite, with 3.3 wt.% Al_2O_3 and an Mg number of 0.893 (Table 1), which is interpreted to represent the prevalent lithospheric mantle beneath the Canadian Cordillera (Shi et al., 1998). Os model ages for these Cordilleran spinel lherzolite xenoliths (Peslier et al., 2000a,b, 2002) indicate that the lithospheric mantle along the western margin to the North American craton stabilized in the mid-Proterozoic (~ 1.5 Ga), in agreement with a recent compilation of Os isotopic results (Meisel et al., 2001) from a variety of off-craton basalt-hosted

mantle xenoliths that has demonstrated the Proterozoic age of lithospheric mantle peripheral to many of the Earth's continental cratons.

Three Cordilleran xenolith sites clustered near the Yukon–British Columbia border (Fig. 1) are characterized by bimodal populations (Fig. 2), with one mode corresponding to the fertile spinel lherzolite observed in the 13 other Cordilleran xenolith suites, while the other mode corresponds to relatively refractory spinel harzburgite (Shi et al., 1998), with lower Al contents (~ 0.8 wt.%) and higher Mg numbers (0.912) (Fig. 2 and Table 1). The spinel harzburgite xenoliths of both the bimodal and unimodal suites are more refractory in composition than harzburgites that dominate the

Table 1
Average composition and normative mineralogy of mantle xenoliths

	Kaapvaal	Siberia	Tanzania	Slave	Somerset	MORB	Cordillera harzburgite	Cordillera lherzolite	Primitive mantle
Number	55	19	29	30	16	79	70	158	–
Major elements in wt.% oxide:									
SiO_2	46.20	44.56	43.63	44.82	43.55	44.69	43.76	44.60	45.62
TiO_2	0.09	0.04	0.11	0.05	0.05	0.02	0.03	0.11	0.22
Al_2O_3	1.31	0.89	0.65	0.95	1.87	1.45	0.78	3.28	4.73
Cr_2O_3	0.31	0.00	0.00	0.42	0.51	0.42	0.41	0.40	0.37
MgO	44.03	46.14	47.51	45.30	44.45	43.98	45.86	39.53	36.37
FeO	6.49	7.23	7.36	6.95	7.61	8.42	7.92	8.46	8.18
MnO	0.11	0.13	0.13	0.11	0.12	0.13	0.13	0.14	0.14
NiO	0.28	0.00	0.00	0.30	0.33	0.32	0.34	0.28	0.24
CaO	0.93	0.82	0.43	0.78	1.32	0.51	0.64	2.95	3.75
Na_2O	0.11	0.07	0.07	0.18	0.10	0.04	0.10	0.22	0.35
K_2O	0.15	0.12	0.10	0.14	0.08	0.02	0.02	0.05	0.03
LOI	2.81	11.32	0.00	2.75	3.73	11.72	0.21	0.21	0.10
Mg number	0.924	0.919	0.920	0.921	0.912	0.903	0.912	0.893	0.888
Calculated mode in the garnet stability field in oxygen units:									
Cpx	3.0	3.1	1.6	3.9	4.8	0.0	1.9	12.7	16.4
Opx	29.5	18.7	14.5	20.6	12.9	24.3	16.3	14.8	19.2
Oliv	62.4	74.7	81.3	71.7	74.0	68.5	77.5	59.8	47.3
Garn	5.2	3.4	2.7	3.9	8.3	7.2	4.4	12.8	17.2

Average composition and normative mineralogy of low-temperature mantle xenoliths in kimberlites from the Kaapvaal, Siberian, Tanzanian, and Slave cratons, as well as those from Somerset Island, MORB peridotites dredged from the MARK region of the mid-Atlantic ridge, the harzburgite, and lherzolite modes of Canadian Cordilleran mantle xenolith suites, and a recent estimate of the composition of primitive mantle composition (O'Neill and Palme, 1998). The modes were calculated in the garnet stability field at a temperature of 1200 °C (data sources as in Fig. 3).

dredge hauls from the MARK region (Casey, 1997) of the mid-Atlantic ridge ($Al_2O_3 \sim 1.5$ wt.% and Mg number ~ 0.90). The array from fertile lherzolite to depleted harzburgite that characterizes all the Cordilleran mantle xenolith suites, and the majority of the Earth's off-craton mantle xenolith suites hosted by alkaline basalts, is equivalent to the "oceanic" trend of Boyd (1989) and is well modeled (Fig. 3) as a series of residues produced by progressive melting (0–25%) of a Pyrolitic primitive mantle source (Francis, 1987; Shi et al., 1998). The anomalous abundance of harzburgites in the three xenolith suites near the Yukon–northern British Columbia border overlie a teleseismic S wave slowness anomaly (Frederiksen et al., 1998) in the underlying asthenospheric mantle, and has been proposed to reflect melting due the ingress of volatiles and heat into the overlying lithospheric mantle (Shi et al., 1998).

Peridotite mantle xenolith suites from the Jericho pipe (Fig. 1) in the northern Slave Province (Kopy-

lova and Russell, 2000) are dominated by garnet harzburgites with low Al contents (~ 1 wt.% Al_2O_3), similar to those of Cordilleran harzburgites (Fig. 2 and Table 1), but with distinctly higher Mg numbers (0.92), and olivine forsterite (Fo) contents, as do the garnet harzburgites in kimberlites from other Archean cratons, such as those of the Kaapvaal, Siberian, and Tanzanian cratons. The Nikos kimberlite xenoliths from Somerset Island (Schmidberger and Francis, 1999) are also dominated by garnet harzburgites, but they are relatively less refractory ($Al_2O_3 \sim 1.9$ wt.%, Mg number ~ 0.91) than those of the Jericho pipe in the Slave Province and the harzburgites of the adjacent Canadian Cordillera. The relatively more refractory nature of cratonic mantle xenoliths is particularly characteristic of the low-temperature mantle xenoliths, which preserve equilibration pressures that are shallower than the "kink" that commonly characterizes cratonic paleogeotherms defined by kimberlite-hosted mantle xenoliths, and

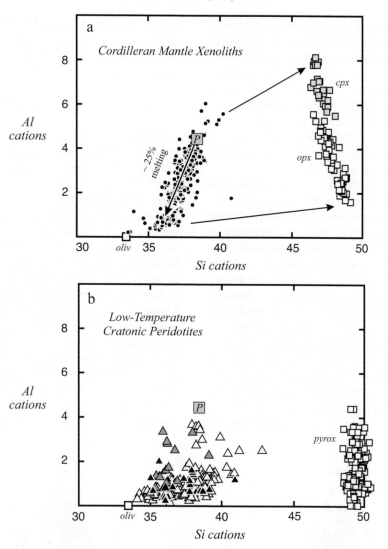

Fig. 3. Al versus Si in cation units showing the whole-rock compositions of (a) Canadian Cordilleran mantle xenoliths and (b) low-temperature cratonic mantle xenoliths in kimberlites. Symbols: Cordilleran mantle xenoliths—black circles; low-temperature cratonic mantle xenoliths: Jericho—black triangles, Somerset Island—gray triangles, Kaapvaal, Siberian, and Tanzanian cratons—open triangles (data sources as in Fig. 2).

thus sample the shallowest upper mantle beneath the cratons. Rather than defining a trend towards a Pyrolitic primitive mantle in Al–Si space, low-temperature cratonic harzburgites trend from those with Si contents similar to the spinel harzburgites associated with mid-ocean ridges and those of the Canadian Cordillera towards anomalously Si-rich compositions (Fig. 3). A few harzburgites dredged from mid-ocean ridges do have high Si contents, but they are characterized by low Mg numbers (< 0.89) and are probably cumu-

lates. In contrast, all low temperature cratonic harzburgites, regardless of Si content, have higher Mg numbers than harzburgites with similarly low Al contents from the Canadian Cordillera or mid-ocean ridges.

Richardson (Richardson et al., 1984) first demonstrated that garnet inclusions in South African diamonds yielded a 3.5-Ga Nd isochron, indicating an Archean age for the Kaapvaal craton mantle roots. More recently, a number of Os isotopic studies have demonstrated the likely 3^+-Ga age of low-temperature garnet peridotites in kimberlites from the Kaapvaal, Siberian, and Slave cratons (Chesley et al., 1999; Irvine et al., 1999; Pearson et al., 1995a,b), and have established the antiquity of the shallow portions of cratonic mantle roots in general. Most recently, a 2.8-Ga whole-rock Lu–Hf isochron obtained on low-temperature garnet harzburgites from the Nikos pipe has confirmed the validity of the Archean Os-depletion ages (Schmidberger et al., 2002).

The contrast between Proterozoic Cordilleran mantle xenoliths and Archean low-temperature harzburgites from cratonic kimberlites is particularly striking when their normative mineralogy is calculated at similar conditions of pressure and temperature (Fig. 4). While the Proterozoic Cordilleran mantle xenoliths range from depleted harzburgites to fertile lherzolites similar to primitive mantle in composition, Archean low-temperature cratonic mantle xenoliths trend from depleted harzburgites towards increasing orthopyroxene, with little increase in clinopyroxene or garnet, and no evident trend towards the composition of primitive mantle (Fig. 4). This trend towards increasing orthopyroxene is most clearly developed in the low temperature harzburgites of the Kaapvaal, Siberia, and Slave cratons, whereas low temperature harzburgites of Somerset Island and the Tanzanian and Greenland cratons have relatively low Si contents, similar to the harzburgites of the Canadian Cordillera or mid-ocean ridges.

2.2. Tertiary versus Archean high-Mg lavas

In addition to the generally higher Mg contents of Archean komatiites (Arndt et al., 1997; Herzberg and O'Hara, 1997), there are systematic differences between Phanerozoic and Archean high-Mg lavas. The array of Archean high-Mg lavas, including both komatiites and ferropicrites, appears to be shifted to higher Fe contents and lower Al/Si ratios than Tertiary high-Mg lavas (Figs. 5 and 6). Some would attribute these differences solely to higher mantle temperatures in the Archean, and thus higher pressures and degrees of partial melting that yielded liquids with both higher Mg and Fe contents compared to Tertiary picrites. This explanation does not, however, account for the systematic mismatch between Fe and incompatible minor or trace elements (e.g. Ti and Zr) when comparing Archean and Tertiary high-Mg lavas. For example, although Archean Al-undepleted komatiites have Fe contents and Al/Si ratios similar to those of modern Hawaiian picrites, their low Ti and Zr contents are more

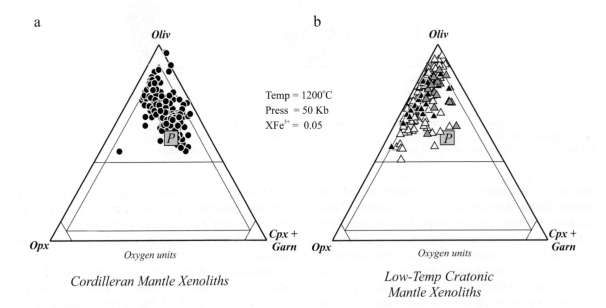

Fig. 4. Calculated modes in the garnet stability field for (a) Cordilleran mantle xenoliths and (b) low temperature cratonic mantle xenoliths (symbols and data sources as in Fig. 3).

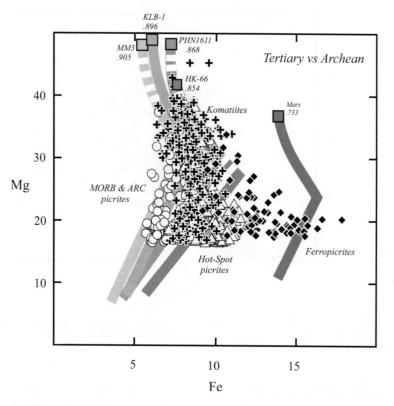

Fig. 5. Mg versus Fe in cation units for high-Mg (12$^+$ wt.% MgO) Tertiary and Archean lavas along with the spectra of experimental melts produced from a variety of model mantle compositions. Symbols: Tertiary MORB and Arc picrites—open circles, Tertiary hot-spot picrites—open triangles, Archean komatiites—black crosses, Archean ferropicrites—black diamonds, model mantle compositions MM3, KLB-1, PHN1611, HK-66, and Mars—shaded squares, with their corresponding spectra of experimental partial melts indicated by the shaded paths (data sources: Berka and Holloway, 1994; Francis et al., 1999; Hirose and Kushiro, 1993; Kushiro, 1998; Takahashi, 1986; Takahashi et al., 1994; Walter, 1998).

typical of relatively Fe-poor MORB picrites (Fig. 7a,b). In contrast, Archean ferropicrites that are similar to Hawaiian picrites and mildly alkaline ankaramites in terms of Ti and Zr have systematically higher Fe (Fig. 7a,b) and lower Al/Si ratios (Fig. 6) than both Tertiary hotspot picrites and rarer Phanerozoic ferropicrites (Gibson et al., 2000). Furthermore, although Al-depleted komatiites have been documented together with Al-undepleted komatiites in the late Archean (Cattell and Arndt, 1987), there appears to be a general temporal evolution in the composition of komatiitic lavas, from the Al-depleted (Al/Ti ~ 10) komatiites that characterize 3$^+$ Ga greenstone belts to Al-undepleted (Al/Ti ~ 20) komatiites that characterize circa 2.6–2.9 Ga greenstone belts to the Al-rich Mesozoic komatiites of Gorgona Island, which exhibit Al–Si systematics similar to those of modern MORB (Fig. 8a).

3. Discussion

A fundamental mass balance problem arises when attempts are made to relate high-Mg Archean lavas to a refractory restite equivalent to the low-temperature garnet harzburgites of lithospheric mantle roots beneath Archean cratons (Herzberg, 1993). If a Pyrolitic primitive mantle source is assumed, the majority of high-Mg Archean magmas are too low in Al and high in Si to leave behind a refractory residue with the composition of the harzburgite xenoliths that constitute the Archean mantle roots beneath continental cratons (Fig. 6). This problem is particularly acute for 3$^+$ Ga Al-depleted komatiites and the Si-rich harzburgites of the Kaapvaal, Siberian, and Slave cratons. A number of proposals have been made to explain the orthopyroxene-rich nature

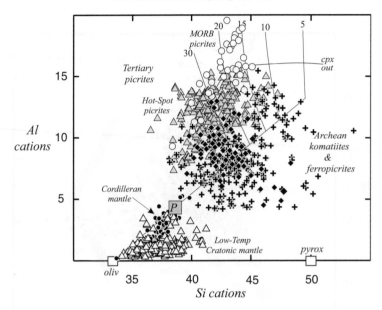

Fig. 6. Al versus Si in cation units for high-Mg (12$^+$ wt.% MgO) Tertiary and Archean lavas, along with Cordilleran and low-temperature cratonic mantle xenoliths, and a grid indicating the pressures in kilobars of experimental partial melts of primitive mantle. Symbols as in Figs. 3 and 5, except hot-spot picrites—gray triangles (data sources as in Figs. 3 and 5).

of some low-temperature harzburgite xenoliths in cratonic kimberlites:

(1) Prograde metamorphism of a residual MORB peridotite protolith that had been relatively enriched in silica during serpentinization of the oceanic lithosphere by the preferential leaching of other cations such as Ca (Helmstaedt and Schulze, 1989; Schulze, 1986). Actual analyses of serpentinized MORB peridotites with Mg numbers greater than 0.89, however, are not anomalous in Si (Fig. 8b), but resemble those of mantle harzburgites found in alkaline basalts.
(2) Floatation of residual harzburgite on top of komatiitic melts followed by metamorphic differentiation into orthopyroxene-rich and orthopyroxene-poor layers and cooling (Boyd, 1989).
(3) Tectonic mixing of orthopyroxene-rich cumulates derived from komatiitic melt or a high Si terrestrial magma ocean (Boyd, 1989; Herzberg, 1993).
(4) Reaction of residual harzburgite with fluids or siliceous melts rising from underlying subduction zones, as a result of which olivine is converted to orthopyroxene (Kelemen et al., 1999).

Although each of these proposals has some merit, with the exception of Herzberg (1993), most have interpreted the origin of the orthopyroxene enrichment in terms of the action of a secondary process, following the generation of komatiitic magmas from a Pyrolitic mantle source. The unstated assumption is that the Earth's asthenospheric mantle has been in a compositional steady state since the Hadean, with the composition of fertile mantle having always been approximately that of primitive mantle Pyrolite (McDonough and Sun, 1995; O'Neill and Palme, 1998). With the recent recognition that cratonic mantle xenoliths from Tanzania (Chesley et al., 1999), Greenland (Bernstein et al., 1998), and Somerset Island (Schmidberger and Francis, 1999) are not enriched in orthopyroxene compared to oceanic lithospheric mantle, interest in the Si-enrichment of cratonic mantle roots has waned. It is important to realize, however, that the mass balance problem in relating komatiites to cratonic mantle roots persists even for Archean mantle restites with the relatively Si-poor compositions of younger harzburgites in alkaline basalt suites and beneath mid-ocean ridges. The derivation of liquid compositions ranging from Al-depleted to Al-undepleted Archean komatiites from a

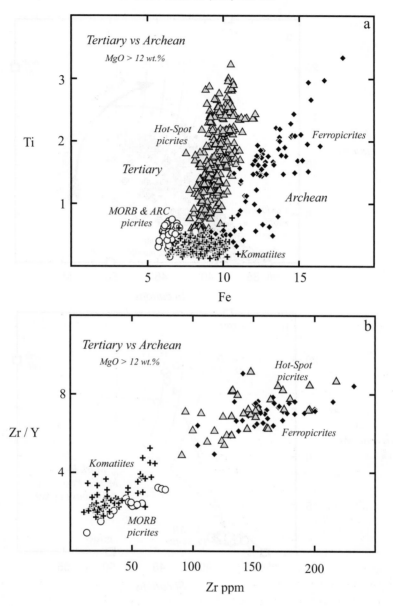

Fig. 7. (a) Fe versus Ti in cation units and (b) Zr versus Zr/Y in ppm for high-Mg (12+ wt.% MgO) Tertiary and Archean lavas. Symbols as in Fig. 6 (data sources as in Fig. 5).

primitive mantle source would leave refractory residues that range in composition from garnetiferrous dunite to dunite (Fig. 6). Neither of these two lithologies are well represented in cratonic mantle xenolith populations. Only komatiites with Al contents significantly above the extension of line joining olivine to primitive mantle in a plot of Al versus Si (Fig. 8) are capable of leaving a refractory residue with significant orthopyroxene. The fact that a number of experimenters have obtained komatiitic partial melts from model Pyrolitic primitive mantle sources (Walter, 1998) does not invalidate the foregoing argument. The phase rule requires that the partial melt of any peridotite will have a similar major element composition at similar

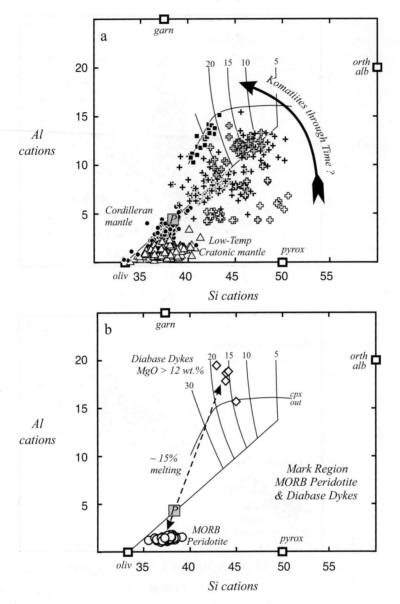

Fig. 8. Al versus Si in cation units for (a) komatiites through time and (b) dredged peridotites and picritic dykes that cut them from the MARK region of the mid-Atlantic ridge. Symbols: 3⁺ Ga komatiites—open crosses, 2.6–2.9 Ga komatiites—black crosses, 1.0–2.5 Ga komatiites—gray crosses, Mesozoic Gorgona komatiites—black squares, MARK region MORB peridotites—open circles, and picritic dykes—open diamonds. Other symbols as in Fig. 3 (data sources: Casey, 1997, others as in Fig. 5).

P–T conditions, it is rather the composition of the refractory restite that is sensitive to the original source composition. Furthermore, experimental melts of primitive mantle compositions that have pyroxene in their coexisting refractory restite have relatively high Al/Si ratios, and pyroxene is absent in the restites of experimental melts that approach the lower Al/Si ratios typical of most Archean komatiites (Walter, 1998). A more complex explanation for the low Al/Si ratios of Al-depleted komatiites involving a derivation from a mantle source that had been depleted with respect to primitive mantle by a previous melting event(s) makes their relatively high Fe contents even more difficult to rationalize.

The concept of a primitive mantle composition has its underpinnings in the approach of Pyrolite models for mantle source regions of basaltic magmas to the intersection of the compositional array of basalt-hosted mantle xenoliths and the extrapolation of the array of chondritic meteorites in Al/Si, Ca/Si, and Mg/Si spaces (Fig. 9) (Jagoutz et al., 1979). The presently accepted major element composition of fertile Pyrolite mantle (McDonough and Sun, 1995) is essentially that of the most Al-rich spinel lherzolite xenoliths hosted by off-craton alkaline basalts. Estimates for the major element composition of Earth's primitive mantle are essentially indistinguishable from fertile Pyrolite, in that they are obtained by a small correction ($\sim 1\%$) for what is thought to have been extracted to form the continental crust (O'Neill and Palme, 1998). Pyrolitic mantle sources appear to work well in Phanerozoic volcanic suites, as can be seen in Fig. 8b where a

Pyrolite source can relate picritic MORB dykes to the residual MORB harzburgite they cut by $\sim 15\%$ partial melting (Casey, 1997). Archean komatiites that could coexist with olivine of Fo 92.5–94 composition, however, cannot be linked to the harzburgites of cratonic mantle roots with a line that passes through the composition of primitive mantle, but such a join does pass through the compositions of chondritic meteorites in Mg/Si versus Al/Si space (Fig. 9). This suggests that Archean komatiitic magmas could be simply related to the lithospheric mantle roots beneath Archean cratons if they were derived from mantle sources that were more chondritic in terms of Mg/Si than the presently accepted composition for fertile Pyrolitic mantle. This interpretation would require that the Earth's fertile convecting upper mantle has evolved since the Hadean to its present Pyrolite composition. Further support for such a possibility can be found in the compositions of chondritic meteorites. The intersection between the compositional

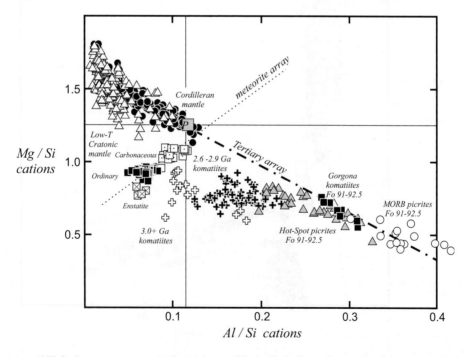

Fig. 9. Mg/Si versus Al/Si for low-temperature cratonic mantle xenoliths in kimberlites, off-craton basalt-hosted mantle xenoliths from the Canadian Cordillera, along with the compositions of chondritic meteorites, high-Mg MORB picrites, and Tertiary hot-spot picrites that would coexist with an olivine of Fo 91–92.5 composition, and komatiites that would coexist with an olivine of Fo 92.5–94 composition. Symbols: carbonaceous chondrites—squares with dots, ordinary chondrites—black squares, enstatite chondrites—crossed squares. Other symbols as in Figs. 3, 6, and 8 (data sources: Jarosewich, 1990, others as in Figs. 3 and 5).

arrays of chondritic meteorites and mantle xenoliths is also evident a simple plot of Fe versus Mg (Fig. 10), in which the silicate portions of the chondritic meteorites form a linear array trending directing away from Fe. Here, however, the point of intersection with the mantle array is further from the composition of primitive mantle than in Al/Si or Ca/Si versus Mg/Si space. When corrected to an Mg number (~ 0.86) that would make them approximately collinear with the compositions of Archean komatiites that would coexist with olivine of Fo 92.5–94 composition and the compositions of low-temperature cratonic mantle xenoliths in Mg–Fe space (Fig. 10) by mathematically

removing additional FeO, both the carbonaceous and the ordinary chondrite meteorites fall to more Si-rich compositions than primitive mantle, and would provide much more suitable sources with which to link komatiitic magmas and refractory cratonic mantle xenoliths in Si–Al space (Fig. 11). A mantle with a composition of the Mg number 86-normalized silicate portions of the chondritic meteorites would melt at the same pseudo-invariant point as a Pyrolitic mantle, with the initial liquid coexisting with olivine, orthopyroxene, clinopyroxene, and a pressure-dependent aluminous phase, but at a slightly lower solidus temperature (~ 20°) because of its lower Mg number

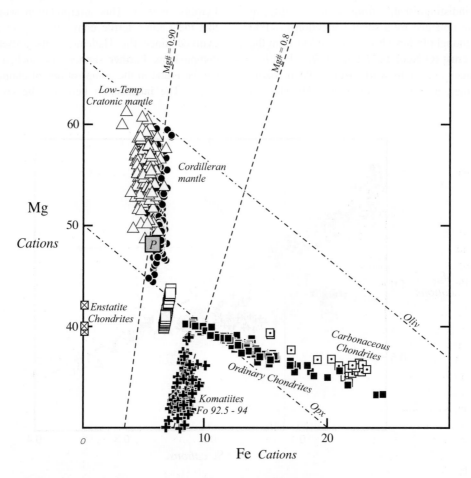

Fig. 10. Mg versus Fe in cation units for low-temperature cratonic mantle xenoliths in kimberlites, off-craton basalt-hosted mantle xenoliths from the Canadian Cordillera, the silicate portions of chondritic meteorites, the silicate portions of chondritic meteorites normalized to an Mg number of 0.86 by the removal of FeO, and komatiites that would coexist with an olivine of Fo 92.5–94 composition. Symbols: Mg number 0.86 normalized meteorites—open squares, komatiites that would coexist with an olivine of Fo 92.5–93 composition—black crosses. Other symbols as in Figs. 3 and 8 (data sources as in Figs. 3, 5, and 9).

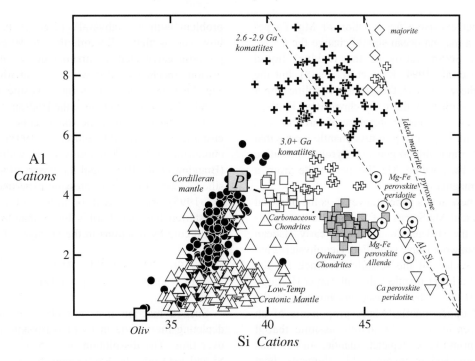

Fig. 11. Al versus Si in cation units for low-temperature cratonic mantle xenoliths in kimberlites, off-craton basalt-hosted mantle xenoliths from the Canadian Cordillera, chondritic meteorites normalized to an Mg number of 0.86, as well as the compositions of experimentally determined higher pressure phases in mantle peridotite. Symbols: carbonaceous chondrites—open squares, ordinary chondrites—shaded squares, Mg–Fe perovskite—dotted circles, Ca perovskite—inverted triangles, majorite—open diamonds. Other symbols as in Figs. 3 and 9 (data sources: Collerson et al., 2000; Wood, 2000, others as in Figs. 3, 5, and 9).

(Hirschmann, 2000). The removal of this pseudo-invariant melt will, however, leave residues with higher orthopyroxene contents than off-cratonic harzburgites, and the extent of melting at any given temperature will be greater than that for a Pyrolitic source, leading to higher Mg numbers in the residue. These are exactly the characteristics of many low-temperature cratonic mantle xenoliths.

The high Mg/Si ratio of the Earth's upper mantle compared to chondritic meteorites has been a persistent cosmochemical enigma (O'Neill and Palme, 1998). Either bulk silicate Earth is not chondritic or the composition of the Earth's mantle has changed, and a Si-rich reservoir must exist within the Earth to balance the observed high Mg/Si ratio of the Earth's upper mantle. The two most likely candidates for such a reservoir are the recently hypothesized "abyssal layer" (Kellogg et al., 1999) in the lower mantle and the Earth's liquid outer core. The approximate alignment of primitive mantle, chondritic meteorites,

and mantle perovskites suggests that the fractionation of Mg–Fe perovskite might be a mechanism to produce the proposed change in the composition of the mantle. The lever rule in Al–Si space is consistent with modern Pyrolite mantle being derived by ~ 20–30% fractionation of Mg–Fe perovskite from the composition of carbonaceous chondrites (Fig. 11), after the loss of metal, sulfide, and/or magnesiowustite to raise their Mg number to 0.86.

The fact that experimental perovskites have higher Mg number than their peridotite hosts (Wood, 2000), however, would require that magnesiowustite segregation accompany perovskite fractionation in order to prevent a decrease in Mg number. Numerous authors have discussed the possibility of perovskite fractionation from a magma ocean in the early Hadean, shortly after accretion, to explain the high Mg/Si ratio of the Earth's present upper mantle. This explanation has run into difficulty because experimental data on perovskite/melt element partitioning suggests that extensive fractionation of Mg–Fe perovskite in a magma ocean should change the Ca/Al, Al/Ti, and Sc/Sm ratios of the residual liquid (McFarlane et al., 1994; Kato et al., 1988), but the values of these ratios in the present Pyrolitic upper mantle are similar to those of chondrites. Furthermore, the lower melting temperature of carbonaceous chondrite with respect to primitive mantle means that it would be impossible to generate Pyrolitic primitive mantle by fractionating Mg–Fe perovskite from a chondritic magma ocean.

What is new in this paper, however, is the proposal that the present Mg/Si ratio of the mantle has developed since the Hadean, long after an early Hadean magma ocean would have solidified. Geophysicists are presently divided on the seismic evidence for the existence of a lower mantle "abyssal" layer (Kellogg et al., 1999) that is both denser and hotter than the convecting upper mantle. Many proponents of an abyssal layer in the lower mantle assume that it represents primitive undepleted mantle, and is thus relatively enriched in incompatible elements, heat producing elements, and primordial noble gases (Kellogg et al., 1999) compared to the asthenospheric upper mantle we observe today. Could the abyssal layer, however, represent an Mg–Fe perovskite enriched separate removed from the upper mantle to produce the latter's present Pyrolitic composition? The similarity of komatiitic Ca/Al ratios, as well as ratios of refractory lithophile incompatible elements in general, to those of chondrites and modern picrites, indicates that the Ca- and Al-rich perovskites and majorites that have been reported (Collerson et al., 2000; Wood, 2000) have not been preferentially removed from the upper mantle. If it is assumed that the higher Ca and Al contents of the present Pyrolite upper mantle reflect the preferential loss of Mg–Fe perovskite to a dense "abyssal" layer in the lower mantle (assuming a bulk silicate Earth that resembles carbonaceous chondrites), then the minimum mass of the lower layer must also be on the order of 20%, assuming it contained little Ca or Al. Given the low incompatible trace element contents of Mg–Fe perovskites compared to Ca–Al perovskites, the general absolute enrichment in incompatible elements in Pyrolite over carbonaceous chondrites (~ 1.2) (McDonough and Sun, 1995) is also consistent with a mantle reservoir poor in incompatible trace elements constituting 20% of the mantle. The major

problem with an "abyssal" layer reservoir model, however, is that following the solidification of a possible early Hadean magma ocean, there is no obvious mechanism that will preferentially sequester Mg–Fe perovskite and magnesiowustite in a lower mantle "abyssal" layer since the Hadean.

The presence of metallic silicon in highly reducing enstatite chondrites (O'Neill et al., 1998), the suprachondritic $^{186}Os/^{188}Os$ ratios of Hawaiian picrites (Brandon et al., 1998), and the observation of metal-silicate reactions in high pressure experiments (Ito et al., 1995) suggest that both Si and Fe may be transferred between Earth's lower mantle and outer liquid core by reactions in the D'' layer of the type:

$$(Mg_x, Fe_{1-x})SiO_{3Pv} + 3(1-x)Fe_{metal} \rightarrow xMgSiO_{3Pv}$$
$$+ (1-x)SiFe_{3metal} + (1-x)FeO_{metal}$$

Such a reaction may provide a mechanism for depleting the mantle in its Fe-perovskite component over time. The dissolution of ~ 5–6 wt.% each of Si and FeO into the Earth's core by such a reaction mechanism in the D'' layer could have produced the difference between the Mg/Si ratios of the chondritic meteorites and Pyrolitic primitive mantle (O'Neill et al., 1998) and the increase in the mantle Mg number since the Hadean (Francis et al., 1999). The fact that the present upper mantle has relatively high and unfractionated highly siderophile trace element contents (e.g. platinum group elements) does not invalidate this possibility because current estimates for the partitioning of siderophile elements into FeO (Ohtani et al., 1997) indicate that the loss of FeO to the core might have a negligible effect on the siderophile element abundances of the remaining mantle.

Perhaps the best estimate for the starting composition of the Earth's silicate mantle is obtained after the manner of Larimer and Anders (1970) (O'Neill and Palme, 1998), in which the co-variation of Ni/Mg with Fe/Mg in the chondritic meteorites can be interpreted to indicate that bulk silicate Earth had an Mg number of ~ 0.80 at the end of the Hadean (Fig. 12), a value that corresponds closely with the lowest Mg numbers of the silicate portions of ordinary chondrites (Fig. 10). The compositions of Archean komatiites suggest that the Mg number of the fertile mantle had increased to ~ 0.86 by the early Archean,

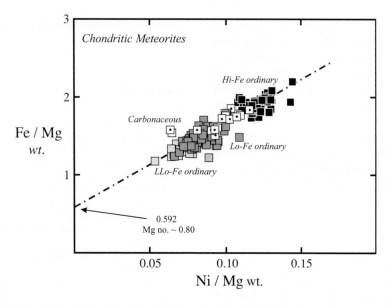

Fig. 12. Fe/Mg versus Ni/Mg in weight units for chondritic meteorites. Symbols: high-Fe chondrites—black squares, low-Fe chondrites—gray squares, very low-Fe chondrites—open squares, carbonaceous chondrites—dotted squares (data source as in Fig. 9).

but since then there appears to have been a further decrease in Si and increase in Al, Ca, and Mg numbers to the present Pyrolitic mantle values (Mg number ~ 0.89). The increase in the Al content of komatiitic magmas from the early Archean through to the Mesozoic (Fig. 8b) would, in this scenario, reflect the increase in Al and decrease in Si of the fertile upper mantle caused by the preferential loss of Fe perovskite and FeO to the core. The lever rule would require that Al-depleted 3^+ Ga komatiites represent very large degrees of partial melting (70–80%), which are more easily reconciled with magma ocean-type models, rather than mantle plume models (Fig. 11). The estimated degree of partial melting in Al–Si space falls to ~ 50% for Al-undepleted komatiites of the late Archean and 30–40% for the Mesozoic komatiites of Gorgona (Fig. 11).

Although the exact mechanism(s) that has produced the proposed decrease in the Fe and Si contents of fertile mantle since the Hadean remains to be established, one of the most intriguing aspects of this proposal lies in its implications for the origin of the distinctive chemical character of the picritic magmas of ocean island basalt (OIB) suites, which are enriched in Fe and Si, and depleted in Al compared to MORB picrites (Fig. 6) (Francis, 1995). If MORB and Ha-

waiian primitive magmas were derived from the same mantle source, then the lower Al content of Hawaiian picrites would require that they represent a greater degree of partial melting. Yet, the fractionated trace element profiles of Hawaiian picrites suggest smaller degrees of partial melting of the source that gave rise to MORB, leaving a significant proportion of garnet in the residue. This paradox, combined with the absence of coexisting olivine and garnet on the liquidus of Hawaiian picrites at any pressure (Eggins, 1992), requires that the major element composition of Hawaiian mantle source(s) differs from that of the convecting upper mantle that produces MORB. The enrichments in Fe and Si contents that characterize Hawaiian, and OIB picrites in general, mimic the characteristics of Archean high-Mg magmas, and these are exactly the two elements that the proposed model requires to be sequestered in the lower mantle or outer core.

The most fertile mantle xenoliths in off-craton alkaline volcanic suites, those that constrain the presently accepted composition of fertile Pyrolite mantle, have Os isotopic compositions that are similar to those of the majority of modern MORB samples that have appreciable Os contents ($^{187}Os/^{188}Os \sim 0.13$) (Meisel et al., 2001). This, in combination with their common light rare-earth depleted character (Roden et al., 1984) and the complementarity of MORB picrites and the array of Proterozoic mantle xenoliths, suggests that fertile lherzolite xenoliths may represent actual fragments of the convecting upper mantle source for MORB, trapped and preserved along the margins of continental cratons. The consistent Proterozoic model ages obtained for such fertile lherzolite xenoliths would imply that the Pyrolitic convecting upper mantle that produces modern MORB is a Proterozoic or younger feature of the Earth.

4. Conclusions

The mass balance problem that results when Archean komatiitic magmas are related to cratonic mantle roots may reflect current assumptions about the composition of fertile mantle sources in the Archean. The low Al/Si ratios of komatiites may indicate a more chondritic composition for the Earth's early Archean mantle. The presently accepted major element composition of primitive mantle is based on Proterozoic mantle xenoliths. The compatible major element, trace element, and Os isotopic compositions of fertile Proterozoic mantle xenoliths and MORB suggest that the present Pyrolitic convecting upper mantle that produces MORB may be a Proterozoic or younger feature of the Earth. According to this view, the depleted mantle roots beneath continental cratons would represent the refractory relicts of a Si- and Fe-rich early Archean mantle. Forsaking conventional geodynamic wisdom that the Earth has been in a compositional steady state since the Hadean, and accepting the possibility that the Earth's convecting upper mantle has become poorer in Fe and Si over geologic time, not only provides a simpler way of relating Archean high-Mg lavas to the lithospheric mantle roots that underlie Archean cratons, but could also lead to new models for the nature Archean magmatism and the lower mantle sources of modern hot-spot volcanism.

Acknowledgements

I am indebted to the excellent critical reviews of Claude Herzberg, Bill McDonough, and Bill White that have forced me to clarify my thoughts, as well as numerous discussions with colleagues and graduates students.

References

Arndt, N.T., Albarède, F., Nisbet, E.G., 1997. Mafic and ultramafic magmatism. In: de Wit, M.J., Ashwal, L.D. (Eds.), Greenstone Belts. Clarendon Press, Oxford, pp. 233–264.

Berka, C.M., Holloway, J.R., 1994. Anhydrous partial melting of an iron-rich mantle II: primary melt compositions at 15 kbars. Contrib. Mineral. Petrol. 115, 323–338.

Bernstein, S., Kelemen, R.B., Brooks, C.K., 1998. Depleted spinel harzburgite xenoliths in Tertiary dykes from East Greenland: restites from high degree melting. Earth Planet. Sci. Lett. 154, 221–235.

Boyd, F.R., 1987. High- and low-temperature garnet peridotite xenoliths and their possible relation to the lithosphere–asthenosphere boundary beneath southern Africa. In: Nixon, P.H. (Ed.), Mantle Xenoliths. Wiley, Chichester, pp. 403–412.

Boyd, F.R., 1989. Compositional distinction between oceanic and cratonic lithosphere. Earth Planet. Sci. Lett. 96, 15–26.

Boyd, F.R., Pokhilenko, N.P., Pearson, D.G., Mertzman, S.A., Sobolev, N.V., Finger, L.W., 1997. Composition of the Siberian cratonic mantle: evidence from Udachnaya peridotite xenoliths. Contrib. Mineral. Petrol. 128, 228–246.

Brandon, A.D., Walker, R.J., Morgan, J.W., Norman, M.D., Prichard, H.M., 1998. Coupled ^{186}Os and ^{187}Os evidence for core–mantle interaction. Science 280, 1570–1573.

Carswell, D.A., Clarke, D.B., Mitchell, R.H., 1979. The petrology and geochemistry of ultramafic nodules from Pipe 200, northern Lesotho. In: Boyd, F.R., Meyers, H.O.A. (Eds.), Proc. Sec. Inter. Kimberlite Conf. 2. AGU, Washington, DC, pp. 127–144.

Casey, J.F., 1997. Comparison of major and trace element geochemistry of abyssal peridotites and mafic plutonic rocks with basalts from the MARK region of the mid-Atlantic Ridge. Proc. Ocean Drill. Prog., Sci. Results 153, 181–240.

Cattell, A., Arndt, N., 1987. Low- and High-alumina komatiites from a late Archean sequence, Newton Township, Ontario. Contrib. Mineral. Petrol. 97, 218–227.

Chesley, J.T., Rudnick, R.L., Lee, C.T., 1999. Re–Os systematics of mantle xenoliths from the East African Rift: age, structure, and history of the Tanzanian craton. Geochim. Cosmochim. Acta 63, 1203–1217.

Collerson, K.D., Hapugoda, S., Kamber, B.S., Williams, Q., 2000. Rocks from the mantle transition zone: majorite-bearing xenoliths from Malaita, southwest Pacific. Science 288, 1215–1223.

Cox, K.G., Smith, M.R., Beswetherick, S., 1987. Textural studies of garnet lherzolites: evidence of exsolution origin from high-temperature harzburgites. In: Nixon, P.H. (Ed.), Mantle Xenoliths. Wiley, Chichester, pp. 537–550.

Danchin, R.V., 1979. Mineral and bulk chemistry of garnet lherzolite and garnet harzburgite xenoliths from the Premier Mine, South Africa. In: Boyd, F.R., Meyers, H.O.A. (Eds.), Proc. Sec. Inter. Kimberlite Conf. 2. AGU, Washington, DC, pp. 104–126.

Eggins, S.M., 1992. Petrogenesis of Hawaiian tholeiites: 1. Phase constraints. Contrib. Mineral. Petrol. 110, 387–397.

Francis, D., 1987. Mantle-melt interactions recorded in spinel lherzolite xenoliths from the Alligator Lake volcanic complex, Yukon, Canada. J. Petrol. 28, 569–597.

Francis, D., 1995. Implications of picritic lavas for the mantle sources of terrestrial volcanism. Lithos 34, 89–105.

Francis, D., Ludden, J., Johnstone, R., Davis, W., 1999. Picrite evidence for more Fe in Archean mantle reservoirs. Earth Planet. Sci. Lett. 167, 197–213.

Frederiksen, A.W., Bostock, M.G., VanDecar, J.C., Cassidy, J.F., 1998. Seismic structure of the upper mantle beneath the northern Canadian Cordillera from teleseismic travel-time inversion. Tectonophysics 294, 43–55.

Gibson, S.A., Thompson, R.N., Dickin, A.P., 2000. Ferropicrites: geochemical evidence for Fe-rich streaks in upwelling mantle plumes. Earth Planet. Sci. Lett. 174, 355–374.

Helmstaedt, H., Schulze, D.J., 1989. South African kimberlites and their mantle sample: implications for Archean tectonics and lithospheric evolution. Kimberlites Relat. Rocks 1, 358–368.

Herzberg, C., 1993. Lithosphere peridotites of the Kaapvaal craton. Earth Planet. Sci. Lett. 120, 13–29.

Herzberg, C., O'Hara, M., 1997. Phase equilibria constraints on the origin of basalts, picrites, and komatiites. Earth Sci. Rev. 44, 39–79.

Hirose, K., Kushiro, I., 1993. Partial melting of dry peridotites at high pressures: determination of compositions of melts segregated from peridotite using aggregates of diamonds. Earth Planet. Sci. Lett. 114, 477–489.

Hirschmann, M.M., 2000. Mantle solidus: experimental constraints and the effects of peridotite composition. G³ Geochem. Geophys. Geosys. 1, # 2000GC000070.

Irvine, G.J., Kopylova, M.G., Carlson, R.W., Pearson, D.G., Shirey, S.B., Kjarsgaard, B.A., 1999. Age of the lithospheric mantle beneath and around the Slave craton: a Re–Os isotopic study of peridotite xenoliths from the Jericho and Somerset Island Kimberlites. Ninth Annual V.M. Goldschmidt Conference. Lunar Planet. Sci. Contrib., vol. 971. Lunar and Planetary Science Institute, Houston, TX, pp. 134–135.

Ito, E., Morooka, K., Ujike, O., Katsura, T., 1995. Reactions between molten iron and silicate melts at high pressures: implications for the chemical evolution of Earth's core. J. Geophys. Res. 100, 5901–5910.

Jagoutz, E., Palme, H., Baddenhausen, H., Blum, K., Cendales, M., Dreibus, G., Spettel, B., Lorenz, V., Wänke, H., 1979. The abundances of major, minor and trace elements in the earth's mantle as derived from primitive ultramafic nodules. Proc. 10th Lunar Planet. Sci. Conf. Lunar and Planetary Science Institute, Houston, TX, pp. 2031–2050.

Jarosewich, E., 1990. Chemical analyses of meteorites: a compilation of stony and iron meteorite analyses. Meteor 25, 323–337.

Jordan, T.H., 1988. Structure and formation of the continental tectosphere. J. Petrol. Spec. Lithosphere, 11–37.

Kato, T., Ringwood, A.E., Irifune, T., 1988. Constraints on element partitioning coefficients between MgSiO₃ perovskite and liquid determined by direct measurement. Earth Planet. Sci. Lett. 90, 65–68.

Kelemen, P.B., Hart, S.R., Berstein, S., 1999. Silica enrichment in the continental upper mantle via melt/rock reaction. Earth Planet. Sci. Lett. 164, 387–406.

Kellogg, L.H., Hager, B.H., van der Hilst, R.D., 1999. Compositional stratification in the deep mantle. Science 283, 1881–1884.

Kopylova, M.G., Russell, J.K., 2000. Chemical stratification of cratonic lithosphere: constraints from the northern Slaver craton, Canada. Earth Planet. Sci. Lett. 181, 71–87.

Kushiro, I., 1998. Compositions of partial melts formed in mantle peridotites at high pressures and their relation to those of primitive MORB. Phys. Earth Planet. Inter. 107, 103–110.

Larimer, J.W., Anders, E., 1970. Chemical fractionation in meteorites: III. Major element fractionations in chondrites. Geochim. Cosmochim. Acta 34, 367–387.

Lee, C.T., Rudnick, R.L., 1999. Compositionally stratified cratonic lithosphere: petrology and geochemistry of peridotite xenoliths from the Labait volcano, Tanzania. Proc. 7th Inter. Kimberlite. Red Roof Design, Cape Town, South Africa, pp. 503–521.

McDonough, W.F., Sun, S.S., 1995. The composition of Earth. Chem. Geol. 120, 223–253.

McFarlane, E.A., Drake, M.J., Rubie, D.C., 1994. Element partitioning between Mg-perovskite, magnesiowustite, and silicate melt at conditions of the Earth's mantle. Geochim. Cosmochim. Acta 58, 5161–5172.

Meisel, T., Walker, R.J., Irving, A.J., Lorand, J.P., 2001. Osmium isotopic compositions of mantle xenoliths: a global perspective. Geochim. Cosmochim. Acta 65, 1311–1323.

Nixon, P.H., Boyd, F.R., 1973. Petrogenesis of the granular and sheared ultrabasic nodule suite in kimberlites. Lesotho Kimberlites, 48–56.

Ohtani, E., Yurimoto, H., Seto, S., 1997. Element partitioning between metallic liquid, silicate liquid, and lower mantle minerals: implications for core formation of the Earth. Phys. Earth Planet. Inter. 100, 97–114.

O'Neill, H.S.C., Palme, H., 1998. Composition of the silicate Earth: implications for accretion and core formation. In: Jackson, I. (Ed.), The Earth's Mantle Composition, Structure, and Evolution. Cambridge Univ. Press, Cambridge, UK, pp. 3–126.

O'Neill, H.S.C., Canil, D., Rubie, D.C., 1998. Oxide–metal equilibria to 2500 °C and 25 GPa: implications for core formation and the light component in the Earth's core. J. Geophys. Res. 103, 12239–12260.

Pearson, D.G., Carlson, R.W., Shirey, S.B., Boyd, F.R., Nixon, P.H., 1995a. Stabilization of Archean lithospheric mantle: Re–Os isotope study of peridotite xenoliths from the Kaapvaal craton. Earth Planet. Sci. Lett. 134, 341–357.

Pearson, D.G., Shirey, S.B., Carlson, R.W., Boyd, F.R., Pokhilenko, N.P., Shimizu, N., 1995b. Re–Os, Sm–Nd, and Rb–Sr isotope evidence for thick Archaean lithospheric mantle beneath the Siberian craton modified by multi-stage metasomatism. Geochim. Cosmochim. Acta 59, 959–977.

Peslier, A., Francis, D., Ludden, J., 2002. The lithospheric mantle beneath continental margins: melting and melt-rock reaction in Canadian Cordilleran xenoliths. J. Petrol. 43, 13–48.

Peslier, A., Reisberg, L., Ludden, J., Francis, D., 2000a. Re–Os constraints on harzburgite and lherzolite formation in the litho-

spheric mantle: a study of northern Canadian Cordilleran xenoliths. Geochim. Cosmochim. Acta 64, 3061–3071.

Peslier, A., Reisberg, L., Ludden, J., Francis, D., 2000b. Os isotopic
systematics in mantle xenoliths; age constraints in the Canadian
Cordilleran lithosphere. Chem. Geol. 166, 85–101.

Richardson, S.H., Gurney, J.J., Erlank, A.J., Harris, J.W., 1984. Origin of diamonds in old enriched mantle. Nature 310, 198–202.

Roden, M.F., Frey, F.A., Francis, D., 1984. An example of consequent mantle metasomatism in peridotite inclusions from Nuni-
vak Island, Alaska. J. Petrol. 25, 546–577.

Schmidberger, S.S., Francis, D., 1999. Nature of the mantle roots
beneath the North American Craton: mantle xenolith evidence
from Somerset Island kimberlites. Lithos 48, 195–216.

Schmidberger, S.S., Simonetti, A., Francis, D., Gariepy, C., 2002.
Probing Archean lithosphere using the Lu–Hf isotope systematics of garnet periditite xenoliths from Somerset Island kimberlites, Canada. Earth Planet. Sci. Lett. 197, 245–259.

Schulze, D., 1986. Calcium anomalies in the mantle and a subducted metaserpentinite origin for diamonds. Nature 319,
433–452.

Shi, L., Francis, D., Ludden, J., Frederilsen, A., Bostock, M., 1998.
Xenolith evidence for lithospheric melting above anomalously
hot mantle under the northern Canadian Cordillera. Contrib.
Mineral. Petrol. 131, 39–53.

Takahashi, E., 1986. Melting of a dry peridotite KLB-1 up to 14
GPa: implications on the origin of peridotite upper mantle.
J. Geophys. Res. 91, 9367–9382.

Takahashi, E., Shimazaki, T., Tsuzaki, Y., Yoshida, H., 1994. Melting
study of a peridotite KLB-1 to 6.5 GPa, and the origin of basaltic
magmas. Philos. Trans. R. Soc. Lond., A 342, 105–120.

Walter, M.J., 1998. Melting of garnet peridotite and the origin of
komatiite and depleted lithosphere. J. Petrol. 39, 29–60.

Wood, B.J., 2000. Phase transformations and partitioning relations
in peridotite under lower mantle conditions. Earth Planet. Sci.
Lett. 174, 341–354.

Available online at www.sciencedirect.com

SCIENCE ⒹDIRECT°

LITHOS

Lithos 71 (2003) 153–184

www.elsevier.com/locate/lithos

ELSEVIER

The timing of kimberlite magmatism in North America: implications for global kimberlite genesis and diamond exploration

L.M. Heaman[a,*], B.A. Kjarsgaard[b], R.A. Creaser[a]

[a]Department of Earth and Atmospheric Sciences, University of Alberta, 4–18 Earth Sciences Building, Edmonton, Alberta, Canada T6G 2E3
[b]Geological Survey of Canada, Ottawa, Ontario, Canada K1A OE8

Abstract

Based on a compilation of more than 100 kimberlite age determinations, four broad kimberlite emplacement patterns can be recognized in North America: (1) a northeast Eocambrian/Cambrian Labrador Sea province (Labrador, Québec), (2) an eastern Jurassic province (Ontario, Québec, New York, Pennsylvania), (3) a Cretaceous central corridor (Nunavut, Saskatchewan, central USA), and (4) a western mixed (Cambrian-Eocene) Type 3 kimberlite province (Alberta, Nunavut, Northwest Territories, Colorado/Wyoming). Ten new U–Pb perovskite/mantle zircon and Rb–Sr phlogopite age determinations are reported here for kimberlites from the Slave and Wyoming cratons of western North America. Within the Type 3 Slave craton, at least four kimberlite age domains exist: I-a southwestern Siluro-Ordovician domain (~ 450 Ma), II-a SE Cambrian domain (~ 540 Ma), III-a central Tertiary/Cretaceous domain (48–74 Ma) and IV-a northern mixed domain consisting of Jurassic and Permian kimberlite fields. New U–Pb perovskite results for the 614.5 ± 2.1 Ma Chicken Park and 408.4 ± 2.6 Ma Iron Mountain kimberlites in the State Line field in Colorado and Wyoming confirm the existence of at least two periods of pre-Mesozoic kimberlite magmatism in the Wyoming craton.

A compilation of robust kimberlite emplacement ages from North America, southern Africa and Russia indicates that a high proportion of known kimberlites are Cenozoic/Mesozoic. We conclude that a majority of these kimberlites were generated during enhanced mantle plume activity associated with the rifting and eventual breakup of the supercontinent Gondwanaland. Within this prolific period of kimberlite activity, there is a good correlation between North America and Yakutia for three distinct short-duration (~ 10 my) periods of kimberlite magmatism at 48–60, 95–105 and 150–160 Ma. In contrast, Cenozoic/Mesozoic kimberlite magmatism in southern Africa is dominated by a continuum of activity between 70–95 and 105–120 Ma with additional less-prolific periods of magmatism in the Eocene (50–53 Ma), Jurassic (150–190) and Triassic (~ 235 Ma). Several discrete episodes of pre-Mesozoic kimberlite magmatism variably occur in North America, southern Africa and Yakutia at 590–615, 520–540, 435–450, 400–410 and 345–360 Ma. One of the surprises in the timing of kimberlite magmatism worldwide is the common absence of activity between about 250 and 360 Ma; this period is even longer in southern Africa. This >110 my period of quiescence in kimberlite magmatism is likely linked to relative crustal and mantle stability during the lifetime of the supercontinent Gondwanaland.

Economic diamond deposits in kimberlite occur throughout the Phanerozoic from the Cambrian (Venetia, South Africa; Snap Lake and Kennady Lake, Canada) to the Tertiary (Mwadui, Tanzania; Ekati and Diavik in Lac de Gras, Canada). There are clearly some discrete periods when economic kimberlite-hosted diamond deposits formed globally. In contrast, the Devonian

* Corresponding author. Tel.: +1-780-492-3265; fax: +1-780-492-2030.
E-mail address: larry.heaman@ualberta.ca (L.M. Heaman).

0024-4937/$ - see front matter © 2003 Elsevier B.V. All rights reserved.
doi:10.1016/j.lithos.2003.07.005

event, which is such an important source of diamonds in Yakutia, is notably absent in the kimberlite record from both southern Africa and North America.

Keywords: Kimberlite; Geochronology; North America

1. Introduction

Kimberlites are small potassic ultrabasic intrusions that occur in virtually every Archean craton worldwide. The mineralogy of kimberlites is quite variable (Skinner and Clement, 1979; Mitchell, 1986) but often include macro- and megacrysts of olivine, clinopyroxene, phlogopite, garnet, and ilmenite encapsulated in a finer-grained matrix of calcite, perovskite, phlogopite, and spinel (± apatite, monticellite, rutile, serpentine, sulphide). Kimberlites are volatile- and crystal-rich magmas, derived at great depth within the mantle; some would suggest at the core/mantle boundary (e.g. Haggerty, 1994) or mantle transition zone (e.g. Ringwood et al., 1992). Kimberlitic magmas have a high carrying capacity for mantle and crustal xenoliths/xenocrysts and are most renown for the fact that they are host to the majority of primary diamond deposits worldwide. Despite the large number of kimberlites that have been discovered and mined, there are several outstanding questions that relate to both the origin of kimberlite magmatism in general and, more specifically, to what controls the occurrence of economic diamond deposits.

In order to constrain better the origin and cause of kimberlite magmatism it is important to have a solid understanding of certain aspects of kimberlite genesis such as: (1) what is the likely mantle source material capable of producing an ultrabasic magma with the requisite mineralogy and geochemistry of a kimberlite, (2) what is the depth of melting in the mantle required for kimberlite genesis, and (3) what are the possible triggers responsible for mantle melting to produce kimberlite? To address the latter question in detail it is imperative to understand both the local and global timing/distribution of kimberlite magmatism and the relationship of this magmatism to contemporaneous tectonic processes (e.g. Dawson, 1989). For example, many of the models postulated for kimberlite genesis (see discussion) imply a quite specific temporal–spa-

tial relationship and predict that the timing of kimberlite magmatism is linked to some large-scale tectonic process such as subduction of oceanic lithosphere (e.g. McCandless, 1999), rifting of continents (e.g. Phillips et al., 1998) or impact of mantle plumes (e.g. Heaman and Kjarsgaard, 2000). We agree with the view of Helmstaedt and Gurney (1997) that the structural control of final kimberlite emplacement within individual fields (e.g. proximity to faults, mafic dyke intersections, etc.) may be unrelated to the tectonic forces responsible for triggering mantle melting to produce kimberlite. On the other hand, larger scale patterns of kimberlite emplacement (e.g. craton wide) could help identify subcontinental mantle processes that are linked to kimberlite formation.

However, for many Archean cratons worldwide it is difficult to assess in detail the various models for kimberlite formation because there is a paucity of precise and accurate kimberlite emplacement ages. Part of the reason for this is that many kimberlites do not contain minerals amenable to precise radiometric dating, such as phlogopite (Rb–Sr, ^{40}Ar/^{39}Ar), perovskite (U–Pb), mantle zircon (U–Pb), or mantle rutile (U–Pb). An additional concern is that kimberlites are mineralogically quite complex, containing a mixture of primary minerals crystallizing directly from the kimberlite magma (e.g. phlogopite and perovskite) and a variety of minerals that are entrained in the form of mantle and crustal xenoliths and xenocrysts (e.g. phlogopite and zircon). It is not always easy to decipher whether a particular kimberlite mineral is primary or not.

Compared to the number of kimberlite age determinations available from southern Africa and Russia, until recently there have been very few published age dates for kimberlites in North America. The main reason for this was the fact that relatively few kimberlites were known. Prior to 1988, the total number of kimberlites in North America numbered around 50, and since this time there has been at least a tenfold increase in the number of discovered kimberlites. In this paper we report new U–Pb and Rb–Sr age determinations for a total of ten kimberlites in western North America; eight from the Slave craton, Northwest Territories and Nunavut, Canada and two from the Wyoming craton, Colorado/Wyoming. Together with other published kimberlite age dates, we evaluate the emplacement history of kimberlite magmatism, with emphasis on searching for kimberlite emplacement patterns in North America, southern Africa and Russia that may help elucidate their origins. We further examine whether there is any relationship between the timing of kimberlite magmatism and diamond productivity in these three areas.

1.1. Previous age studies of North America kimberlite magmatism

Phanerozoic kimberlite magmatism in North America can be broadly subdivided into four regions based on the distribution of clusters/fields and emplacement history (Fig. 1). These include: (1) a NE Eocambrian Labrador Sea province (east coast of Labrador and eastern Québec), (2) an eastern dominantly Jurassic (180–140 Ma) province, including many of the kimberlites in Ontario, western Québec, New York, Pennsylvania (Heaman and Kjarsgaard, 2000 and references therein), (3) a Cretaceous (103–94 Ma) central province or corridor (Nunavut, Saskatchewan and central USA), which includes the 103–94 Ma Somerset Island and 101–95 Ma Fort à la Corne fields (Kjarsgaard, 1996b; Leckie et al., 1997a; Heaman and Kjarsgaard, 2002), and (4) a western mixed (Eocene-Cambrian) province, which includes amongst others, kimberlite fields in Alberta, Northwest Territories, Nunavut, and Wyoming/Colorado. The western mixed age province contains two Type 3 kimberlite provinces (as defined by Mitchell, 1986, p. 106) where multiple kimberlite fields with a variety of emplacement ages occur in a small geographic area. In the Slave craton alone, several ages of kimberlite emplacement between ~ 50 and ~ 540 Ma were reported by Heaman et al. (1997). Based on a small number of previous age determinations, the Slave craton contains kimberlite fields of Eocene, Cretaceous, Jurassic, Permian, Siluro-Ordovician and Cambrian age; some of these occur in relatively close spatial proximity (Kjarsgaard, 1996a; Davis and

Kjarsgaard, 1997; Heaman et al., 1997). Likewise, in the Wyoming craton kimberlites have a large range in emplacement age from Devonian (386–400 Ma; Smith, 1979; Carlson and Marsh, 1989) to Eocambrian (e.g. ~ 600 Ma George Creek and < 780 Ma Green Mountain kimberlites; Carlson and Marsh, 1989; Lester and Larson, 1996). Within the western mixed province there are a few isolated kimberlite fields and clusters that appear to have discrete periods of kimberlite magmatism such as the 86–88 Ma Buffalo Head Hills field and 70–78 Ma Birch Mountains cluster in Alberta (Aravanis, 1999; Carlson et al., 1999a,b; Doyle, personal communication), a Permian (256–286 Ma) kimberlite field on Victoria Island, Nunavut Territory (Kahlert, personal communication), and a Devonian (386–400 Ma) cluster in the State Line field, Colorado/Wyoming (Smith, 1979).

From a geochronology perspective, the eastern Jurassic province (Fig. 1) has been studied in most detail (Basu et al., 1984, Barnett et al., 1984; Heaman, 1989; Bikerman et al., 1997; Heaman and Kjarsgaard, 2000). A NE–SW corridor of kimberlite magmatism can be traced for more than 2000 km from as far northwest as Rankin Inlet (196 Ma) through to the Attawapiskat (180 and 155 Ma), Kirkland Lake (157–152 Ma), and Timiskaming (155–134 Ma) fields. This corridor of kimberlite magmatism generally follows a pattern of southeast younging that is geographically coincident with independent estimates for the timing and location of the continental extension of the Great Meteor hotspot track (Heaman and Kjarsgaard, 2000). There are other Jurassic kimberlite occurrences in North America that are apparently not related to this hotspot track, such as the 173 Ma Jericho kimberlite, northern Slave craton (Heaman et al., 2002).

A growing number of Precambrian kimberlites have been discovered in North America. For many years the only known Precambrian kimberlite was the 1.1 Ga Bachelor Lake kimberlite, Québec (Watson, 1967). More recently, the following Precambrian kimberlites and ultramafic lamprophyres have been identified; the 1076 Ma Kyle Lake #5 kimberlite in the James Bay Lowlands (Sage, 1996; Heaman, unpublished data), the 1097 Ma Whitefish Lake ultramafic lamprophyre in the Wawa area (Kaminsky et al., 2000) and a number of ~ 1140 Ma ultramafic lamprophyres in the Lake Superior region (Queen et al., 1996).

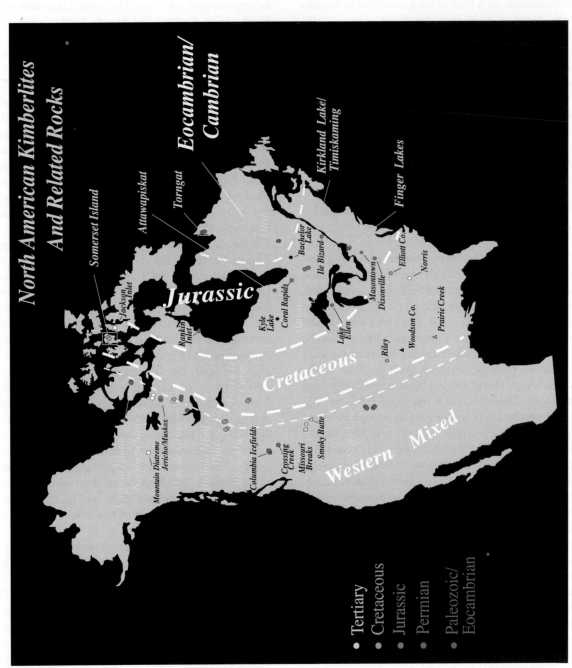

2. U–Pb results

New U–Pb results are presented here for perovskite and mantle zircon isolated from seven Slave craton kimberlites (C13, Anne, Cross, Drybones Bay, Orion, Ursa, and Snowy Owl) and two kimberlites from the Wyoming craton (Chicken Park and Iron Mountain). The U–Pb data are reported in Table 1 and displayed on concordia diagrams in Figs. 3 and 4. The starting material for each sample varied from small pieces of drill core to mantle zircon grains remaining after caustic fusion digestion (e.g. Drybones Bay, Cross and Ursa). The drill core samples were crushed using a jaw crusher and Bico disk mill equipped with hardened steel plates. Perovskite was concentrated using a Wilfley Table, heavy liquids and a Frantz Isodynamic separator. All mantle zircon grains (typically quite large crystals or parts of crystals in the mm–cm size range) investigated in this study were selected from caustic fusion residues. In the case of the Drybones Bay samples (residues of core from three separate drill holes), the surface texture of the zircon was pitted with a slightly frosty appearance, a feature likely generated during the fusion process. The mineral fractions analysed in this study were selected using a stereomicroscope and generally only grains devoid of visible alteration, fractures or inclusions were selected. In some perovskite populations (e.g., State Line kimberlite samples), a chalky white alteration rim occurs on most crystals and could not be entirely excluded.

The selected mineral fractions were washed in 2N HNO_3, Millipore H_2O and distilled acetone prior to weighing (UTM2 ultra-microbalance) and dissolution in TFE Teflon digestion vessels. A detailed description of the analytical procedures for U–Pb perovskite dating used at the University of Alberta is presented in Heaman and Kjarsgaard (2000; see EPSL On-Line Background Information). All isotopic data reported in Table 1 were determined with a VG354 thermal ionization mass spectrometer operating in single Faraday or analogue Daly mode. Details of corrections applied to the isotopic data are given in the Table 1 footnotes. All uncertainties in the isotopic data are reported at two-sigma. Age calculations consist of either weighted average $^{206}Pb/^{238}U$ dates or a two-error linear regression treatment performed with the program Isoplot (Ludwig, 1992). The uranium decay constants (^{238}U—1.55125×10^{-10} year^{-1}; ^{235}U—9.8485×10^{-10} year^{-1}) are those determined by Jaffey et al. (1971) and recommended by Steiger and Jager (1977).

2.1. Slave craton kimberlites

Numerous kimberlite samples from the Slave craton were investigated as part of this study but a relatively high proportion of the investigated samples do not contain minerals amenable to radiometric dating (e.g. Camsell Lake, Nicholas Bay, Ranch Lake, Tli Kwi Cho, Torrie). The emplacement ages for a total of eight Slave Province kimberlites (kimberlite locations are shown in Fig. 2) were investigated in this study including Anne, C13, Orion and Snowy Owl (U–Pb perovskite), Cross, Ursa and Drybones Bay (U–Pb mantle zircon) and AK5034 from the Kennady Lake cluster (Rb–Sr phlogopite).

2.1.1. C13 (Lac de Gras field)

A 0.5-kg piece of 2 in. kimberlite drill core from C13, provided by C. Jennings, yielded less than 100 perovskite crystals. Most of these were dark orange to brown with a cubic habit. The U–Pb results for two multi-grain perovskite fractions consisting of 36 and 31 crystals, respectively, are presented in Table 1 (#1,2). Perovskite in C13 contains moderate uranium (100–111 ppm) and quite high Th (9432–9665 ppm) contents, which accounts for the very high Th/U (85.1–96.4). These two fractions yielded similar $^{206}Pb/^{238}U$ dates of 75.2 ± 2.5 and 72.5 ± 2.2 Ma, respectively. A weighted average $^{206}Pb/^{238}U$ date of 73.7 ± 3.2 Ma is considered the best estimate for the emplacement age of the C13 pipe.

Fig. 1. Location map of kimberlite pipes, clusters and fields in North America as mentioned in the text. Modified from Heaman and Kjarsgaard (2000). Rectangular boxes and bold yellow labels indicate kimberlite clusters or fields. A smaller black font indicates individual kimberlite bodies. Circles represent kimberlites, triangles indicate lampröites, and squares indicate other kimberlite-like units such as olivine melilitite or lamprophyre. The symbols are colour coded by age (see legend). Symbols colour coded black indicates Mesoproterozoic kimberlites and white symbols indicate unknown age.

Table 1
U–Pb results for Slave and Wyoming craton kimberlites

Mineral[a] Analysed	Wt. (μg)	U (ppm)	Th (ppm)	Pb (ppm)	Th/U	Atomic Ratios[b]					Age (Ma)			%Disc
						$^{206}Pb/^{204}Pb$	$^{238}U/^{204}Pb$	$^{206}Pb/^{238}U$	$^{207}Pb/^{235}U$	$^{207}Pb/^{206}Pb$	$^{206}Pb/^{238}U$	$^{207}Pb/^{235}U$	$^{207}Pb/^{206}Pb$	
Slave Craton														
Pipe C13														
1 P	11	100	9665	46	96.44	24.27 ± 44	484.8 ± 356	0.01173 ± 78	–	–	75.2 ± 5.0	–	–	–
2 P	13	111	9432	45	85.04	24.82 ± 42	550.8 ± 350	0.01131 ± 68	–	–	72.5 ± 4.4	–	–	–
Anne (HL-10)														
3 P	40	43	1901	13	43.99	22.86 ± 166	387.4 ± 116	0.01100 ± 98	–	–	70.5 ± 6.2	–	–	–
Orion														
4 P	37	192	1638	60	8.55	94.87 ± 68	1099.7 ± 104	0.06988 ± 46	0.601 ± 22	0.06233 ± 224	435.4 ± 2.8	478 ± 14	686 ± 76	37.7
Drybones Bay 943985 (Hole 95-7)														
5 Z	539	5.8	1.4	0.42	0.25	546	–	0.06902 ± 26	0.540 ± 6	0.05679 ± 66	430.3 ± 1.6	439 ± 5	486 ± 26	11.4
6 Z	256	9.9	2.67	0.62	0.27	373	–	0.05661 ± 26	0.444 ± 8	0.05690 ± 84	354.9 ± 1.6	373 ± 5	488 ± 32	27.9
7 Z	162	5.1	1.85	0.41	0.36	280	–	0.07174 ± 38	0.566 ± 20	0.05718 ± 196	446.6 ± 2.4	455 ± 13	499 ± 74	10.8
8 Z	165	9.6	13.6	0.93	1.42	142	–	0.05441 ± 42	0.434 ± 16	0.05787 ± 212	341.5 ± 2.6	366 ± 12	525 ± 79	35.8
9 Z	362	9.6	2.9	0.65	0.31	350	–	0.05898 ± 26	0.464 ± 6	0.05701 ± 68	369.5 ± 1.6	387 ± 4	492 ± 26	25.6
10 Z	240	4.7	1.4	0.38	0.29	268	–	0.07065 ± 38	0.562 ± 16	0.05765 ± 162	440.1 ± 2.2	453 ± 11	517 ± 61	15.3
11 Z	212	3.0	0.82	0.37	0.28	105	–	0.07333 ± 52	0.618 ± 34	0.06117 ± 318	456.2 ± 3.2	489 ± 21	645 ± 110	30.3
Drybones Bay 943902 (Hole 95-8)														
12 Z	4627	4.7	1.3	0.33	0.29	2272	–	0.07085 ± 36	0.545 ± 4	0.05576 ± 16	441.3 ± 2.2	441.5 ± 2.2	442.7 ± 6.6	0.3
13 Z	254	4.6	1.2	0.44	0.26	153	–	0.07032 ± 42	0.573 ± 18	0.05908 ± 174	438.1 ± 2.4	460 ± 11	569.9 ± 63.0	23.9
14 Z	539	8.3	3.6	0.71	0.43	358	–	0.07179 ± 34	0.556 ± 6	0.05613 ± 58	446.9 ± 2.0	448.7 ± 4.2	457.7 ± 23.0	2.4
15 Z	200	4.0	0.85	0.36	0.22	190	–	0.07520 ± 42	0.609 ± 22	0.05872 ± 194	467.4 ± 2.4	483 ± 13	556.8 ± 71.4	16.7
16 Z	421	4.9	1.2	0.41	0.25	248	–	0.07011 ± 32	0.553 ± 10	0.05723 ± 98	436.8 ± 2.0	447.1 ± 6.6	500 ± 37	13.1

Sample														
Drybones Bay 943930 (Hole 95-9)														
17 Z	875	4.9	1.3	0.68	88	—	—	0.07288 ± 54	0.600 ± 24	0.05972 ± 242	453.5 ± 3.2	477 ± 16	593 ± 86	24.4
Cross														
18 Z	390	38	10	2.8	5273	—	—	0.07237 ± 36	0.568 ± 4	0.05692 ± 14	450.4 ± 2.2	456.7 ± 2.0	488.6 ± 5.4	8.1
Ursa														
19 Z	974	95	38	7.4	2230	—	—	0.0744 ± 28	0.581 ± 22	0.05668 ± 22	463 ± 16	465 ± 14	479.3 ± 8.6	3.6
20 Z	5	20	4.0	1.4	88	—	—	0.0735 ± 24	—	—	457 ± 14	—	—	—
Snowy Owl (MGJ99-01C)														
21 P	28	112	10291	58.7	91.80	56.70 ± 73	893.3 ± 16.9	0.04296 ± 22	0.285 ± 13	—	271.2 ± 2.6	254 ± 20	—	—
Wyoming Craton														
Chicken Park (K33/1)														
22 P	251	96	525	29	5.46	120.61 ± 68	1020.7 ± 72	0.10084 ± 56	0.848 ± 24	0.06102 ± 164	619.3 ± 3.0	624 ± 13	640 ± 57	3.4
23 P	220	95	560	29	5.89	121.19 ± 88	1050.3 ± 90	0.09854 ± 54	0.825 ± 24	0.06074 ± 170	605.9 ± 3.2	611 ± 13	630 ± 59	4.0
24 P	212	88	263	21	2.99	116.41 ± 74	986.6 ± 78	0.10006 ± 50	0.848 ± 24	0.06146 ± 168	614.8 ± 3.0	624 ± 13	656 ± 58	6.5
25 P	174	97	435	26	4.47	131.57 ± 106	1139.4 ± 11	0.09996 ± 50	0.848 ± 20	0.06151 ± 148	614.2 ± 3.0	623 ± 11	657 ± 51	6.9
Iron Mountain (K17/4)														
26 P	101	117	1380	36	11.78	70.14 ± 28	787.5 ± 52	0.06614 ± 60	0.480 ± 30	0.05267 ± 324	412.9 ± 3.6	398 ± 20	315 ± 137	-32.2
27 P	383	117	799	34	6.81	59.44 ± 16	632.3 ± 30	0.06546 ± 64	0.527 ± 36	0.05836 ± 402	408.8 ± 3.8	430 ± 24	543 ± 147	25.6
28 P	249	114	796	33	7.01	59.44 ± 14	633.2 ± 26	0.06535 ± 62	0.517 ± 36	0.05734 ± 398	408.1 ± 3.8	423 ± 24	505 ± 149	19.8

Perovskite $^{238}U/^{204}Pb$ and $^{206}Pb/^{204}Pb$ ratios corrected for fractionation, blank and spike.

Th concentration determined from amount of ^{208}Pb and estimated age.

All errors in this table are reported at 2 sigma and correspond to the last significant digits (e.g. 120.61 ± 34 means 120.61 ± 0.34).

[a] P—perovskite, Z—zircon.

[b] Atomic ratios corrected for mass spectrometer fractionation (Pb—0.088%/amu; U—0.155%/amu), blank (5pg Pb; 1pg U), spike, and initial common Pb (Stacey and Kramers, 1975). $^{206}Pb/^{204}Pb$ zircon ratios corrected for fractionation and spike only.

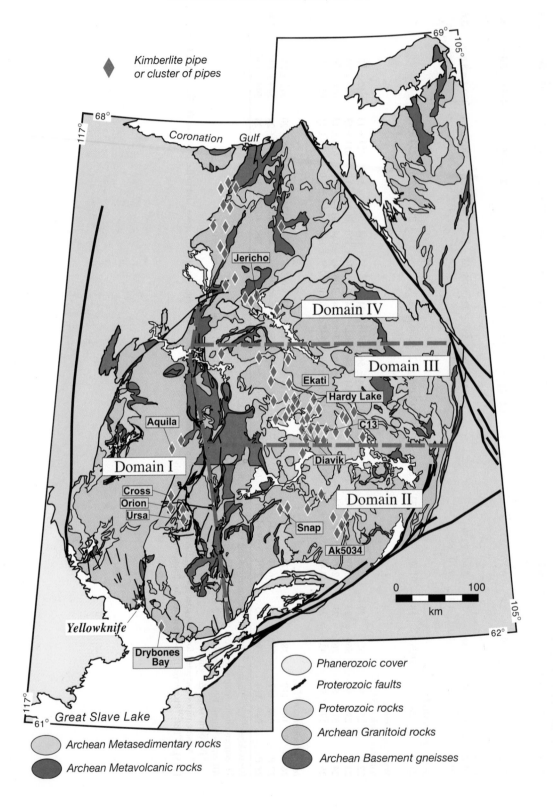

2.1.2. Anne (Hardy Lake cluster, Lac de Gras field)

The Anne kimberlite is one of several kimberlites in the Hardy Lake cluster approximately 25 km east-northeast of the Ekati diamond mine (McKinlay et al., 1997). Abundant tiny (<40 µm) perovskite cubes were recovered from a sample of hypabyssal kimberlite provided by Barbara Scott-Smith. The U–Pb results for one fraction consisting of ~ 250 dark brown cubes (Table 1) indicates the Anne perovskite has relatively low U contents (43 ppm) and higher than average Th/U (44.0) for kimberlite perovskite (typically in the 5–20 range). The 70.5 ± 6.2 Ma $^{206}Pb/^{238}U$ date for perovskite fraction #3 listed in Table 1 is considered a good estimate for the emplacement age of the Anne kimberlite. The U–Pb perovskite age obtained here is in excellent agreement with the $^{40}Ar/^{39}Ar$ phlogopite dates reported for four kimberlites in this cluster by Scott Smith and McKinlay (2002); Anne (72 ± 2 Ma), Drew (73 ± 1 Ma), HL1 (71 ± 2 Ma) and HL 2 (73 ± 1 Ma). These $^{40}Ar/^{39}Ar$ phlogopite dates were determined at the DeBeers Geoscience Center in Johannesburg (Scott Smith, personal communication). Both the Hardy Lake and C13 kimberlites have similar Cretaceous emplacement ages.

2.1.3. Orion (lower Carp Lake cluster, SW slave)

A small piece of kimberlite drill core from Orion yielded a moderate amount of tiny dark brown perovskite cubes. The U–Pb results for a multi-grain (i.e. 80 cubes) perovskite fraction (#4 in Table 1) indicate that Orion perovskite has high uranium contents (192 ppm) and a typical Th/U (8.6) for kimberlitic perovskite. The $^{206}Pb/^{238}U$ date of 435.3 ± 2.8 Ma is interpreted as the best estimate for the emplacement age of the Orion pipe and is the first documented Ordovician kimberlite intrusion in North America.

2.1.4. Cross and Ursa (lower Carp Lake cluster, SW slave)

Large mantle zircon crystals were analysed from the Cross (#18 in Table 1) and Ursa (#19 in Table 1)

kimberlites. A much smaller zircon fragment from the Ursa kimberlite was also analysed (#20) but the low radiogenic ^{207}Pb content (0.3 pg) precluded calculation of $^{207}Pb/^{235}U$ or $^{207}Pb/^{206}Pb$ model dates. The mantle zircons isolated from these two kimberlites contain considerably higher U (38–95 ppm) and Th (10–38 ppm) contents than the Drybones Bay mantle zircon. The U–Pb results are listed in Table 1, however these analyses are not shown on a concordia plot. Interestingly, the U–Pb mantle zircon results for these two kimberlites are somewhat similar to the Drybones Bay mantle zircon data described below. They are discordant (3.6–8.1%) with $^{206}Pb/^{238}U$ dates that vary between 450.4 ± 2.2 and 462.6 ± 8.2 Ma. At face value these $^{206}Pb/^{238}U$ dates could be interpreted to reflect the emplacement age of the Cross (450.4 ± 2.2 Ma) and Ursa (459 ± 10 Ma; weighted average of two zircon analyses) pipes. However, considering the possible presence of an inherited Pb component in these grains, as appears to be the case in some Drybones Bay zircon fragments, these dates should be considered maximum emplacement age estimates. The Cross and Ursa U–Pb zircon results can be interpreted in a similar way to the Drybones Bay zircon data (described below) and it is possible that all three have an identical emplacement age of ~ 440 Ma.

2.1.5. Drybones Bay (SW slave)

Mantle zircon fragments are abundant in the caustic fusion residues of Drybones Bay kimberlite. Three separate samples of Drybones Bay kimberlite (943985, 943902, and 943930) taken from different drill holes (95-7, 95-8, 95-9, respectively) were provided by U. Kretschmar and investigated in this study. The U–Pb results for seven single zircon fragments from sample 943985 are reported in Table 1 (#5–11) and presented on a concordia diagram in Fig. 3a. The Drybones Bay zircon is typical of mantle zircon from elsewhere (e.g. Davis, 1977, 1978) in forming large colourless transparent crystals (up to two centimeters in diameter) that have quite

---

Fig. 2. Location of kimberlites in the Slave craton, Canada. Kimberlites locations are indicated with small red diamonds Existing or planned diamond mines occur at Ekati, Diavik, Snap Lake and Jericho. Also shown is a subdivision of the Slave craton into four kimberlite domains (denoted I–IV) separated by heavy dashed lines; (I) a western Siluro-Ordovician domain, (II) a southeastern Cambrian domain, (III) and a central Tertiary/Cretaceous domain centered around Lac de Gras and (IV) a northern domain dominated by Jurassic and Permian kimberlites.

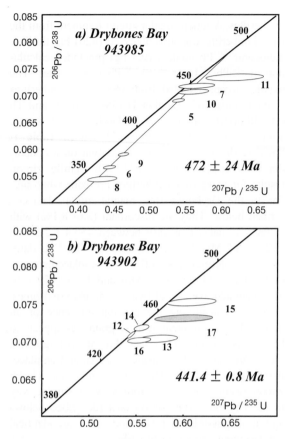

Fig. 3. U–Pb concordia diagrams displaying the mantle zircon results for two kimberlite samples from the Drybones Bay kimberlite near Yellowknife. (a) U–Pb results from sample 943985 and (b) U–Pb results from samples 943902 (open ellipses) and 973930 (shaded ellipse).

analyses. This translates to an enormous uncertainty in the $^{207}Pb/^{235}U$ dates and reflected in the large elongate error ellipsoids in Fig. 3. Most of the zircon data form a collinear array and a regression treatment of the three most precise analyses (#5,6,9) yields a relatively imprecise upper intercept date of 472 ± 24 Ma (Fig. 3a) and a near-zero lower intercept date. It is important to evaluate the significance of the discordant zircon analyses (such as #11 in Fig. 3a), as mantle zircon typically does not display such profound discordance unless there is an inherited Pb component preserved in the grains (e.g. Kinny et al., 1989). If a small amount of inherited Pb component is present in fragment #11 then the model $^{207}Pb/^{206}Pb$ date of 645 Ma could provide a minimum estimate for the age of this older zircon component. Another possibility is that an improper common Pb correction is being applied to these analyses. The common Pb isotopic composition in these analyses is estimated from the average terrestrial Pb evolution model of Stacey and Kramers (1975). This would have the greatest impact on the analyses with the lowest $^{206}Pb/^{204}Pb$ ratios such as grain #11. If analysis #11 plots to the right of the reference discordia line due to an improper common Pb correction then the upper intercept date of 472 Ma is best considered as a maximum estimate for the age of zircon crystallization.

The U–Pb results for five zircon fractions from sample 943902 (#12–16) and one single zircon fragment from sample 943930 (shaded ellipse #17) are listed in Table 1 and displayed on a concordia diagram in Fig. 3b. The zircon analyses from these two samples display a similar pattern as described above for sample 943985; the grains are typically large colourless fragments, have low U (4.0–8.3 ppm) and Th (0.9–3.6 ppm) contents, intermediate Th/U (0.33–0.68) and many grains display considerable discordance (up to 24%). The two most precise analyses (#12,14) are concordant adding some credence to the notion that the discordance pattern observed for many fractions (#13,15–17) could be due to an improper common Pb correction. Fraction #12 is the only multi-grain analysis (32 fragments) and has a sufficiently high $^{206}Pb/^{204}Pb$ as to be relatively insensitive to the common Pb correction. A reference line constructed to pass through the two most precise analyses (#12,14) yields an upper intercept date of 1004 ± 470 Ma and a lower intercept date of 439.7 ± 5.8 Ma. The upper intercept date

low U (3.0–9.9 ppm) and Th (0.8–13.6 ppm) contents. The Th/U ratios for these fractions are generally quite similar (0.25–0.36) with values most consistent with an igneous origin. One exception to this low to moderate range in Th/U is fraction #8 with a Th/U of 1.42. Fraction #10 was given an abrasion treatment (Krogh, 1982) and, although this fraction has one of the older model ages, it has a discordance pattern similar to most zircon analyses from this sample. All zircon analyses from this sample are relatively discordant (11–35%) and display a large range in model $^{207}Pb/^{206}Pb$ dates between 484 and 645 Ma. As a consequence of the very low U contents in these zircon fragments and the longer ^{235}U half-life, there is very little radiogenic ^{207}Pb (4–17 pg) present in these could be interpreted as indicating either the age of original zircon crystallization followed by almost complete Pb-loss until the time of kimberlite magmatism or the presence of Mesoproterozoic inheritance. The lower intercept date is interpreted as a good estimate for the emplacement age of the Drybones Bay kimberlite. This date is also similar to the $^{206}Pb/^{238}U$ dates for discordant fractions #13 and 16 (438.1 and 436.8 Ma, respectively). The best estimate for the time of Drybones Bay kimberlite emplacement is 441.4 ± 0.8 Ma and is derived from the weighted average $^{206}Pb/^{238}U$ and $^{207}Pb/^{235}U$ ages for concordant fraction #12.

The $^{206}Pb/^{238}U$ date of 453.5 Ma for a single zircon fragment (#17) from sample 943930 could indicate that this part of the kimberlite complex is slightly older than 440 Ma; the best estimate for the emplacement age of sample 943902. It is more likely that this one analysis contains a small-inherited Pb component, like other zircon fractions from sample 943902, so the 453.5 Ma date should be considered a maximum estimate for the emplacement age of this sample.

It is clear that the zircon discordance pattern in all three Drybones Bay kimberlite samples is complex and our preferred interpretation is that original zircon crystallization occurred in the Mesoproterozoic and that these mantle zircon megacrysts remained at great depths (>150 km) until entrained in the Drybones Bay kimberlite. The majority of the zircon fragments remained open to radiogenic Pb diffusion until entrainment in the kimberlite magma and transport to the crust, and these crystals provide a good estimate for the timing of kimberlite magmatism and emplacement. Other zircon fragments such as #15 and 17 which have $^{206}Pb/^{238}U$ dates as old as 593 Ma did not completely lose all vestiges of radiogenic Pb. The significance of some zircon analyses with large uncertainties is masked by the substantial common Pb corrections required. The U–Pb data presented here support the possibility of two intrusion ages for the Drybones Bay kimberlite (474 ± 24 and 441.4 ± 0.8 Ma). Although we cannot rule out the possibility of multiple intrusions, we interpret the 441.4 ± 0.8 Ma concordant analysis #12 from sample 943902 to provide the most robust constraint for the timing of Drybones Bay kimberlite emplacement.

2.1.6. Snowy owl MGJ99-01C (Victoria Island cluster, northern Slave)

Perovskite was isolated from a 15-cm long piece of 1 in. kimberlite core taken from the Snowy Owl kimberlite, Victoria Island. One perovskite fraction consisting of 33 dark brown cubes has been analysed (#21 in Table 1). The Snowy Owl perovskite has similar geochemical traits to the C13 pipe with moderate uranium (112 ppm) and very high Th (10,291 ppm). The $^{206}Pb/^{238}U$ date of 271.2 ± 2.6 Ma is interpreted as the current best estimate for the crystallization age of perovskite in this sample. This is somewhat older than the previously reported Late Permian 257 ± 3 Ma $^{40}Ar/^{39}Ar$ phlogopite date for this pipe (Kahlert, personal communication). However, it is still within the range of previously reported Permian dates from this Victoria Island cluster (256–286 Ma; Table 3).

2.1.7. AK5034 (Kennady Lake cluster, SE Slave)

The AK5034 kimberlite occurs within the Kennady Lake (or Gacho Kue) cluster in the SE Slave craton together with the Hearne, Tesla, Tuzo, and Wallace kimberlites. A summary of the mineralogy and petrology of this kimberlite has been presented by Cookenboo (1995). The Rb–Sr results for six phlogopite megacryst analyses are presented in Table 2 and on an isochron plot in Fig. 4. All phlogopite fractions were leached with dilute HCl prior to analyses (Brown et al., 1989) except fraction 022866C. This analysis has the lowest Rb/Sr (5.52) and highest Sr concentration (279 ppm), likely due to the presence of carbon-

Table 2
Rb–Sr Phlogopite Data, Kennady Lake Pipe AK5034

Sample	HCl leach	Rb (ppm)	Sr (ppm)	$^{87}Rb/^{86}Sr$	$^{87}Sr/^{86}Sr$	± 2sm
022866A	Y	429.13	5.07	300.17	3.01562	0.00012
022866B	Y	433.61	5.58	271.36	2.81903	0.00080
022866C	N	529.73	279.07	5.5155	0.75120	0.00005
022943	Y	309.36	15.40	60.782	1.17594	0.00001
022945	Y	330.56	19.17	51.859	1.10974	0.00004
022949	Y	352.15	19.07	55.676	1.14020	0.00002

External reproducibility in $^{87}Rb/^{86}Sr$ ratios is 1% (2 sigma).
Uncertainty in concentrations is 2%.
Analytical techniques and leaching follow the procedures described by Holmden et al. (1996) and Brown et al. (1989).
^{87}Rb decay constant used is 1.42×10^{-11} year^{-1}.

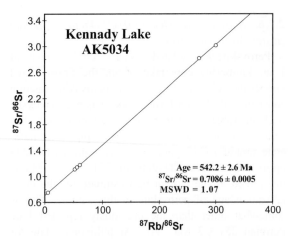

Fig. 4. Rb–Sr isochron diagram for six phlogopite megacrysts from the Kennady Lake AK5034 kimberlite. The phlogopite analysis with the lowest $^{87}Rb/^{86}Sr$ was not given an HCl leaching treatment. The regression of all data indicate a date of 542.2 ± 2.6 Ma.

ate along cleavage planes in the phlogopite. The leached phlogopite fractions contain moderate Rb (309–434 ppm), low Sr (5.1–19.2 ppm) and variable but high $^{87}Rb/^{86}Sr$ (51.86–300.17). A two-error linear regression treatment of all phlogopite data yields an isochron (MSWD = 1.07) corresponding to a date of 542.2 ± 2.6 Ma and an initial strontium ratio of 0.70858 ± 0.00049. Considering only the leached phlogopite analyses, an identical but slightly less precise date of 542 ± 5 Ma with a much larger uncertainty in the initial strontium ratio (0.7085 ± 0.0048). We interpret the relatively precise Rb–Sr phlogopite isochron date of 542.2 ± 2.6 Ma as the best estimated for the emplacement age of the AK5034 kimberlite. This is a slight revision to the date we reported previously (Heaman et al., 1997). It is slightly older, but similar to the 523 and 535 Ma Rb–Sr dates reported for the Cambrian Snap Lake dyke further west (Agashev et al., 2001). The relatively high initial strontium isotopic composition of 0.7085 may not reflect the primary composition of the AK5034 kimberlite magma but could be strongly influenced by the isotopic composition of carbonate in this sample.

2.2. Wyoming craton kimberlites

Several kimberlites from the State Line kimberlite field in Colorado/Wyoming were investigated in this

study including samples from Chicken Park, Iron Mountain, Sloan and George Creek (kimberlite locations shown in Fig. 5a). Unfortunately the Sloan sample was not amenable to Rb–Sr phlogopite or U–Pb perovskite/zircon dating and the George Creek sample did not contain perovskite.

2.2.1. Iron Mountain

Three perovskite analyses from the Iron Mountain kimberlite are reported in Table 1 (#26–28) and displayed on a concordia diagram in Fig. 5b. Iron Mountain perovskite consists of small (~ 50 μm) dark brown fragments and cubes with pervasive white rims. This perovskite population is geochemically similar to Chicken Park perovskite in having moderate contents of U (114–117 ppm) and Th (796–1380 ppm) and average perovskite Th/U (7.0–11.8). As can be seen on Fig. 5b, all three multi-grain (150–200 crystals per fraction) perovskite fractions plot within error of the concordia curve with similar $^{206}Pb/^{238}U$ dates that vary between 408.1 and 412.9 Ma. The weighted average $^{206}Pb/^{238}U$ date for all three fractions is 410.0 ± 6.5 Ma (MSWD = 2). We interpret the slightly more precise weighted average $^{206}Pb/^{238}U$ date of 408.4 ± 2.6 Ma (MSWD = 0.06) obtained for fractions #27 and 28 to be the best estimate for the emplacement age of the Iron Mountain kimberlite. The U–Pb perovskite age obtained here is similar to a ca. 400 Ma Rb–Sr phlogopite age determined by Smith (1979) for Iron Mountain, and is slightly older than other kimberlites in the State Line field such as the Estes Park kimberlite dyke (386 ± 9 Ma; Smith, 1979).

2.2.2. Chicken Park

Abundant dark brown, rounded perovskite cubes were recovered from a sample of Chicken Park kimberlite. A large proportion of the perovskite crystals contains a white chalky rim and these grains were generally avoided in all fractions except #24. The U–Pb results for four multi-grain perovskite analyses consisting of between 140 and 300 crystals are reported in Table 1 (#22–25) and shown on a concordia diagram in Fig. 5c. Chicken Park perovskite contains moderate U (88–97 ppm) and Th (263–560 ppm) and typical perovskite Th/U (3.0–5.9). All four U–Pb perovskite analyses from the Chicken Park kimberlite overlap within error of the

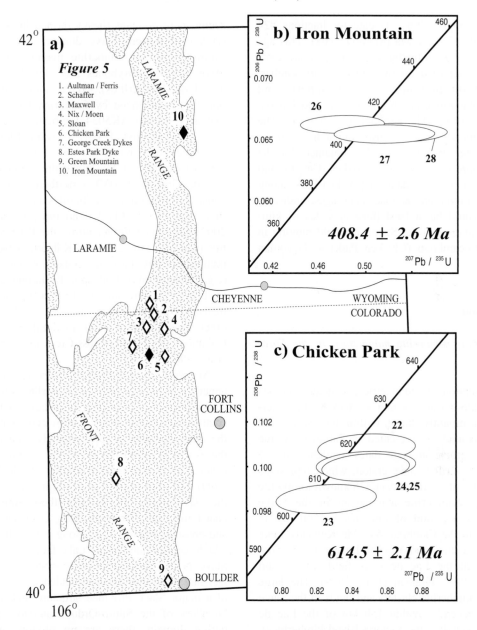

Fig. 5. U–Pb results for kimberlites from the Wyoming craton. (a) The location of kimberlites in the State Line field of Colorado and Wyoming. (b) U–Pb perovskite results for three perovskite fractions from the Iron Mountain kimberlite yields a date of 408.4 ± 2.6 Ma. (c) U–Pb results for four perovskite fractions from the Chicken Park kimberlite. The weighted average $^{206}Pb/^{238}U$ date for the two identical analyses indicate a date of 614.5 ± 2.1 Ma.

concordia curve but have a slight range of model $^{206}Pb/^{238}U$ dates that vary between 606 and 619 Ma (Fig. 5c). The weighted average $^{206}Pb/^{238}U$ date for all analyses is 613.8 ± 8.7 Ma with a relatively high MSWD of 13. The best estimate for the emplacement age of the Chicken Park kimberlite is the weighted average $^{206}Pb/^{238}U$ date for fractions #24 and 25 of 614.5 ± 2.1 Ma (MSWD=0.08). This age is similar to, but distinctly younger than, the circa 640 Ma age reported for the Chicken Park kimberlite based on preliminary paleomagnetic and $^{40}Ar/^{39}Ar$ phlogopite age results (Lester and Larson, 1996). The 614.5 Ma age obtained here for the Chicken Park kimberlite is also similar to a Rb–Sr phlogopite age of circa 600 Ma obtained for the nearby George Creek kimberlite dykes (Carlson and Marsh, 1989). The age data for Chicken Park, along with other kimberlite emplacement ages, indicates that there could be at least three episodes of kimberlite magmatism in the State Line field ranging in age from Eocambrian (Chicken Park) to Devonian (Estes Park).

3. Discussion

3.1. Timing of kimberlite magmatism in the Slave craton

A compilation of radiometric age determinations for kimberlites located within western North America, but dominantly those within the Slave and Wyoming cratons, is presented in Table 3. At the present time there are age dates available for 25 kimberlites from the Slave craton, which represents less than 10% of the more than 326 known kimberlite occurrences. Kimberlite magmatism in the Slave craton spans a period of time nearly 500 my in duration from the Cambrian 542 Ma Kennady Lake AK5034 pipe (this study) to the Eocene (47.5–56.0 Ma) pipes in the Lac de Gras field (Davis and Kjarsgaard, 1997; Graham et al., 1999). The large numbers of kimberlite clusters and fields in the Slave craton, all generally within 150 km of the Lac de Gras area (excluding the Victoria Island kimberlites), have discrete emplacement ages spanning most of the Phanerozoic. In this respect, the Slave craton is identical to the highly diamondiferous Type 3 kimberlite provinces of southern Africa and Yakutia, as previously defined by Mitchell (1986). The Slave Province kimberlites depicted in Fig. 2 demonstrate a distinct spatial correlation with certain periods of kimberlite magmatism. On this basis the Slave craton has been crudely subdivided into four kimberlite age

segments or domains (heavy dashed lines in Fig. 2): (I) a southwestern Slave domain characterized by Siluro-Ordovician kimberlite magmatism (435 Ma Orion, 441 Ma Drybones Bay, 450 Ma Cross and 459 Ursa kimberlites), (II) a southeastern Slave domain characterized by Cambrian kimberlite magmatism (542 Ma AK5034 kimberlite and 523–535 Ma Snap Lake dyke), (III) a central Slave domain dominated by Cretaceous and Eocene kimberlite magmatism (48–74 Ma kimberlites centered around Lac de Gras) and, (IV) a northern mixed domain consisting of Jurassic kimberlites in the Contwoyto field (e.g. ~ 173 Ma Jericho pipe; Heaman et al., 2002) and, much further north, the Permian kimberlite field (~ 256–286 Ma; Kahlert, personal communication, this study) recognized on Victoria Island (e.g. Golden Plover, Longspur, Phalarope, and Snowy Owl; Fig. 1). Permian and Triassic kimberlite magmatism is generally rare in North America but one other example does exist in southeastern British Columbia; the 241 Ma Cross Creek kimberlite (Smith et al., 1988).

At present there is no obvious explanation for this apparent domain pattern of kimberlite emplacement. The inferred N–S trending boundary separating the southwestern Slave Siluro-Ordovician domain from the other domains is approximately coincident with the exposed eastern margin of the Anton terrane (or Central Slave Basement Complex of Bleeker et al., 1999); the geologically ancient (pre-2.8 Ga) part of the Slave craton. Interestingly, these Siluro-Ordovician kimberlites (Drybones Bay, Ursa, Orion, Cross and possibly Aquila) align along a N–S linear trend, which parallels the Beniah Lake fault—a major intra-Slave N–S crustal-scale fault system. Despite the fact that there may be a coincidence in the position of a major Slave Province structure and the eastern boundary of the Siluro-Ordovician Domain I kimberlite domain, there are no obvious geological controls that might coincide with the other kimberlite age domains.

The Eocene to Cretaceous kimberlite magmatism in the central Slave Domain III is focused within a 50-km^2 area in the Lac de Gras region. Four kimberlites from the Diavik cluster have Rb–Sr phlogopite dates in the range 54.8 and 56.0 Ma. One of these (A154S) has been dated in two different laboratories yielding two identical Rb–Sr age determinations within analytical error

Table 3
Compilation of kimberlite emplacement ages, western North America

Pipe name	Province/State	Field	Age	Method	Reference
Cross Creek	BC	Cross Creek	241 ± 5	Rb–Sr phlogopite	Smith et al., 1988
BHP/92-F	NWT	Lac de Gras	47.5 ± 1.6	Rb–Sr phlogopite	Amstrong and Moore, 1998
BHP/92-C	NWT	Lac de Gras	52.1 ± 1.0	Rb–Sr phlogopite	Amstrong and Moore, 1998
North Paul	NWT	Lac de Gras	47.5 ± 0.5	Rb–Sr phlogopite	Davis and Kjarsgaard, 1997
Panda	NWT	Lac de Gras	53.2 ± 3.8	Rb–Sr phlogopite	Carlson et al., 1999a,b
Leslie	NWT	Lac de Gras	53.9 ± 2.0	Rb–Sr phlogopite	Berg and Carlson, 1998
A154S	NWT	Lac de Gras	54.8 ± 0.3	Rb–Sr phlogopite	Graham et al., 1999
A418	NWT	Lac de Gras	55.2 ± 0.3	Rb–Sr phlogopite	Graham et al., 1999
A154S	NWT	Lac de Gras	55.5 ± 0.5	Rb–Sr phlogopite	Graham et al., 1999
A21	NWT	Lac de Gras	55.7 ± 2.1	Rb–Sr phlogopite	Graham et al., 1999
A154N	NWT	Lac de Gras	56.0 ± 0.7	Rb–Sr phlogopite	Graham et al., 1999
C13	NWT	Lac de Gras	73.7 ± 3.2	U–Pb perovskite	this study
Anne (HL-10)	NWT	Hardy Lake	70.5 ± 6.2	U–Pb perovskite	this study
Anne	NWT	Hardy Lake	72 ± 2	Ar/Ar phlogopite	Scott Smith and McKinlay, 2002
Drew	NWT	Hardy Lake	73 ± 1	Ar/Ar phlogopite	Scott Smith and McKinlay, 2002
HL1	NWT	Hardy Lake	71 ± 2	Ar/Ar phlogopite	Scott Smith and McKinlay, 2002
HL2	NWT	Hardy Lake	73 ± 1	Ar/Ar phlogopite	Scott Smith and McKinlay, 2002
DryBones Bay	NWT	Great Slave	441.4 ± 0.8	U–Pb zircon 943902	this study
Cross	NWT	Great Slave	450.4 ± 2.2	U–Pb zircon	this study
Ursa	NWT	Great Slave	459 ± 10	U–Pb zircon	this study
Orion	NWT	Great Slave	435.4 ± 2.8	U–Pb perovskite	this study
AK5034	NWT	Kennady Lake	542.2 ± 2.6	Rb–Sr phlogopite	this study
Snap Lake	NWT	Snap Lake	522.9 ± 6.9	Rb–Sr phlogopite	Agashev et al., 2001
Snap Lake	NWT	Snap Lake	535 ± 11	Rb–Sr phlogopite	Agashev et al., 2001
Jericho JD-1/3	Nunavut	Contwoyto	173.1 ± 1.3	Rb–Sr phlogopite	Heaman et al., 2002
Phalarope	Nunavut	Victoria Island	256.0 ± 3.0	Ar/Ar phlogopite	Kahlert, personal communication, 2002
Snowy Owl	Nunavut	Victoria Island	257.0 ± 3.0	Ar/Ar phlogopite	Kahlert, personal communication, 2002
Snowy Owl	Nunavut	Victoria Island	271.2 ± 2.6	U–Pb perovskite	this study
Longspur	Nunavut	Victoria Island	276.0 ± 7.0	Ar/Ar phlogopite	Kahlert, personal communication, 2002
Golden Plover	Nunavut	Victoria Island	286.0 ± 4.0	Ar/Ar phlogopite	Kahlert, personal communication, 2002
Mountain Lake	Alberta	Mountain Lake	75 ± 10	apatite fission track	Leckie et al., 1997b
Phoenix	Alberta	Birch Mountains	70.3 ± 1.6	U–Pb perovskite	Aravanis, 1999
Phoenix	Alberta	Birch Mountains	70.9 ± 0.4	Rb–Sr phlogopite	Aravanis, 1999
Dragon	Alberta	Birch Mountains	72.4 ± 0.9	Rb–Sr phlogopite	Aravanis, 1999
Xena	Alberta	Birch Mountains	72.6 ± 2.1	Rb–Sr phlogopite	Aravanis, 1999
Valkyrie	Alberta	Birch Mountains	75.8 ± 2.7	U–Pb perovskite	Aravanis, 1999
Legend	Alberta	Birch Mountains	77.6 ± 0.8	Rb–Sr phlogopite	Aravanis, 1999
K7A	Alberta	Buffalo Hills	86 ± 3	U–Pb perovskite	Carlson et al., 1999a,b
K5	Alberta	Buffalo Hills	87 ± 3	U–Pb perovskite	Carlson et al., 1999a,b
K14	Alberta	Buffalo Hills	88 ± 5	U–Pb perovskite	Carlson et al., 1999a,b
Williams	MO	Missouri Breaks	48.0 ± 2.5	K–Ar phlogopite	Marvin et al., 1980
Estes Park	CO	State Line	386 ± 9	Rb–Sr phlogopite	Smith, 1979
Iron Mountain	WY	State Line	400	Rb–Sr phlogopite	Smith, 1979
Iron Mountain	WY	State Line	408.4 ± 2.6	U–Pb perovskite	this study
George Creek	CO	State Line	600	Rb–Sr phlogopite	Carlson and Marsh, 1989
Chicken Park	CO	State Line	614.5 ± 2.1	U–Pb perovskite	this study
Chicken Park	CO	State Line	640	Ar/Ar phlogopite	Lester and Larson, 1996
Green Mountain	CO	State Line	<780	Ar/Ar phlogopite	Lester and Larson, 1996

$(54.8 \pm 0.3$ and 55.5 ± 0.5 Ma). The five dated kimberlites from the Ekati cluster have slightly younger Rb–Sr phlogopite ages of 47.5–53.9 Ma. Eocene magmatism is common in western North America, especially in central Montana, and includes the 48 Ma Williams kimberlite (Marvin et al., 1980).

3.2. Timing of kimberlite magmatism in the Wyoming craton

More than 90 kimberlite pipes and dykes occur in the State Line field of Colorado/Wyoming (Lester and Larson, 1996) however very few of these have any type of robust age constraint. The Eocene (48 Ma) Williams kimberlite (Marvin et al., 1980) is part of the Missouri Breaks alkaline magmatism and is significantly younger than any State Line field kimberlites. It is located off the northern edge of the Wyoming craton, so is not discussed further in this section. Based on radiometric dating of five intrusions, the kimberlite magmatism in the Wyoming craton is Paleozoic or older and spans a period from 615 to 386 Ma, forming two age clusters (Devonian and Eocambrian). The oldest confidently dated kimberlite in the State Line field is Chicken Park (614.5 ± 2.1 Ma U–Pb perovskite, this study). A slightly older ^{40}Ar/^{39}Ar groundmass phlogopite date of 640 Ma was previously reported for Chicken Park by Lester and Larson (1996). An even older ^{40}Ar/^{39}Ar 780 Ma megacryst phlogopite "pseudo-plateau" date has been reported for the Green Mountain intrusion (Lester and Larson, 1996). However it is likely that the age spectrum obtained for this sample indicates the presence of excess argon (e.g. see discussion on excess argon in kimberlitic phlogopite megacrysts by Phillips, 1991) so the 780 Ma date should be considered a maximum estimate for the emplacement age of the Green Mountain intrusion.

The youngest period of kimberlite magmatism currently recognized in the Wyoming craton is Devonian (386–408 Ma) and includes the 386 ± 9 Ma Estes Park (Rb–Sr phlogopite; Smith, 1979) and the 408.4 ± 2.6 Ma Iron Mountain kimberlites (Smith, 1979; this study). Based on a study of fossiliferous limestone xenoliths, Chronic et al. (1969) documented five other kimberlites (Aultman, Ferris, Nix, Shaffer, and Sloan) that have Ordovician to Silurian maximum age constraints so these bodies could also have Devonian emplacement ages. The Devonian kimberlite magmatism in the State Line field is 30–70 my younger than Siluro-Ordovician kimberlites in the western Slave craton (435–460 Ma; Drybones Bay, Cross, Orion, and Ursa).

3.3. Constraints on the origin of kimberlite-general

There are basically two aspects of kimberlite formation that most researchers do agree on; (1) the geochemistry of kimberlites, in particular the pronounced LREE enrichment, requires some form of metasomatism in the source region prior to melt formation and (2) the ultrabasic nature of kimberlites requires generation, by small degree partial melting, of fertilized peridotite. There is no consensus regarding the depth at which kimberlite magmas are generated or what triggers kimberlite melt generation. Numerous origins have been proposed for kimberlite genesis and include; (1) decarbonation/dehydration of subducted oceanic crust and partial melting of overlying mantle (e.g. Sharp, 1974; Helmstaedt and Schulze, 1979; Helmstaedt and Gurney, 1984; McCandless, 1999), 2) small volume melts associated with mantle plume hotspot magmatism (e.g. Crough et al., 1980; Crough, 1981; England and Houseman, 1984; le Roex, 1986; Haggerty, 1994; Gibson et al., 1995; Heaman and Kjarsgaard, 2000; Schissel and Smail, 2001), (3) continental extension of oceanic transform faults (e.g. Sykes, 1978; Taylor, 1984), (4) melting of plume-enriched subcontinental mantle along zones of failed intercontinental rifts (e.g. Phipps, 1988) or propagating rifts associated with continental break-up (Phillips et al., 1998), (5) lower mantle volatile fluxing, asthenosphere partial melting and lithosphere fracture propagation (e.g. Wyllie, 1980, 1989), (6) melting of a mixed mantle source characterized by pre-kimberlite veins that host carbonate and hydrous mineral assemblages (Foley, 1991), and (7) transition zone (~ 410–650 km) partial melting of refertilized pyrolite in a mixed boundary layer (Ringwood et al., 1992) or peridotite (Edgar and Charbonneau, 1993).

Of the many models that have been proposed for the origin of kimberlite, the first two listed above (i.e. a link to subduction of oceanic lithosphere or mantle plume hotspot tracks) have received the most attention. The essence of the subduction hypothesis has been summarized by McCandless (1999). During subduction beneath continents, the subducted oceanic crust reaches conditions where entrapped fluids are released from the slab and this fluid fluxing will promote small degrees of partial melting in the overlying mantle and generation of kimberlite magmas. As the velocity of subduction is reduced or ceases then heat migrates progressively up the slab, releasing more fluid and generating a temporal pattern of kimberlite magmatism that gets progressively younger towards the trench. Two examples were used to support this hypothesis; (1) the general younging of North American kimberlite magmatism from Jurassic in the east to Eocene/Cretaceous in the west, which was interpreted to reflect a link between kimberlite magmatism and subduction of the Farallon plate beginning at about 200 my ago and (2) the general westward younging of kimberlites in southern Africa from about 140–60 Ma that could be linked to a foundered east-dipping subduction zone beneath Gondwanaland (west Gondwanaland megalith of Helmstaedt and Gurney, 1997) prior to the opening of the South Atlantic Ocean at about 130 Ma (McCandless, 1999). Further support for the subduction hypothesis in North America was derived from global seismic tomography, which shows a large tabular anomaly of relatively fast seismic velocity descending eastward beneath North America that Grand et al. (1997) interpret as the relict Farallon plate.

There are two major difficulties in accepting this hypothesis for the origin of the majority of Cenozoic/Mesozoic North American kimberlites. The first is the timing and location of the oldest Jurassic kimberlites (e.g. ∼ 200 Ma Rankin Inlet and ∼ 180 Ma Attawapiskat fields). The emplacement of these kimberlite fields would correspond to the timing of the initial stages of subduction of the Farallon plate but are located more than 2000 km from the site of subduction (i.e. at 200–180 Ma the subducted Farallon plate is only beginning to have an impact on the tectonics of western North America, not in eastern North America). The second difficulty relates to the origin of the geochemical signature of kimberlites. It is now well established that kimberlite magmas must be derived from peridotite source regions that have had a multistage history involving initial depletion (removal of basalt or komatiite melt), metasomatism and final melting of a phlogopite-garnet peridotite (Wyllie, 1980; Eggler, 1989; Tainton and McKenzie, 1994). Although release of fluids from subducted oceanic crust might initiate melting of previously metasomatized peridotite in subcontinental lithosphere regions (Helmstaedt and Gurney, 1997; McCandless, 1999), such fluids will not be sufficiently enriched in incompatible or high field strength elements to represent the requisite precursor metasomatic agent necessary to explain many of the geochemical features of kimberlite magmas, such as high Nb contents (100–300 ppm). As pointed out by Green and Pearson (1987), the Nb content of fluids or small degree partial melts derived from a mafic/ultramafic source will be quite low, especially if a residual Ti-bearing mineral such as rutile remains in the source. It is unlikely therefore that a one-stage dewatering or melting of oceanic crust containing approximately 2 ppm Nb (typical MORB) will enrich the overlying mantle wedge sufficiently in high field strength elements to account for the HFSE composition of kimberlites.

In addition to the arguments presented above for the Cenozoic/Mesozoic kimberlite magmatism, there is no temporal link between the timing of pre-Mesozoic kimberlite magmatism and known periods of pre-Mesozoic subduction. For example, Devonian and Siluro-Ordovician kimberlites are abundant in the Slave and Wyoming cratons of the western mixed domain (Fig. 1) but are far removed from the site of Taconic subduction (∼ 550–450 Ma; Waldron et al., 1998) in eastern North America. In addition, Archean cratons (e.g. Slave, Superior, Wyoming) now comprising the core of the Precambrian Shield in North America were all subjected to major episodes of Paleoproterozoic (1.9–1.7 Ga) subduction. If there is a genetic relationship between subduction and kimberlite formation then there should be abundant Paleoproterozoic kimberlites in North America—none are known at present. In fact, there is evidence of Paleoproterozoic metasomatised eclogite xenoliths in the Jericho kimberlite, N.W.T. that could be a relict piece of east-dipping subducted oceanic crust beneath the Slave craton (Heaman et al., 2002), however there is currently no indication of Paleoproterozoic kimberlite magmatism in the Slave Province.

An origin for kimberlite magmas linked to mantle plume hot spots has been proposed by many researchers but gained popularity following the studies of Crough et al. (1980) and Crough (1981) where it was shown that many Mesozoic kimberlites in North America, South America and South Africa are located within five degrees of predicted hotspot tracks. Evidence in support of this hypothesis derives largely from the observation that kimberlite clusters and fields (primarily in southern Africa, Brazil and eastern North America) are located along continental extensions of aseismic ridges and seamount chains that are produced as a result of magmatism associated with the trace of a current oceanic hotspot track (Crough et al., 1980; le Roex, 1986; Haggerty, 1994). Proposals for the actual site of kimberlite magma formation associated with mantle plumes is quite variable and could be at the core/mantle boundary (e.g. Haggerty, 1994), mantle transition zone (e.g. Ringwood et al., 1992) or sub-continental lithospheric mantle (in the case of the Group II kimberlites, e.g. le Roex, 1986). It is clear that if most diamonds are extricated near their site of formation then some kimberlite magmas must originate or pass through the mantle transition zone as rare majorite garnet (an ultrahigh pressure form of garnet stable at depths >400 km; Ringwood and Major, 1971) inclusions occur in diamonds (e.g. Moore et al., 1991; Sautter et al., 1991; Stachel et al., 2000). The arguments put forward against a hotspot model for kimberlite genesis are the fact that oceanic mantle plumes such as Hawaii produce volumetrically large amounts of magma, quite in excess of the amount of kimberlite magmatism preserved in the geological record, and that "no presently known hotspot tracks correlate with the temporal-spatial patterns of kimberlites in southern Africa or in North America" (McCandless, 1999). The 'volume of magma' argument of McCandless (1999) is difficult to assess because it is unknown what proportion of plume related magmatism in continental settings (basaltic, kimberlitic or otherwise) is underplated or solidifies in the subcontinental mantle lithosphere.

One of the major obstacles in addressing the temporal-spatial relationship of kimberlite magmatism has been the paucity of precise and accurate age determinations for kimberlites (especially in North America). However, over the past decade there have been a number of detailed studies of temporal-spatial patterns of kimberlite emplacement and at least three areas do show age progression corridors of kimberlite magmatism that correlate with independent estimates for the continental extensions of current oceanic mantle plume hotspot tracks: (1) Cretaceous (140–110 Ma) Group II kimberlites in South Africa (Crough et al., 1980; Haggerty, 1994), linked to Shona, Bouvet or Tristan plumes (Hartnady and le Roex, 1983; le Roex, 1986; Duncan et al., 1978), (2) Cretaceous (90–50 Ma) Alto Paranaíba Igneous Province in Brazil, linked to the Trindade plume (Gibson et al., 1995), and (3) Jurassic (200–140 Ma) kimberlite

fields in eastern North America, linked to the Great Meteor plume (Heaman and Kjarsgaard, 2000). Although we do not propose that all kimberlites are linked to mantle plume hotspot tracks, there is evidence that at least some can be explained in this way.

In the subduction and hotspot hypotheses, the duration of kimberlite magmatism is on the order of 100–150 and 30–60 my, respectively, but the spatial relationships are quite different. Kimberlites linked to hotspot tracks will form along a relatively narrow path (100–300 km wide) that youngs towards the rifted plate margin, consistent with the pattern of most modern oceanic plume hotspot tracks (Morgan, 1983). In contrast, kimberlite magmatism linked to subduction (e.g. as proposed by McCandless, 1999) will display corridors of kimberlite magmatism of similar age that can span distances of continental proportion and will young towards the trench and convergent plate margin.

In order to evaluate whether there is any correlation between the timing of kimberlite magmatism in North America, we have indicated on Fig. 6 the timing of known major tectonic processes, such as subduction or rifting events. It is difficult to do a similar comparison for the Russian database because details of the dated kimberlites (locations, etc.) are often not given in the data summaries (e.g. Brakhfogel, 1995). However, it is interesting to note that the Triassic kimberlite magmatism in Russia coincides with the timing of Siberian flood basalt volcanism (e.g. Baksi and Farrar, 1991) and the coeval Noril'sk ultramafic intrusion at 251.2 ± 0.3 Ma (Kamo et al., 1996).

During the Eocambrian and Phanerozoic, major rifting events include opening of the Iapetus Ocean (initiation of this event occurred at about 615 Ma in NE Laurentia based on U–Pb dating of the Long Range dyke swarm; Kamo et al., 1989), and the North Atlantic (~ 200 Ma; Dunning and Hodych, 1990). Not shown in Fig. 6 is the rifting of western Rodinia at about 780 Ma based on Gunbarrel igneous events (LeCheminant and Heaman, 1994; Park et al., 1995). In addition, the opening of the South Atlantic (~ 133 Ma; Renne et al., 1992) is shown for comparison in Fig. 6. The timing of major periods of subduction in North America are also shown in Fig. 6 and comprise Taconian orogenesis along the east coast (~ 550–450 Ma; Waldron et al., 1998) and subduction of the Farallon plate between 200 Ma to present beneath western North America.

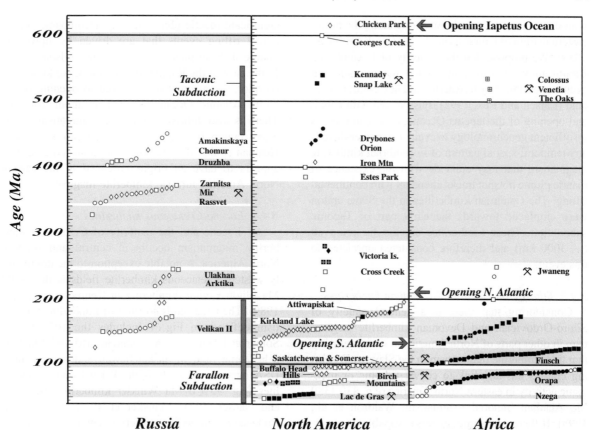

Fig. 6. Compilation of kimberlite age determinations from North America, southern Africa and Russia based primarily on Rb–Sr phlogopite (squares), ^{40}Ar/^{39}Ar phlogopite (square with cross), U–Pb perovskite (diamonds) and U–Pb mantle zircon (circles). The shaded bands indicate major periods of kimberlite magmatism (i.e. six or more known kimberlites emplaced in that period). Also shown are major periods of Phanerozoic/Eocambrian rifting (large arrows) and fields where existing or planned diamond mining occurs (mining symbol). The Slave and Kaapvaal craton kimberlites are denoted with solid symbols in the North America and Africa panels, respectively. The majority of kimberlites in the 105–120 Ma band in the African compilation are Group II kimberlites. Data sources: *Africa*—Allsopp and Barrett (1975), Davis (1977), MacIntyre and Dawson (1976), Davis (1978a,b), Raber (1978), Fitch and Miller (1983), Kramers and Smith (1983), Kramers et al. (1983), Allsopp and Roddick (1984), Allsopp and Hargraves (1985), Allsopp et al. (1985), Brown (1985), Smith et al. (1985), Kinny et al. (1989), Allsopp et al. (1989), Gobba (1989), Smith et al. (1989), Schärer et al. (1992), Smith et al. (1994), Phillips et al. (1999). *North America*—Smith (1979), Marvin et al. (1980), Barnett et al. (1984), Smith et al. (1988), Carlson and Marsh (1989), Heaman (1989), Paces et al. (1991), Duskin and Jarvis (1993), Kjarsgaard and Heaman (1995), Hegner et al. (1995), Lester and Larson (1996), Davis and Kjarsgaard (1997), Bikerman et al. (1997), (Davis and Kjarsgaard, 1997; Graham et al., 1999), Armstrong and Moore (1998), Berg and Carlson (1998), Carlson et al. (1995, 1999a,b), Agashev et al. (2001), Heaman and Kjarsgaard (2000), Heaman et al. (2002), Heaman and Kjarsgaard, (2002, unpublished data). *Russia*—Davis et al. (1980), Komarov and Ilupin (1990), Heaman and Mitchell (1995), Pearson et al. (1996), Kinny et al. (1995, 1997) plus data summary by Brakhfogel, 1995.

3.4. Constraints on the origin of North American kimberlites

3.4.1. Eocambrian/Cambrian kimberlites

North American Eocambrian and Cambrian kimberlite magmatism is currently recognized in the Wyoming Craton (615–600 Ma; Chicken Park, George Creek and possibly Green Mountain kimberlites), Slave Craton (542–523 Ma; Snap and Kennady Lake kimberlites) and in the Labrador Sea Province (580–550 Ma; Beaver Lake kimberlite and Abloviak dykes). The Eocambrian period of kimberlite magmatism predates estimates for the timing of Taconic subduction (e.g. 550–450 Ma; Waldron et al., 1998) but overlaps with the timing of Eocambrian rifting along the north-eastern margin of Laurentia at 615 Ma (e.g. Kamo et al., 1989). We propose that the majority of Eocambrian/ Cambrian kimberlite and related magmatism that occurs in eastern North America and in western Green-land (Larsen and Rex, 1992) is associated with rifting and opening of the Iapetus Ocean. At present there is insufficient geochronology to evaluate whether there is any temporal-spatial pattern of Eocambrian kimberlite magmatism that may coincide with the existence of mantle plume hotspot tracks attendant with continental rifting. The Cambrian kimberlites in the Slave craton were emplaced towards the early part of Taconic subduction. These kimberlites are too far removed (~ 3000 km) and therefore considered unrelated to this event.

3.4.2. Siluro-Ordovician and Devonian kimberlites

Considering that there is a general paucity of Siluro-Ordovician and Devonian kimberlite magma-tism in other parts of North America, it seems unlikely that this kimberlite magmatism in the Slave or Wyom-ing cratons has any link to coeval tectonic or mag-matic activity in eastern North America (e.g. 430–380 Ma Acadian bimodal magmatism; Waldron et al., 1998). If there was some processes capable of gener-ating kimberlite magmas linked to Acadian magma-tism, which is not subduction related, then it might be anticipated that Paleozoic kimberlite magmatism would be much more prevalent in some of the recent fields discovered in Québec and Labrador. There are currently no known kimberlites of this age in eastern North America (e.g. Digonnet et al., 2000; Moorhead et al., 2002).

3.4.3. Jurassic kimberlites

The North American Jurassic kimberlite magma-tism provides some insight into one possible cause of protracted periods of kimberlite magmatism. The majority of Jurassic kimberlites in North America occur in Ontario and follow a NW–SE trending corridor between the Rankin Inlet and Kirkland Lake fields. An important observation for this kimberlite magmatism is that it follows a pattern of southeast-ward younging, which is coincident with the previ-ously hypothesized continental trace of the Great Meteor mantle plume hotspot track (Heaman and Kjarsgaard, 2000). In such cases, it is possible that

significant mantle plume activity accompanies conti-nental rifting events that are driven by upwelling mantle asthenosphere (Weinstein and Olson, 1989). In the case of the Rankin to Kirkland Lake kimberlite corridor, the magmatism is linked to opening of the North Atlantic Ocean, initiated at about 200 Ma. There is also Jurassic kimberlite magmatism in the Slave craton (e.g. the 173 Ma Jericho kimberlite, Heaman et al., 2002; Heaman, unpublished data) that appears to have an origin unrelated to the eastern North America Jurassic kimberlite magmatism.

3.4.4. Eocene-Cretaceous kimberlites

All Eocene and the majority of Cretaceous kim-berlite magmatism occurs in central and western North America. A notable exception to the dominant-ly western Cretaceous kimberlite fields is the ~ 90 Ma Elliott County kimberlite in Kentucky (Heaman, 1989). The Cretaceous corridor of kimberlite magma-tism defined on Fig. 1 includes the 103–94 Ma Somerset Island field in Nunavut, 101–95 Ma Fort à la Corne field in Saskatchewan and some central USA occurrences such as the Riley County kimber-lites (e.g. 95 ± 6 Ma Winkler kimberlite; Brookins and Naeser, 1971; Mansker et al., 1987). Several Cretaceous kimberlite clusters occur in the Western Mixed Domain (Fig. 1; Table 3) and include the 88–86 Ma Buffalo Head Hills field and 78–70 Ma Birch Mountains cluster in Alberta and some kimberlites in the Lac de Gras field, such as the 73–70 Ma Hardy Lake cluster. Eocene kimberlite magmatism is preva-lent in the Lac de Gras field, N.W.T. and in the Missouri Breaks region of Montana (e.g. 48 Ma Williams kimberlite; Marvin et al., 1980).

There are some noticeable temporal-spatial patterns of Cenozoic/Mesozoic kimberlite magmatism in west-ern North America. The most notable is the Creta-ceous corridor (Fig. 1), where kimberlites in a number of fields extending from Somerset Island to Kansas have emplacement ages in the narrow time range between 103 and 94 Ma. Equally striking is the corridor-like pattern of Maastrichtian/Campanian kim-berlites that occur in the Lac de Gras field (e.g. 73–71 Ma Hardy Lake cluster) and in Alberta (e.g. 78–70 Ma Birch Mountains cluster). Although it is more difficult to define a corridor of Eocene kimberlite magmatism, the 56–48 Ma Lac de Gras and much of the Eocene Missouri Breaks magmatism does occur west of most Cretaceous kimberlite occurrences (i.e. west of the thin white dashed line in Fig. 1) and interestingly coincides with a major flare-up of Eocene volcanic activity in the Cordillera (Armstrong, 1988).

The origin of the Cretaceous/Eocene kimberlite magmatism in western North America cannot be linked to a mantle plume hotspot track. The only possible track that could be proposed involves a NNE–SSW Permian to Jurassic track segment extending from Victoria Island to the Contwoyto cluster then a change to a more SSE direction and ending in the Cretaceous/Eocene approximately in the Lac de Gras region (a total distance of approximately 500 km). Considering only the Jurassic to Eocene segment of this possible track, a relative northwestern directed plate motion for North America in the period 173–48 Ma of approximately 1–2 mm/year is calculated. This rate is considerably slower than other estimates for the North American plate motion during the Mesozoic (20–60 mm/year) based on the Great Meteor Hotspot Track (e.g. Sleep 1990, Heaman and Kjarsgaard, 2000 and references therein). The proposed Labrador hotspot track (Morgan, 1983) does pass through Saskatchewan and Alberta in mid-Cretaceous times but the track location (south of Great Slave Lake) and orientation makes it unlikely that it is linked in any way to the Slave craton kimberlites. In addition, this tracks youngs towards the east, opposite to the westward younging of kimberlite magmatism in Saskatchewan and Alberta. We therefore conclude that there is no convincing evidence at present that the Cenozoic/Mesozoic kimberlite magmatism in the western North America is linked to a mantle plume hotspot track.

One pattern of kimberlite emplacement revealed from the western North American database is the progressive younging of ~ N–S trending corridors of Cretaceous to possibly Eocene kimberlite magmatism (Fig. 1). At least one of these corridors, containing the Cretaceous Somerset Island, Fort à la Corne and Riley County clusters, transects most of North America. We observe that slightly younger corridors of kimberlite magmatism parallel this Cretaceous corridor (e.g. 80–70 Ma Hardy Lake and Birch Mountains clusters and 56–48 Ma Missouri Breaks and Lac de Gras clusters) and these corridors become progressively younger westward. The thin dashed

white line on Fig. 1 denotes the approximate eastern limit of known Tertiary magmatism. Such corridors of kimberlite magmatism that generally mirror the margin of the continent and young towards the convergent margin represent a pattern consistent with the subduction hypothesis (Sharp, 1974; Helmstaedt and Gurney, 1984; McCandless, 1999). However, the geochemical arguments presented above are inconsistent with a single-stage derivation of kimberlite magmas from melting of mantle wedge material triggered by dewatering of subducted oceanic crust.

One possible explanation for the observed pattern of Cretaceous/Tertiary kimberlite magmatism in western North America is that during this time the margin was subjected to large-scale extension. We note that Eocene extension and associated bimodal magmatism is prevalent throughout the Cordillera of North America (e.g. Lipman et al., 1972; Armstrong, 1988), and has been attributed to a variety of tectonic processes. For example, an origin linked to lithospheric decompression (Morris et al., 2000), back-arc spreading (Ewing, 1981) and asthenosphere upwelling (O'Brien et al., 1995) possibly linked to the Kula-Farallon slab window (Thorkelson and Taylor, 1989; Breitsprecher and Thorkelson, 2001) have been proposed. At present it is unclear whether the origin of the Eocene kimberlite magmatism in the Missouri Breaks and Lac de Gras regions could be linked to Eocene crustal extension caused by changes in geometry to Kula-Farallon plate subduction or upwelling of asthenosphere.

3.5. Temporal patterns of kimberlite magmatism in North America, Southern Africa and Russia

A compilation of kimberlite emplacement ages from North America, Russia (mostly from Yakutia) and southern Africa (but mostly from South Africa) are presented in Fig. 6. A number of radiometric dating techniques have been used and are denoted with different symbols: Rb–Sr and $^{40}Ar/^{39}Ar$ phlogopite (square and square with cross), U–Pb perovskite (diamond), and U–Pb mantle zircon (circle). The Slave craton kimberlites represented in the North America panel and the Kaapvaal craton kimberlites in the Africa panel are denoted by solid symbols. In addition, kimberlite fields that are particularly noted for their diamond production are indicated with a mining symbol. The total number of robust published age determinations for kimberlites from Africa and North America are similar (~ 100) and greater than that available for Russia. Several summaries of Russian kimberlite age dates have been published based on about 500 age determinations for approximately 280 kimberlites (Brakhfogel, 1995). Therefore, in addition to the individual published ages, the yellow shaded bands in Fig. 6 also indicate significant periods of kimberlite magmatism in Russia reported in these data summaries. For example, there are apparently abundant Tertiary, Cretaceous and Eocambrian kimberlites in Russia, but the details of the dating have not been published.

There are many interesting features of this kimberlite age database presented graphically in Fig. 6. The majority of dated Phanerozoic kimberlites in both Africa (95%) and North America (83%) are Cenozoic/Mesozoic (i.e. younger than ~ 250 Ma). The youngest kimberlite magmatism in all three panels is Eocene and includes the diamondiferous 53–52 Ma Mwadui kimberlite, Tanzania (Davis, 1977) and the 56–48 Ma Lac de Gras area kimberlites in the Slave Province, Canada (e.g. Davis and Kjarsgaard, 1997; Graham et al., 1999). The oldest known kimberlite magmatism is not represented in Fig. 6. Examples include Mesoproterozoic to Paleoproterozoic kimberlites such as the 2188 ± 11 Ma Turkey Well kimberlite, Australia (Kiviets et al., 1998), the 1.7–1.6 Ga Kuraman Province kimberlites, Northwest Cape, South Africa (Shee et al., 1989), the ~ 1.73 Ga Guaniamo intrusions, Venezuala (Nixon et al., 1994), the ~ 1.18 Ga Premier and National kimberlites, South Africa (Smith, 1983; Allsopp et al., 1989), the 1.1 Ga kimberlites in south India (Kumar et al., 1993), and the ~ 1.1 Ga Bachelor Lake, Whitefish Lake and Kyle Lake #5 kimberlites/ultramafic lamprophyres in North America (Sage, 1996; Kaminsky et al., 2000; Heaman, unpublished data).

For the following discussion a distinction is made between currently recognized major periods of kimberlite magmatism, where six or more dated pipes have been identified from one or more fields in a particular craton within a <40 my time interval (denoted with a yellow shaded band in Fig. 6). It is possible to recognize two global patterns of kimberlite emplacement worldwide. The first involves multiple discrete periods (duration of 20 my or less) of kim-

berlite magmatism and is best exemplified in the Russian database. In Yakutia, at least seven main kimberlite events of relatively short duration are recognized (50–60, 95–105, 150–170, 220–240, 350–370, 400–410, and 590–605 Ma). In North America at least six periods of Cenozoic/Mesozoic kimberlite can be discerned: (1) 48–56 Ma (Lac de Gras), (2) 70–78 Ma (Lac de Gras, Birch Mountains), (3) 86–88 Ma (Buffalo Head Hills), (4) 94–103 Ma (Saskatchewan and Somerset Island), (5) 140–160 Ma (Kirkland Lake, Timiskaming, Finger Lakes) and (6) 170–180 Ma (Attawapiskat). Major periods of pre-Mesozoic kimberlite magmatism are difficult to identify at the moment because there is insufficient geochronology. However, there could be major periods of kimberlite activity in North America in the Permian/Triassic (Cross Creek, Victoria Island kimberlites), Devonian (Estes Park, Iron Mountain), Siluro-Ordovician (Drybones Bay, Orion, Cross, Ursa), Cambrian (Kennady and Snap Lake) and Eocambrian (Chicken Park, George Creek, Beaver Lake and Abloviak dykes).

The second prominent kimberlite emplacement pattern is best portrayed in the Mesozoic kimberlite magmatism of South Africa where there appears to be a continuum of kimberlite magmatism spanning large periods of time (30–50 my or more). It is much more difficult to make a case for discrete periods of kimberlite magmatism in southern Africa as any possible hiatus in kimberlite activity is quite short (i.e. 10–15 my at most). Apart from a possible short hiatus at between 95 and 100 Ma, 69 of the 95 dated kimberlites in South Africa have emplacement ages in the range 70–116 Ma. Although the total number of accurately dated kimberlites in South Africa and North America as a whole is similar, there simply is not enough geochronology available from the Slave craton at present to evaluate whether the notion of discrete "periods" of kimberlite activity noted for some kimberlite fields in North America (e.g. Somerset Island, Fort à la Corne, Kirkland Lake) is applicable to the western mixed domain. For example, very few kimberlites in the Lac de Gras field have been dated so it remains to be seen whether discrete Cretaceous and Tertiary kimberlite "events" truly exist or whether in fact there is a continuum of kimberlite magmatism between 48 and 75 Ma.

The kimberlite age compilation in Fig. 6 provides an ideal opportunity to evaluate 'active' versus 'inactive' periods of kimberlite magmatism during the Phanerozoic and whether "global" kimberlite events are in fact a common feature on Earth. Within the plethora of Cenozoic/Mesozoic kimberlite magmatism noted above, there are possibly four periods of kimberlite activity that could be considered global in nature. These occur in the Tertiary (45–60 Ma), Cretaceous (95–105 Ma), Jurassic (140–160 Ma), and perhaps a less convincing event in the Triassic (215–240 Ma). The Tertiary event is well represented in the Slave craton (kimberlites in the Lac de Gras field emplaced between 48 and 56 Ma), in Yakutia (based on age data summaries; Brakhfogel, 1995) and the Eocene kimberlite pipes in the Singida and Nzega fields, Tanzania (Gobba, 1989). Discrete Cretaceous events in the period 95–105 Ma are best recognized in Russia (based on age data summaries; e.g. Brakhfogel, 1995) and North America (e.g. Fort à la Corne and Somerset Island fields). The correlation of this Cretaceous event to the Kaapvaal craton is rather weak, as there are apparently only four kimberlites in southern Africa emplaced during this interval. There is a remarkable synchronicity of Jurassic kimberlite magmatism in Russia (e.g. 154.7 Ma Velikan II, 156.2 Ma Krisoliovaya, 159.3 Ma Marichka; Davis et al., 1980; Heaman and Mitchell, 1995; age summary of Brakhfogel, 1995) and North America (e.g. 140–160 Ma Kirkland Lake, Timiskaming and Finger Lakes fields; Heaman and Kjarsgaard, 2000). Two Group I (e.g. Ramatseliso, Mzongwana) and the following Group II (Middleputts, Lace, Swartruggens, Melton Wold, Klipfontein; Smith et al., 1985, 1994) kimberlites in South Africa were also emplaced during this interval.

There are very few Triassic kimberlites currently recognized worldwide, however there is a period of Triassic kimberlite activity in Russia (e.g. 245 Ma Ulakhan and 232 Ma Arkitika alnöite; Kinny et al., 1997) and in Botswana, the richly diamondiferous 235 Ma Jwaneng kimberlite is a noteworthy example. The only known Triassic kimberlite in North America is the 241 Ma Cross Creek diatreme in British Columbia (Smith et al., 1988). Slightly older Permian kimberlite magmatism (~ 256–286 Ma) occurs on Victoria Island, Canada.

The age database for pre-Mesozoic kimberlites is sparse and may reflect the fact that kimberlite magmatism has been less prolific earlier in earth history compared to the Cenozoic/Mesozoic activity. Alternatively, it may be more difficult to preserve (or identify) older kimberlites. If the latter is true then there may be no significance to the observation from Fig. 6 that global episodes of kimberlite magmatism are harder to recognize prior to about 240 Ma. Despite the sparse data, there are some interesting correlations in the timing of Eocambrian and Paleozoic kimberlite magmatism. A good age correlation in kimberlite magmatism that is apparent from Fig. 6 is the Siluro-Ordovician event recognized in both North America and Yakutia. There is possibly a contemporaneous Devonian event in Russia (400–410 Ma; West Ukukit, Middle Olenek; Davis et al., 1980; Kumar et al., 1995) and in North America (408 Ma Iron Mountain; this study) and possibly again in the Siluro-Ordovician at 435–460 Ma (e.g. 450 Ma Amakinskaya, Russia; Davis et al., 1980; Brakhfogel, 1995: 435 Ma Orion and 441 Ma Drybones Bay, Slave craton, this study). Another interesting correlation is the similarity in the emplacement ages for the 533 Ma Colossus and 520 Ma Venetia kimberlite pipes in southern Africa (Allsopp and Smith, in press) with the 542 Ma Kennady Lake pipe AK5034 (this study) and 523–535 Ma Snap Lake dyke (Agashev et al., 2001) in the Slave craton. It is even more intriguing that, despite the apparent paucity of Cambrian kimberlite magmatism worldwide, Venetia, Snap Lake and Kennady Lake are all highly diamondiferous (see Fig. 7).

Another interesting global correlation is the Eocambrian kimberlite magmatism in North America, Greenland and Yakutia. Eocambrian kimberlite magmatism in North America occurs in the Wyoming craton (e.g. 600 Ma George Creek and 615 Ma Chicken Park kimberlites in the State Line field; Carlson and Marsh, 1989; this study) and in the Labrador Sea domain (e.g. 550–580 Ma); including the 550.9 ± 3.5 Ma Beaver Lake kimberlite, Québec (Moorhead et al., 2002) and the Abloviak ultrabasic lamprophyres and aillikites in NE Labrador (550 ± 15 Ma ^{40}Ar/^{39}Ar phlogopite date; Digonnet et al., 2000; circa ~ 580 Ma U–Pb perovskite age for K35, Heaman unpublished data). In addition, slightly older lamprophyres, lamproïtes and kimberlites occur in western Greenland (583–607 Ma, Larsen and Rex, 1992) and in Yakutia (590–605 Ma; age summary of Brakhfogel, 1995). Considering the

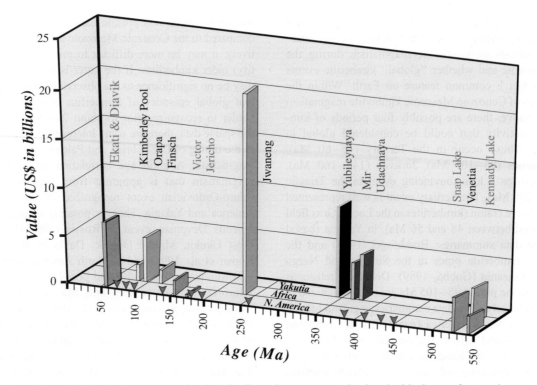

Fig. 7. Compilation of kimberlites that are currently mined for diamonds or are expected to be mined in the near future and corresponding emplacement ages. The value of each deposit is based on average grades, estimated diamond value and tonnage if mined to a depth of 120 m. Red arrows indicate other periods of kimberlite magmatism in North America that are currently not considered to have economic significance. The Triassic Jwaneng pipe has by far the greatest value (~ 20 billion US$). Interestingly, Cambrian kimberlites are not that common worldwide but Venetia, Snap Lake and Kennady Lake have similar emplacement ages and are quite economic with respect to diamond deposit value.

close proximity of Eocambrian kimberlites and related magmatism in Labrador and Greenland, we propose that the Labrador Sea domain identified in Fig. 1 be expanded to include the Eocambrian magmatism in Greenland.

In summary, there are some very active periods of kimberlite magmatism that occurred simultaneously in quite different regions on Earth. This may attest to the global nature of the triggering mechanism for kimberlite formation at these times (e.g. Haggerty, 1994). Furthermore, there is a good correlation between periods of increased kimberlite activity and supercontinent demise and associated enhanced mantle plume activity. One of the best examples of this is the good temporal correlation between increased Mesozoic kimberlite activity recorded in many continents with the breakup of Gondwanaland. However, there are also clearly periods of kimberlite magmatism that are not global in nature. For example, the period of

dominantly Devonian kimberlite magmatism (350–370 Ma) that is prolific in Yakutia, is apparently non-existent in Africa and North America (see Fig. 6).

Just as there are some global periods of kimberlite magmatism, there are also global periods of kimberlite quiescence (gray shaded areas in Fig. 6). The longest time period when kimberlite magmatism appears to be absent occurs during a span of time greater than 100 my centered on the Carboniferous period. Contemporaneous global "quiet" periods of kimberlite magmatism occur between 345–245 Ma in Russia, 380–280 Ma in Canada and from 500–240 Ma in southern Africa. It is not entirely clear what the significance of global "quiet" periods in kimberlite magmatism might represent, but it is worth noting that the circa 350–250 Ma period coincides with a time of relative tectonic stability during the lifetime of the supercontinent Gondwanaland. If the existence of supercontinents has directly or indirectly some influence on the production of kimberlite magmas or their transport to the surface then it can be anticipated that there will be a dearth of kimberlites during other supercontinent stability cycles, such as Pangea and Rodinia.

3.6. Is there a relationship between the timing of kimberlite emplacement and diamond potential?

Kimberlites that contain economic diamond deposits are indicated on Fig. 6 with a mining symbol and it is clear from this diagram that economically important diamond mines and deposits occur in kimberlites that have a variety of ages: 52–56 Ma (Ekati and Diavik, Lac de Gras; Mwadui, Tanzania), 84–94 Ma (e.g. Kimberley Pool, Orapa), 118–124 Ma (e.g. Group II kimberlites Finsch and Roberts Victor), 235 Ma (Jwaneng), 360–376 Ma (e.g. Mir, Udachnaya, Zarnitsa, International) and 523–542 Ma (Venetia, Snap Lake, Kennady Lake). An abbreviated compilation of kimberlites that are currently mined for diamonds, or are expected to be mined in the near future, and corresponding emplacement ages is presented in Fig. 7. The value of each deposit is based on average grades, estimated diamond value and tonnage if mined to a depth of 120 m. In terms of the most valuable deposits, it can be seen on Fig. 7 that in Canada there are many periods of kimberlite magmatism (indicated with red arrows) but the current most economic deposits are the Eocene Ekati and Diavik pipes in the Lac de Gras field, followed by the Cambrian Snap Lake dyke and kimberlites in the Kennady Lake cluster and possibly the Jurassic Victor kimberlite in the Attawapiskat cluster in Ontario. Although at first glance there is a striking correlation between Cambrian kimberlite magmatism and enormous diamond potential in both North America (Snap Lake, Kennady Lake) and South Africa (Venetia), there are several Cambrian kimberlites that are currently not economic (e.g. River Ranch and Colossus). Apart from this excellent correlation between economic diamond deposits and emplacement age for the Cambrian examples listed above, the economic Cretaceous kimberlites in southern Africa do not have temporal economic equivalents in Yakutia or North America. Furthermore, the richly diamondiferous Devonian kimberlite event is only observed in Yakutia (e.g. Mir, Udachnaya), but not in North America or southern Africa.

There is evidence that a certain proportion of diamonds in the Slave subcontinental mantle are much older that the emplacement ages of many kimberlites in the Slave craton. The fact that the 542 Ma Kennady Lake and 523–535 Ma Snap Lake kimberlites are richly diamondiferous indicates that some diamonds in the Slave subcontinental mantle are older than the Cambrian emplacement ages of their kimberlite hosts. In a separate study, Stachel et al. (2002) deduced that the mantle residence times for Panda diamonds is on the order of 10–50 Ma, based on very low aggregation levels of nitrogen impurities so these diamonds could easily have crystallized prior to the majority of Lac de Gras kimberlite magmatism (48–74 Ma). These authors also report that diamond mineral inclusions record a higher geothermal gradient (40–42 mW/m^2) compared to estimates from mantle xenoliths (e.g. Kopylova et al., 1999), consistent with diamond formation prior to establishing the cooler geotherm (38 mW/m^2) recorded in the xenoliths. In addition, eclogite xenoliths (including a small proportion of diamond-bearing eclogite) entrained in the Jericho kimberlite have Paleoproterozoic U–Pb zircon dates, interpreted to reflect a minimum estimate for the eclogite protolith age (Heaman et al., 2002). Some of these eclogites have experienced Precambrian metasomatism and partial melting so the likelihood of Proterozoic diamond formation in these samples is great.

It is clear from the above discussion that at least some diamonds were present in the Slave subcontinental mantle prior to Cenozoic/Mesozoic kimberlite magmatism. The fact that only the Eocene kimberlites in the Lac de Gras field appear to be of economic significance, even though both Eocene and Cretaceous kimberlite pipes occur in close proximity (within less than 2 km) at the present erosional level, indicates that other factors must control the distribution of diamonds in this kimberlite field. The strong correlation between kimberlite emplacement age and diamond potential for the Lac de Gras field is presented in Fig. 8. It is clear from this diagram that the most economic kimberlites in this field, including pipes at the Ekati and Diavik mines, are of Eocene age.

We conclude that, except for an exceptionally high proportion of Cambrian kimberlites, diamond potential is not specifically related to the global patterns of

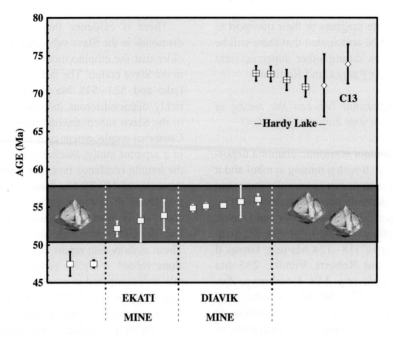

Fig. 8. Summary of kimberlite emplacement ages from the Cretaceous/Eocene central Slave domain illustrating the ~ 28 my time span of kimberlite emplacement in the Lac de Gras area. Note that all economic kimberlites were emplaced within a narrow time window (56–52 Ma) Modified from Kjarsgaard et al. (2002).

kimberlite magmatism. For example, the most economic diamond deposits are hosted in Devonian kimberlites in Russia, Cretaceous and Permian kimberlites in southern Africa and in Eocene kimberlites in North America (Fig. 7). The very fact that old diamonds (>3 Ga) exist beneath Archean cratons (e.g. Richardson et al., 1984) but are only sampled by a specific period of kimberlite magmatism indicates that the control on diamond content and grade must be related to other factors such as the depth of magma generation, magma oxidation state, magma ascent rates, composition and corrosiveness of the magma, and nature of the Archean mantle roots through time.

4. Conclusions

The U–Pb and Rb–Sr age dates obtained in this study for ten kimberlites from the Slave and Wyoming cratons confirms that each craton is a Type-3 kimberlite province with multiple fields emplaced over a long period of time (in excess of 200 and 500 my, respectively). Perhaps most revealing from this study is that the Slave craton can be divided into at least

four age domains of kimberlite magmatism. Although we admit that the kimberlite age database for the Slave craton is small by comparison to other cratons, it is intriguing that Cambrian magmatism occurs in the SE quadrant, Siluro-Ordovician magmatism follows a N–S trending distribution in the western Slave, which is characterized by the most ancient crust, the central Slave is dominated by Tertiary and Cretaceous magmatism in the Lac de Gras field and the northern Slave consists of Jurassic and Permian kimberlites. The proximity of diamond-rich Eocene and diamond-poor Cretaceous pipes in the Lac de Gras field likely reflects a fundamental difference in the depth of kimberlite magma genesis. If true then the Eocene pipes in this field will have the greatest potential for entraining material from the greatest mantle depths.

Based on a compilation of more than 250 published kimberlite age dates (and perhaps double this number in published data summaries from Yakutia) the following conclusions can be drawn:

(1) The majority of known kimberlites on Earth are Cenozoic to Mesozoic in age.

(2) The "quiescent period" of kimberlite activity between ca. 250 and 400 Ma possibly reflects the

relative crustal (and mantle?) stability attendant with creation of the supercontinent Gondwanaland.

(3) Multiple discrete episodes of kimberlite magmatism (i.e. Type 3 kimberlite provinces) spanning the entire Phanerozoic are typical of Archean cratons most noted for their diamond potential (Anabar, Kaapvaal, Slave).

(4) There are no known temporal equivalents in Africa or North America to the Devonian highly diamondiferous kimberlites in Yakutia.

(5) Some kimberlite magmatism can be directly linked to mantle plume hotspot tracks. For example, Cenozoic/Mesozoic kimberlites in Brazil, eastern North America, and Group II kimberlites in South Africa follow a relatively narrow linear unidirectional age progression consistent with independent estimates for the timing and location of continental portions of oceanic hotspot tracks, contemporaneous with the breakup and rifting of the supercontinent Gondwanaland between 180 and 50 Ma. If true, then prolific kimberlite magmatism should be attendant with the breakup of other supercontinents during Earth history. The continuum of Mesozoic Group I kimberlite magmatism in southern Africa could also be linked to mantle plume hotspot tracks but the large number of plumes that passed beneath the Kaapvaal craton during this relatively short period of geological time (70–130 Ma) has conspired to disguise individual tracks.

Acknowledgements

We greatly appreciate the generosity of the following individuals who over the years have donated samples for this study: Chris Jennings, Ulrich Kretschmar, Wolf Skublak and Howard Coopersmith. We also appreciate the cooperation of Ashton Mining of Canada, Pure Gold and Tennajon in supplying samples from the Cross, Orion and Ursa pipes and for permission to publish the dating results. We would like to thank Barbara Scott-Smith for supplying a perovskite-bearing sample from the Anne kimberlite. Joe Joyce (DeBeers Canada Exploration), Buddy Doyle (Kennecott), and Bernard Kahlert (Diamonds North Resources) kindly gave us permission to report previously unpublished age dates for kimberlites in the Hardy Lake, Birch Mountains and Victoria Island clusters, respectively. In addition we acknowledge the help of Mike Waldman in locating samples from the reclaimed Chicken Park and Sloan pipes, Wyoming. LMH wishes to acknowledge the assistance of A. Berggren, S. Hagen, L. Raynor, S. Ross, R. Stefaniuk, K. Toope, the geochronology staff who collectively contributed to the effective and smooth operation of the University of Alberta radiogenic isotope facility U–Pb labs, and financial support from NSERC MFA and Research grants. The comments by T. Stachel on an early version of the manuscript are greatly appreciated. Journal reviews by T. McCandless and D. Phillips helped improve the final version of the manuscript and their efforts are greatly appreciated. This is geological Survey of Canada contribution #2002135 and Lithoprobe publication #1315.

References

Agashev, A.M., Pokhilenko, N.P., McDonald, J.A., Takazawa, E., Vavilov, M.A., Sobolev, N.V., Watanabe, T., 2001. A unique kimberlite–carbonatite primary association in the Snap Lake dyke system, Slave craton: evidence from geochemical and isotopic studies. The Slave—Kaapvaal Workshop, Program with Abstracts, pp. 42–44.

Allsopp, H.L., Barrett, D.R., 1975. Rb–Sr age determinations on South African kimberlite pipes. Physics and Chemistry of the Earth 9, 605–617.

Allsopp, H.L., Roddick, J.C., 1984. Rb–Sr and $^{40}Ar–^{39}Ar$ age determinations on phlogopite micas from the pre-Lebombo Group Dokolwayo kimberlite pipe. In: Erlank, A.J. (Ed.), Petrogenesis of the Volcanic Rocks of the Karoo Province. Special Publication, vol. 13. The Geological Society of South Africa, Johannesburg, pp. 267–271.

Allsopp, H.L., Hargraves, R.B., 1985. Rb–Sr ages and paleomagnetic data for some Angolan alkaline intrusives. Transactions Geological Society of South Africa 88, 295–300.

Allsopp, H.L., Smith, C.B., 2002. The emplacement age and geochemical character of the Venetia kimberlite bodies, Limpopo belt, Northern Transvaal. South African Journal of Geology (in press).

Allsopp, H.L., Bristow, J.W., Skinner, E.M.W., 1985. The Rb–Sr geochronology of the Colossus kimberlite pipe, Zimbabwe. Transactions Geological Society of South Africa 88, 245–248.

Allsopp, H.L., Bristow, J.W., Smith, C.B., Brown, R., Gleadow, A.J.W., Kramers, J.D., Garvie, O.G., 1989. A summary of radiometric dating methods applicable to kimberlite and related rocks. In: Ross, J. (Ed.), Proceedings of the Fourth International Kimberlite Conference, v. 1, Kimberlites and Related Rocks: Their Composition, Occurrence, Origin and Emplacement. Geological Society of Australia Special Publication, vol. 14. Blackwell Scientific Publications, Oxford, pp. 343–357.

Armstrong, R.L., 1988. Mesozoic and early Cenozoic magmatic evolution of the Canadian Cordillera. In: Clark, S.P., Burchfiel, B.C., Suppe, J. (Eds.), Processes in Continental Lithospheric Deformation. Geological Society of America Special Paper. vol. 218, pp. 55–91.

Armstrong, R.A., Moore, R.O., 1998. Rb–Sr ages on kimberlites from the Lac de Gras area, Northwest Territorries, Canada. South African Journal of Geology 101 (2), 155–158.

Aravanis, T., 1999. Legend property assessment report, Birch Mountains area, Alberta. Alberta Energy and Utilities Board/ Alberta Geological Survey Assessment File Report 20000003. 23 pp.

Baksi, A.K., Farrar, E., 1991. ^{40}Ar/^{39}Ar dating of the Siberian Traps, USSR: evaluation of the ages of the two major extinction events relative to episodes of flood-basalt volcanism in the USSR and the Deccan Traps, India. Geology 19, 461–464.

Barnett, R.L., Arima, M., Blackwell, J.D., Winder, C.G., Palmer, H.C., Hayatsu, A., 1984. The Picton and Varty Lake ultramafic dikes: Jurassic magmatism in the St. Lawrence Platform near Belleville, Ontario. Canadian Journal of Earth Sciences 21, 1460–1472.

Basu, A.R., Rubury, E., Mehnert, H., Tatsumoto, M., 1984. Sm–Nd, K–Ar and petrologic study of some kimberlites from eastern United States and their implications for mantle evolution. Contributions to Mineralogy and Petrology 86, 35–44.

Berg, G.W., Carlson, J.A., 1998. The Leslie kimberlite pipe of Lac de Gras, Northwest Territories, Canada: evidence for near surface hypabyssal emplacement. Seventh International Kimberlite Conference Extended Abstracts, Capetown, pp. 81–83.

Bikerman, M., Prellwitz, H.S., Dembosky, J., Simonetti, A., Bell, K., 1997. New phlogopite K–Ar dates and the age of southwestern Pennsylvania kimberlite dikes. Northeastern Geology and Environmental Sciences 19, 302–308.

Bleeker, W., Ketchum, J.W.F., Davis, W.J., 1999. The Central Slave Basement Complex: Part II. Age and tectonic significance of high strain zones along the basement-cover contact. Canadian Journal of Earth Sciences 36, 1111–1130.

Brakhfogel, F.F., 1995. The age division of the kimberlitic and related magmatites in the N.E. of the Siberian platform (methods and results). Sixth International Kimberlite Conference, Extended Abstracts, Novosibirsk, pp. 60–62.

Breitsprecher, K., Thorkelson, D.J., 2001. Spatial coincidence of the Kula-Farallon slab window with trench-distal volcanism of the Eocene Magmatic Belt in the southern Cordillera. SNORCLE Workshop. Lithoprobe Report, vol. 79, pp. 178–183.

Brookins, D.G., Naeser, C.W., 1971. Age of emplacement of Riley County, Kansas, kimberlites and possible minimum age for the Dakota sandstone. Geological Society of America Bulletin 82, 1723–1726.

Brown, R.W., 1985. An application of the 'leaching technique' of Rb–Sr dating to phlogopite micas from the Makganyene kimberlite, northern Cape Province, South Africa. Unpublished BSc Thesis, Rhodes University. 28 pp.

Brown, R.W., Allsopp, H.L., Bristow, J.W., Smith, C.B., 1989. Improved precision of Rb–Sr dating of kimberlitic micas: an assessment of a leaching technique. Chemical Geology 79, 125–136.

Carlson, J.A., Marsh, S.W., 1989. Discovery of the kimberlite dikes, Colorado. In: Ross, J. (Ed.), Proceedings of the Fourth International Kimberlite Conference 2, Kimberlites and Related Rocks: Their Mantle/Crustal Setting, Diamonds and Diamond Exploration. Geological Society of Australia Special Publication, vol. 14. Blackwell Scientific Publications, Oxford, pp. 1169–1178.

Carlson, J.A., Kirkley, M.B., Ashley, R.M., Moore, R.O., Kolebaba, M.R., 1995. Geology and exploration of kimberlites on the BHP/Dia Met claims, Lac de Gras region, Northwest Territories, Canada. Sixth International Kimberlite Conference, Extended Abstracts, Novosibirsk, pp. 98–100.

Carlson, J.A., Kirkley, M.B., Thomas, E.M., Hillier, W.D., 1999a. Recent Canadian kimberlite discoveries. Seventh International Kimberlite Conference, Capetown, vol. 1, pp. 81–89.

Carlson, S.M., Hillier, W.D., Hood, C.T., Pryde, R.P., Skelton, D.N., 1999b. The Buffalo Hills Kimberlites: a newly discovered diamondiferous kimberlite province in north-central Alberta, Canda. Seventh International Kimberlite Conference, Capetown, vol. 1, pp. 109–116.

Chronic, J., McCallum, M.E., Ferris Jr., C.S., Eggler, D.H., 1969. Lower Paleozoic rocks in diatremes, southern Wyoming and northern Colorado. Geological Society of America Bulletin 80, 149–156.

Cookenboo, H.O., 1995. Mineralogy, petrology and possible rock types of the 5034 kimberlite at Kennady Lake, N.W.T. Exploration Overview, Northwest Territories Mining, Exploration and Geological Investigations, p. 37. DIAND.

Crough, S.T., 1981. Mesozoic hotspot epeirogeny in Eastern North America. Geology 9, 2–6.

Crough, S.T., Morgan, W.J., Hargraves, R.B., 1980. Kimberlites: their relation to mantle hotspots. Earth and Planetary Science Letters 50, 260–274.

Davis, G.L., 1977. The ages and uranium contents of zircons from kimberlites and associated rocks. Carnegie Institute of Washington. Yearbook, vol. 76, pp. 631–635.

Davis, G.L., 1978a. Zircons from the mantle. Carnegie Institute of Washington. Yearbook, vol. 77, pp. 805–807.

Davis, G.L., 1978b. Zircons from the mantle. United States Geological Survey Open File Report 78-701, 86–88.

Davis, W.D., Kjarsgaard, B.A., 1997. A Rb–Sr isochron age for a kimberlite from the recently discovered Lac de Gras field, Slave Province, Northwest Territories, Canada. Journal of Geology 105, 503–509.

Davis, G.L., Sobalev, N.V., Kharkiv, A.D., 1980. New data on the age of Yakutian kimberlites obtained by the uranium–lead method on zircon. Geol. Dokl. Akad. Nauk S.S.S.R. 254, 53–57.

Dawson, J.B., 1989. Geographic and time distribution of kimberlites and lampröites: relationship to tectonic processes. In: Ross, J. (Ed.), Proceedings of the Fourth International Kimberlite Conference 1, Kimberlites and Related Rocks: Their Composition, Occurrence, Origin and Emplacement. Geological Society of Australia Special Publication, vol. 14. Blackwell Scientific Publications, Oxford, pp. 323–342.

Digonnet, S., Goulet, N., Stevenson, R., 2000. Petrology of the Aillikite dykes, Abloviak New Quebec: New exploration target and evidence for a Cambrian diamondiferous province in Northeastern America. GeoCanada 2000, abstract #1079.

Duncan, R.A., Hargraves, R.B., Brey, G.P., 1978. Age, palaeomagnetism and chemistry of melilite basalts in the Southern Cape, South Africa. Geological Magazine 115, 317–327.

Dunning, G.R., Hodych, J.P., 1990. U–Pb zircon and baddeleyite ages for the Palisades and Gettysburg sills of the northeastern United States: implications for the age of the Triassic/Jurassic boundary. Geology 18, 795–798.

Duskin, D.J., Jarvis, W., 1993. Kimberlites in Michigan. In: Dunne, K.P.E., Grant, B. (Eds.), Mid-Continent Diamonds. GAC Mineral Deposits Division, Vancouver, pp. 105–106.

Edgar, A.D., Charbonneau, H.E., 1993. Melting experiments on a SiO_2 poor, CaO-rich aphanitic kimberlite from 5–10 GPa and their bearing on the sources of kimberlite magma. American Mineralogist 78, 132–142.

Eggler, D.H., 1989. Kimberlites: How do the form? In: Ross, J. (Ed.), Proceedings of the Fourth International Kimberlite Conference1 Kimberlites and Related Rocks: Their Composition, Occurrence, Origin, and Emplacement. Geological Society of Australia Special Publication, vol. 14. Blackwell Scientific Publications, Oxford, pp. 489–504.

England, P., Houseman, G., 1984. On the geodynamic setting of kimberlite genesis. Earth and Planetary Science Letters 67, 109–122.

Ewing, T.E., 1981. Petrology and geochemistry of the Kamloops Group volcanics, British Columbia. Canadian Journal of Earth Sciences 18, 1478–1491.

Fitch, F.J., Miller, J.A., 1983. K–Ar age of the east peripheral kimberlite at De Beers Mine, Kimberley, R.S.A. Geological Magazine 120, 505–512.

Foley, S.F., 1991. The origin of kimberlite and lamproites in veined lithospheric mantle. Fifth International Kimberlite Conference, Extended Abstracts, Araxá, pp. 109–111.

Gibson, S.A., Thompson, R.N., Leonardos, O.H., Dickin, A.P., Mitchell, J.G., 1995. The Late Cretaceous impact of the Trindade mantle plume: evidence from large-volume, mafic, potassic magmatism in SE Brazil. Journal of Petrology 36, 189–229.

Gobba, J.M., 1989. Kimberlite exploration in Tanzania. Journal of African Earth Sciences 9 (3/4), 565–578.

Graham, I., Burgess, J.L., Bryan, D., Ravenscroft, P.J., Thomas, E., Doyle, B.J., Hopkings, R., Armstrong, K.A., 1999. Exploration history and geology of the Diavik kimberlites, Lac de Gras. Northwest Territories, Canada. Seventh International Kimberlite Conference, vol. 1, pp. 262–279.

Grand, S.P., van der Hilst, R.D., Widiyantoro, S., 1997. Global seismic tomography: a snapshot of convection in the Earth. GSA Today 7, 1–7.

Green, T.H., Pearson, N.J., 1987. An experimental study of Nb and Ta partitioning between Ti-rich minerals and silicate liquids at high pressure and temperature. Geochimica et Cosmochimica Acta 51, 55–62.

Haggerty, S., 1994. Superkimberlites: a geodynamic window to the Earth's core. Earth and Planetary Science Letters 122, 57–69.

Hartnady, C.J., le Roex, A.P., 1983. Southern Ocean hotspot tracks and the Cenozoic absolute motions of the African, Antarctic and South American plates. Earth and Planetary Science Letters 75, 245–257.

Heaman, L.M., 1989. The nature of the subcontinental mantle from Sr–Nd–Pb isotopic studies on kimberlitic perovskite. Earth and Planetary Science Letters 92, 323–334.

Heaman, L.M., Mitchell, R.H., 1995. Constraints on the emplacement age of Yakutian Province kimberlites from U–Pb perovskite dating. Sixth International Kimberlite Conference, Extended Abstracts, Novosibirsk, pp. 223–224.

Heaman, L.M., Kjarsgaard, B.A., 2000. Timing of eastern North American kimberlite magmatism: continental extension of the Great Meteor hotspot track? Earth and Planetary Science Letters 178, 253–268.

Heaman, L.M., Kjarsgaard, B.A., 2002. A Cretaceous corridor of kimberlite magmatism: U–Pb results from the Fort à la Corne field, central Saskatchewan. Geological Association of Canada/ Mineralogical Association of Canada Meeting in Saskatoon, Saskatchewan, 47.

Heaman, L.M., Kjarsgaard, B.A., Creaser, R.A., Cookenboo, H.O., Kretschmar, U., 1997. Multiple episodes of kimberlite magmatism in the Slave Province, North America. Lithoprobe Workshop Report, vol. 56, pp. 14–17.

Heaman, L.M., Creaser, R.A., Cookenboo, H.O., 2002. Extreme high-field-strength element enrichment in Jericho eclogite xenoliths: A cryptic record of Paleoproterozoic subduction, partial melting and metasomatism beneath the Slave craton, Canada. Geology 30, 507–510.

Hegner, E., Roddick, J.C., Fortier, S.M., Hulbert, L., 1995. Nd, Sr, Pb, Ar and O isotopic systematics of Sturgeon Lake kimberlite, Saskatchewan, Canada: constraints on emplacement age, alteration and source composition. Contributions to Mineralogy and Petrology 120, 212–222.

Helmstaedt, H., Gurney, J.J., 1984. Kimberlites of Southern Africa—are they related to subduction processes? Third International Kimberlite Conference, vol. 1, pp. 425–434.

Helmstaedt, H., Gurney, J.J., 1997. Geodynamic controls of kimberlites—What are the roles of hotspot and plate tectonics? Russian Geology and Geophysics 38, 492–508.

Helmstaedt, H., Schulze, D.J., 1979. Type A-Type C eclogitic transition in xenoliths from the Moses Rock diatreme-further evidence for the presence of metamorphosed ophiolites beneath the Colorado Plateau. Second International Kimberlite Conference, vol. 2, pp. 357–365.

Holmden, C.E., Creaser, R.A., Muehlenbachs, K., Bergstrom, S.M., Leslie, S.A., 1996. Isotopic and elemental systematics of Sr and Nd in 454 Ma biogenic apatites: implications for paleoseawater studies. Earth and Planetary Science Letters 142, 425–437.

Jaffey, A.H., Flynn, K.F., Glendenin, L.E., Bentley, W.C., Essling, A.M., 1971. Precision measurements of half-lives and specific activities of ^{235}U and ^{238}U. Physics Review C4, 1889–1906.

Kaminsky, F.V., Sablukov, S.M., Sablukova, L.I., Shchukin, V.S., 2000. Petrology of kimberlites from the newly discovered Whitefish Lake field in Ontario. GeoCanada 2000, abstract #1203.

Kamo, S.L., Gower, C.F., Krogh, T.E., 1989. Birthdate for the Iapetus Ocean? A precise U–Pb zircon and baddeleyite age for the Long Range dikes, southeast Labrador. Geology 17, 602–605.

Kamo, S.L., Czamanske, G.K., Krogh, T.E., 1996. A minimum U–

Pb age for Siberian flood-basalt volcanism. Geochimica et Cosmochimica Acta 60, 3505–3511.

Kinny, P.D., Compston, W., Bristow, J.W., Williams, I.S., 1989. Archean mantle xenocrysts in a Permian kimberlite: two generations of kimberlitic zircon in Jwaneng DK2, southern Botswana. In: Ross, J. (Ed.), Proceedings of the Fourth International Kimberlite Conference 2, Kimberlites and Related Rocks: Their Mantle/Crustal Setting, Diamonds and Diamond Exploration. Geological Society of Australia Special Publication, vol. 14. Blackwell Scientific Publications, Oxford, pp. 833–842.

Kinny, P.D., Griffin, B.J., Brakhfogel, F.F., 1995. SHRIMP U–Pb ages of perovskite and zircon from Yakutian kimberlites. Sixth International Kimberlite Conference, Extended Abstracts, Novosibirsk, p. 275.

Kinny, P.D., Griffin, B.J., Heaman, L.M., Brakhfogel, F.F., Spetsius, Z.V., 1997. SHRIMP U–Pb ages of perovskite from Yakutian kimberlites. Russian Geology and Geophysics 38, 97–105.

Kiviets, G.B., Phillips, D., Shee, S.R., Vercoe, S.C., Barton, E.S., Smith, C.B., Fourie, L.F., 1998. $^{40}Ar/^{39}Ar$ dating of yimengite from the Turkey Well kimberlite, Australia: the oldest and the rarest. Seventh International Kimberlite Conference Extended Abstracts, Capetown, pp. 432–433.

Kjarsgaard, B.A., 1996a. Slave province kimberlites, NWT. In: LeCheminant, A.N., Richardson, D.G., DiLabio, R.N.W., Richardson, K.A. (Eds.), Searching for Diamonds in Canada. Geological Survey of Canada Open File, vol. 3228, pp. 55–60.

Kjarsgaard, B.A., 1996b. Somerset Island kimberlite field, District of Franklin, NWT. In: LeCheminant, A.N., Richardson, D.G., DiLabio, R.N.W., Richardson, K.A. (Eds.), Searching for Diamonds in Canada. Geological Survey of Canada Open File, vol. 3228, pp. 61–66.

Kjarsgaard, B.A., Heaman, L.M., 1995. Distinct emplacement periods of Phanerozoic kimberlites in North America, and implications for the Slave Province. Exploration Overview 1995, Northwest Territories (Igboji, E.I., compiler). Department of Indian and Northern Affairs, Yellowknife, NWT, pp. 3–22.

Kjarsgaard, B.A., Wilkinson, L., Armstrong, J.A., 2002. Geology, Lac de Gras kimberlite field, central Slave Province, Northwest Territories-Nunavut. Geological Survey of Canada Open File 3238, scale 1:250,000.

Komarov, F.N., Ilupin, I.P., 1990. Geochronology of kimberlites of the Siberian Platform. Geokhimiya 3, 365.

Kopylova, M.G., Russell, J.K., Cookenboo, H., 1999. Petrology of peridotite and pyroxenite xenoliths from the Jericho kimberlite: implications for the thermal state of the mantle beneath the Slave craton, northern Canada. Journal of Petrology 40, 79–104.

Kramers, J.D., Smith, C.B., 1983. A feasibility study of U–Pb and Pb–Pb dating of kimberlites using groundmass mineral fractions and whole-rock samples. Chemical Geology (Isotope Geoscience) 1, 23–38.

Kramers, J.D., Roddick, J.C., Dawson, J.B., 1983. Trace element and isotopic studies on veined, metasomatic and 'marid' xenoliths from Bultfontein, South Africa. Earth and Planetary Science Letters 65, 90–106.

Krogh, T.E., 1982. Improved accuracy of U–Pb zircon ages using an air abrasion technique. Geochimica et Cosmochimica Acta 46, 637–649.

Kumar, A., Padma Kumari, V.M., Dayal, A.M., Murthy, D.S.N., Gopalan, K., 1993. Rb–Sr ages of Proterozoic kimberlites of India: evidence for contemporaneous emplacement. Precambrian Research 62, 227–237.

Kumar, A., Gopalan, K., Padmakumari, V.M., Kornilova, V.P., Oleinikov, O.B., Safronov, A.F., 1995. Precise Rb–Sr ages of Siberian kimberlites. Sixth International Kimberlite Conference, Extended Abstracts, Novosibirsk, p. 307.

Larsen, L.M., Rex, D.C., 1992. A review of the 2500 Ma span of alkaline-ultramafic, potassic and carbonatitic magmatism in West Greenland. Lithos 28, 367–402.

LeCheminant, A.N., Heaman, L.M., 1994. 779 Ma mafic magmatism in the northwestern Canadian Shield and northern Cordillera: a new regional time-marker. Eighth International Conference on Geochronology and Isotope Geology. U.S. Geological Survey Circular, vol. 1107, p. 197.

Leckie, D.A., Kjarsgaard, B.A., Bloch, J., McIntyre, D., McNeil, D., Stasiuk, L.S., Heaman, L.M., 1997a. Emplacement and reworking of Cretaceous, diamond-bearing, crater facies kimberlite of central Saskatchewan, Canada. Geological Society of America Bulletin 109, 1000–1020.

Leckie, D.A., Kjarsgaard, B., Pierce, J.W., Grist, A.M., Collins, M., Sweet, A., Stasiuk, L., Tomica, M.A., Eccles, R., Dufresne, M., Fenton, M.M., Pawlowicz, J.G., Balzer, S.A., McIntyre, D.J., McNeil, D.H., 1997b. Geology of a late Cretaceous possible kimberlite at Mountain Lake, Alberta-chemistry, petrology, indicator minerals, aeromagnetic signature, age, stratigraphic position and setting. Geological Survey of Canada, Open File, vol. 3441.

Le Roex, A.P., 1986. Geochemical correlation between southern African kimberlites and South Atlantic hotspots. Nature 324, 243–245.

Lester, A.P., Larson, E.E., 1996. New geochronologic evidence for Late Proterozoic emplacement in the Colorado–Wyoming kimberlite belt. EOS Transactions 77 (46), F821 (American Geophysical Union abstract).

Lipman, P.W., Prostka, J.J., Christiansen, R.L., 1972. Cenozoic volcanism and plate tectonic evolution of the western United States. Philosophical Transactions, Royal Society of London Series A 271, 217–248.

Ludwig, K.R., 1992. ISOPLOT—a plotting and regression program for radiogenic-isotope data, version 2.57. U.S. Geological Survey Open File Report, vol. 91-445. 40 pp.

MacIntyre, R.M., Dawson, J.B., 1976. Age significance of some South African kimberlites. Fourth European Colloquium of Geochronology, Cosmochronology and Isotope, p. 66.

Mansker, W.L., Richards, B.D., Cole, G.P., 1987. A note on newly discovered kimberlites in Riley County, Kansas. Geological Society of America Special Paper, vol. 215, pp. 197–204.

Marvin, R.F., Hearn Jr., B.C., Mehnert, H.H., Naeser, C.W., Zartman, R.E., Lindsay, D.A., 1980. Late Cretaceous–Paleocene–Eocene igneous activity in north central Montana. Isochron West 29, 5–25.

McCandless, T.E., 1999. Kimberlites: mantle expressions of deep-seated subduction. In: Gurney, J.J., Gurney, J.L., Pacsoe, M.D.,

Richardson, S.H. (Eds.), Proceedings of the Seventh International Kimberlite Conference, vol. 2, pp. 545–549.

McKinlay, F.T., Williams, A.C., Kong, J., Scott-Smith, B.I., 1997. An integrated exploration case history for diamonds, Hardy Lake project, NWT. Proceedings of Exploration 97, International Conference on Mineral, pp. 1029–1038.

Mitchell, R.H., 1986. Kimberlites. Plenum Press, New York. 442 pp.

Moore, R.O., Gurney, J.J., Griffin, W.L., Shimizu, N., 1991. Ultra-high pressure garnet inclusions in Monastery diamonds: trace element abundance patterns and conditions of origin. European Journal of Mineralogy 3, 213–230.

Moorhead, J., Girard, R., Heaman, L., 2002. Caracterisation de kimberlites au Quebec. Quebec Geological Survey Open House.

Morgan, W.J., 1983. Hotspot tracks and the early rifting of the Atlantic. Tectonophysics 94, 123–139.

Morris, G.A., Larson, P.B., Hooper, P.R., 2000. 'Subduction style' magmatism in a non-subduction setting: the Colville Igneous Complex, NE Washington State, USA. Journal of Petrology 41, 43–67.

Nixon, P.H., Griffin, B.L., Davies, G.R., Condliffe, E., 1994. Cr garnet indicators in Venezuela kimberlites and their bearing on the evolution of the Guyana craton. Proceedings of the Fifth International Kimberlite Conference, Araxá, Brazil, pp. 378–387.

O'Brien, H.E., Irving, A.J., MaCallum, I.S., Thirwall, M.F., 1995. Strontium, neodymium and lead isotopic evidence for interaction of post-subduction asthenospheric potassic mafic magmas of the Highwood Mountains, Montana, USA, with ancient Wyoming craton lithospheric mantle. Geochimica et Cosmochimica Acta 59, 4539–4562.

Paces, J.B., Zartman, R.E., Taylor, L.A., Futa, K., Kwak, L.M., 1991. Pb isotopic evidence for multiple episodes of lower crustal growth and modification in granulite nodules from the Superior Province, Michigan. Geological Society of America (abstracts), Dallas, vol. A119.

Park, J.K., Buchan, K.L., Harlan, S.S., 1995. A proposed giant radiating dyke swarm fragmented by the separation of Laurentia and Australia based on paleomagnetism of ca. 780 Ma mafic intrusions in western North America. Earth and Planetary Science Letters 132, 129–139.

Pearson, D.G., Harris, J.W., Shirey, S.B., Carlson, R.W., Boyd, F.R., 1996. Re–Os study of sulfide diamond inclusions from S. Africa: constraints on timing of diamond formation and lithosphere evolution. EOS Transactions 77 (46), F816.

Phillips, D., 1991. Argon isotope and halogen chemistry of phlogopite from South African kimberlites: a combined step-heating, laser probe, electron microprobe and TEM study. Chemical Geology 87, 71–98.

Phillips, D., Machin, K.J., Kiviets, G.B., Fourie, L.F., Roberts, M.A., Skinner, E.M.W., 1998. A petrographic and $^{40}Ar/^{39}Ar$ geochronological study of the Voorspoed kimberlite, South Africa: implications for the origin of Group II kimberlite magmatism. South African Journal of Geology 101, 299–306.

Phillips, D., Kiviets, G.B., Barton, E.S., Smith, C.B., Viljoen, K.S., Fourie, L.F., 1999. $^{40}Ar/^{39}Ar$ dating of kimberlites and related rocks: problems and solutions. In: Gurney, J.J., Gurney, J.L., Pacsoe, M.D., Richardson, S.H. (Eds.), Proceedings of the Seventh International Kimberlite Conference, vol. 2, pp. 677–688.

Phipps, S.P., 1988. Deep rifts as sources for alkaline intraplate magmatism in eastern North America. Nature 334, 27–31.

Raber, E., 1978. Zircons from diamond bearing kimberlites: oxide reactions, fission track dating and a mineral inclusion study. Unpublished MSc thesis, University of Massachusetts.

Renne, P.R., Emestro, M., Pacca, I.G., Coe, R.S., Glen, J.M., Prévot, M., Perrin, M., 1992. The age of Paraná flood volcanism, rifting of Gondwanaland, and the Jurassic–Cretaceous boundary. Science 258, 975–979.

Richardson, S.H., Gurney, J.J., Erlank, A.J., Harris, J.W., 1984. Origin of diamonds in old enriched mantle. Nature 310, 198–202.

Ringwood, A.E., Major, A., 1971. Synthesis of majorite and other high pressure garnets and perovskites. Earth and Planetary Science Letters 12, 411–418.

Ringwood, A.E., Kesson, S.E., Hibberson, W., Ware, N., 1992. Origin of kimberlites and related magmas. Earth and Planetary Science Letters 113, 521–538.

Queen, M., Heaman, L.M., Hanes, J.A., Archibald, D.A., Farrar, E., 1996. $^{40}Ar/^{39}Ar$ and U–Pb perovskite dating of lamprophyre dykes from the eastern Lake Superior region: evidence for a 1.14 Ga magmatic precursor to Midcontinent Rift volcanism. Canadian Journal of Earth Sciences 33, 958–965.

Sage, R.P., 1996. Kimberlites of the Lake Timiskaming Structural Zone. Ontario Geological Survey, Open File, vol. 5937. 435 pp.

Sautter, V., Haggerty, S.E., Field, S., 1991. Ultra-deep (>300 km) ultramafic xenoliths: new petrologic evidence from the transition zone. Science 252, 827–830.

Schärer, U., Corfu, F., DeMaiffe, D., 1992. U–Pb and Lu–Hf isotopes in baddelyite and zircon megacrysts from the Mbuji-Mayi kimberlite: constraints on the subcontinental mantle. Chemical Geology 143, 1–16.

Schissel, D., Smail, R., 2001. Deep mantle plumes and ore deposits. In: Ernst, R.E., Buchan, K.L. (Eds.), Mantle Plumes: Their Identification Through Time. Geological Society of America Special Publication, vol. 352, pp. 291–322.

Scott Smith, B.H., McKinlay, T., 2002. Emplacement of the Hardy Lake kimberlites, NWT, Canada. Geological Association of Canada/Mineralogical Association of Canada Meeting in Saskatoon, Saskatchewan, p. 106.

Sharp, W.E., 1974. A plate tectonic origin for diamond-bearing kimberlite. Earth and Planetary Science Letters 21, 351–354.

Shee, S.R., Bristow, J.W., Bell, B.R., Smith, C.B., Allsopp, H.L., Shee, P.B., 1989. The petrology of kimberlites, related rocks and associated mantle xenoliths from the Kuruman Province, South Africa. In: Ross, J. (Ed.), Kimberlites and Related Rocks, Their Composition, Occurrence, Origin and Emplacement. Geological Society of Australia Special Publication, vol. 14, pp. 60–82.

Skinner, E.M.W., Clement, C.R., 1979. Mineralogical classification of Southern African kimberlites. Second International Kimberlite Conference, vol. 1, pp. 129–139.

Sleep, N.H., 1990. Monteregian hotspot track: a long-lived mantle plume. Journal of Geophysical Research 95, 21983–21990.

Smith, C.B., 1979. Rb–Sr age determinations on kimberlites from the State Line field, Colorado. Cambridge Kimberlite Symposium II, Extended Abstracts, Cambridge, UK, pp. 61–66.

Smith, C.B., 1983. Pb, Sr and Nd isotopic evidence for sources of southern African Cretaceous kimberlites. Nature 304, 51–54.

Smith, C.B., Allsopp, H.L., Kramers, J.D., Hutchinson, G., Roddick, J.C., 1985. Emplacement ages of Jurassic–Cretaceous South African kimberlites by the Rb–Sr method on phlogopite and whole-rock samples. Transactions of the Geological Society of South Africa 88 (2), 249–266.

Smith, C.B., Colgan, E.A., Hawthorne, J.B., Hutchinson, G., 1988. Emplacement age of the Cross kimberlite, southesastern British Columbia, by the Rb–Sr phlogopite method. Canadian Journal of Earth Sciences 25, 790–793.

Smith, C.B., Allsopp, H.L., Garvie, O.G., Kramers, J.D., Jackson, P.F.S., Clement, C.R., 1989. Note on the U–Pb perovskite method for dating kimberlites; examples from the Wesselton and DeBeers mines, South Africa, Somerset Island, Canada. Chemical Geology (Isotope Geoscience) 79, 137–145.

Smith, C.B., Clark, T.C., Barton, E.A., Bristow, J.W., 1994. Emplacement ages of kimberlite occurrences in the Prieska region, southwest border of the Kaapvaal Craton, South Africa. Chemical Geology (Isotope Geoscience) 113, 149–169.

Stacey, J.S., Kramers, J.D., 1975. Approximation of terrestrial lead isotope evolution by a two-stage model. Earth and Planetary Science Letters 26, 207–221.

Stachel, T., Brey, G.P., Harris, J.W., 2000. Kankan kiamonds (Guinea) I: from the lithosphere down to the transition zone. Contributions to Mineralogy and Petrology 140, 1–15.

Stachel, T., Harris, J.W., Tappert, R., Brey, G.P., 2002. Peridotitic inclusions in diamonds from the Slave and the Kaapvaal cratons—similarities and differences based on a preliminary data set. Lithos (this volume).

Steiger, R.H., Jager, E., 1977. Subcommission of geochronology: convention on the use of decay constants in geo- and cosmo-chronology. Earth and Planetary Science Letters 36, 359–362.

Sykes, L.R., 1978. Intraplate seismicity, reactivation of pre-existing zones of weakness, alkaline magmatism and other tectonism post-dating continental fragmentation. Reviews of Geophysics and Space Physics 16, 621–688.

Tainton, K.M., McKenzie, D., 1994. The generation of kimberlites, lampröites, and their source rocks. Journal of Petrology 35, 787–817.

Taylor, L.A., 1984. Kimberlitic magmatism in the eastern United States: relationship to mid-Atlantic tectonism. In: Kornprobst, J. (Ed.), Proceedings of the Third International Kimberlite Conference 1. Kimberlites I, Kimberlites and Related Rocks, pp. 417–424.

Thorkelson, D.J., Taylor, R.P., 1989. Cordilleran slab windows. Geology 17, 833–836.

Waldron, F.W.F., Anderson, S.D., Cawood, P.A., Goodwin, L.B., Hall, J., Jamieson, R.A., Palmer, S.E., Stockmal, G.S., Williams, P.F., 1998. Evolution of the Appalachian Laurentian margin: lithoprobe results in western Newfoundland. Canadian Journal of Earth Sciences 35, 1271–1287.

Watson, K.D., 1967. Kimberlites of eastern North America. In: Wyllie, P.J. (Ed.), Ultramafic and Related Rocks. Wiley, New York. 464 pp.

Weinstein, S.A., Olson, P.L., 1989. The proximity of hotspots to convergent and divergen plate boundaries. Geophysical Research Letters 16, 433–436.

Wyllie, P.J., 1980. The origin of kimberlite. Journal of Geophysical Research 85, 6902–6910.

Wyllie, P.J., 1989. The genesis of kimberlites and some low SiO_2, high-alkali magmas. In: Ross, J. (Ed.), Proceedings of the Fourth International Kimberlite Conference 1, Kimberlites and Related Rocks: Their Composition, Occurrence, Origin and Emplacement. Geological Society of Australia Special Publication, vol. 14. Blackwell Scientific Publications, Oxford, pp. 603–615.

Available online at www.sciencedirect.com

SCIENCE @ DIRECT°

Lithos 71 (2003) 185–193

LITHOS

www.elsevier.com/locate/lithos

Thermal and chemical variations in subcrustal cratonic lithosphere: evidence from crustal isostasy

Walter D. Mooney[a,*], John E. Vidale[b]

[a] Earth Quake Hazards Team, United States Geological Survey, MS 977, 345 Middlefield Road, Menlo Park, CA 94025, USA
[b] Earth and Space Sciences Department, UCLA, Los Angeles, CA 90095-1567, USA

Abstract

The Earth's topography at short wavelengths results from active tectonic processes, whereas at long wavelengths it is largely determined by isostatic adjustment for the density and thickness of the crust. Using a global crustal model, we estimate the long-wavelength topography that is not due to crustal isostasy. Our most important finding is that cratons are generally depressed by 300 to 1500 m in comparison with predictions from pure crustal isostasy. We conclude that either: (1) cratonic roots may be 50 to 300 °C colder than previously suggested by thermal models, or (2) cratonic roots may be, on average, less depleted than suggested by studies of shallow mantle xenoliths. Alternatively, (3) some combination of these conditions may exist. The thermal explanation is consistent with recent geothermal studies that indicate low cratonic temperatures, as well as seismic studies that show very low seismic attenuation at long periods (150 s) beneath cratons. The petrologic explanation is consistent with recent studies of deep (>140 km) mantle xenoliths from the Kaapvaal and Slave cratons that show 1–2% higher densities compared with shallow (<140 km), highly depleted xenoliths.
© 2003 Elsevier B.V. All rights reserved.

Keywords: Isostasy; Crustal structure; Cratons; Xenoliths

1. Introduction

The long-wavelength topography of the Earth's surface is generally due to variations in crustal thickness, in combination with the large density contrast between crust and mantle (Turcotte and Schubert, 1982; Schubert et al., 2001; Watts, 2001). Topography can also be supported by the flexural rigidity of the lithosphere, but only on length scales much shorter than we investigate here. Topography that is not compensated by the crust is compensated by lateral

density contrasts below the crust. Here, we take advantage of recent improvements in our understanding of crustal structure to look deeper, at the lithosphere below the crust. The essence of our analysis is to make an isostatic correction for the known crustal part of the lithosphere to reveal the part of the topography that can only result from lateral density variations below the crust. Local dynamic uplifts associated with mantle convection, and mantle plumes in particular, may be important in some regions, but are not modeled in this global study.

Lateral density contrasts in the lithosphere arise from variations in both composition and temperature. The small magnitude of topographic and geoid anomalies, coupled with geochemical evidence, has

---

* Corresponding author. Tel.: +1-650-329-4764; fax: +1-650-329-5163.
E-mail address: mooney@usgs.gov (W.D. Mooney).

0024-4937/$ - see front matter © 2003 Elsevier B.V. All rights reserved.
doi:10.1016/j.lithos.2003.07.004

led to the recognition that the uppermost mantle beneath continents is likely to be colder, but of lighter composition than mantle beneath oceans (Jordan, 1975; Boyd, 1989). The isopicnic hypothesis states that these two effects nearly cancel, leaving only small lateral differences in density (Jordan, 1978). In this paper we show that cratons are generally depressed by about 300 to 1500 m in comparison with predictions from pure crustal isostasy. We interpret this observation in terms of excess density within craton roots, either due to colder temperatures or less chemical depletion, or both (cf. Kaban et al., 2003).

2. Data and analysis

We use ETOPO5 (National Geophysical Data Center, 1993), which consists of the Earth's surface elevation averaged over a 5 min grid, filtered as shown in Fig. 1a. The same filter is applied to all the data presented in this paper. This degree of smoothing removes the short-wavelength uncompensated topography that results from the elastic strength of the lithosphere, and emphasizes the long-wavelength patterns of uncompensated topography. Thus, we model lithospheric properties at wavelengths comparable to models derived from global surface wave tomography (i.e., >2000 km).

The signature of the crust, which is the upper 5 to 80 km of the Earth, must be removed to reveal subcrustal anomalies. Lateral variations in crustal thickness and density dominate the topography. For example, the high Andes and the Tibetan plateau are buoyed by the thickest crust, whereas the low-lying oceans have the thinnest crust. We use the global model Crust 5.1 (Mooney et al., 1998), which is based on measurements of crustal thickness and seismic velocity from seismic refraction experiments, and densities from laboratory studies (Christensen and Mooney, 1995). Where refraction data are not available, the model extrapolates based on tectonic affinity.

In order to avoid model artifacts, we emphasize only the conclusions that are derived from the regions of the densest refraction coverage.

Previous attempts to model the Earth's topography differ from this study in the following respects: (1) LeStunff and Ricard (1995) and Simons and Hager (1997) used less comprehensive models of crustal structure; (2) Forte (1995) and Forte et al. (1993) divided the Earth into oceanic and continental regions, each with an average ocean or continent structure and (3) Pari and Peltier (1996) defined cratons from seismic tomography maps rather than geologic maps, as is done here. Studies that focus on shorter-wavelength structure often use gravity data (Pari and Peltier, 1996; Simons and Hager, 1997). Other studies have focused on the properties of the deep mantle (Forte and Mitrovica, 2001). The approach taken in our study is most similar to that of Forte and Perry (2000); however, they also use global seismic tomographic models to infer density variations in the mantle. We do not model the geoid here because lower mantle density variations cause larger geoid signals than lithospheric density variations (Simons and Hager, 1997).

The topography estimated from Crust 5.1 is shown in Fig. 1b. Airy compensation is assumed at the base of the crust, with a constant mantle density of 3350 kg/m^3 (Dziewonski and Anderson, 1981). In other words, when calculating pure crustal isostasy, the crust in each cell is assumed to float isostatically on the upper mantle (Turcotte and Schubert, 1982).

The sinking that accompanies the cooling of oceanic crust with age is well known from the observation that mid-ocean ridges stand significantly higher than the old ocean floor. This topography has been explained by a half-space model of a lithosphere that cools by conduction (Parsons and Sclater, 1977). The topography predicted by the age of the seafloor, as derived from magnetic anomalies (Mueller et al., 1993), is shown in Fig. 1c, with zero depth assigned to 150 Ma ocean floor. There are alternative models for the dependence of topography on seafloor age that

---

Fig. 1. (a) The topography of the Earth. ETOPO5 (National Geophysical Data Center, 1993) is averaged over 5° by 5° cells. This and the following figures are resampled by bilinear interpolation to 2.5° by 2.5° cells, then smoothed by five convolutions with a 7.5° by 7.5° boxcar function. The result is that only long-wavelength features are resolved. (b) Topography predicted by Crust 5.1 (Mooney et al., 1998) floating on mantle with density 3350 kg/m^3. (c) Topography predicted by a half space model of cooling oceanic lithosphere (Parsons and Sclater, 1977). Note that 150 Ma oceanic crust is arbitarily assigned zero topography.

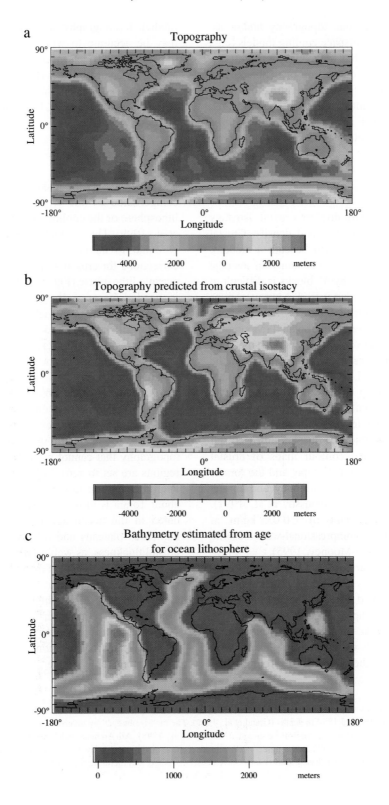

correct for inferred dynamic topography and/or assume plate rather than halfspace cooling models (e.g.,
Stein and Stein, 1996), but our conclusions for cratonic lithosphere do not depend on our choice of an
oceanic cooling model. Thermal perturbations from
hotspots are not taken into account in our modeling,
and the anomalies that appear to be due to these
features are identified below.

3. Uncertainties

Topography (Fig. 1b) due to crustal isostasy
depends upon crustal thickness and density. Crustal
thickness is accurate to ± 2 km beneath a modern,
reversed refraction line. The uncertainty in average
crustal thickness across 5° by 5° blocks is difficult to
quantify, but we estimate that it is ± 3 km in regions
with seismic refraction profiles and ± 7 km in regions
for which crustal structure is extrapolated (Mooney et
al., 1998; 2002). However, the multi-cell smoothed
values in all our figures average out much of the
variation from cell to cell, so we subjectively estimate
that the long-wavelength crustal thickness is accurate
to ± 2 km in regions with seismic refraction profiles
and ± 4 km in regions for which crustal structure is
extrapolated. A comparison of our smoothed crustal
thickness with published contour maps of crustal
thickness for North America, Europe, and the former
USSR support these estimates (Mooney and Braile,
1989; Meissner et al., 1987; Beloussov et al., 1991).
There is an average uncertainty of ± 0.035 kg/m^3 in
estimating density from compressional-wave seismic
velocity (Christensen and Mooney, 1995).

Topography due to crustal isostasy is best determined in North America, Eurasia, and Australia (Fig.
1d). In these well-studied areas, the uncertainty in the
smoothed model topography (Fig. 1b) is 200 m. The
crustal structure in Africa, Greenland, and South America, in contrast, is largely extrapolated from areas with
similar geology. Using the estimated uncertainties, the

predicted topography in these extrapolated regions
may be accurate to within 400 m. In view of this
variability in model uncertainty, we base our conclusions only on results for well-studied areas, and include
other regions for completeness and for comparison
with previous work (e.g., Forte and Perry, 2000).

4. Results

Fig. 2a shows the residual topography that is not
compensated by the crustal part of the continental
lithosphere or the cooling of oceanic plates (cf. Kaban
et al., 1999). The residual topography is approximately equal to actual topography (Fig. 1a) with the
corrections for crustal structure (Fig. 1b) and cooling
oceanic lithosphere (Fig. 1c) subtracted. (A correction
is needed because the oldest oceanic floor was
assigned zero depth in Fig. 1c, rather than the actual
average bathymetry of 150 Ma crust.).

Three regions in Fig. 2a have been filled with zeroes
prior to smoothing. (1) The Ross Ice Sheet, on the edge
of Antarctica south of New Zealand, was given zero
elevation in ETOPO5, rather than the true ocean depth.
(2) Near the North Pole, our ocean age model did not
have ages for crust in several ocean basins. Both of
these unknowns result in several km of apparent
topography that cannot be interpreted, so these small
regions are set to zero. (3) The area in the southwest
Pacific extending from the east coast of Australia to
about Tonga is also set to zero because it also is not
dated in the ocean age model. This area contains
numerous plateaus and back-arc basins of uncertain
crustal thickness, as well as several subduction zones.

Fig. 2b shows a recent map of variations in seismic
shear velocity from 100 to 175 km depth (Grand et al.,
1997), and Fig. 2c is a highly smoothed map of the
distribution of cratons that has been filtered in the
same manner as the geophysical maps (Fig. 2a and b).
The well-known correlation of high seismic velocities
with cratons, and low seismic velocities with young

Fig. 2. (a) Residual topography unexplained by pure crustal isostasy and the cooling plate oceanic lithosphere model (Fig. 1c).(b) Shear wave
velocity variation between 100- and 175-km depths (Grand et al., 1997). The map is obtained by inversion of seismic body waves (Grand and
Helmberger, 1984). (c) Highly smoothed map of the cratons (Mooney et al., 1998). All Archean and Proterozoic shields and platforms are
considered cratons. The initial, unsmoothed map agrees well with a recent compilation (Goodwin, 1996). (d) Distribution of the refraction
profiles on which the model Crust 5.1 is based. Darker areas have denser coverage, and therefore the crustal structure is better resolved than in
lighter areas.

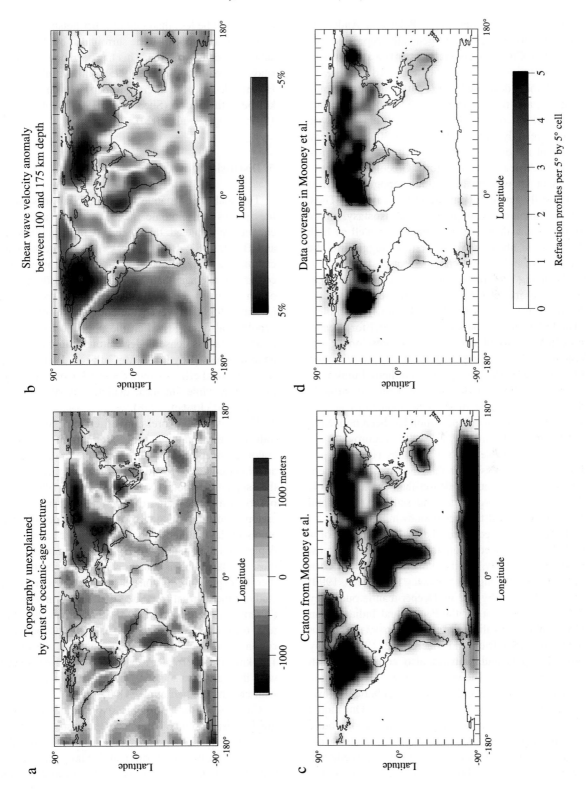

oceanic crust, is evident (e.g., Polet and Anderson, 1995). Cratonic regions are generally depressed 300 to 1500 m. Cratons where refraction data are available to constrain crustal properties (Fig. 1d) appear, in general, more depressed. Large parts of Eurasia, South America, and the cratonic part of North America are depressed by 600 to 1500 m. Australia and west Africa are only depressed by about 300 m. Several continent regions, including western North America, southern Africa, western Europe, Greenland and eastern Asia show little anomalous topography. Most tectonically active regions are in this category. Thus, when compared with pure crustal isostasy, many cratons show depressed topography, as well as high lithospheric shear wave velocities, suggesting the presence of cold, dense cratonic roots.

Changing the two most loosely constrained features of our model, (1) making the reference age of the oceanic crust younger than 150 My, or (2) assuming a plate rather than halfspace cooling model for oceanic lithosphere, would enhance the depression of the cratonic regions. A region with abundant refraction control, the cratons comprising northern Eurasia including the Caspian Sea and the Urals, is strongly depressed. The crust of the Ural mountains is well studied; this lack of compensation has been noticed in local studies (Artemjev et al., 1994; Yegorova et al., 1995) and is visible in a similar study (LeStunff and Ricard, 1995). It is clear that the crust of northern Eurasia is underlain by denser-than-average mantle lithosphere. This region is composed of Archean and Proterozoic crustal terranes (Goodwin, 1996).

In several well-constrained areas, the ocean floor is higher than predicted by our model. Iceland is the most prominent anomaly, but the Indian and Pacific Ocean hotspots with the highest buoyancy flux (Sleep, 1990) are also elevated by 300 m. Deeper mantle buoyancy forces may account for the Pacific and Indian ocean topographic anomalies, since these hotspots appear in older crust (Sandwell and MacKenzie, 1989). Many other hotspots and plateaus also correspond with topographic highs. The positive topographic anomaly associated with Iceland requires a subcrustal low-density anomaly (i.e., positive buoyancy). Our model already corrects for the thermal structure of young oceanic lithosphere, but the region surrounding Iceland still stands 500 to 1000 m higher than predicted. Although the hotspot buoyancy flux under Iceland is

estimated to be less than that of other hotspots (Sleep, 1990), Iceland appears to be surrounded by the largest subcrustal density deficit. This has been noted in local studies (Smallwood et al., 1995; White et al., 1995). Hot mantle has been inferred to persist as deep as 660 km beneath Iceland (Shen et al., 1998; Wolfe et al., 1997; Foulger et al., 2000).

5. Discussion

The analysis of mantle xenoliths found in kimberlites shows that cratonic mantle is colder and more chemically depleted than oceanic mantle (Finnerty and Boyd, 1987; Boyd, 1989). The mantle beneath cratons has been estimated to be approximately 200° colder on average than the mantle beneath mature ocean basins in the depth range from 40 to 200 km (Jordan, 1988). However, the composition of mantle xenoliths also suggests that the uppermost mantle is roughly 1% less dense under cratons than under non-cratonic crust (Boyd and McCallister, 1976). Thus, purely thermal effects would cause 1 km of depression of cratons, while chemical depletion would cause 1 km of uplift (Jordan, 1988).

The observed depression of cratons in Fig. 2a indicates that cratonic roots are denser than non-cratonic upper mantle. Therefore, the cratonic roots are either colder or less depleted than is suggested by the isopicnic hypothesis (Jordan, 1978; 1988). The observed average value of 500 m of depression of cratons can be explained by 100 °C colder temperatures in the cratonic roots or, alternatively, by 0.7% depletion, rather than 1% depletion. A combination of these explanations is also a viable possibility. The thermal explanation is consistent with a recent global study of the thermal regime of continental lithosphere (Artemieva and Mooney, 2001) that shows temperature differences of as much as 500 °C between cratonic and Phanerozoic upper mantle at a depth of 150 km. The thermal explanation is also consistent with the very low seismic attenuation that correlates with cratons (Billien et al., 2000).

Conversely, the petrologic explanation for topographic depression of cratons is consistent with recently reported data from mantle xenoliths from the Kaapvaal (southern Africa) and Slave (northwest Canada) cratons that indicate less chemical depletion in deep lithospheric roots. Boyd et al. (1999) report densities for Kaapvaal garnet lherzolites, and shows that the deeper (>140 km), high-temperature samples are 1–2% denser than the highly depleted samples from shallower (<140 km) depth. In accord with Boyd et al. (1999), xenolith samples from the Slave craton indicate that the lithosphere is layered, with a denser, more iron-rich layer below 145 km (Griffin et al., 1999). Thus, evidence from the Kaavaal and Slave cratons points to dense lithospheric roots below 140 km. These observations suggest that the magnitude of negative lithospheric buoyancy is proportional to the thickness of the lithospheric root below 140 km.

Dense cratonic roots, such as we infer, are consistent with observed geoid anomalies only if restricted to roughly the upper 100–250 km of the subcrustal lithosphere. The geoid, whose variations have an amplitude of about 100 m, is dominated by the effects of slabs, hotspots, and glacial rebound (Simons and Hager, 1997). The average geoid low over platforms and shields has been estimated to be only 5 to 10 m (LeStunff and Ricard, 1995; Shapiro et al., 1999). Compensation at 100-km depth of 1 km of depressed topography would produce a 10 m geoid low (Haxby and Turcotte, 1978).

Our model of dense cratonic roots differs from previous estimates of density variation in the subcrustal lithosphere. The model of Jordan (1988) has no density contrast between cratonic and non-cratonic lithosphere. The model of LeStunff and Ricard (1995), based on geoid and topography modeling, indicates different density variations below the crust, although similar anomalous topography is observed. Mantle convection modeling of the geoid and gravity also suggests that there is excess density beneath cratons, but that it is deep-seated and descending with mantle convection (Pari and Peltier, 1996; Pari, 2001).

Part of the residual topography in Fig. 2a may be due to the density variations deep in the mantle that drive mantle convection (Forte and Perry, 2000). For example, Africa has the only large craton that is not depressed in Fig. 2a. The low shear velocities at the base of the mantle beneath Africa (Woodhouse and Dziewonski, 1984) may be due to a low-density structure that can elevate Africa by more than a kilometer (Lithgow-Bertelloni and Silver, 1998). This anomalous elevation of Africa has been previously noted from bathymetric data (Nyblade and Robinson,

1994). Further crustal data are needed to evaluate our African results because there is little control on crustal thickness and density (Fig. 1d).

We may compare our map with predictions of topography calculated from estimates of mantle density variations based on the history of subduction in the past few hundred million years and the Earth's geoid anomaly (Ricard et al., 1993; Lithgow-Bertelloni and Richards, 1995). Such modeling of the geoid and topography predicts that subduction at the western edge of the Americas and along the western margin of the Pacific would produce up to 1 km of depressed topography. Our estimates of residual topography, however, are significantly smaller in amplitude and more localized than the predictions of flow modeling of the geoid.

It has been noted that whole-mantle models of circulation could explain the dynamic surface topography data if there is a large downward increase in viscosity (by a factor of 50 or more) across the 660 km seismic discontinuity (Pari, 2001). Pari (2001) also noted that a perfectly layered model of the circulation led to accurate descriptions of the long-wavelength dynamic surface topography constraints. Forte and Mitrovica (2001) argue for a high-viscosity layer near a depth of 2000 km by modeling surface topography, gravity, and plate velocity data. Radial variations in viscosity may also be important in the uppermost mantle: thick cratonic roots appear to encounter higher viscosity, with basal drag slowing plate motion (Stoddard and Abbott, 1996; Artemieva and Mooney, 2002). Thus, deep lithospheric roots appear to play an important role in determining both surface topography and the velocity of plate motions.

6. Conclusions

We have used a recent global crustal model to calculate crustal isostasy, which we compare with observed topography, upper mantle shear-wave tomography maps, and the global distribution of cratonic crust. The calculation of crustal isostasy is most reliable for those regions of the Earth with good seismic control on crustal structure (North America, Eurasia, Australia, and oceanic regions), and we base our conclusions on those regions. After subtracting pure crustal isostasy from observed topography, we find that cratons are generally depressed by about 300 to 1500 m due to negative buoyancy (excess density) in the sub-crustal lithosphere (cf. Kaban et al., 2003). This negative buoyancy may be due to either: (1) lithospheric temperatures that are about 100 °C colder than is commonly assumed, or (2) a lesser degree of chemical depletion (approximately 0.7% vs. 1.0%) in cratonic lithosphere. Combinations of these two factors will also satisfy our observations. The thermal explanation is consistent with high seismic velocities and very low seismic attenuation in cratonic lithosphere. The petrologic explanation is consistent with recent studies of xenoliths from the Kaapvaal and Slave cratons that indicate that the density of the lithosphere increases 1–2% below 140 km. The petrologic explanation of our results also implies that the deeper a cratonic root extends below 140 km, the greater the negative buoyancy, and the larger the topographic depression.

Acknowledgements

This study was made possible by the provision of data and models from Steve Grand, Gabi Laske, Guy Masters, Carolina Lithgow-Bertelloni, David Sandwell, and William Moore. Discussions with Thomas Jordan, Art Lachenbruch, Misha Kaban, Wayne Thatcher, Yu-Shen Zhang, Brad Hager, Irina Artemieva, Paul Tackley, and the late Ron Girdler are appreciated. Irina Artemieva, Wouter Bleeker, Walter R. Roest, and Alan Jones provided reviews that significantly improved the text.

References

Artemieva, I.M., Mooney, W.D., 2001. Thermal thickness and evolution of Precambrian lithosphere: a global study. J. Geophys. Res. 106, 16387–16414.

Artemieva, I.M., Mooney, W.D., 2002. On the relations between cratonic lithosphere thickness, plate motions, and basal drag. Tectonophysics 358, 211–231.

Artemjev, M.E., Kaban, M.K., Kucherinenko, V.A., Demyanov, G.V., Taranov, V.A., 1994. Subcrustal density inhomogeneities of Northern Eurasia as derived from the gravity data and isostatic models of the lithosphere. Tectonophysics 240, 249–280.

Beloussov, V.V., Pavlenkova, N.I., Egorkin, A.V., 1991. Deep Structure of the Territory of the USSR Nauka, Moscow. 224 pp.

Billien, M., Leveque, J.-J., Trampert, J., 2000. Global maps of Rayleigh wave attenuation for periods between 40 and 150 seconds. Geophys. Res. Lett. 27, 3619–3622.

Boyd, F.R., 1989. Compositional distinction between oceanic and cratonic lithosphere. Earth Planet. Inter. 96, 15–26.

Boyd, F.R., McCallister, R.H., 1976. Densities of fertile and sterile garnet peridotites. Geophys. Res. Lett. 3, 509–512.

Boyd, F.R., Pearson, D.G., Mertzman, S.A., 1999. Spinel-facies peridotites from the Kaapvaal root. In: Gurney, J.J., Gurney, J.L., Pascoe, M.D., Richardson, S.H. (Eds.), Proceedings of the VII International Kimberlite Conference. Red Roof Design, Cape Town, pp. 40–48.

Christensen, N.I., Mooney, W.D., 1995. Seismic velocity structure and composition of the continental crust. J. Geophys. Res. 100, 9761–9788.

Dziewonski, A.M., Anderson, D.L., 1981. Preliminary reference Earth model. Phys. Earth Planet. Inter. 25, 297–356.

Finnerty, A.A., Boyd, F.R., 1987. Thermobarometry for garnet peridotites: basis for the determination of thermal and compositional structure of the upper mantle. In: Nixon, P.H. (Ed.), Mantle Xenoliths. Wiley, Chichester, pp. 381–402.

Forte, A.M., 1995. Continent ocean chemical heterogeneity in the mantle based on seismic tomography. Science 268, 789.

Forte, A.M., Mitrovica, J.X., 2001. Deep-mantle high-viscosity flow and thermochemical structure inferred from seismic and geodynamic data. Nature 410, 1049–1056.

Forte, A.M., Perry, H.K.C., 2000. Geodynamic evidence for a chemically depleted continental tectosphere. Science 290, 1940–1944.

Forte, A.M., Peltier, W.R., Dziewonski, A.M., Woodward, R.L., 1993. Dynamic surface topography—a new interpretation based upon mantle flow derived from seismic tomography. Geophys. Res. Lett. 20, 225–228.

Foulger, G., 2000. The seismic anomaly beneath Iceland extends down to the mantle transition zone and no deeper. Geophys. J. Int. 142, F2–F5.

Goodwin, A.M., 1996. Principles of Precambrian Geology. Academic Press, San Diego. 327 pp.

Grand, S.P., Helmberger, D.V., 1984. Upper mantle shear structure of North America. Geophys. J. R. Astron. Soc. 76, 399–438.

Grand, S.P., vanderHilst, R., Widiyantoro, S., 1997. Global seismic tomography: a snapshot of convection in the Earth. GSA Today 7, 1–7.

Griffin, W.L., Doyle, B.J., Ryan, C.G., Pearson, N.J., O'Reilly, S.Y., Davies, R., Kivi, K., Van Achterbergh, E., Natapov, L.M., 1999. Layered mantle lithosphere in the Lac de Gras, Slave Craton: composition, structure and origin. J. Petrol. 40, 705–727.

Haxby, W.F., Turcotte, D.L., 1978. On isostatic geoid anomalies. J. Geophys. Res. 83, 5473–5478.

Jordan, T.H., 1975. The continental tectosphere. Rev. Geophys. 13, 1–12.

Jordan, T.H., 1978. Composition and development of continental tectosphere. Nature 274, 544–548.

Jordan, T.H., 1988. Structure and formation of continental tectosphere. J. Petrol., 11–37.

Kaban, M.K., Schwintzer, P., Tikhotsky, S.A., 1999. A global isostatic gravity model of the Earth. Geophys. J. Int. 136, 519–538.

Kaban, M.K., Schwintzer, P., Artemieva, I.M., Mooney, W.D., 2003. Density of the continental roots: compositional and thermal contributions. Earth Planet. Sci. Lett. 209, 53–69.

LeStunff, Y., Ricard, Y., 1995. Topography and geoid due to lithospheric mass anomalies. Geophys. J. Int. 122, 982–990.

Lithgow-Bertelloni, C., Richards, M.A., 1995. Cenozoic plate driving forces. Geophys. Res. Lett. 22, 1317–1320.

Lithgow-Bertelloni, C., Silver, P.G., 1998. Dynamic topography, plate driving forces and the African superswell. Nature 395, 269–272.

Meissner, R., Wever, T., Flueh, E.R., 1987. The Moho in Europe—implications for crustal development. Ann. Geophys., Ser. B 5, 357–364.

Mooney, W.D., Braile, L.W., 1989. The seismic structure of the continental crust and upper mantle of North America. In: Bally, A.W., Palmer, A.R. (Eds.), The Geology of North America—An overview. Geological Society of America, Boulder, CO, pp. 39–52.

Mooney, W.D., Laske, G., Masters, T.G., 1998. Crust 5.1: a global crustal model at $5° \times 5°$. J. Geophys. Res. 103, 727–748.

Mooney, W.D., Prodehl, C., Pavlenkova, N.I., 2002. Seismic velocity structure of the continental lithosphere from controlled source data. In: Lee, W.H.K., Kanamori, H., Jennings, P.C., Kisslinger, C. (Eds.), International Handbook of Earthquake and Engineering Seismology, vol. 81A. Academic Press, San Diego, CA, pp. 887–910.

Mueller, R.D., Roest, W.R., Royer, J.-Y., Gahagan, L.M., Sclater, J.G., 1993. A digital age map of the ocean floor.

Nyblade, A.A., Robinson, S.W., 1994. The African Superswell. Geophys. Res. Lett. 21, 765–768.

Pari, G., 2001. Crust 5.1-based inference of the Earth's dynamic surface topography: geodynamic implications. Geophys. J. Int. 144, 501–516.

Pari, G., Peltier, W.R., 1996. The free-air gravity constraint on subcontinental mantle dynamics. J. Geophys. Res. 101, 28105–28132.

Parsons, B., Sclater, J.G., 1977. An analysis of ocean floor bathymetry and heat flow with age. J. Geophys. Res. 82, 803–827.

Polet, J., Anderson, D.L., 1995. Depth extent of cratons as inferred from tomographic studies. Geology 23, 205–208.

Ricard, Y., Richards, M.A., Lithgow-Bertelloni, C., LeStunff, Y., 1993. A geodynamic model of mantle density heterogeneity. J. Geophys. Res. 98, 21895–21909.

Sandwell, D.T., MacKenzie, K.R., 1989. Geoid height versus topography for oceanic plateaus and swells. J. Geophys. Res. 94, 7403–7418.

Schubert, G., Turcotte, D.L., Olsen, P., 2001. Mantle Convection in the Earth and Planets. Cambridge Univ. Press, Cambridge, UK. 940 pp.

Shapiro, S.S., Hager, B.H., Jordan, T.H., 1999. Stability and dynamics of the continental tectosphere, in composition, deep structure, and evolution of continents. In: van der Hilst, R.D., McDonough, W.F. (Eds.), Lithos 48, pp. 115–133.

Shen, Y., Solomon, S.C., Bjarnason, I.T., Wolfe, C.J., 1998. Seismic evidence for lower mantle origin of the Iceland plume. Nature 395, 62–65.

Simons, M., Hager, B.H., 1997. Localization of the gravity field and the signature of glacial rebound. Nature 390, 500–504.

Sleep, N.H., 1990. Hotspots and mantle plumes: some phenomenology. J. Geophys. Res. 95, 6715–6736.

Smallwood, J.R., White, R.S., Minshull, T.A., 1995. Sea-floor spreading in the presence of the Iceland plume: the structure of the Reykjanes Ridge at $61°40'$N. J. Geol. Soc. (Lond.) 152, 1023–1029.

Stein, S., Stein, C.A., 1996. Thermo-mechanical evolution of oceanic lithosphere: Implication for the subduction process and deep earthquakes. In: Bebout, G.E., Scholl, D.W., Kirby, S.H., Platt, J.P. (Eds.), In Subduction: Top to Bottom. Geophysical Monograph, vol. 96. American Geophysical Union, Washington, pp. 1–17.

Stoddard, P.R., Abbott, D., 1996. Influence of the tectosphere upon plate motion. J. Geophys. Res. 101, 5425–5433.

Turcotte, D.L., Schubert, G., 1982. Geodynamics; Applications of Continuum Physics to Geological Problems. Wiley, New York. 450 pp.

Watts, A.B., 2001. Isostasy and Flexure of the Lithosphere. Cambridge Univ. Press, Cambridge, UK. 458 pp.

White, R.S., Brown, J.W., Smallwood, J.R., 1995. The temperature of the Iceland plume and origin of outward-propagating V-shaped ridges. J. Geol. Soc. (Lond.) 152, 1039–1045.

Wolfe, C.J., Bjarnason, I.T., Vandercar, J.C., Solomon, S.C., 1997. Seismic structure of the Iceland mantle plume. Nature 385, 245–247.

Woodhouse, J.H., Dziewonski, A.M., 1984. Mapping the upper mantle: three-dimensional modeling of earth structure by inversion of seismic waveforms. J. Geophys. Res. 89, 5953–5986.

Yegorova, T.P., Kozlenko, V.G., Pavlenkova, N.I., Starostenko, V.I., 1995. 3-D density model for the lithosphere of Europe: construction method and preliminary results. Geophys. J. Int. 121, 873–892.

Available online at www.sciencedirect.com

SCIENCE @ DIRECT°

Lithos 71 (2003) 195–213

LITHOS

www.elsevier.com/locate/lithos

A classification scheme for mantle-derived garnets in kimberlite: a tool for investigating the mantle and exploring for diamonds

Daniel J. Schulze

Department of Geology, University of Toronto, Erindale College, 3359 Mississauga Road North, Mississauga, Ontario, Canada L5L 1C6

Abstract

A new empirical method has been devised for classification of mantle-derived garnets in kimberlite. Simple chemical screens have been developed to distinguish between garnets from different parageneses, based on Mg, Fe, Ca, Cr, Ti and Na values of published analyses of garnets from >2000 ultramafic xenoliths in kimberlite. Although crustal garnets are typically uncommon as xenocrysts in kimberlite, the first step in the classification is to screen these from the mantle population, using data from >600 garnet-bearing crustal rocks. Such a screen may also prove useful in evaluating the source (crust vs. mantle) of garnet in kimberlite exploration samples. Subsequent steps divide mantle garnets into eclogite, peridotite and Cr-poor megacryst groupings, and sub-groups of the peridotite (lherzolite, harzburgite, wehrlite) and eclogite (Groups I and II and A, B, C and grospydite) populations. Important features of this classification include the fact that it is based on distinctions between groups of fundamental geological significance (e.g., peridotite vs. eclogite) and it is based on a large, well-documented and well-understood xenolith database. As it utilizes oxide values and molar ratios of major and minor elements, the rationale for the screens is readily understood and it is simple to use.
© 2003 Elsevier B.V. All rights reserved.

Keywords: Garnet; Kimberlite; Upper mantle; Eclogite; Peridotite

1. Introduction

Our knowledge of the constitution of the inaccessible parts of the interior of Earth comes from a variety of sources, including cosmochemical modelling and geophysical studies. Direct study of actual samples of the deep interior of Earth, such as the upper mantle, is made possible by the occurrence of such material as tectonically emplaced slices in orogenic belts (e.g., Alpine peridotites) and as accidental inclusions of mantle material brought to the surface by some kimberlites and other alkaline volcanic rocks. Intact rocks (xenoliths) are ideal samples to study, but these are not as abundant as are single crystals (xenocrysts) derived by the disaggregation of mantle rocks during the typically violent transport to the surface in the host magmas. Great progress has been made in the study of the composition and structure of the upper mantle through xenocryst investigations, especially studies of xenocrystal garnet and chromite (e.g., Gurney and Switzer, 1973; Sobolev et al., 1973, 1975; Boyd and Gurney, 1982; Schulze, 1989b; Griffin et al., 1999; Grutter et al., 1999; Pokhilenko et al., 1999). Most attention has been turned towards garnet, as it occurs in a variety of rock types, such as eclogites and peridotites, rocks prevalent in most

E-mail address: dschulze@credit.erin.utoronto.ca (D.J. Schulze).

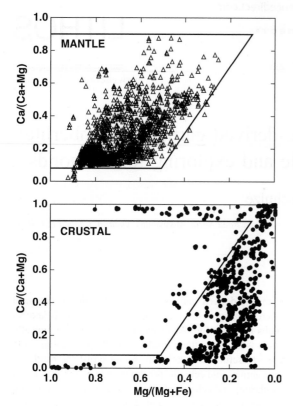

Fig. 1. Distinction between mantle-derived garnets and those from crustal rocks, in terms of Mg/(Mg + Fe) and Ca/(Ca + Mg). As discussed in the text, mantle-derived garnets are from peridotites (lherzolites and harzburgites), eclogites, alkremites and Cr-poor megacrysts. Crustal sources of garnets are dominantly amphibolite to granulite grade meta-pelites and meta-basites, in which garnets are rich in the almandine and spessartine components, but also included are less common garnets rich in the grossular, uvarovite, andradite and pyrope end members.

upper mantle xenolith suites in kimberlite and thus thought to dominate the upper mantle.

A serious drawback of dealing with xenocrysts is that, as they are single crystals, the nature of their parent rock prior to disaggregation is uncertain. One must make inferences about the characteristics of the intact parent rocks based on chemical, and to some extent physical, characteristics of the xenocrysts. To this end, a number of classification schemes have been devised to assist in deciphering more about the pre-disaggregation host of the xenocryst garnets.

Chemical screens have been proposed to help distinguish between garnets from various parageneses,

although most workers have dealt with a restricted range of rock types. Sobolev et al. (1973), for example, outlined the field of compositions of garnets from lherzolites, relative to those from harzburgites and wehrlites, in terms of their CaO and Cr_2O_3 contents and Gurney (1984) proposed a screen to distinguish between garnets from lherzolites and low-Ca garnet harzburgites, also based on CaO and Cr_2O_3 contents. Jago and Mitchell (1989) devised a cluster analysis scheme to group garnets from individual kimberlites based on chemical similarities, although the linkage of the various groups with xenolith types was tenuous. Ramsay (1992) presented graphical screens to subdivide garnets from various rock types.

The most comprehensive and widely used classification scheme for mantle-derived garnets is the cluster analysis of Dawson and Stephens (1975), in which

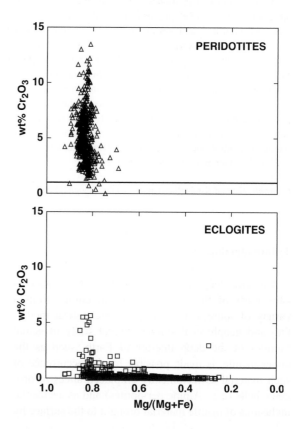

Fig. 2. Variation in Cr_2O_3 and Mg/(Mg + Fe) in garnets from eclogites and peridotites. A value of 1 wt.% Cr_2O_3 is used to discriminate between the two groups, with peridotite garnet values above this and those of eclogite garnets below.

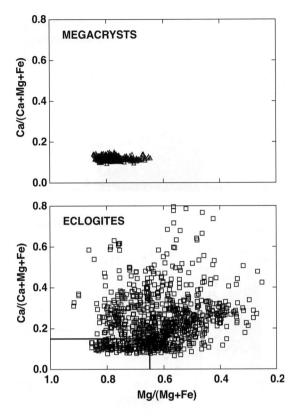

Fig. 3. Variation in Mg/(Mg + Fe) and Ca/(Ca + Mg + Fe) that assist in discrimination between garnets from the Cr-poor megacryst suite and those from eclogites. The solid lines enclose the range of eclogite garnet compositions that overlap with those of Cr-poor megacrysts. Eclogite garnets within this range are distinguished from megacrysts using Ti content in Fig. 4.

garnets from kimberlite are classified into 12 groups based on FeO, MgO, CaO, TiO_2 and Cr_2O_3 contents. Drawbacks of this scheme, however, include the facts that some groups contain garnets from more than one rock type, some rock types fall into two different groups (e.g., Cr-poor megacrysts can be classified into both Groups 1 and 2) and some garnets used to set up the classification are apparently of crustal derivation. Furthermore, the database was relatively small, utilizing only 136 xenoliths, and a substantial number of garnets (163) were single crystals of unknown, or at least uncertain, paragenesis. Danchin and Wyatt (1979) used a similar technique, although their method divided garnets into an unwieldy 52 groups.

A new classification scheme for mantle garnets is presented in this paper. It is based on chemical

analyses of 2073 garnets in mantle xenoliths from kimberlite, together with analyses of 624 garnets from crustal rocks. Figs. 1–7 illustrate the chemical screens that allow distinction between garnets from different parageneses. Figs. 8–11 are flow charts outlining steps using the chemical screens, first to discriminate crustal from mantle garnets (Fig. 8), followed by subdivision of mantle garnets into groups representing those from peridotite, eclogite and the Cr-poor megacryst suite (Fig. 9). Figs. 10 and 11 illustrate subdivision of the eclogite and peridotite groups, respectively. By connecting to the University of Toronto web site at http://www.geology.utoronto.ca/faculty/schulze.html, an Excel macro that executes this classification can be downloaded.

The screens are based on Ca, Mg, Fe, Cr, Ti and Na contents of the garnets. Routine modern electron microprobe procedures should typically provide data of high enough quality that analytical accuracy should not be a concern in applying this classification to new garnet data. The one possible exception, however, is the quality of sodium analyses. If garnets classified as eclogite-derived are to be divided into Groups I and II on the basis of their Na_2O contents (as described in a later section), care must be taken to ensure that appropriate WDS microprobe methods have been

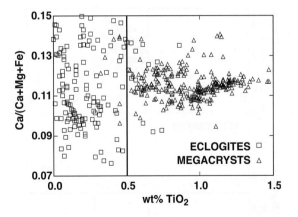

Fig. 4. Variation in TiO_2 and Ca/(Ca + Mg + Fe) in garnets from Cr-poor megacrysts with $Cr_2O_3 < 1$ wt.% and those from eclogites with Mg/(Mg + Fe) > 0.65 and Ca/(Ca + Mg + Fe) < 0.15. A cut-off value of 0.5 wt.% TiO_2 is used to discriminate between the two groups. Most of the eclogite garnets with $TiO_2 > 0.5$ wt.% are from a single locality, the Kaalvallei kimberlite, as discussed in the text.

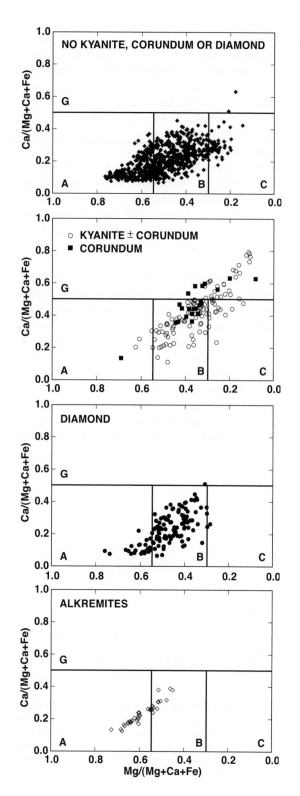

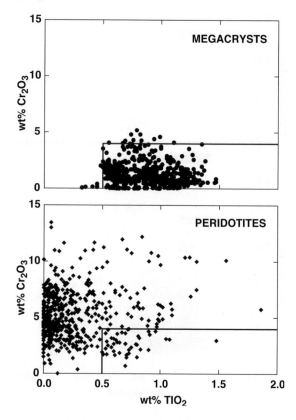

Fig. 6. Variation in Ti and Cr contents of garnets from the Cr-poor megacryst suite and peridotites. In the classification, Cr-poor megacryst garnets are defined as those with $TiO_2>0.5$ wt.% and $Cr_2O_3<4.0$ wt.%, and peridotite garnets have values outside these ranges.

used to obtain low Na detection limits (e.g., ~ 0.01 wt.% Na_2O).

2. The database

The garnet data used in this study have been taken primarily from the published literature and theses,

Fig. 5. Variation in $Mg/(Mg+Ca+Fe)$ and $Ca/(Ca+Mg+Fe)$ used to subdivide eclogite garnet compositions. "Grospydite" garnets (field G) have $Ca/(Ca+Mg+Fe)>0.50$, and those with lower calcium contents are subdivided into groups "A", "B" and "C" at $Mg/(Mg+Ca+Fe)$ values of 0.55 and 0.30, as discussed in the text. Figure is subdivided into garnets from eclogites lacking diamond or the aluminous phase kyanite or corundum, those with kyanite (with or without corundum) and those with corundum (but lacking kyanite), and those containing diamond. The compositional range of garnets from alkremites is also shown.

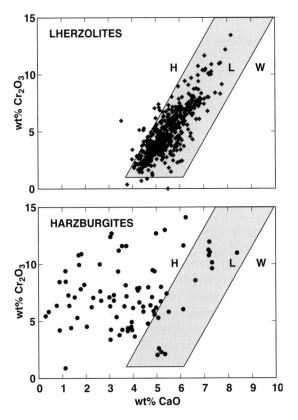

Fig. 7. Variation in CaO and Cr_2O_3 in garnets from lherzolites and harzburgites. Garnets from lherzolites are dominantly in the shaded area within the solid lines, with the points at CaO values significantly below the lower lherzolite limit likely not to be in equilibrium with both clinopyroxene and orthopyroxene. Garnets from harzburgites (lacking modal clinopyroxene) fall both within the lherzolite field, indicating equilibration with clinopyroxene, and at CaO values below the lherzolite field. The latter are typical of peridotite–suite garnets in diamonds. Garnets from wehrlites, not in equilibrium with orthopyroxene, will plot at CaO values above those of the lherzolite field. Fields defined in this classification are harzburgite (H), lherzolite (L, shaded field) and wehrlite (W).

and some previously unpublished microprobe analyses have also been used. These new WDS data (obtained at the University of Toronto) consist of analyses of 56 eclogite garnets from Kaalvallei and Bobbejaan and 94 lherzolite garnets from Kimberley dumps, Hamilton Branch and Liqhobong. By connecting to the University of Toronto web site at http://www.geology.utoronto.ca/faculty/schulze.html, these new data can be downloaded. Although some published analyses included calculated Fe_2O_3 values, all iron values used in this study have been recalculated

to total Fe as FeO, the typical method used to report electron microprobe data.

2.1. Crustal garnets

The most common occurrence of garnet in crustal rocks is in medium to high-grade meta-pelites and meta-basites. These garnets are rich in the almandine end-member, $Fe_3Al_2Si_3O_{12}$, and can have significant Mn contents (spessartine end-member, $Mn_3Al_2Si_3O_{12}$). Such garnets dominate the crustal component of the database as they are readily obtainable from the published literature. A specific search was also made for less common compositions of crustal garnets, including those rich in the grossular ($Ca_3Al_2Si_3O_{12}$), andradite ($Ca_3Fe_2Si_3O_{12}$), uvarovite ($Ca_3Cr_2Si_3O_{12}$), and pyrope ($Mg_3Al_2Si_3O_{12}$) end-members. Crustal garnet compositions in the database are from Dunn (1978), Nixon (1979), Percival (1981), Pattison et al. (1982), Chopin (1984), Cortesogno and Lucchetti (1986), von Knorring et al. (1986), Chopin et al. (1991), Gordon et al. (1991, 1994), Pattison (1991), Bégin and Pattison (1994), Fitzsimons and Harley (1994), Pattison and Bégin (1994), Nyman et al. (1995), Santos de Lima et al. (1995), Fitzsimons (1996), Guiraud et al. (1996), Hartel and Pattison (1996), Knudsen (1996), Kryza et al. (1996), Shaw and Arima (1996), Thöni and Miller (1996), Whitney et al. (1996, 2001), Azor and Ballèvre (1997), Cartwright et al. (1997), Simon et al. (1997), Willner et al. (1997), Rosenberg et al. (1998), Vance and Mahar (1998), Parthasarathy et al. (1999), Abd El-Naby et al. (2000), Bose et al. (2000), Clarke et al. (2000), Cooke et al. (2000), Fraser et al. (2000), Gayk and Kleinschrodt (2000), Gupta et al. (2000), Jones and Strachan (2000), Moraes and Fuck (2000), Parkinson (2000), Zhao et al. (2000), Bruno et al. (2001), Compagnoni and Hirajima (2001), Garcia-Casco et al. (2001), Habler and Thoni (2001), Harangi et al. (2001), Lang and Gilotti (2001), Pita and de Waal (2001), Rolland et al. (2001), Rotzler and Romer (2001), Satish-Kumar et al. (2001), Schmadicke et al. (2001), Scrimegour et al. (2001), Stowell et al. (2001), White et al. (2001) and Zeh and Millar (2001).

2.2. Mantle-derived garnets

All of the garnets in this category are from ultramafic xenoliths from kimberlites, specifically lherzo-

CRUSTAL vs. MANTLE GARNETS:

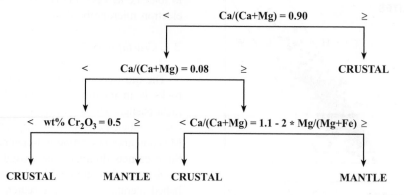

Fig. 8. Flow chart illustrating the steps used to distinguish between garnets derived from crustal rocks and those of mantle origin.

lites, harzburgites, eclogites, and Cr-poor megacrysts. Garnet inclusions in diamonds were not included as in some instances their compositions are well outside of the range of compositions of minerals from ultramafic xenoliths (e.g., Gurney et al., 1984; Schulze, 1997). Table 1 lists the regional distribution of the kimberlite sources of the 2073 mantle xenoliths in the database. Approximately half of the samples are from kimberlites on the Kaapvaal Craton. Specific locations can be obtained from the references cited.

Eclogite garnet data are from Sobolev et al. (1968, 1994), Meyer and Brookins (1971), Chinner and Cornell (1974), Reid et al. (1976), Smyth and Hatton (1977), Shee (1978), Shee and Gurney (1979), Boyd and Danchin (1980), Ater (1982), Tollo (1982), McGee and Hearn (1984), Robinson et al. (1984), Smyth and Caporuscio (1984), Spetsius et al. (1984), Hall (1985), MacGregor and Manton (1986), McCandless and Gurney (1986), Ford (1987), Schulze and Helmstaedt (1988), Sommerville (1988), de Bruin (1989), Hills and Haggerty (1989), Smith et al. (1989), Taylor and Neal (1989), Schulze (1992, 1997), Jerde et al. (1993a,b), Jacob et al. (1994), Viljoen (1994, 1995), Fung and Haggerty (1995), Beard et al. (1996), Schulze et al. (1996, 1997, 2000), Taylor et al. (1996), Viljoen et al. (1996),

MANTLE GARNETS:

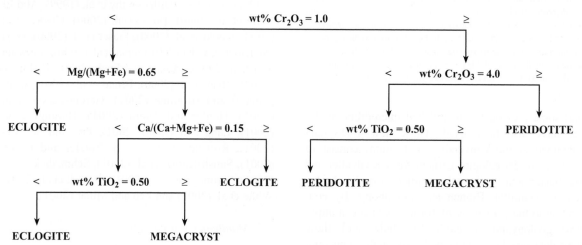

Fig. 9. Flow chart illustrating the steps used to subdivide mantle garnets into groups representing those from eclogites, peridotites and Cr-poor megacrysts.

ECLOGITE GARNETS:

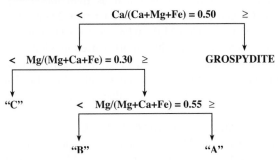

Fig. 10. Flow chart illustrating the steps used to subdivide eclogite garnets. Eclogite garnets are further subdivided into Groups I and II using sodium content, if reliable values for Na$_2$O values are available. See text for discussion.

Snyder et al. (1997), El Fadili and Demaiffe (1999), Kopylova et al. (1999) and this paper. Alkremite garnets, used for comparison with other rock types but not specifically used to define the chemical screens, are from Nixon et al. (1978), Shee (1978), Exley et al. (1983), Mazzone and Haggerty (1989a,b), and Viljoen (1994).

Harzburgite garnet data were taken from Pokhilenko et al. (1977, 1991, 1993), Dawson et al. (1978), Sobolev et al. (1984) Nixon et al. (1987), Boyd and Nixon (1988), Hops (1989), Skinner (1989), Viljoen et al. (1992), Boyd et al. (1993), Schulze (1995), Pearson et al. (1994, 1999), Schulze et al. (1997) and Burgess and Harte (1999). Lherzolite garnet data are from Boyd (1973), Cox et al. (1973), Nixon and Boyd (1973b), Hearn and Boyd (1975), Boyd et al. (1976), Danchin and Boyd (1976), Pokhilenko et al. (1977, 1991), Smith (1977), Sobolev (1977), Bishop et al. (1978), Boyd and Nixon (1978), Carswell et al.

(1979), MacGregor (1979), Boyd and Danchin (1980), Robey (1981), Shee et al. (1982), Hearn and McGee (1984), Mitchell (1984), Sobolev et al. (1984), Eggler et al. (1987), Hops (1989), Nehru and Reddy (1989), Skinner (1989), Hall (1991), Pearson et al. (1994, 1999), Viljoen et al. (1994), Vicker (1997), Burgess and Harte (1999), Kopylova et al. (1999) and this paper.

Analyses of garnets belonging to the Cr-poor megacryst suite are from Nixon and Boyd (1973a), Jakob (1977), Robey (1981), Hops (1989), de Bruin (1991) and Schulze (1997).

3. Distinction between crustal and mantle garnets

As kimberlites pass through both the uppermost upper mantle and the entire thickness of the continental crust on their rise to Earth's surface, they are likely to contain garnets derived from both mantle and crustal regions. In most kimberlites, mantle-derived xenoliths appear to be far more abundant than xenoliths that belong to the crustal portion of the crystalline basement, although locally garnet-bearing (and other) crystalline crustal rocks may be relatively abundant (e.g., Dawson and Smith, 1987). In studies of garnet xenocrysts from 32 kimberlites from South Africa and North America (4500 garnets), Schulze (1993, 1994) found that 2/3 of the kimberlites contained some crustal garnets and 3% of all the garnet xenocrysts in these kimberlites were derived from crustal rocks (using an earlier formulation of this classification scheme). Approximately one-quarter of the garnet xenocrysts in the Lace kimberlite are crustal (Schulze, 1994). Thus, it is important to be able to

PERIDOTITE GARNETS:

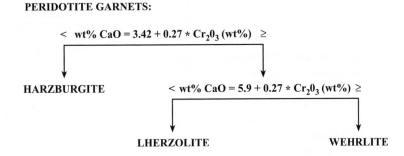

Fig. 11. Flow chart illustrating the steps used to subdivide peridotite garnets into harzburgite, lherzolite and wehrlite subgroups.

Table 1
Geographic–tectonic distribution of samples in database

	Lherzolite	Harzburgite	Megacryst	Eclogite
Kaapvaal Craton	208	55	422	458
Kaapvaal margin	14		27	
Zimbabwe Craton				117
West African Craton				87
Congo Craton	12			16
Tanzanian Craton	3			
Namaqua Belt	55			
Siberian Platform	18	29		116
Dharwar Craton	6			
Slave Craton	26	5		28
Superior Craton	82			9
Grenville Province	43			
Canadian Cordillera	2			
Wyoming Craton	15			
Yavapai-Mazatzal Belt, USA	14		98	96
Adeladian Belt, Australia			12	
Total	498	89	559	927

screen out crustal garnets from kimberlite xenocryst populations in order to interpret properly the mantle population. Furthermore, in kimberlite and diamond exploration work, in which most garnets from sediment samples are likely to be derived from exposed crustal rocks, it is essential to be able to screen out the crustal garnet population in the search for the kimberlite-derived (i.e., mantle) garnets.

Garnets from crustal and mantle sources are compared in Fig. 1, in terms of $Ca/(Ca+Mg)$ and $Mg/(Mg+Fe)$ values (mole proportions). In general, crustal garnets are characterized by lower $Mg/(Mg+Fe)$ values relative to those from mantle rocks, but many crustal garnets with very high or low values of $Ca/(Ca+Mg)$ overlap mantle garnets in terms of $Mg/(Mg+Fe)$.

With the exception of some Ca-poor Cr-pyropes from harzburgites, mantle garnets do not extend to the very low $Ca/(Ca+Mg)$ values of some of the pyrope-rich garnets in high-grade quartzites and whiteschists, such as those described from coesite-bearing rocks in the Alps (e.g., Chopin, 1984) and a cut-off of $Ca/(Ca+Mg)=0.08$ discriminates between the Mg-rich, Ca-poor crustal garnets and most of the Mg-rich mantle garnets that have higher values of $Ca/(Ca+Mg)$. The crustal Ca-poor pyropes are essentially Cr-free, however, whereas similarly magnesian and

Ca-poor mantle garnets (such as diamond-bearing low-Ca garnet harzburgites) are Cr-rich. Mantle garnets with $Ca/(Ca+Mg)<0.08$ have Cr_2O_3 contents above 0.5 wt.%, and this cutoff is used in addition to distinguish between Ca-poor, Mg-rich garnets from the crust and those from the mantle (Fig. 1). The upper limit of $Ca/(Ca+Mg)$ values for mantle garnets is 0.90.

For garnets of intermediate $Ca/(Ca+Mg)$ values $(0.08<Ca/(Ca+Mg)<0.90)$, the two garnet groups are divided at $Ca/(Ca+Mg)=1.1-2Mg/(Mg+Fe)$, with mantle garnets having $Mg/(Mg+Fe)$ values above this and those from crustal rocks below (Fig. 1). Five mantle garnets, all from eclogites, and 27 crustal garnets fall outside of the defined boundaries for these rock types, representing $<1\%$ and 4% of the database for each rock type, respectively (Table 2). The discriminant line has been chosen to favour correct classification of mantle garnets, as few crustal garnets are expected to occur in most kimberlites and the classification is primarily intended for use with garnets in kimberlite. If this discriminant is used for garnets from exploration samples, some Mg-rich crustal garnets could be incorrectly classified as mantle-derived (specifically eclogitic). In the absence of a significantly larger number of additional clearly mantle-derived garnets in such an exploration sample (e.g., Cr-pyropes and Cr-poor megacryst garnets), as predicted by garnet populations from known kimberlites (Schulze, 1997), such a result should be viewed with suspicion.

Table 2
Comparison of garnet classifications: classification proposed in this paper vs. rock name given in the literature for garnets in the database

Name in proposed classification	Literature rock name				
	Crustal (624)[a]	Lherzolite (498)	Harzburgite (89)	Megacryst (559)	Eclogite (927)
Crustal	597	0	0	0	5
Wehrlite	0	0	1	0	2
Lherzolite	0	438	15	12	22
Harzburgite	0	12	72	0	4
Megacryst	0	45	0	541	21
Eclogite	27	3	1	6	873
% Agreement	96	88	81	97	94

[a] Number in parentheses indicates number of garnets from that rock type in the database.

4. Subdivision of mantle garnets

4.1. Eclogite vs. peridotite

The three general categories of mantle garnets considered in this paper are eclogite, peridotite, and Cr-poor megacryst. Eclogites are essentially biminer-alic garnet–clinopyroxene rocks that may contain a variety of accessory minerals, such as rutile, kyanite, diamond, graphite and coesite. With few exceptions, the silicates in rocks termed "eclogite" have low Cr_2O_3 contents, and cut-off values at modest levels of Cr_2O_3 (0.5–4.0 wt.%; e.g., Gurney, 1984; Schulze, 1989b) have been proposed to distinguish eclogite-derived garnets from those from peridotites. A cut-off of 1.0 wt.% Cr_2O_3 is used in this classification. In the database, garnets from 29 of 927 rocks described as eclogite have $Cr_2O_3 > 1.0$ wt.% (3% of eclogite garnets) and only 5 of 541 peridotitic garnets have $Cr_2O_3 < 1.0$ wt.% (less than 1% of peridotitic garnets) (Fig. 2).

4.2. Pyroxenites

One group that is not distinguished in this classi-fication is "pyroxenite". Rocks rich in pyroxene and poor in, or free of, olivine are common, though typically not abundant, in many xenolith suites in kimberlites. They undoubtedly have a variety of ori-gins. The Cr-rich eclogites in the database would be called garnet clinopyroxenites by many workers, and some garnet pyroxenites, especially those containing primary spinel (e.g., Eggler et al., 1987), are probably more closely related to peridotites than to eclogites. Two-pyroxene garnet websterites may be also be modal variants of peridotites (e.g., Cox et al., 1973) and others may be closely related to eclogites (e.g., Hatton, 1978). Some Ti-rich pyroxenites may be cumulates related to the Cr-poor megacryst suite (e.g., Kopylova et al., 1999). In the classification scheme proposed here, such garnets will be classified as the variety of peridotite, eclogite or Cr-poor megacryst that they would most closely resemble chemically.

4.3. Eclogite vs. Cr-poor megacryst

Although peridotite garnets are distinct from eclo-gite garnets in terms of Cr_2O_3, the other important type of garnet from the mantle, from the Cr-poor megacryst suite (e.g., Eggler et al., 1979; Gurney et al., 1979; Schulze, 1987) overlaps both eclogitic and peridotitic garnets in Cr_2O_3 content. Unlike the large variation in $Ca/(Ca+Mg+Fe)$ and $Mg/(Mg+Fe)$ exhibited by eclogite garnets, however, Cr-poor meg-acryst garnets show only a small range in these values, illustrated in Fig. 3. Compositions of only two mega-cryst garnets of 559 in the database ($< 1\%$) are outside of the ranges $0.09 < Ca/(Ca+Mg+Fe) < 0.15$ and $Mg/(Mg+Fe) > 0.65$. Cr-poor megacryst garnets typically have elevated Ti contents relative to eclogite garnets, however, and the two groups can be effectively separated on the basis of TiO_2, $Mg/(Mg+Fe)$ and $Ca/(Ca+Mg)$. Only 150 eclogite garnets in the data-base overlap Cr-poor megacryst garnets in terms of $Ca/(Ca+Mg)$ and $Mg/(Mg+Fe)$. Comparison of Ti contents of these two groups (megacrysts with $Cr_2O_3 < 1.0$ wt.% and the sub-set of 150 eclogitic garnets) in Fig. 4 shows little overlap, and a distinc-tion between the two groups is made at 0.5 wt.% TiO_2. Seven of the 265 Cr-poor megacryst garnets with Cr_2O_3 below 1.0 wt.% have less than 0.5 wt.% TiO_2 (3% of this megacryst sub-group and 1% of the whole megacryst suite). Eighteen eclogite garnets have $TiO_2 > 0.5$ wt.%, which would cause them to be mis-classified as garnets belonging to the Cr-poor megacryst suite. Fourteen of those are from the Kaalvallei kimberlite, however, which is anomalous in the Ti-rich eclogites that occur there (Viljoen, 1994). Aside from the Kaalvallei Ti-rich eclogite garnets, only four other eclogite garnets have TiO_2 values corresponding to the Cr-poor megacryst field (less than 1% of the 927 eclogites in the database). Even including the Kaalvallei Ti-rich eclogites, only 2% of the eclogites overlap garnets from the Cr-poor megacryst suite (Table 2).

4.4. Subdivision of eclogite garnets

Although mantle eclogites contain a wide variety of primary accessory minerals (e.g., rutile, kyanite, co-rundum, diamond, graphite, coesite, sanidine, ortho-pyroxene, ilmenite, apatite, zircon, phlogopite, titanite, spinel, amphibole, sulphides), and their garnets and clinopyroxenes range widely in composition, there are few paragenetic or chemical screens that allow mean-ingful subdivision of this group. One classification that has gained wide acceptance, however, is based on the Na$_2$O content of eclogitic garnet. Many eclogite suites have garnets with bimodal distribution of Na contents, which has been especially well documented in samples from the Roberts Victor Mine (e.g., Mac-Gregor and Carter, 1970; Hatton, 1978; McCandless and Gurney, 1989; Schulze et al., 2000). Based on Roberts Victor and other eclogite suites (e.g., Gurney, 1984), eclogite garnets are divided into two groups at 0.07 wt.% Na$_2$O. Group II eclogite garnets have < 0.07 wt.% Na$_2$O and those from Group I have ≥0.07 wt.% Na$_2$O.

The high sodium content of Group I eclogite garnets has been proposed as a diamond exploration tool, as the garnets in most diamond-bearing eclogites have Na$_2$O>0.07 wt.% and are classified as Group I. Of the 145 diamond-bearing eclogites in the database, only seven have Na$_2$O < 0.07 wt.%. Group I eclogite garnets are not completely correlative with diamonds, however, as garnets from many graphite-bearing eclogites have Na$_2$O>0.07 wt.% (e.g., Grutter and Quadling, 1999), and most Group I eclogites do not contain any carbon polymorph.

Not all published eclogite garnet data have reliable sodium values. Sodium is typically not reported for analyses obtained using energy dispersive microprobe methods, and some older wavelength dispersive microprobe data are imprecise. In this classification scheme, sub-division into Groups I and II based on sodium content of garnet is applied at the end of the routine in appropriate cases, modifying the other eclogite sub-groupings suggested below.

Kyanite-bearing eclogites are well known and fairly common and corundum-bearing examples also occur, but are less abundant. Kyanite eclogites with high Ca garnets (Ca/(Ca + Mg + Fe)>0.50) are referred to as grospydites (e.g., Sobolev et al., 1968). In Fig. 5, eclogites containing kyanite and/or corundum are compared with eclogites lacking these aluminous minerals, and it is clear that most eclogites in which garnets have Ca/(Ca + Mg + Fe)>0.50 also contain kyanite and/or corundum. Thus, garnets with Ca/(Ca + Mg + Fe)>0.50 are classified as G-type (for grospydite), although not all such garnets necessarily represent disaggregated kyanite eclogites, which is part of the true definition of grospydite. Some garnets with Ca/(Ca + Mg + Fe)>0.50 are from kyanite-free, corundum-bearing eclogites and a few contain neither

corundum nor kyanite (Fig. 5). Furthermore, it is clear from the wide variation in Ca/(Ca + Mg + Fe) contents of garnets from corundum/kyanite eclogites that the presence of these aluminous minerals is not restricted to eclogites with high calcium garnets.

Eclogitic garnets with Ca/(Ca + Mg + Fe) < 0.50 have been subdivided into three groups, based on their Mg/(Mg + Ca + Fe) values (Fig. 5). These three groups, termed A, B and C have ranges of Mg/(Mg + Ca + Fe) as illustrated in Fig. 5, corresponding to the compositional breaks suggested by Coleman et al. (1965) in their study of eclogites from a variety of geologic settings. The original three Groups A, B and C were established in order to "subdivide the various types [of eclogite] into geologically similar occurrences" (Coleman et al., 1965, p. 485) and it was *not* their intention to use these groupings to classify eclogites based on garnet composition. [At that time, the mineral chemical database for mantle eclogites was so small that all mantle eclogite garnets were thought to be pyropic and correspond to Group A. It has since been shown that mantle eclogites have an extremely wide compositional range (e.g., MacGregor and Carter, 1970; Hatton, 1978; Ater, 1982; Fung and Haggerty, 1995). The A, B and C designations are used here solely to divide the large compositional range into smaller, but arbitrary, fields (with no particular petrologic significance), to facilitate discussion of various eclogite populations. For example, diamond-bearing eclogites (Fig. 5) can be described as primarily belonging to Group B, with Group A examples being less abundant, and Group C and grospydite varieties being extremely rare. The compositional ranges and A, B, C designations have been chosen for historical purposes, consistent with continued use of the Coleman et al. (1965) terminology (e.g., Fung and Haggerty, 1995; Snyder et al., 1997).

Also illustrated in Fig. 5 are compositions of garnets from alkremites (Ponomarenko, 1975), which are garnet–spinel rocks, some of which contain corundum (e.g., Nixon et al., 1978). Ponomarenko (1975) considered alkremites to belong to an "ultramafic" paragenesis. They are much less abundant and less widely distributed than are either peridotites or eclogites. Although their mineral assemblage is distinct from eclogites (alkremites lack clinopyroxene and primary spinel is extremely rare in mantle eclogites; Viljoen, 1994), and they have only a small range in composition, in terms of Ca/Mg/Fe values (Fig. 5), as well as Ti, Cr and Mn contents, their garnets are similar in composition to some in eclogites. In garnet composition, alkremites form a sub-group of the eclogites and thus in the classification scheme alkremite garnets are grouped with those from eclogites.

4.5. Cr-poor megacryst vs. peridotite

As noted above, eclogites with $Cr_2O_3>1.0$ wt.% are uncommon, and most garnets with $Cr_2O_3>1.0$ wt.% are from peridotites and the Cr-poor megacryst suite. As the name implies, Cr-poor megacryst garnets do not have very high Cr_2O_3 contents, and those with $Cr_2O_3 < 1.0$ wt.% were distinguished from eclogites above. Garnets from most Cr-poor megacryst suites have $Cr_2O_3 < 2-3$ wt.%, especially those from Group I kimberlites in southern Africa (e.g., Nixon and Boyd, 1973a,b; Gurney et al., 1979), although those from North America (e.g., Eggler et al., 1979; Schulze, 1997), and from Group II kimberlites in southern Africa (e.g., Moore and Gurney, 1991), can reach Cr_2O_3 values above 4 wt.% (Fig. 6).

Garnets in most peridotites are considerably richer in Cr_2O_3. Cr_2O_3 values above 5 wt.% are typical and those above 10 wt.% are not uncommon (Fig. 6).

Titanium contents are also useful in distinguishing between garnets that belong to the Cr-poor megacryst suite and those from peridotites. Megacryst garnets are mostly above 0.5 wt.% TiO_2 whereas those from peridotites, though occupying a large range, have TiO_2 contents primarily below that value. The smallest overlap between these two groups occurs by defining megacryst garnets as having $TiO_2>0.50$ wt.% and $Cr_2O_3 < 4.0$ wt.%, and peridotite garnets having compositions outside this range, as illustrated in Fig. 6. Using these values as screens, 12 of 559 Cr-poor megacryst garnets and 45 of 587 peridotites are incorrectly classified, 5% of the summed populations of the two groups. Within some individual kimberlite populations, there is no overlap between garnets from the megacryst suite and those from "sheared" peridotites, such as at Jagersfontein (Hops, 1989).

Most of the peridotite garnets that correspond in composition to Cr-poor megacrysts occur in the population of high-temperature, deformed ("sheared") peridotites (e.g., Nixon and Boyd, 1973a,b). This is not surprising, as many workers have concluded that

the parent magma of the megacrysts has infiltrated the sheared peridotites and metasomatised them (e.g., Gurney and Harte, 1980) and so the similarities in composition between the two groups reflect genetic links. As many of the Ti-rich garnets in the high-temperature peridotites are zoned (e.g., Smith and Boyd, 1992), whereas garnet megacrysts are typically homogeneous, a test for homogeneity in suspect garnets might assist in distinguishing one type of Ti-rich, Cr-poor garnet from the other.

4.6. Subdivision of peridotite garnets

Pyroxene-bearing peridotites (i.e., not dunites) are subdivided into lherzolite, harzburgite and wehrlite, based on the presence or absence of modal orthopyroxene and clinopyroxene. Harzburgites contain orthopyroxene and lack clinopyroxene, wehrlites contain clinopyroxene but lack orthopyroxene and lherzolites contain both clinopyroxene and orthopyroxene. Sobolev et al. (1973) showed that Ca and Cr correlate positively in garnets from lherzolites, and the field of garnets from lherzolites in the database is shown in terms of wt.% CaO and wt.% Cr_2O_3 (Fig. 7). Garnets that plot within the field defined by the lherzolite garnets are classified as lherzolite, those with CaO values above the lherzolite field are termed wehrlite, and those with CaO values below the lherzolite field are termed harzburgite. The lower limit of the lherzolite field in this figure ($CaO = 3.42 + 0.27Cr_2O_3$, with oxide values in weight percent) is taken from Fig. 3 of Gurney (1984), dividing the garnets termed G9 and G10 in his study of peridotite–suite garnets included in diamonds. Garnets termed G9 correspond to those from lherzolites and those termed G10 correspond to harzburgite garnets. The upper CaO limit of the lherzolite field in Fig. 7 is drawn to enclose the lherzolite garnets in the database, and be parallel to the lower CaO limit. Four lherzolite garnets have Cr_2O_3 contents below 1.0 wt.%, and 12 plot below the lower CaO limit. Of the latter, two significantly within the harzburgite field are in a zoned garnet from Jagersfontein described by Burgess and Harte (1999). The composition of the core of this garnet is so far into the harzburgite field that it is probably not in equilibrium with the clinopyroxene in the rock and thus is not considered in defining the lherzolite field (also see Pokhilenko et al., 1999). A similar situation was documented in a lherzolite from the Roberts Victor mine by Viljoen et al. (1994), who concluded that a low-CaO garnet in a lherzolite was out of equilibrium with the clinopyroxene in the peridotite, which probably represented late, metasomatic introduction of clinopyroxene into a low-Ca garnet harzburgite.

Fig. 7 also illustrates compositions of garnets from rocks described as in the literature as harzburgites. As this type of garnet does not have much influence on the screens used for classification, only a few sources were used for data, simply to illustrate how harzburgite garnets might be classified. A significant proportion (~ 17%) of these garnets plot in the lherzolite field. This suggests that the minerals in these rocks are in equilibrium with clinopyroxene, though it does not appear in the mode of the rock. Such garnets would be classified as lherzolite-derived, and are common in harzburgites (Schulze, 1995).

Most peridotite–suite garnets that occur as inclusions in diamonds correspond to compositions in the harzburgite field (the G10 garnets of Gurney, 1984), and this chemical characteristic has proven to be one of the most valuable features of garnets used in diamond exploration. Further subdivisions of the harzburgite field have been used by Gurney (1980) to give "scores" to garnet xenocryst populations, "scores" used in ranking the diamond potential of individual kimberlites. Versions of these harzburgite subdivisions have been illustrated by Hill (1989), Griffin et al. (1992), and Lee (1993), but are not utilized in this classification scheme.

Garnets that plot above the lherzolite field, in the wehrlite field, are not common in kimberlites. Some true garnet wehrlites have been described (e.g., Sobolev et al., 1973; Schulze, 1989a) and garnets from rocks referred to as Cr-rich eclogites also fall within the wehrlite field. Most of the unusual green garnet xenocrysts, rich in the uvarovite component, such as those described by Sobolev et al. (1973), Clarke and Carswell (1977) and Schulze (1989) correspond to wehrlite garnets.

5. Evaluation of the classification scheme

Table 2 is cross-tabulation comparing original rock names in the literature for garnets in the database with names that would be assigned to them using the new classification. The original name is returned in over 93% of the cases.

Three suites of mantle garnets not included in the database have also been used to independently evaluate the new classification. They are garnets from peridotites from Letseng-la-Terae, Lesotho (new data from Moore and Lock, 2001), garnet megacrysts from Orapa, Botswana (Shee, 1978) and garnets from eclogites from the Jagersfontein kimberlite in South Africa (Dawson and Smith, 1986; Pyle and Haggerty, 1998). Garnets from these populations have been classified using the new scheme in this paper, and results are compared with the names given to the rocks in the original studies.

Of the 17 clinopyroxene-bearing peridotites from Letseng, all are classified as peridotite. Garnets from 16 of these are classified as lherzolite and one as harzburgite (though very near the lherzolite field). All 20 of the Orapa garnet megacrysts are classified as megacrysts by the new classification scheme. Of the 76 eclogites from Jagersfontein, one Cr-rich ($Cr_2O_3 = 1.59$ wt.%) garnet would be classified as a lherzolite garnet. One Ti-rich eclogite garnet (0.90 wt.% TiO_2) would be classified as derived from the Cr-poor megacryst suite. As the clinopyroxene with which this garnet coexists (Pyle and Haggerty, 1998) is anomalous relative to other eclogite clinopyroxenes at Jagersfontein, but similar to Cr-poor clinopyroxene megacrysts from this location (Hops, 1989), it is likely that the sample was originally mis-classified, but is correctly classified in the new scheme. The remaining 74 Jagersfontein eclogite garnets are classified as eclogite-derived. Eclogites with Na_2O values of both Groups I and II occur, although there is not a clear separation between the two groups at $Na_2O = 0.07$ wt.%. In terms of Mg/(Mg + Ca + Fe) values, Group C eclogites are absent and the ratio of Group B to Group A is about 2:1.

Thus, the new classification scheme yields satisfactory results in classifying garnets from mantle xenoliths of known paragenesis for the three suites used here, and for the original data in the database.

6. Concluding remarks

The mantle garnet classification scheme proposed in this paper is thought to represent a significant improvement over those currently available in the published literature. It is based on a large body of published data of garnets from mantle xenoliths in kimberlites worldwide. The chemical screens utilized to distinguish between populations typically only "mis-classify" a few percent of garnets from any given rock type, giving a high degree of confidence in their general applicability to garnets from unknown sources.

Although the primary intention of the classification is to understand better garnets from the upper mantle, to aid in furthering our knowledge of the constitution of this inaccessible part of Earth, several features of the scheme have direct applicability to the more practical pursuit of exploration for kimberlites and diamonds. With this scheme, garnets from crustal sources can be screened out from exploration samples, and in a single sequence of steps those that have a high probability of indicating the presence of diamonds (e.g., Group I eclogites and low-Ca garnet harzburgites) can be identified.

Acknowledgements

I thank Claudio Cermignani and Kimberley Scully for assistance with electron microprobe analysis, and Mike Kolcun for producing the classification macro. Reviews by Nick Pokhilenko, Bruce Wyatt and Herman Grutter improved the manuscript, and I also benefited from discussions with Herman and Bruce. Alison Dias and Jennifer Storer-Folt are thanked for their assistance with figure and manuscript preparation. Financial assistance was provided by N.S.E.R.C.

References

Abd El-Naby, H., Frisch, W., Hegner, E., 2000. Evolution of the Pan-African Wadi Haimur metamorphic sole, Eastern Desert, Egypt. J. Metamorph. Geol. 18, 639–651.

Ater, P.C., 1982. Petrology and geochemistry of mantle eclogite xenoliths from Colorado–Wyoming kimberlites, MS thesis, Colorado State University.

Azor, A., Ballèvre, M., 1997. Low-pressure metamorphism in the Sierra Albarrana Area (Variscan Belt, Iberian Massif). J. Petrol. 38, 35–64.

Beard, B.L., Fraracci, K.N., Taylor, L.A., Snyder, G.A., Clayton, R.A., Mayeda, T.K., Sobolev, N.V., 1996. Petrography and geo-

chemistry of eclogites from the Mir kimberlite, Yakutia, Russia. Contrib. Mineral. Petrol. 125, 293–310.

Bégin, N.J., Pattison, D.R.M., 1994. Metamorphic evolution of granulites in the Minto Block, northern Québec: extraction of peak P–T conditions taking account of late Fe–Mg exchange. J. Metamorph. Geol. 12, 411–428.

Bishop, F.C., Smith, J.V., Dawson, J.B., 1978. Na, K, P and Ti in garnet, pyroxene and olivine from peridotite and eclogite xenoliths from African kimberlites. Lithos 11, 153–173.

Bose, A., Fukuoka, M., Sengupta, P., Dasgupta, S., 2000. Evolution of high-Mg–Al granulites from Sunkarametta, eastern Ghats, India: evidence for a lower crustal heating-cooling trajectory. J. Metamorph. Geol. 18, 223–240.

Boyd, F.R., 1973. Appendix of mineral analyses. In: Nixon, P.H. (Ed.), Lesotho Kimberlites. Lesotho National Development, Maseru, pp. 33–36.

Boyd, F.R., Danchin, R.V., 1980. Lherzolites, eclogites and megacrysts from some kimberlites of Angola. Am. J. Sci. 280-A, 528–549.

Boyd, F.R., Gurney, J.J., 1982. Low calcium garnets: keys to craton structure and diamond crystallization. Carnegie Inst. Wash. Yrbk. 81, 261–267.

Boyd, F.R., Nixon, P.H., 1978. Ultramafic nodules from the Kimberley pipes. Geochim. Cosmochim. Acta 42, 1267–1282 (Appendix).

Boyd, F.R., Nixon, P.H., 1988. Low-Ca garnet harzburgites: origin and role in craton structure. Annu. Rep. Dir. Geophys. Lab. 2102, 8–13.

Boyd, F.R., Fujii, T., Danchin, R.V., 1976. A noninflected geotherm for the Udachnaya kimberlite pipe, USSR. Carnegie Inst. Wash. Yrbk. 75, 523–531.

Boyd, F.R., Pearson, D.G., Nixon, P.H., Mertzman, S.A., 1993. Low-calcium garnet harzburgites from southern Africa: their relations to craton structure and diamond crystallization. Contrib. Mineral. Petrol. 113, 352–366.

Bruno, M., Compagnoni, R., Rubbo, M., 2001. The ultra-high pressure coronitic and pseudomorphous reactions in a metagranodiorite from the Brossasco-Isasca Unit, Dora-Maira Massif, western Italian Alps: a petrographic study and equilibrium thermodynamic modeling. J. Metamorph. Geol. 19, 33–43.

Burgess, S.R., Harte, B., 1999. Tracing lithosphere evolution through the analysis of heterogeneous G9/G10 garnets in peridotite xenoliths: I. Major element chemistry. In: Gurney, J.J., Gurney, J.L., Pascoe, M.D., Richardson, S.H. (Eds.), Proceedings of the VIIth International Kimberlite Conference, vol. 1. Red Roof Design, Cape Town, pp. 66–79.

Carswell, D.A., Clarke, D.B., Mitchell, R.H., 1979. The petrology and geochemistry of ultramafic nodules from Pipe 200, northern Lesotho. In: Boyd, F.R., Meyer, H.O.A. (Eds.), The Mantle Sample: Inclusions in Kimberlites and Other Volcanics. Amer. Geophys. Union, Washington, pp. 127–144.

Cartwright, I., Buick, I.S., Maas, R., 1997. Fluid flow in marbles at Jervois, central Australia: oxygen isotope disequilibrium and zoning produced by decoupling of mineralogical and isotopic resetting. Contrib. Mineral. Petrol. 128, 335–351.

Chinner, G.A., Cornell, D.H., 1974. Evidence of kimberlite–grospydite reaction. Contrib. Mineral. Petrol. 45, 153–160.

Chopin, C., 1984. Coesite and pure pyrope in high-grade blues-chists of the Western Alps: a first record and some consequences. Contrib. Mineral. Petrol. 86, 107–118.

Chopin, C., Henry, C., Michard, A., 1991. Geology and petrology of the coesite-bearing terrain, Dora Maira massif, Western Alps. Eur. J. Mineral. 3, 263–291.

Clarke, D.B., Carswell, D.A., 1977. Green garnets from the New-lands kimberlite, Cape Province, South Africa. Earth Planet. Sci. Lett. 34, 30–38.

Clarke, G.L., Klepeis, A., Daczko, N.R., 2000. Cretaceous high-*P* granulites at Milford Sound, New Zealand: metamorphic history and emplacement in a convergent margin setting. J. Metamorph. Geol. 18, 359–374.

Coleman, R.G., Lee, D.E., Beatty, L.B., Brannock, W.W., 1965. Eclogites and eclogites: their similarities and differences. Geol. Soc. Amer. Bull. 76, 483–508.

Compagnoni, R., Hirajima, T., 2001. Superzoned garnets in the coesite-bearing Brossasco-Isasca Unit, Dora-Maira massif, Western Alps, and the origin of the whiteschists. Lithos 57, 219–236.

Cooke, R.A., O'Brien, P.J., Carswell, D.A., 2000. Garnet zoning and the identification of equilibrium mineral compositions in high-pressure–temperature granulites from the Moldanubian Zone, Austria. J. Metamorph. Geol. 18, 551–569.

Cortesogno, L., Lucchetti, G., 1986. Andradites and chromian an-dradites from Northern Apennine ophiolites (Italy). Neues Jahrb. Mineral. Abh. 155, 165–184.

Cox, K.G., Gurney, J.J., Harte, B., 1973. Xenoliths from the Mat-soku Pipe. In: Nixon, P.H. (Ed.), Lesotho Kimberlites. Lesotho National Development, Maseru, pp. 76–92. Plus Appendix.

Danchin, R.V., Boyd, F.R., 1976. Ultramafic nodules from the Pre-mier kimberlite pipe, South Africa. Carnegie Inst. Wash. Yrbk. 75, 531–538.

Danchin, R.V., Wyatt, B.A., 1979. Statistical Cluster Analysis of Garnets from Kimberlites and Their Xenoliths. Kimberlite Sym-posium II, Cambridge.

Dawson, J.B., Smith, J.V., 1986. Relationship between eclogites and certain megacrysts from the Jagersfontein kimberlite, South Africa. Lithos 19, 325–330.

Dawson, J.B., Smith, J.V., 1987. Reduced sapphirine granulite xen-oliths from the Lace kimberlite, South Africa: implications for the deep structure of the Kaapvaal Craton. Contrib. Mineral. Petrol. 95, 376–383.

Dawson, J.B., Stephens, W.E., 1975. Statistical classification of garnets from kimberlites and associated xenoliths. J. Geol. 83, 589–607.

Dawson, J.B., Smith, J.V., Delaney, J.S., 1978. Multiple spinel-garnet peridotite transitions in upper mantle: evidence from a garnet harzburgite xenolith. Nature 273, 741–743.

de Bruin, D., 1989. Mantle eclogites from the Schuller kimberlite, Transvaal, South Africa. S. Afr. J. Geol. 92, 134–145.

de Bruin, D., 1999. The megacryst suite from the Schuller kimber-lite, South Africa. PhD thesis, University of Cape Town.

Dunn, P.J., 1978. On the composition of some Canadian green garnets. Can. Mineral. 16, 205–206.

Eggler, D.H., McCallum, M.E., Smith, C.B., 1979. Megacryst as-semblages in kimberlites from northern Colorado and southern

Wyoming. In: Boyd, F.R., Meyer, H.O.A. (Eds.), The Mantle Sample: Inclusions in Kimberlites and Other Volcanics. AGU, Washington, pp. 213–226.

Eggler, D.H., McCallum, M.E., Kirkley, M.B., 1987. Kimberlite-transported nodules from Colorado–Wyoming: a record of en-richment of shallow portions of an infertile lithosphere. In: Mor-ris, E.M., Pasteris, J.D. (Eds.), Mantle Metasomatism and Alkaline Magmatism. Geol. Soc. Amer. Spec. Paper, vol. 215, pp. 77–90.

El Fadili, S., Demaiffe, D., 1999. Petrology of eclogite and gran-ulite nodules from the Mbuji Mayi kimberlites (Kasai, Congo): significance of kyanite–omphacite intergrowths. In: Gurney, J.J., Gurney, J.L., Pascoe, M.D., Richardson, S.H. (Eds.), Pro-ceedings of the VIIth International Kimberlite Conference, vol. 1. Red Roof Design, Cape Town, pp. 205–213.

Exley, R.A., Smith, J.V., Dawson, J.B., 1983. Alkremite, garnetite and eclogite xenoliths from Bellsbank and Jagersfontein. Am. Mineral. 68, 512–516.

Fitzsimons, I.C.W., 1996. Metapelitic migmatites from Brattstrand Bluffs, East Antarctica—metamorphism, melting and exhuma-tion of the mid crust. J. Petrol. 37, 395–414.

Fitzsimons, I.C.W., Harley, S.L., 1994. The influence of retrograde cation exchange on granulite *P–T* estimates and a convergence technique for the recovery of peak metamorphic conditions. J. Petrol. 35, 543–576.

Ford, F.D., 1987. Petrology and geochemistry of xenoliths from the Blaauwbosch kimberlite pipe. R.S.A. BSc thesis, Queen's University.

Fraser, G., Worley, B., Sandiford, M., 2000. High-precision geother-mobarometry across the High Himalayan metamorphic sequence, Langtang Valley, Nepal. J. Metamorph. Geol. 18, 665–681.

Fung, A., Haggerty, S.E., 1995. Petrography and mineral composi-tions of eclogites from the Koidu kimberlite complex, Sierra Leone. J. Geophys. Res. 100, 20451–20473.

Garcia-Casco, A., Torres-Roldan, R.L., Millan, G., Monie, P., Hais-sen, F., 2001. High-grade metamorphism and hydrous melting of metapelites in the Pinos terrane (W Cuba): evidence for crus-tal thickening and extension in the northern Caribbean colli-sional belt. J. Metamorph. Geol. 19, 699–715.

Gayk, T., Kleinschrodt, R., 2000. Hot contacts of garnet peridotites in middle/upper crustal level: new constraints on the nature of the late Variscan high-*T*/low-*P* event in the Moldanubian (cen-tral Voges/NE France). J. Metamorph. Geol. 18, 293–305.

Gordon, T.M., Ghent, E.D., Stout, M.Z., 1991. Algebraic analysis of the biotite–sillimanite isograd in the File Lake area, Mani-toba. Can. Mineral. 29, 673–686.

Gordon, T.M., Aranovich, L.Ya., Fed'kin, V.V., 1994. Exploratory data analysis in thermobarometry: an example from the Kissey-new Sedimentary Gneiss Belt, Manitoba, Canada. Am. Mineral. 79, 973–982.

Griffin, W.L., Gurney, J.J., Ryan, C.G., 1992. Variations in trapping temperatures and trace elements in peridotite–suite inclusions from African diamonds: evidence for two inclusion suites, and implications for lithosphere stratigraphy. Contrib. Mineral. Pet-rol. 110, 1–15.

Griffin, W.L., Fisher, N.I., Friedman, J., Ryan, C.G., O'Reilly, S.Y., 1999. Cr-pyrope garnets in the lithospheric mantle: I. Compositional systematics and relations to tectonic setting. J. Petrol. 40, 679–704.

Grutter, H.S., Quadling, K.E., 1999. Can sodium in garnet be used to monitor eclogitic diamond potential? In: Gurney, J.J., Gurney, J.L., Pascoe, M.D., Richardson, S.H. (Eds.), Proceedings of the VIIth International Kimberlite Conference, vol. 1. Red Roof Design, Cape Town, pp. 314–320.

Grutter, H.S., Apter, D.B., Kong, J., 1999. Crust–mantle coupling: evidence from mantle-derived xenocrystic garnets. In: Gurney, J.J., Gurney, J.L., Pascoe, M.D., Richardson, S.H. (Eds.), Proceedings of the VIIth International Kimberlite Conference, vol. 1. Red Roof Design, Cape Town, pp. 307–313.

Guiraud, M., Powell, R., Cottin, J.-Y., 1996. Hydration of orthopyroxene-cordierite-bearing assemblages at Laouni, Central Hoggar, Algeria. J. Metamorph. Geol. 14, 467–476.

Gupta, S., Bhattacharya, A., Raith, M., Nanda, J.K., 2000. Contrasting pressure–temperature–deformation history across a vestigal craton mobile belt boundary: the western margin of the Eastern Ghats Belt at Deobhog, India. J. Metamorph. Geol. 18, 683–697.

Gurney, J.J., 1980. Variations in garnet, chromite and ilmenite chemistry and in diamond and micro diamond content from some well characterised kimberlites: An orientation survey. Report published by Mineral Services.

Gurney, J.J., 1984. A correlation between garnets and diamonds. In: Glover, J.E., Harris, P.G. (Eds.), Kimberlite Occurrence and Origin: A Basis for Conceptual Models in Exploration. University of Western Australia Publication, vol. 8, pp. 376–383.

Gurney, J.J., Harte, B., 1980. Chemical variations in upper mantle nodules from southern African kimberlites. Philos. Trans. R. Soc. Lond., A 297, 273–293.

Gurney, J.J., Switzer, G.S., 1973. The discovery of garnets closely related to diamonds in the Finsch pipe, South Africa. Contrib. Mineral. Petrol. 39, 103–116.

Gurney, J.J., Jakob, W.R.O., Dawson, J.B., 1979. Megacrysts from the Monastery kimberlite pipe, South Africa. In: Boyd, F.R., Meyer, H.O.A. (Eds.), The Mantle Sample: Inclusions in Kimberlites and Other Volcanics. Amer. Geophys. Union, Washington, pp. 227–243.

Gurney, J.J., Harris, J.W., Rickard, R.S., 1984. Minerals associated with diamonds from the Roberts Victor Mine. In: Kornprobst, J. (Ed.), Kimberlites I: Kimberlites and Related Rocks. Elsevier, Amsterdam, pp. 25–32.

Habler, G., Thoni, M., 2001. Preservation of Permo Triassic low pressure assemblages in the Cretaceous high pressure metamorphic Saualpe crystalline basement (Eastern Alps, Austria). J. Metamorph. Geol. 19, 679–697.

Hall, D.C., 1985. The petrology of xenoliths from the Orapa AK1 kimberlite pipe, Botswana. MSc thesis, Queen's University.

Hall, D.C., 1991. A petrological investigation of the Cross kimberlite occurrence, southeastern British Columbia, Canada. PhD Thesis, Queen's University.

Harangi, S.Z., Downes, H., Kosa, L., Szabo, C.S., Thirwall, M.F., Mason, P.R.D., Mattey, D., 2001. Almandine garnet in calc alkaline volcanic rocks of the northern Pannonian basin (Eastern Central Europe): geochemistry, petrogenesis and geodynamic considerations. J. Petrol. 42, 1813–1843.

Hartel, T.H.D., Pattison, D.R.M., 1996. Genesis of the Kapuskasing (Ontario) migmatitic mafic granulites by dehydration melting of amphibolite: the importance of quartz to reaction progress. J. Metamorph. Geol. 14, 591–611.

Hatton, C.J., 1978. The geochemistry and origin of xenoliths from the Roberts Victor mine. PhD thesis, University of Cape Town.

Hearn Jr., B.C., Boyd, F.R., 1975. Garnet peridotite xenoliths in a Montana, U.S.A. kimberlite. Phys. Chem. Earth 9, 247–255.

Hearn Jr., B.C., McGee, E.S., 1984. Garnet peridotites from Williams kimberlites, north central Montana, U.S.A. In: Kornprobst, J. (Ed.), Kimberlites II: The Mantle and Crust–Mantle Relationships. Elsevier, Amsterdam, pp. 57–70.

Hill, S.J., 1989. A study of the diamonds and xenoliths from the Star kimberlite, Orange Free State, South Africa, MSc Thesis, University of Cape Town.

Hills, D.V., Haggerty, S.E., 1989. Petrochemistry of eclogites from the Koidu kimberlite complex, Sierra Leone. Contrib. Mineral. Petrol. 103, 397–422.

Hops, J.J., 1989. Some aspects of the geochemistry of high-temperature peridotites and megacrysts from the Jagersfontein kimberlite pipe. South Africa, PhD Thesis, University of Cape Town.

Jacob, D., Jagoutz, E., Lowry, D., Mattey, D., Kudrjavtseva, G., 1994. Diamondiferous eclogites from Siberia: Remnants of Archean oceanic crust. Geochim. Cosmochim. Acta 58, 5191–5207.

Jago, B.C., Mitchell, R.H., 1989. A new garnet classification technique: divisive cluster analysis applied to garnet populations from Somerset Island kimberlites. In: Ross, J. (Ed.), Kimberlites and Related Rocks, vol. 1: Their Composition, Occurrence, Origin and Emplacement. Geol. Soc. Austr. Spec. Pub., vol. 14, pp. 298–310.

Jakob, W.R.O., 1977. Geochemical aspects of the megacryst suite from the Monastery kimberlite pipe. MSc Thesis, University of Cape Town.

Jerde, E.A., Taylor, L.A., Crozaz, G., Sobolev, N.V., Sobolev, V.N., 1993a. Diamondiferous eclogites from Yakutia, Siberia: evidence for a diversity of protoliths. Contrib. Mineral. Petrol. 114, 189–202.

Jerde, E.A., Taylor, L.A., Crozaz, G., Sobolev, N.V., 1993b. Exsolution of garnet within clinopyroxene of mantle eclogites: major- and trace-element chemistry. Contrib. Mineral. Petrol. 114, 148–159.

Jones, K.A., Strachan, R.A., 2000. Crustal thickening and ductile extension in the NE Greenland Caledonides: a metamorphic record from anatectic pelites. J. Metamorph. Geol. 18, 719–735.

Knudsen, T.-L., 1996. Petrology and geothermobarometry of granulite facies metapelites from the Hisøy-Torungen area, south Norway: new data on the Sveconorvegian $P–T–t$ path of the Bamble sector. J. Metamorph. Geol. 14, 267–287.

Kopylova, M.G., Russell, J.K., Cookenboo, H., 1999. Petrology of peridotite and pyroxenite xenoliths from the Jericho kimberlite: implications for the thermal state of the mantle beneath the slave Craton, northern Canada. J. Petrol. 40, 79–104.

Kryza, R., Pin, C., Vielzeuf, D., 1996. High-pressure granulites from the Sudetes (south-west Poland): evidence of crustal subduction and collisional thickening in the Variscan Belt. J. Metamorph. Geol. 14, 531–546.

Lang, H.M., Gilotti, J.A., 2001. Plagioclase replacement textures in partially eclogitised gabbros from the Sanddal mafic-ultramafic complex, Greenland Caledonides. J. Metamorph. Geol. 19, 497–517.

Lee, J.E., 1993. Indicator mineral techniques in a diamond exploration program at Kokong, Botswana. Diamonds: Exploration, Sampling and Evaluation, Short Course Proceedings, Pros. Dev. Assoc. Can., pp. 213–235.

MacGregor, I.D., 1979. Mafic and ultramafic xenoliths from the Kao kimberlite pipe. In: Boyd, F.R., Meyer, H.O.A. (Eds.), The Mantle Sample: Inclusions in Kimberlites and Other Volcanics. Amer. Geophys. Union, Washington, pp. 156–172.

MacGregor, I.D., Carter, J.L., 1970. The chemistry of clinopyroxenes and garnets of eclogite and peridotite xenoliths from the Roberts Victor Mine, South Africa. Phys. Earth Planet. Inter. 1, 391–397.

MacGregor, I.D., Manton, W.I., 1986. Roberts Victor eclogites: ancient oceanic crust. J. Geophys. Res. 91, 14063–14079.

Mazzone, P., Haggerty, S.E., 1989a. Corganites and corgaspinites: two new types of aluminous assemblages from the Jagersfontein pipe. In: Ross, J. (Ed.), Kimberlites and Related Rocks: vol. 2. Their Mantle/Crust Setting, Diamonds and Diamond Exploration. Geol. Soc. Austr. Spec. Pub., vol. 14. Blackwell, Carlton, pp. 795–808.

Mazzone, P., Haggerty, S.E., 1989b. Peraluminous xenoliths in kimberlite: metamorphosed restites produced by partial melting of pelites. Geochim. Cosmochim. Acta 53, 1551–1561.

McCandless, T.E., Gurney, J.J., 1986. Sodium in garnet and potassium in clinopyroxene: criteria for classifying mantle eclogites. Univ. Cape Town, Kimberlite Research Group, Data Appendix, Internal Report, vol. 10. 60 pp.

McCandless, T.E., Gurney, J.J., 1989. Sodium in garnet and potassium in clinopyroxene: criteria for classifying mantle eclogites. In: Ross, J. (Ed.), Kimberlites and Related rocks: vol. 2. Their Mantle/Crust Setting, Diamonds and Diamond Exploration. Blackwell, Carlton, Australia, pp. 827–832.

McGee, E.S., Hearn Jr., B.C., 1984. The Lake Ellen kimberlite, Michigan, U.S.A. In: Kornprobst, J. (Ed.), Kimberlites I: Kimberlites and Related Rocks. Elsevier, Amsterdam, pp. 143–154.

Meyer, H.O.A., Brookins, D.G., 1971. Eclogite xenoliths from Stockdale kimberlite, Kansas. Contrib. Mineral. Petrol. 34, 60–72.

Mitchell, R.H., 1984. Garnet lherzolites from the Hanaus-I and Lowrensia kimberlites of Namibia. Contrib. Mineral. Petrol. 86, 178–188.

Moore, R.O., Gurney, J.J., 1991. Garnet megacrysts from Group II kimberlites in southern Africa. Ext. Abstr. 5th Int. Kimb. Conf., Araxa, Brasil, 298–300.

Moore, A.E., Lock, N.P., 2001. The origin of mantle-derived megacrysts and sheared peridotites—evidence from kimberlites in the northern Lesotho-Orange Free State (South Africa) and Botswana pipe clusters. S. Afr. J. Geol. 104, 23–38.

Moraes, R.D., Fuck, R.A., 2000. Ultra-high-temperature metamorphism in central Brazil: the Barro Alto complex. J. Metamorph. Geol. 18, 345–358.

Nehru, C.E., Reddy, A.K., 1989. Ultramafic xenoliths from Vajrakarur kimberlites, India. In: Ross, J. (Ed.), Kimberlites and Related Rocks: vol. 2. Their Mantle/Crust Setting, Diamonds and Diamond Exploration. Geol. Soc. Austr. Spec. Pub., vol. 14. Blackwell, Carlton, pp. 745–758.

Nixon, P.H., 1979. Chromium garnet, uvarovite, from eastern Papua New Guinea. Sci N. Guin. 6, 16–18.

Nixon, P.H., Boyd, F.R., 1973a. The discrete nodule (megacryst) association in kimberlites from northern Lesotho. In: Nixon, P.H. (Ed.), Lesotho Kimberlites. Lesotho National Development, Maseru, pp. 67–75.

Nixon, P.H., Boyd, F.R., 1973b. Petrogenesis of the granular and sheared ultrabasic nodule suite in kimberlites. In: Nixon, P.H. (Ed.), Lesotho Kimberlites. Lesotho National Development, Maseru, pp. 48–56.

Nixon, P.H., Chapman, N.A., Gurney, J.J., 1978. Pyrope-spinel (alkremite) xenoliths from kimberlite. Contrib. Mineral. Petrol. 65, 341–346.

Nixon, P.H., van Calsteren, P.W.C., Boyd, F.R., Hawkesworth, C.J., 1987. Harzburgites with garnets of diamond facies from southern African kimberlites. In: Nixon, P.H. (Ed.), Mantle Xenoliths. Wiley, New York, pp. 524–533.

Nyman, M.W., Pattison, D.R.M., Ghent, E.D., 1995. Melt extraction during formation of k-feldspar + sillimanite migmatites, west of Revelstoke, British Columbia. J. Petrol. 36, 351–372.

Parkinson, C.D., 2000. Coesite inclusions and prograde compositional zonation of garnet in whiteschist of the HP-UHPM Kokchetav massif, Kazakhstan: a record of progressive UHP metamorphism. Lithos 52, 215–233.

Parthasarathy, G., Balaram, V., Srinivasan, R., 1999. Characterization of green garnets from an Archean calc-silicate rock, Bandihalli, Karnatacka, India: evidence for a continuous solid solution between uvarovite and grandite. J. Asian Earth Sci. 17, 345–352.

Pattison, D.R.M., 1991. Infiltration-driven dehydration and anatexis in granulite facies metagabbro, Grenville Province, Ontario, Canada. J. Metamorph. Geol. 9, 315–332.

Pattison, D.R.M., Bégin, N.J., 1994. Zoning patterns in orthopyroxene and garnet in granulites: implications for geothermometry. J. Metamorph. Geol. 12, 387–410.

Pattison, D.R.M., Carmichael, D.M., St-Onge, M.R., 1982. Geothermometry and geobarometry applied to Early Proterozoic "S-type" granitoid plutons, Wopmay Orogen, Northwest Territories, Canada. Contrib. Mineral. Petrol. 79, 394–404.

Pearson, D.G., Boyd, F.R., Haggerty, S.E., Pasteris, J.D., Field, S.W., Nixon, P.H., Pokhilenko, N.P., 1994. The characterisation and origin of graphite in cratonic lithospheric mantle: a petrological carbon isotope and Raman spectroscopic study. Contrib. Mineral. Petrol. 115, 449–466.

Pearson, N.J., Griffin, W.L., Doyle, B.J., O'Reilly, S.Y., van Achterbergh, E., Kivi, K., 1999. Xenoliths from kimberlite pipes of the Lac de Gras area, Slave Craton, Canada. In: Gurney, J.J., Gurney, J.L., Pascoe, M.D., Richardson, S.H. (Eds.), Proceedings of the VIIth International Kimberlite Conference, vol. 2. Red Roof Design, Cape Town, pp. 644–658.

Percival, J., 1981. Geological evolution of part of the central Superior Province based on relationships among the Abitibi and Wawa sub-provinces and the Kapuskasing Structural Zone. PhD thesis, Queen's University.

Pita, P., de Waal, S.A., 2001. High-temperature, low-pressure metamorphism and development of prograde symplectites, Marble Hall Fragment, Bushveld Complex (South Africa). J. Metamorph. Geol. 19, 311–315.

Pokhilenko, N.P., Sobolev, N.V., Lavrent'ev, Y.G., 1977. Xenoliths of diamondiferous ultramafic rocks from Yakutian kimberlites. Ext. Abstr. 2nd Kimb. Conf., Santa Fe, USA. unpaged.

Pokhilenko, N.P., Pearson, D.G., Boyd, F.R., Sobolev, N.V., 1991. Megacrystalline dunites: sources of Siberian diamonds. Ann. Rev. Dir. Geophys. Lab. Carneg. Inst. Wash. 90, 11–18.

Pokhilenko, N.P., Sobolev, N.V., Boyd, F.R., Pearson, D.G., Shimizu, N., 1993. Megacrystalline pyrope peridotites in the lithosphere of the Siberian Platform: mineralogy, geochemical features and the problem of their origin. Russ. J. Geol. Geophys. 34, 56–67.

Pokhilenko, N.P., Sobolev, N.V., Kuligin, S.S., Shimizu, N., 1999. Peculiarities of distribution of pyroxenite paragenesis garnets in Yakutian kimberlites and some aspects of evolution of Siberian Craton lithospheric mantle. In: Gurney, J.J., Gurney, J.L., Pascoe, M.D., Richardson, S.H. (Eds.), Proceedings of the VIIth International Kimberlite Conference, vol. 2. Red Roof Design, Cape Town, pp. 689–698.

Ponomarenko, A.I., 1975. Alkremite, a new variety of aluminous ultramafic rock in xenoliths from the Udachnaya kimberlite pipe. Dokl. Earth Sci. 225, 155–157.

Pyle, J.M., Haggerty, S.E., 1998. Eclogites and the metasomatism of eclogites from the Jagersfontein kimberlite: punctuated transport and implications for alkali magmatism. Geochim. Cosmochim. Acta 62, 1207–1231.

Ramsay, R.R., 1992. Geochemistry of diamond indicator minerals. PhD Thesis, University of Western Australia.

Reid, A.M., Brown, R.W., Dawson, J.B., Whitfield, G.G., Siebert, J.C., 1976. Garnet and pyroxene compositions in some diamondiferous eclogites. Contrib. Mineral. Petrol. 58, 203–220.

Robey, J.v.A., 1981. Kimberlites of the central Cape Province, R.S.A. PhD thesis, University of Cape Town.

Robinson, D.N., Gurney, J.J., Shee, S.R., 1984. Diamond eclogite and graphite eclogite xenoliths from Orapa, Botswana. In: Kornprobst, J. (Ed.), Kimberlites II: Their Mantle and Crust–Mantle Relationships. Elsevier, Amsterdam, pp. 11–24.

Rolland, Y., Maheo, G., Guillot, S., Pecher, A., 2001. Tectonometamorphic evolution of the Karakorum metamorphic complex (Dassu-Askole area, NE Pakistan): exhumation of mid-crustal HT-MP gneisses in a convergent context. J. Metamorph. Geol. 19, 717–737.

Rosenberg, J.L., Spry, P.G., Jacobson, C.E., Cook, N.J., Vokes, F.M., 1998. Thermobarometry of the Bleikvassli Zn–Pb–(Cu) deposit, Nordland, Norway. Miner. Depos. 34, 19–34.

Rotzler, J., Romer, R.L., 2001. P–T–t evolution of ultrahigh-temperature granulites from the Saxon granulite massif, Germany: Part I. Petrology. J. Petrol. 42, 1995–2013.

Santos de Lima, E., Vannucci, R., Bottazzi, P., Ottolini, L., 1995. Reconnaissance study of trace element zonation in garnet from the Central Structural Domain, Northeastern Brazil: an example of polymetamorphic growth. J. South Am. Earth Sci. 8, 315–324.

Satish-Kumar, M., Wada, H., Santosh, M., Yoshida, M., 2001. Flu-

id-rock history of granulite-facies humite-marbles from Ambasamudram, southern India. J. Metamorph. Geol. 19, 395–410.

Schmadicke, E., Okrusch, M., Schubert, W., Elwart, B., Gorke, U., 2001. Phase relations of calc-silicate assemblages in the Auerbach marble, Odenwald Crystalline Complex, Germany. Mineral. Petrol. 72, 77–111.

Schulze, D.J., 1987. Megacrysts in alkaline volcanic rocks. In: Nixon, P.H. (Ed.), Mantle Xenoliths. Wylie, London, pp. 433–451.

Schulze, D.J., 1989a. Green garnets from South African kimberlites and their relationship to wehrlites and crustal uvarovites. In: Ross, J. (Ed.), Kimberlites and Related Rocks: vol. 2. Their Mantle/Crust Setting, Diamonds, and Diamond Exploration. Blackwell, Carlton, Australia, pp. 820–826.

Schulze, D.J., 1989b. Constraints on the abundance of eclogite in the upper mantle. J. Geophys. Res. 94, 4205–4212.

Schulze, D.J., 1992. Diamond eclogite from Sloan Ranch, Colorado, and its bearing on the diamond grade of the Sloan kimberlite. Econ. Geol. 87, 2175–2179.

Schulze, D.J., 1993. Garnet xenocryst populations in North American kimberlites. Diamonds: Exploration, Sampling and Evaluation, Short Course Proceedings, Pros. Dev. Assoc. Can., pp. 359–377.

Schulze, D.J., 1994. Abundance and distribution of low-Ca garnet harzburgites in the subcratonic lithosphere of southern Africa. In: Meyer, H.O.A., Leonardos, O. (Eds.), Vol. I: Kimberlites, Related Rocks, and Mantle Xenoliths. CPRM (Brasil) Spec. Pub., vol. 1-A/94, pp. 327–335.

Schulze, D.J., 1995. Low-Ca garnet harzburgites from Kimberley, South Africa: abundance and bearing on the structure and evolution of the lithosphere. J. Geophys. Res. 100, 12513–12526.

Schulze, D.J., 1997. The significance of eclogite and Cr-poor megacryst garnets in diamond exploration. Explor. Min. Geol. 6, 349–366.

Schulze, D.J., Helmstaedt, H., 1988. Coesite-sanidine eclogites from kimberlite: products of mantle fractionation or subduction? J. Geol. 96, 435–443.

Schulze, D.J., Wiese, D., Steude, J., 1996. Abundance and distribution of diamonds in eclogite revealed by volume visualization of CT x-ray scans. J. Geol. 104, 109–114.

Schulze, D.J., Valley, J.W., Viljoen, K.S., Stiefenhofer, J., Spicuzza, M., 1997. Carbon isotope composition of graphite in mantle eclogites. J. Geol. 105, 379–386.

Schulze, D.J., Valley, J.W., Spicuzza, M.J., 2000. Coesite eclogites from the Roberts Victor kimberlite, South Africa. Lithos 54, 23–32.

Scrimegour, I., Smith, J.B., Raith, J.G., 2001. Paleoproterozoic high-T, low-P metamorphism and dehydration melting in metapelites from the Mopunga Range, Arunta Inlier, central Australia. J. Metamorph. Geol. 19, 739–757.

Shaw, R.K., Arima, M., 1996. High-temperature metamorphic imprint on calc-silicate granulites on Rayagada, Eastern Ghats, India: implication for the isobaric cooling path. Contrib. Mineral. Petrol. 126, 169–180.

Shee, S.R., 1978. The mineral chemistry of xenoliths from the Orapa kimberlite pipe, Botswana. MSc thesis, University of Cape Town.

Shee, S.R., Gurney, J.J., 1979. The mineralogy of xenoliths from

Orapa, Botswana. In: Boyd, F.R., Meyer, H.O.A. (Eds.), The Mantle Sample: Inclusions in Kimberlites and Other Volcanics. A.G.U., Washington, pp. 37–49.

Shee, S.R., Gurney, J.J., Robinson, D.N., 1982. Two diamond-bearing peridotite xenoliths from the Finsch kimberlite, South Africa. Contrib. Mineral. Petrol. 81, 79–87.

Simon, G., Chopin, C., Schenk, V., 1997. Near-end-member magnesiochloritoid in prograde-zoned pyrope, Dora-Maira massif, western Alps. Lithos 41, 37–57.

Skinner, C.P., 1989. The petrology of peridotite xenoliths from the Finsch kimberlite, South Africa. S. Afr. J. Geol. 92, 197–206.

Smith, C.B., 1977. Kimberlite and mantle derived xenoliths at Iron Mountain, Wyoming. MS thesis, Colorado State University.

Smith, D., Boyd, F.R., 1992. Compositional zonation in garnets in peridotite xenoliths. Contrib. Mineral. Petrol. 112, 134–147.

Smith, C.B., Gurney, J.J., Harris, J.W., Robinson, D.N., Shee, S.R., Jagoutz, E., 1989. Sr and Nd isotope systematics of diamond-bearing eclogite xenoliths and eclogitic inclusions in diamond from southern Africa. In: Ross, J. (Ed.), Kimberlites and Related Rocks: vol. 2. Their Mantle/Crust Setting, Diamonds, and Diamond Exploration. Blackwell, Carlton, Australia, pp. 853–863.

Smyth, J.R., Caporuscio, F., 1984. Petrology of a suite of eclogite inclusions from the Bobbejaan kimberlite: II. Primary phase compositions and origin. In: Kornprobst, J. (Ed.), Kimberlites II: Their Mantle and Crust–Mantle Relationships. Elsevier, Amsterdam, pp. 121–131.

Smyth, J.R., Hatton, C.J., 1977. A coesite-sanidine grospydite xenolith from the Roberts Victor kimberlite. Earth Planet. Sci. Lett. 34, 284–290.

Snyder, G.A., Taylor, L.A., Crozaz, G., Halliday, A.N., Beard, B.L., Sobolev, V.N., Sobolev, N.V., 1997. The origins of Yakutian eclogite xenoliths. J. Petrol. 38, 85–113.

Sobolev, N.V., 1977. Deep-Seated Inclusions in Kimberlites and the Problem of the Composition of the Upper Mantle. AGU, Washington, DC.

Sobolev, N.V., Kuznetsova, I.K., Zyuzin, N.I., 1968. The petrology of grospydite xenoliths from the Zagodachnaya kimberlite pipe in Yakutia. J. Petrol. 9, 253–280.

Sobolev, N.V., Lavrent'ev, Y.G., Pokhilenko, N.P., Usova, L.V., 1973. Cr-rich garnets from kimberlites of Yakutia and their parageneses. Contrib. Mineral. Petrol. 40, 39–52.

Sobolev, N.V., Pokhilenko, N.P., Lavrent'ev, Y.G., Usova, L.V., 1975. Peculiarities of composition of chrome-spinels from diamonds and kimberlites of Yakutia. Geol. Geofiz. 11, 7–24.

Sobolev, N.V., Pokhilenko, N.P., Efimova, E.S., 1984. Diamond-bearing peridotite xenoliths in kimberlites and the problem of the origin of diamonds. Sov. Geol. Geophys. 25 (12), 62–76.

Sobolev, V.N., Taylor, L.A., Snyder, G.A., 1994. Diamondiferous eclogites from the Udachnaya kimberlite pipe, Yakutia. Int. Geol. Rev. 36, 42–64.

Sommerville, T.A., 1988. Petrography and geochemistry of xenoliths from Newlands kimberlite pipes. R.S.A., BSc Thesis, Queen's University.

Spetsius, Z.V., Nikishov, K.N., Makhotko, V.F., 1984. Kyanite eclogite with sanidine from the Udachnaya kimberlite pipe. Dokl. Earth Sci. Sect. 279, 138–141.

Stowell, H.H., Taylor, D.L., Tinkham, D.L., Goldberg, S.A., Ou-

derkirk, K.A., 2001. Contact metamorphic $P–T–t$ paths from Sm–Nd garnet ages, phase equilibria modelling and thermobarometry: garnet Ledge, south-eastern Alaska, USA. J. Metamorph. Geol. 19, 645–660.

Taylor, L.A., Neal, C.R., 1989. Eclogites with oceanic crustal and mantle signatures from the Bellsbank kimberlite, South Africa: Part I. Mineralogy, petrography and whole rock chemistry. J. Geol. 97, 551–567.

Taylor, L.A., Snyder, G.A., Crozaz, G., Sobolev, V.N., Yefimova, E.S., Sobolev, N.V., 1996. Eclogitic inclusions in diamonds: evidence of complex mantle processes over time. Earth Planet Sci. Lett. 142, 535–551.

Thöni, M., Miller, Ch., 1996. Garnet Sm–Nd data from the Saualpe and the Koralpe (Eastern Alps, Austria): chronological and $P–T$ constraints on the thermal and tectonic history. J. Metamorph. Geol. 14, 453–466.

Tollo, R.P., 1982. Petrography and mineral chemistry of ultramafic and related inclusions from the Orapa A/K-1 kimberlite pipe, Botswana. PhD thesis, University of Massachusetts.

Vance, D., Mahar, E., 1998. Pressure–temperature paths from $P–T$ pseudosections and zoned garnets: potential, limitations and examples from the Zanskar Himalaya, NW India. Contrib. Mineral. Petrol. 132, 225–245.

Vicker, P.A., 1997. Garnet peridotite xenoliths from kimberlites near Kirkland Lake, Canada. MSc Thesis, University of Toronto.

Viljoen, K.S., 1994. The petrology and geochemistry of a suite of mantle-derived eclogite xenoliths from the Kaalvallei kimberlite, South Africa. PhD thesis, University of Witwatersrand.

Viljoen, K.S., 1995. Graphite- and diamond-bearing eclogite xenoliths from the Bellsbank kimberlites, Northern Cape, South Africa. Contrib. Mineral. Petrol. 121, 414–423.

Viljoen, K.S., Swash, P.M., Otter, M.L., Schulze, D.J., Lawless, P.J., 1992. Diamondiferous garnet harzburgites from the Finsch kimberlite, Northern Cape, South Africa. Contrib. Mineral. Petrol. 110, 133–138.

Viljoen, K.S., Robinson, D.R., Swash, P.M., Griffin, W.L., Otter, M.L., Ryan, C.G., Win, T.T., 1994. Diamond- and graphite-bearing peridotite xenoliths from the Roberts Victor kimberlite, South Africa. In: Meyer, H.O.A., Leonardos, O. (Eds.), Kimberlites, Related Rocks and Mantle Xenoliths. C.P.R.M. Special Publication 1/A, Rio de Janeiro, pp. 285–303.

Viljoen, K.S., Smith, C.B., Sharp, Z.D., 1996. Stable and radiogenic isotope study of eclogite xenoliths from the Orapa kimberlite, Botswana. Chem. Geol. 131, 235–255.

von Knorring, O., Condliffe, E., Tong, Y.L., 1986. Some mineralogical and geochemical aspects of chromium-bearing skarn minerals from northern Karelia, Finland. Bull. Geol. Soc. Finl. 58, 277–292.

White, C.E., Barr, S.M., Jamieson, R.A., Reynolds, P.H., 2001. Neoproterozoic high-pressure/low-temperature metamorphic rocks in the Avalon terrane, southern New Brunswick, Canada. J. Metamorph. Geol. 19, 519–530.

Whitney, D.L., Mechum, T.A., Huehner, S.M., Dilek, Y.R., 1996. Progressive metamorphism of pelitic rocks from protolith to granulite facies, Dutchess County, New York, USA: constraints on the timing of fluid infiltration during regional metamorphism. J. Metamorph. Geol. 14, 163–181.

Whitney, D.L., Teyssier, C., Dilek, Y., Fayon, A.K., 2001. Metamorphism of the central Anatolian Crystalline Complex, Turkey: influence of orogen-normal collision vs. wrench-dominated tectonics on $P-T-t$ paths. J. Metamorph. Geol. 19, 411–432.

Willner, A.P., Rötzler, K., Maresch, W.V., 1997. Pressure–temperature and fluid evolution of quartzo-feldspathic metamorphic rocks with a relic high-pressure, granulite-facies history from the Central Erzgebirge (Saxony, Germany). J. Petrol. 38, 307–336.

Zeh, A., Millar, I.L., 2001. Metamorphic evolution of garnet–epidote–biotite gneiss from the Moine Supergroups, Scotland, and geotectonic implications. J. Petrol. 42, 529–554.

Zhao, G.C., Wilde, S.A., Cawood, P.A., Lu, L.Z., 2000. Petrology and $P-T$ path of the Fuping mafic granulites: implications for tectonic evolution of the central zone of the North China craton. J. Metamorph. Geol. 18, 375–391.

Available online at www.sciencedirect.com

SCIENCE DIRECT°

Lithos 71 (2003) 215–241

LITHOS

www.elsevier.com/locate/lithos

The evolution of lithospheric mantle beneath the Kalahari Craton and its margins

W.L. Griffin[a,b,*], Suzanne Y. O'Reilly[a], L.M. Natapov[a], C.G. Ryan[b]

[a] ARC National Key Centre for Geochemical Evolution and Metallogeny of Continents, Department of Earth and Planetary Sciences, Macquarie University, Sydney, NSW 2109, Australia
[b] CSIRO Exploration and Mining, North Ryde, NSW 2113, Australia

Abstract

The compositional structure and thermal state of the subcontinental lithospheric mantle (SCLM) beneath the Kalahari Craton and the surrounding mobile belts have been mapped in space and time using >3400 garnet xenocrysts from >50 kimberlites intruded over the period 520–80 Ma. The trace-element patterns of many garnets reflect the metasomatic refertilisation of originally highly depleted harzburgites and lherzolites, and much of the lateral and vertical heterogeneity observed in the SCLM within the craton is the product of such metasomatism. The most depleted, and possibly least modified, SCLM was sampled beneath the Limpopo Belt by early Paleozoic kimberlites; the SCLM beneath other parts of the craton may represent similar material modified by metasomatism during Phanerozoic time. In the SW part of the craton, the SCLM sampled by "Group 2" kimberlites (>110 Ma) is thicker, cooler and less metasomatised than that sampled by "Group 1" kimberlites (mostly ≤ 95 Ma) in the same area. Therefore, the extensively studied xenolith suite from the Group 1 kimberlites probably is not representative of primary Archean SCLM compositions. The relatively fertile SCLM beneath the mobile belts surrounding the craton is interpreted as largely Archean SCLM, metasomatised and mixed with younger material during Paleoproterozoic to Mesoproterozoic rifting and compression. This implies that at least some of the observed secular evolution in SCLM composition worldwide may reflect the reworking of Archean SCLM. There are strong correlations between mantle composition and the lateral variations in seismic velocity shown by detailed tomographic studies. Areas of relatively low Vp within the craton largely reflect the progressive refertilisation of the Archean root during episodes of intraplate magmatism, including the Bushveld (2 Ga) and Karroo (ca. 180 Ma) events; areas of high Vp map out the distribution of relatively less metasomatised Archean SCLM. The relatively low Vp of the SCLM beneath the mobile belts around the craton is consistent with its fertile composition. The seismic data may be used to map the lateral extent of different types of SCLM, taking into account the small lateral variations in the geotherm identified using the techniques described here.
© 2003 Elsevier B.V. All rights reserved.

Keywords: Kalahari Craton; Kaapvaal Craton; Lithospheric mantle; Cr-pyrope garnets; Seismic tomography; Kimberlites

1. Introduction

* Corresponding author. ARC National Key Centre for Geochemical Evolution and Metallogeny of Continents, Department of Earth and Planetary Sciences, Macquarie University, Sydney, NSW 2109, Australia. Tel.: +61-2-9850-8954; fax: +61-2-9850-8943.
E-mail address: bill.griffin@mq.edu.au (W.L. Griffin).

The 4D lithosphere mapping methodology (O'Reilly and Griffin, 1996) uses geochemical information from xenoliths and xenocrysts in volcanic rocks to map the compositional structure and thermal state

0024-4937/$ - see front matter © 2003 Elsevier B.V. All rights reserved.
doi:10.1016/j.lithos.2003.07.006

of the subcontinental lithospheric mantle (SCLM), in time slices represented by the eruption ages of the volcanic rocks. Geophysical data can in principle be used to extend this mapping between the "boreholes" represented by volcanic vents, with the caveat that geophysical data represent the present-day situation, while the volcanic eruptions sampled the mantle in another time slice.

This combination of techniques allows the detailed investigation of the deep structure of the continents and provides insights into the formation and evolution of the crust and mantle. For example, lithosphere mapping along a 1000-km traverse in NE Siberia (Griffin et al., 1999a) revealed a marked change in the nature of the SCLM beneath Archean and Proterozoic portions of the Siberian Craton, and showed that different crustal terranes within the Archean core are underlain by distinctly different types of SCLM. This observation implies that crustal terrane boundaries are near-vertical and translithospheric, and that

Archean crust

I - Zimbabwe craton:
Ia -Tokwe terrane
Ib - North-Western terrane

II - Kaapvaal craton:
IIa - South-Eastern terrane
IIb - Central terrane
IIc - Pietersburg terrane
IId - Western terrane

III - Limpopo microcontinent
IIIa - Central zone
IIIb - Northern marginal zone
IIIc - Southern marginal zone

IV -Angolan craton

Early Proterozoic Crust (Passive margin of Kalahari Continent)

Va - Kheis fold belt
Vb - Okwa inlier
Vc - Makondi fold belt

Early-Middle Proterozoic Crust (Accretionary fold belts)

VI - Namaqua-Natal belt
VIa - Namaqua province
VIb - Natal province
VII - Rehobothian subprovince

Late Proterozoic Crust (Pan-African Orogeny)

VIII - Damara province
IX - Saldanian province

Other Symbols

⊕ ancient gneisses
🦇 greenstone belts
ᵥ ᵛ volcanics: active margin?
⤢ shear zones
ɪɪ rift zones

↗ thrusts
— "terrane" boundaries
···· inferred boundaries
≡ Lebombo-Sabi monoclines
𝄐 Kalahari suture zone
⸲⸲⸲ Political boundaries

Fig. 1. Structural units of southern Africa, as discussed in this paper, superimposed on a shaded map of smoothed aeromagnetic anomalies.

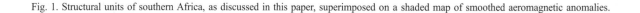

each terrane carried its own "root" at the time of craton amalgamation. In the Slave Craton, Griffin et al. (1999b) mapped an unusual two-layered SCLM, in which an ultradepleted upper layer is separated by a sharp boundary (145 ± 5 km) from a deeper less depleted layer, more typical of Archean SCLM. This layered SCLM has been mapped over an area of 12,000 km^2 in the central part of the craton (Griffin et al., 1999c); its distribution coincides with that of an anomalously conductive layer in the upper mantle (Jones et al., 2001).

The Kalahari Craton is a key area for studies of the SCLM. In many respects, it is the "type" Archean craton, by virtue of the extensive studies of xenoliths from its many productive diamond mines and from other kimberlites. More data probably are available on the composition of the upper mantle beneath this area than on the rest of the world's cratons combined. The recent addition of detailed geophysical data, including seismic tomography (James et al., 2001), has made this area even more important as a natural laboratory.

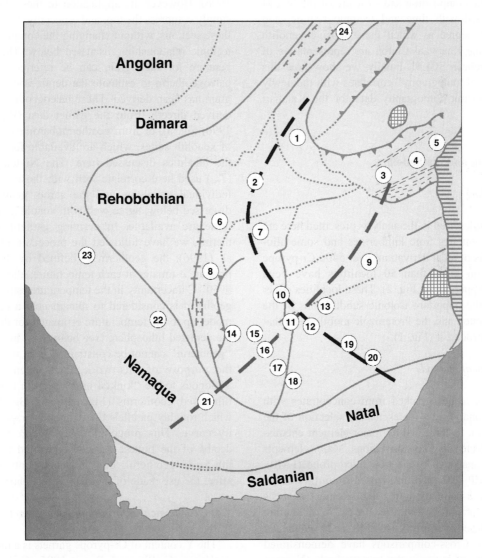

Fig. 2. Location map showing kimberlite fields (Appendix A) discussed in the text, and the boundaries of structural units from Fig. 1. Dashed lines show two SCLM traverses discussed in the text.

Most of the xenolith data on the Kalahari Craton come from its SW corner, but the cratonic crust is a complex collage of tectonic units (Fig. 1), and equivalent complexity might be expected in its SCLM. The large number of kimberlites provides mantle-derived xenocryst suites across most of the craton, and in several time slices. In this study, we use these xenocryst suites, including >3500 Cr-pyrope garnets from >50 kimberlites (Fig. 2), to examine the composition of the SCLM beneath the different terranes that make up the craton and its margins. We investigate how the composition and structure of the SCLM have changed with time and tectonic regime, and discuss the degree to which the extensive xenolith data from the Kaapvaal Craton are representative of primary Archean SCLM. Finally, we show how this "geochemical tomography" correlates with the newly available seismic tomography data for the Kalahari Craton.

2. Sampling and methodology

2.1. Sampling

The samples used in the analysis presented here are garnet concentrates from kimberlites and some alluvial deposits of local derivation; over 3400 Cr-pyrope garnets from more than 50 localities have been analysed (Appendix A; Fig. 2). These localities cover several of the important tectonic subdivisions of the Kalahari Craton and the Proterozoic mobile belts that partially surround it (Fig. 1).

2.2. Analytical methods

Grains have been picked from concentrates with the aim of analysing a representative selection of the types present, as indicated by major-element chemistry and reflected in colour variations. Major elements have been analysed by electron microprobe (EMP), and the EMP data have been used to further select representative populations for trace-element analysis by proton microprobe (Ni, Zn, Ga, Sr, Y, Zr; before 1995) or laser-microprobe ICPMS (>25 elements; after 1995). Cross-comparisons have demonstrated good agreement between the two methods (Norman et al., 1998). Detailed discussions of the garnet

database, including data quality and detection limits, are given by Griffin et al. (1999d, submitted).

2.3. Thermometry and barometry

The key technique used in constructing the mantle sections presented here is the determination of the equilibration temperature of each garnet grain, using its Ni content and the Ni thermometer as calibrated by Ryan et al. (1996). An alternative calibration by Canil (1994) is regarded as invalid (Griffin and Ryan, 1996). However, its application to these data would simply compress the top, and especially the bottom, of these sections, without changing the compositional or tectonic relationships discussed below. The Ni temperature for each grain can be referred to a local paleogeotherm to estimate the depth from which the grain has been derived. These paleogeotherms can be derived directly from the garnet data (Ryan et al., 1996; Fig. 3), or from geothermobarometric analyses of xenolith suites, which are available for several of the localities discussed here. The Ni temperatures (T_{Ni}) used here correlate well with those derived for individual xenoliths, and the garnet geotherms, as discussed below, agree well with xenolith data where these are available. In deriving garnet-based geotherms, we have followed the procedure of Ryan et al. (1996); the geotherm is defined by the highest pressure estimates at each temperature, allowing for a ± 50 °C uncertainty in the temperature estimates. The geotherm is considered to remain near a conductive model up to the temperature estimated for the base of the depleted lithosphere (see below). At higher T, the "geotherm" cannot be constrained from the data. For the purpose of this work, it is taken as a line analogous to the "kinked limb" seen in many xenolith-based geotherms (Finnerty and Boyd, 1987), which roughly parallels the diamond–graphite stability curve. This procedure may underestimate the depths of the hottest garnets, compared to a xenolith-based geotherm, but this uncertainty does not affect the use made of the data in this paper.

2.4. Geochemical information from garnets

The Y content of Cr-pyrope garnets is a measure of depletion (Griffin and Ryan, 1995; Griffin et al., 1999d), and plots of Y vs. T_{Ni} (Fig. 4) show the

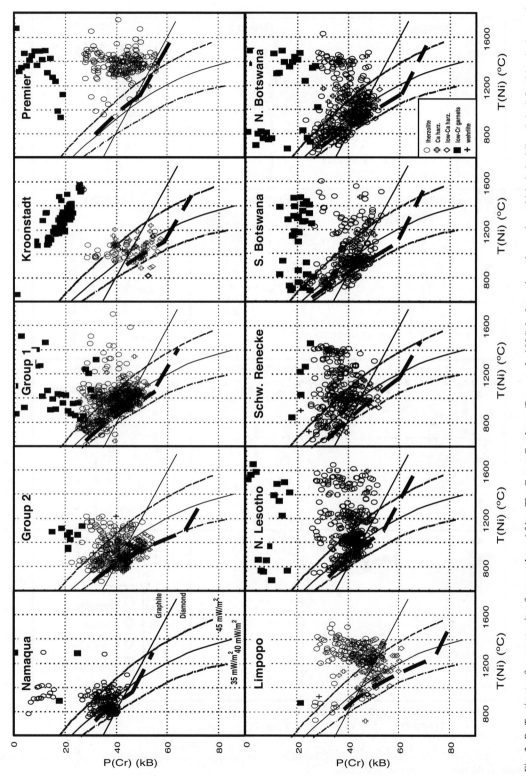

Fig. 3. *P–T* estimates for garnet suites from southern Africa. The Garnet Geotherm (Ryan et al., 1996) is shown in each case by a thick dashed line. It is defined by the locus of maximum P_{Cr} at each T_{Ni}, and extrapolated up to the temperature given by a "Y edge" (Fig. 4), representing the base of the depleted lithosphere. Above this temperature, it is generalised as a line parallel to the diamond–graphite stability curve. Model conductive geotherms of Pollack and Chapman (1977) are shown as a reference frame.

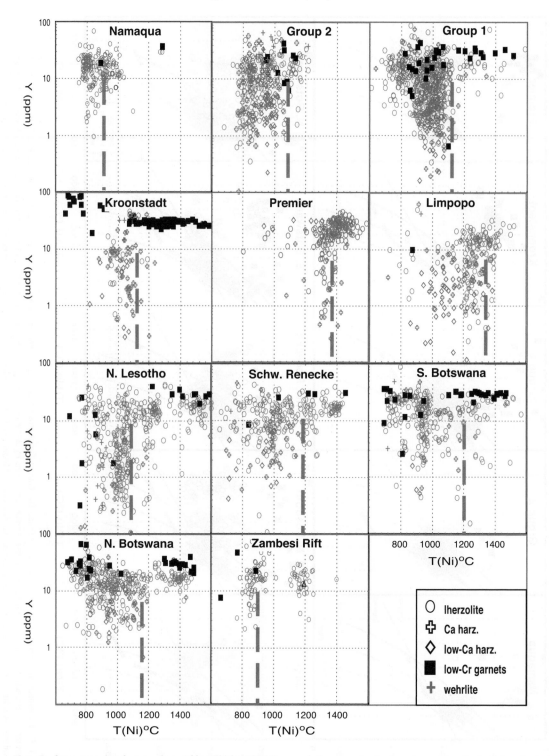

Fig. 4. Y vs. T_{Ni} for garnet suites from southern Africa. Thick dashed lines mark the approximate upper temperature limit of strongly depleted garnets, interpreted as the base of the depleted lithosphere.

distribution of more- and less-depleted peridotites with depth. Estimates of the depth to the base of the lithosphere can be derived from these plots, using the assumption that depleted peridotites are diagnostic of the lithosphere. The median Y content of Cr-pyrope garnets from Archean SCLM is ca. 12 ppm (Griffin et al., 1999d, 2002a), and values less than 8–10 ppm can be regarded as evidence of depleted peridotites. Many of these plots (Fig. 4) show a relatively sharp high-temperature limit to the distribution of such Y-depleted garnets, which can be regarded as representing the temperature at the base of the depleted lithosphere. Some (e.g., S. Botswana, N. Lesotho) show a further scatter of Y-depleted grains to quite high temperatures, which may reflect local heating near magma conduits; in these cases, the definition of lithosphere thickness is more ambiguous. These plots also illustrate the relative degree of depletion of different SCLM sections.

The interelement correlations in a large database of mantle-derived Cr-pyrope garnets have been described by Griffin et al. (1999d), and used by Griffin et al. (2002a) to evaluate several approaches to the definition of populations using multivariate statistics.

The technique described as Cluster Analysis by Recursive Partitioning (CARP) recognised 16 distinctive populations, which show significant variations in relative abundance and depth distribution in the SCLM across different tectonic settings, and can be correlated with specific types of xenoliths found in kimberlites and other volcanic rocks. Fourteen of these CARP classes are represented in the Kalahari dataset, and they can be grouped into four major categories (Table 1). *Depleted harzburgites* contain subcalcic (CaO <4%) garnets depleted in Y, Ga, Zr, Ti and HREE; *depleted lherzolites* have garnets with Ca–Cr relationships indicating equilibration with clinopyroxene (Griffin et al., 1999d), but depleted in HREE, HFSE and Ga. The garnets of *depleted/ metasomatised* lherzolites contain garnets that are depleted in Y and HREE, but enriched in Zr and LREE, suggesting that they were subjected to depletion and subsequent re-fertilisation; xenoliths of this type commonly contain phlogopite ± amphibole. The garnets of *fertile lherzolites* have high contents of HREE and near-median contents of HFSE; they retain no evidence of a depletion event. The garnets of *melt-metasomatised peridotites* show a character-

Table 1
Mean compositions of olivine and garnet in xenoliths with garnets of selected CARP classes

	Olivine	Garnet										Xenolith types
	%Fo	Cr_2O_3	FeO	MnO	MgO	CaO	Ti	Ga	Y	Zr	Ni	
Depleted classes												
CARP-H2	93.6	6.9	5.5	0.3	23.1	3.0	335	2.6	3	41	59	subcalcic harzburgites ± phlogopite
CARP-L3	91.6	5.8	7.3	0.3	19.2	6.5	215	3.4	2	6	38	lherzolites to Ca-harzburgites
CARP-L5	92.9	5.5	6.5	0.3	20.9	5.3	220	3.0	3	62	46	depleted lherzolite/Ca-harzburgite ± phlogopite
Depleted/metasomatised classes												
CARP-H3	93.2	4.8	6.2	0.3	22.6	3.3	520	3.0	29	111	48	Subcalcic harzburgites ± phlogopite
CARP-L15	92.4	4.3	7.0	0.3	20.2	5.7	230	4.7	15	70	38	lherzolites, mostly with phlogopite
CARP-L18	92.8	7.0	6.3	0.3	21.0	4.9	1320	4.0	10	71	56	lherzolites/Ca-harzburgites with phlogopite
CARP-L19	92.6	7.2	6.7	0.3	19.9	6.0	655	3.8	8	120	47	lherzolites with phlogopite
CARP-L21	92.0	6.4	6.8	0.3	20.1	5.9	2695	6.0	20	81	62	lherzolites ± phlogopite; melt metasomatism
Fertile lherzolite classes												
CARP-L9	87.4	2.5	10.5	0.4	18.6	4.8	1080	6.3	18	54	29	Fe-metasomatised lherzolites ± ilmenite
CARP-L10a	92.1	3.5	7.8	0.4	20.3	4.9	450	3.8	18	42	27	lherzolites with phlogopite; Ca-Al metasomatism
CARP-L10b	89.6	2.0	7.3	0.3	20.6	5.2	1315	5.7	41	46	72	very fertile lherzolites, high Cpx/Gnt
Melt-metasomatised classes												
CARP-L13	90.3	3.5	7.4	0.3	21.0	5.0	4085	11	20	62	88	many sheared lherzolites
CARP-L25	91.7	6.4	7.4	0.3	20.1	5.7	1900	7.8	9	59	65	many sheared lherzolites
CARP-L27	90.4	4.6	8.0	0.3	20.2	5.5	5765	12	23	89	91	mostly sheared lherzolites

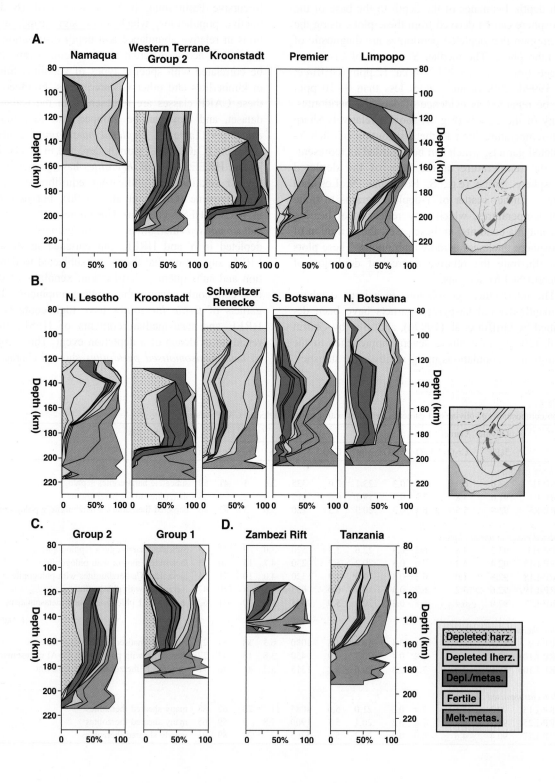

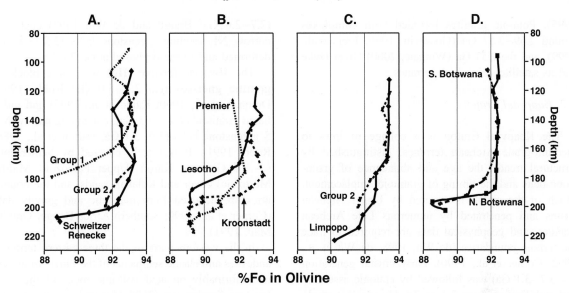

Fig. 6. Distribution of calculated olivine composition (method of Gaul et al., 2000) with depth for selected SCLM sections.

istic enrichment in Zr, Ti, Y and Ga, and correspond to the sheared and enriched lherzolite xenoliths found in many kimberlites. In Fig. 5, we have plotted the relative abundances of the CARP classes (not numbered) and the major categories, against depth. The data have been averaged in 100 °C windows, overlapped by 50 °C, to smooth local variations.

Given the major-element composition and T_{Ni} of a Cr-pyrope garnet, and an estimate of its depth of origin, it is possible to calculate the mg (100 Mg/ (Mg + Fe), or %Fo) of the coexisting olivine (Gaul et al., 2000). We have calculated this composition for each garnet grain, and present the data in terms of the mean olivine composition at each depth, averaged over windows ranging from 50 to 150 °C wide, depending on data density (Fig. 6). Olivine composition is an important parameter in determining the physical properties of ultramafic rocks, and these plots allow a qualitative assessment of the relationship between composition and depth along teleseismic ray paths.

3. Regional tectonic setting

The Kalahari Craton (Fig. 1) is made up of the Archean Kaapvaal and Zimbabwe cratons, separated by the Limpopo Belt. It is partially surrounded by Proterozoic fold belts and is bounded on the east by the Phanerozoic Lebombo and Sabi monoclines, formed during the break-up of Gondwana (Brandl and de Wit, 1997).

3.1. Zimbabwe Craton

Kusky (1998) recognised two main structural blocks: the Tokwe continental terrane (~ 3.5–2.82 Ga) (Fig. 1; **Ia**) and a younger (2.8–2.6 Ga) accretionary terrane (**Ib**) in the northwest. The older complex consists dominantly of tonalitic gneisses with relics of older greenstone successions, and is unconformably overlain by younger granite–greenstone complexes (Petters, 1991). The two complexes are divided by 2.75–2.65 Ga greenstone belts of the Northern Magmatic Belt (Wilson, 1979, Wilson et al.,

Fig. 5. SCLM sections for southern Africa, showing the relative abundances of different broad garnet classes (Table 1) with depth. (A) Traverse 1 along path shown in inset. (B) Traverse 2 along path shown in inset. (C) Comparison between SCLM sampled by "Group 2" kimberlites (>110 Ma) and "Group 1" kimberlites (≤ 95 Ma) in the SW part of the Western Terrane (see Appendix A for sample sites). (D) Comparison of SCLM beneath the Zambesi Rift and the East African Rift in Tanzania.

1995). Potassic granites intruded both complexes around 2.69–2.57 Ga (Goodwin, 1996; Frei et al., 1999), and the 2.57 Ga (Wingate, 2000) Great Dyke and its satellites cross both terranes.

3.2. Kaapvaal Craton

The Kaapvaal Craton is a collage of low- to medium-grade Archean terranes, distinguished by structural trends, the age and abundance of greenstone belts and the timing of granitoid emplacement. Much of the craton is covered by Upper Archean basins and penetrated by numerous Late Archean granites, and geophysical data are required to trace the terrane boundaries (de Wit, 1998; de Wit et al., 1992; Corner et al., 1990). Early crustal generation (~ 3.7–3.1 Ga) was followed by cratonic assembly and stabilization from 3.1 to 2.6 Ga (de Wit et al., 1992). The first collision events, marked by the intrusion of potassic granites, ended by 3.0 Ga. The first volcano-sedimentary cover (Pongola basin) was deposited on this basement between 3.0 and 2.9 Ga, and the boundaries of the eastern terranes were covered by sedimentary and volcanic rocks of the Witwatersrand and Ventersdorp Supergroups between 2.9 and 2.7 Ga. The collision of the Western Terrane with the amalgamated "eastern" terranes occurred about 2.5 Ga.

The *Southeastern Terrane* (**IIa**) contains the oldest gneisses (including the Swaziland Block) and greenstone belts with a NE–SW trend. The dominant TTG gneisses were emplaced between 3.68 and 3.2 Ga (Kröner and Tegtmeyer, 1994) and intrude remnants of greenstone belts. The Barberton greenstone belt (3.57–3.08 Ga; Kamo and Davies, 1994; de Ronde and de Wit, 1994) stretches along the northern boundary of the terrane, which is marked by a belt of northward-directed thrusts. Granites intruded inside and along the boundaries of this block at ~ 3.0–3.2 and ~ 2.5–2.75 Ga.

The *Central Terrane* (**IIb**) is a narrow block devoid of greenstone belts. It consists of granitic gneisses and migmatites with granitoid domes, some intruded at ~ 3.1–3.2 Ga and metamorphosed at ca. 2.6 Ga (Allsopp, 1964; Anhaeusser and Burger, 1982; Anhaeusser, 1999). The *Pietersburg Terrane* (**IIc**) consists largely of tonalitic–trondhjemitic gneisses (2.9–2.8 Ga) and massive granodiorites and granites

(2.7–2.6 Ga; Brandl and de Wit, 1997) enclosing narrow NE-trending greenstone belts (3.5–3.4 Ga) deformed and metamorphosed at ca. 2.8 Ga.

The *Western Terrane* (**IId**) is a large block of granitic gneisses (ca. 2.8–3.0 Ga; Robb, 1991; Drennan et al., 1990; Kamo et al., 1995) and unfoliated granitoids (ca. 2.65 Ga), containing narrow N–S greenstone belts (3.0–3.1 Ga; Barton et al., 1986; Robb, 1991). It is largely covered by Ventersdorp (~ 2.7 Ga) and Karoo (Upper Paleozoic–Middle Mesozoic) cover and Kalahari sands, and the boundaries are deduced from magnetic and gravity data (Corner et al., 1990; Southern Africa Bouger Gravity Map, 1991).

In the western parts of the three eastern terranes, andesites and clastic sediments of the Pinel Group rest unconformably on acid volcanic rocks of the Ventersdorp Supergroup (2620 Ma) and are overlain by early horizons of the Transvaal basin (Buffelsfontein Group, 2460 Ma; Visser, 1984). This belt may represent an active continental margin with an eastward-dipping subduction zone, suggesting that the Western Terrane collided with the rest of the craton near the end of Archean time.

The mafic–ultramafic Bushveld complex covers an area of >60,000 km^2 ($3-4 \times 10^5$ km^3) in the Central and Pietersburg terranes. It was intruded within the time span 2070–2055 Ma (Tankard et al., 1982; Cawthorn and Webb, 2001), with the final granitic phases dated to 2052 Ma (Walraven et al., 1990).

3.3. Limpopo Belt

This terrane contrasts in deformational style with the Kaapvaal and Zimbabwe Cratons and appears to be an ancient microcontinent. However, it also has been regarded as a "mobile belt" (Petters, 1991; Goodwin, 1996 and others), and the high metamorphic grade of the exposed rocks has been attributed to crustal thickening as a direct result of collision between the Kaapvaal and Zimbabwe cratons 2.7–2.6 Ga ago (Limpopo Orogeny). It commonly is divided into a Central Zone with N-trending structures and Northern and Southern Marginal Zones of dominantly ENE-trending structures.

The *Central Zone* (**IIIa**) may contain the oldest basement in Africa. This consists of granulite- and amphibolite-facies gneisses of granodiorite–diorite–tonalite composition with ages of 3.8 Ga (Sand River Gneiss; Tankard et al., 1982; Barton and Key, 1981; Petters, 1991) or 3.9–3.2 Ga (Harris et al., 1987; Retief et al., 1990). The anorthositic Messina Suite (~ 3.1 Ga; Watkeys, 1983) intruded near the contact between this basement and its cover of high-grade metasedimentary rocks (Beitbridge Sequence). Granites of 2.7–2.65 Ga age coincide with the Limpopo Orogeny. On the north side of the block, collision had ended before 2.57 Ga when Great Dyke satellites were intruded (Wingate, 2000).

The *Northern* (**IIIb**) and *Southern* (**IIIc**) *Marginal Zones* comprise reworked basement of the Limpopo terrane, separated from the Central Zone by systems of shear zones (Cahen et al., 1984; McCourt and Vearncombe, 1992; Treloar et al., 1992). The maximum age recorded in the basement of the Southern Zone is 3.15 Ga (Barton et al., 1994a,b). In the Northern zone, low-pressure granulite metamorphism took place either before 2.9 Ga (Petters, 1991), or ca. 2.87 Ga ago (Goodwin, 1996), and porphyritic char-nockitic–enderbitic gneisses were intruded at 2.52–2.69 Ga (Frei et al., 1999).

High-grade metamorphism (2.7 Ga) and the intrusion of voluminous porphyritic granites and grano-diorites (ca. 2.60 Ga; Goodwin, 1996; Frei et al., 1999) marked the final joining of the Zimbabwe Craton, the Limpopo microcontinent and the Kaapvaal Craton into the unified Kalahari Craton. The Limpopo terrane was thrust over the Zimbabwe Craton, and the Kaapvaal Craton was thrust under the Limpopo terrane along the Southern Marginal Zone (Durrheim et al., 1992); the resulting crustal thickening produced the granulite-facies rocks exposed in the Limpopo Belt today (McCourt and Vearncombe, 1992; Treloar et al., 1992; Roering et al., 1992; de Beer and Stettler, 1992). Numerous 2.05- to 1.95 Ga Rb–Sr ages on micas from the Central and Southern Marginal Zones record late uplift and cooling (Barton et al., 1994a,b). The western end of the Limpopo Belt is cut off by the Proterozoic Kheis–Okwa–Makondi fold belt, containing the 2.0 Ga Mahalapy granite. The pattern of magnetic anomalies suggests a tectonic boundary between these granites and the South Marginal Zone of the Limpopo terrane.

3.4. The proterozoic Kheis–Okwa–Makondi belt

The *Kheis Belt* (**Va**) is an east-vergent thin-skinned fold and thrust belt composed largely of sediments and volcanic rocks. It is bounded by the Blackridge thrust on the northeast, the Doringberg fault on the southeast and the Trooilapspan shear zone and Brackbosch fault on the west (Moen, 1999, Visser, 1984). The northwestern boundary of the belt (the "Kalahari Suture Zone" (KSZ)) coincides with a strong magnetic anomaly, interpreted as the unexposed basic–ultrabasic Tshane Complex (1.02–1.07 Ga; Key and Ayres, 2000).

The oldest sedimentary rocks in the eastern part of the Kheis Belt are dated at 1.93–1.89 Ga (Armstrong, 1987). Mafic dykes and sills intruded the Olifantshoek sedimentary sequence at 1.75 Ga (Cornell et al., 1998). The deformed sedimentary rocks of the belt are intruded by numerous syn- and post-tectonic granite plutons and 1180 Ma felsic dykes along the northeastern boundary of the belt (Moen, 1999), and the 1.27 Ga Kalkwerf augen gneisses in the central part (Barton and Burger, 1983). It is clear that the main orogeny cannot be Eburnean (~ 1.8 Ga) in age, as has been suggested previously (e.g., Beukes and Smit, 1987; Altermann and Hälbich, 1990; Goodwin, 1996; Cornell et al., 1998). The Kheis Belt therefore is regarded as the Mid-Proterozoic passive margin of the Kaapvaal Craton, deformed in Late Proterozoic time.

The *Okwa inlier* (**Vb**) is a well-defined magnetic feature at the north end of the Kheis Belt in central Botswana (Key and Ayres, 2000). A 2.0–2.1 Ga basement (Ramokate et al., 1996; Botswana Geological Survey, 2000) is overlain by, or in tectonic contact with, low-grade clastic sediments and limestones (Key and Ayres, 2000). The age of deformation of the Okwa inlier may be ca. 1.7–1.2 Ga.

The *Makondi (Magondi) Belt* (**Vc**) contains a thick sequence of metasediments and metavolcanics (Makondi Supergroup), deformed at 2.1–1.8 Ga (Stowe et al., 1984; Treloar and Kramers, 1989; Petters, 1991). These are interpreted as deposits on the passive margin of the Zimbabwe Craton. The belt has a SE-directed thin-skinned fabric, and the grade of regional metamorphism increases towards the NW. Undeformed granite and older migmatites in the Sua Pan area (2.0 Ga; Key and Ayres, 2000; Majaule et al., 1998) and the Mahalapy granite–migmatite complex

(2.0 Ga; van Beerman and Dodson, 1972; McCourt and Armstrong, 1998) confirm a major Proterozoic thermal event.

The *Namaqua–Natal Orogenic Belt* (**VI**) commonly is divided into two parts. *The Namaqua* (or *Orange River–Namaqua*) *Province* (**VIa**) is comprised of the SE trending *Bushmanland* (southern) *subprovince*, dominated by 2.1–1.9 Ga gneisses, granulites and granitoids deformed ca. 1.2 Ga ago, and the *Gordonia* (northern) *subprovince*, which is comprised of supracrustal rocks in a zone of convergent thrusts against the adjoining Kheis Belt. The Namaqua province may be a Proterozoic terrane accreted to the Kalahari Craton about 1.2 Ga. The *Natal Province* (**VIb**) contains a complex of granitic and migmatitic gneisses, and minor amphibolite occurs in a tectonic melange beneath metamorphic rocks (1.2–0.9 Ga) that have been transported northwards onto the Kaapvaal Craton. Matthews (1981) suggests that the Natal granite–gneiss complex and the infolded supracrustal sequence represent the former margin of a continental plate, deformed during collision with the Kaapvaal Craton.

3.5. Rehobothian Subprovince and Damara Province

The *Rehobothian Subprovince* (**VII**) west of the Kheis–Okwa–Makondi belt is hidden beneath the relatively undeformed Late Proterozoic Nama and Karoo volcano-sedimentary cover sequence (Hartnady et al., 1985). The "Kalahari Line", a distinctive geophysical lineament along the 22°E meridian, separates deep (~ 15 km) magnetic basement to the west from shallow basement to the east (Hartnady et al., 1985, Reeves and Hutchins, 1982). In the northern part of this subprovince the *Northwest Botswana Rift* contains 1.1–1.25 Ga bimodal volcanic rocks (Schwartz et al., 1996; Ramokate and Mapeo, 1998).

The *Damara Orogen* (**VIII**) in Namibia and NW Botswana experienced major deformation ca. 950 Ma, and a second episode 650–540 Ma. The orogen can be divided into several tectonic belts, located between the Kalahari Craton and the Angolan terrane (**IV**) of the Zaire (Congo) craton.

3.6. Eastern edge of the Kalahari Craton

Between Zululand and the Limpopo River, the Kalahari Craton is bounded by the east-dipping Leb-ombo monocline, which rotates into the Sabi monocline north of the Limpopo River. The third branch of this rift system is formed by the Tuli and Nuanetsi synclines, which extend along the Limpopo terrane. The Karoo volcanics, a typical bimodal rift association of Late Triassic–Early Jurassic age (205–180 Ma), reach a thickness of 10 km in the Lebombo monocline.

4. Results: mantle petrology

The garnet data from the individual areas are discussed with reference to two generalised traverses running diagonally across the craton (Fig. 2).

Traverse 1 (Fig. 5A) extends from the Namaqua–Natal belt in the SW to the Limpopo Belt in the NE. The section for the *Namaqua Fold Belt* includes data from three areas spread along the belt from Uintjiesberg in the SE (Fig. 2; area 21) through Rietfontein (area 22) to the Gibeon area (area 23) in the NW. The mantle sections from all three areas are very similar. The garnet geotherm (Fig. 3) is not well-defined for any of the areas, due to the small T range covered by the garnets. Taken together, the data suggest a geotherm that lies between the 35 and 40 mW/m² conductive models, lower than the 39–44 mW/m² derived from xenoliths in several of these areas by Finnerty and Boyd (1987). However, the garnets that constrain the low geotherm are from Rietfontein; the data from Uintjiesberg and Namibia are consistent with a geotherm of ≥ 40 mW/m². In each case, the base of the lithosphere is poorly constrained, because Y-depleted garnets are only present in the upper parts of the section (Fig. 4); our best estimate for the temperature at the base of the lithosphere is ca. 900 °C, corresponding to ca. 150 km depth. Subcalcic garnets are absent, and the sections are dominated by fertile lherzolites. Garnets from depleted and depleted/metasomatised lherzolites are most important between 90 and 120 km. Melt-related metasomatism is not pronounced. The mean olivine composition is essentially constant at Fo$_{92.5}$ throughout the section; this is more magnesian than most Proterozoic sections worldwide (Gaul et al., 2000; Griffin et al., 2002a).

The Western Terrane includes kimberlites intruded in two time slices (Appendix A) that show significant differences in SCLM composition and must be considered separately. The "Group 2" section includes data from a series of pipes with ages >110 Ma, spread across Western Terrane from its western edge (Finsch; area 14) through Barkley West (area 15) to Roberts Victor (area 11) on its eastern edge. Most of these are Group 2 kimberlites as defined by Smith (1983). Some data from the Star pipe (area 12) are included, although it lies within the margin of the SE Terrane, because its SCLM is very similar to that of the other pipes. The geotherm for each locality, and for the group as a whole, is well-defined between 800 and 1000 °C, and lies near the 35 mW/m^2 conductive model (Fig. 3); this is consistent with limited xenolith data from Finsch and West End, and from Frank Smith, an older Group 1 kimberlite (Finnerty and Boyd, 1987). A pronounced Y edge at 1100 °C (ca. 190 km; Fig. 4) is consistent with the position of the geotherm inflection in the xenolith data. The section shows little stratification; depleted and depleted/metasomatised lherzolites make up 50–60% of the section between 120 and 180 km; melt-related metasomatism increases with depth and becomes dominant below ca. 190 km. The mean olivine composition is nearly constant at Fo$_{93}$ down to 170–180 km, then decreases slowly with depth (Fig. 6A).

The "Group 1" section from the same area includes data from kimberlites in the Kimberley area (area 16), the Barkley West area to the NW and from the Koffiefontein pipe (area 17). All except Frank Smith are ≤ 95 Ma in age (Appendix A). Jagersfontein (area 18; not included) lies on the edge of the SE Terrane, but its SCLM is similar in composition and structure to the Group 1 section shown here (Griffin et al., unpublished data). The geotherm is well-defined between 700 and 1050 °C, and lies near a 37–38 mW/m^2 conductive model (Fig. 3); a large body of xenolith data from these pipes gives a mean geotherm near the 39 mW/m^2 model (Finnerty and Boyd, 1987), and the two estimates appear to be within their mutual uncertainties. A well-defined Y edge at ca. 1100 °C (Fig. 4) suggests that the base of the depleted lithosphere lies near 160–170 km. The proportion of garnets from depleted harzburgites and lherzolites is significantly lower than in the "Group 2" section from the same area. Garnets from fertile lherzolites make up most of the sample at depths < 110 km, and the degree of melt-metasomatism increases rapidly at depths >160 km. The strong influence of metasoma-

tism between 100 and 120 km is reflected in a drop in the mean olivine composition from Fo$_{93}$ to < Fo$_{92}$ at this level (Fig. 6A). The mean Fo content of olivine drops rapidly below 150 km, reaching Fo$_{90}$ at 175 km; this suggests that the depleted lithosphere does not extend below ca. 170–175 km.

The *Kroonstadt area* (Fig. 2; area 13) includes several kimberlites that sample the SCLM in the northern part of the SE Terrane. The geotherm is not well-defined except in the range 900–1100 °C; it lies near the 35 mW/m^2 conductive model (Fig. 3). The data show a pronounced Y edge at ca. 1100 °C (Fig. 4). Harzburgites and depleted lherzolites are abundant from 130 to 190 km depth, and the section is overall more depleted than the "Group 2" section from the SW Terrane. The mean olivine composition (Fig. 6B) increases downward from Fo$_{92}$ at 140 km to Fo$_{93.5}$ at 180 km, then decreases rapidly with depth to reach asthenospheric values (Fo$_{89-90}$) at 190–195 km. Below this depth nearly all garnets are low-Cr types with the high-Ti, high-Zr signature of melt-related metasomatism.

The Premier kimberlite (1.25 Ga; Barrett and Allsopp, 1973; area 9 on Fig. 2) intruded through the 2.06 Ga Bushveld complex. It sampled SCLM within the Central Terrane of the Kaapvaal Craton, but is expected to reflect the intraplate magmatic episode of which the Bushveld is a part. The garnet geotherm is poorly defined, because most of the garnets in the sample are from the high-T lower part of the section. The few lower-T data are broadly consistent with the xenolith-derived geotherm of Danchin (1979), which lies above the 40 mW/m^2 conductive model to ca. 1100 °C, then rises sharply (see also Finnerty and Boyd's (1987) analysis of the same data). We have used Danchin's xenolith-based geotherm to derive the depths of derivation for the garnet grains. The Y edge is at 1350–1400 °C, indicating that some depleted garnets have been heated to high T without being metasomatised (Fig. 4). The high mean T_{Ni} indicates that most of the sample is derived from depths of >180–190 km, and the upper part of the section was poorly sampled. The available data indicate that it is strikingly different from other sections in the Archean part of craton. Strongly subcalcic garnets are absent, depleted/metasomatised or "fertile" lherzolites are more common, and melt-metasomatised garnets are abundant throughout the sampled part of the section.

The mean calculated olivine composition of the Premier SCLM section varies from $Fo_{91.5}$ to $Fo_{92.5}$ in the 130–180 km depth range; this is significantly lower than other sections (Fig. 6B). The mean Fo content decreases rapidly below 180 km, suggesting that this is the maximum thickness of the depleted SCLM. Despite the small number of samples in the upper part of the section, these estimates are consistent with xenolith data reported by Danchin (1979). The mean olivine composition in coarse lherzolites, which occur at depths of 140–170 km, is $Fo_{92.1}$. In Danchin's Group II harzburgites, which occur at depths of 180–190 km, the mean olivine composition is $Fo_{92.6}$; in his Group I (sheared) harzburgites and deformed garnet lherzolites, which occur mainly at depths >190 km, the mean values are $Fo_{91.3}$ and $Fo_{90.7}$, respectively.

The *Limpopo Belt* section includes data from three localities (areas 3–5) in the Central Zone. The Venetia pipe is dated at ca. 520 Ma (Phillips et al., 1998), and the other pipes are believed to be of similar age. All show similar sections, which are taken as representative of this microcontinent's SCLM at that time. The garnet geotherm (Fig. 3) lies near a 37 mW/m² conductive model up to >1200 °C; depleted lherzolites are present up to 1400 °C, implying a very thick section (Fig. 4). From 140–180 km, the section consists mainly of harzburgites and depleted lherzolites. The mean olivine composition (Fig. 6C) is constant at $Fo_{93.5}$ down to ca. 170 km. Below this, the section contains a higher proportion of lherzolites, but is still depleted; the mean olivine composition remains around $Fo_{92.5}$ down to 200 km, then becomes more iron-rich as the effects of melt-related metasomatism increase with depth. The depleted lithosphere is at least 200 km thick.

Traverse 2 (Fig. 5B) extends from northern Lesotho in the SE and into the Kheis Belt in the NW. The *N. Lesotho* section includes data from ca. 90 Ma kimberlites in the NE highlands of Lesotho (area 20). They sampled the SCLM of the SE Terrane of the Kaapvaal Craton, probably modified by continental-margin processes during the Late Proterozoic emplacement of Natal Province mobile belt (Irvine et al., 2001). The kimberlites erupted through a thick section of Karoo lavas, and the SCLM may also have been affected by Karoo plume activity. Data from the Monastery kimberlite to the NW (area 19) are not included because the SCLM is highly anomalous (heavily melt-metaso-

matised) relative to the Lesotho SCLM. The garnet geotherm lies near a 37 mW/m² conductive model, and is well-defined from 800 to 1000 °C (Fig. 3); this is parallel to, but lower than, the N. Lesotho geotherm of Finnerty and Boyd (1987); the inflection point of the xenolith-based geotherm at 1100 °C correlates with a pronounced Y edge (Fig. 4). The proportion of garnets from subcalcic harzburgites is low, but depleted lherzolites are abundant in the 140–160 km depth range. Fertile lherzolites dominate the upper part of the section. Melt-related metasomatism increases rapidly below 160 km, affecting most of the section below 170 km. The mean Fo content of the olivine (Fig. 6B) decreases from Fo_{93} at 100 km depth to Fo_{92} at 170 km, then drops rapidly to asthenospheric values ($\leq Fo_{90}$). The data indicate that the depleted SCLM is not more than 170–175 km thick.

The *Schweizer Renecke* (area 10; Fig. 2) samples are from many small alluvial deposits on a highland area; a detailed study of the indicator minerals indicated derivation from local sources (Griffin et al., 1994). The kimberlites from which these alluvial deposits are derived sampled the SCLM near the eastern margin of the Western Terrane. Their ages are not known, but the abundance of ilmenite macrocrysts in association with the diamonds in these deposits suggests that they are Group 1 kimberlites, and thus probably have ages < 100 Ma (Smith, 1983).

The garnet geotherm is well-defined between 700 and 1000 °C (Fig. 3). It appears to be steeper than the conductive models, rising from near the 35 mW/m² model at 700 °C to near a 38 mW/m² model at 1000 °C. There is a weak Y edge near 1200 °C (Fig. 4). The section is generally similar to that sampled by the Group 1 kimberlites to the south, but with a higher proportion of subcalcic harzburgite at shallow depths (100–120 km). Fertile peridotites make up a smaller portion of the section between 100 and 160 km depth than in the Group 1 section, and the degree of melt-related metasomatism is generally lower down to ca. 200 km. The mean olivine composition (Fig. 6A) varies irregularly around Fo_{93} from 100 to 200 km depth, then decreases rapidly to asthenospheric values at ca. 210 km. The data suggest that the depleted lithosphere is ca. 200 km thick.

Most of the data in the *southern Botswana* section are from the Tsabong kimberlites (area 8; Fig. 2) and the Jwaneng pipe (area 7); a few samples from the

Matahebuse and Kokong kimberlites (area 6) are very similar. These kimberlites have sampled the SCLM beneath the Kheis Belt and the outer margin of the Western Terrane, subjected to extension and then collision in late Proterozoic time. Jwaneng lies near the buried Molopo Farm Complex, and the SCLM also may have been affected by the ca. 2 Ga intraplate magmatism. The Jwaneng pipe is Permian in age (235 Ma; Kinney et al., 1989); the ages of the other kimberlites are not well known. The garnet geotherm is reasonably well-defined from 700 to 1000 °C by data from most of the pipes, and lies near the 37 mW/m^2 conductive model. However, some Jwaneng garnets lie below this model, suggesting an even lower geotherm (Fig. 3). There is a poorly defined Y edge near 1200 °C; some Y-depleted garnets give higher temperatures, but have high Ti contents, suggesting that they have been affected by heating and metasomatism (Fig. 4).

The SCLM section (Fig. 5B) is clearly different from most of the Archean ones. Garnets from subcalcic harzburgites and depleted lherzolites are very scarce. Depleted/metasomatised lherzolites make up 20–30% of most of the section; garnets from fertile lherzolites are most abundant at depths < 130 km, as in the Western Terrane Group 1 section. Melt-related metasomatism increases with depth, becoming dominant below ca. 170–180 km. The mean olivine composition is Fo$_{92}$ from 130 to 170 km, and is lower at the shallow depths where fertile peridotites dominate. The mean Fo content decreases below Fo$_{92}$ at 170–180 km, suggesting a rapid transition from depleted SCLM to asthenosphere-like compositions (Fig. 6D).

The data for the *northern Botswana* mantle section come from two main areas of the Kheis Belt. Orapa, Letlhakane and several small kimberlites to the west (area 1; Fig. 1) lie within the Makondi Belt, and the Gope cluster (area 2) to the SW lies within the Okwa inlier. There are no obvious differences between the garnet suites from these two areas, so they are considered together here. The kimberlites have sampled SCLM affected by Proterozoic rifting and collision along the passive margin of the Kalahari Craton. The garnet geotherm is well-defined from 700 to 1000 °C, and lies near the 37 mW/m^2 conductive model (Fig. 3). A pronounced Y edge at 1150 °C (Fig. 4) corresponds to a depth of ca. 190 km. The section is similar to the S. Botswana section; it contains a low

proportion (20–30%) of depleted rocks, and fertile lherzolites are the dominant rock type at depths < 120 km. The degree of melt-related metasomatism is generally lower; it increases slowly with depth, but sharply at ca. 190 km. The mean olivine composition (Fig. 6D) is constant around Fo$_{92}$ down to 190 km, then drops rapidly with depth.

The *Zambesi Rift* (area 24) lies along the extension of this traverse to the NE. The section (Fig. 3D) contains data from three kimberlites in the Quest cluster on shores of Lake Kariba. The rift reflects Mesozoic extension of the Proterozoic crust of the Makondi Belt on the rifted margin of the Zimbabwe Craton. The garnet geotherm is not well-defined because the garnets represent only a small *T* range (Fig. 3), but lies between the 35 and 40 mW/m^2 conductive models. There is a Y edge at ca. 900 °C (Fig. 4). The short section is dominated by lherzolites, with relatively more depleted rocks in the upper part. Melt-related metasomatism increases sharply below ca. 140 km, defining the base of the depleted SCLM. The mean Fo content of olivine decreases steadily with depth from depleted values (Fo$_{93}$) at the top, to asthenospheric values (Fo$_{90}$) at 150 km. The upper part of the section is very similar to the upper parts of the N. Botswana and S. Botswana sections; this is consistent with thinning and metasomatism of a thicker Proterozoic SCLM section, inferred to have been present before rifting.

5. Discussion

5.1. Intra-terrane differences in SCLM

Comparisons of the SCLM beneath the different structural provinces of the Kalahari Craton (Fig. 1) are complicated because different kimberlite fields have sampled the SCLM at different times, while comparisons of different time slices show that late events have modified the SCLM at least in the SW part of the craton. However, some idea of differences and similarities can be gained by comparing the sections sampled by the oldest kimberlites in each area.

The SCLM of the Limpopo Belt may represent the most primitive section available. The 520 Ma kimberlites sampled the mantle prior to the Karoo and Cretaceous magmatic activity, which may have affected much of the craton (see below). The Limpopo SCLM is extremely depleted, with a high proportion of subcalcic harzburgites and depleted lherzolites compared to the other sections studied here. Re–Os data are available for seven xenoliths from Venetia (Carlson et al., 1999); four of these give T_{RD} model ages of 2.7–3.7 Ga, and T_{MA} ages back to 3.3–3.6 Ga, consistent with very early formation of this lithosphere. The section has a distinct lower layer of less depleted material at depths >170 km; its sharp boundary to the rest of the section suggests it may have been added from below. Two xenoliths from this layer have T_{RD} ages of 1.3–1.5 Ga (Re–Os not measured). These ages do not correspond to known crustal ages in the area, and they may represent mixtures related to later modification of the original SCLM.

The Kroonstadt section (SE Terrane) was sampled mainly by ca. 145 Ma kimberlites, and has a lower proportion of depleted rocks than the Limpopo section. However, the high proportion of depleted/metasomatised rocks suggests that it may originally have been nearly as depleted as the Limpopo section, and has been more strongly modified by later metasomatism. The depleted part of the section extends to greater depth than beneath the Limpopo Belt, and has a very sharp lower limit between 190 and 200 km.

The section sampled by the Group 2 kimberlites (>110 Ma) in the Western Terrane is less depleted than the Kroonstadt section, but much of the difference may relate to metasomatic modification. The section contains a higher proportion of fertile rocks at shallow depth, and has experienced more melt-related metasomatism at depth, giving a more transitional lithosphere–asthenosphere boundary.

The differences between these sections may indicate that each terrane was carrying its own lithospheric keel at the time of craton amalgamation (2.6–2.7 Ga), as occurred in the Siberian craton (Griffin et al., 1999a). However, ambiguity remains because of the age differences among the kimberlites, which would allow longer time for post-accretion modification. This ambiguity may only be resolved by detailed dating of mantle depletion and enrichment events.

5.2. Premier section: Fe metasomatism of SCLM

The SCLM section sampled by the 1.25 Ma Premier kimberlite is markedly different from the

others studied here. The most striking compositional difference is the lower mean Fo content of the olivine. The section is short; despite an intensive search of concentrates, relatively few peridotitic garnets have been found from depths <160 km, and most of the section is strongly affected by melt-related metasomatism. Danchin (1979) has described Fe enrichment in peridotite xenoliths from Premier, and noted that harzburgitic rocks are extremely scarce at depths <180 km; he also found few garnet peridotites derived from depths <150 km. Hoal et al. (this volume) describe a suite of Fe-rich pyroxenites and peridotites from shallower depths. These observations, combined with the data presented here, suggest that the entire section has been extensively Fe-metasomatised during pyroxenite intrusion (cf. Pokhilenko et al., 1999). If the section originally was similar to the Kroonstadt section to the SW, the metasomatism has removed garnet from much of the section above ca. 160 km, converted harzburgites to lherzolites, and reduced the mean Fo content of olivine in the remaining garnet peridotites.

The Premier kimberlite erupted through the Bushveld mafic–ultramafic complex ca. 800 Ma after its intrusion. Half of the analysed Premier peridotite xenoliths give Proterozoic, rather than Archean, Re–Os model ages (T_{RD}) (Carlson et al., 1999). It seems apparent that the metasomatism of the Premier SCLM section is related to the passage of large volumes of magma through the lithosphere in connection with the intrusion of the Bushveld complex. The elevated geotherm produced by this event would have relaxed back to a conductive regime over much less than 800 Ma. However, the xenolith-derived geotherm for the Premier SCLM (Danchin, 1979) is slightly higher than the geotherms for the least-metasomatised sections studied here (Limpopo, Kroonstadt, Group 2; Fig. 3), which suggests that the K, U and Th introduced by the metasomatism (Hoal et al., this volume) also has resulted in a higher mantle heat production.

The presence of diamonds in the Premier pipe requires either that the diamonds survived the metasomatic event and the inferred associated heating, or that the diamonds grew in the mantle after the event. We have found 11 analyses of peridotitic garnets included in Premier diamonds (Tsai et al., 1979; Gurney et al., 1985; Moore et al., 1989). Of these, seven (64%) are lherzolitic or very weakly subcalcic, compared to 5–10% in other Kaapvaal diamond-inclusion suites (Gurney, 1984; Griffin et al., 1992). Five have high Ti contents (0.17–0.68% TiO_2) compared to other peridotitic garnets included in diamonds, suggesting that they grew in a metasomatised protolith. We are not aware of model ages for diamond inclusions from Premier that are older than 2 Ga. We therefore suggest that few diamonds survived the Bushveld igneous event, and that many of the diamonds in the Premier pipe have grown in the mantle after this event.

5.3. Mantle chemistry and seismic tomography

James et al. (2001, this volume) have provided detailed seismic tomography along a SW–NE swath through the area studied here. The tomographic images show a compact "core" of high-Vp SCLM beneath the SW part of the Western Terrane and the adjacent part of the SE Terrane, and another, more fragmentary, core under the Limpopo Belt and the southern part of the Zimbabwe Craton (Fig. 7). The two cores are separated by a large "hole" of lower Vp, corresponding to the Bushveld complex, the related Molopo Farm complex in Botswana, and the dike swarm between them (Fig. 7A). This area of lower Vp also extends under the Kheis Belt and other areas interpreted here as the passive margin of the ancient Kalahari Craton.

The lateral variations in Vp are on the order of 1–1.5% and could in principle be explained by variations in either the temperature or the bulk composition of the SCLM. The available data from xenoliths and garnet concentrates do not indicate large variations in the geotherm across the region. It therefore is probable that the regional variations in seismic velocity mainly reflect differences in the bulk composition of the SCLM.

At constant T, the seismic velocity of peridotite is a function of fertility (Griffin et al., 1998, 1999e). The more iron-rich olivine of fertile peridotites produces lower Vp; an increase in garnet content will raise Vp, but this is more than outweighed by the decrease in Vp that results from increasing clinopyroxene, orthopyroxene or phlogopite contents. For the typical modal composition of Archean lherzolites, a decrease in mg from Fo_{93} to Fo_{90} produces an decrease in Vp of ca. 0.6% (Griffin et al., 1999e). At constant Fo content, a 20% decrease in the modal proportion of

olivine (with the relative proportions of other phases being held constant) leads to a 0.8% decrease in Vp. The metasomatic modification of peridotite at 120–150 km depth typically leads to a correlated decrease in olivine content (increase in pyroxene ± phlogopi-phlogopite ± amphibole) and mg (Erlank et al., 1987; Griffin et al., 1999f; van Achterbergh et al., 2001); this interplay of modal composition and bulk mg provides sufficient variation to explain most of the range of seismic velocities observed by James et al. (2001).

There is a good correlation between the lateral distribution of Vp across the craton and the composition of the SCLM, as illustrated by "geochemical tomography" maps showing the distribution of different mantle types and mean Fo content of olivine at 150 km depth (Fig. 8). The high-Vp core beneath the SW part of the craton corresponds to the depleted SCLM mapped in this area, whereas the large "hole" in the high-Vp core beneath the central part of the craton corresponds to the marked iron enrichment documented in the SCLM sampled by the Premier pipe. The relatively low Vp along the western margin of the craton is consistent with the more fertile mantle mapped beneath the Kheis Belt and the adjacent margin of the craton in Botswana, and beneath the Namaqua–Natal Belt.

The most depleted mantle section mapped here, and one of the thickest, is beneath the Limpopo Belt, whereas the high-Vp core seen beneath this area in the tomographic images is fragmentary. The discrepancy probably reflects the different time slices sampled by the kimberlites (ca. 500 Ma) and the geophysical study (today). We suggest that the Mesozoic Karoo dike swarm that cuts across this area, related to the Lebombo–Sabi monoclines (Fig. 7B), has affected the SCLM of the Limpopo Belt in the same way that the formation of the Bushveld complex affected the SCLM of the Central Terrane, but less severely. Similarly, the dike swarm extending to the SSW separates the main root from a smaller high-Vp volume to the E. In general, there is a good correlation between the lower-Vp areas within the cratonic root, as seen on the seismic tomography images, and the distribution of intraplate magmatism (Fig. 7). The correlation, coupled with our observations on the section beneath the Premier pipe, suggests that a once laterally continuous craton root has been progressive-

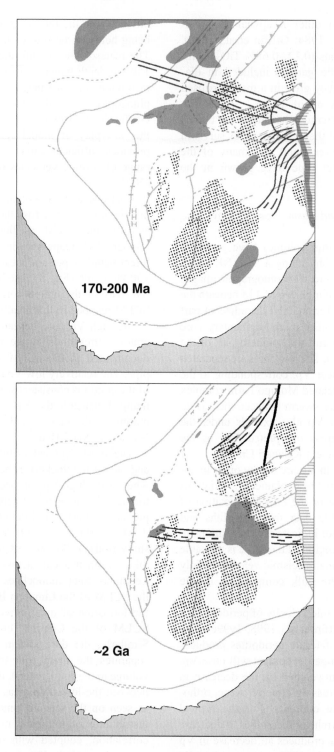

Fig. 7. Distribution of intraplate magmatism in two time slices; heavy shading is surface or buried lavas or igneous complexes; dashed fields are known dike swarms. Stippled pattern shows approximate extent of high-Vp mantle at 150 km, after James et al. (2001).

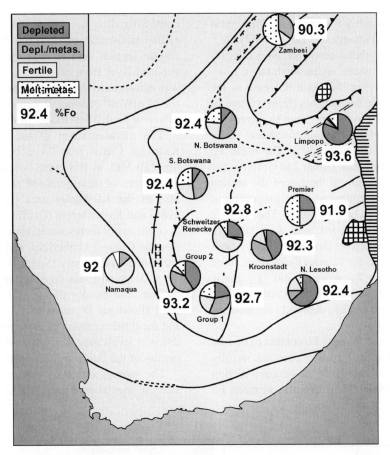

Fig. 8. "Chemical tomography" map showing relative abundances of different garnet classes, and calculated mean olivine composition, at 150 km depth. Note that areas of high Vp (Fig. 7) correspond to areas with high mean Fo content in olivine, and high proportions of depleted peridotites; low-Vp areas have more Fe-rich olivine and high proportions of metasomatised and fertile lherzolites.

ly "eroded" by metasomatism related to intraplate and craton-margin magmatism.

There is no clear lower boundary to the high-Vp cores mapped by James et al. (2001); the depth to the velocity value chosen as the limit between higher-Vp and lower-Vp mantle varies from 150 to 300 km. The boundaries of the high-Vp volumes outlined in this way are irregular and show considerable vertical relief. Some of this relief may be an artefact of the seismic data processing, related to the use of vertical ray paths, but some may be real and related to metasomatic processes. Magma conduits affected by high degrees of metasomatism would become lower-velocity volumes, and appear as "pipes" through the high-Vp keel, while less metasomatised volumes appear as "roof pendants" of higher Vp. A resolution

of these alternatives will have to await more detailed processing of the seismic data.

5.4. Temporal evolution of SCLM: Western Terrane

The comparison between the Group 2 (>110 Ma) and Group 1 (mostly ≤ 95 Ma) sections from the SW part of the Kaapvaal Craton shows a major temporal change in the compositional and thermal structure of the SCLM through this time range. The geotherm derived from the garnet data rises from near the 35 mW/m^2 conductive model in the older kimberlites to closer to 37–38 mW/m^2 in the younger ones. As noted above, this rise is supported by the lower temperatures at each pressure seen in xenoliths suites from the Finsch, Frank Smith and West End kimberlites, relative to estimates for xenoliths from most younger kimberlites (Finnerty and Boyd, 1987).

This rise in the geotherm accompanied a refertilisation of the depleted mantle. In the depth range 130–170 km, this is seen (Fig. 5C) as a decrease in the proportion of subcalcic harzburgites (from a mean of 24% in Group 2 to 17% in Group 1) and an increase in the proportion of depleted/metasomatised and fertile lherzolites from ca. 30% to ca. 50%. The mean Fo content of olivine drops from Group 2 to Group 1, but the decrease occurs mainly in the top of the section (corresponding to a large increase in the proportion of "fertile" lherzolites) and below 150 km. The Group 2 kimberlites contain few garnets with $T_{Ni} < 800$ °C, suggesting that this represents the upper stability limit of garnet at that time. In the younger kimberlites, there are many garnets with T_{Ni} down to 700 °C; this is consistent with an expansion of the garnet stability field to shallower depths in less depleted (metasomatically refertilised) rocks.

The marked decrease in mean Fo content of olivine with depth in the Group 1 section contrasts sharply with the relatively constant Fo in the Group 2 section between 110 and 200 km. This decrease in mean Fo

content parallels an increase in the degree of melt-related metasomatism in the lower part of the Group 1 mantle section, and a consequent thinning of the depleted layer from ca. 200 km to ca. 170 km. This was accompanied by a rise in the geotherm from near the 35 mW/m^2 conductive model in the older section, to near a 38 mW/m^2 model in the younger section.

The metasomatism of the SCLM in the SW Kaapvaal Craton may have begun in Karoo time (ca. 180 Ma), as this event is reflected in the U–Pb systematics of metasomatised xenoliths from kimberlites in the Kimberley area (Hawkesworth et al., 1983) and Jagersfontein (Griffin et al., 1999g). However, the major metasomatic episode must be younger than the Group 2 kimberlites, and it may be related to the plume-type activity (Smith, 1983) responsible for the widespread Cretaceous Group 1 kimberlite magmatism. The density of the SCLM increases downward (Poudjom Djomani et al., 2001; Fig. 9 (left)) and the different styles of metasomatism expressed at different levels may simply be related to the different density of the fluids or melts involved (Fig. 9 (right)).

The overall effect of this refertilisation was to increase the mean density (at standard temperature

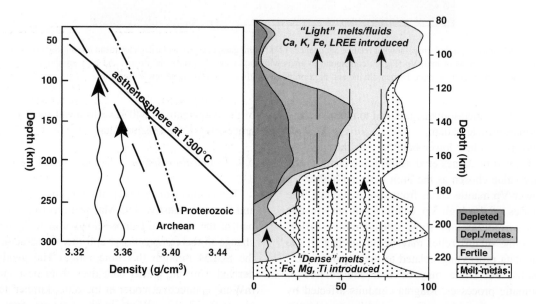

Fig. 9. Cartoon illustrating the relationship between lithospheric density and metasomatism. (Left) Density vs. depth for typical Archean and Proterozoic SCLM sections (after Poudjom Djomani et al., 2001). Arrows represent rising mafic/ultramafic melts of different densities, which will "die" at the depth where their density is equivalent to that of the mantle wall rocks. (Right) Typical distribution of metasomatic styles with depth (Group 1 SCLM section, from Fig. 5C).

and pressure) of the lithospheric section, both by decreasing the mg of the relatively depleted SCLM and raising its pyroxene and garnet content, and by decreasing the thickness of the depleted layer (Figs. 5C and 6A). However, these effects are more than offset by the density decrease resulting from the rise in the geotherm (Fig. 3). The predicted result of the changes in SCLM thermal state and composition from Group 2 to Group 1 time is a surface uplift of at least 1 km over the affected area. Such uplift has been documented by fission-track studies in the southern Kaapvaal Craton, and is reflected in a dramatic increase in sedimentation rates in the Mozambique Channel, beginning ca. 90 Ma ago (Brown et al., 1998); the southern Kaapvaal Craton still maintains a high elevation (mean ca. 1500 m o.s.l.).

Much of our picture of the composition of Archean SCLM is based on studies of xenoliths from Kaapvaal Craton kimberlites, and these data sets are dominated by material from the Group 1 kimberlites in the SW part of the craton. The data presented here indicate that the xenoliths from Group 1 kimberlites represent a strongly modified version of the SCLM, and do not give a reliable picture of the composition, structure or thermal state of primary Archean SCLM, or even of the SCLM across the rest of the Kaapvaal Craton.

5.5. Modification of SCLM during extension

The modification of SCLM during rifting events is demonstrated by sections from the Tanzania and Zambesi Rift zones (Fig. 3D). The Tanzanian rift is developed on an Archean shield, and the underlying SCLM is at least partly Archean in age (Chesley et al., 1999); this is confirmed by the presence of subcalcic harzburgites and relatively abundant depleted lherzolites. The Zambesi Rift affects an area of Proterozoic crust, and the underlying SCLM may have resembled that of the sections from northern and southern Botswana. Although the sections have different thicknesses of SCLM (160 for Tanzania vs. 140 km for the Zambesi Rift), they also show strong similarities. Fertile lherzolites are abundant at intermediate depths, there is a strong increase in metasomatic effects downward through the section, and the base of the depleted lithosphere is relatively sharp and marked by an abrupt increase in the degree of melt-related metasomatism.

These patterns are similar to those seen in the Botswana sections, and especially in northern Botswana. They suggest that the Archean and Proterozoic SCLM can be thinned in extensional situations, because progressive infiltration of asthenospheric melts into the base of the section is assisted by the extension of the lithosphere. The new SCLM formed in the lower parts of the rift sections is very fertile. Analyses of refertilised sheared peridotites derived from a similarly melt-metasomatised zone suggest that it is only mildly depleted in bulk composition relative to the asthenospheric mantle (Smith and Boyd, 1987; Smith et al., 1991); in this respect, it is more like Phanerozoic SCLM worldwide (Griffin et al., 1999e).

5.6. Archean vs. proterozoic SCLM—secular evolution by metasomatism?

Griffin et al. (1998, 1999e) have documented a secular evolution in the composition of the SCLM, correlated with the age of the last tectothermal event in the overlying crust. SCLM beneath areas with Archean crust, unaffected by later events, is more depleted than that beneath areas that experienced Proterozoic thermal events; SCLM beneath many areas of Phanerzoic thermal and magmatic activity is still less depleted. This overall pattern is seen in the data from the Kalahari Craton as well; the SCLM beneath the Proterozoic crust of the Kheis Belt and beneath the Proterozoic Bushveld Complex is significantly less depleted than that beneath the central Archean areas of the craton.

The Botswana sections are similar to many Proterozoic sections worldwide (Griffin et al., 2002a, and unpublished data). They contain a low proportion of depleted rocks, and especially of harzburgites; fertile lherzolites are abundant at depths < 130 km. However, high proportions of depleted/metasomatised lherzolites in the lower portions of the sections suggest that the whole SCLM was originally more depleted. Eighty percent of Re–Os T_{RD} model ages of xenoliths from the Letlhakane pipe in northern Botswana are Archean (2.5–2.8 Ga; Carlson et al., 1999), and indicate that at least the N. Botswana SCLM is Archean in origin.

Comparison with the time-related differences between the Group 2 and Group 1 SCLM in the Western

Terrane (Fig. 5C) suggests that the Botswana sections may be the result of similar processes of metasomatism, carried closer to completion. The metasomatism that affected the Group 1 SCLM section has been studied in detail in xenoliths from the Wesselton pipe in the Kimberley area (Griffin et al., 1999f). The progress of this style of metasomatism converts harzburgite to depleted lherzolite, and depleted lherzolite to fertile rocks (CARP class L10A; Table 1). In the Kimberley area, the timing of the metasomatism suggests a relation to plume-related magmatism. However, kimberlite magmatism in S. Botswana, at least, precedes both the Karoo plume and Group 1 kimberlite magmatism. We therefore suggest that these sections reflect modification of Archean SCLM during rifting of the passive margin of the craton, and perhaps during subsequent collisional events, in Proterozoic time.

Modelling based on the mean composition and typical thermal states of Archean SCLM (Griffin et al., 1999e; Poudjom Djomani et al., 2001; O'Reilly et al., 2001) shows that Archean SCLM is so buoyant relative to the underlying asthenosphere that it cannot be removed through gravitational forces. This buoyant SCLM would tend to persist during rifting (see above) and collision events, but be progressively transformed to more fertile compositions. In extreme cases, the refertilised lower parts of some sections might become gravitationally unstable. Does all of the secular evolution in SCLM composition from Archean to Proterozoic time represent progressive modification of Archean SCLM, or can SCLM that was newly generated in Proterozoic time be recognised?

The SCLM beneath the Namaqua–Natal Belt is thinner and more fertile than the Botswana SCLM, although its overall structure is generally similar to the top of the N. Botswana section. Re–Os studies of xenoliths from the Namaqua Belt have produced several Proterozoic model ages, but no Archean ones (Hoal et al., 1995). Thus, the Namaqua Belt SCLM might have been generated in Proterozoic time. However, detailed in situ Re–Os analysis will be required to tell whether the Proterozoic Re–Os model ages reflect Proterozoic formation of the SCLM, or overprinting of Archean ages by several generations of sulfide phases, as has been observed in the Siberian SCLM (Griffin et al., 2002b).

6. Conclusions

The data presented here reveal considerable lateral heterogeneity in the bulk composition and lithological structure of the SCLM beneath the Kalahari Craton and its surrounding mobile belts. Much of the heterogeneity within the craton is interpreted to be the product of metasomatism related to intraplate magmatism, and this makes it difficult to know whether the individual terranes had distinctive SCLM roots at the time of craton amalgamation. The most depleted SCLM section is the one sampled beneath the Limpopo Belt by early Paleozoic kimberlites; the SCLM beneath the southern parts of the craton may represent similar material modified by metasomatism during Phanerozoic time, and sampled by late Mesozoic kimberlites. The extensively studied xenolith suite from the Group 1 kimberlites in the SW part of the craton area represents a strongly modified mantle sample, and is not a good model for primary Archean SCLM compositions. The relatively fertile SCLM beneath the mobile belts surrounding the craton is interpreted as largely Archean SCLM, metasomatised and possibly mixed with younger material during Paleoproterozoic to Mesoproterozoic rifting and compression. This implies that at least some of the observed secular evolution in SCLM composition worldwide may reflect the reworking of Archean SCLM that has persisted because of its intrinsic initial buoyancy.

There are strong correlations between the mantle composition and structure mapped in this work, and the distribution of seismic velocities shown by the tomographic work of James et al. (2001). Areas of relatively low Vp within the craton largely reflect the progressive refertilisation of the Archean root during episodes of intraplate magmatism, including the Bushveld (2.05 Ga) and Karoo (ca. 180 Ma) events. Areas of high Vp map out the distribution of relatively less metasomatised Archean SCLM.

Acknowledgements

This study uses data accumulated over a period of 13 years, through basic research and applied research carried out in collaboration with diamond exploration companies. We thank the companies and individuals that have provided samples, including John Gurney and his students, DeBeers, MPH Botswana, Auridiam, Reunion Mining and many others. We are grateful to DeBeers for arranging collecting trips. Funding has been provided by a number of exploration companies, and by ARC grants to WLG/SYO, Macquarie University internal research funds and GEMOC. We are grateful to Tin Tin Win, Ashwini Sharma, Norm Pearson, Carol Lawson and Oliver Gaul for assistance with analytical work, and Oleg Belousov for designing the GeoSpeed software used to analyse the data. The shaded magnetic map used in Fig. 1 was kindly provided by Bruce Wyatt of DeBeers Australia Exploration. The conclusions presented here have benefited from discussions with many colleagues, especially Joe Boyd, Buddy Doyle, John Gurney, Bram Janse, Kevin Kivi, Simon Shee, Craig Smith and Bruce Wyatt. We thank Hermann Grutter, Martin Menzies and Russell Sweeney for critical reviews of the original manuscript. Special thanks go to Barry Hawthorne and Bobby Danchin for their invaluable moral support in the early stages of this work. This is contribution no. 294 from the ARC National Key Centre for Geochemical Evolution and Metallogeny of Continents (GEMOC).

Appendix A. Sampling

Grouping	Locality	Age, Ma	Key (Fig. 2)	Gnts, LAM	Gnts, PMP
Group 1 (n = 638)	Boshoff Road	86–94	16	55	
	Dutoitspan	86–94	16	71	
	Kamfersdam	86–94	16	76	58
	Pulsator Dump	86–94	16	72	
	Wesselton	88.6	16	74	
	Koffiefontein	<90	17		77
	Frank Smith	116	15		47
	Leicester	<90	15		28
	Snyder	<90	15		14
	Balmoral	<90	15	36	
	Kouewater	<90	15		30
Group 2 (n = 363)	Star	124	12		37
	Sover	120	15		13
	Newlands	116	15		74
	Roberts Victor	125	11		174
	Finsch	124	14		65
N. Lesotho (n = 391)	Letseng-le-Terai	88	20	93	
	Thaba Putsoa	80–90	20	92	

Appendix A (*continued*)

Grouping	Locality	Age, Ma	Key (Fig. 2)	Gnts, LAM	Gnts, PMP
N. Lesotho (n = 391)	Liqhobong	80–90	20	34	
	Liqhobong satellite	80–90	20	109	
	Kao	89	20	63	
Kroonstadt area (n = 232)	Lace	145	13		61
	Kaalvallei	85	13		20
	K4	145?	13	51	
	K4 satellite	145?	13	50	
	K6	145?	13	50	
S. Botswana (n = 301)	MD24	ca. 80?	8	26	
	M1	78	8	53	
	T-6	ca. 80?	8	49	
	MD19	ca. 80?	8	23	
	MD35	ca. 80?	8	28	
	MD41	ca. 80?	8	10	
	MD73	ca. 80?	8	60	
	Jwaneng	235	7		52
N. Botswana (n = 393)	Orapa	93	1	45	78
	Letlhakane	93	1		23
	BK16	80–90	1		11
	BK15	80–90	1		32
	BK2	80–90	1		84
	BK4	80–90	1		45
	BK5	80–90	1		38
	Gope 25	80–90	2		37
Limpopo Belt (n = 251)	Venetia	500	3	92	
	River Ranch	500	4	46	83
	Fort Victoria	500	5	30	
Namaqua Fold Belt (n = 198)	Rietfontein	72	22	89	
	Uintjiesberg	99	21	17	34
	Deutsches Erde	66	23		48
	Gruen Dorner	71.5	23		10
Zambesi Rift (n = 105)	Quest 1, 2, 3	86?	24		105
Premier (n = 215)	Premier mine	1200	9	215	
Schweizer Renecke (n = 398)	Alluvial deposits	80–90?	10		398
Total = 3485				1709	1776

References

Allsopp, H.L., 1964. Rubidium–strontium ages from the western Transvaal. Nature 204, 361–363.

Altermann, W., Hälbich, I.W.H., 1990. Thrusting, folding and stratigraphy of the Ghaap Group along the south-western margin of the Kaapvaal Craton. S. Afr. J. Geol. 93, 553–567.

Anhaeusser, C.R., 1999. Archaean crustal evolution of the central Kaapvaal craton, South Africa: evidence from the Johannesburg Dome. S. Afr. J. Geol. 102, 303–322.

Anhaeusser, C.R., Burger, A.J., 1982. An interpretation of U–Pb zircon age for Archaean tonalite gneisses from the Johannesburg–Pretoria granite dome. Trans. Geol. Soc. S. Afr. 85, 111–116.

Armstrong, R.A., 1987. Geochronological studies on Archaean and Proterozoic formations of the foreland of the Namaqua front and possible correlates on the Kaapvaal Craton. PhD Thesis, Witwatersrand University, Johannesburg. 274 pp.

Barrett, D.R., Allsopp, H.L., 1973. Rubidium–strontium age determinations on South African kimberlite pipes. Extended Abstracts of the 1st International Kimberlite Conference. University of Cape Town, Cape Town, pp. 23–25.

Barton, E.S., Burger, A.J., 1983. Reconnaissance isotopic investigations in the Namaqua Mobile Belt and implications for Protozoic crustal evolution—Upington Geotraverse. In: Botha, B.J.V. (Ed.), Namaqualand Metamorphic Complex, vol. 10. Spec. Publ. Geol. Soc. S. Afr., Johannesburg, pp. 173–191.

Barton, E.S., Key, R.M., 1981. The tectonic development of the Limpopo mobile belt and the evolution of the cratons of southern Africa. In: Kröner, A. (Ed.), Precambrian Plate Tectonics. Elsevier, Amsterdam, pp. 21–185.

Barton, E.S., Armstrong, R.A., Cornell, D.H., Welke, H.I., 1986. Feasibility of total-rock Pb–Pb dating of metamorphosed banded iron-formation: the Marydale Group, South Africa. Chem. Geol. 59, 255–271.

Barton Jr, J.M., Holzer, L., Kamber, B., 1994a. Discrete metamorphic events in the Limpopo belt, southern Africa: implications for application of P–T paths in complex metamorphic terrains. Geology 22, 1035–1038.

Barton, J.M., Holzer, L., Kamber, B., Doig, R., Kramers, J.D., Nyfeler, J.D., 1994b. Discrete metamorphic events in the Limpopo belt, Southern Africa: implications for the application of P–T path in complex metamorphic terrains. Geology 22, 1035–1038.

Beukes, N.J., Smit, C.A., 1987. New evidence for thrust faulting in Griqualand West, South Africa: implication for stratigraphy and the age of the red beds. S. Afr. J. Geol. 90, 378–394.

Botswana Geological Survey, 2000. Report of Botswana Ongoing Project on Regional Geology, Gaborone, 1–3. http://www.gov.bw/government/regional_geology.htm.

Brandl, G., de Wit, M.J., 1997. The Kaapvaal Craton. In: de Wit, M.J., Ashwal, L.D. (Ed.), Greenstone Belts. Clarendon Press, Oxford, pp. 581–607.

Brown, R., Gallagher, K., Griffin, W.L., de Wit, M., Belton, D., Harman, R., 1998. Kimberlites, accelerated erosion and evolution of the lithospheric mantle beneath the Kaapvaal craton during the mid-Cretaceous. Extended Abstracts of the 7th International Kimberlite Conference. University of Cape Town, Cape Town, pp. 105–107.

Cahen, L., Snelling, N.J., Delhal, J., Vail, J.R., 1984. The Geochronology and Evolution of Africa. Clarendon Press, Oxford. 184 pp.

Canil, D., 1994. An experimental calibration of the "Nickel in Garnet" geothermometer with applications. Contrib. Mineral. Petrol. 117, 410–420.

Carlson, R.W., Pearson, D.G., Boyd, F.R., Shirey, S.B., Irvine, G., Menzies, A.H., Gurney, J.J., 1999. Re–Os systematics of lithospheric peridotites: implications for lithosphere formation and preservation. Proceedings of the 7th International Kimberlite Conference. Red Roof Design, Cape Town, pp. 99–108.

Cawthorn, R.G., Webb, S.J., 2001. Connectivity between the western and eastern limbs of the Bushveld Complex. Tectonophysics 330, 195–209.

Chesley, J.T., Rudnick, R.L., Lee, C.-T., 1999. Re–Os systematics of mantle xenoliths from the East African Rift: age, structure, and history of the Tanzanian craton. Geochim. Cosmochim. Acta 63 (7/8), 1203–1217.

Cornell, D.H., Armstrong, R.A., Walraven, F., 1998. Geochronology of the Proterozoic Hartley Basalt Formation, South Africa: constraints on the Kheis tectogenesis and the Kaapvaal Craton's earliest Wilson Cycle. J. Afr. Earth Sci. 26, 5–27.

Corner, B., Durrheim, R.J., Nicolaysen, L.O., 1990. Relationship between the Vredefort structure and the Witwatersrand basin within the tectonic framework of the Kaapvaal Craton as interpreted from regional gravity and aeromagnetic data. Tectonophysics 171, 49–61.

Danchin, R.V., 1979. Mineral and bulk chemistry of garnet lherzolite and garnet harzburgite xenoliths from the Premier Mine, South Africa. In: Boyd, F.R., Meyer, H.O.A. (Eds.), The Mantle Sample: Inclusions in Kimberlites and Other Volcanics. American Geophysical Union, Washington, DC, pp. 104–126.

de Beer, J.H., Stettler, E.H., 1992. The deep structure of the Limpopo Belt from geophysical studies. Precambrian Res. 55, 173–186.

de Ronde, C.E.J., de Wit, M.J., 1994. The tectonothermal evolution of the Barberton Greenstone Belt, South Africa: 490 million years of crustal evolution. Tectonics 13, 983–1005.

de Wit, M.J., 1998. On Archean granites, greenstones, cratons and tectonics: does the evidence demand a verdict? Precambrian Res. 91, 181–224.

de Wit, M.J., Roering, Ch., Hart, R.G., Armstrong, R.A., de Ronde, C.E.J., Green, R.W.E., Tredoux, M., Peberdy, E., Hart, R.A., 1992. Formation of an Archean continent. Nature 357, 553–562.

Drennan, G.R., Robb, L.J., Meyer, F.M., Armstrong, R.A., de Bruin, H., 1990. The nature of the Archaean basement in the hinterland of the Witwatersrand basin: II. A crustal profile west of the Welkom goldfield and comparisons with the Vredefort crustal profile. S. Afr. J. Geol. 93, 41–53.

Durrheim, R.J., Barker, W.H., Green, R.W.E., 1992. Seismic studies in the Limpopo Belt. Precambrian Res. 55, 187–200.

Erlank, A.J., Waters, F.G., Hawkesworth, C.J., Haggerty, S.E., Allsopp, H.L., Rickard, R.S., Menzies, M.A., 1987. Evidence for mantle metasomatism in peridotite nodules from the Kimberley pipes, South Africa. In: Menzies, M.A., Hawkesworth, C.J. (Eds.), Mantle Metasomatism. Academic Press, London, pp. 221–311.

Finnerty, A.A., Boyd, F.R., 1987. Thermobarometry for garnet peridotite xenoliths: a basis for mantle stratigraphy. In: Nixon, P.H. (Ed.), Mantle Xenoliths. Wiley, New York, pp. 381–402.

Frei, R., Blenkinsop, T.G., Shönberg, R., 1999. Geochronology of the late Archean Razi and Chilimanzi suites of granites in Zimbabwe: implications for the late Archean tectonics of the Limpopo belt and Zimbabwe craton. S. Afr. J. Geol. 102, 55–63.

Gaul, O.F., Griffin, W.L., O'Reilly, S.Y., Pearson, N.J., 2000. Mapping olivine composition in the lithospheric mantle. Earth Planet. Sci. Lett. 182, 223–235.

Goodwin, A.M., 1996. Precambrian Geology. Academic Press, London. 327 pp.

Griffin, W.L., Ryan, C.G., 1995. Trace elements in indicator minerals: area selection and target evaluation in diamond exploration. J. Geochem. Explor. 53, 311–337.

Griffin, W.L., Ryan, C.G., 1996. "An experimental calibration of the "Nickel in Garnet" geothermometer, with applications" by D. Canil: discussion. Contrib. Mineral. Petrol. 124, 216–218.

Griffin, W.L., Gurney, J.J., Ryan, C.G., 1992. Variations in trapping temperatures and trace elements in peridotite-suite inclusions from African diamonds: evidence for two inclusion suites, and implications for lithosphere stratigraphy. Contrib. Mineral. Petrol. 110, 1–15.

Griffin, W.L., Ryan, C.G., Win, T.T., 1994. Trace elements in indicator minerals: application to paleodrainage analysis. CSIRO Exploration and Mining Report 50R, North Ryde. 120 pp.

Griffin, W.L., O'Reilly, S.Y., Ryan, C.G., Gaul, O., Ionov, D., 1998. Secular variation in the composition of subcontinental lithospheric mantle. In: Braun, J., Dooley, J.C., Goleby, B.R., van der Hilst, R.D., Klootwijk, C.T. (Eds.), Structure and Evolution of the Australian Continent, Geodynamics, vol. 26. American Geopyhysical Union, Washington, DC, pp. 1–26.

Griffin, W.L., Ryan, C.G., Kaminsky, F.V., O'Reilly, S.Y., Natapov, L.M., Win, T.T., Kinny, P.D., Ilupin, I.P., 1999a. The Siberian Lithosphere Traverse: mantle terranes and the assembly of the Siberian Craton. Tectonophysics 310, 1–35.

Griffin, W.L., Doyle, B.J., Ryan, C.G., Pearson, N.J., O'Reilly, S.Y., Davies, R.M., Kivi, K., van Achterbergh, E., Natapov, L.M., 1999b. Layered Mantle Lithosphere in the Lac de Gras Area, Slave Craton: composition, structure and origin. J. Petrol. 40, 705–727.

Griffin, W.L., Doyle, B.J., Ryan, C.G., Pearson, N.J., O'Reilly, S.Y., Natapov, L., Kivi, K., Kretschmar, U., Ward, J., 1999c. Lithosphere Structure and Mantle Terranes: Slave Craton, Canada. Proceedings of the 7th International Kimberlite Conference, Red Roof Design, Cape Town, pp. 299–306.

Griffin, W.L., Fisher, N.I., Friedman, J.H., Ryan, C.G., O'Reilly, S.Y., 1999d. Cr-pyrope garnets in the lithospheric mantle: I. Compositional systematics and relations to tectonic setting. J. Petrol. 40, 679–704.

Griffin, W.L., O'Reilly, S.Y., Ryan, C.G., 1999e. The composition and origin of subcontinental lithospheric mantle. In: Fei, Y., Bertka, C.M., Mysen, B.O. (Eds.), Mantle Petrology: Field Observations and High-Pressure Experimentation: A Tribute to Francis R. (Joe) Boyd. Geochem. Soc. Spec. Publ., vol. 6. The Geochemical Society, Houston, pp. 13–45.

Griffin, W.L., Shee, S.R., Ryan, C.G., Win, T.T., Wyatt, B.A., 1999f. Harzburgite to lherzolite and back again: metasomatic processes in ultramafic xenoliths from the Wesselton kimberlite, Kimberley, South Africa. Contrib. Mineral. Petrol. 134, 232–250.

Griffin, W.L., Pearson, N.J., Jackson, S.E., Zhang, M., O'Reilly, S.Y., Wang, Z., 1999g. Hf, Pb and Sr isotopes in LIMA from the Jagersfontein kimberlite: in-situ analysis by LAM-MC-ICPMS. Abstracts of the Goldschmidt Conference, Boston.

Griffin, W.L., Fisher, N.I., Friedman, J.H., O'Reilly, S.Y., Ryan, C.G., 2002a. Cr-pyrope garnets in the lithospheric mantle: II. Compositional populations and their distribution in time and space. Geochem. Geophys. Geosyst. 3, 1073.

Griffin, W.L., Spetsius, Z., Pearson, N.J., O'Reilly, S.Y., 2002b. In-situ Re–Os analysis of sulfide inclusions in kimberlitic olivine: new constraints on depletion events in the Siberian lithospheric mantle. Geochem. Geophys. Geosyst. 3, 1069.

Gurney, J.J., 1984. A correlation between garnets and diamonds. In: Glover, J.E., Harris, P.G. (Eds.), Kimberlite Occurrence and Origins: A Basis for Conceptual Models in Exploration, vol. 8. Geol. Dept. Univ. Ext., Univ. Western Austr. Publ., Perth, pp. 143–166.

Gurney, J.J., Harris, J.W., Rickard, R.S., Moore, R.O., 1985. Inclusions in Premier Mine diamonds. Trans. Geol. Soc. S. Afr. 88, 301–310.

Harris, N.B.W., Hawkesworth, C.G., Van Calstern, P., McDermott, F., 1987. Evolution of continental crust in southern Africa. Earth Planet. Sci. Lett. 83, 85–93.

Hartnady, C.H., Joubert, P., Stowe, C., 1985. Proterozoic crustal evolution in southwestern Africa. Episodes 8, 236–243.

Hawkesworth, C.J., Erlank, A.J., Kempton, P.D., Waters, F.G., 1983. Mantle metasomatism: isotope and trace-element trends in xenoliths from Kimberley, South Africa. Chem. Geol. 85, 19–34.

Hoal, B.G., Hoal, K.E.O., Boyd, F.R., Pearson, D.G., 1995. Age constraints on crustal and mantle lithosphere beneath the Gibeon kimberlite field, Namibia. S. Afr. J. Geol. 85, 19–34.

Irvine, G.J., Pearson, D.G., Carlson, R.W., 2001. Lithospheric mantle evolution of the Kaapvaal Craton: a Re–Os isotope study of peridotite xenoliths from Lesotho kimberlites. Geophys. Res. Lett. 28, 2505–2508.

James, D.E., Fouch, M.J., Van Decar, J.C., van der Lee, S., the Kaapvaal Seismic Group, 2001. Tectospheric structure beneath southern Africa. Geophys. Res. Lett. 28, 2485–2498.

Jones, A.G., Ferguson, I.J., Chave, A.D., Evans, R.L., McNeice, G.W., 2001. The electric lithosphere of the Slave craton. Geology 29, 423–426.

Kamo, S.L., Davies, D.W., 1994. Geochronology and tectonic evolution of the Barberton greenstone belt. Tectonics 13, 167–192.

Kamo, S.L., Key, R.M., Daniels, L.R.M., 1995. New evidence for Neoarcheaean hydrothermally altered granites in south-central Botswana. J. Geol. Soc. Lond. 152, 747–750.

Key, R.M., Ayres, N., 2000. The 1998 edition of the National Geological Map of Botswana. J. Afr. Earth Sci. 30, 427–451.

Kinney, P.D., Compston, W., Bristow, J.W., Williams, I.S., 1989. Archean mantle xenocrysts in a Permian kimberlite: two generations of kimberlitic zircon in Jwaneng DK2, southern Botswana. Geol. Soc. Austr. Spec. Publ. 14, 833–842.

Kröner, A., Tegtmeyer, A., 1994. Gneiss–greenstone relationships in the Archean Gneiss Complex of south-western Swaziland, southern Africa and implications for early crustal evolution. Precambrian Res. 67, 63–71.

Kusky, T.M., 1998. Tectonic setting and terrane accretion of the Archean Zimbabwe craton. Geology 2, 153–156.

Majaule, T., Key, R.N., Hanson, R., 1998. The south-western margin of the Magondi Belt in north-east Botswana: new geological and geochronological information from the Sua Pan area, Botswana. Abstracts of the International Conference on the Role of BNGS in Sustainable Development, Gaborone, pp. 48–50.

Matthews, P.E., 1981. Eastern or Natal sector of the Namaqua–Natal mobile belt in Southern Africa. In: Hantner, D.R. (Ed.), Precambrian of the Southern Hemisphere. Elsevier, Amsterdam, pp. 705–715.

McCourt, S., Armstrong, R.A., 1998. SHRIMP U–Pb zircon geochronology of granites from the central Zone, Limpopo belt, southern Africa: implications for the age of Limpopo Orogeny. S. Afr. J. Geol. 101, 329–338.

McCourt, S., Vearncombe, J.R., 1992. Shear zones of Limpopo Belt and adjacent granitoid–greenstone terranes: implications for the late Archaean collision tectonics in southern Africa. Precambrian Res. 55, 553–570.

Moen, H.F.G., 1999. The Kheis Tectonic Subprovince, southern Africa: a lithostratigraphic perspective. S. Afr. J. Geol. 102, 27–42.

Moore, R.O., Gurney, J.J., Fipke, C.E., 1989. The development of advanced technology to distinguish between diamondiferous and barren diatremes. Geological Survey of Canada Open File Report 2124, 4 volumes.

Norman, M.D., Griffin, W.L., Pearson, N.J., Garcia, M.O., O'Reilly, S.Y., 1998. Quantitative analysis of trace element abundances in glasses and minerals: a comparison of laser ablation ICPMS, solution ICPMS, proton microprobe, and electron microprobe data. J. Anal. At. Spectrosc. 13, 477–482.

O'Reilly, S.Y., Griffin, W.L., 1996. 4-D lithospheric mapping: a review of the methodology with examples. Tectonophysics 262, 3–18.

O'Reilly, S.Y., Griffin, W.L., Poudjom Djomani, Y., Morgan, P., 2001. Are lithospheres forever? Tracking changes in subcontinental lithspheric mantle through time. GSA Today 11, 4–9.

Petters, S.W., 1991. Regional Geology of Africa. Springer-Verlag, Berlin. 722 pp.

Phillips, D., Kiviets, G.B., Barton, E.S., Smith, C.B., Fourie, L.F., 1998. 40Ar/39Ar dating of kimberlites and related rocks: problems and solutions. Extended Abstracts of the 7th International Kimberlite Conference. University of Cape Town, Cape Town.

Pokhilenko, N.P., Sobolev, N.V., Kuligin, S.S., Shimizu, N., 1999. Pecularities of distribution of pyroxenitic paragenesis garnets in Yakutian kimberlites and some aspects of the evolution of the Siberian Craton lithospheric mantle. Proceedings of the 7th International Kimberlite Conference, Red Roof Design, Cape Town, pp. 689–698.

Pollack, H.N., Chapman, D.S., 1977. On the regional variation of heat flow, geotherms and lithospheric thickness. Tectonophysics 38, 279–296.

Poudjom Djomani, Y.H., O'Reilly, S.Y., Griffin, W.L., Morgan, P., 2001. The density structure of subcontinental lithosphere: constraints on delamination models. Earth Planet. Sci. Lett. 184, 605–621.

Ramokate, L.V., Mapeo, R.B.M., 1998. Palaeoproterozoic to Neoproterozoic geology of western Botswana; the Ganzi–Makunda area, a case study. Abstracts of the International Conference on the Role of BNGS in Sustainable Development. Botswana Geological Survey, Gaborone, pp. 60–64.

Ramokate, L.V., Mapeo, R.B.M., Davis, D.W., Corfu, F., Kampunzu,

A.B., 1996. The geology, geochronology and regional correlation of the Palaeoproterozoic Okwa Inlier, western Botswana. Abstracts IGCP 358 Meeting (Palaeoproterozoic of Sub-Equatorial Africa), Geological Survey Department, Lusaka, pp. 7–8.

Reeves, C.V., Hutchins, D.G., 1982. A progress report on the geophysical exploration of the Kalahari in Botswana. Geoexploration 20, 209–224.

Retief, E.A., Compston, W., Armstrong, R.A., Williams, I.S., 1990. Characteristics and preliminary U–Pb ages of zircons from Limpopo belt lithologies. The Limpopo Belt: A Field Workshop on Granulites and Deep Crustal Tectonics. Rand Afrikaans University, Johannesburg, pp. 95–98.

Robb, L.J., 1991. The Schweitzer–Reneke dome. In: Anhaeusser, C.R. (Ed.), The Archaean Kraaipan Group Volcano-Sedimentary Rocks and Associated Granites and Gneisses of South-Western Transvaal, North-Western Cape Province and Bophuthatswana, Guidebook. Inform. Circ. Econ. Geol. Res. Unit. Univ. Witwatersrand, Johannesburg, vol. 244, pp. 7–13.

Roering, C., van Reenen, D.D., Smit, C.A., Barton, J.M., de Beer, J.H., de Wit, M.J., Stettler, E.H., Stevens, J., Pretorius, S., 1992. Tectonic model for the evolution of the Limpopo Belt. Precambrian Res. 55, 539–552.

Ryan, C.G., Griffin, W.L., Pearson, N.J., 1996. Garnet geotherms: a technique for derivation of P–T data from Cr-pyrope garnets. J. Geophys. Res. 101, 5611–5625.

Schwartz, M.O., Kwok, Y.Y., Davis, D.W., Akanyang, P., 1996. Geology, geochronology and regional correlation of the Ganzi Ridge, Botswana. S. Afr. J. Geol. 99, 245–250.

Smith, C.B., 1983. Pb, Sr and Nd isotopic evidence for sources of southern African Cretaceous kimberlites. Nature 304, 51–54.

Smith, D., Boyd, F.R., 1987. Compositional heterogeneities in a high-temperature lherzolite nodule and implications for mantle processes. In: Nixon, P.H. (Ed.), Mantle Xenoliths. Wiley, Chichester, pp. 551–562.

Smith, D., Griffin, W.L., Ryan, C.G., Cousens, D.R., Sie, S.H., Suter, G.F., 1991. Trace-element zoning of garnets from The Thumb: a guide to mantle processes. Contrib. Mineral. Petrol. 107, 60–79.

Southern Africa Bouger Gravity Map, 1991. Scale 1:2,500,000. IGSN-71. Anglo-American Prospecting Services.

Stowe, C.W., Hartnady, C.G.H., Joubert, P., 1984. Proterozoic tectonic provinces of Southern Africa. Precambrian Res. 25, 229–241.

Tankard, A.J., Jackson, M.P.A., Eriksson, K.A., Hobday, D.K., Hunter, D.R., Minter, W.E.L., 1982. Crustal Evolution of Southern Africa. 3.8 Billion Years of Earth History. Springer-Verlag, Berlin. 523 pp.

Treloar, P.J., Kramers, J.D., 1989. Metamorphism and geochronology of granulites and migmatic granulites from the Mogondi Mobile Belt, Zimbabwe. Precambrian Res. 38, 55–73.

Treloar, P.J., Coward, M.P., Harris, N.B.W., 1992. Himalayan–Tibetian analogies for the evolution of the Zimbabwe Craton and Limpopo Belt. Precambrian Res. 55, 571–587.

Tsai, H.M., Meyer, H.O.A., Moreau, J., Milledge, H., 1979. Mineral inclusions in diamond: premier, jagersfontein, and finsch kimberlites, South Africa, and Williamson Mine, Tanzania. In: Boyd, F.R., Meyer, H.O.A. (Eds.), Kimberlites, Diatremes, and Diamonds: Their Geology, Petrology, and Geochemistry.

Proc. 2nd Int. Kimb. Conf., vol. 1. American Geophysical Union, Washington, DC, pp. 16–26.

van Achterbergh, E., Griffin, W.L., Stiefenhofer, J., 2001. Metasomatism in mantle xenoliths from the Letlhakane kimberlites: estimation of element fluxes. Contrib. Mineral. Petrol. 141, 397–414.

van Beerman, O., Dodson, M.H., 1972. Metamorphic chronology of the Limpopo Belt, southern Africa. Geol. Soc. Amer. Bull. 83, 2005–2018.

Visser D.J.L. (Compiler), 1984. Geological Map of the republics of South Africa, Transkei, Bophuthatswana, Venda and Ciskei and the Kingdoms of Lesotho and Swaziland. Scale 1:1,000,000. The Government Printer, Pretoria, 4 sheets.

Walraven, F., Armstrong, R.A., Kruger, F.J., 1990. A chronostratigraphic framework for the north-central Kaapvaal craton, the Bushveld complex and the Vredefort structure. Precambrian Res. 171, 23–48.

Watkeys, M.K., 1983. Brief explonatery notes on the provisional geological map of the Limpopo Belt and environs. Geol. Soc. S. Afr. Spec. Publ. 8, 5–8.

Wilson, J.F., 1979. A preliminary reappraisal of the Rodesian basement complex. Geol. Soc. S. Afr. Spec. Publ. 5, 1–23.

Wilson, J.F., Nesbitt, R.W., Fanning, S.M., 1995. Zircon geochronology of Archean felsic sequences in the Zimbabwe craton: a revision of greenstone stratigrphy and a model for crustal growth. In: Coward, M.p., Ries, A.C. (Eds.), Early Precambrian Processes. Geol. Soc. Lond. Spec. Publ., vol. 92, pp. 109–126.

Wingate, M.T.D., 2000. Ion microprobe U–Pb zircon and baddeleyite ages from the Great Dyke and its sattelite dykes, Zimbabwe. S. Afr. J. Geol. 103, 74–80.

Available online at www.sciencedirect.com

SCIENCE  DIRECT°

Lithos 71 (2003) 243–258

www.elsevier.com/locate/lithos

Regional patterns in the paragenesis and age of inclusions in diamond, diamond composition, and the lithospheric seismic structure of Southern Africa

Steven B. Shirey[a,*], Jeffrey W. Harris[b], Stephen H. Richardson[c], Matthew Fouch[d], David E. James[a], Pierre Cartigny[e], Peter Deines[f], Fanus Viljoen[g]

[a] Department of Terrestrial Magnetism, Carnegie Institution of Washington, 5241 Broad Branch Road, NW, Washington, DC 20015, USA
[b] Division of Earth Sciences, University of Glasgow, Glasgow G12 8QQ, UK
[c] Department of Geological Sciences, University of Cape Town, Rondebosch 7701, South Africa
[d] Department of Geological Sciences, Arizona State University, Box 871404, Tempe, AZ 85287-1404, USA
[e] Laboratoire de Geochemie des Isotopes Stables, Universite de Paris VII, Institut de Physique du Globe de Paris (IPGP),
4 Place Jussieu, 75251 Paris Cedex 05, France
[f] Department of Mineral Sciences, The Pennsylvania State University, University Park, PA, USA
[g] De Beers GeoScience Centre, PO Box 82232, Southdale, 2135, South Africa

Abstract

The Archean lithospheric mantle beneath the Kaapvaal–Zimbabwe craton of Southern Africa shows ± 1% variations in seismic P-wave velocity at depths within the diamond stability field (150–250 km) that correlate regionally with differences in the composition of diamonds and their syngenetic inclusions. Seismically slower mantle trends from the mantle below Swaziland to that below southeastern Botswana, roughly following the surface outcrop pattern of the Bushveld-Molopo Farms Complex. Seismically slower mantle also is evident under the southwestern side of the Zimbabwe craton below crust metamorphosed around 2 Ga. Individual eclogitic sulfide inclusions in diamonds from the Kimberley area kimberlites, Koffiefontein, Orapa, and Jwaneng have Re–Os isotopic ages that range from circa 2.9 Ga to the Proterozoic and show little correspondence with these lithospheric variations. However, silicate inclusions in diamonds and their host diamond compositions for the above kimberlites, Finsch, Jagersfontein, Roberts Victor, Premier, Venetia, and Letlhakane do show some regional relationship to the seismic velocity of the lithosphere. Mantle lithosphere with slower P-wave velocity correlates with a greater proportion of eclogitic versus peridotitic silicate inclusions in diamond, a greater incidence of younger Sm–Nd ages of silicate inclusions, a greater proportion of diamonds with lighter C isotopic composition, and a lower percentage of low-N diamonds whereas the converse is true for diamonds from higher velocity mantle. The oldest formation ages of diamonds indicate that the mantle keels which became continental nuclei were created by middle Archean (3.2–3.3 Ga) mantle depletion events with high degrees of melting and early harzburgite formation. The predominance of sulfide inclusions that are eclogitic in the 2.9 Ga age population links late Archean (2.9 Ga) subduction-accretion events involving an oceanic lithosphere component to craton stabilization. These events resulted in a widely distributed younger Archean generation of eclogitic diamonds in the lithospheric mantle. Subsequent Proterozoic tectonic and magmatic events

---

* Corresponding author. Fax: +1-202-478-8821.
  E-mail addresses: shirey@dtm.ciw.edu (S.B. Shirey), jwh@earthsci.gla.ac.uk (J.W. Harris), shr@geology.uct.ac.za (S.H. Richardson), fouch@asu.edu (M. Fouch), cartigny@ipgp.jussieu.fr (P. Cartigny), p7d@psu.edu (P. Deines).

0024-4937/$ - see front matter © 2003 Elsevier B.V. All rights reserved.
doi:10.1016/j.lithos.2003.07.007

altered the composition of the continental lithosphere and added new lherzolitic and eclogitic diamonds to the already extensive Archean diamond suite.

Keywords: Diamond; Eclogite; Peridotite; Inclusion; P-wave; Craton; Lithosphere

1. Introduction

A worldwide association between ancient lithospheric mantle keels beneath cratons and diamond occurrences has long been known (Boyd and Gurney, 1986; Janse, 1992). Seismic imaging of the lithospheric mantle beneath the Kaapvaal and Zimbabwe cratons and the Limpopo mobile belt known as the Southern Africa Seismic Experiment (SASE, James et al., 2001; James and Fouch, 2002) carried out during the multidisciplinary, multinational Kaapvaal Lithosphere Project (Carlson et al., 1996, 2000) has produced the first detailed picture of the lithospheric mantle at depths within the diamond stability field. It reveals prominent variations in the present seismic velocity structure of this mantle keel. Within the south African cratonic keel, Archean mantle peridotite and eclogite host multiple generations of diamonds that are both Archean and Proterozoic (Kramers, 1979; Richardson et al., 1984, 1993, 2001; Navon, 1999; Pearson and Shirey, 1999; Shirey et al., 2002) as well as less common occurrences of younger diamonds (Boyd et al., 1987; Akagi and Masuda, 1988; Navon et al., 1988; Schrauder and Navon, 1994; Pearson et al., 1998; Izraeli et al., 2001). Xenoliths of peridotite and eclogite and xenocrysts of diamond have been brought to the surface from depths as great as 150–180 km in kimberlites whose ages are typically young (65–150 Ma) but can be significantly older (240–1600 Ma; Davis et al., 1976; Davis, 1977; Smith et al., 1985). A major goal of Kaapvaal lithosphere studies is to relate current lithospheric structure to previous episodes of kimberlite eruption and to the distribution of Archean and Proterozoic peridotite, eclogite, and diamonds within the lithosphere. For the present study, we wish to put two decades of research on the geochronological age and composition of inclusion-bearing diamonds from southern Africa's major diamond deposits (Harris, 1992; Pearson and Shirey, 1999), comprising some 4000 individual diamond

specimens, into regional geologic context at diamond source depths. The goal is to uncover any connection between southern African diamonds and the composition of their mantle source. We are specifically looking for any regional control of lithospheric mantle seismic structure on the formation of diamond or whether geological processes associated with diamond formation produced the extant mantle seismic structure.

2. Methods and sources of data

2.1. Seismic data

Seismic velocities in the lithospheric mantle beneath the Kaapvaal–Zimbabwe cratons have been mapped using teleseismic broadband waveform data from the period 1997–1999, gathered during the SASE (James et al., 2001; James and Fouch, 2002). For constructing tomographic images of the lithosphere used in this study, P-wave data have been used because the available data set is nearly twice as large as that available for S-waves (James and Fouch, 2002). To produce quantitative P-wave anomaly data for each diamond source area, a 50-km radius cylinder of mantle extending from 150 to 225 km depth has been averaged. In the tomographic image and the data table, seismic velocity anomalies for P-waves are represented as % deviation of this average from a cratonic reference model.

2.2. Ages on inclusions in diamond

The compositions of silicate inclusions in diamond can be grouped according to their mineralogical similarity to eclogite or peridotite xenoliths in kimberlite. Diamonds are classified as peridotitic or 'P-type' (or U-type by some workers, e.g. Gurney and Switzer, 1973; Sobolev et al., 1998) when they contain Cr-pyrope, diopside, enstatite, chromite, or olivine (Meyer and Boyd, 1972; Harris and Gurney, 1979). P-type diamonds can be further subdivided into harzburgitic or lherzolitic by the degree of depletion indicated by the Cr_2O_3 and CaO content of their included garnet and when applicable, the composition of included olivine or orthopyroxene (e.g. Sobolev et al., 1973; Harris and Gurney, 1979). Diamonds with silicate inclusions are classified as eclogitic or 'E-type' when they contain pyrope-almandine garnet, omphacite, coesite, or kyanite.

Sulfide inclusions in diamonds are chiefly inter-growths of pyrrhotite and pentlandite with subordinate chalcopyrite that have exsolved from a monosulfide solid solution. A similar E versus P paragenetic distinction, as seen for silicate inclusions, has been applied to sulfide inclusions. P-type sulfides from the Siberian craton typically contain Ni contents greater than 22.8 wt.% whereas E-type sulfides typically have less than 8 wt.% (Yefimova et al., 1983). Such a clear break between E- and P- type sulfides has not yet been established for inclusions in Kaapvaal craton diamonds (Harris and Gurney, 1979; Deines and Harris, 1995) and with the advent of Re–Os isotopic studies on individual sulfides, it has been suggested that Os content of the sulfide is a more accurate discriminant (Pearson et al., 1998; Pearson and Shirey, 1999). Thus, for the present work, if bulk sulfide inclusions have high pentlandite content (e.g. Ni equivalent to >16 wt.% in monosulfide solid solution) and 2–30 ppm Os, they are referred to as P-type and if they have low pentlandite content (e.g. Ni equivalent to <9 wt.% in monosulfide solid solution) and less than 200–300 ppb Os, they are referred to as E-type.

Over the past two decades, ages of diamonds from the major mines have been derived from Rb–Sr and Sm–Nd studies of silicate inclusions by Richardson and coworkers (Richardson (1986); Richardson et al. (1984, 1990, 1993, 1999)). Due to the low concen-trations of Rb, Sr, Sm, and Nd in garnet and clinopyr-oxene and the small size of such inclusions, these studies added together many inclusions of similar mineral composition to produce isochrons that repre-sent many tens to hundreds of diamonds from any one kimberlite. Early attempts to date sulfide inclusions relied on rare large grains and the Pb–Pb isotope system (e.g. Kramers, 1979). But within the last 5 years, improvements in Re–Os analytical sensitivity

have made the dating of individual sulfide inclusions in diamonds routine (Pearson et al., 1998; Pearson and Shirey, 1999; Richardson et al., 2001).

Data used in this study come from this literature and the ongoing work of Shirey et al. (Orapa) and Richardson et al. (Jwaneng). It is important to recog-nize the large variability of inclusion suites between kimberlites and that there may be some bias in the total data set from sample selection. For example, while E-type silicates and E-type sulfides have clear eclogitic affinities, a connection to the eclogite host rocks in the lithosphere can best be established at kimberlites where both sulfide and silicate E-type inclusions (in separate diamonds) occur commonly (e.g. Jwaneng, Orapa, and Premier). An ideal situation would be to have silicate and sulfide E-type inclusions in the same diamond, but such diamonds with inclu-sions large enough for analysis are exceedingly rare. In addition, to get sulfides large enough for single grain analyses of Re and Os, the larger stones are routinely sought. The abundance of E-type diamonds is observed to increase with increasing diamond size for both sulfide and silicate inclusions (Gurney, 1989; Sobolev et al., 2001; Viljoen, unpublished data), thus, in seeking larger stones, there is a greater chance that they will be E-type. This may provide an explanation for why, with the exception of one P-type diamond from Koffiefontein, all sulfide inclusions analyzed to date from southern Africa have been E-type. All available Kaapvaal inclusion ages prior to 1999 have been summarized recently in Pearson and Shirey (1999) which also contains detailed discussion of inclusion syngenecity and paragenesis.

2.3. C isotopic composition

The carbon isotopic composition of diamonds from the major mines in southern Africa has been deter-mined over the last two decades by total combustion of inclusion-bearing diamond chips and analysis of the resultant CO_2 by Deines (1980), Deines et al. (1984, 1987, 1989, 1991a,b, 1993, 1995, 1997, 2001) and Cartigny (1998), Cartigny et al. (1998, 1999). Sample sizes for the chips in these studies ranged from 0.3 to 1.3 mg for the work of Deines et al. (e.g. 1984) and from 1 to 4 mg for that of Cartigny et al. (e.g. 1998). Based on the type of C isotopic variability of growth horizons seen in ion microprobe analyses of diamond plates (Hauri et al., 1999, 2002), bulk analyses of chips of this size suggest that any C isotopic variability associated with diamond growth zonation would have been mixed together in these isotopic analyses.

2.4. N abundance and aggregation

Fourier transform infrared spectroscopy (FTIR) studies of the abundance and crystal–chemical aggregation of trace nitrogen in the diamond lattice were also carried out on diamond chips from the same samples that were used for carbon isotopic study by Deines et al. (1984, 1987, 1989, 1991a, 1991b), Deines and Harris (1995), Cartigny (1998) and Cartigny et al. (1998, 1999). Diamonds with no detectable N are termed Type II and their percentage in a diamond population reflects low partitioning of N into diamonds during their growth in the mantle. Sufficiently old diamonds with sufficiently high N will show clustering or aggregation of N into optically active centers (see Wilks and Wilks, 1991, pp. 67–77 for a discussion of diamond types and optical centers based on N). Diamonds with enough N to detect optically (termed Type Ia diamonds) will typically progress from C-centers (Ib; single N) through A-centers (IaA; paired N) to B-centers (IaB; clusters of four N and a vacancy) in proportion with geological time (hundreds of Ma), temperature, and nitrogen content (Evans and Qi, 1982; Evans and Harris, 1989; Taylor et al., 1990; Navon, 1999). Of these three variables, N aggregation will be sensitive chiefly to temperature (Navon, 1999) but still is a function of N content. Thus, diamond populations with a lesser percentage of highly aggregated nitrogen (e.g. low % Type IaB) are more likely to have resided in cooler lithosphere whereas diamonds with a higher Type IaB

nitrogen aggregation state are more likely to have resided in warmer lithosphere. As mentioned above for C isotopic studies, the diamond chips analyzed by FTIR for N are large enough to include multiple diamond growth zones.

3. Results

3.1. Seismic velocity variations

Fig. 1, showing the tomographic inversion for P-wave data, presents a picture of the current lithospheric seismic velocity structure at a depth of 150 km. The Kaapvaal–Zimbabwe craton is marked by relatively high P-wave velocity lithospheric mantle that occurs in two prominent but irregularly shaped lobes separated by a broad west-northwest trending band of relatively lower velocity mantle. The seismic array coverage crosses the margin of the craton only in the southwest (Fig. 1). There, the craton boundary as mapped at the surface is marked by a change in the regional magnetic fabric (Ayres et al., 1998), an abrupt change in Re–Os model age of mantle xenoliths (Janney et al., 1999; Carlson et al., 2000; Irvine et al., 2001), and a sharp decrease of about 1% in seismic velocity. Very low mantle velocities are evident off-craton in the far southwest underneath the western Cape Foldbelt. Kimberlites distributed across southern Africa have diamonds that derive from mantle with large differences in seismic velocity (Table 1, Fig. 1). Jwaneng, Letlhakane, Orapa, and Premier diamonds were hosted in slower lithospheric mantle with relative P-wave velocity perturbations that vary from -0.209% to -0.006% whereas diamonds in the Roberts Victor, Jagersfontein, Finsch,

Fig. 1. Tomographic image of the lithospheric mantle derived from seismic P-wave data at a depth of 150 km (James et al., 2001; James and Fouch, 2002). The color scheme depicts % deviation from an average cratonic lithosphere velocity model. Areal coverage spans the lithospheric mantle of the Kaapvaal (K) and Zimbabwe (Z) cratons and the Limpopo mobile belt (L; see inset, left). Bold green line indicates the outermost boundary of the Archean cratons as defined by a sharp break between Archean and Proterozoic Re–Os ages on peridotite xenoliths (Carlson et al., 1999; Janney et al., 1999; Irvine et al., 2001). The location of diamond mines is shown by colored squares. Red squares are localities whose silicate inclusion in diamond suites are predominately eclogitic (Jagersfontein = JA, Jwaneng = JW, Letlhakane = LE, Orapa = O, Premier = P) and green squares are localities whose silicate inclusion suites are predominately peridotitic (Kimberley area mines of Wesselton, Bultfontein, and Dutoitspan termed De Beers Pool = D, Finsch = F, Koffiefontein = KO, Roberts Victor = R, and Venetia = V). Mines located above red-yellowish areas referred to in text as having diamonds derived from seismically slower mantle; mines located above greenish-blue areas referred to in text as having diamonds derived from seismically faster mantle. Reprinted with permission from Shirey et al. (2002). Diamond genesis, seismic structure, and evolution of the Kaapvaal–Zimbabwe craton. Science, 297: 1683–1686. © 2002 American Association for the Advancement of Science.

Venetia, Koffiefontein, and the Kimberley area kimberlites (Bultfontein, De Beers, Dutoitspan, Wesselton; known collectively as the *De Beers Pool*) were hosted in seismically faster mantle with relative P-wave velocity perturbations that vary from +0.084% to +0.357%.

3.2. Sulfide inclusions in diamond

Data on single E-type sulfides from the De Beers Pool, Jwaneng, Koffiefontein, and Orapa form a $^{187}Re/^{188}Os$ versus $^{187}Os/^{188}Os$ array (Fig. 2) that has a circa 2.9 Ga age slope (Pearson et al., 1998;

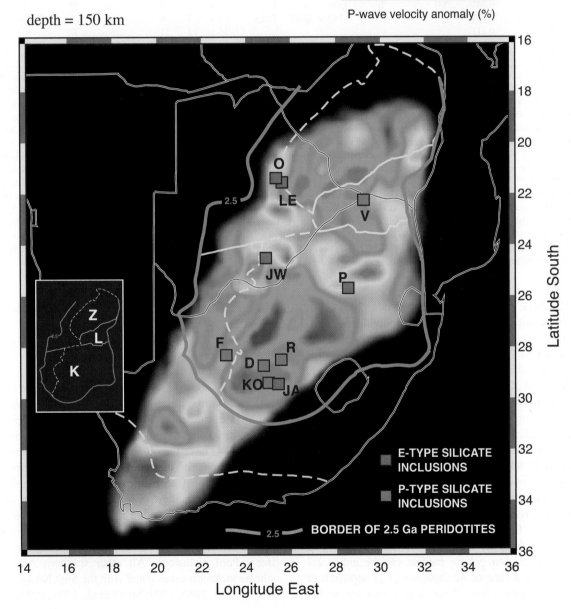

Table 1

Seismic velocity of the lithospheric mantle, inclusion Re–Os or Sm–Nd age (Ma), diamond carbon isotopic composition, nitrogen abundance and aggregation, and paragenesis compiled for Southern African diamonds and their localities

Location	Seismic velocity	Sulfide inclusion		Silicate inclusion		$\delta^{13}C$		N and N aggregation		
		Re–Os Age (Ma)	Para	Sm–Nd Age (Ma)	Para	Range (‰)	%< −9	N (ppm)	% Type Ia	Para
Jwaneng	−0.006	1500 to 2900	E	1540	E	−19 to −2	52	400	91	E,P
Letlhakane	−0.008							345	87	E,P
Orapa	−0.010	1000 to 2900	E	990	E	−26 to −3	65	478	94	E,P,(W)
Premier	−0.209			1150 to 1930	E,(P)	−12 to −2	1	413	90	E,P
Venetia	0.194					−18 to −2	11	259	77	P, (E,W)
De Beers Pool	0.245	2900	E	3200	P	−7 to −3	9	170	55	P,(E)
Finsch	0.084			1580 to 3200	P,(E)	−8 to −3	0	199	74	P,(E)
Roberts Victor	0.211					−16 to −3	71	260	73	P,(E)
Jagersfontein	0.357					−21 to −3	83	291	81	E,P
Koffiefontein	0.327	990 to 2900	E			−17 to −2	10	201	85	P,E

Average seismic velocity for P-waves in % deviation from a cratonic reference model for a 50 km radius cylinder of mantle extending from 150 to 225 km depth (James et al., 2001; James and Fouch, 2002). Parageneses (P = peridotitic, E = eclogitic, W = websteritic) are listed in relative order of abundance; subordinate parageneses are in parentheses. About 100 individual sulfide inclusions have been analyzed for Re–Os ages; about 3000 silicate inclusions (mostly composites but a few individual grains) have been analyzed for Sm–Nd ages. C isotopic and N aggregation studies have been carried out on diamonds enclosing silicate, oxide and sulfide inclusions; more than 900 individual diamonds have been studied. C isotopic composition is represented with the $\delta^{13}C$ notation in ‰ relative to the PDB standard. The percentage of diamonds with $\delta^{13}C$ less than −9‰ out of the total number of diamonds studied is shown under the column labeled '%< −9'. Nitrogen concentration is the average of the total diamond population as measured by Fourier transform infrared spectroscopy (FTIR); De Beers Pool data is by mass spectrometry. '%Type Ia' data are the percentage of diamonds in the studied population with aggregated nitrogen >20 ppm. Sources of data are from the literature as follows: De Beers Pool (Richardson et al., 1984, 2001; Cartigny, 1998; Cartigny et al., 1998); Finsch (Deines et al., 1984, 1989; Richardson et al., 1984, 1990; Smith et al., 1991); Jagersfontein (Deines et al., 1991a), Jwaneng (Cartigny et al., 1998; Richardson, unpublished data; Richardson et al., 1999); Koffiefontein (Deines et al., 1991a; Pearson et al., 1998); Orapa (Richardson et al., 1990; Shirey, unpublished data; Deines et al., 1991b, 1993); Premier (Milledge et al., 1983; Deines et al., 1984, 1989; Richardson, 1986; Richardson et al., 1993); Roberts Victor (Deines et al., 1987); Venetia (Deines et al., 2001; Viljoen, 2002). Note that the sulfide, silicate, and C–N studies were all carried out on separate suites of inclusion-bearing diamonds.

Richardson et al., 2001; Shirey et al., 2001). For each of these four inclusion suites individually, most of the inclusions have Re–Os systematics that approximate a 2.9 Ga age slope. De Beers Pool sulfide inclusions shows the tightest array whereas Jwaneng, Orapa, and Koffiefontein have a significant percentage of sulfide inclusions that plot at clearly younger ages ranging from 1 to 1.5 Ga. There is no obvious correspondence of sulfide inclusion age with lithospheric seismic velocity as both fast and slow mantle apparently hosted diamonds with circa 2.9 Ga ages. However, younger ages may be more frequent for diamonds from localities in slower mantle (e.g. Jwaneng for which more radiogenic data is not shown in Fig. 2). The circa 2.9 Ga age is not resolvable from the Re–Os model ages obtained on mantle peridotites whose median can be either 2.7 Ga if using a time of *Re depletion* (T_{rd}) approach or 3.1 Ga if using a time of *mantle reservoir separation*

(T_{ma}) approach (Carlson et al., 1999, 2000; Irvine et al., 2001).

3.3. Silicate inclusions in diamond

Sm–Nd isochron and mantle model ages for silicate inclusion suites from the Kaapvaal–Zimbabwe craton are shown in Fig. 3. P-type (harzburgitic) garnets from Finsch and De Beers Pool (Richardson et al., 1984) have the oldest model ages yet recorded for inclusions in diamond from southern Africa. Premier P-type (lherzolitic) garnets have a much younger isochron age (1.9 Ga, Richardson et al., 1993) that, with reasonable assumptions of protolith Sm/Nd (see Fig. 3, caption) would lead to a younger 3.0 Ga depleted mantle model age compared to the De Beers Pool inclusions. All other southern African silicate inclusion suites dated with the Sm–Nd system (Richardson, 1986, 1990; Smith et al., 1991, 1999) are

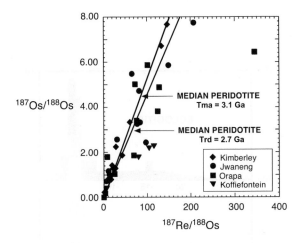

Fig. 2. Re–Os isotopic array for individual sulfide inclusions in single diamonds compared to typical Re–Os model ages on peridotites from the Kaapvaal–Zimbabwe craton. Figure modified after Shirey et al. (2001). Data sources are as follows: De Beers Pool (Richardson et al., 2001), Jwaneng (Richardson et al., unpublished), Koffiefontein (Pearson et al., 1998), and Orapa (Shirey et al., unpublished). Peridotite Re–Os model ages from Irvine et al. (2001) and Carlson et al. (1999).

E-types and yield Proterozoic isochron or model ages. P-type silicate inclusion model ages in the 3.2–3.3 Ga range overlap some of the older crustal U–Pb ages obtained (Moser et al., 2000; Schmitz, 2002), but Fig. 3 shows that they predate the Re–Os model ages of many peridotite xenoliths (Carlson et al., 1999; Irvine et al., 2001), sulfide inclusions in diamond (see Fig. 2), diamondiferous eclogite xenoliths (Shirey et al., 2001; Menzies et al., 2003), and the U–Pb ages of crustal metamorphism and plutonism (Schmitz, 2002). A common history for these diverse lithospheric components could be postulated from circa 2.9 Ga onwards.

Most diamondiferous kimberlites have diamond populations with silicate inclusions from both P-type and E-type parageneses (Table 1). It is whether the majority of the inclusions are eclogitic or peridotitic that shows a correspondence with slow versus fast lithospheric mantle (Fig. 1). Jwaneng, Letlhakane, Orapa, and Premier have inclusion-bearing diamond populations where E-type inclusions predominate; these kimberlites have penetrated seismically slower mantle. De Beers Pool, Finsch, Koffiefontein, Roberts Victor, and Venetia have inclusion-bearing diamond populations where P-type inclusions predom-

inate; these kimberlites have penetrated seismically fast mantle. Only Jagersfontein does not fit this pattern, having penetrated seismically faster mantle but containing an E-type dominant silicate inclusion suite. Jagersfontein diamonds may be exceptional because of the evident fertilization and metasomatism of Jagersfontein xenoliths (Haggerty, 1983; Winterburn et al., 1990; Pyle and Haggerty, 1998) and the presence of sublithospheric xenoliths (Haggerty and Sautter, 1990) and diamonds (Deines et al., 1991b).

3.4. C isotopic compositions

It has long been known (Sobolev et al., 1979) that diamonds of both P-type and E-type paragenesis show a prevalent mantle-like carbon isotopic composition ($\delta^{13}C = -3\%$ to -7%) with an isotopically light sub-population ($\delta^{13}C = -10\%$ to -34%) dominated

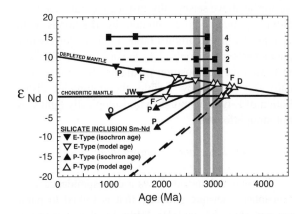

Fig. 3. Sm–Nd isochron and model ages of silicate inclusions in diamond (sources, Table 1 caption) compared to the ages of major crust forming events in the Kaapvaal craton (1; Schmitz, 2002), peridotite xenolith Re–Os model ages (2; sources, Fig. 2 caption), diamondiferous eclogite xenoliths (3; Shirey et al., 2001; Menzies et al., 2003), and sulfide inclusions in diamonds (4; sources, Fig. 2 caption). Growth curves from the isochron age (crystallization) of silicate inclusions to their reservoir separation from depleted mantle are estimated by assuming that the measured Sm/Nd of the clinopyroxene associated with lherzolitic garnet at Premier and the clinopyroxene associated with eclogitic garnet at Finsch, Jwaneng and Orapa represent a maximum (and perhaps typical) Sm/Nd of the protolith. These assumptions can be supported because most of the light REE in an eclogite will reside in clinopyroxene (e.g. Taylor et al., 1996) and lherzolitic garnet and associated clinopyroxene show regular REE patterns (e.g. non-sinusoidal; Stachel and Harris, 1997; Stachel et al., 1998).

by diamonds of E-type paragenesis (e.g. Gurney, 1989; Galimov, 1991; Kirkley et al., 1991). If the C isotopic composition of all southern African diamonds studied is broken down by either paragenesis or the seismic velocity of the source mantle, it can be seen that the distribution of C isotope composition for E-type diamonds (Fig. 4A) appears nearly identical to that for diamonds from the seismically slow lithospheric mantle (Fig. 4B). But if the eclogitic diamonds with a C isotopic composition less than $-9‰$ (e.g. those that comprise the isotopically light tail of the $\delta^{13}C$ distribution) are treated mine by mine (Table 1), it can be seen that the large number of specimens from Jwaneng and Orapa dominate the isotopically light population and its correlation with seismically slow mantle. Premier, Jagersfontein, and Roberts Victor also directly contraindicate such a correlation: Premier lies above seismically slow mantle but has no isotopically light diamonds in its E-type population whereas Jagersfontein and Roberts Victor lie above seismically fast mantle but have their E-type populations dominated by diamonds with isotopically light carbon (Table 1). Therefore, while paragenesis is the controlling factor in the differences in C isotopic composition of diamonds from seismically fast versus slow mantle, the cratonwide distribution of E-type diamonds with isotopically light C is not straightforward.

3.5. N abundance and N aggregation

Nitrogen abundance shows a correspondence with lithospheric seismic structure that is linked to paragenesis (Fig. 5A and B; Table 1), the statistically significant higher N content of E-type diamonds noted previously for Finsch, Jwaneng, Orapa, and Premier by Deines et al. (1989, 1993, 1997). Diamonds from slower lithospheric mantle (Jwaneng, Letlhakane, Orapa, and Premier localities) have a higher percentage of Type Ia diamonds and an average N content above 300 ppm to accompany their greater percentage of E-type inclusions (Fig. 5A and B). Except for Jagersfontein, diamonds from faster lithospheric mantle (Koffiefontein, Finsch, Roberts Victor, Venetia, and the De Beers Pool) have a lower percentage of Type Ia diamonds and an average N content below 300 ppm to accompany their greater percentage of P-type inclusions. The aggregation state

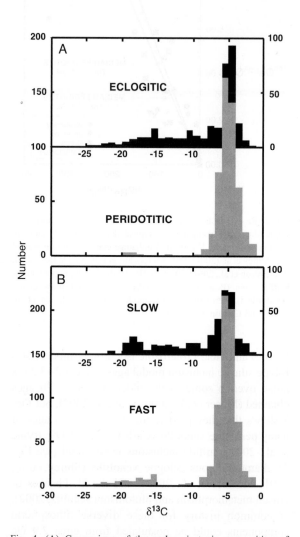

Fig. 4. (A) Comparison of the carbon isotopic composition of individual diamond analyses grouped according to E-type or P-type paragenesis. Note that both E-type and P-type histograms include diamonds from all nine localities. (B) Comparison of the carbon isotopic composition of southern African diamonds, grouped according to their derivation from a locality in seismically slower (Jwaneng, Letlhakane, Orapa, and Premier) or seismically faster lithospheric mantle (Venetia, De Beers Pool, Finsch, Roberts Victor, Jagersfontein and Koffiefontein). The similarities in the histograms in A and B occur because a greater proportion of diamonds of E-type paragenesis and isotopically light carbon derive from localities occurring in seismically slower lithospheric mantle. See text and Table 1 for sources of data. Reprinted with permission from Shirey et al. (2002). Diamond genesis, seismic structure, and evolution of the Kaapvaal–Zimbabwe craton. Science, 297: 1683–1686. © 2002 American Association for the Advancement of Science.

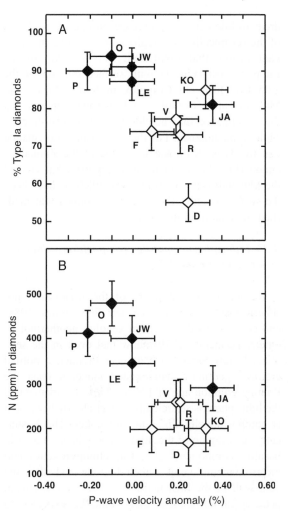

Fig. 5. Percentage of Type Ia diamonds (5A) and average N abundance in diamonds (5B) versus P-wave velocity anomaly. Error bars have been set at the ± 5 % level for percentage of Type Ia diamonds, ± 50 ppm for average N abundance, and ± 0.1 % for P-wave velocity anomaly. Type Ia diamonds are the total of IaA, IaA/B and IaB diamonds (Table 1). Same lettering scheme as in Fig. 1. Peridotitic = open symbols; eclogitic = closed symbols. Reprinted with permission from Shirey et al. (2002). Diamond genesis, seismic structure, and evolution of the Kaapvaal–Zimbabwe craton. Science, 297: 1683–1686. © 2002 American Association for the Advancement of Science.

of N in the Ia diamonds (not shown) displays no systematic variation with lithospheric seismic velocity although De Beers Pool, Finsch, and Roberts Victor diamonds from seismically fast mantle display the lowest percentages of aggregation to B-centers.

4. Discussion

4.1. Cratonic keel structure and diamond age distribution

With the advent of new age information on mantle lithologies and diamonds from the Re–Os system, it is interesting to re-evaluate the timing of diamond formation in the mantle beneath southern Africa and its relationship to seismic structure. There was a long-standing suspicion that cratonic mantle keels are old (Holmes and Paneth, 1936; Kramers, 1979). This initially was confirmed for the Kaapvaal craton by Richardson et al. (1984) using the Sm–Nd system on harzburgitic garnets included in diamonds from Finsch and the De Beers Pool that gave mid-Archean ages. Subsequent work of Richardson et al. (1986), Richardson et al. (1990, 1993, 1999) on diverse eclogitic and lherzolitic silicate inclusions in diamonds produced Proterozoic ages. Recent Re–Os data shows a preponderance of old eclogitic sulfide inclusion ages and cratonic peridotite mantle model ages of late Archean age (Carlson et al., 1999; Irvine et al., 2001; Shirey et al., 2001). These ages confirm that widespread diamond formation took place in a cratonic root that was being created or stabilized in the late Archean.

In detail, harzburgitic silicate inclusions in diamonds from the De Beers Pool and Finsch that give the oldest mid-Archean model ages (3.2–3.3 Ga; Richardson et al., 1984) currently have been found only above seismically faster lithospheric mantle. All other silicate inclusion suites dated so far are either lherzolitic or eclogitic, Proterozoic in age (Richardson, 1986, 1990, 1993, 1999) and occur above either seismically fast or slow mantle. Eclogitic sulfide inclusions from the four localities studied so far have all given late Archean ages (Pearson et al., 1998; Richardson et al., 2001; Shirey et al., 2001) as well as occasional Proterozoic ages that match the Proterozoic ages of the eclogitic silicate inclusions. Eclogitic sulfide inclusions are found above either seismically slow or fast mantle.

The connection between inclusion paragenesis and lithospheric seismic structure (Fig. 1) is clear only for silicate inclusion suites in diamonds. Most of these suites have not had age determinations because their diamonds were studied chiefly for C isotopic composition, N abundance and N aggregation (Table 1). It is noted that dated P-type silicate inclusions yield Archean (model) ages (Finsch, DeBeers Pool, Premier; Richardson et al., 1984, 1993, Fig. 3) and E-type silicate inclusions give Proterozoic model ages (Jwaneng, Orapa, Premier, Finsch; Richardson, 1986; Richardson et al., 1990, 1999). The best explanation for the complexities in the range and distribution of diamond ages and the distribution of diamond inclusion types lies in the uneven make-up of the overall diamond sample and the non-overlapping nature of the extant diamond studies in addition to the episodic nature by which the cratonic keel was generated, stabilized, and modified. The picture emerging from the diamond data is of a keel that has retained distinct populations of diamonds that were formed during each of the major tectonothermal episodes to have affected the cratonic keel (Shirey et al., 2002).

4.2. Age of the South African lithospheric seismic velocity structure

Several lines of evidence suggest that the current lithospheric seismic structure of the craton is a mid-Proterozoic overprint to this predominantly 2.9–3.3 Ga keel. Proterozoic E-type, silicate inclusion-bearing diamonds from seismically slow lithosphere (Figs. 1 and 3; Table 1) occur in the same kimberlites that contain Archean E-type sulfide inclusions (e.g. Orapa and Jwaneng) or peridotites with Archean Re–Os model ages (e.g. Letlhakane, Irvine et al., 2001). The Premier kimberlite, penetrating seismically slow mantle, not only contains peridotite xenoliths with Archean Re–Os model ages but also the clearest example of peridotite overprinted in the Proterozoic (Carlson et al., 1999; Irvine et al., 2001). Premier P-type silicate inclusions (lherzolitic garnet and clinopyroxene) that were equilibrated at the 1.9 Ga isochron age which represents the time of encapsulating diamond growth (Richardson et al., 1993) have Sm–Nd mantle model ages that are Archean (Fig. 3).

Modification of the cratonic root, which was originally thought to affect chiefly the Premier locality on the basis of Re–Os studies of lithospheric peridotite (Carlson et al., 1999; Irvine et al., 2001), is apparently more widespread than originally thought, extending across the northern Kaapvaal craton and to the south of the Zimbabwe craton–Limpopo Belt some hun-

dreds of kilometers from the craton edge. Correlation of the seismically slow regions of the northern Kaapvaal Craton that trend ESE–WNW south of the Limpopo belt (Fig. 1) with surface outcrop of the 2.05 Ga Bushveld Complex in South Africa and Molopo Farms Complex in Botswana suggests that the modification may be closely related to Bushveld–Molopo magmatism. For the seismically slow mantle that trends N–S on the west side of the Zimbabwe craton (Fig. 1), regional metamorphism that created the Magondi–Okwa terranes was likely to have been the surface manifestation of the tectonism that modified the craton on this margin.

4.3. Lithospheric mantle composition, diamond growth and storage

Recent thermal structure of the Kaapvaal lithospheric mantle from surface heat flow (Jones, 1988) and cratonic geotherms from xenolith geothermobarometry (Danchin, 1979) show that even in the seismically slow areas, such as near Premier, a normal cratonic geotherm has existed since at least the Premier eruption age of 1.2 Ga. This is evidence that the lithosphere is seismically slower chiefly because it is compositionally different, not hotter. The seismically slow region of the lithosphere is likely higher in basaltic components (Fe, Ca, clinopyroxene) and metasomatizing veins which hydrate and alter the vein wall of the host peridotite. These are the main petrological differences that would be expected to account for the 1% difference in P-wave velocity seen in Fig. 1.

Differences in silicate inclusion paragenesis and their diamond host composition correlate with the compositional differences recorded in the seismic structure of the lithosphere. Diamond suites from seismically slower lithosphere have a greater percentage of eclogitic inclusions which is in direct agreement with regions of the lithosphere that would have a higher proportion of basaltic components. Also, diamond suites from these regions have the eclogitic subset with isotopically light $\delta^{13}C$, the highest percentage of Type Ia stones and a higher average N content (>300 ppm). It is not clear whether higher average N incorporation during growth of these diamonds was due to a diamond-forming fluid with higher N content or due to faster growth. The lack of an obvious difference in N aggregation characteristics between diamonds from seismically slow and fast mantle (Table 1) indicates that diamonds from both types of mantle were stored in the lithosphere at hot enough temperatures (e.g. 1150 ± 50 °C) for long enough to allow substantial aggregation of N to B centers (Evans and Harris, 1989; Navon, 1999). In this case, the low IaB % of De Beers Pool, Finsch, and Roberts Victor diamonds (Table 1) would be related perhaps more to the lower average N content of these diamond populations than to temperature. Simple heating of the lithosphere in the seismically slow regions, as might be suggested by the resetting of U–Pb ages in low-closure-temperature minerals such as rutile found in lower crustal granulites elsewhere (Schmitz, 2002) is not recorded in the N aggregation data. This could be because any thermal pulse was too short-lived (Danchin, 1979) for substantial N aggregation to occur in a short period (Richardson and Harris, 1997; Navon, 1999). The current N aggregation data also are not detailed enough to resolve any systematic temperature or depth differences between E-type and P-type diamonds.

Diamonds and their inclusions examined on the regional scale of this study are surprisingly good at retaining evidence of lithospheric processes and craton-scale compositional effects. Craton-wide patterns in xenolith and megacryst suites that correlate with lithospheric seismic structure are subtler perhaps because silicate mineral assemblages not protected by encapsulating diamond continue to equilibrate and are subject to infiltrating metasomatic fluids. Some Premier peridotites, for example, have younger Re–Os model ages attributable to interaction with Bushveld age melts (Carlson et al., 1999) and the $Cr/Cr + Al$ of spinel megacrysts is lower (Schulze, 2001) in kimberlites from seismically slower lithosphere. These examples provide further evidence that the current physiochemical features that produce the observed seismic structure were created during Proterozoic modification of Archean lithosphere.

4.4. Creation, stabilization, and modification of lithospheric mantle

Constraints on lithosphere creation come from those diamonds that are Archean: the four suites with E-type sulfides that have been dated using the Re–Os system at around 2.9 Ga (Pearson and Shirey, 1999; Richardson et al., 2001, Richardson et al., unpublished data) and two suites with P-type silicates that have been dated using the Sm–Nd system at 3.2–3.3 Ga (Richardson et al., 1984). Until the advent of Re–Os analyses on sulfide inclusions, the prevalent model for lithosphere creation and stabilization involved degrees of melting high enough to create komatiite (e.g. Richardson et al., 1984; Walker et al., 1989; Boyd et al., 1999). Although this model failed to account for the high orthopyroxene (e.g. silica content) of the lithospheric mantle, it could simultaneously account for the high Mg# of peridotitic olivine, the bouyancy (and hence preservation) of the lithospheric mantle, its low heat production, its highly unradiogenic Os isotopic composition, and the harzburgitic composition of the 3.2–3.3 Ga garnet inclusions. Surprisingly, *none* of the Archean sulfide inclusions analyzed yet for Re–Os are P-type, they are all E-type. As seen in Figs. 2 and 3, typical ages do not cluster around the oldest P-type silicate age of 3.2–3.3 Ga (Richardson et al., 1984) or the dominant crust-forming age of 3.1 Ga (Moser et al., 2000; Schmitz, 2002) but have a typical age varying around a younger 2.9 Ga age. Nowhere is this age distinction clearer than with the De Beers Pool samples from the Kimberley area kimberlites. Here, 3.2–3.3 Ga P-type, depleted (harzburgitic) garnets coexist in the same lithospheric mantle section sampled by kimberlitic volcanism as 2.9 Ga E-type sulfides that have elevated, enriched initial Os isotopic composition.

All indications are that diamond formation in Archean cratonic mantle is episodic and such episodicity may apply to the formation and stabilization of the cratonic mantle itself. The occurrence of 3.2–3.3 Ga diamonds with depleted harzburgitic silicate inclusions (Sm–Nd, Rb–Sr model ages; Richardson et al., 1984) and 2.9 Ga diamonds with enriched eclogitic sulfide inclusions (Re–Os isochron age; Richardson et al., 2001) in the same Kimberley kimberlites (De Beers Pool) indicates that creation and assembly of the craton was *at least* a two-stage process (Shirey et al., 2002). Considering Re–Os, Sm–Nd, and Rb–Sr model age relationships, a time gap of 300 ± 200 million years is required between the two Archean diamond formation events recorded in De Beers Pool diamonds. Cratonic nuclei were first created by mantle-melting processes that produced severe depletion.

This could have occurred by vertical underplating (Haggerty, 1986) above a mantle plume (Arndt et al., 1997) although recent experimental and trace element work suggests that such depletion might have occurred initially at shallower depths (Canil and Wei, 1992; Stachel et al., 1998; Walter, 1998) or in a hot and wet subduction setting (Parman et al., 2001). Melting within a thickening volcanic pile led to crustal differentiation, the production of sial, and the first phase of crustal preservation. Preservation of the first-stage depleted mantle in the lithosphere may have been partial due to the magmatism and delamination required by the crustal differentiation process, an idea supported by the restricted distribution of 3.2–3.3 Ga harzburgitic silicate inclusions that thus far have only been documented from the Kimberley–Finsch area. It must be noted that the spatial distribution of extensively studied diamond-bearing kimberlites in southern Africa limit the above model for episodic creation and stabilization of the Kaapvaal–Zimbabwe craton chiefly to its western part. However, mid-Archean Re–Os model ages on lithospheric mantle peridotites, which have a slightly wider geographic distribution than the diamonds presently studied, are proposed similarly to reflect the existence of early depleted cratonic nuclei (Pearson et al., 2002).

The existence of these early cratonic nuclei provided a locus against which to accrete cratonic lithosphere created in the second step of the process (Shirey et al., 2002). In this phase, subduction of depleted oceanic harzburgite, (e.g. Helmstaedt and Schulze, 1989) which may have included the roots of Archean oceanic plateaus, built the rest of the lithospheric mantle. This second phase of lithosphere creation could account for widespread circa 2.9 Ga age Archean E-type sulfide inclusions in diamonds (Table 1; Shirey et al., 2001), their enriched initial Os isotopic composition (Richardson et al., 2001), the presence of enriched and depleted Archean inclusion chemistry in the same kimberlite, the younger age of these E-type diamonds compared to harzburgitic P-Type diamonds, the typical Re–Os model age for mantle peridotite (Carlson et al., 1999; Irvine et al., 2001), and the occurrence of 2.9 Ga diamondiferous eclogite with basaltic komatiitic compositions (Shirey et al., 2001; Menzies et al., 2003). This two-step lithosphere creation model also fits the detailed geochronological record for the lower and upper crust of the western Kaapvaal craton in which the earliest crustal components are formed from 3.20 to 3.26 Ga and the craton is sutured together by subduction convergence at 2.88–2.94 Ga (Schmitz, 2002).

Subduction for the western craton is supported by stable isotopic data from Orapa and Jwaneng. Orapa E-Type diamonds have light C and heavy N isotopic compositions (Cartigny et al., 1999) and sulfide inclusions with mass-independent sulfur isotopic fractionations (Farquhar et al., 2002) that are consistent with incorporation of C, N, and S from surficial sedimentary endmember reservoirs (Navon, 1999; Farquhar et al., 2002). Also, some Archean Jwaneng megacrystic zircon has low O isotope compositions suggesting that the zircon host lithologies were once hydrothermally altered as oceanic crust (Valley et al., 1998). Subduction is likely not to be the only source of C and N isotopic variability, however, because of the difficulties of finding subducted materials with appropriately high C/N ratios (e.g. Cartigny et al., 1998, 1999). Furthermore, intra-mantle processes (e.g. Cartigny et al., 2001) that must have been changing C and N isotopic composition of fluids during diamond growth are required by the isotopic composition differences observed in the growth horizons of some diamonds (Hauri et al., 1999; Hauri et al., 2002). The lack of complete C and N isotopic data sets from all but a few of south Africa's diamond suites means that currently it is not possible to discuss a cratonwide picture of intra-mantle isotopic fractionation of C and N in diamonds.

The near one-to-one correspondence of Proterozoic diamond suites having a majority of E-type silicate inclusions with seismically slow mantle suggests that craton modification and Proterozoic diamond formation were part of the same process (Shirey et al., 2002). Proterozoic craton modification that reduced the seismic velocity of the lithosphere did not apparently involve the addition of new mantle from the asthenosphere to the lithosphere. This is indicated by a lack of dominant Proterozoic Re–Os model ages on peridotites from the seismically slower portions of the lithosphere. Since the paragenesis of the Proterozoic silicate inclusions is chiefly E-type, it is likely that this process involved basaltic components generated in conjunction with Bushveld-Molopo Farms magmatism under the center of the northern Kaapvaal craton or in conjunction with some form of subduction along the western Kaapvaal–Zimbabwe craton margin. Both sublithospheric magmatism and western craton margin subduction were tectonothermal events that altered the composition of the lithosphere and added new diamonds to an already extensive Archean diamond population resident in the lithosphere.

5. Conclusions

Regional patterns of diamond composition, inclusion age, and paragenesis were derived from a data set comprising measurements on some 4000 individual stones or their inclusions. When these patterns are compared to the seismic velocity structure of the lithosphere in the diamond stability field, they lead to new insights on the creation of the lithospheric seismic velocity anomaly, the formation of various diamond generations, and the formation of continental lithosphere itself. Mantle lithosphere with slow P-wave velocity correlates with a greater proportion of eclogitic versus peridotitic silicate inclusions in diamond, a greater range in Sm–Nd age of silicate inclusions, and diamond suites that have a greater proportion of lighter C isotopic composition diamonds, a higher average N content, and more Type Ia diamonds. Mantle lithosphere with high P-wave velocity is typified by a higher proportion of peridotitic versus eclogitic silicate inclusions, the oldest Sm–Nd model ages, and diamond suites that have a lower average N content and fewer Type Ia diamonds.

The oldest formation ages of diamonds indicate that the mantle keels which became continental nuclei were created by middle Archean (3.2–3.3 Ga) mantle depletion events with high degrees of melting and early harzburgite formation. The predominance of sulfide inclusions that are eclogitic in the 2.9 Ga age population links late Archean subduction-accretion events involving an oceanic lithosphere component to craton stabilization which resulted in a widely distributed, late-Archean generation of eclogitic diamonds. Subsequent Proterozoic tectonic and magmatic events altered the composition of the continental lithosphere, produced the seismic velocity variations, and added new lherzolitic and eclogitic diamonds to the already extensive Archean diamond suite.

Tests of these ideas will come with future studies, including geographically more complete, overlapping data sets from individual mines, studies of sulfide inclusion composition (Fe, Ni, and Cu content) and C and N isotopic and FTIR studies on individual diamonds that have been dated with the Re–Os system. These need to be pursued in order to establish the relative importance of intra-mantle processes and regional variability on the isotopic composition of diamonds and the ultimate source of diamond-forming fluids.

Acknowledgements

Discussions with D. Bell, R.W. Carlson, I. Chinn, D.G. Pearson, and K. Westerlund during the preparation are greatly appreciated as are reviews of K. Burke, F. R. Boyd, R.W. Carlson, E. H. Hauri, and M. D. Schmitz on earlier versions of these ideas. The authors appreciate the constructive criticism of H. Grütter, O. Navon, and D.G. Pearson. The authors and all researchers working on inclusions in diamond are grateful to V. Anderson, E. van Blerk, R. Ferraris, R. Hamman, W. Moore, A. Ntidisang, G. Parker, and others at Harry Oppenheimer House, Kimberley for their skill in selecting specimens and to the De Beers Diamond Trading Company for making them available. M. Horan and T. Mock are thanked for their help in the DTM chemistry and mass spectrometry labs, respectively. This work was supported chiefly by the NSF EAR Continental Dynamics Grant 9526840.

References

Akagi, T., Masuda, A., 1988. Isotopic and elemental evidence for a relationship between kimberlite and Zaire cubic diamonds. Nature 336, 665–667.

Arndt, N.T., Kerr, A.C., Tarney, J., 1997. Dynamic melting in plume heads; the formation of Gorgona komatiites and basalts. Earth and Planetary Science Letters 146, 289–301.

Ayres, N.P., Hatton, C.J., Quadling, K.E., Smith, C.B., 1998. An update on the distribution in time and space of southern African kimberlites, 1:5,000,000 scale maps. DeBeers GeoScience Centre, Southdale (South Africa).

Boyd, F.R., Gurney, J.J., 1986. Diamonds and the African lithosphere. Science 232, 472–477.

Boyd, S.R., Mattey, D.P., Pillinger, C.T., Milledge, H.J., Mendelssohn, M., Seal, M., 1987. Multiple growth events during diamond genesis; an integrated study of carbon and nitrogen isotopes and nitrogen aggregation state in coated stones. Earth and Planetary Science Letters 86, 341–353.

Boyd, F.R., Pearson, D.G., Mertzman, S.A., 1999. Spinel-facies peridotites from the Kaapvaal root. In: Gurney, J.J., Gurney, J.L., Pascoe, M.D., Richardson, S.H. (Eds.), The J.B. Dawson Volume—Proceedings of the Seventh International Kimberlite Conference, Cape Town. Red Roof Design, Cape Town, pp. 40–48.

Canil, D., Wei, K., 1992. Constraints on the origin of mantle-derived low-Ca garnets. Contributions to Mineralogy and Petrology 109, 421–430.

Carlson, R.W., Grove, T.L., de Wit, M.J., Gurney, J.J., 1996. Program to study crust and mantle of the Archean craton in southern Africa. EOS, Transactions of the American Geophysical Union 77, 273–277.

Carlson, R.W., Pearson, D.G., Boyd, F.R., Shirey, S.B., Irvine, G., Menzies, A.H., Gurney, J.J., 1999. Re–Os systematics of lithospheric peridotites: implications for lithosphere formation and preservation. In: Gurney, J.J., Gurney, J.L., Pascoe, M.D., Richardson, S.H. (Eds.), The J.B. Dawson Volume—Proceedings of the Seventh International Kimberlite Conference, Cape Town. Red Roof Design, Cape Town, pp. 99–108.

Carlson, R.W., Boyd, F.R., Shirey, S.B., Janney, P.E., Grove, T.L., Bowring, S.A., Schmitz, M.D., Dann, J.C., Bell, D.R., Gurney, J.J., Richardson, S.H., Tredoux, M., Menzies, A.H., Pearson, D.G., Hart, R.J., Wilson, A.H., Moser, D., 2000. Continental growth, preservation, and modification in southern Africa. GSA Today 10, 1–7.

Cartigny, P., 1998. Carbon isotopes in diamond. PhD Thesis, Universite de Paris VII.

Cartigny, P., Harris, J.W., Javoy, M., 1998. Eclogitic diamond formation at Jwaneng; no room for a recycled component. Science 280, 1421–1424.

Cartigny, P., Harris, J.W., Javoy, M., 1999. Eclogitic, peridotitic and metamorphic diamonds and the problems of carbon recycling—the case of Orapa (Botswana). In: Gurney, J.J., Gurney, J.L., Pascoe, M.D., Richardson, S.H. (Eds.), The J.B. Dawson Volume—Proceedings of the Seventh International Kimberlite Conference, Cape Town. Red Roof Design, Cape Town, pp. 117–124.

Cartigny, P., Harris, J.W., Javoy, M., 2001. Diamond genesis, mantle fractionations and mantle nitrogen content: a study of $\delta^{13}C$–N concentrations in diamonds. Earth and Planetary Science Letters 185, 85–98.

Danchin, R.V., 1979. Mineral and bulk chemistry of garnet lherzolite and garnet harzburgite xenoliths from the Premier Mine, South Africa. In: Boyd, F.R., Meyer, H.O.A. (Eds.), The Mantle Sample: Inclusions in Kimberlites and Other Volcanics, Proceedings of the Second International Kimberlite Conference, Santa Fe. American Geophysical Union, Washington, pp. 104–126.

Davis, G.L., 1977. The ages and uranium contents of zircons from kimberlites and associated rocks. Year Book Carnegie Institution of Washington 76, 631–635.

Davis, G.L., Krogh, T.E., Erlank, A.J., 1976. The ages of zircons from kimberlites from South Africa. Year Book Carnegie Institution of Washington 75, 821–824.

Deines, P., 1980. The carbon isotopic composition of diamonds; relationship to diamond shape, color, occurrence and vapor composition. Geochimica et Cosmochimica Acta 44, 943–962.

Deines, P., Harris, J.W., 1995. Sulfide inclusion chemistry and carbon isotopes of African diamonds. Geochimica et Cosmochimica Acta 59, 3173–3188.

Deines, P., Gurney, J.J., Harris, J.W., 1984. Associated chemical and carbon isotopic composition variations in diamonds from Finsch and Premier kimberlite, South Africa. Geochimica et Cosmochimica Acta 48, 325–342.

Deines, P., Harris, J.W., Gurney, J.J., 1987. Carbon isotopic composition, nitrogen content and inclusion composition of diamonds from the Roberts Victor Kimberlite, South Africa; evidence for $\delta^{13}C$ depletion in the mantle. Geochimica et Cosmochimica Acta 51, 1227–1243.

Deines, P., Harris, J.W., Spear, P.M., Gurney, J.J., 1989. Nitrogen and ^{13}C content of Finsch and Premier diamonds and their implications. Geochimica et Cosmochimica Acta 53, 1367–1378.

Deines, P., Harris, J.W., Gurney, J.J., 1991a. The carbon isotopic composition and nitrogen content of lithospheric and asthenospheric diamonds from the Jagersfontein and Koffiefontein Kimberlite, South Africa. Geochimica et Cosmochimica Acta 55, 2615–2625.

Deines, P., Harris, J.W., Robinson, D.N., Gurney, J.J., Shee, S.R., 1991b. Carbon and oxygen isotope variations in diamond and graphite eclogites from Orapa, Botswana, and the nitrogen content of their diamonds. Geochimica et Cosmochimica Acta 55, 515–524.

Deines, P., Harris, J.W., Gurney, J.J., 1993. Depth-related carbon isotope and nitrogen concentration variability in the mantle below the Orapa Kimberlite, Botswana, Africa. Geochimica et Cosmochimica Acta 57, 2781–2796.

Deines, P., Harris, J.W., Gurney, J.J., 1997. Carbon isotope ratios, nitrogen content and aggregation state, and inclusion chemistry of diamonds from Jwaneng, Botswana. Geochimica et Cosmochimica Acta 61, 3993–4005.

Deines, P., Viljoen, F., Harris, J.W., 2001. Implication of the carbon isotope and mineral inclusion record for the formation of diamonds in the mantle underlying a mobile belt: Venetia, South Africa. Geochimica et Cosmochimica Acta 65, 813–838.

Evans, T., Harris, J.W., 1989. Nitrogen aggregation, inclusion equilibration temperatures and the age of diamonds. In: Ross, J., Jaques, A., Ferguson, J., Green, D., O'Reilly, S., Danchin, R., Janse, A. (Eds.), Kimberlites and Related Rocks—Proceedings of the Fourth International Kimberlite Conference, Perth. Blackwells, Melbourne, pp. 1001–1006.

Evans, T., Qi, Z., 1982. The kinetics of the aggregation of nitrogen atoms in diamond. Proceedings of the Royal Society of London, Series A: Mathematical and Physical Sciences 381, pp. 159–178.

Farquhar, J., Wing, B.A., McKeegan, Harris, J.W., Cartigny, P., Thiemens, M., 2002. Mass-independent sulfur in inclusions in diamond and sulfur recycling on early Earth. Science 298, 2369–2372.

Galimov, E.M., 1991. Isotope fractionation related to kimberlite magmatism and diamond formation. Geochimica et Cosmochimica Acta 55, 1697–1708.

Gurney, J.J., 1989. Diamonds. In: Ross, J., Jaques, A., Ferguson, J., Green, D., O'Reilly, S., Danchin, R., Janse, A. (Eds.), Kimberlites and Related Rocks—Proceedings of the Fourth International Kimberlite Conference, Perth. Blackwells, Melbourne, pp. 935–965.

Gurney, J.J., Switzer, G.S., 1973. The Discovery of Garnets Closely Related to Diamonds in the Finsch Pipe, South Africa. Contributions to Mineralogy and Petrology 39, 103–116.

Haggerty, S.E., 1983. The mineral chemistry of new titanates from the Jagersfontein kimberlite, South Africa; implications for metasomatism in the upper mantle. Geochimica et Cosmochimica Acta 47, 1833–1854.

Haggerty, S.E., 1986. Diamond genesis in a multiply-constrained model. Nature 320, 34–38.

Haggerty, S.E., Sautter, V., 1990. Ultradeep (greater than 300 kilometers), ultramafic upper mantle xenoliths. Science 248, 993–996.

Harris, J.W., 1992. Diamond geology. In: Field, J.E. (Ed.), The Properties of Natural and Synthetic Diamond. Academic Press, New York, pp. 345–393.

Harris, J.W., Gurney, J.J., 1979. Inclusions in diamond. In: Field, J.E. (Ed.), The Properties of Diamond. Academic Press, New York, pp. 555–591.

Hauri, E.H., Pearson, D.G., Bulanova, G.P., Milledge, H.J., 1999. Microscale variations in C and N isotopes within mantle diamonds revealed by SIMS. In: Gurney, J.J., Gurney, J.L., Pascoe, M.D., Richardson, S.H. (Eds.), The J.B. Dawson Volume—Proceedings of the Seventh International Kimberlite Conference, Cape Town. Red Roof Design, Cape Town, pp. 341–347.

Hauri, E.H., Wang, J., Pearson, D.G., Bulanova, G.P., 2002. Microanalysis of $\delta^{13}C$, $\delta^{15}N$, and N abundances in diamonds by secondary ion mass spectrometry. Chemical Geology 185, 149–163.

Helmstaedt, H., Schulze, D.J., 1989. Southern African kimberlites and their mantle sample: implications for Archean tectonics and lithosphere evolution. In: Ross, J., Jaques, A., Ferguson, J., Green, D., O'Reilly, S., Danchin, R., Janse, A. (Eds.), Kimberlites and Related Rocks—Proceedings of the Fourth International Kimberlite Conference, Perth. Blackwells, Melbourne, pp. 358–368.

Holmes, A., Paneth, F.A., 1936. Helium-ratios of rocks and minerals from the diamond pipes of South Africa. Proceedings of the Royal Society of London, Series A: Mathematical and Physical Sciences, vol. 154, pp. 385–413.

Irvine, G.J., Pearson, D.G., Carlson, R.W., 2001. Lithospheric mantle evolution of the Kaapvaal craton: a Re–Os isotope study of peridotite xenoliths from Lesotho kimberlites. Geophysical Research Letters 28, 2505–2508.

Izraeli, E.S., Harris, J.W., Navon, O., 2001. Brine inclusions in diamonds; a new upper mantle fluid. Earth and Planetary Science Letters 187, 323–332.

James, D.E., Fouch, M.J., VanDecar, J.C., van der Lee, S., Kaapval Seismic, 2001. Tectospheric structure beneath southern Africa. Geophysical Research Letters 28, 2485–2488.

James, D.E., Fouch, M.J., 2002. Formation and evolution of Archean cratons: insights from Southern Africa. In: Ebinger, C., Fowler, M., Hawkesworth, C.J. (Eds.), The Early Earth: Physical, Chemical and Biological Development. Geological Society of London, London, Special Publication 199, pp. 1–26.

Janney, P.E., Carlson, R.W., Shirey, S.B., Bell, D.R., le Roex, A.P.,

1999. Temperature, pressure, and rhenium–osmium age systematics of off-craton peridotite xenoliths from the Namaqua-Natal Belt, western Southern Africa. Ninth Annual V.M. Goldschmidt Conference. Lunar and Planetary Institute, Houston, p. 139.

Janse, A.J.A., 1992. New ideas in subdividing cratonic areas. Russian Geology and Geophysics 33, 9–25.

Jones, M.Q.W., 1988. Heat flow in the Witwatersrand Basin and environs and its significance for the south African shield geotherm and lithosphere thickness. Journal of Geophysical Research 93, 3243–3260.

Kirkley, M.B., Gurney, J.J., Levinson, A.A., 1991. Age, origin, and emplacement of diamonds; scientific advances in the last decade. Gems and Gemology 27, 2–25.

Kramers, J.D., 1979. Lead, uranium, strontium, potassium and rubidium in inclusion-bearing diamonds and mantle-derived xenoliths from southern Africa. Earth and Planetary Science Letters 42, 58–70.

Menzies, A.H., Carlson, R.W., Shirey, S.B., Gurney, J.J., 2003. Re–Os systematics of diamond-bearing eclogites from the Newlands kimberlite. In: Carlson, R.W., Grutter, H., Jones, A. (Eds.), Tale of two cratons: the Slave-Kaapvaal Workshop. Lithos, vol. 71, pp. 323–336. this issue.

Meyer, H.O.A., Boyd, F.R., 1972. Composition and origin of crystalline inclusions in natural diamonds. Geochimica et Cosmochimica Acta 36, 1255–1273.

Milledge, H.J., Mendelssohn, M.J., Seal, M., Rouse, J.E., Swart, P.K., Pillinger, P.T., 1983. Carbon isotopic variation in spectral type II diamonds. Nature 303, 791–792.

Moser, D.E., Flowers, R., Hart, R.J., 2000. Birth of the Kaapvaal lithosphere at 3.08 Ga; implications for the growth of ancient continents. Science 291, 465–468.

Navon, O., 1999. Diamond formation in the Earth's mantle. In: Gurney, J.J., Gurney, J.L., Pascoe, M.D., Richardson, S.H. (Eds.), The P.H. Nixon Volume—Proceedings of the Seventh International Kimberlite Conference, Cape Town. Red Roof Design, Cape Town, pp. 584–604.

Navon, O., Hutcheon, I.D., Rossman, G.R., Wasserburg, G.J., 1988. Mantle-derived fluids in diamond micro-inclusions. Nature 335, 784–789.

Parman, S.W., Dann, J.C., Grove, T.L., 2001. The production of Barberton komatiites in an Archean subduction zone. Geophysical Research Letters 28, 2513–2516.

Pearson, D.G., Shirey, S.B., 1999. Isotopic dating of diamonds. In: Lambert, D.D., Ruiz, J. (Eds.), Application of Radiogenic Isotopes to Ore Deposit Research and Exploration. Reviews in Economic Geology. Society of Economic Geologists, Denver, pp. 143–172.

Pearson, D.G., Shirey, S.B., Harris, J.W., Carlson, R.W., 1998. Sulfide inclusions in diamonds from the Koffiefontein kimberlite, S. Africa: Constraints on diamond ages and mantle Re–Os systematics. Earth and Planetary Science Letters 160, 311–326.

Pearson, D.G., Irvine, G.J., Carlson, R.W., Kopylova, M.G., Ionov, D.A., 2002. The development of lithospheric mantle keels beneath the earliest continents: time constraints using PGE and Re–Os isotope systematics. In: Ebinger, C., Fowler, M., Hawkesworth, C.J. (Eds.), The Early Earth: Physical, Chemical and

Biological Development. Geological Society of London, London, Special Publication, pp. 665–690.

Pyle, J.M., Haggerty, S.E., 1998. Eclogites and the metasomatism of eclogites from the Jagersfontein Kimberlite; punctuated transport and implications for alkali magmatism. Geochimica et Cosmochimica Acta 62, 1207–1231.

Richardson, S.H., 1986. Latter-day origin of diamonds of eclogitic paragenesis. Nature 322, 623–626.

Richardson, S.H., Harris, J.W., 1997. Antiquity of peridotitic diamonds from the Siberian craton. Earth and Planetary Science Letters 151, 271–277.

Richardson, S.H., Gurney, J.J., Erlank, A.J., Harris, J.W., 1984. Origin of diamonds in old enriched mantle. Nature 310, 198–202.

Richardson, S.H., Erlank, A.J., Harris, J.W., Hart, S.R., 1990. Eclogitic diamonds of Proterozoic age from Cretaceous kimberlites. Nature 346, 54–56.

Richardson, S.H., Harris, J.W., Gurney, J.J., 1993. Three generations of diamonds from old continental mantle. Nature 366, 256–258.

Richardson, S.H., Chinn, I.L., Harris, J.W., 1999. Age and origin of eclogitic diamonds from the Jwaneng kimberlite, Botswana. In: Gurney, J.J., Gurney, J.L., Pascoe, M.D., Richardson, S.H. (Eds.), The P.H. Nixon Volume—Proceedings of the Seventh International Kimberlite Conference, Cape Town. Red Roof Design, Cape Town, pp. 734–736.

Richardson, S.H., Shirey, S.B., Harris, J.W., Carlson, R.W., 2001. Archean subduction recorded by Re–Os isotopes in eclogitic sulfide inclusions in Kimberley diamonds. Earth and Planetary Science Letters 191, 257–266.

Schmitz, M.D., 2002. Geology and thermochronology of the lower crust of southern Africa. PhD Thesis, Massachusetts Institute of Technology, Cambridge, 269 pp.

Schrauder, M., Navon, O., 1994. Hydrous and carbonatitic mantle fluids in fibrous diamonds from Jwaneng, Botswana. Geochimica et Cosmochimica Acta 58, 761–771.

Schulze, D.J., 2001. Origins of chromian and aluminous spinel macrocrysts from kimberlites in southern Africa. Canadian Mineralogist 39, 361–376.

Shirey, S.B., Carlson, R.W., Richardson, S.H., Menzies, A.H., Gurney, J.J., Pearson, D.G., Harris, J.W., Wiechert, U., 2001. Archean emplacement of eclogitic components into the lithospheric mantle during formation of the Kaapvaal craton. Geophysical Research Letters 28, 2509–2512.

Shirey, S.B., Harris, J.W., Richardson, S.H., Fouch, M.J., James, D.E., Cartigny, P., Deines, P., Viljoen, F., 2002. Diamond genesis, seismic structure, and evolution of the Kaapvaal–Zimbabwe craton. Science 297, 1683–1686.

Smith, C.B., Allsopp, H.L., Kramers, J.D., Hutchinson, G., Roddick, J.C., 1985. Emplacement ages of Jurassic-Cretaceous South African kimberlites by the Rb–Sr method on phlogopite and whole-rock samples. Transactions of the Geological Society of South Africa 88, 249–266.

Smith, C.B., Gurney, J.J., Harris, J.W., Robinson, D.N., Kirkley, M.B., Jagoutz, E., 1991. Neodymium and strontium isotope systematics of eclogite and websterite paragenesis inclusions from single diamonds. Geochimica et Cosmochimica Acta 55, 2579–2590.

Sobolev, N.V., Laurent'ev Yu, G., Pokhilenko, N.P., 1973. Chrome-rich garnets from the kimberlites of Yakutia and their paragenesis. Contributions to Mineralogy and Petrology 40, 39–52.

Sobolev, N.V., Galimov, E.M., Ivanovskaya, I.N., Yefimova, E.S., 1979. The carbon isotope composition of diamonds containing crystalline inclusions. Doklady Akademy Nauk USSR 189, 133–136.

Sobolev, N.V., Yefimova, E.S., Channer, D.M.D., Anderson, P.F.N., Barron, K.M., 1998. Unusual upper mantle beneath Guaniamo, Guyana Shield, Venezuela: evidence from diamond inclusions. Geology 26, 971–974.

Sobolev, N.V., Yefimova, E.S., Loginova, A.M., Sukhodol'skaya, O.V., Solodova, Y.P., 2001. Abundance and composition of mineral inclusions in large diamonds from Yakutia. Doklady Earth Sciences 376, 34–38.

Stachel, T., Harris, J.W., 1997. Diamond precipitation and mantle metasomatism; evidence from the trace element chemistry of silicate inclusions in diamonds from Akwatia, Ghana. Contributions to Mineralogy and Petrology 129, 143–154.

Stachel, T., Viljoen, K.S., Brey, G., Harris, J.W., 1998. Metasomatic processes in lherzolitic and harzburgitic domains of diamondiferous lithospheric mantle; REE in garnets from xenoliths and inclusions in diamonds. Earth and Planetary Science Letters 159, 1–12.

Taylor, W.R., Jaques, A.L., Ridd, M., 1990. Nitrogen-defect aggregation characteristics of some Australasian diamonds: Time–temperature constraints on the source regions of pipe and alluvial diamonds. American Mineralogist 75, 1290–1310.

Taylor, L.A., Snyder, G.A., Crozaz, G., Sobolev, V.N., Yefimova, E.S., Sobolev, N.V., 1996. Eclogitic inclusions in diamonds; evidence of complex mantle processes over time. Earth and Planetary Science Letters 142, 535–551.

Valley, J.W., Kinny, P.D., Schulze, D.J., Spicuzza, M.J., 1998. Zircon megacrysts from kimberlite: oxygen isotope variability among mantle melts. Contributions to Mineralogy and Petrology 133, 1–11.

Viljoen, F., 2002. An infrared investigation of inclusion-bearing diamonds from the Venetia kimberlite, Northern Province, South Africa: implications for diamonds from craton-margin settings. Contributions to Mineralogy and Petrology 144, 98–108.

Walker, R.J., Carlson, R.W., Shirey, S.B., Boyd, F.R., 1989. Os, Sr, Nd, and Pb isotope systematics of Southern African peridotite xenoliths; implications for the chemical evolution of subcontinental mantle. Geochimica et Cosmochimica Acta 53, 1583–1595.

Walter, M.J., 1998. Melting of garnet peridotite and the origin of komatiite and depleted lithosphere. Journal of Petrology 39, 29–60.

Wilks, E., Wilks, J., 1991. Properties and Applications of Diamond. Butterworth-Heinemann, Boston. 525 pp.

Winterburn, P.A., Harte, B., Gurney, J.J., 1990. Peridotite xenoliths from the Jagersfontein kimberlite pipe; I, Primary and primary-metasomatic mineralogy. Geochimica et Cosmochimica Acta 54, 329–341.

Yefimova, E.S., Sobolev, N.V., Pospelova, L.N., 1983. Sulfide inclusions in diamonds and specific features of their paragenesis. Zapiski Vsesoyuznogo Mineralogicheskogo Obshchestva 112, 300–310.

Available online at www.sciencedirect.com

SCIENCE @ DIRECT°

Lithos 71 (2003) 259–272

LITHOS

www.elsevier.com/locate/lithos

Samples of Proterozoic iron-enriched mantle from the Premier kimberlite

K.O. Hoal*

Hazen Research, Inc., 4601 Indiana Street, Golden, CO 80401, USA

Abstract

Two populations of mantle xenoliths from the Proterozoic Premier kimberlite show an absence of potassic metasomatism common in Phanerozoic kimberlites. The Premier samples are relatively enriched in Fe and Ti, and contain Fe mica and aluminous amphibole instead of Mg-phlogopite and K-richterite. These features are consistent with a recently identified ρ-wave anomaly beneath this part of the Kaapvaal craton ascribed to refertilization of the mantle. Upwelling of sublithospheric mantle to produce the Bushveld Igneous Complex is considered to be the source of silicate melt available for metasomatism. The resultant refertilized Fe-, Ti-, and Al-enriched mantle composition resembles that which is required to form Proterozoic troctolitic magmas.
© 2003 Elsevier B.V. All rights reserved.

Keywords: Proterozoic; Lithosphere; Iron-enriched; Xenolith; Amphibole

1. Introduction

Mantle xenoliths offer field access to a region otherwise accessible by experimental study, geochemical modeling, and geophysical measurement (see for example, Nixon, 1987; Fei et al., 1999). Most xenoliths from subcratonic regions come to the surface via Phanerozoic volcanic eruptions and show evidence of potassic metasomatism (e.g., Erlank et al., 1987). This paper describes mantle xenoliths that are samples of relatively iron-rich and aluminous Proterozoic lithospheric mantle. Such a mantle composition has been used to explain, for example, a large seismic anomaly beneath the Proterozoic Bushveld Igneous Complex (James et al., 2001) and the gener-

ation of certain Proterozoic magmatic suites (Olson and Morse, 1990).

Kimberlites from the Kaapvaal craton, southern Africa, have provided many of the mantle samples that have been available for study, largely through over a century of diamond mining. Studies of mantle peridotites from southern African kimberlites, and associated diamond inclusions, reveal a mantle lithosphere that was broadly chemically depleted in the past, and subsequently re-enriched prior to being incorporated in kimberlite (e.g., Gurney et al., 1975; Rhodes and Dawson, 1975; Gurney and Harte, 1980; Hawkesworth et al., 1984, 1990; Boyd and Mertzman, 1987; Erlank et al., 1987). The xenolith record from the Kaapvaal craton indicates that chemical depletion resulted from widespread extraction of basalt and komatiite, probably during craton building in the Archean (e.g., Boyd and Gurney, 1982; De Wit et

---

* Fax: +1-303-278-1528.

E-mail address: hoalko@hazenusa.com (K.O. Hoal).

0024-4937/$ - see front matter © 2003 Elsevier B.V. All rights reserved.
doi:10.1016/S0024-4937(03)00116-6

al., 1992; Walter, 1999). Re-enrichment of coarse peridotites is evident in modal and chemical metasomatic features reflecting the interaction of mantle peridotite with potassic, aluminum-poor, and trace-element-enriched (Mg-phlogopite- and K-richterite amphibole-producing) mantle fluids (e.g., Wyllie, 1987; Erlank et al., 1987). The alteration effects of these fluids are apparent in most southern African Phanerozoic mantle xenolith suites, and the age of the process may have been anytime up to emplacement age (largely Cretaceous). For example, Hawkesworth et al. (1990) implicated Jurassic Karoo magmatism in the metasomatism of peridotites entrained in mid-Cretaceous kimberlites. Hoal et al. (1994) described peridotite mineral disequilibrium resulting from metasomatism prior to kimberlite emplacement. Shimizu (1999) underscored the implications of chemical disequilibrium and nonoriginal compositions induced by melt–rock interactions potentially throughout the history of the lithosphere and also at the time of eruption. The simple model of Archean depletion and Cretaceous eruption leaves open a large time span in which to modify mantle materials.

There are few examples of Proterozoic or Archean mantle xenolith-bearing kimberlite pipes in the world, and thus information on re-enrichment processes in older lithospheric mantle is less widely available. The 1198 ± 32 m.y. old (Sm/Nd isochron age on megacrysts in kimberlite; Jones, 1987) Premier kimberlite, in Cullinan, South Africa contains well preserved mantle xenoliths (Danchin and Boyd, 1976; Danchin, 1979). Emplacement of the kimberlite formed part of a regional alkaline event from about 1400 to 1100 Ma (Harmer, 1985; Bartlett, 1990) that included syenites, carbonatites, and alkaline basalts. The Premier pipe was emplaced through the 2050 ± 22 m.y. old (Rb/Sr; Von Gruenewaldt et al., 1985) Bushveld Igneous Complex and is currently exposed at the southern margin of the larger body (Fig. 1).

Seismic experiments conducted across the Kaapvaal craton as part of the Kaapvaal Project (Carlson et al., 1996) show anomalously low ρ-wave velocities in the mantle region beneath the Bushveld Igneous Complex with a distribution beyond the current outcrop of the intrusion (James et al., 2001). These results were interpreted by James et al. (2001) to reflect refertiliza-

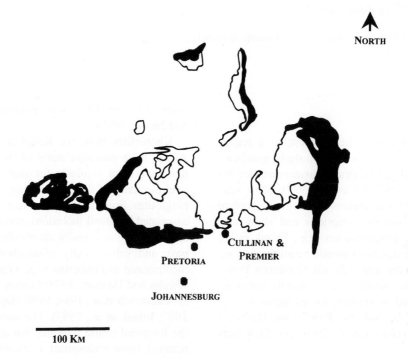

Fig. 1. Sketch map (after Von Gruenewaldt et al., 1985) showing the outline of the Bushveld intrusion, with the mafic Rustenburg Layered Suite in black, and location of the Premier kimberlite, at Cullinan, South Africa.

tion of lithospheric mantle, particularly in iron, related to the Bushveld magmatic event. The extensive seismic anomaly may reflect the effect on the lithospheric mantle of sublithospheric mantle upwelling, whose surface expression, the Bushveld Igneous Complex, required an estimated 1×10^6 km^3 or more of magma (Cawthorn and Walraven, 1998) to produce. The Premier kimberlite sampled this mantle approximately 800 m.y. after emplacement of the Bushveld intrusion, thus providing samples of the regional iron-rich refertilization observed in the seismic record.

2. Premier xenoliths

Samples of Premier xenoliths were collected from 1990 to 1992 as part of a broader study of mantle metasomatism by A.J. Erlank at the University of Cape Town, and with the assistance of P. Barlett of De Beers. The rocks, thin sections, photographs, and geochemical data currently reside in the Mantle Room Collection of the Department of Geological Science, University of Cape Town. Whole rock (major elements, trace elements, rare

Table 1
Compositions of some Premier peridotite xenoliths

	Garnet harzburgite, Prem9004	Garnet phlogopite lherzolite, Prem9009	Garnet phlogopite lherzolite, Prem9006	Phlogopite lherzolite, PDPK1829	Phlogopite lherzolite, Prem8970
SiO$_2$, wt.%	45.36	45.26	49.43	45.37	44.73
TiO$_2$	0.02	0.10	0.07	0.16	1.36
Al$_2$O$_3$	1.91	2.49	1.75	0.67	2.90
FeO$_{total}$	7.14	7.75	7.36	6.70	8.88
MnO	0.10	0.12	0.10	0.11	0.14
MgO	44.81	41.99	39.20	44.76	35.71
CaO	0.63	2.15	1.29	1.75	4.49
Na$_2$O	<0.02	<0.02	0.17	0.08	0.22
K$_2$O	0.04	0.13	0.61	0.37	1.51
P$_2$O$_5$	0.01	0.01	0.02	0.03	0.06
Sc, ppm	6.6	11.5	4.3	5.5	12.6
V	29	53	37	36	147
Cr	2270	3341	2919	4091	2934
Mn	752	862	737	866	1045
Co	127	123	127	117	112
Ni	2641	2387	2606	2385	1948
Cu	3.3	11.9	9.2	7.9	20.8
Zn	35.2	39.6	31.5	35.5	58.2
Ga	0.7	2.1	1.1	1.0	9.9
Rb	2.4	10.1	20.4	25.7	128.6
Sr	8.6	28.5	64.3	46.5	131.2
Y	<0.7	2.0	0.8	1.0	1.3
Zr	1.2	7.6	11.2	13.4	24.6
Nb	1.4	2.1	10.6	2.6	19.8
Cs	<4.4	<4.6	<4.6	<4.4	8.0
Ba	16	240	715	158	428
La	0.47	1.07	5.57	3.3	26.6
Ce	1.15	2.31	7.29	5.2	16.3
Pr	0.14	0.29	0.67		1.5
Nd	0.70	1.29	2.85	2.8	5.6
Sm	0.23	0.45	0.57		1.08
Eu	0.065	0.11	0.16		0.29
Gd	0.26	0.37	0.37		0.80
Tb	0.05	0.055	0.045		0.105
Dy	0.36	0.34	0.21		1.525
Er	0.26	0.20	0.075		0.21
Yb	0.27	0.21	0.05		0.13

earth elements, and radiogenic isotopes) and mineral major and trace-element analyses were performed by X-ray fluorescence, ion chromatography, mass spectrometry, and electron microprobe, respectively, at the University of Cape Town.

Two mantle xenolith populations from Premier were examined during the course of this study: peridotitic and igneous. The first population includes coarse (i.e. not sheared) garnet- and aluminous spinel-bearing harzburgite and lherzolite, and web-

Table 2
Compositions of Premier igneous xenoliths and kimberlite

	PREM8908	PREM8960[a]	PREM8955	PREM8999[a]	PREM89112	PREM89116	PREM9071
SiO_2, wt.%	47.74	47.24	49.89	46.12	50.96	49.94	43.18
TiO_2	1.89	4.37	1.72	1.57	1.80	2.04	2.78
Al_2O_3	4.05	5.20	7.04	3.74	5.19	5.85	3.95
FeO_{total}	10.41	11.40	8.86	10.03	8.02	9.03	14.24
MnO	0.18	0.23	0.16	0.15	0.14	0.12	0.26
MgO	27.25	19.82	21.44	32.23	21.94	21.92	26.77
CaO	6.85	9.59	8.61	4.23	8.91	8.78	6.85
Na_2O	0.70	0.70	0.81	0.45	0.55	0.86	0.40
K_2O	0.72	0.84	1.21	1.32	2.17	1.10	1.40
P_2O_5	0.21	0.63	0.27	0.16	0.34	0.36	0.16
Sc, ppm	17.45	25.92	19.39	11.35	13.50	13.56	20.53
V	122.6	211.4	132.4	93.0	116.3	111.8	101.3
Cr	1194	531	901	1453	1029	1065	465
Mn	1338	1688	1213	1080	1011	855	1847
Co	95	83	84	109	88	105	78
Ni	1328	675	959	1632	1225	1452	479
Cu	38	39	58	64	41	25	77
Zn	90.7	107.4	80.0	68.4	66.2	78.8	128.9
Ga	5.5	7.5	9.0	5.4	7.6	9.3	7.9
Rb	34.1	52.2	70.3	62.0	124.4	57.0	81.8
Sr	210.0	187.7	210.4	159.3	371.4	255.2	223.5
Y	9.4	14.6	13.4	8.2	12.5	12.4	8.3
Zr	92.3	145.0	114.3	84.5	119.9	96.3	105.0
Nb	60	97	67	58	73	66	71
Mo	0.8	0.7	1.1	<0.591	1.0	0.7	1.4
Cs	6.7	10.8	<5.7	<5.27	6.3	<5.76	6.6
Ba	486.2	740.1	724.7	616.2	964.1	1199.7	1057.0
Pb	22.1	5.6	6.2	2.7	7.7	7.1	11.3
La	28.9	44.3	32.8	21.4	35.6	30.1	29.5
Ce	57.4	70.0	64.1	35.7	72.8	61.8	63.3
Pr		7.8		4.3			
Nd	25.9	30.7	27.4	15.5	32.6	27.9	27.4
Sm		5.58		2.67			
Eu		1.36		0.66			
Gd		4.35		2.06			
Tb		0.52		0.29			
Dy		2.84		1.51			
Er		1.41		0.72			
Yb		1.12		0.65			
$(^{87}Sr/^{86}Sr)_p$		0.71631		0.71771			
$(^{87}Sr/^{86}Sr)_i$		0.70508		0.69633			
$(^{143}Nd/^{144}Nd)_p$		0.51191		0.51184			
$(^{143}Nd/^{144}Nd)_i$		0.51110		0.51108			
ε_{Nd}		−2.69		−2.40			

(p) indicates present-day, measured isotopic values; (i) indicates initial values calculated at 1180 Ma.

[a] Rare earth elements analyzed by isotope dilution.

sterite, as described by Danchin and Boyd (1976), Danchin (1979), and Olson and Erlank (1993), among others. Whole-rock compositions of some garnet- and phlogopite-bearing xenoliths are listed in Table 1. Where possible, study of the peridotites was focused on samples showing little to no visual effects of minerals having precipitated from the host kimberlite. The relative proportions of olivine, orthopyroxene, clinopyroxene, garnet, and spinel vary from sample to sample, as is common in mantle

PREM9107	PREM9108	PREM89111	PREM9105	PREM9106[a]	SHR202, Kimb	SHR206, Kimb	SHR207, Kimb
47.59	48.39	49.05	50.49	48.75	39.69	34.36	33.33
2.00	1.36	2.19	1.51	2.14	2.89	2.56	2.52
4.11	3.88	4.30	5.31	3.85	2.37	2.06	2.18
9.76	9.14	9.00	8.76	9.17	11.47	10.85	10.24
0.16	0.17	0.17	0.16	0.16	0.18	0.22	0.20
29.44	28.24	28.34	23.66	29.18	39.38	37.68	36.69
5.00	6.78	5.79	7.34	3.70	3.56	11.02	14.42
0.82	0.74	0.41	1.13	1.34	<0.023	<0.024	<0.023
0.88	1.06	0.55	1.43	1.56	0.03	0.73	0.22
0.24	0.23	0.20	0.20	0.16	0.46	0.55	0.24
11.37	13.09	11.61	17.17	10.66	17.65	15.62	15.66
97.5	96.9	105.4	108.3	103.7	107.1	81.5	65.7
1574	1215	1533	982	1368	2191	1718	1764
1209	1325	1075	1245	1212	1303	1757	1579
108	105	87	85	103	123	122	114
1517	1447	1383	1130	1481	1509	1587	2025
30	34	43	47	47	56	97	124
88.3	97.6	60.3	79.4	72.3	63.4	65.9	82.0
5.3	5.9	6.5	7.9	5.6	4.9	5.7	5.1
33.4	55.9	47.1	74.5	78.4	5.2	45.6	38.9
161.1	230.0	238.1	263.2	186.2	330.8	440.5	466.3
10.9	10.3	8.3	11.8	8.7	11.0	11.7	11.4
90.2	78.4	103.3	97.0	85.5	94.2	83.8	84.0
61	44	78	60	72	159	140	299
1.0	<0.587	0.7	0.9	<0.582	1.1	1.8	<0.665
6.2	6.3	9.3	6.7	8.7	<5.58	<6.19	<6.59
268.7	707.1	194.8	847.5	471.9	249.8	182.7	1280.1
8.3	8.8	6.4	11.4	3.4	<2.58	11.0	4.6
21.1	23.8	25.7	33.4	26.3	57.1	50.8	147.5
47.3	46.2	36.6	59.0	48.3	81.6	76.4	214.9
				5.5			
23.2	24.0	14.3	25.6	18.0	31.7	27.5	70.2
				3.41			
				0.85			
				2.43			
				0.29			
				1.59			
				0.75			
				0.66			
				0.72071			0.70642
				0.69207			0.70228
				0.51180			0.51187
				0.51104			0.51131
				−4.99			3.87

Table 3
Representative mica compositions from Premier peridotite and igneous xenoliths

	Micas from peridotite xenoliths				Micas from igneous xenoliths			
	9036-5	9040-4	KI829-1	9046-8	8960-2	8999-8	9071-2	9101-2
SiO_2	39.37	38.16	41.46	37.46	38.35	39.67	37.55	39.32
TiO_2	7.19	6.36	1.61	2.73	5.38	4.98	5.54	4.13
Al_2O_3	14.36	13.98	12.50	17.40	14.84	13.78	13.94	12.79
Cr_2O_3	0.32	0.19	0.68	0.27	0.00	0.00	0.00	0.13
FeO_{total}	5.63	5.48	2.92	6.35	9.04	7.00	7.24	6.73
MnO	0.00	0.00	0.00	0.00	0.00	0.00	0.00	0.00
MgO	20.98	20.96	25.78	20.88	19.73	21.10	20.82	23.16
CaO	0.00	0.00	0.00	0.00	0.00	0.00	0.00	0.00
Na_2O	0.23	0.27	0.09	0.00	0.75	0.66	0.75	0.61
K_2O	8.83	8.27	9.61	9.82	7.04	5.82	7.83	6.79
F	0.00	0.00	0.00	0.00	0.00	0.00	0.00	0.00
NiO	0.00	0.00	0.00	0.00	0.00	0.00	0.00	0.00
Total	97.42	94.29	95.48	94.91	96.81	93.01	94.19	93.66
Oxygen	22.00	22.00	22.00	22.00	22.00	22.00	22.00	22.00
Si	5.52	5.52	5.88	5.41	5.51	5.75	5.48	5.74
Ti	0.76	0.69	0.17	0.30	0.58	0.54	0.61	0.45
Al	2.37	2.39	2.09	2.96	2.51	2.35	2.40	2.20
Cr	0.04	0.02	0.08	0.03	0.00	0.00	0.00	0.02
Fe	0.66	0.66	0.35	0.77	1.09	0.85	0.88	0.82
Mn	0.00	0.00	0.00	0.00	0.00	0.00	0.00	0.00
Mg	4.39	4.52	5.45	4.50	4.23	4.56	4.53	5.04
Ca	0.00	0.00	0.00	0.00	0.00	0.00	0.00	0.00
Na	0.06	0.08	0.02	0.00	0.21	0.19	0.21	0.17
K	1.58	1.53	1.74	1.81	1.29	1.08	1.46	1.26
F	0.23	0.29	0.37	0.00	0.76	0.00	0.24	0.00
Ni	0.00	0.00	0.00	0.00	0.00	0.00	0.00	0.00
Sum	15.60	15.70	16.15	15.78	16.19	15.32	15.81	15.71

peridotites. Mica is a common accessory mineral that, by analogy with similar examples reported in the literature, represents part of the metasomatic effects of secondary melts or fluids. These micas are commonly coarse and euhedral, and in many instances appear to have coexisted with secondary diopside, as the two minerals are intergrown or in contact in most samples. Amphibole, a key indicator of metasomatism in peridotites from Cretaceous kimberlites (Erlank et al., 1987), is rare and has been observed only in websterite. Examples of these mineral assemblages are shown in Fig. 2.

The second xenolith population includes a group of small, rounded igneous xenoliths containing abundant mica and amphibole (Olson and Erlank, 1993). Whole-rock compositions of these rocks are shown in Table 2. A variety of igneous textures is shown in these xenoliths. Some samples contain euhedral to subhedral amphibole and mica phenocrysts in a matrix dominated by mica, amphibole, and diopside. Others contain autoliths and brecciated fragments of similar rocks or of individual crystals. Amphibole, mica, and diopside commonly enclose rounded olivine microcrysts poikilitically; rims of amphibole occur on mica and diopside, and rims of mica occur on amphibole and diopside. Diopsides occur as oikocrysts, as long bladed crystals, or are poikilitically enclosed in amphibole or mica. Magnetite is present as a product of oxyexsolution from mica, and ilmenite and apatite are abundant as small grains. Some of these textures are illustrated in Fig. 3. A few samples contain abundant ilmenite and no amphibole. They are not included in the tables and figures below.

A whole-rock plot of FeO against TiO_2 (Fig. 4) illustrates the relatively iron-rich nature of the Premier peridotite and igneous xenoliths compared to representative samples from Kimberley (Erlank et al.,

Table 4
Representative amphibole compositions from Premier peridotite (websterite) and igneous xenoliths

	Websterite	Igneous xenoliths														
	AMPH 2 9046	AMPH 1 8960	AMPH 2 8960	AMPH 1 8999	AMPH 4 8999	AMPH 6 8999	AMPH 9 8999	AMPH 10 8999	AMPH 2 9071	AMPH 3 9106	AMPH 4 9106	AMPH 1 9051	AMPH 1 9055	AMPH 1 9105	AMPH 1 9108	AMPH 3 9108	AMPH 4 9108
SiO$_2$	41.05	44.59	48.45	46.01	43.15	50.96	50.87	51.32	48.78	51.88	49.49	50.78	53.53	49.55	51.84	44.29	44.29
TiO$_2$	1.45	2.75	0.42	2.76	4.16	0.75	1.28	3.40	2.61	3.18	2.04	4.32	0.94	0.34	0.78	3.34	3.34
Al$_2$O$_3$	15.83	9.33	7.17	8.71	9.82	4.26	5.09	3.71	5.84	3.44	5.48	2.95	2.13	7.63	4.28	10.28	10.28
Cr$_2$O$_3$	0.31	0.00	0.00	0.00	0.15	0.39	0.38	0.36	0.00	0.29	0.93	0.00	0.00	0.00	0.00	0.00	0.00
FeO$_{total}$	7.67	9.73	8.69	6.96	8.38	9.72	7.72	7.10	8.33	4.67	6.47	10.87	6.93	7.57	5.43	7.33	7.33
MnO	0.00	0.13	0.00	0.00	0.00	0.00	0.00	0.00	0.18	0.00	0.13	0.15	0.14	0.13	0.15	0.00	0.00
MgO	15.17	16.80	19.29	18.49	16.62	17.66	18.04	18.70	18.23	20.62	18.88	15.95	20.22	19.81	21.35	17.92	17.92
CaO	12.17	11.30	10.80	11.24	11.49	11.89	11.84	8.59	9.01	9.14	9.59	5.30	11.11	11.13	10.94	11.52	11.52
Na$_2$O	2.80	2.79	3.27	3.27	3.45	1.24	1.35	3.42	3.60	4.44	4.55	6.41	1.53	3.17	2.53	2.86	2.86
K$_2$O	1.31	0.61	0.35	0.42	0.55	0.10	0.06	0.33	0.35	0.86	0.49	1.27	0.24	0.39	0.14	0.68	0.68
Total	97.76	98.03	98.43	97.86	97.77	96.97	96.63	96.93	96.93	98.52	98.05	98.78	96.77	99.72	97.67	98.22	98.22
Oxygen	23.00	23.00	23.00	23.00	23.00	23.00	23.00	23.00	23.00	23.00	23.00	23.00	23.00	23.00	23.00	23.00	23.00
Si	5.96	6.45	6.86	6.57	6.28	7.33	7.28	7.29	7.02	7.27	7.09	7.32	7.61	6.87	7.30	6.32	6.32
Al^{IV}	2.04	1.55	1.14	1.43	1.68	0.68	0.72	0.62	0.98	0.57	0.91	0.50	0.36	1.13	0.70	1.68	1.68
Al (total)	2.71	1.59	1.20	1.47	1.68	0.72	0.86	0.62	0.99	0.57	0.93	0.50	0.36	1.25	0.71	1.73	1.73
Al^{VI}	0.67	0.04	0.06	0.03	0.00	0.05	0.14	0.00	0.01	0.00	0.01	0.00	0.00	0.12	0.01	0.05	0.05
Ti	0.16	0.30	0.05	0.30	0.46	0.08	0.14	0.36	0.28	0.34	0.22	0.47	0.10	0.04	0.08	0.36	0.36
Fe^3	0.18	0.22	0.34	0.20	0.04	0.28	0.18	0.09	0.08	0.16	0.07	0.18	0.04	0.44	0.05	0.29	0.29
Mg	3.28	3.62	4.07	3.93	3.60	3.78	3.85	3.96	3.91	4.31	4.03	3.43	4.28	4.09	4.48	3.81	3.81
Fe^2	0.75	0.96	0.69	0.64	0.85	0.89	0.74	0.47	0.92	0.30	0.71	0.97	0.79	0.44	0.59	0.59	0.59
Mn	0.00	0.02	0.00	0.00	0.00	0.00	0.00	0.00	0.02	0.00	0.02	0.02	0.02	0.02	0.02	0.00	0.00
Sum (FM)	13.04	13.15	13.21	13.10	12.91	13.08	13.05	12.79	13.23	12.94	13.05	12.88	13.19	13.14	13.23	13.09	13.09
Ca	1.89	1.75	1.64	1.72	1.79	1.83	1.82	1.31	1.39	1.37	1.47	0.82	1.69	1.65	1.65	1.76	1.76
Na (M4)	0.07	0.10	0.16	0.19	0.30	0.09	0.13	0.90	0.38	0.69	0.48	1.30	0.12	0.21	0.12	0.14	0.14
Na (total)	0.79	0.78	0.90	0.91	0.97	0.35	0.38	0.94	1.01	1.21	1.26	1.79	0.42	0.85	0.69	0.79	0.79
Na (A)	0.72	0.68	0.74	0.72	0.67	0.26	0.24	0.04	0.63	0.52	0.79	0.49	0.30	0.64	0.57	0.65	0.65
K	0.24	0.11	0.06	0.08	0.10	0.02	0.01	0.06	0.06	0.15	0.09	0.23	0.04	0.07	0.03	0.12	0.12

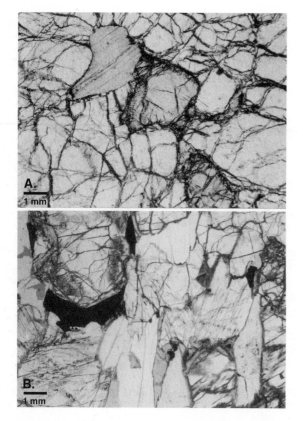

Fig. 2. Photomicrographs of metasomatized Premier peridotite xenoliths. (A) Lherzolite with phlogopite and diopside surrounded by olivine. (B) Garnet websterite with phlogopite and ferroan pargasite (center bottom) between diopside grains.

element patterns suggests a magmatic affinity among these samples.

The isotope compositions of the igneous xenoliths (Table 2; initial ratios calculated for 1180 Ma) reflect an enriched mantle composition (Sm/Nd) that differs from the depleted mantle composition of the host Group I kimberlite (last column in Table 2). The samples also show a range of values for $(^{87}Sr/^{86}Sr)_i$ that may reflect variably low Rb in the source, combined with possible altering effects of the kimberlite. While the group is isotopically distinct from the host kimberlite (Table 2), further study using other isotopic systems would help clarify the mantle history of these rocks.

Mineral compositions show that major element variations in whole rocks are not a result of interstitial

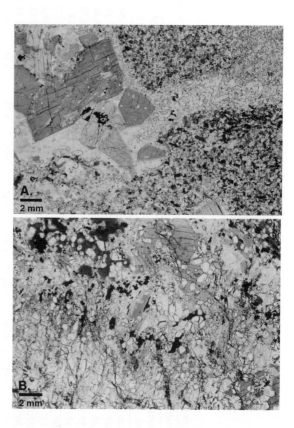

Fig. 3. Photomicrographs of Premier igneous xenoliths. (A) Euhedral pargasitic amphibole and phlogopite in fine-grained matrix dominated by pargasite, phlogopite, and diopside. (B) Rounded olivine microcysts in diopside, pargasite, and phlogopite oikocrysts.

1987); mineral compositions such as Fo in olivine also reflect whole-rock compositions (Boyd, 1987) but whole rocks are used in this case because parageneses differ among the xenolith populations. For the peridotites, Fe enrichment was noted by Danchin (1979) in xenoliths lacking metasomatic mineral phases such as Mg-phlogopite and K-richterite. The igneous xenoliths considerably extend the range of iron and titanium compositions observed in Premier peridotites. Trace-element compositions in the igneous xenoliths (Fig. 5) are consistent with this group representing a distinct compositional population. The group is enriched in incompatible elements relative to primitive mantle (Hoffmann, 1988), and displays comparatively high levels of Ti, Ca, and Fe, as well as Nb, Nd, Ba, and Rb. The coherency of the trace-

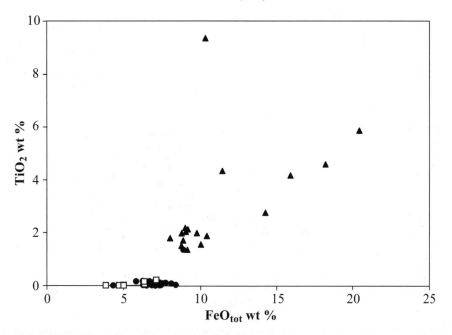

Fig. 4. Whole rock FeO_{tot} vs. TiO_2 for representative Kimberley peridotite xenoliths (open squares), Premier peridotite xenoliths (closed circles), and the Premier igneous xenoliths (triangles).

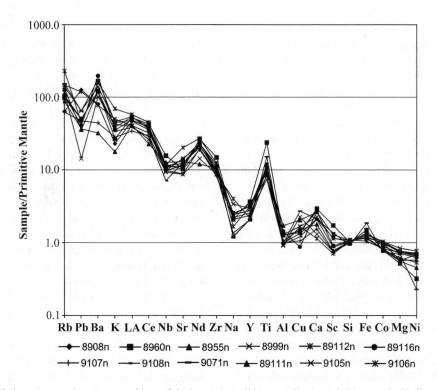

Fig. 5. Whole rock trace element compositions of the igneous xenoliths, normalized to primitive mantle (Hoffmann, 1988).

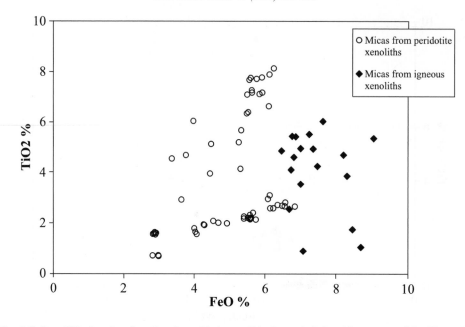

Fig. 6. FeO vs. TiO$_2$ for micas from Premier peridotite xenoliths (open circles) and igneous xenoliths (diamonds).

alteration, veins, or late-stage kimberlite alteration, but are related to mineralogy. Among the more distinctive metasomatic minerals are mica and amphibole. Phanerozoic kimberlites generally contain Mg-phlogopite mica and K-richterite amphibole, low in Al and Fe and siliceous. (Jagersfontein, a Cretaceous kimberlite showing Proterozoic metasomatism of peridotite (Field et al., 1989; Winterburn et al., 1990), contains both K-richterite and aluminous amphibole parageneses.)

Representative mica and amphibole compositions from Premier peridotite and igneous xenoliths are listed in Tables 3 and 4. Micas from Premier xenoliths are generally Fe–Ti-rich, and micas from igneous xenoliths define a population of higher Fe content than those from peridotite xenoliths (Fig. 6). Where present, compositional zoning in micas is toward relatively Fe–Ti-poor rims. Amphiboles from Premier are aluminous (Fig. 7), unlike amphiboles from Cretaceous kimberlites and related mantle peridotites, which are generally K-richterite. Premier amphiboles range in composition from ferroan pargasite (field defined by amphiboles in websterite, Fig. 7) and magnesiohastingsite to edenite, with the most silicic compositions being that of silicic edenite (terminology of Leake, 1978). Zoning extends from pargasitic cores to edenitic rims (Fig. 7).

3. Discussion

The presence of Fe–Ti metasomatism of depleted mantle xenoliths reflects local interaction with silicate melt (as opposed to hydrous fluid) such as basaltic melt, as observed for xenoliths from oceanic lithosphere and orogenic belts (Boyd, 1987, 1989). For example, high-temperature garnet peridotites have relatively low Mg numbers, enrichment in Fe and Ti, and fabrics indicative of stresses induced by magma contact, possibly within the asthenosphere (Boyd, 1987). The samples described here, however, are coarse peridotite and igneous xenoliths. In this case, it is possible that these rocks also are the products of silicate melt interaction, and subsequently have either fully recrystallized (in the case of the peridotites) or quenched (in the case of the igneous xenoliths). One needs to look no further than the local Bushveld Igneous Complex for the source of the silicate melt. The amount of sublithospheric magmatic upwelling to form the intrusion was considerable (e.g., Cawthron and Walraven, 1998), and the effects of this event are still observed in the recently identified widespread ρ-wave anomaly (James et al., 2001) beneath the region. In addition, unlike Kaapvaal Archean ages, diamond inclusion ages from Premier

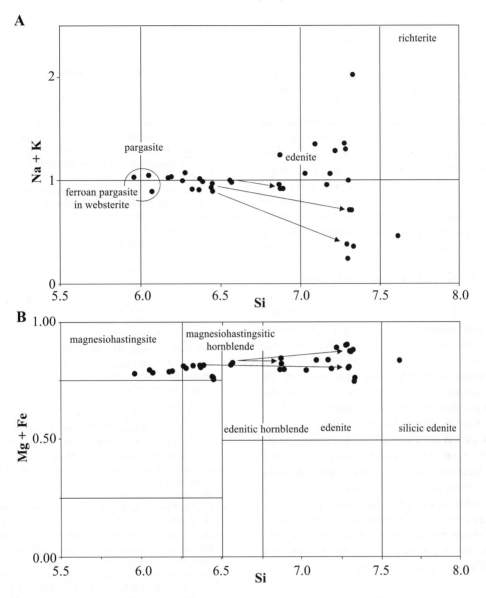

Fig. 7. Compositions of amphiboles from Premier websterite xenolith and igneous xenoliths (fields of Leake, 1978).

are Proterozoic (2.2–1.9 Ga; Richardson et al., 1993; Carlson et al., 1999), broadly conforming to the age of the Bushveld event and indicating possible isotopic resetting at that time (Carlson et al., 2000).

In contrast to the local effects of silicate melt or fluid metasomatism in mantle peridotites that commonly vary between samples and even minerals (Shimizu, 1999), the geochemical and geophysical evidence from this part of the Kaapvaal craton indi-

cate a regional influence on the lithosphere by sub-lithospheric mantle upwelling. The Premier xenoliths are direct samples of that mantle.

Whether this widespread silicate melt metasomatism of the lithosphere is a Proterozoic feature, not having been observed in Phanerozoic xenolith suites, or whether it is a one-time effect of the unusual circumstances of emplacement of the Bushveld Igneous Complex, is beyond the scope of this paper.

However, it is important to recognize that Fe–Al mantle of this type is necessary to form the troctolitic magmas that were emplaced only in the Proterozoic (2120–920 Ma; Hamilton et al., 1998; Scoates and Mitchell, 2000), largely in Laurentia and Fennoscandia (Olson and Morse, 1990). Some 114 examples of these magmas have been identified worldwide (Scoates and Mitchell, 2000). While it is not yet possible to effectively model the origin of troctolitic magmas and their associated noritic and anorthositic 6w>rocks, experimental work by Scoates and Lindsley (2000) shows that olivine-normative high-Al basalt can produce anorthosite and related rocks under polybaric fractionation and require an Fe-rich mantle source. The observations that parental magma compositions to the Bushveld Complex were in part boninitic (Sharpe and Hulbert, 1985) and in part anorthositic–noritic (e.g., Kruger and Marsh, 1985), and that the Rustenburg Layered Suite contains layered norite, noritic anorthosite, and gabbronorite (Vermaak and Von Gruenewaldt, 1986), may ultimately provide the petrogenetic link between Fe–Al mantle present at Premier and that required for troctolitic magmas elsewhere. Finally, it may also be observed that all of these occurrences were emplaced prior to, during, or after Rodinia assembly (1700 to 1300 Ma until about 750 Ma; Torsvik et al., 1996; Weil et al., 1998; Karlstrom et al., 1999), when the Kaapvaal craton probably faced Laurentia (Hoffman, 1999), but not after breakup of the supercontinent and its lithosphere (Hoal and Hoal, 1999). Further study of Fe–Al sublithospheric mantle beneath the Bushveld Igneous Complex, and sampled at Premier, may thus provide useful information for a number of diverse geological subjects.

Acknowledgements

This research was sponsored by the Foundation for Research and Development of South Africa, a University of Cape Town research postdoctoral fellowship, and De Beers Consolidated Mines (Pty) Geology Department. The manuscript was significantly improved by the reviews and comments of Tony Morse, Brian Hoal, Jock Harmer, Russell Sweeney, Herman Grütter, and an anonymous reviewer.

References

Bartlett, P.D., 1990. Unpublished report on Premier mine geology. Premier Mine, DeBeers Consolidated Mines, Cullinan. 53 pp.
Boyd, F.R., 1987. High- and low-temperature garnet peridotite xenoliths and their possible relation to the lithosphere–asthenosphere boundary beneath southern Africa. In: Nixon, P.H. (Ed.), Mantle Xenoliths. Wiley, New York, pp. 403–412.
Boyd, F.R., 1989. Compositional distinction between oceanic and cratonic lithosphere. Earth and Planetary Science Letters 96, 15–26.
Boyd, F.R., Gurney, J.J., 1982. Low-calcium garnets: keys to craton structure and diamond crystallization. Carnegie Institution of Washington Yearbook 81, 261–267.
Boyd, F.R., Mertzman, S.A., 1987. Composition and structure of the Kaapvaal lithosphere, southern Africa. In: Mysen, B.O. (Ed.), Magmatic Processes: Physicochemical Principles. Geochemical Society Special Publication, vol. 1. The Geochemical Society, Houston, pp. 3–12.
Carlson, R.W., Grove, T.L., de Wit, M.J., Gurney, J.J., 1996. Program to study the crust and mantle of the Archean craton in southern Africa. EOS, Transactions of the American Geophysical Union 77 (273), 277.
Carlson, R.W., Pearson, D.G., Boyd, F.R., Shirey, S.B., Irvine, G., Menzies, A.H., Gurney, J.J., 1999. Re–Os systematics of lithospheric peridotites: implications for lithosphere formation and preservation. In: Proceedings of the 7th International Kimberlite Conference, Red Roof Design, Cape Town, pp. 99–108.
Carlson, R.W., Boyd, F.R., Shirey, S.B., Janney, P.E., Grove, T.L., Bowring, S.A., Schmitz, M.D., Dann, J.C., Bell, D.R., Gurney, J.J., Richardson, S.H., Tredoux, M., Menzies, A.H., Pearson, D.G., Hart, R.J., Wilson, A.H., Moser, D., 2000. Continental growth, preservation, and modification in southern Africa. GSA Today 10 (2), 1–4.
Cawthorn, R.G., Walraven, F., 1998. Emplacement and crystallization time for the Bushveld complex. Journal of Petrology 39, 1669–1687.
Danchin, R.V., 1979. Mineral and bulk chemistry of garnet lherzolite and garnet harzburgite xenoliths from the Premier mine, South Africa. In: Boyd, F.R., Meyer, H.O.A. (Eds.), The Mantle Sample: Inclusions in Kimberlites and Other Volcanics. Proceedings of the Second International Kimberlite Conference. The American Geophysical Union, Washington, DC, pp. 104–126.
Danchin, R.V., Boyd, F.R., 1976. Ultramafic nodules from the Premier kimberlite pipe, South Africa. Carnegie Institution of Washington Yearbook 75, 531–538.
De Wit, M.J., Roering, C., Hart, R.J., Armstrong, R.A., De Ronde, C.E.J., Green, R.W.E., Tredoux, M., Peberdy, E., Hart, R.A., 1992. Formation of an Archaean continent. Nature 357, 553–562.
Erlank, A.J., Waters, F.J., Hawkesworth, C.J., Haggerty, S.E., Allsopp, H.L., Rickard, R.S., Menzies, M.A., 1987. Evidence for mantle metasomatism in peridotite nodules from the Kimberley pipes, South Africa. In: Menzies, M.A., Hawkesworth, C.J. (Eds.), Mantle Metasomatism. Academic Press, London, pp. 221–311.

Fei, Y., Bertka, C.M., Mysen, B.O. (Eds.), 1999. Mantle Petrology: Field Observations and High-pressure Experimentation, A Tribute to Francis R. (Joe) Boyd. Special Publication, vol. 6. The Geochemical Society, Houston. 322 pp.

Field, S.W., Haggerty, S.E., Erlank, A.J., 1989. Subcontinental metasomatism in the region of Jagersfontein, South Africa. In: Ross, J., Jaques, A.L., Ferguson, J., Green, D.H., O'Reilly, S.Y., Danchin, R.V., Janse, A.J.A. (Eds.), Kimberlites and Related Rocks, Proceedings of the Fourth International Kimberlite Conference. Geological Society of Australia Special Publication, vol. 14(1). Blackwell, Melbourne, pp. 771–783.

Gurney, J.J., Harte, B., 1980. Chemical variations in upper mantle nodules from southern African kimberlites. Philosophical Transactions of the Royal Society of London, A 297, 273–293.

Gurney, J.J., Harte, B., Cox, K.G., 1975. Mantle xenoliths in the Matsoku kimberlite pipe. Physics and Chemistry of the Earth 9, 507–529.

Hamilton, M.A., Ryan, A.B., Emslie, R.F., Ermanovics, I.F., 1998. Identification of Paleoproterozoic anorthositic and monzonitic rocks in the vicinity of the Mesoproterozoic Nain plutonic suite, Labrador: U-Pb evidence. In: Radiogenic Age and Isotope Studies, Report 11, Geological Survey of Canada, Current Research 1998-F, 23–40.

Harmer, R.E., 1985. Rb–Sr isotopic study of units of the Pienaars River alkaline complex, north of Pretoria, South Africa. Transactions of the Geological Society of South Africa 88, 215–223.

Hawkesworth, C.J., Rogers, N.W., van Calsteren, P., 1984. Mantle enrichment processes. Nature 322, 356–359.

Hawkesworth, C.J., Erlank, A.J., Kempton, P.D., Waters, F.G., 1990. Mantle metasomatism: isotope and trace-element trends in xenoliths from Kimberley, South Africa. Chemical Geology 85, 19–34.

Hoal, K.O., Hoal, B.G., 1999. Contrasting styles of mantle metasomatism with time: implications for Al–Fe magmatism in Rodinia. Geological Society of America - Abstracts with Programs 31 (7), A318.

Hoal, K.E.O., Hoal, B.G., Erlank, A.J., Shimizu, N., 1994. Metasomatism in the mantle lithosphere recorded by rare earth elements in garnets. Earth and Planetary Science Letters 126, 303–313.

Hoffman, P.F., 1999. The break-up of Rodinia, birth of Gondwana, true polar wander and the snowball Earth. Journal of African Earth Sciences 26 (1), 9–26.

Hoffmann, A.W., 1988. Chemical differentiation of the Earth: the relationship between mantle, continental crust, and oceanic crust. Earth and Planetary Science Letters 90, 297–314.

James, D.E., Fouch, M.J., VanDecar, J.C., van der Lee, S., Kaapvaal Seismic Group, 2001. Tectospheric structure beneath southern Africa. Geophysical Research Letters 28 (13), 2485–2488.

Jones, R.A., 1987. Strontium and neodymium isotopic and rare earth element evidence for the genesis of megacrysts in kimberlites of southern Africa. In: Nixon, P.H. (Ed.), Mantle Xenoliths. Wiley, New York, pp. 711–724.

Karlstrom, K.E., Harlan, S.S., Williams, M.L., McLelland, J., Geissman, J.W., Åhäll, K.-I., 1999. Refining Rodinia: Geolog-

ical evidence for the Australia–Western U.S. connection for the Proterozoic. GSA Today 9 (10), 1–7.

Kruger, F.J., Marsh, J.S., 1985. The mineralogy, petrology, and origin of the Merensky cyclic unit in the western Bushveld complex. Economic Geology 80, 958–974.

Leake, B.E., 1978. Nomenclature of amphiboles. Canadian Mineralogist 16, 501–515.

Nixon, P.H. (Ed.), 1987. Mantle Xenoliths. Wiley, New York. 844 pp.

Olson, K.E., Erlank, A.J., 1993. Magmas and metasomites from the Proterozoic Premier kimberlite, South Africa. In: Dunne, K.P.E., Grant, B. (Eds.), GAC-MAC Symposium on Mid-continent Diamonds, May 17–18. Geological Association of Canada Mineral Deposits Division, Edmonton, p. 133.

Olson, K.E., Morse, S.A., 1990. Regional Al–Fe mafic magmas associated with anorthosite-bearing terranes. Nature 344, 760–762.

Rhodes, J.M., Dawson, J.B., 1975. Major and trace element chemistry of peridotite inclusions from the Lashaine volcano, Tanzania. Physics and Chemistry of the Earth 9, 545–557.

Richardson, S.H., Harris, J.W., Gurney, J.J., 1993. Three generations of diamonds from old continental mantle. Nature 366, 256–258.

Scoates, J.S., Lindsley, D.H., 2000. New insights from experiments on the origin of anorthosite. EOS Transactions, American Geophysical Union 81 (suppl. 48), F1300.

Scoates, J.S., Mitchell, J.N., 2000. The evolution of troctolitic and high-Al basaltic magmas in Proterozoic anorthosite plutonic suites and implications for Voisey's Bay massive Ni–Cu sulfide deposits. In: Naldrett, A.J., Li, C. (Eds.), Special Issue on Voisey's Bay Ni–Cu–Co Deposits. Economic Geology 95 (4), 677–701.

Sharpe, M.R., Hulbert, L.J., 1985. Ultramafic sills beneath the eastern Bushveld complex: mobilized suspensions of early lower zone cumulates in a parental magma with boninitic affinities. Economic Geology 80, 849–871.

Shimizu, N., 1999. Young geochemical features in cratonic peridotites from southern Africa and Siberia. In: Fei, Y., Bertka, C.M., Mysen, B.O. (Eds.), Mantle Petrology: Field Observations and High-pressure Experimentation, A Tribute to Francis R. (Joe) Boyd. Special Publication, vol. 6. The Geochemical Society, Houston, pp. 47–55.

Torsvik, T.H., Smethurst, M.A., Meert, J.G., Van der Voo, R., McKerrow, W.S., Brasier, M.D., Sturt, B.A., Walderhaug, H.J., 1996. Continental break-up and collision in the Neoproterozoic and Palaeozoic—a tale of Baltica and Laurentia. Earth-Science Reviews 40, 229–258.

Vermaak, C.F., Von Gruenewaldt, G., 1986. Introduction to the Bushveld complex. In: Anhaeusser, C.R., Maske, S. (Eds.), Mineral Deposits of Southern Africa. Geological Society of South Africa, Johannesburg, pp. 1021–1029.

Von Gruenewaldt, G., Sharpe, M.R., Hatton, C.J., 1985. The Bushveld complex: introduction and review. Economic Geology 80, 1049–1061.

Walter, M.J., 1999. Melting residues of fertile peridotite and the origin of cratonic lithosphere. In: Fei, Y., Bertka, C.M., Mysen, B.O. (Eds.), Mantle Petrology: Field Observations and High-pressure Experimentation, A Tribute to Francis R. (Joe) Boyd. Special Publication, vol. 6. The Geochemical Society, Houston, pp. 225–239.

Weil, A.B., Van der Voo, R., Mac Niocaill, C., Meert, J.G., 1998. The Proterozoic supercontinent Rodinia: paleomagnetically derived reconstructions for 1100 to 800 Ma. Earth and Planetary Science Letters 154, 13–24.

Winterburn, P.A., Harte, B., Gurney, J.J., 1990. Peridotite xenoliths from the Jagersfontein kimberlite pipe: I. Primary and primary-metasomatic mineralogy. Geochimica et Cosmochimica Acta 54, 329–341.

Wyllie, P.J., 1987. Metasomatism and fluid generation in mantle xenoliths. In: Nixon, P.H. (Ed.), Mantle Xenoliths. Wiley, New York, pp. 609–621.

Available online at www.sciencedirect.com

SCIENCE @ DIRECT°

Lithos 71 (2003) 273–287

LITHOS

www.elsevier.com/locate/lithos

Mesozoic thermal evolution of the southern African mantle lithosphere

David R. Bell[a,b,*], Mark D. Schmitz[b,1], Philip E. Janney[c,2]

[a]*Department of Geological Sciences, University of Cape Town, Rondebosch 7700, South Africa*
[b]*Department of Earth, Atmospheric and Planetary Sciences, Massachusetts Institute of Technology, Cambridge, MA 02139, USA*
[c]*Department of Terrestrial Magnetism, Carnegie Institution of Washington, 5241 Broad Branch Road, N.W., Washington, DC 20015, USA*

Abstract

The thermal structure of Archean and Proterozoic lithospheric terranes in southern Africa during the Mesozoic was evaluated by thermobarometry of mantle peridotite xenoliths erupted in alkaline magmas between 180 and 60 Ma. For cratonic xenoliths, the presence of a 150–200 °C isobaric temperature range at 5–6 GPa confirms original interpretations of a conductive geotherm, which is perturbed at depth, and therefore does not record steady state lithospheric mantle structure.

Xenoliths from both Archean and Proterozoic terranes record conductive limb temperatures characteristic of a "cratonic" geotherm (~ 40 mW m^{-2}), indicating cooling of Proterozoic mantle following the last major tectonothermal event in the region at ~ 1 Ga and the probability of thick off-craton lithosphere capable of hosting diamond. This inference is supported by U–Pb thermochronology of lower crustal xenoliths [Schmitz and Bowring, 2003. Contrib. Mineral. Petrol. 144, 592–618].

The entire region then suffered a protracted regional heating event in the Mesozoic, affecting both mantle and lower crust. In the mantle, the event is recorded at ~ 150 Ma to the southeast of the craton, propagating to the west by 108–74 Ma, the craton interior by 85–90 Ma and the far southwest and northwest by 65–70 Ma. The heating penetrated to shallower levels in the off-craton areas than on the craton, and is more apparent on the southern margin of the craton than in its western interior. The focus and spatial progression mimic inferred patterns of plume activity and supercontinent breakup 30–100 Ma earlier and are probably connected.

Contrasting thermal profiles from Archean and Proterozoic mantle result from penetration to shallower levels of the Proterozoic lithosphere by heat transporting magmas. Extent of penetration is related not to original lithospheric thickness, but to its more fertile character and the presence of structurally weak zones of old tectonism. The present day distribution of surface heat flow in southern Africa is related to this dynamic event and is not a direct reflection of the pre-existing lithospheric architecture.
© 2003 Elsevier B.V. All rights reserved.

Keywords: Lithosphere; Mantle; Craton

* Corresponding author. Current address: Department of Chemistry and Biochemistry, Arizona State University, Main Campus P.O. Box 871604 Tempe, AZ 85287-1604 USA. Fax: +1-480-965-2747.
E-mail address: david.r.bell@asu.edu (D.R. Bell).
[1] Current address: Department of Terrestrial Magnetism, Carnegie Institution of Washington, 5241 Broad Branch Road, N.W., Washington, DC 20015, USA.
[2] Current address: Field Museum of Natural History, 1400 S. Lake Shore Drive, Chicago, IL 60615, USA.

1. Introduction

The continental crust is a complex assemblage of terranes of varying age and geologic history, of which the compositions and internal structures reflect the changing nature of crust and mantle processes through geologic time. These differences persist through the

0024-4937/$ - see front matter © 2003 Elsevier B.V. All rights reserved.
doi:10.1016/S0024-4937(03)00117-8

crust and into the subcontinental lithospheric mantle (Griffin et al., 1999). In studying the evolution of continental lithosphere, we seek to understand the varying responses of these diverse continental lithospheric packages to external geologic processes acting upon them. The present contribution examines how various measurements pertaining to the thermal structure of the southern African lithosphere may be used to infer its internal response to large scale tectonic processes (e.g., mantle upwelling and continental breakup), and thereby lead to better understanding of its internal material and compositional structure.

Archean cratons, of which the Kaapvaal craton of southern Africa is a well-studied example (De Wit et al., 1992), are characterized globally by low average surface heat flow of 41 ± 12 mW m^{-2} (Nyblade and Pollack 1993) that is commonly attributed to the insulating effects of a thick, non-convecting mantle root beneath (Jordan, 1978; Ballard and Pollack 1987). The surface heat flow from Proterozoic terranes that lie adjacent to Archean continental nuclei is similar to that of the cratons, but those lying more distal ($> \sim 400$ km) exhibit a somewhat higher global average of 55 ± 17 mW m^{-2} (Nyblade and Pollack 1993). This difference is commonly attributed to some combination of increased crustal heat production (Morgan 1984) and a thinner mantle lithosphere in the Proterozoic domains (Ballard and Pollack 1987; Jones 1988; Nyblade and Pollack 1993).

In the regional case of southern Africa, the difference is accentuated to the extent that the Proterozoic mobile belts exhibit elevated heat flow averaging ~ 65 mW m^{-2} (Jones, 1987, 1998; Ballard et al., 1987; Nyblade et al., 1990). Thinner lithospheric mantle beneath these regions has been generally favored for explaining the enhanced heat flow of these regions, in part from the apparent agreement with certain steeper P–T arrays calculated from the mineral compositions in mantle-derived inclusions in craton-margin and "off-craton" kimberlites piercing the Proterozoic mobile belts (Boyd, 1973; MacGregor, 1975; Boyd and Gurney, 1986), this despite a long history of controversy over the interpretation of the latter data (e.g., Finnerty, 1989).

Whereas detailed, self-consistent local models of thermal structure based on heat flow and xenolith thermobarometry have been constructed for both Archean and Proterozoic terranes in southern Africa

(Jones 1987, 1988), there is some question as to whether or not these models have general significance. The existence of a canonical "Proterozoic" thermal structure has been questioned by Jaupart and Mareschal (1999) who demonstrated a wide range of thermal structures for the Proterozoic terranes of the Canadian shield, with a lack of correlation of heat flow with tectonic age for Archean, Proterozoic and Paleozoic terranes with similar crustal heat production. These observations cast doubt on the notion that Proterozoic terranes possess a characteristic component of reduced heat flow and by implication a particular lithospheric architecture at depth.

In this study, we have re-examined the xenolith-based evidence for mantle thermal structure in southern Africa and find that there is a significant temporal component to the variability that has important consequences for its interpretation. These inferences are also consistent with observational constraints from lower crustal thermochronology and previous inferences from garnet xenocryst studies. Together, these approaches provide a dynamic view of lithospheric thermal structure that has significant implications for present day surface heat flow, the nature of continental lithosphere, and models for diamond exploration.

2. Peridotite thermobarometry and geotherms: a brief discussion

Boyd (1973) presented a geotherm for the mantle beneath Lesotho that was characterized by a conductive limb corresponding to geophysical predictions, inflected at depth to a steeper thermal gradient indicative of non-steady-state conditions. Whether or not the inflection is an artifact of the calculation method and, if it is not, what its geophysical significance might be, have subsequently been debated at length. The former controversy has to a large degree centered on thermobarometer accuracy, in particular the capacity of the formulation to deal adequately with bulk compositional changes that frequently accompany the proposed geotherm inflection (Nixon and Boyd, 1973; Finnerty, 1989; Carswell 1991). Taylor (1998) addressed this question directly and proposed a new thermobarometer based on experiments in a range of bulk compositions, including fertile ones appropriate to the high-temperature nodules. Nimis and Taylor (2000) reformulated the pyroxene solvus thermometer so that it could be applied to clinopyroxene alone, and developed a new barometer based on the Cr content of clinopyroxene. This provides an alternative to the Al-in-opx barometry and can be used as an independent check. Nimis and Taylor (2000) noted that the Cr-in-cpx barometer produced a smaller depth range for the high temperature xenoliths than Al-in-opx barometer, suggesting that the Cr-in-cpx is less affected by the chemical disequilibrium that may exist in these rocks due to transient heating effects.

Despite the frustrations of this protracted debate, it remains an extremely important one to resolve. Xenolith-derived *P–T* arrays are potentially the most quantitative way to test models for the thermal structure and evolution of the continental mantle. As discussed by Rudnick et al. (1998), the interpretation of the xenolith *P–T* array as a steady state or perturbed phenomenon produces substantially different constraints on lithospheric thickness and composition.

It is commonly the case that xenolith suites in kimberlites (or in the currently accessible portion of a kimberlite intrusion) do not represent the entire column of mantle through which the magma has ascended. It was suggested by Dawson et al. (1975) and Harte (1978) that the process of collating data from different kimberlite pipes produced interpretations different from those drawn from the individual occurrences. Here an attempt has been made to examine the xenolith *P–T* arrays on several geographic scales, with further attention to tectonic setting, host-rock type, and xenolith composition.

3. Methods

3.1. Mantle peridotite thermobarometry

Compositional data on coexisting minerals from xenoliths were compiled from the literature, and from unpublished data collections, including our own recent studies. Pressures and temperatures were computed using three modern thermobarometers: Brey and Köhler (1990) [BKN90], Taylor (1998) [T98] and Nimis and Taylor (2000) [NT00]. The raw data were not screened before performing *PT* calculations,

but were re-examined in the few cases that generated extreme *PT* outliers. In such cases, the anomalous results could always be attributed to typographical errors or obvious instances of chemical disequilibrium between phases, and were eliminated from the plots. We attach relatively low significance to the *PT* results for the lowest temperature (shallowest) xenoliths because of the decreasing sensitivity of the pyroxene composition to temperature as temperature decreases (Smith, 1999), and because of the in-

Table 1
Xenolith suites used in this study

Locality name	Age (Ma)	Host rock type
Kaapvaal Craton		
Kimberley cluster (Bultfontein, Wesselton, Du Toitspan, De Beers, Kamfersdam pipes)	84[1]	kimberlite
Frank Smith/Weltevreden	114[2]	kimberlite
Finsch	118[2]	orangeite
Jagersfontein	86[2]	kimberlite
Monastery	88[1]	kimberlite
Lesotho cluster (Kao, Letseng, Liquobong, Matsoku, Mothae, Pipe 200, Thaba Putsoa)	~ 90[3]	kimberlite
Zimbabwe craton/Magondi Belt		
Letlhakane AK1, AK	~ 90[3]	kimberlite
Proterozoic Namaqua–Natal terrane		
East Griqualand cluster (Abbotsford, Clarkton, Mzongwana, Ramatseliso, Zeekoegat)	150[2,3]	kimberlite
Melton Wold*	140/174[4,5]	orangeite
Uintjiesberg*	101[2]	kimberlite
Hebron*	74[5]	kimberlite
Gansfontein*	? ~ 74[6]	kimberlite
Hoedkop*	80[7]	melilitite
Rehoboth terrane		
Gibeon cluster* (Anis Kubub, Gibeon Townlands, Hanaus, Louwrensia)	65–75[8,9]	kimberlite
Damara Belt (Pan African)		
Swakopmund	75[10]	nephelinite

Age references: (1) Allsopp and Barrett (1975), Kramers et al. (1983), (2) Smith et al. (1985), (3) Davis (1977), (4) Skinner et al. (1994), (5) Smith et al. (1994), (6) Age estimate based on petrographic similarity to nearby Hebron kimberlite, (7) G. Kiviets, unpublished data, (8) Spriggs (1988), (9) Allsopp et al. (1989), (10) Whitehead et al. (2002).

creased possibility of frozen mineral equilibria at these low temperatures.

The three calibrations produce somewhat different absolute results that are not critically important to the conclusions drawn here, which rely on relative comparisons within and between suites of xenoliths. However, the absolute data will influence quantitative estimates of lithospheric thickness, temperature and so on. These significant issues demand a more detailed assessment of thermobarometer accuracy than can be addressed here, and their accurate evaluation may require the application of different thermobarometers for different conditions of origin, rather than the present application of a single technique throughout the entire *PT* range.

The general discussion and data presented follow the BKN90 method for two reasons: (1) this is a commonly used method and we wish to illustrate that these interpretations do not result from a recalibration of the more established thermobarometers and (2)

because the BKN90 calibration tends to suppress the thermal perturbation signal (i.e., the "kink" in the geotherm) and is therefore the most conservative method. No general superiority of this method over the others is implied. In this context, it should be expected that for each thermobarometer there will be a realm where it offers superior accuracy. For example, it is quite conceivable that a thermobarometer producing the most accurate *P–T* estimates for fertile, high-temperature nodules may be unable to satisfy the *P–T* constraints of diamond or graphite-bearing xenoliths of refractory composition.

3.2. Samples

Table 1 lists the sample host intrusion localities, their petrologic affinity, age and tectonic setting and indicates where new data are reported in this study. Their geographic positions and relation to tectonic and crustal age provinces are shown in Fig. 1.

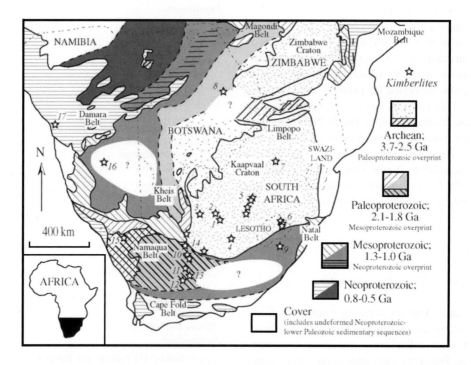

Fig. 1. Crustal age provinces of southern Africa, with location of the xenolith-bearing volcanics yielding data discussed in this study. Locality key: (1) Kimberley cluster, (2) Frank Smith, (3) Finsch, (4) Jagersfontein, (5) Free State province, (6) Northern Lesotho province, (7) Premier cluster, (8) Orapa–Letlhakane cluster, (9) East Griqualand province, (10) Uintjiesberg, (11) Hebron (Hartebeestfontein), (12) Gansfontein, (13) Melton Wold, (14) Eastern Namaqualand province, (15) Hoedkop melilitite, (16) Gibeon province, (17) Swakopmund nephelinites. Modified after Hanson (2003).

4. Results of mantle thermobarometry

4.1. The regional data set and perturbed geotherms

Results from the entire data set, including both cratonic and non-cratonic xenoliths are plotted in Fig. 2, using the BKN90 formulation. The upper part of the diagram consists of a low $P-T$ array with some overlap, but also some distinction, between the on- and off-craton samples, which is discussed later in the paper. At higher P and T (lower part of the diagram), this broad array appears to splay into three branches. The upper (lowest pressure) branch corresponds to the off-craton xenoliths, while the two lower (deeper) branches are both comprised of cratonic xenoliths. Also shown are $P-T$ data for xenoliths from Udachnaya, Siberia (Boyd, 1976, 1984), that lie together with a number of Kaapvaal xenoliths on the lowermost branch. The presence of these two branches of cratonic xenoliths demonstrates that, over the integrated time of sampling (120–80 Ma) there was substantial thermal heterogeneity (~ 150–200°) beneath the Kaapvaal craton. Thus, the combined regional data set supports the perturbed geotherm hypothesis of Boyd (1973), as noted previously by Harte (1978).

The lowermost branch is interpreted as the unperturbed limb of a typical continental shield geotherm, with the middle and upper branches representing perturbations at two broadly different levels in the mantle (e.g., MacGregor, 1975; Boyd and Gurney 1986). While the Siberian PT data are not directly relevant to the southern African mantle, they are included in Fig. 2 because it was concluded (Boyd 1984) that the Siberian xenoliths show no $P-T$ inflection. In Fig. 2, they appear to coincide with a trend that is rather weakly established in the Kaapvaal xenolith suite at pressures greater than 60 kbar, but is well populated at pressures between 50 and 60 kbar. They thus provide additional circumstantial evidence that the deepest $P-T$ data points on the lowermost Kaapvaal $P-T$ branch represent thermally undisturbed mantle lithosphere, and help to define better the suggested trend.

Fig. 2b and c shows the same mineral data recalculated according to the T98 and NT00 thermobarometers. The former barometer, formulated expressly to account for the high-Fe, Ti compositions

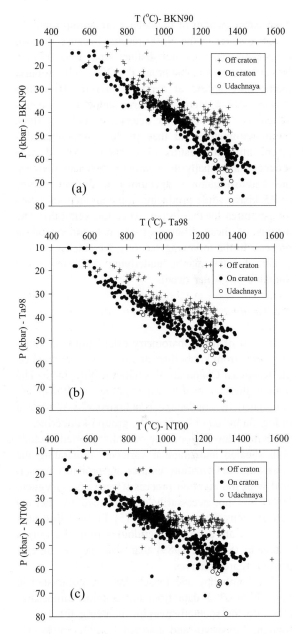

Fig. 2. A comparison of three modern thermobarometers applied to on-craton (solid circles) and off-craton (crosses) mantle xenoliths from southern Africa, with a suite from Udachnaya (Siberia) for reference. (a) BKN, (b) T98, (c) NT00.

commonly encountered in the deeper xenoliths, produces somewhat lower pressure and temperature estimates for these xenoliths that result in an enhanced pattern of thermal perturbation. The thermal disturbance pattern of the deeper on-craton xenoliths is less evident in the NT00 results, because the two deeper branches are not distinct from one another. Nevertheless, a substantial (~ 200 °C) isobaric temperature spread at depth is present. It is thus appears that a degree of lateral temperature heterogeneity of 150–200° is detectable with all three thermobarometers, but they produce somewhat different absolute results. The NT00 method also reduces substantially the relative differences between on- and off-craton temperatures at shallow depths (<35 kbar), while producing some anomalously low temperatures for the deeper off-craton xenoliths. The origins of these differences are not understood at present. In the following sections, we discuss the *P–T* data for the mantle underlying different crustal domains in greater detail.

4.2. Cratonic xenolith suites

Results of thermobarometry calculations on xenolith suites from kimberlites and orangeites erupted on the Kaapvaal Craton are illustrated in Fig. 3a–c. The typical thermal profile expected for a lithosphere dominated by conductive heat transfer is illustrated in Fig. 3a by xenoliths from the group I kimberlites of the Kimberley region. The samples plotted include both sheared and granular varieties, illustrating the points that deformation textures are probably associated with the eruption process (Green and Gueguen, 1974) and not necessarily accompanied by thermal perturbations (Dawson et al., 1975). Further illustrated in Fig. 3a are a suite of peridotites from the Finsch orangeite (Skinner, 1989) that also include diverse textural representatives.

The Kimberley and Finsch data are contrasted in Fig. 3b by *P–T* data from the cratonic kimberlites closer to the southern margin, including the Jagersfontein, Monastery and assorted Lesotho intrusions, and from the Frank Smith kimberlite NW of Kimberley. These xenolith suites all contain substantial proportions of high-temperature xenoliths, that give rise to the central branch of the *P–T* array in Fig. 2a. The highest temperatures of all are found in the Lesotho suite. In detail, with benefit of the large sample suite, it appears that Kimberley, Jagersfontein and Lesotho all carry representatives of both the low-temperature and high-temperature limbs at depth, but it is the

Finsch data that present the strongest argument for an unperturbed geotherm at depth. All three thermobarometers place the Finsch data at lower temper-

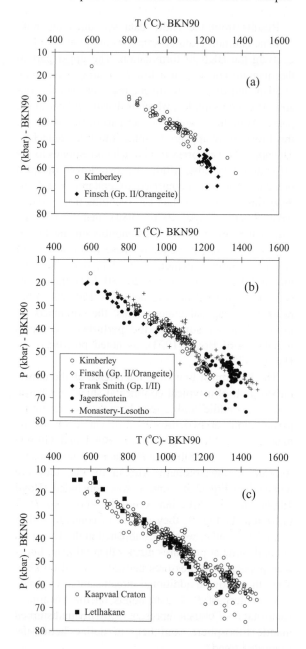

Fig. 3. Cratonic *P–T* arrays calculated according to the BKN90 thermobarometer: (a) Kimberley, Finsch (west central Kaapvaal) (b) southern margin of Kaapvaal compared with Kimberley–Finsch array and the Frank Smith Mine, (c) Letlhakane (Zimbabwe Craton) compared with Kaapvaal array.

atures than other xenoliths derived from similar depths.

Fig. 3c shows the *P–T* array for xenoliths from Letlhakane (Stiefenhofer, 1993), situated in the Proterozoic crust of the Magondi Belt, but assumed to be underlain by mantle of the Zimbabwe craton at its northwest margin. In contrast to those of the southern Kaapvaal margin, the Letlhakane xenoliths show no inflection from the presumably conductive array of the central western Kaapvaal. This is somewhat surprising given the craton margin setting, although the robustness of this observation is compromised by the restricted number of high-pressure samples.

There is some evidence for lateral heterogeneity in the lower *P–T* limb of the cratonic geotherm. Shallow seated (~ 20–30 kbar), commonly amphibole-bearing xenoliths, from the Jagersfontein kimberlite, record lower temperatures than those from Kimberley and the Lesotho kimberlites at similar depths. Furthermore, several xenoliths from the Frank Smith kimberlite lie on the low-temperature side of the Kimberley array. The Frank Smith kimberlite is an unusual body that contains characteristics of both group I and group II kimberlites (Wagner, 1914; Smith, 1983; Bell and Mofokeng, 1998). An age of 114 Ma (Smith et al 1985) is similar to other group II kimberlite (orangeite) intrusions in the area. Similarly low mantle geothermal gradients were reported by Menzies et al. (1999) at the nearby and contemporaneously erupted Newlands kimberlite, and have been recently confirmed with a more extensive sample suite (Bell, 2002). Data from garnet xenocrysts also suggests some differences between orangeite and later kimberlite geotherms (Brown et al., 1998; Menzies and Baumgartner, 1998).

In Fig. 4, the same data as in Fig. 2b (i.e., BKN90 thermobarometer) are plotted, with samples classified according to Ti content of the garnet. This illustrates that the apparent thermal perturbation using this thermobarometer is not an artifact of the high-Ti mineral compositions, because high-Ti samples from Finsch have lower temperatures of origin. However, it does indicate that a great many of the deep seated cratonic xenoliths are Ti-rich. This feature has previously been explained as due to a transition to asthenospheric mantle (Boyd, 1973; Nixon and Boyd, 1973), or as a result of metasomatism by asthenosphere-

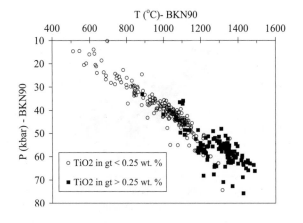

Fig. 4. *P–T* array for cratonic xenoliths classified according to the Ti content of their garnets.

derived melts (Ehrenberg 1979; Gurney and Harte, 1980).

4.3. The Gibeon xenoliths

P–T data from xenoliths from the Gibeon kimberlites, some 500 km west of the craton margin in Namibia are illustrated in Fig. 5a. These xenoliths present a rather different picture from the typical cratonic situation illustrated in Fig. 2. The data are bounded on their low-temperature side by temperatures that correspond to those defined by the cratonic array as at Kimberley, but extend over a broad temperature range at any given pressure, with a general tendency towards increased temperature heterogeneity at higher pressures. This gives the general impression of a scattered shallow geotherm. As in the case of many cratonic xenolith suites, the higher temperature xenoliths are mostly of the high-Ti compositional variety, and those recording highest temperatures at a given pressure are commonly the most intensely deformed (Fig. 5b). However, both sheared, low temperature xenoliths and undeformed, high temperature xenoliths exist. The NT00 and T98 thermobarometers that are formulated to take more explicit account of fertile mineral compositions have the effect of compressing the pressure range, so that high temperature xenoliths occur over essentially the same depth range as the lower temperature xenoliths. The Ti-rich mineral compositions of the highest temperature peridotites are identical to those in Cr-

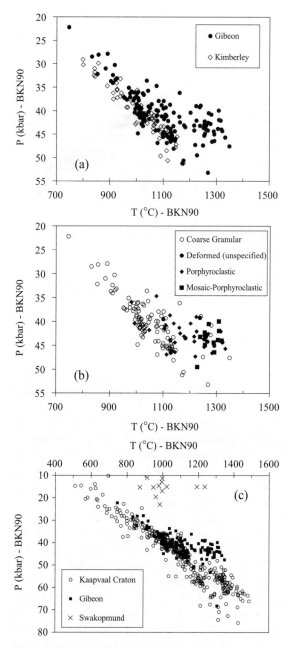

Fig. 5. *P–T* data for Namibian xenoliths. (a) Gibeon province demonstrating low-temperature overlap with Kimberley array. (b) Textural correlation in the Gibeon xenoliths. (c) Gibeon province *P–T* array compared with cratonic array and Swakopmund spinel lherzolites (Damara belt).

poor megacrysts, indicating equilibration with a magma. We concur with previous studies that have interpreted the Gibeon xenolith *P–T* data in terms of

disturbance from a steady state profile by metasomatism and magmatic activity (Mitchell, 1984; Franz et al., 1996a,b). However, we emphasize here the low-temperature overlap with the cratonic xenolith thermal profile.

The Swakopmund xenoliths (Fig. 5c) are spinel peridotites many of which have compositions typical of Phanerozoic mantle (Boyd, 1989; Griffin et al., 1999), while a subset of more refractory harzburgites appear to be compositionally similar to the Gibeon xenoliths (Whitehead et al., 2002). All samples indicate high temperatures at shallow depths: a lower temperature group at 870–1020 °C lie well above temperatures at similar pressures (12–22 kbar) for Proterozoic and cratonic domains. These samples show no signs of deformation or high-Ti metasomatism. Another even higher temperature group (~ 1200 °C) may have experienced magmatic heating (Whitehead et al., 2002).

4.4. Xenoliths peripheral to the Kaapvaal craton

Xenoliths from these areas are sampled by kimberlites and orangeites of a wide range of ages, from ~ 175 to 70 Ma. The *P–T* data are illustrated in Fig. 6. The oldest Mesozoic group I kimberlites are the ~ 150 Ma East Griqualand intrusions to the SE of the craton. *P–T* data for peridotite xenoliths from these kimberlites (Fig. 6a) consist of a few points lying close to the cratonic Kimberley array, but a majority are displaced to higher temperatures, including both very shallow level high-*T* samples and a deep (~ 40 kbar) high-temperature cluster. In contrast, xenoliths from the Melton Wold orangeite, located to the southwest of the craton, fall mostly on the cratonic array (Fig. 6a and b). The age of Melton Wold is somewhat uncertain, but is distinctly older than group I kimberlites in this region. Imprecise U–Pb perovskite ages suggest an age in the region of 140–145 Ma (Smith et al 1994), but a more precise Rb–Sr age of 174 ± 2 Ma has been determined for the closely located Droogfontein orangeite that has the same "transitional" isotopic composition (Smith et al., 1994).

Xenoliths from the younger kimberlites and melilites that intrude Proterozoic Namaqualand basement to the southwest of the craton from 103 to 68 Ma all record temperatures above those of the Melton Wold

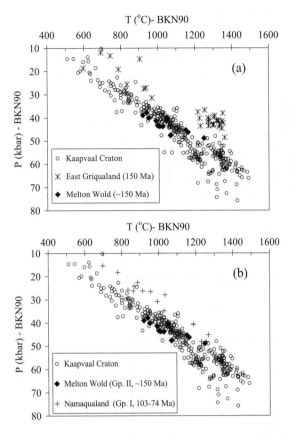

Fig. 6. P–T arrays for peridotites from the Namaqua–Natal mobile belt, adjacent to (<400 km from) the Kaapvaal craton. (a) Jurassic kimberlites and orangeites: East Griqualand province and Melton Wold. (b) Comparison of P–T array for Cretaceous kimberlites and melilitites with the closely situated, but older Melton Wold orangeite array.

suite (Fig. 6c). Most of the xenoliths derive from kimberlites in Eastern Namaqualand and include samples from the Hebron and Gansfontein kimberlites that are within a few tens of kilometers of Melton Wold (Fig. 1). The shallower xenoliths are similar to the East Griqualand suite in being displaced to higher T throughout their depth range, showing no instances of overlap with the cratonic geotherm. However, there is no group of deeper, highly thermally perturbed samples, such as occurs at East Griqualand (or Gibeon). The Eastern Namaqualand kimberlite and Hoedkop melilitite xenoliths are uniformly displaced above the cratonic array throughout their depth range and appear to represent a suite that has equilibrated to a higher geotherm.

5. Discussion

5.1. Significance of low-temperature xenoliths off craton

The presence of mantle temperatures characteristic of Archean lithosphere in xenoliths from Melton Wold situated off the craton to the southwest, in many of the Gibeon xenoliths some 500 km to the northwest of the craton, and in a few East Griqualand xenoliths to the southeast of the craton is strong evidence that a craton-like geotherm existed at one time in the pericratonic Proterozoic lithosphere. The implication is that after Namaqua orogenesis, terminating near 1.0 Ga, the entire interior of the subcontinent, including Archean craton and surrounding Proterozoic belts through which the abovementioned kimberlites were erupted, cooled to a craton-like thermal gradient of ~ 40 mW m^{-2}. This accords with the expectation from global heat flow studies (Nyblade and Pollack, 1993), but not from local observations in southern Africa. These conclusions are, therefore, at odds with standard models of lithospheric structure in southern Africa that propose steeper steady state thermal gradients in the mantle beneath Proterozoic crustal terranes.

How can an apparent craton-like geotherm be reconciled with the higher surface heat flow for the Proterozoic mobile belts within a steady-state framework? In fact, it remains plausible that the contrast in present-day surface heat flow between Proterozoic and Archean lithosphere is a function of differential crustal heat production (Morgan, 1984; Jones, 1987). Jones (1987) noted that surface heat flow in the Proterozoic crust adjacent to the Kaapvaal craton is generally lower than that westward away from the craton. On a local scale of ≤ 100 km, heat flow measurements exhibit considerable dispersion and are well correlated with both borehole heat production and thermal conductivity measurements (Fig. 7a), supporting a shallow crustal control on local thermal gradient. A variety of crustal radiogenic heat production models consistent with surface measurements as well as Eastern Namaqualand lower crustal xenolith compositions can reproduce the modestly elevated surface heat flow data for the region with the same basal heat flux into the Proterozoic crust as that inferred for the Archean craton (~ 16 mW/m^2) (Fig. 7b). The resultant model geothermal gradients

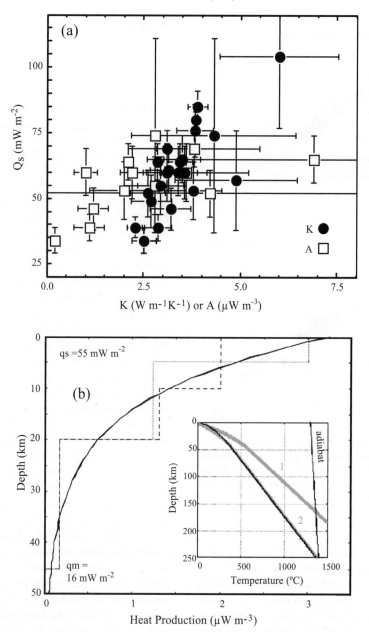

Fig. 7. (a) Measured average thermal conductivity (K) and radiogenic heat production (A) for borehole lithologies at heat flow measurement sites of Jones (1987), illustrating their correlation with surface heat flow. (b) Several crustal heat production models (thin black lines) satisfying an average surface heat flow of 55 mW m^{-2} for Eastern Namaqualand while maintaining a basal (mantle-derived) heat flow of 16 mW m^{-2}, and a craton-like lithospheric thermal gradient (thin black line of insert). Reference geotherms in the inset for the (1) Namaqua mobile belt and (2) Kaapvaal craton are from Jones (1988); the former may be derived from the crustal heat production model drawn as a thick gray line.

in the deep Proterozoic and Archean lithospheric sections are indistinguishable, in contrast to the model for the Namaquan lithosphere of Jones (1988). Al-

though it was pointed out by Jaupart and Mareschal (1999) that differences in crustal heat production can lead to divergent lithospheric mantle geotherms at depth, this divergence is minimized if the elevated heat production is concentrated in the upper to middle crust. A shallow concentration of heat producing elements is supported by Eastern Namaqualand lower crustal xenolith compositions, which exhibited very low heat production values averaging $\sim 0.2 \, \mu W \, m^{-3}$ (Schmitz, unpublished data).

Although we consider that inferences of a pre-Jurassic craton-like thermal state in the Proterozoic mobile belt lithosphere may be reconciled with present-day surface heat flow measurements simply through differential crustal heat production, the mantle xenolith data obviously lead us to question the more fundamental assumption that the present-day surface heat flow of the southern African continent represents a thermal steady-state condition, despite the antiquity of the constituent crust. Whether some component of the elevated heat flow beneath the Proterozoic mobile belts is a remnant transient response to changes in the subcontinental lithosphere over the past 200 Ma is explored in the following discussion.

The presence of an approximately cratonic thermal gradient west of the craton suggests that the thickness of mantle lithosphere in this region would have been similar to that beneath the craton, unless there was significantly lower heat production at the deeper levels of off-craton, compared with on-craton, lithosphere. The potential of Proterozoic terranes to exhibit a thermal structure equivalent to that of Archean cratonic lithosphere (documented by heat flow measurements on the Canadian Shield by Jaupart and Mareschal, 1999) has potentially important implications for diamond exploration models and suggests that the general paucity of primary diamond deposits in Proterozoic lithospheric domains is probably not a consequence of their thermal state.

5.2. Thermal disturbance of Proterozoic lithosphere in space and time

The large range of temperatures at pressures in the 30–40 kbar range, extending from craton-like temperatures near 900 to 1200–1300 °C is taken to indicate that this region of the mantle was being heated at the time of sampling and therefore represents a non-steady state thermal condition, as proposed by Mitchell (1984). It is shown in Fig. 5b that this heating is commonly associated with the presence of deformation and there are also correlations of temperature with Ti-enriched mineral compositions, of which the extreme examples are very similar to the compositions of the Cr-poor megacryst suite from the region. This xenolith suite therefore records a perturbed geotherm that represents passage from a close-to-cratonic, purely conductive state, to a condition where heat is being advected to high levels in the lithosphere. The thermal heterogeneity, textural evidence and the presence of chemical zoning in a few samples indicates that some of this heating is occurring together with kimberlite eruption, but it may also have been going on for some time prior to eruption. This situation is similar to that observed in the cratonic xenoliths from the southern margin of the Kaapvaal, but occurs at shallower levels and is also more extensive. MacGregor (1975) and Boyd and Gurney (1986) proposed that this depth difference was an indication of contrasting lithospheric thickness, but this conclusion is difficult to reconcile with the $P-T$ coincidence of the thermally unperturbed xenoliths from Gibeon and the Kaapvaal craton. Rather than a "kink" in the geotherm, the Namibian mantle lithosphere appears to be extensively infiltrated and heated over a depth range of about 50 km. This is particularly apparent if the T98 thermobarometer is applied.

The East Griqualand xenolith suite represents a somewhat similar thermal situation to Gibeon, but shows less evidence of the original craton-like geotherm, and a more restricted zone of magmatic reaction. In this region, the situation is however recorded at 150 Ma, about 80 Ma earlier than Gibeon. At a similar time, to the southwest of the craton at Melton Wold, only low temperatures, similar to those on the craton, are recorded. However, by 100–70 Ma, the geotherm recorded by kimberlites in this area is displaced to higher temperature over its entire depth range, yet does not attain the very shallow thermal gradient and magmatic-type temperatures and compositions of Gibeon or East Griqualand. There is, thus, less direct evidence for large scale local magma incursion into the lithosphere in this region, and the heating may be of a more regional nature.

Finally, the most extreme lithospheric heating is observed further to the northwest of Gibeon, in the Swakopmund nephelinite plugs (Whitehead et al.,

2002), where this heating had penetrated to the spinel stability field, i.e., ~ 40 km depth by 75 Ma. However, in this region, the prior lithospheric structure is likely to have been different, because these lavas erupt through the Pan African age Damara belt, the development of which is proposed to include a phase of intra-continental rifting (Kröner, 1982).

It thus appears that the regional heating event swept in a general westward direction across the region. The first sign of a major thermal perturbation to the southern African mantle in the Mesozoic is the outpouring at ~ 183 Ma of Karoo flood basalts sequences preserved in greatest thickness along the eastern margin of the craton in the Lebombo and Lesotho. At 150 Ma, a thermal event was recorded in the East Griqualand xenoliths but was not yet apparent at Melton Wold, approximately contemporaneous, but 300 km to the west. When sampled in this region at 103 Ma and again at 74 Ma (Uintjiesberg and Hebron, respectively), the entire mantle in this region had been heated by ~ 100°. Two hundred kilometers further NW, at Hoedkop (80 Ma), a similar situation prevailed. However, at approximately the same time (~ 70 Ma) the Gibeon transition from a craton-like thermal regime was still in progress.

Whether the migration was continuous or episodic cannot be determined from the sparse sampling. This pattern however mimics broadly that of continental breakup in the region, with opening of the Indian ocean close to the time of Karoo volcanism at ~ 180 Ma followed by the South Atlantic opening closer to 130 Ma coincident with the Parana–Etendeka volcanism. A continuous migration of the thermal pulse is suggested by a compilation of ages of Mesozoic alkaline eruptive rocks in southern Africa. It is tempting to seek the origins of the observed thermal perturbations in these large scale events.

5.3. Supporting evidence from crustal xenoliths

Schmitz and Bowring (2003) have recently traced the thermal evolution of the lower crust of southern Africa by U–Pb thermochronometry of accessory minerals in kimberlite-hosted lower crustal xenoliths. These studies show that by ~ 700 Ma, the Proterozoic lower crust sampled by the 115 to 74 Ma Eastern Namaqualand, Northern Lesotho, and northern Bot-

swana (Orapa) kimberlites (i.e., southwest, southeast, and north of the craton, respectively) had cooled to below the limit of ~ 450 °C required for closed system behavior of the U–Pb system in rutile. This temperature corresponds to that of a cratonic geotherm for the ~ 1.2 GPa pressure of origin of the samples. Due to the lower radiogenic heat production and higher thermal conductivity of the mantle relative to crustal materials, the cratonic state recorded in the lower crust necessarily extends to the underlying lithospheric mantle. This result and those derived above from mantle xenolith thermobarometry and heat flow modeling are thus mutually consistent.

In addition, these craton-margin and off-craton lower crustal samples record a protracted late Mesozoic heating event that raised the lower crust up to ~ 550 °C for tens of millions of years, inducing substantial Pb-loss in the rutile and apatite U–Pb thermochronometers. By contrast, the thermochronological manifestations of heating are absent or only weakly manifested in the cratonic interior (e.g., Newlands and Free State kimberlites and orangeites). We propose that this event corresponds to that recorded by the thermal perturbations of the various mantle geotherms that are seen to propagate to high levels in the off-craton lithosphere, but remain confined to the deepest xenolith samples in most cratonic suites. Simple one-dimensional thermal diffusion calculations show that the time scales and magnitude of the requisite craton-margin lower crustal heating are consistent with the conductive relaxation of a transient Middle to Late Jurassic advective magmatic heating event penetrating to the 30–40 kbar depths indicated by perturbed mantle xenolith thermobarometry (Figs. 2 and 6).

5.4. Implications for the thermal structure of continental lithosphere

Our interpretation of the results of mantle thermobarometry is that the kimberlite-hosted xenoliths of southern Africa record a snapshot of a thermally evolving system and not a steady state picture that is generally representative of the various types of subcontinental mantle. The time scale for thermal diffusion over distances characterizing the thickness of the thermally conductive lithosphere implies relaxation times of hundreds of Ma for major mantle heating events. Thus, in the southern African case where heating in the past 200 Ma is recorded, there is likely to be little correspondence between the present day measured surface heat flow and the steady state thermal structure of the mantle. In fact, it may be expected that the present day surface heat flow contains a significant component residual from this event and that the higher average heat flow of the off-craton areas reflects in part the advection of heat to shallower depth in these areas.

This hypothesis may explain the anomalously high reduced heat flow calculated for southern African Proterozoic terranes compared with others worldwide (Artemieva and Mooney, 2001). This susceptibility to thermal incursion may be due to the less refractory character of Proterozoic mantle lithosphere (Boyd and Nixon, 1979; Pearson et al., 1994; Hoal et al., 1995; Griffin et al., 1998), as well as the possible penetration of magmas along zones of weakness near craton boundaries that represent old loci of tectonic activity. The advection of heat to high levels of the lithosphere does not necessarily imply large scale erosion of the thermal boundary layer and may be accomplished by local channeling of melt. Seismic tomography of the southern African mantle (James et al., 2001) indicates that pervasive erosion has not occurred to the shallow depths at which xenolith compositions indicate the presence of "fertile" components (Fig. 4).

The present study has been concerned primarily with some of the gross qualitative features of xenolith thermobarometry, but there is an indication of further information in the quantitative and finer scale details. Large magnitude thermal perturbation is confined to the deepest cratonic xenoliths, suggesting that the cratons are relatively impervious to magmatic infiltration. However, small degrees of advective heat transfer by more ephemeral metasomatic melts may play a role in their thermal evolution at shallower levels. We have noted above some variation for craton-derived xenoliths in what has been inferred here to represent the unperturbed (steady state) limb of the cratonic geotherm. Whether this is truly unperturbed from the previous steady state, or represents an adjustment to new conditions, requires further investigation. Variation may also be due to local differences in mantle heat production as a result of different intensities of metasomatism. Possible perturbation has been suggested previously in studies of garnet

xenocrysts from Kaapvaal kimberlites and orangeites (Griffin and Ryan, 1995; Brown et al., 1998; Menzies and Baumgartner, 1998). The range of characteristic diffusive equilibration rates for various chemical equilibria used in thermometry and barometry of peridotites provides, in theory, a means to investigating the time scale of this process. The quantitative constraints that the form of this array places on lithospheric thickness and composition, and the implications of these variations for lithospheric processes are the subject of a separate study.

Acknowledgements

This research was supported by the National Science Foundation Continental Dynamics Program grant NSF-9526840. DRB gratefully acknowledges support from the Crosby Fellowship Fund during residence at MIT, that made this collaboration possible, as well as support from the Carnegie Institution of Washington. Thanks to Joe Boyd at the Geophysical Laboratory for hospitality, discussion and access to unpublished data, and to Chris Hadidiacos and David George for the benefit of their expertise and assistance with the electron microprobe. Work at the University of Cape Town was hosted by J.J. Gurney and A.P. le Roex and supported by De Beers and the National Research Foundation of South Africa. Margaret Merry of Westerford High School is thanked for assistance with the thermobarometry calculations. We thank Mike Jones for enlightening discussion and J.K. Russell and W. Davis for constructive reviews of the manuscript.

References

Allsopp, H.L., Barrett, D.R., 1975. Rb-Sr age determinations on South African kimberlite pipes. Phys. Chem. Earth 9, 605–617.

Allsopp, H.L., Bristow, J.W., Smith, C.B., Brown, R., Gleadow, A.J.W., Kramers, J.D., Garvie, O.G., 1989. A summary of radiometric dating methods applicable to kimberlites and related rocks. In: Ross, J. (Ed.), Kimberlites and Related Rocks, Vol. 1. Geol. Soc. Australia Spec. Publ., vol. 14, pp. 343–357.

Artemieva, I.M., Mooney, W.D., 2001. Thermal thickness and evolution of Precambrian lithosphere: a global study. J. Geophys. Res. 106, 16387–16414.

Ballard, S., Pollack, H.N., 1987. Diversion of heat by Archean cratons: a model for southern Africa. Earth Planet. Sci. Lett. 85, 253–264.

Ballard, S., Pollack, H.N., Skinner, N.J., 1987. Terrestrial heat flow in Botswana and Namibia. J. Geophys. Res. 92, 6291–6300.

Bell, D.R., 2002. Mesozoic thermal evolution of the Kaapvaal Craton mantle root. Eos Trans. AGU 83 (47), Fall Meeting Suppl., Abstr. T61A-1243.

Bell, D.R., Mofokeng, S.W., 1998. Cr-poor megacrysts from the Frank Smith Mine and the source region of transitional kimberlites. Extd. Abstrs., 7th Internat. Kimberlite Conf., Cape Town, pp. 64–66.

Boyd, F.R., 1973. A pyroxene geotherm. Geochim. Cosmochim. Acta 37, 2533–2546.

Boyd, F.R., 1976. Inflected and noninflected geotherms. Carnegie Inst. Washington Yearb. 75, 521–523.

Boyd, F.R., 1984. A Siberian geotherm based on lherzolite xenoliths from the Udachnaya kimberlite. U.S.S.R. Geology 12, 528–530.

Boyd, F.R., 1989. Compositional distinction between oceanic and cratonic lithosphere. Earth Planet. Sci. Lett. 96, 15–26.

Boyd, F.R., Gurney, J.J., 1986. Diamonds and the African lithosphere. Science 232, 472–477.

Boyd, F.R., Nixon, P.H., 1979. Garnet lherzolite xenoliths from the kimberlites of East Griqualand, South Africa. Carnegie Inst. Washington Yearb. 78, 488–492.

Brey, G.P., Köhler, T., 1990. Geothermobarometry in four phase lherzolites: II. New thermobarometers and practical assessment of existing thermobarometers. J. Petrol. 31, 1353–1378.

Brown, R.W., Gallagher, K., Griffin, W.L., Ryan, C.G., de Wit, M.C.J., Belton, D.X., Harman, R., 1998. Kimberlites, accelerated erosion and evolution of the lithospheric mantle beneath the Kaapvaal Craton during the mid-Cretaceous. Extd. Abstrs, 7th Internat. Kimberlite Conf., Cape Town, pp. 105–107.

Carswell, D.A., 1991. The garnet–orthopyroxene Al barometer: problematic application to natural garnet lherzolite assemblages. Mineral. Mag. 55, 19–31.

Davis, G.L., 1977. The ages and uranium contents of zircons from kimberlites and associated rocks. Carnegie Inst. Washington Yearb. 76, 631–635.

Dawson, J.B., Gurney, J.J., Lawless, P.J., 1975. Palaeogeothermal gradients derived from xenoliths in kimberlite. Nature 257, 299–300.

De Wit, M.J., Roering, C., Hart, R.J., Armstrong, R.A., De Ronde, C.E.J., Green, R.W.E., Tredoux, M., Peberdy, E., Hart, R.A., 1992. Formation of an Archaean continent. Nature 357, 553–562.

Ehrenberg, S.N., 1979. Garnetiferous ultramafic inclusions in minette from the Navajo Volcanic Field. In: Boyd, F.R., Meyer, H.O.A. (Eds.), The Mantle Sample: Inclusions in Kimberlites and other Volcanics. Proceedings of the 2nd International Kimberlite Conference, Santa Fe. AGU, Washington, pp. 330–344.

Finnerty, A.A., 1989. Xenolith-derived mantle geotherms: whither the inflection? Contrib. Mineral. Petrol. 102, 367–375.

Franz, L., Brey, G.P., Okrusch, M., 1996a. Re-equilibration of ultramafic xenoliths from Namibia by metasomatic processes at the mantle boundary. J. Geol. 104, 599–615.

Franz, L., Brey, G.P., Okrusch, M., 1996b. Steady state geotherm, thermal disturbances, and tectonic development of the lower lithosphere underneath the Gibeon kimberlite province, Namibia. Contrib. Mineral. Petrol. 126, 181–198.

Green II, H.W., Gueguen, Y. 1974. Origin of kimberlite pipes by diapiric upwelling in the upper mantle. Nature 249, 617–620.

Griffin, W.L., Ryan, C.G., 1995. Trace elements in indicator minerals: area selection and target evaluation in diamond exploration. J. Geochem. Explor. 53, 311–337.

Griffin, W.L., O'Reilly, S.Y., Ryan, C.G., Gaul, O., Ionov, D.A., 1998. Secular variation in the composition of subcontinental lithospheric mantle: geophysical and geodynamic implications. In: Braun, J., Dooley, J.C., Goleeby, B.R., Van der Hilst, R.D., Klootwijk, C.T. (Eds.), Structure and Evolution of the Australian Continent. AGU Geodynamics Series, vol. 26, pp. 1–26.

Griffin, W.L., O'Reilly, S.Y., Ryan, C.G., 1999. The composition and origin of subcontinental lithospheric mantle. In: Fei, Y., Bertka, C., Mysen, B.O. (Eds.), Mantle Petrology: Field Observations and High-pressure Experimentation: A Tribute to Francis R. (Joe) Boyd. Geochemical Soc. Spec., vol. 6. The Geochemical Society, San Antonio, pp. 13–45.

Gurney, J.J., Harte, B., 1980. Chemical variations in upper mantle nodules from southern African kimberlites. Philos. Trans. R. Soc. Lond., A 297, 273–293.

Hanson, R.E., 2003. Proterozoic geochronology and tectonic evolution of southern Africa. In: Yoshida, M., Windley, B.F., Dasgupta, S. (Eds.), Proterozoic East Gondwana: supercontinent assembly and breakup. Geological Society of London Spec. Publ. 206, pp. 427–463.

Harte, B., 1978. Kimberlite nodules, upper mantle petrology and geotherms. Philos. Trans. R. Soc. Lond., Ser. A 288, 487–500.

Hoal, B.G., Hoal, K.E.O., Boyd, F.R., Pearson, D.G., 1995. Tectonic setting and mantle composition inferred from peridotite xenoliths, Gibeon kimberlite field. Namibia Extd. Abstrs., Sixth International Kimberlite Conference, Novosibirsk, pp. 239–241.

James, D.E., Fouch, M.J., VanDecar, J.C., van der Lee, S., 2001. Tectospheric structure beneath Southern Africa. Geophys. Res. Lett. 28, 2485–2488.

Jaupart, C., Mareschal, J.C., 1999. The thermal structure and thickness of continental roots. Lithos 48, 93–114.

Jones, M.Q.W., 1987. Heat flow and heat production in the Namaqua Mobile Belt, South Africa. J. Geophys. Res. 92, 6273–6289.

Jones, M.Q.W., 1988. Heat flow in the Witwatersrand Basin and environs and its significance for the South African shield geotherm and lithospheric thickness. J. Geophys. Res. 93, 3243–3260.

Jones, M.Q.W., 1998. A review of heat flow in southern Africa and the thermal structure of the lithosphere. South. Afr. Geophys. Rev. 2, 115–122.

Jordan, T.H., 1978. Composition and development of the continental tectosphere. Nature 274, 544–548.

Kramers, J.D., Roddick, J.C.M., Dawson, J.B., 1983. Trace element and isotopic studies on veined, metasomatic and "MARID" xenoliths from Bultfontein, South Africa. Earth Planet. Sci. Lett. 65, 90–106.

Kröner, A., 1982. Rb–Sr geochronology and tectonic evolution of the Pan African Damara belt of Namibia, southwestern Africa. Am. J. Sci. 82, 1471–1507.

MacGregor, I.D., 1975. Petrologic and thermal structure of the upper mantle beneath South Africa in the Cretaceous. Phys. Chem. Earth 9, 455–466.

Menzies, A.H., Baumgartner, M.C., 1998. Application of garnet geothermobarometry to southern African kimberlites. Extd. Abstrs., 7th Internat. Kimberlite Conf., Cape Town, pp. 570–572.

Menzies, A.H., Carlson, R.W., Shirey, S.B., Gurney, J.J., 1999. Re-Os systematics of Newlands peridotite xenoliths; implications for diamond and lithosphere formation. In: Gurney, J.J., et al., (Eds.), The P.H. Nixon Volume. Proceedings of the 7th International Kimberlite Conference, Vol. 2. Red Roof Design, Cape Town, pp. 566–573.

Mitchell, R.H., 1984. Garnet lherzolites from the Hanaus-I and Louwrensia kimberlites of Namibia. Contrib. Mineral. Petrol. 86, 178–188.

Morgan, P., 1984. The thermal structure and thermal evolution of the continental lithosphere. Phys. Chem. Earth 15, 107–193.

Nimis, P., Taylor, W.R., 2000. Single clinopyroxene thermobarometry for garnet peridotites: Part 1. Calibration and testing of a Cr-in-cpx barometer and an enstatite-in-cpx thermometer. Contrib. Mineral. Petrol. 139, 541–554.

Nixon, P.H., Boyd, F.R., 1973. Petrogenesis of the granular and sheared ultrabasic nodule suite in kimberlite. In: Nixon, P.H. (Ed.), Lesotho Kimberlites. Lesotho National Development, Maseru, pp. 48–56.

Nyblade, A.A., Pollack, H.N., 1993. A global analysis of heat flow from Precambrian terrains: implications for the thermal structure of Archean and Proterozoic lithosphere. J. Geophys. Res. 98, 12207–12218.

Nyblade, A.A., Pollack, H.N., Jones, D.L., Podmore, F., Mushayandebvu, M., 1990. Terrestrial heat flow in east and southern Africa. J. Geophys. Res. 95, 17371–17384.

Pearson, D.G., Boyd, F.R., Hoal, K.E.O., Hoal, B.G., Nixon, P.H., Rogers, N.W., 1994. A Re-Os isotopic and petrological study of Namibian peridotites: contrasting petrogenesis and composition of on- and off-craton lithospheric mantle. Mineral. Mag. 58A, 703–704.

Rudnick, R.L., McDonough, W.F., O'Connell, R.J., 1998. Thermal structure, thickness and composition of continental lithosphere. Chem. Geol. 145, 395–411.

Schmitz, M.D., Bowring, S.A., 2003. Constraints on the thermal evolution of continental lithosphere from U–Pb accessory mineral thermochronology of lower crustal xenoliths, southern Africa. Contrib. Mineral. Petrol. 144, 592–618.

Skinner, C.P., 1989. The petrology of peridotite xenoliths from the Finsch kimberlite, South Africa. Trans. Geol. Soc. South. Afr. 92, 197–206.

Skinner, E.M.W., Viljoen, K.S., Clark, T.C., Smith, C.B., 1994. The petrography, tectonic setting and emplacement ages of kimberlites in the southwestern border region of the Kaapvaal craton, prieska area, South Africa. In: Meyer, H.O.A., Leonardos, O. (Eds.), Kimberlites, related rocks and mantle xenoliths: Proceedings of the 5th International Kimberlite Conference. Special Publ. 92/1, CPRM, Brasilia, pp. 80–97.

Smith, C.B., 1983. Pb- Sr and Nd-isotopic evidence for the source regions of southern African Cretaceous kimberlites. Nature 304, 51–54.

Smith, D., 1999. Temperatures and pressures of mineral equilibration in peridotite xenoliths: review, discussion, and implications. In: Fei, Y., Bertka, C.M., Mysen, B.O. (Eds.), Mantle petrology: field observations and high pressure experimentation: a tribute to Francis R. (Joe) Boyd. Geochemical Soc. Spec. Publ. 6, pp. 171–188.

Smith, C.B., Allsopp, H.L., Kramers, J.D., Hutchinson, G., Roddick, J.C., 1985. Emplacement ages of Jurassic–Cretaceous kimberlites by the Rb–Sr method on phlogopite and whole rock samples. Trans. Geol. Soc. South. Afr. 88, 249–266.

Smith, C.B., Clark, T.C., Barton, E.S., Bristow, J.W., 1994. Emplacement ages of kimberlite occurrences in the Prieska region, southwest border of the Kaapvaal craton, South Africa. Chem. Geol. 113, 149–169.

Spriggs, A.J., 1988. An isotopic and geochemical study of kimberlites and associated alkaline rocks from Namibia. Unpublished. PhD Thesis. University of Leeds, UK.

Stiefenhofer, J., 1993. The petrography, mineral chemistry and isotope geochemistry of a mantle xenolith suite from the Letlhakane DK 1 and DK 2 kimberlite pipes, Botswana. PhD thesis, Rhodes Univ., South Africa.

Taylor, W.R., 1998. An experimental test of some geothermometer and geobarometer formulations for upper mantle peridotites with application to the thermobarometry of fertile lherzolite and garnet websterite. Neues Jahrb. Miner. Abh. 172, 381–408.

Wagner, P.A., 1914. The Diamond Fields of Southern Africa. The Transvaal Leader, Johannesburg. Reprinted by C. Struik, Cape Town, 1971.

Whitehead, K., le Roex, A.P., Class, C., Bell, D.R., 2002. Composition and Cretaceous thermal structure of the upper mantle beneath the Damara Mobile Belt: evidence from nephelinites-hosted peridotite xenoliths, Swakopmund, Namibia. J. Geol. Soc. Lond. 159, 307–321.

Available online at www.sciencedirect.com

SCIENCE @ DIRECT°

ELSEVIER

Lithos 71 (2003) 289–322

LITHOS

www.elsevier.com/locate/lithos

The origin of garnet and clinopyroxene in "depleted" Kaapvaal peridotites

Nina S.C. Simon [a,*], Gordon J. Irvine [b,1], Gareth R. Davies [a],
D. Graham Pearson [b], Richard W. Carlson [c]

[a] Faculty of Earth and Life Sciences, Vrije Universiteit Amsterdam, De Boelelaan 1085,
1181 HV Amsterdam, The Netherlands
[b] Department of Geological Sciences, Durham University, South Road, Durham DH1 3LE, UK
[c] Department of Terrestrial Magnetism, Carnegie Institution of Washington, 5241 Broad Branch Road N.W.,
Washington, DC 20015, USA

Abstract

A detailed petrographic, major and trace element and isotope (Re–Os) study is presented on 18 xenoliths from Northern Lesotho kimberlites. The samples represent typical coarse, low-temperature garnet and spinel peridotites and span a P–T range from ~ 60 to 150 km depth. With the exception of one sample (that belongs to the ilmenite–rutile–phlogopite–sulphide suite (IRPS) suite first described by [B. Harte, P.A. Winterburn, J.J. Gurney, Metasomatic and enrichment phenomena in garnet peridotite facies mantle xenoliths from the Matsoku kimberlite pipe, Lesotho. In: Menzies, M. (Ed.), Mantle metsasomatism. Academic Press, London 1987, 145–220.]), all samples considered here have high Mg# and show strong depletion in CaO and Al_2O_3. They have bulk rock Re depletion ages (T_{RD}) >2.5 Ga and are therefore interpreted as residua from large volume melting in the Archaean. A characteristic of Kaapvaal xenoliths, however, is their high SiO_2 concentrations, and hence, modal orthopyroxene contents that are inconsistent with a simple residual origin of these samples. Moreover, trace element signatures show strong overall incompatible element enrichment and REE disequilibrium between garnet and clinopyroxene. Textural and subtle major element disequilibria were also observed. We therefore conclude that garnet and clinopyroxene are not co-genetic and suggest that (most) clinopyroxene in the Archaean Kaapvaal peridotite xenoliths is of metasomatic origin and crystallized relatively recently, possibly from a melt precursory to the kimberlite.

Possible explanations for the origin of garnet are exsolution from a high-temperature, Al- and Ca-rich orthopyroxene (indicating primary melt extraction at shallow levels) or a majorite phase (primary melting at >6 GPa). Mass balance calculations, however, show that not all garnet observed in the samples today is of a simple exsolution origin. The extreme LREE enrichment (sigmoidal REE pattern in all garnet cores) is also inconsistent with exsolution from a residual orthopyroxene. Therefore, extensive metasomatism and probably re-crystallization of the lithosphere after melt-depletion and garnet exsolution is required to obtain the present textural and compositional features of the xenoliths. The metasomatic agent that modified or perhaps even precipitated garnet was a highly fractionated melt or fluid that might have been derived from the asthenosphere or from recycled oceanic crust. Since, to date, partitioning of trace elements between orthopyroxene and garnet/clinopyroxene is poorly constrained, it was impossible to assess if orthopyroxene is in chemical equilibrium with garnet or clinopyroxene. Therefore, further trace element and isotopic studies are required to

---

* Corresponding author. Tel.: +31-20-4447403; fax: +31-20-6462457.
  E-mail address: simn@geo.vu.nl (N.S.C. Simon).
[1] Present address: Department of Earth Sciences, University of Cardiff, PO Box 914, Cardiff CF14 3YE, UK.

0024-4937/$ - see front matter © 2003 Elsevier B.V. All rights reserved.
doi:10.1016/S0024-4937(03)00118-X

constrain the timing of garnet introduction/modification and its possible link with the SiO_2 enrichment of the Kaapvaal lithosphere.

Keywords: Kaapvaal craton; Peridotite; Trace elements; Re–Os; Disequilibrium

1. Introduction

The formation and evolution of cratonic lithospheric mantle has long been a matter of debate. The high Mg# in olivine (~ 92–94) and strong depletions in CaO and Al_2O_3 in cratonic peridotites indicate an origin as residua from high degrees of partial melting (e.g., Boyd and Mertzman, 1987). Their high SiO_2 contents and enrichment in other magmaphile components (e.g., incompatible trace elements) compared to oceanic peridotites (Boyd, 1989), however, suggest a more complex history.

A number of authors (e.g., Takahashi, 1986; Falloon and Green, 1987, 1988; Falloon et al., 1988; Hirose and Kushiro, 1993; Herzberg and Zhang, 1996; Kinzler, 1997; Walter, 1998) carried out high $P–T$ melting experiments on pyrolite and other model mantle compositions to constrain the nature of the melts and residua produced at different depths. The results show that melting at pressure ≪3 GPa, similar to a modern MOR environment, as well as deep (>5 GPa) komatiite melt extraction can account for many of the observed major element characteristics of low-T cratonic peridotites. No experiment, however, reproduced the high Si-contents of most low-T peridotites from the Kaapvaal craton, unless the source was initially significantly more Si-rich than generally assumed for primitive mantle compositions (see summary by Kelemen et al., 1998). In contrast, the major element composition of many other non-silica-enriched cratonic peridotites (Siberia, Tanzania), including Kaapvaal high-T xenoliths, can be explained by melt extraction (Walter, 1999).

Three different models have been proposed to account for the high orthopyroxene (opx) contents in most Kaapvaal low-T peridotites: (1) extensive melting followed by cooling and metamorphic un-mixing of opx- and olivine-rich layers (Boyd et al., 1997); (2) mixing of the residue with an opx-rich cumulate (Herzberg, 1999); (3) re-enrichment of the residue by infiltrating Si-rich fluids or melts (trondhjemites) derived from low degree melting of subducted oceanic crust (Kesson and Ringwood, 1989; Kelemen et al., 1992, 1998; Rudnick et al., 1994; Kelemen and Hart, 1996). The reader is referred to Walter (1998) for a detailed discussion of all three models.

Another important observation is that many Kaapvaal low-T xenoliths contain abundant garnet and clinopyroxene (cpx), which often are spatially related. Cox et al. (1987) first suggested that the spatial relation of garnet, cpx and opx might be explained by cooling of a harzburgitic residue and subsequent exsolution of garnet and cpx from high-T, Al and Cr-rich opx. This model was experimentally demonstrated by Canil (1991) and statistically developed further by Saltzer et al. (2001). Exsolution of garnet (± cpx) from opx is certainly a process that takes place in the lithosphere and is preserved in the texture of some rocks (Fig. 1a). However, exsolution from opx is unlikely to account for the extreme incompatible trace element enrichment and trace element zonation in garnet (e.g., Griffin et al., 1999; Shimizu, 1999) nor the major and trace element disequilibrium between garnet and cpx that is abundant in many xenoliths (e.g., Shimizu, 1975; Brey, 1989; Günther and

---

Fig. 1. Backscattered electron images of selected samples. (a) Exsolution of garnet and spinel from opx in sample J8 from Jagersfontein; (b) LET2 garnet with inner fine grained and outer coarse grained kelyphite rim; (c) fractured garnet with kelyphitization starting in the core in LQ8; (d) spinel–cpx–opx "fingerprint" texture in TP5; (e) garnet–spinel–cpx–opx symplectite with kelyphite and cpx on the rim (TP9); (f) garnet inclusion in olivine in M2; (g) cpx porphyroblast with inclusions, "spongy" overgrowth rim and vein-like extensions showing melt-like textures (M2); (h) trapped kimberlite melt with microphenocrysts and serpentine-filled vugs in LET64.

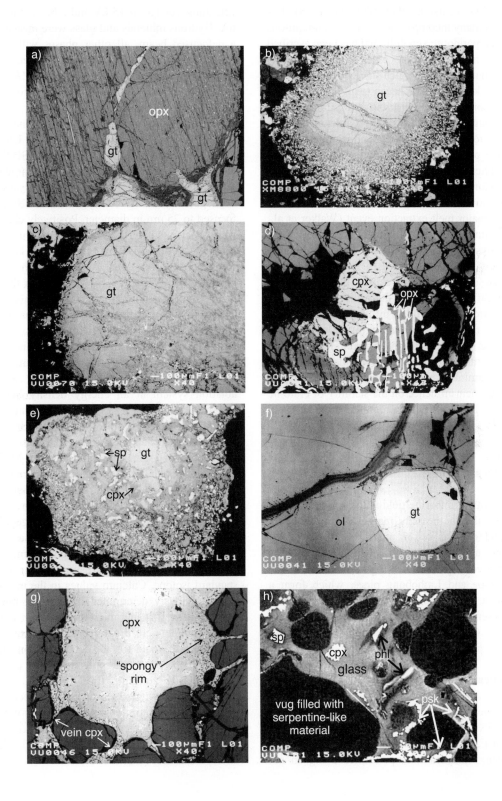

Jagoutz, 1994; Pearson, 1999; Shimizu, 1999). Enrichment in many incompatible elements is ubiquitous in Kaapvaal peridotite xenoliths and has been ascribed to different metasomatic processes (e.g., Menzies et al., 1987) that have taken place since formation of the lithospheric keel in the Archaean (e.g., Pearson et al., 1995; Carlson et al., 1999; Irvine et al., 2001).

In this ongoing project, we report a detailed petrographic, major and trace element and isotope study to address the problem of the temporal modification of the lithosphere. Re–Os isotope systematics provide constraints on the age of the bulk xenoliths (Irvine et al., 2001 and this work) that is believed to reflect the age of the major melt extraction event (Walker et al., 1989). Textural relations and major and trace element compositions of the phases are assessed to obtain information about the processes that affected the lithospheric peridotites and caused their formation and/or modification. The density of lithospheric peridotites is strongly controlled by their bulk compositions and resulting mineralogies. One of our objectives is to determine the formation age of the constituent minerals of cratonic peridotites with a view to examining possible temporal variations in density and their influence on the stability of cratonic lithosphere. This contribution represents our first step towards this aim.

2. Geological setting

The Lesotho kimberlite field is located in northern Lesotho on the southeastern margin of the Kaapvaal Craton and consists of 17 pipes (Nixon, 1973) of Cretaceous age (Davis, 1977). Peridotite xenoliths were sampled from the Letseng-la-Terai, Matsoku, Thaba Putsoa and Liqhobong pipes, which form a cluster of ca. 15 km radius. The 18 samples examined here represent a sub-set of the xenoliths described in Irvine et al. (2001) and Irvine (2002).

3. Analytical techniques

Major elements in minerals were analyzed using the Jeol JXA-8800M electron microprobe at the Faculty of Earth and Life Sciences, Vrije Universiteit Amsterdam. Analyses were performed using an acceleration voltage of 15 kV and a beam current of 25 nA. Hydrous minerals and glass were measured using a defocused beam, all other phases with a focused beam. Standards are natural and synthetic minerals, corrections were performed using the ZAF method. Relative errors are usually 1–2% for major and about 5% for minor elements.

Trace element analyses were performed by laser ablation ICP-MS at the University of Utrecht and using the Cameca IMS 6f ion probe (SIMS) at the Department of Terrestrial Magnetism, Carnegie Institution of Washington. For ion probe analyses, a beam of negatively charged oxygen ions (10 nA) was focused to 25 μm in diameter. Positively charged ions were accelerated at 10 kV and analyzed using energy filtering, as described by Shimizu and Hart (1982). The energy offset was -70 eV with an energy bandpass of ± 25 eV. ^{30}Si was used as the normalizing isotope. Calibration curves were checked in each session by analyzing mafic glasses, pyroxenes and garnets described by Jochum et al. (2000), Norman et al. (1996), Norman (1998) and Eggins et al. (1998). The laser ablation ICP-MS analyses were carried out using a 193-nm GeoLas 200Q laser ablation system at an energy density between 10 and 15 J cm^{-2} and a pulse repetition rate of 10 Hz resulting in craters of 40 or 120 μm in diameter and < 50 μm depth. The chemistry of the ablated particles was determined using a quadrupole ICP-MS (Micromass Platform ICP) under standard operating conditions (see de Hoog et al., 2001 for more details). Analysis was performed with respect to time using the data acquisition protocols of Longerich et al. (1996). NIST SRM 612 glass was used as a calibration standard with the preferred concentration values of Pearce et al. (1997) and using ^{44}Ca as an internal standard isotope. Calcium values in the minerals had been previously determined by electron microprobe analysis. The first few microns ablated at the surface of the sample were discarded when calculating concentrations to avoid surface contamination from thin-section preparation and polishing. The presence of inclusions or zoning within minerals was easily observed and avoided or taken into account when calculating concentrations. Accuracy was within 10% of the recommended values for most elements for USGS basaltic glass reference materials BCR2-G, BIR1-G and BHVO2-G.

Table 1
Sample classification and calculated equilibrium pressures and temperatures for mineral cores

Sample	Locality	Type	Paragenesis	Pressures and temperatures calculated iteratively										Preset P (4 GPa for gt-, 2 for sp-peridotites)				
				BKN/BKN		BKN/KB		Krogh/BKN		O'Neill/BKN		Harley/BKN		Ca(opx)	Na(cpx)	OW	WS	Balhaus
				T	P	T	P	T	P	T	P	T	P	T	T	T	T	T
LET2*	LET	gt-harzburgite	ol,opx,(cpx),gt,(sp)							936	3.1	929	3.0					
LET8	LET	sp-harzburgite	ol,opx,(cpx),(sp),gr											837	730	1036	855	1001
LET28	LET	sp-harzburgite	ol,opx,sp,(cpx)											899	956	905	874	887
LET29+	LET	gt-sp-lherzolite	ol,opx,cpx,gt,sp	1081	4.0	1110	5.6	1099	4.1	957	3.4	937	3.2	986	1066			
LET38*+	LET	gt-lherzolite	ol,opx,cpx,gt	1027	4.1	1064	6.2	1004	3.9	1013	4.0	971	3.7	997	1070			
LET64*+	LET	gt-lherzolite	ol,opx,cpx,gt,sp,mi	1123	4.5	1189	8.1	1151	4.7	1057	4.1	987	3.8	996	1061			
LQ5	LQ	grt-lherzolite	ol,opx,cpx,gt	1084	4.4	1100	5.3	1092	4.5	1012	4.0	980	3.8	994	1035			
LQ6*	LQ	grt-lherzolite	ol,opx,cpx,gt	1150	4.5	1216	7.9	1128	4.4	974	3.5	951	3.4	1002	1058			
LQ8*	LQ	grt-lherzolite	ol,opx,cpx,gt	1085	3.8	1121	5.7	1090	3.8	901	2.8	894	2.8	1090	1105			
M1+	M	IRPS	ol,opx,cpx,gt,il,sul,mi	1081	4.4	1039	2.5	1135	4.8	1034	4.1	1002	3.9	1030	1090			
M2*	M	grt-lherzolite	ol,opx,cpx,gt,mi	928	3.5	931	3.7	1013	4.0	911	3.4	902	3.3	923	942	616		614
M5	M	sp-lherzolite	ol,opx,cpx,sp				0.0		0.0		0.0		0.0	798	694			
M9	M	grt-lherzolite	ol,opx,cpx,gt	1075	4.4	1106	6.1	1144	4.9	1020	4.1	984	3.9	942	1022			
M13*	M	grt-sp-lherzolite	ol,opx,cpx,gt,sp, mi	1026	4.1	1035	4.6	1109	4.5	996	3.9	909	3.5	962	992	853		786
GP404*	M	grt-lherzolite	ol,opx,cpx,gt	1028	4.0	1038	4.5	1087	4.3	1032	4.0	985	3.7	995	1050			
TP5	TP	grt-sp-lherzolite	ol,opx,cpx,gt,sp	907	3.2	922	4.1	986	3.5	867	3.0	859	3.0	881	894	767		
TP6	TP	grt-sp-lherzolite	ol,opx,cpx,gt,sp	972	3.7	1000	5.4	1091	4.3	929	3.5	889	3.4	907	826	985		
TP9*	TP	grt-sp-lherzolite	ol,opx,cpx,gt,sp	943	3.6	955	4.4	1116	4.3	875	3.3	854	3.2	863	879	898		
J8*	JAG	grt-sp-lherzolite	ol,opx,cpx,gt,sp	759	2.9	712	neg.	633	2.3	695	2.6	789	3.1	780	859	729		688

Sample names with a * indicates that these samples were analyzed by ion probe, + indicates LA-ICP-MS. Localities: LET = Letseng-la-Terai, LQ = Liqhobong, M = Matsoku, TP = Thaba Putsoa, JAG = Jagersfontein. ol = olivine, opx = orthopyroxene, cpx = clinopyroxene, gt = garnet, sp = spinel, mi = mica (phlogopite), il = ilmenite, gr = graphite. Minerals in brackets indicate secondary origin of this phase.
T(BKN): 2-px thermometer of Brey and Köhler (1990), P(BKN): Al-in-opx barometer of Brey and Köhler (1990), P(KB): Ca-in-ol barometer of Köhler and Brey (1990), T(Krogh): gt-cpx thermometer of Krogh (1988), T(O'Neill): gt-ol thermometer of O'Neill and Wood (1979), T(Harley): gt-opx thermometer of Harley (1984), T(Ca-in-opx): T(Na-in-cpx): Brey and Köhler (1990), T(OW): ol-sp thermometer of O'Neill and Wall (1987), T(WS): Al-in-opx thermometer of Witt-Eickschen and Seck (1991), T(Balhaus): ol-sp thermometer of Ballhaus et al. (1991). P–T calculations for LQ6 and LQ8 are from rim compositions. See text for explanation.

Table 2
Major and trace element analyses of whole rocks, mineral modes and Re–Os isotope compositions and model ages

Sample	LET2	LET8	LET28	LET29	LET38	LET64	LQ5	LQ6	LQ8	M1	M2	M5	M9	M13	GP404	TP5	TP6	TP9
Major elements																		
SiO_2	45.45	46.20	46.02	46.32	48.18	47.96	48.14	47.38	47.63	44.98	46.81	46.45	48.01	46.18	45.95	46.31	45.82	46.36
TiO_2	0.04	0.01	0.01	0.02	0.02	0.08	0.01	0.03	0.01	0.58	0.04	0.03	0.11	0.07	0.01	0.02	0.01	0.01
Al_2O_3	0.60	0.58	0.60	1.05	1.57	1.80	1.76	1.48	1.22	0.82	1.17	1.49	1.46	0.86	1.59	1.23	0.67	0.89
FeO	3.82	4.85	3.95	4.42	5.17	4.35	4.33	5.79	5.03	8.17	4.83	5.68	5.99	5.47	5.76	5.63	5.59	5.65
Fe_2O_3	2.67	1.16	2.80	2.26	0.88	2.02	1.92	0.64	1.10	1.96	1.77	1.27	0.48	1.40	0.31	0.68	1.27	1.08
MnO	0.10	0.10	0.12	0.11	0.12	0.13	0.12	0.12	0.11	0.14	0.12	0.12	0.13	0.12	0.12	0.12	0.12	0.12
MgO	46.82	46.86	46.15	44.92	43.13	42.56	42.66	43.37	43.99	42.55	41.28	43.86	42.65	45.06	42.61	45.16	45.99	45.25
CaO	0.38	0.18	0.27	0.78	0.81	0.90	0.91	0.94	0.72	0.58	0.97	0.88	0.93	0.63	0.74	0.72	0.45	0.55
Na_2O	0.07	0.04	0.05	0.08	0.08	0.10	0.09	0.13	0.11	0.12	0.13	0.10	0.14	0.11	0.10	0.08	0.07	0.07
K_2O	0.02	0.01	0.00	0.01	0.01	0.08	0.03	0.10	0.05	0.07	0.05	0.10	0.08	0.05	0.05	0.02	0.00	0.07
P_2O_5	0.03	0.02	0.03	0.02	0.02	0.03	0.02	0.02	0.02	0.02	0.03	0.03	0.02	0.03	0.30	0.02	0.02	0.00
LOI	4.9	4.7	5.4	4.5	3.8	4.9	2.8	3.1	1.8	3.0	2.1	1.6	2.6	1.9	1.7	3.1	3.2	3.0
FeO(total)	6.22	5.89	6.46	6.45	5.96	6.17	6.06	6.36	6.02	9.93	6.42	6.82	6.42	6.74	6.03	6.25	6.73	6.63
Mg# (ol)	0.93	0.94	0.93	0.93	0.93	0.93	0.93	0.93	0.93	0.88	0.93	0.92	0.92	0.92	0.92	0.93	0.93	0.93
Modes																		
olivine	76	70	68	70	56	54	54	60	61	69	53	59	53	69	61	67	72	67
opx	20	29	30	26	36	37	37	32	34	27	38	35	38	26	33	27	24	28
cpx	0.9	0.5	0.8	2.4	1.7	1.6	1.8	2.9	2.2	1.6	2.9	3.3	3.9	1.4	1.7	1.2	0.8	0.4
garnet	3.1			0.1	5.7	7.7	7.1	5.5	3.6	2.5	5.7		5.1	3.7	4.9	5.2	2.9	3.8
spinel		0.0	0.5	1.6								2.7		0.0		0.2	0.2	0.4
RMS	0.059	0.238	0.003	0.006	0.011	0.027	0.005	0.007	0.030	0.001	0.294	0.156	0.084	0.002	0.212	0.009	0.064	0.005

Trace elements

La	2.41	0.93	2.98	1.17	1.35	1.73	1.18	1.00	1.07	0.75		2.39	1.38	2.50		1.35	0.49	0.54
Ce	4.93	1.72	5.57	2.25	2.63	3.39	2.37	2.09	2.19	1.52		4.36	2.77	4.95		2.85	1.06	1.25
Pr	0.64	0.23	0.68	0.30	0.35	0.44	0.32	0.29	0.29	0.20		0.51	0.36	0.60		0.41	0.15	0.18
Nd	2.27	0.86	2.43	1.19	1.42	1.78	1.35	1.19	1.17	0.80		1.82	1.43	2.22		1.70	0.65	0.83
Sm	0.32	0.16	0.35	0.24	0.26	0.33	0.26	0.24	0.23	0.15		0.30	0.29	0.33		0.28	0.13	0.18
Eu	0.09	0.04	0.08	0.07	0.07	0.10	0.08	0.08	0.07	0.05		0.08	0.09	0.09		0.08	0.04	0.06
Gd	0.23	0.11	0.21	0.23	0.19	0.26	0.20	0.25	0.20	0.17		0.23	0.27	0.23		0.21	0.12	0.18
Tb	0.03	0.01	0.02	0.03	0.02	0.03	0.03	0.04	0.02	0.03		0.03	0.04	0.02		0.03	0.02	0.03
Dy	0.16	0.05	0.09	0.14	0.09	0.14	0.10	0.23	0.12	0.17		0.12	0.20	0.09		0.09	0.09	0.14
Ho	0.03	0.01	0.01	0.02	0.01	0.02	0.02	0.04	0.02	0.04		0.02	0.03	0.02		0.01	0.02	0.02
Er	0.09	0.01	0.03	0.04	0.02	0.05	0.03	0.09	-0.03	0.09		0.04	0.08	0.03		0.03	0.03	0.05
Tm	0.015	0.001	0.004	0.005	0.003	0.007	0.004	0.013	0.004	0.014		0.006	0.012	0.004		0.004	0.005	0.006
Yb	0.102	0.010	0.016	0.028	0.017	0.043	0.024	0.067	0.026	0.081		0.040	0.059	0.023		0.024	0.018	0.032
Lu	0.017	0.002	0.002	0.004	0.003	0.007	0.004	0.010	0.004	0.012		0.007	0.009	0.003		0.004	0.003	0.005
Cr [ppm]	2025	2140	1935	2295	2710	2865	2725	2435	2260	1040	2738	2570	2650	2515	1832	3080	2165	2700
Ni [ppm]	2647	2601	2683	2508	2215	2091	2161	2367	2338	2316	2597	2404	2198	2419	2750	2317	2621	2508

Re–Os

Re [ppb]	0.330	0.251	0.096	0.260	0.090	0.598	0.012	0.113	0.010	0.164	0.002	0.090	0.015	0.031	0.041		0.021	0.024
Os [ppb]	5.71	1.60	3.88	3.44	13.73	6.42	6.56	2.28	3.82	2.21	3.01	3.16	9.55	5.36	5.13		5.46	4.57
$^{187}Re/^{188}Os$	0.277	0.753	0.119	0.362	0.031	0.448	0.009	0.239	0.012	0.357	0.000	0.137	0.007	0.028	0.040		0.018	0.025
$^{187}Os/^{188}Os(m)$	0.1086	0.1086	0.1079	0.1101	0.1090	0.1100	0.1091	0.1102	0.1080	0.1220	0.1100	0.1118	0.1085	0.1095	0.1100		0.1073	0.1077
±	12	9	10	8	9	10	12	11	12	10		17	9	15			12	12
$^{187}Os/^{188}Os(i)$	0.1082	0.1075	0.1078	0.1095	0.1090	0.1093	0.1091	0.1098	0.1080	0.1215	0.1100	0.1116	0.1085	0.1095	0.1100		0.1073	0.1076
T_{RD}(erup)	2.8	2.9	2.9	2.7	2.7	2.7	2.7	2.6	2.9	1.0	2.8	2.4	2.8	2.7	2.7		3.0	2.9
T_{MA}	7.7	-3.8	3.9	15.8	2.9	-97.0	2.8	5.7	2.9	5.7	2.8	3.4	2.8	2.8	3.0		3.1	3.1

Duplicate analyses were performed on some samples using both facilities, and results from LA-ICP-MS and ion probe are in relatively good agreement (within 30% for REE, Sr, Y, Zr and Hf, 50% for Nb and 85% for Rb).

4. Petrography and chemical composition

4.1. Petrography

Hand specimens are between 10 and 25 cm in diameter. They were split in half and one half was then analyzed for Os isotopic composition, platinum group elements (PGEs) and whole rock major and trace element composition. The results are discussed in detail in Irvine (2002) and Irvine et al. (2001). The other half of the hand specimen was used in this work. Based on an initial petrographic study and the data of Irvine et al. (2001), a sub-set of samples was selected for a detailed petrographic study and in situ major and trace element analyses on thin sections and polished thick sections, respectively.

A short description of all samples studied here, together with the Re–Os ages of Irvine et al. (2001) is given in Tables 1 and 2. With the exception of M1 (from Matsoku), all samples are coarse peridotites (following the nomenclature of Harte, 1977) and show no deformation apart from occasional undulose extinction in olivine. Sample M1 is texturally and chemically distinct and seems to belong to the IRPS suite described by Harte et al. (1987). Interestingly, M1 is also much younger than the other xenoliths, giving a T_{RD} (Re-depletion age) of 1 Ga, compared to an average T_{RD} of 2.5 Ga for the other 17 samples (Irvine et al., 2001, see Table 2). Schmitz and Bowring (2001) reconstructed the thermal evolution of the Kaapvaal lithosphere from U–Pb rutile and titanite thermochronology of lower crustal xenoliths and identified a major thermal pulse that affected the lower crust underneath northern Lesotho at ~ 1 Ga. This thermal event in the crust might have been linked to the magmatic event in the mantle recorded by the IRPS suite from Matsoku.

Mineral modes were calculated from whole rock composition and electron microprobe data using a least-squares method. The range of modal compositions is 53–76% olivine, 20–38% opx, 0–3.9% cpx

and 0–7.7% garnet ± < 1–2.4% spinel ± < 1% phlogopite (Table 2). Thus, the samples represent typical low-T peridotites with relatively low modal olivine (relative to olivine Mg#), are enriched in opx (average mode 31%) and have relatively low modal garnet and cpx contents.

Garnet in all samples is kelyphitized, but to different extents. Some garnets (especially in the samples from Letseng-la-Terai) are surrounded by up to three distinct kelyphite rims (Fig. 1b). The inner rim is very fine grained, while the outer rim consists of coarser (5–100 μm) spinel, cpx, opx and phlogopite. Most garnets have fractures, which are often connected to the coarse kelyphite rim and filled with the same material (Fig. 1b). In samples GP404 and M2, some very fresh and almost unaltered garnets are preserved. Clinopyroxene is often spatially related to garnet but also occurs as distinct grains.

The majority of the Lesotho samples show evidence for melt infiltration and interaction with the kimberlite. This is most obvious for LET64, where pools of remnant kimberlitic glass and microphenocrysts can be found (Simon et al., 2000). In all samples, a second generation of cpx (in addition to the high-Al kelyphitic cpx) could be identified, having distinctly higher Ti contents and highly variable major and trace element composition. This secondary cpx occurs on grain boundaries and in cracks and have a vein-like appearance (Fig. 1g). Large cpx porphyroblasts often show spongy overgrowth rims of up to 100 μm that have compositions similar to the vein-like cpx. This occurrence was first described as a cloudy "porous" zone around clinopyroxene porphyroblasts in kimberlite xenoliths by Carswell (1975).

All samples from Thaba Putsoa are garnet-spinel lherzolites and contain clusters of garnet symplectites that are very complexly intergrown with spinel, cpx and minor opx (Fig. 1d and e). Minerals within the symplectites are generally strongly zoned in major elements. Clusters are surrounded by thick kelyphite rims and some clusters are entirely kelyphitized. Apart from the symplectites, spinel can also be found as relatively big (1 mm) single grains. Cpx occurs on garnet rims but also as single grains with curved grain boundaries, displaying textural equilibrium with opx and olivine. Cpx is often associated with spinel. As with most other Lesotho samples, late stage secondary phlogopite and spinel occur as small interstitial grains, as constituents in the coarse outer kelyphite rims and in fractures.

4.2. Whole rock chemistry

Fig. 2 shows that all selected xenoliths from our Lesotho sample set, with the exception of M1, have compositions typical for Kaapvaal peridotites. They have a modal olivine content between 53% and 76% and Mg# in olivine from 91.8 to 93.7. The entire sample set of Irvine (2002) includes three more samples similar to M1 (low Mg#, high Fe contents, very low Cr, younger Re–Os ages, different trace element signature). We interpret these rocks as cumulates or wall rock adjacent to pyroxenite veins, in accordance with the interpretation of Harte et al. (1987).

If the cratonic samples originated on a trend parallel to the melt depletion trend defined by oceanic peridotites (Boyd, 1989; Fig. 2), the garnet and spinel lherzolites have experienced the lowest degree of partial melting. The garnet–spinel lherzolites have undergone intermediate degrees of melting. The most

depleted (highest degree of melt extraction) rocks are a spinel harzburgite and a garnet harzburgite from Letseng-la-Terai, with an olivine Mg# close to 94. Experiments of Walter (1998) have shown that Mg# in residual olivine is almost insensitive to the pressure and temperature of melting. Hence, the observation made here might be interpreted in two ways: either the high-Mg#, high modal olivine samples underwent the most extensive melting, or the lherzolitic samples with the lower Mg# are extensively modified by re-fertilization (re-introduction of Fe ± cpx ± opx). Overall, there seems to be no clear correlation between degree of melting and depth. It is important to remember, however, that geothermobarometry yields pressures and temperatures of the most recent major element equilibration event, which does not necessarily represent the $P–T$ conditions at the time of melt extraction.

Bulk major element contents (Table 2) vary with abundance of the constituent minerals as expected. Garnet lherzolites have the highest bulk SiO_2, Al_2O_3, CaO, MnO, Na_2O, Cr and lowest MgO, FeO and Ni contents. Minor elements such as TiO_2, K_2O and P_2O_5 vary independently from the abundance of the main phases and are most likely controlled by phlogopite content and secondary phases related to kimberlite magmatism. There is no correlation evident between SiO_2 and FeO.

Fig. 3 shows that, in general, all samples are LREE enriched with La_N between 2 and 10.5, Lu_N between 0.08 and 0.8 and $(LREE/HREE)_N$ of ~ 10–100. There is no variation of REE content with modal mineralogy or whole rock major element composition, with the exception of the alkalies, which correlate positively with HREE contents. Garnet harzburgite LET2, however, does not follow this trend and has too low a concentration of alkalies relative to its very high HREE content. M1 (IRPS) has a very distinct chondrite normalized REE pattern compared to all other samples. It is much flatter with LREE contents comparable to the garnet–spinel lherzolites TP6 and TP9 and HREE similar to the garnet harzburgite LET2, which has the highest HREE (Fig. 3).

Compared to oceanic peridotites with comparably high Mg# in olivine (e.g., Ross and Elthon, 1997), the Lesotho samples are enriched in incompatible elements by 3–4 orders of magnitude. Hence, the xenoliths are not simple residua from large degrees of

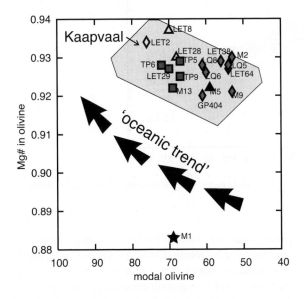

Fig. 2. Mg# in olivine vs. modal olivine content. All Lesotho peridotites (with the exception of IRPS-sample M1 with a very low Mg#) plot in the field defined for Kaapvaal peridotite xenoliths (Boyd, 1989). Solid symbols are cpx-bearing (lherzolites), open symbols are cpx-free (harzburgites). Diamonds stand for garnet peridotites, squares are garnet–spinel peridotites and triangles are spinel peridotites.

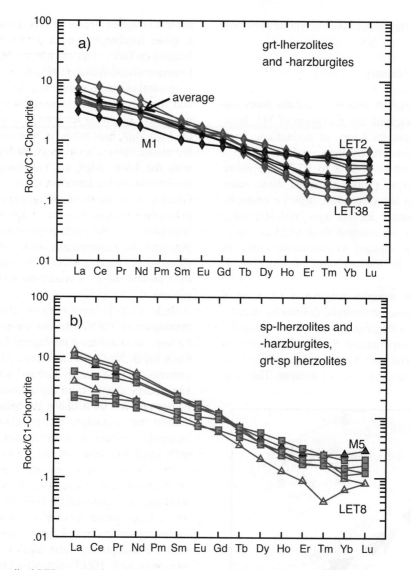

Fig. 3. Chondrite normalized REE concentrations (McDonough and Sun, 1995) of whole rocks. Note the distinctively flatter REE pattern of IRPS-sample M1. Symbols as in Fig. 2.

melting but have undergone metasomatism and re-enrichment in incompatible elements (e.g., Menzies et al., 1987).

4.3. Mineral chemistry

In the following section, a distinction is made between "primary" phases (porphyroblasts) and secondary minerals. The latter are derived from decomposition of primary phases (e.g., cpx and spinel on

garnet rims) or formed by crystallization from the kimberlite or by reaction between kimberlitic magma and primary phases (e.g., vein-like cpx, interstitial phlogopite and overgrowth rims on primary phases). Alteration products such as serpentine or magnetite are also present. In the context of this paper, we focus on the composition of the porphyroblasts. However, distinction between primary porphyroblasts and phases originated from, or modified by, late stage processes is not always clear. Therefore, the composition of non-primary phases will also be discussed briefly to aid comprehension of the genesis of the xenolith samples.

Pieces of rock were cut to 1-in. diameter slabs and polished to 1–2 mm thickness. Specimens were used for electron microprobe, ion probe and Laser-ICP-MS. Thoroughly characterized mineral grains with known composition and textural relationship will be separated for Hf, Nd and Sr isotope studies in a later stage of this project. Cores and rims of all mineral phases were analyzed for major and minor elements by electron microprobe. Where elemental zonation was detected, a more detailed concentration profile was measured through selected grains. In addition, major element maps for selected elements were obtained for some garnets, cpx and opx. Trace elements were measured on cores and rims of garnet, cpx, opx and phlogopite. Trace element contents in olivine were too low to obtain useful data, but occasionally ion probe analyses of olivine were performed to provide an indication of detection limits and reproducibility of the method. Where trace element zonation was detected in garnet or cpx, quantitative trace element profiles through grains were measured with the ion probe to provide information about the process that produced trace element variation. Wherever possible, major and trace element analyses were carried out on the same grains. Selected results for mineral cores and rims are presented in Table 3. Complete analyses of microprobe and ion probe transects can be obtained from the senior author on request.

4.3.1. Olivine

Olivine in all samples has homogeneous major element contents. Only the outermost rims (10–100 μm) have elevated Ca and Fe and variable Ni contents. Fe-rich overgrowth rims are found in samples (e.g. LET64) where there are numerous indications for interaction with the kimberlite. Thus, we ascribe the olivine overgrowth rims to reaction of the xenolith with the infiltrating kimberlite during ascent. Since all rims are very thin, this must have been a very recent process of short duration, consistent with the results of Canil and Fedortchouk (1999) who carried out experiments on dissolution of garnet in kimberlite and concluded that the ascent of kimberlite is very fast, probably on the scale of minutes or hours.

4.3.2. Orthopyroxene

Orthopyroxene porphyroblasts are also generally homogeneous in major and trace elements and only show variation in the outermost rims (max. 100 μm). Mg# in opx are very similar to olivine and vary between 92 and 94 for the whole sample set (excluding M1). Cr# ($100 \times Cr/Cr + Al$) in opx cores shows a positive correlation with Mg# and varies from 10 in spinel lherzolite M5 to 25 in garnet-spinel lherzolite M13. There is some core-rim variation in Ca, Al, Cr, Fe, Na, Ti and Mn, but it is not systematic and is restricted to the outermost rims. Light and heavy REE concentrations in opx vary by a factor of 30 between the samples. MREE contents are more uniform (Table 3b). Two types of patterns can be recognized (Fig. 4a): (i) a smooth slope decreasing from LREE to HREE; (ii) a slightly curved pattern with a maximum at Nd–Eu, and only slightly lower chondrite-normalized HREE than LREE concentrations. Some samples show extreme LREE enrichment in opx rims compared to cores (e.g., M13, LET38; Fig. 4a). Since LREE enrichment is accompanied by high Ba, Sr and Ti contents, this is most probably due to small melt inclusions or other impurities in the outermost rim, and probably indicates melt-rock interaction during transport to the surface in the kimberlite.

Opx derived from breakdown of garnet has a very different major element composition: high Al_2O_3 (~9.7 wt.%), Cr_2O_3, FeO, and low MgO and SiO_2 (51 wt.%). They are distinct from opx porphyroblasts that have homogeneous cores with low Al (<1 wt.% Al_2O_3), and slightly higher Al, Fe, Ca, Na and Ti in the outermost rims. In sample LET64, opx microphenocrysts can be found in melt pools together with cpx, spinel and phlogopite (Fig. 1h), which are believed to have crystallized from the kimberlite.

4.3.3. Clinopyroxene

The distinct generations of cpx identified by textural relationships also have different major and trace element compositions. Vein-cpx often have similar composition to overgrowth rims on the big porphyroblasts (elevated Ti, different Ca, Fe, Na, Cr and Al, higher REE compared to porphyroblast cores). Cpx derived from breakdown of garnet can easily be identified by very high (up to 10 wt.%) Al_2O_3 contents. Compositionally distinct rims sometimes extend

Table 3
Selected analyses of phases

(a) Olivine

Sample	M1			M2			M5		M9		M13		GP404			LET2		LET8	
	Core	Rim	Incl.	Core	Rim	Incl.	Core	Rim	Core	Rim	Core	Rim	Core	Rim	Incl.	Core	Rim	Core	Rim
SiO_2	40.7	40.7	40.3	41.2	41.4	41.8	41.9	41.8	41.9	41.9	41.8	41.9	41.9	41.9	42.1	41.9	41.9	41.6	41.6
Al_2O_3	0.03	0.02	0.03	0.00	0.01	0.02	0.00	0.00	0.01	0.01	0.00	0.01	0.02	0.03	0.00	0.03	0.03	0.00	0.01
Cr_2O_3	0.03	0.04	0.04	0.02	0.03	0.05	0.01	0.00	0.02	0.05	0.03	0.03	0.08	0.09	0.10	0.07	0.10	0.08	0.07
FeO	11.4	11.5	11.5	7.12	7.11	7.01	7.73	7.72	7.72	7.73	7.71	7.72	8.08	8.16	8.04	6.44	6.59	6.55	6.51
MnO	0.12	0.12	0.12	0.08	0.10	0.08	0.11	0.09	0.11	0.09	0.10	0.09	0.10	0.09	0.10	0.07	0.10	0.08	0.07
MgO	48.1	48.0	48.4	51.5	51.3	51.4	51.1	50.9	50.8	50.8	51.2	51.4	51.3	51.3	51.4	51.1	51.1	51.9	51.9
NiO	0.41	0.38	0.40	0.43	0.40	0.34	0.45	0.40	0.42	0.44	0.41	0.40	0.45	0.43	0.41	0.40	0.42	0.44	0.43
CaO	0.04	0.05	0.04	0.02	0.02	0.07	0.01	0.02	0.02	0.03	0.02	0.04	0.02	0.03	0.04	0.02	0.02	0.01	0.01
Total	100.9	100.8	100.9	100.4	100.4	100.7	101.2	101.0	101.1	101.0	101.3	101.6	101.8	102.0	102.1	100.0	100.2	100.6	100.6
Mg#	0.88	0.88	0.88	0.93	0.93	0.93	0.92	0.92	0.92	0.92	0.92	0.92	0.92	0.92	0.92	0.93	0.93	0.94	0.94

Sample	LET28		LET29		LET38		LET64		LQ5		LQ6		LQ8		TP5		TP6	TP9		J8
	Core	Rim	Core	Rim	Core	Rim	Core	Rim	Core	Rim	Core	Rim	Core	Rim	Core	Rim	Core	Core	Rim	All
SiO_2	41.2	41.2	42.0	41.8	42.0	42.0	41.4	41.5	41.9	41.8	42.0	42.0	41.9	41.6	41.5	41.2	41.5	41.6	41.7	40.5
Al_2O_3		0.00	0.01	0.02	0.01	0.02	0.01	0.02	0.00	0.01	0.02	0.01	0.01	0.02	0.01	0.01		0.01	0.02	0.01
Cr_2O_3	0.00		0.02	0.05	0.02	0.03	0.02	0.07	0.01	0.03	0.02	0.03	0.02	0.06	0.01	0.01	0.01	0.01	0.02	0.01
FeO	6.87	6.89	7.25	7.46	7.03	7.10	7.13	8.53	7.18	7.10	7.39	7.45	7.18	7.11	6.86	6.9	7.40	7.38	7.46	6.57
MnO	0.07	0.08	0.09	0.12	0.08	0.09	0.09	0.14	0.11	0.09	0.09	0.11	0.10	0.09	0.08	0.10	0.08	0.08	0.10	0.08
MgO	51.0	51.0	51.4	50.9	51.5	51.5	51.1	50.6	51.5	51.5	51.6	51.5	51.7	51.4	51.5	51.2	50.9	51.3	51.2	51.3
NiO	0.44	0.41	0.42	0.44	0.48	0.44	0.44	0.36	0.44	0.47	0.50	0.44	0.45	0.42	0.42	0.42	0.43	0.45	0.44	0.43
CaO	0.01	0.02	0.02	0.06	0.03	0.02	0.02	0.09	0.02	0.04	0.02	0.03	0.02	0.08	0.02	0.02	0.02	0.02	0.03	0.04
Total	99.6	99.6	101.1	100.8	101.2	101.2	100.2	101.3	101.2	101.1	101.6	101.5	101.3	100.8	100.4	99.9	100.3	100.8	100.9	99.0
Mg#	0.93	0.93	0.93	0.92	0.93	0.93	0.93	0.91	0.93	0.93	0.93	0.92	0.93	0.93	0.93	0.93	0.93	0.93	0.92	0.93

(b) Orthopyroxene

Sample	M1		M2			M5		M9		M13			GP404		LET2		LET8		LET28	
	Core	Rim	Core	Rim	Incl.	Core	Rim	Core	Rim	Core	Rim	Incl.	Core	Rim	Core	Rim	Core	Rim	Core	Rim
wt.%																				
SiO_2	57.2	57.3	59.0	59.0	57.5	57.2	57.7	58.2	58.6	58.9	58.5	57.1	57.9	58.2	58.3	58.4	57.6	57.9	56.8	57.2
TiO_2	0.17	0.15	0.00	0.01	0.00	0.04	0.04	0.14	0.14	0.01	0.11	0.13	0.02	0.05	0.01	0.01	0.01	0.00	0.00	0.00
Al_2O_3	0.85	0.85	0.78	0.83	0.86	2.34	1.91	0.80	0.89	0.66	0.86	0.81	0.78	0.83	0.82	0.90	1.42	1.26	1.58	1.45
Cr_2O_3	0.22	0.23	0.32	0.36	0.34	0.32	0.21	0.31	0.36	0.35	0.41	0.70	0.26	0.27	0.39	0.44	0.43	0.34	0.46	0.33
FeO	6.90	6.92	4.33	4.37	4.41	5.26	5.11	4.67	4.64	4.48	4.71	4.75	4.88	4.90	4.01	4.07	4.20	4.26	4.41	4.55
MnO	0.13	0.15	0.10	0.11	0.09	0.14	0.13	0.11	0.13	0.11	0.13	0.11	0.11	0.11	0.08	0.08	0.10	0.10	0.10	0.11
MgO	34.0	34.2	35.9	35.9	35.6	34.8	35.2	35.1	35.2	35.7	35.1	35.9	35.6	35.6	35.6	35.6	35.8	35.9	35.0	35.1
NiO	0.12	0.11	0.12	0.13	0.09	0.09	0.09	0.12	0.15	0.13	0.14				0.10	0.11	0.10	0.09	0.10	0.08
CaO	0.55	0.55	0.34	0.35	0.34	0.20	0.17	0.45	0.47	0.41	0.43	0.43	0.47	0.50	0.29	0.34	0.33	0.29	0.45	0.30
Na_2O	0.18	0.17	0.10	0.10	0.00	0.03	0.03	0.15	0.15	0.11	0.12	0.14	0.12	0.13	0.12	0.15	0.02	0.02	0.02	0.01
Total	100.3	100.6	100.9	101.2	99.3	100.4	100.6	100.0	100.7	100.9	100.5	100.0	100.2	100.6	99.7	100.1	100.0	100.1	98.9	99.2
ppm																				
Sc			20.2	21.1						16.0	13.1		11.9	11.9	15.5	16.1				
Ti			25.7	29.8						201	612		99.2	177	90.9	92.3				
Cr			2644	2817						886	2707		1722	1789	2442	2539				
Rb			0.59	0.63						2.37	0.60		0.45	0.45	0.32	0.36				
Sr			0.80	0.88						1.09	6.14		0.54	0.52	0.55	0.48				
Y			0.05	0.05						0.09	0.10		0.06	0.06	0.12	0.14				
Zr			0.35	0.35						0.74	1.52		0.53	0.59	1.80	1.86				
Nb			0.21	0.22						0.21	0.59		0.10	0.10	0.31	0.32				
Ba			0.06	0.07						0.10	6.82		0.02	0.02	0.04	0.04				
La			0.02	0.02						0.01	0.40		0.02	0.01	0.02	0.02				
Ce			0.11	0.11						0.01	0.83		0.03	0.03	0.07	0.06				
Nd			0.14	0.12						0.02	0.34		0.04	0.04	0.10	0.09				
Sm			0.02	0.03						0.01	0.09		0.02	0.01	0.04	0.04				
Eu			0.00	0.01						0.00	0.01		0.01	0.01	0.01	0.01				
Gd			0.01	0.01						b.d.l.	0.02		0.02	0.02	0.03	0.04				
Dy			0.01	b.d.l.						0.01	0.02		0.01	0.01	0.03	0.03				
Er			0.00	b.d.l.						0.00	0.02		0.01	0.01	0.02	0.03				
Yb			b.d.l.	0.00						0.00	0.01		0.00	0.00	0.02	0.01				
Hf			0.01	0.01						0.02	0.05		0.02	0.02	0.05	0.04				

(continued on next page)

Table 3 (continued)

(b) Orthopyroxene

Sample	LET29		LET38		LET64		LQ5		LQ6		LQ8			TP5		TP6	TP9		J8	
	Core	Rim	Core	Rim	Core	Rim	Core	Rim	Core	Rim	Core	Rim	Rim	Core	Rim	Core	Core	Rim	Core	Rim
wt.% (La-ICP-MS)																				
SiO_2	58.7	58.4	58.7		58.3	58.2	58.7	59.0	58.8	58.7	58.4	57.4	55.7	58.5	58.0	58.2	58.2	58.1	57.2	
TiO_2	0.01	0.01	0.01		0.02	0.04	0.01	0.01	0.03	0.04	0.02	0.05	0.03	0.01	0.01	0.01	0.00	0.00	0.01	
Al_2O_3	0.84	0.86	0.77		0.77	0.86	0.74	0.85	0.86	0.90	0.84	1.46	3.87	0.96	1.09	0.85	0.86	1.20	0.70	
Cr_2O_3	0.36	0.39	0.29		0.30	0.35	0.26	0.32	0.31	0.34	0.33	0.87	1.00	0.34	0.35	0.30	0.31	0.43	0.17	
FeO	4.47	4.48	4.28		4.27	4.35	4.41	4.47	4.56	4.53	4.39	4.23	4.66	4.23	4.26	4.52	4.56	4.59	4.16	
MnO	0.10	0.12	0.10		0.10	0.12	0.09	0.09	0.11	0.11	0.10	0.09	0.14	0.10	0.11	0.10	0.13	0.10	0.10	
MgO	35.6	35.5	35.7		35.5	35.4	35.9	36.4	35.7	35.7	35.4	34.8	33.1	35.8	35.6	35.6	35.8	35.1	36.0	
NiO	0.11	0.13	0.10		0.12	0.12	0.13	0.15	0.12	0.11	0.11	0.12	0.07	0.08	0.11	0.09	0.06	0.11	0.08	
CaO	0.46	0.50	0.48		0.48	0.47	0.47	0.48	0.46	0.49	0.46	0.86	1.15	0.33	0.35	0.33	0.30	0.33	0.18	
Na_2O	0.15	0.14	0.10		0.14	0.14	0.11	0.12	0.17	0.16	0.15	0.19	0.14	0.05	0.06	0.05	0.06	0.11	0.03	
Total	100.9	100.6	100.5		100.0	100.1	100.8	101.9	101.1	101.1	100.2	100.0	99.9	100.4	99.9	100.1	100.3	100.1	98.6	
ppm																				
Sc			20.1	17.5	18.1	18.1			14.8	19.2	17.7	16.5							10.9	8.80
Ti			34.9	84.8	74.5	255			201	211	206	201							8.55	7.32
Cr			2518	4519	2202	2541			2315	2459	2349	914							1045	960
Rb	0.11	1.09	0.59	2.02	0.55	0.56			0.57	1.37	0.61	0.56							0.30	0.23
Sr	1.07	4.00	0.84	19.47	0.72	1.79			0.78	3.35	0.98	0.66							0.57	0.42
Y	0.04	0.07	0.06	0.14	0.11	0.16			0.08	0.11	0.10	0.09							0.01	0.02
Zr	0.59	0.94	0.42	2.22	1.06	1.79			0.91	1.38	0.99	0.95							0.27	0.30
Nb	0.20	0.45	0.18	1.71	0.12	0.28			0.14	0.24	0.18	0.12							0.03	0.02
Ba	0.24	1.90	0.01	9.49	0.06	1.20			0.29	0.59	0.42	0.10							0.48	0.29
La	0.01	0.16	0.01	1.61	0.01	0.12			0.04	0.10	0.06	0.01							0.02	0.01
Ce	0.08	0.36	0.04	2.48	0.03	0.26			0.09	0.20	0.15	0.03							0.03	0.02
Nd	0.07	0.15	0.07	1.04	0.06	0.18			0.06	0.13	0.08	0.05							0.02	0.01
Sm	0.01	0.03	0.02	0.13	0.02	0.04			0.04	0.04	0.04	0.02							0.01	0.01
Eu	0.00	0.01	0.01	0.03	0.01	0.01			0.01	0.01	0.01	0.01							0.00	0.00
Gd	0.02	0.03	0.01	0.10	0.01	0.04			0.02	0.03	0.02	0.02							0.01	0.02
Dy	0.01	0.04	0.01	0.03	0.02	0.03			0.02	0.02	0.02	0.02							0.01	0.01
Er	<0.003	0.01	0.00	0.01	0.01	0.01			0.01	0.01	0.01	0.01							0.01	0.01
Yb	<0.006	<0.004	0.01	b.d.l.	0.01	0.01			0.00	0.00	0.01	0.00							0.01	0.00
Hf	0.02	0.01	0.02	0.06	0.04	0.06			0.04	0.05	0.03	0.04							0.02	0.04

(c) Clinopyroxene

Sample	M1 Core	M1 Rim	M2 Core	M2 Rim	M2 Vein	M5 Core	M5 Rim	M5 Vein	M9 Core	M9 Rim	M13 Core	M13 Rim	GP404 Core	GP404 Rim	LET2 Vein	LET8 Core	LET8 Rim	LET28 Core	LET28 Rim	LET29 Core	LET29 Rim	LET29 Vein
wt.%																						
SiO_2	54.7	55.0	55.2	56.0	56.0	54.1	54.1	54.4	55.0	54.5	54.8	52.9	54.6	55.0	55.5	55.3	54.5	54.4	54.5	55.6	55.5	55.9
TiO_2	0.41	0.44	0.02	0.08	0.07	0.15	0.16	0.18	0.35	0.28	0.26	0.27	0.10	0.09	0.23	0.22	0.32	0.00	0.15	0.01	0.02	0.15
Al_2O_3	3.14	3.29	2.46	4.11	3.26	4.15	4.05	3.65	2.90	1.57	2.32	1.72	2.46	2.54	0.62	0.69	2.51	1.07	1.06	2.55	2.59	1.67
Cr_2O_3	1.34	1.39	2.23	2.58	2.28	1.28	1.17	0.95	2.24	2.46	2.61	2.65	1.53	1.64	0.79	1.25	2.89	0.47	3.51	2.30	2.44	0.85
FeO	3.72	3.82	1.98	2.41	2.33	1.91	1.86	1.91	2.36	2.52	2.17	2.36	2.39	2.44	2.59	2.99	2.30	1.38	1.93	2.25	2.31	2.97
MnO	0.12	0.10	0.08	0.09	0.09	0.07	0.07	0.07	0.10	0.10	0.08	0.09	0.09	0.09	0.11	0.12	0.06	0.08	0.08	0.08	0.08	0.11
MgO	15.7	15.8	15.6	14.4	14.8	15.1	15.1	15.5	15.8	16.6	16.1	16.2	16.4	16.5	20.4	20.2	18.3	17.6	17.8	16.2	16.1	20.1
NiO	0.06	0.05				0.05	0.04	0.06	0.06	0.03	0.04				0.05	0.04	0.03	0.04	0.04	0.03	0.05	0.06
CaO	18.3	18.0	19.7	17.2	18.2	21.6	21.7	22.0	18.6	20.2	19.4	22.3	19.6	19.5	19.0	18.2	17.6	23.8	18.6	18.9	18.8	18.1
Na_2O	2.70	2.80	2.45	3.63	3.19	1.84	1.82	1.74	2.69	1.71	2.35	1.28	2.10	2.17	0.69	0.97	1.86	0.34	1.74	2.46	2.42	0.98
Total	100.2	100.7	99.7	100.5	100.3	100.3	100.1	100.4	100.1	99.9	100.1	99.7	99.3	100.0	99.9	100.1	100.3	99.3	99.3	100.5	100.3	100.9
ppm (La-ICP-MS)	*(La-ICP-MS)*										*(La-ICP-MS)*									*(La-ICP-MS)*		
Sc			15.9	42.7							28.0	31.5	27.2	19.1								
Ti			122	579							1553	1618	656	569								
Cr			14,279	28,688							17,452	18,191	10,680	11,175								
Rb	0.29		0.18	0.67							0.23	0.30	0.53	0.32						0.44		
Sr	172		416	389							338	355	243	191						268		
Y	4.00		0.95	6.85							2.42	2.31	2.36	3.34						1.76		
Zr	77.1		51.4	149							37.7	37.8	48.3	55.5						51.8		
Nb	1.41		1.51	3.43							1.11	1.26	1.30	0.94						1.31		
Ba	2.1		0.30	0.60							0.24	0.26	0.27	0.41						9.40		
La	7.4		10.5	17.4							6.25	6.51	3.54	2.71						3.50		
Ce	25.7		43.8	58.2							26.1	26.7	14.4	11.3						15.1		
Nd	16.4		36.3	50.4							25.1	25.2	15.2	12.7						14.2		
Sm	3.15		4.92	9.82							4.81	4.86	3.48	3.08						3.40		
Eu	0.93		1.06	2.70							1.25	1.27	0.97	0.88						1.02		
Gd	2.5		b.d.l.	7.88							1.69	1.88	2.41	1.76						2.40		
Dy	1.31		0.31	2.84							0.84	0.79	0.90	1.08						0.87		
Er	0.39		0.13	0.83							0.22	0.22	0.26	0.32						0.10		
Yb	0.16		0.38	b.d.l.							0.22	0.21	0.04	0.15						0.06		
Hf	4.5		1.75	4.80							2.03	2.10	2.02	2.20								

(continued on next page)

Table 3 (*continued*)

(c) Clinopyroxene

Sample	M1		M2			M5			M9		M13		GP404		LET2	LET8		LET28		LET29		
	Core	Rim	Core	Rim	Vein	Core	Rim	Vein	Core	Rim	Core	Rim	Core	Rim	Vein	Core	Rim	Core	Rim	Core	Rim	Vein
wt.%																						
SiO_2	55.2	54.7	54.8	54.9	55.5	54.5	54.5	52.6	55.6	55.6	55.8	55.6	55.3	54.6	55.6	54.8	54.9	55.5	55.7	55.2		
TiO_2	0.01	0.01	0.02	0.27	0.11	0.11	0.54	1.30	0.02	0.03	0.09	0.10	0.03	0.10	0.01	0.01	0.02	0.01	0.00	0.02		
Al_2O_3	2.00	1.91	2.01	1.86	2.60	2.63	2.15	5.08	2.43	2.51	3.22	3.22	2.80	3.19	2.81	2.02	2.12	2.31	2.59	3.63		
Cr_2O_3	1.62	1.29	1.64	1.05	1.71	1.75	2.12	1.69	1.57	1.80	2.27	2.28	2.20	2.31	2.22	1.32	1.62	1.06	0.98	2.34		
FeO	1.90	1.89	1.96	2.79	2.23	2.23	2.58	3.36	2.16	2.23	2.37	2.48	2.14	2.33	2.09	1.83	1.70	2.07	2.19	2.16		
MnO	0.09	0.08	0.08	0.13	0.09	0.10	0.10	0.18	0.08	0.09	0.09	0.10	0.08	0.08	0.10	0.09	0.08	0.07	0.06	0.09		
MgO	16.9	17.1	16.9	19.4	16.5	18.4	17.3	16.8	16.7	16.7	15.8	16.1	16.1	15.7	16.1	16.7	16.6	16.5	16.4	15.1		
NiO	0.04	0.02	0.01	0.05	0.05	0.02	0.07	0.06	0.06	0.05	0.06	0.06	0.06	0.05	0.04	0.06	0.04	0.03	0.04	0.05		
CaO	20.4	20.5	20.2	18.6	19.0	17.4	20.4	18.2	19.6	19.3	18.2	18.1	18.9	18.6	18.8	21.5	21.2	20.6	20.6	19.0		
Na_2O	1.69	1.53	1.69	0.98	2.28	2.17	1.36	1.42	2.04	2.12	2.91	2.72	2.48	2.52	2.51	1.42	1.54	1.85	2.00	2.88		
Total	99.9	99.1	99.3	100.0	100.0	99.3	101.1	100.7	100.4	100.5	100.9	100.8	100.2	99.6	100.3	99.8	100.0	100.0	100.7	100.6		
ppm (La-ICP-MS)																						
Sc	22.8	23.0			34.2	35.7					27.7	27.9	28.8	26.4								
Ti	126	134			1889	868					685	644	700	270								
Cr	9799	13,153			13,534	16,937					16,842	14,985	7177	16,984								
Rb	0.42	0.46			0.55	0.54					0.43	3.65	0.54	0.46								
Sr	388	308			185	211		312			202	169	199	251								
Y	0.45	0.48			5.85	5.88		1.13			2.99	3.17	2.99	1.09								
Zr	15.3	18.0			68.0	116		55.3			91.0	84.2	90.9	46.1								
Nb	1.19	1.86			2.48	1.91		8.19			1.68	4.97	1.88	1.83								
Ba	0.28	0.49			3.28	0.85		41.2			1.14	4.11	1.60	0.32								
La	4.77	4.09			5.69	3.20		19.7			3.14	5.11	3.20	3.81								
Ce	20.4	17.8			18.6	13.2		53.5			13.2	16.4	13.3	15.8								
Nd	20.4	19.0			17.0	15.0		49.2			14.6	15.0	14.6	16.4								
Sm	3.04	3.05			4.04	3.93		9.51			3.43	3.40	3.46	3.31								
Eu	0.72	0.71			1.23	1.26		1.59			1.02	1.01	1.04	0.87								
Gd	1.43	1.42			3.40	3.56		6.07			2.38	2.38	2.48	1.99								
Dy	0.29	0.27			1.89	2.01		4.64			1.13	1.16	1.17	0.53								
Er	0.11	0.15			0.62	0.63		1.16			0.30	0.35	0.30	0.15								
Yb	b.d.l.	b.d.l.			0.22	0.21		0.79			0.10	0.13	0.09	0.04								
Hf		2.20			3.50	5.81		2.31			4.59	3.83	4.79	2.18								

(d) Garnet

Sample	M1 Core	M1 Rim	M2 Core	M2 Rim	M2 Overgr.	M9 Core	M9 Rim	M13 Core	M13 Rim	GP404 Core	GP404 Rim	LET2 Core	LET2 Rim	LET29 Core	LET29 Rim	LET38 Core	LET38 Rim
wt.%																	
SiO_2	42.1	42.0	41.9	42.0	42.7	42.6	42.7	42.1	42.5	42.8	42.6	42.1	42.6	42.1	42.6	42.3	42.4
TiO_2	0.40	0.41	0.02	0.03	0.03	0.19	0.38	0.10	0.13	0.07	0.14	0.04	0.03	0.03	0.02	0.02	0.02
Al_2O_3	21.9	21.7	20.9	21.0	21.3	20.3	21.0	17.9	18.8	21.2	21.3	18.3	20.6	19.7	19.8	20.4	20.5
Cr_2O_3	2.33	2.31	4.41	4.37	4.27	5.04	4.13	8.07	7.18	4.05	3.59	8.16	4.83	5.85	5.71	5.07	4.67
FeO	9.88	9.88	6.81	6.90	6.99	7.03	7.08	6.85	6.91	7.21	7.17	6.28	6.22	6.63	6.70	6.31	6.37
MnO	0.41	0.40	0.41	0.40	0.41	0.43	0.38	0.46	0.47	0.39	0.38	0.34	0.34	0.42	0.41	0.37	0.37
MgO	19.8	19.6	20.4	21.0	21.7	20.6	21.0	19.3	19.6	20.9	21.2	22.0	22.4	20.3	20.5	20.3	20.4
CaO	4.32	4.35	4.97	4.16	4.12	5.24	4.78	6.32	6.23	5.06	4.75	4.36	4.51	5.24	5.23	5.49	5.30
Na_2O	0.07	0.07	n.d.	n.d.	0.04	n.d.	n.d.	n.d.	n.d.	n.d.	n.d.	n.d.	n.d.	0.02	0.03	0.01	0.02
Total	101.2	100.8	99.8	99.9	101.5	101.5	101.3	101.1	101.9	101.7	101.2	101.5	101.4	100.4	101.0	100.3	100.2
ppm (La-ICP-MS)														*(La-ICP-MS)*			
Sc			94.7	96.4	94.6			139.9	131.4	96.8	86.1	131.3				98.5	93.0
Ti			116	181	343			202	2825	570	1044	229				173	178
Cr			12,506	32,989	32,049			66,685	61,547	30,360	25,462	52,108				900	10,049
Rb	0.35		1.92	2.27	2.21			2.94	3.16	2.37	2.05	1.31		0.17		2.19	2.01
Sr	3.1		1.41	0.95	0.66			5.89	2.24	0.76	0.58	1.95		0.88		NaN	1.19
Y	26.6		2.92	7.18	26.37			0.64	23.25	5.50	25.63	1.65		1.60		0.93	3.71
Zr	105		43.1	72.5	167			9.69	70.3	37.8	138	12.3		21.4		24.6	58.5
Nb	n.a.		1.33	1.44	1.24			3.71	3.18	1.40	0.93	1.38		0.41		2.19	1.80
Ba	0.5		0.40	0.20	0.16			0.07	0.09	0.08	0.08	0.09		<0.024		0.10	0.13
La	0.04		0.08	0.06	0.03			0.89	0.11	0.04	0.02	0.37		0.06		0.08	0.04
Ce	0.17		1.27	0.99	0.38			5.80	1.48	0.49	0.28	4.54		1.10		1.41	0.70
Nd	0.82		5.74	5.07	2.91			4.45	6.22	2.40	2.10	4.05		2.80		4.27	4.21
Sm	1.03		2.55	2.74	2.89			1.27	4.13	1.38	1.95	0.63		1.08		1.15	2.15
Eu	0.65		0.75	1.00	1.59			0.46	1.75	0.47	0.96	0.17		0.40		0.37	0.78
Gd	2.76		1.89	3.38	7.26			1.57	6.52	1.57	4.50	0.42		0.92		1.10	2.32
Dy	4.8		0.81	1.99	6.63			0.27	4.94	1.10	4.98	0.29		0.44		0.39	1.19
Er	3		0.25	0.56	1.92			0.14	2.03	0.43	2.22	0.18		0.20		0.09	0.30
Yb	3.1		0.40	0.53	1.14			0.17	1.99	0.59	2.02	0.24		0.21		0.11	0.29
Hf	2		0.50	0.92	1.80			0.10	1.56	0.43	1.83	0.21		0.33		0.32	0.94

(continued on next page)

Table 3 (continued)

(d) Garnet

Sample	LET64		LQ5		LQ6		LQ8		TP5		TP6		TP9		J8
	Core	Rim	Core	Rim	Core	Rim	Core	Rim	Core	Rim	Core	Rim	Core	Rim	
wt.%															
SiO_2	42.1	42.8	42.9	42.9	42.7	42.9	42.6	42.2	42.2	43.0	41.4	42.4	41.8	42.7	41.8
TiO_2	0.03	0.03	0.03	0.03	0.12	0.11	0.05	0.15	0.00	0.04	0.00	0.03	0.01	0.01	0.00
Al_2O_3	20.1	20.9	21.0	21.0	20.8	21.3	19.9	20.7	18.5	21.2	17.7	21.5	17.8	21.8	22.8
Cr_2O_3	4.61	4.58	4.39	4.27	4.40	3.79	5.42	4.01	7.31	3.93	8.16	3.59	8.17	3.41	1.91
FeO	6.33	6.50	6.57	6.54	6.91	6.92	6.32	6.72	6.69	6.63	7.02	7.26	7.05	7.14	7.82
MnO	0.35	0.37	0.37	0.38	0.39	0.36	0.39	0.37	0.52	0.46	0.40	0.40	0.55	0.47	0.47
MgO	20.8	20.6	20.8	21.0	21.2	21.4	20.3	21.0	19.1	20.6	18.6	21.4	17.7	20.8	20.2
CaO	5.32	5.25	5.34	5.21	4.70	4.45	5.73	4.45	6.75	5.61	7.88	4.85	8.19	4.52	5.02
Na_2O	0.03	0.03	0.02	0.02	0.04	0.05	0.01	0.04	0.02	0.01	n.d.	n.d.	0.01	0.03	0.01
Total	99.7	101.0	101.5	101.3	101.3	101.3	100.7	99.6	101.1	101.4	101.2	101.5	101.2	100.8	100.1
ppm															
Sc	92.6	96.1			104	93.8	102	92.6					154	138	79.4
Ti	146	218			1034	597	1039	530					73.9	75.5	11.4
Cr	35,729	34,680			36,302	37,235	1520	14,388					55,161	40,674	11,906
Rb	2.21	1.93			2.33	2.17	2.31	2.01					1.46	1.34	1.19
Sr	0.98	1.13			0.72	0.88	0.72	1.02					1.49	0.59	0.51
Y	0.80	1.12			25.4	3.86	31.1	1.46					15.4	41.2	0.29
Zr	43.9	27.5			147	33.0	165	20.6					40.2	89.3	0.81
Nb	1.78	1.50			1.47	1.56	1.20	1.52					1.90	1.04	0.20
Ba	0.13	0.29			0.07	0.06	0.32	0.15					0.96	0.17	0.73
La	0.06	0.10			0.02	0.19	0.02	0.21					0.35	0.04	0.02
Ce	0.92	1.32			0.34	2.34	0.28	2.63					3.30	0.60	0.05
Nd	5.41	6.14			2.40	4.69	2.07	4.73					7.49	5.93	0.10
Sm	2.72	1.99			2.04	0.92	1.98	0.85					2.30	4.12	0.05
Eu	0.80	0.57			0.95	0.31	0.99	0.24					0.84	1.80	0.02
Gd	2.05	1.37			4.34	1.00	4.74	0.68					2.72	6.72	0.04
Dy	0.46	0.38			4.98	0.74	5.91	0.36					2.83	7.27	0.04
Er	0.18	0.13			2.01	0.31	2.65	0.14					0.99	2.64	0.06
Yb	0.26	0.25			1.57	0.32	2.26	0.24					0.65	1.67	0.19
Hf	0.52	0.30			2.04	0.37	2.31	0.25					0.24	0.40	0.03

(e) Spinel

Sample	M2		M5		M9		M13	LET2		LET8		LET28		LET29	
	Core	Rim	Core	Rim	Core	Rim	Core	Core	Rim	Interst.	Incl.	Incl.	Rim	Core	Rim
SiO_2	0.09	0.27	0.03	0.03	0.15	0.11	0.07	0.10	0.18	0.11	0.05	0.03	0.07	0.14	0.13
TiO_2	0.04	0.05	0.06	0.05	0.97	0.24	1.98	2.62	0.13	1.13	0.02	0.01	0.32	0.61	0.08
Al_2O_3	34.5	53.5	46.4	47.3	29.8	49.2	9.07	14.6	54.1	14.4	20.3	26.0	9.31	13.3	52.2
Cr_2O_3	34.7	14.9	23.0	22.7	38.9	20.0	56.1	50.9	14.3	54.4	50.9	44.4	60.0	54.7	16.0
FeO	11.8	11.6	13.0	11.9	13.6	10.7	19.6	16.4	10.0	13.1	13.4	12.7	15.8	16.0	10.5
MnO	0.15	0.23	0.07	0.06	0.08	0.12	0.13	0.24	0.25	0.11	0.12	0.11	0.17	0.05	0.21
MgO	17.8	20.8	18.3	19.2	17.5	20.2	13.0	14.9	20.4	15.6	15.0	16.4	14.0	15.5	20.5
NiO	0.11	0.06	0.24	0.26	0.25	0.09	0.16	0.20	0.14	0.14	0.10	0.14	0.12	0.15	0.08
ZnO			0.21	0.19				0.02	0.00	0.08	0.15	0.15	0.07	0.02	0.05
CaO	0.00	0.06	0.00	0.02	0.02	0.05	0.03	0.03	0.05	0.22	0.01	0.00	0.06	0.06	0.01
Total	99.3	101.5	101.3	101.7	101.2	100.7	100.3	99.9	99.6	99.2	100.0	99.8	100.0	100.6	99.8
Mg#	0.73	0.76	0.73	0.78	0.70	0.77	0.54	0.66	0.80	0.72	0.69	0.70	0.61	0.63	0.78
Cr#	0.40	0.16	0.25	0.14	0.47	0.21	0.81	0.70	0.15	0.72	0.63	0.53	0.81	0.73	0.17

Sample	LET38		LET64		LQ8		TP5		TP6		TP9		J8	
	Core	Rim	Core	Rim	Core	Rim	Core	Rim	Core	Rim	Core	Rim	Rim	Sympl.
SiO_2	0.12	0.12	0.15	0.13	0.41	0.32	0.03	0.03	0.03	0.14	0.04	0.04	0.26	0.51
TiO_2	0.20	0.12	1.97	0.32	0.64	0.17	0.04	0.03	0.02	0.36	0.04	0.04	0.01	0.01
Al_2O_3	47.75	53.1	10.8	54.2	17.5	56.9	13.9	15.8	14.7	53.42	13.4	14.8	18.8	19.1
Cr_2O_3	21.64	15.3	52.2	13.7	50.6	9.67	55.3	53.1	54.9	12.75	55.6	53.7	51.0	50.5
FeO	9.89	10.4	21.0	10.6	15.1	10.1	17.4	17.8	13.2	9.58	14.7	13.1	16.1	16.4
MnO	0.10	0.18	0.05	0.13	0.09	0.16	0.09	0.17	0.11	0.21	0.10	0.06	0.25	0.26
MgO	21.2	21.3	13.8	21.1	15.6	21.6	13.1	13.5	14.0	20.2	12.8	13.9	13.5	12.5
NiO	0.14	0.06	0.18	0.05	0.19	0.03	0.12	0.06	0.12	0.09	0.13	0.16	0.07	0.06
ZnO	0.04	0.03	0.08	0.01	0.11	0.01			0.08	0.00				
CaO	0.01	0.03	0.06	0.07	0.07	0.06	0.02	0.06	0.01	0.09	0.01	0.01	0.13	0.21
Total	101.1	100.5	100.4	100.3	100.4	99.0	99.9	100.6	101.4	99.7	100.8	100.7	100.1	100.2
Mg#	0.79	0.79	0.54	0.78	0.65	0.79	0.57	0.57	0.65	0.79	0.56	0.59	0.60	0.57
Cr#	0.23	0.16	0.76	0.14	0.66	0.10	0.73	0.69	0.72	0.14	0.74	0.71	0.65	0.64

(a) Olivine major elements. (b) Opx major and trace elements. (c) Cpx major and trace elements. (d) Garnet major and trace elements. (e) Spinel major elements. Trace element analyses are SIMS unless otherwise stated.

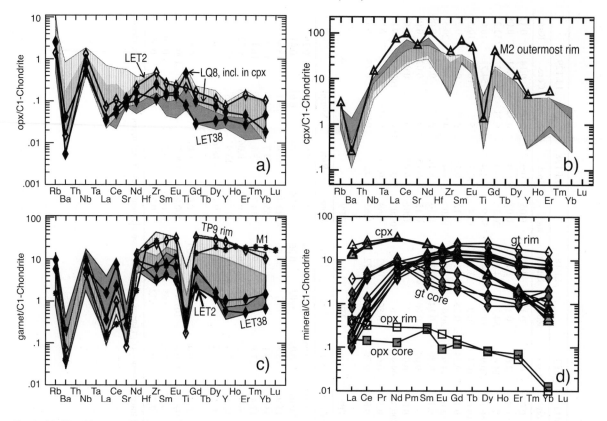

Fig. 4. (a) Chondrite normalized trace elements in opx (McDonough and Sun, 1995). Grey field: opx cores; striped field: opx rims. Three individual patterns for LET2 and LET38 cores and an opx inclusion in cpx in LQ8 are also shown. (b) Chondrite normalized trace elements in cpx. Grey field: cpx cores; striped field: cpx rims. The open triangles represent the outermost rim of a cpx in M2 with very high trace element contents. (c) Chondrite normalized trace elements in garnet. Grey field: garnet cores; striped field: garnet rims. Three individual pattern for LET2 and LET38 cores and TP9 rim are also shown. (d) REE variation in garnet, cpx and opx in LQ6. Note the extreme REE variation in garnet and the rotation of the pattern around Nd–Sm. Sigmoidal pattern occur in the core and unaltered parts of the garnets, pattern with positive (LREE/HREE)$_N$ slope are from outermost rim and close to fractures.

into the crystal and cpx porphyroblasts in many samples contain inclusions (Fig. 1g). Therefore, spots for in situ analyses were selected very carefully. Cpx compositions record systematic elemental variations. Na and Al contents show broad negative correlations with Mg#, which is very similar to Mg# in opx. Ca increases with increasing Mg# in cpx.

On a single sample basis, REE contents within cpx grains vary little (Fig. 4b). In contrast, cpx from different samples have REE contents that differ by an order of magnitude (3–19 ppm La, 0.04–0.4 ppm Yb). If zonation is observed, e.g. in M2, rims show generally higher overall REE abundances than cores. All cpx are LREE enriched, with (La)$_N$ = 10–100, (La/Yb)$_N$ ~ 10–100 and a chondrite normalized max-

imum at Nd. Sample M2 contains by far the highest overall REE concentrations. Cpx in M2 and LET38 have slightly sigmoidal REE patterns with a maximum at Nd and a steep decrease between Nd and Dy. Positive correlation exists between Hf and Sc. Hf is also positively correlated with HREE in cpx. LET64 contains cpx with the highest Hf (up to 5 ppm) and HREE contents, comparable to the levels observed in M2 cpx rims. LET 38 cpx have the lowest Hf contents (0.7 ppm). Sr in cpx varies between 200 and 400 ppm, with M2 cpx having the highest Sr. Sc is sometimes strongly zoned in cpx. M2 cpx have the highest Sc in the rims (45 ppm) and the lowest in the core (15 ppm). Ti contents vary widely from 100 ppm (M2 core, LET38) to 1900 ppm (LET64 core). Cpx contain <1 to 7 ppm Y. Y correlates positively with Zr. Abundances of both elements are highest in M2 cpx rims and LET64.

4.3.4. Garnet

Most garnets display some variations in Cr, Al and Ca, comparable to the trend described as "type I" zonation by Burgess and Harte (1998) for the deformed high-temperature peridotite suite from Jagersfontein. All garnets from samples studied here show this compositional zonation parallel to the "lherzolite trend", except sample M2, which has a distinct low-Ca outer rim. This low-Ca rim is related to fracturing of the garnet and is absent in unfractured grains that occur as inclusions in opx or olivine (Fig. 1f). It is similar to low-Ca overgrowth rims on garnets from Wesselton described by Griffin et al. (1999).

The most noticeable feature of garnet REE patterns (Fig. 4c and d) is the sigmoidal shape of almost all of the cores with a maximum between Nd and Eu and minimum at Dy to Er. These patterns indicate extreme LREE enrichment and are typical of "G10" garnets found as inclusions in diamonds, in heavy mineral separates and rare harzburgite xenoliths (Shimizu and Richardson, 1987; Pearson et al., 1995; Shimizu et al., 1999). In contrast to typical subcalcic G10 garnets, the garnets analyzed here are Ca-rich and relatively low in Cr. Most garnets plot in or near the lherzolite field in the garnet classification diagram (Fig. 5, after Sobolev et al., 1973). Thus, most garnets are saturated in Ca and the samples contain cpx (Table 1). The presence of sigmoidal REE patterns in garnet therefore indicates disequilibrium between garnet and cpx because of the preferred equilibrium partitioning of LREE into cpx.

Cores of garnets are relatively homogeneous and substantial trace element variation is detected only in the outermost rim and where fractures crosscut the grains. It is difficult, however, to determine "real" rim compositions in garnet since almost all of them are surrounded by kelyphite rims of variable thickness. In some samples (M2 and GP404), garnet locally occurs as inclusions in olivine or opx (Fig. 1f). The inclusions were shielded from infiltrating fluids or melts and have almost no kelyphite rims. These garnet inclusions show very limited major or trace element zonation and have the same composition as the cores of the big garnet porphyroblast.

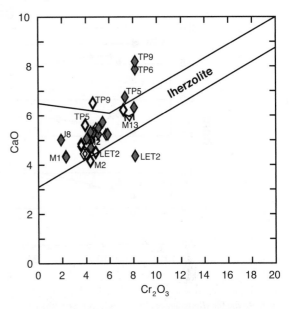

Fig. 5. Garnet classification diagram after Sobolev et al. (1973). Closed diamonds are core open symbols are rim compositions. Garnet rims are generally lower in Ca and Cr than cores. Note the distinct composition of M1, the TP samples and the garnet–spinel lherzolite M13.

Another important observation is that trace element variation towards the rims of garnets depends on the contact mineral. In sample GP404, a trace element profile was measured through a garnet that is in contact with olivine on one side and cpx on the other. The boundary with olivine is strongly kelyphitized and trace element zonation is observed whereas the boundary with cpx is relatively sharp and shows much less trace element variation (Fig. 6). The phase contact dependency of kelyphitisation has been previously observed by Canil and Fedortchouk (1999) for garnet xenocrysts in kimberlites. We agree with their conclusion that kelyphite could only form if olivine was available for the reaction $ol + gt = sp + px$. This is also consistent with the absence of keliphite in olivine-free eclogites. Here, we might be able to directly relate kelyphite formation to trace element zonation in garnet. We, therefore, conclude that the extreme trace element zonation in garnet rims is probably a young feature that might be related to kimberlite volcanism. These conclusions are in contrast to the interpretation of Shimizu et al. (1999) who carried out a very detailed study on trace element zonation in garnets from Siberian harzbur-

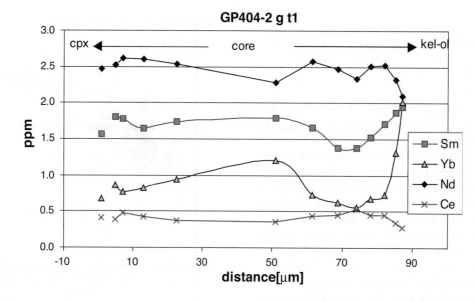

Fig. 6. Profile for selected trace elements (Ce, Nd, Sm, Yb) across a garnet in GP404. The left side of the garnet is in contact with cpx and only slightly kelyphitized. The right edge of the garnet was in contact with olivine and has a broad kelyphite rim. Zonation in REE is much more pronounced where kelyphitization is more advanced. See text for detailed discussion.

gites and dunites. They concluded that trace element zoning of garnets resulted from temporal evolution of kinetic effects during garnet growth, in which element partitioning between garnet and melt/fluid was controlled by rates of trace element incorporation reactions. We agree that whatever caused the zoning in garnet most likely involved complex reaction–precipitation processes and conventional assumptions of equilibrium chemistry and kinetics cannot be applied here. Nevertheless, the zoning observed in the Lesotho samples is different to that observed in the Siberian garnets, and we believe that especially the virtual absence of zonation in the garnet inclusions in M2 and GP404 is an indication for the link between fluid infiltration and zoning of garnet rims.

Where REE zonation in Lesotho garnets is observed (except for the LQ garnets, where the keliphite is in the core—see below and discussion in Section 5.2.2), the pattern always evolves from cores that are strongly sigmoidal (highly LREE enriched) to more "normal" in garnet rims, showing lower (LREE)$_N$ and higher (HREE)$_N$ contents and LREE/HREE < 1 (Fig. 4c and d). Rim compositions are as would be expected if garnet is in REE equilibrium with cpx (e.g., Shimizu, 1975; Zack et al., 1997; Eggins et al.,

1998, and references therein). Some variation in other trace elements (Rb, Sr, Y, Zr, Ti, Cr and Sc) is also observed in garnet. Yb, Zr and Hf are generally low in cores (<1, <50 and <1 ppm, respectively) and elevated only in the very outer rims. Cr correlates positively with Sc, Zr with Y and Nb with Sr. Sample M13, a garnet-spinel lherzolite with very high Cr in garnet (Fig. 5), has the highest LREE, Sr, Y, Rb, Sc, Nb and a low Zr content in garnet.

Garnets in xenoliths from Liqhobong (LQ) are very cracked and appear to have been kelyphitized from the inside (Fig. 1c). This texture is different to all other samples examined here, where kelyphitization was initiated on the margins of the grain. Samples from Thaba Putsoa are also distinct from common garnet lherzolites. These garnet-spinel lherzolites have garnets intergrown as symplectites ("fingerprint" texture) with spinel, cpx and opx (Fig. 1e). Garnets and spinels in the clusters are strongly zoned in major elements, especially in Cr and Al. Garnet-spinel transitions are preserved in several stages of reaction and mineral compositions reflect only very local equilibration of phases. The garnet–spinel reaction probably has been triggered by a change in $P–T$ conditions but it is difficult to establish what process was responsible for initiating the reaction.

4.3.5. Spinel

Spinel in most samples is either derived from breakdown of garnet or infiltration of kimberlite melt. These secondary spinels have high Ti-contents and are strongly zoned, especially in Cr and Al. Cr# varies from l0 to 80. Only samples listed in Table 1 as spinel or garnet-spinel peridotites are likely to contain "primary spinel". In contrast to "secondary" spinels, primary spinels are rather homogeneous in major element composition and have relatively high Cr# (72–80) if coexisting with garnet. Spinel in the only spinel lherzolite within our sample suite (M5) is texturally and compositionally distinct from spinel in garnet-bearing rocks. It has a relatively high Mg# of 73 and a very low Cr# (25).

4.3.6. Phlogopite

Phlogopite can be classified in the same way as spinel. We interpret most as secondary in origin. It is commonly associated with secondary spinel. Secondary phlogopite can be found in garnet kelyphite rims and as microphenocrystic quench crystals in samples that contain kimberlitic melt remnants. Some samples contain large phlogopite porphyroblasts (LET64, M2) which, in the case of LET64, form phlogopite-cpx veins through the xenolith. This textural relation suggests a co-genetic origin for cpx and phlogopite porphyroblasts. In M2, phlogopite crystals mostly occur attached to garnet. Large porphyroblasts are compositionally distinct to secondary phlogopites. The latter contain much less Si (Table 3), indicating an origin in the spinel stability field (Arai, 1984). We therefore conclude that secondary phlogopite most likely formed at a very late stage during ascent of the xenoliths in the kimberlite.

5. Discussion

5.1. Major element equilibrium and pressure–temperature estimates

The complex textural relationships, often involving multiple generations of minerals, imply that chemical equilibrium cannot easily be demonstrated. Therefore, great caution has to be applied when pressure and temperature are calculated from major element exchange equilibria. We tried to avoid these problems by

a combination of very careful selection of microprobe mineral analyses together with the application of a variety of different geothermobarometers (O'Neill and Wood, 1979; O'Neill, 1980; Harley, 1984; Krogh, 1988; Brey and Köhler, 1990; Ballhaus et al., 1991a,b; Witt-Eickschen and Seck, 1991; see Table 1). This method enables us to compare results for different formulations and reactions, involving different mineral pairs with different closure temperatures and different sensitivities to metasomatic interaction. This approach provides a tool for determining which phases are equilibrated in terms of major elements. Since we have demonstrated that zonation in the outer rims is probably caused by late-stage processes and might not represent equilibrium conditions, we used core compositions for P–T calculation (except for LQ6 and LQ8, see discussion below). The results are listed in Table 1.

Pressures and temperatures were calculated iteratively. Temperatures for rim compositions are usually slightly higher, consistent with heating of the xenoliths during transport in the kimberlite. Where minerals show extensive overgrowths, calculated pressures and temperatures become unrealistically high or low.

Following the recommendation of Smith (1999), we employ pressures calculated with the Al-in-opx barometer (P(BKN)) of Brey and Köhler, 1990) for garnet-bearing samples and use the Ca-in-ol barometer of the same authors for garnet-free lherzolites, in combination with several thermometers. The Ca-in-ol barometer is strongly temperature dependent and diffusion of Ca in olivine is rather fast. Moreover, EPMA measurements of Ca in olivine might be affected by secondary fluorescence boundary effects from nearby Ca-rich phases. Therefore, pressures obtained with this barometer have to be considered with caution and only give rough pressure estimates, especially for garnet-free samples. No pressure could be calculated for spinel harzburgites due to the lack of suitable barometers for garnet- and cpx-free peridotites. Temperatures for garnet-bearing samples were calculated with the olivine-garnet Fe–Mg exchange thermometer of O'Neill and Wood (1979 and O'Neill, 1980) in combination with P(BKN). It was shown by Smith (1999) and Brey and Köhler (1990) that, for relatively reduced conditions (low Fe^{3+} content), the O'Neill thermometer reproduces experimental temperatures very well, and it has the advantage of being applicable to garnet harzburgites and lherzolites. Spinel peridotite temperatures are calculated using the thermometers of Witt-Eickschen and Seck (1991: WS) and Ballhaus et al. (1991a,b) and the Ca-in-opx thermometer of Brey and Köhler (1990) for a preset pressure of 2 GPa. Spinel peridotite temperatures calculated with WS are relatively high and do not plot on the Kalahari (Zimbabwe and Kaapvaal craton) geotherm obtained by Rudnick and Nyblade (1999; Fig. 7).

Sample M5 contains large texturally equilibrated spinels and clearly is a spinel facies lherzolite. Temperatures calculated with formulations other than WS (Table 1) are much lower and range from 614 °C (ol-sp of Ballhaus et al., 1991a,b) to 798 °C (Ca-in-opx, Brey and Köhler, 1990). We therefore assume that the WS thermometer significantly overestimates temperature for this sample due to slower diffusivities of Cr and Al compared with divalent cations, and that equilibration conditions more realistically are 600–700 °C and ~ 2–2.5 GPa. For LET8 and LET28, textural relationships do not make it clear that spinel is a primary phase. Boyd et al. (1999) showed that opx in spinel-facies peridotites mostly have Al_2O_3 contents higher than 1.5 wt.%. Opx cores in LET8 and LET28 have 1.42 and 1.58 wt.% Al_2O_3, respectively. Therefore, these samples might either represent spinel peridotites or highly depleted harzburgites that are

void of garnet due to their highly refractory character, but equilibrated in the garnet stability field. The latter option is supported by other thermometer formulations that result in equally high temperatures (Table 1). Boyd et al. (1999) also found high calculated temperatures in Kaapvaal spinel peridotites and concluded that these might reflect incomplete major element (especially Al and Cr) re-equilibration during cooling. They also proposed that there might be a certain metastable overlap in spinel and garnet stability fields in the Kaapvaal lithosphere.

Deviation of the spinel peridotites from the Kalahari geotherm (Fig. 7) due to heating and incomplete re-equilibration is consistent with the heating event at 1 Ga that is recorded in lower crustal xenoliths from northern Lesotho (Schmitz and Bowring, 2001).

In accordance with the results of Brey (1989) for garnet lherzolites from several pipes in northern Lesotho, we find a significant discrepancy between temperatures calculated with thermometers using cpx in equilibrium with another phase (2-px: e.g., Brey and Köhler, 1990; Wells, 1977; Bertrand and Mercier, 1985; Finnerty and Boyd, 1987; Taylor, 1998; cpx-gt: e.g., Krogh, 1988) compared to cpx-free formulations (e.g. opx-gt: Harley, 1984; ol-gt: O'Neill and Wood, 1979). Brey (1989) attributes this difference in calculated temperatures to incomplete equilibration of cpx with garnet and opx. This indicates that recent mantle metasomatism may have either introduced new cpx or completely re-crystallized cpx without significantly affecting garnet. Alternatively, more oxidizing conditions in the mantle compared to experimental conditions (elevated Fe^{3+} in garnet) would reduce temperatures calculated with T(O'Neill) or T(Harley). An increase of ferric iron in garnet with depth was demonstrated by Luth et al. (1990) for low-T peridotite xenoliths from the Kaapvaal craton.

5.2. The origin of garnet and cpx

5.2.1. Major element constraints

There is general agreement that Archaean lithospheric peridotites represent residua from large degrees of melting. The residual opx are saturated with Al, Cr and Ca at high temperatures, but can only host lower amounts of these elements when cooled (Boyd and England, 1964; Wood and Banno, 1973). Therefore, depending on composition and ambient

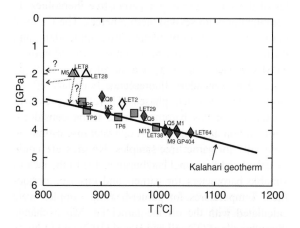

Fig. 7. P–T diagram for the Lesotho xenoliths. Temperature and pressures for garnet-bearing samples were calculated using the olivine–garnet thermometer of O'Neill and Wood (1979) in combination with the barometer of Brey and Köhler (1990). Note the deviations of the spinel peridotites from the Kalahari xenoliths geotherm of Rudnick and Nyblade (1999). See text for discussion.

conditions, garnet, spinel and cpx exsolve from opx in variable proportions. This raises the question: can this process account for all the garnet and cpx in the Lesotho peridotites? To investigate this problem, we followed the procedure described by Canil (1992) who mass-balanced the reaction high-T opx → garnet + cpx + low-T opx and concluded that this reaction potentially can account for the varying amounts of opx, garnet and cpx in Kaapvaal low-T peridotites. If garnet, cpx and opx contents are linked by this reaction, a positive correlation between modal opx and garnet + cpx in the rock is expected (Canil, 1992). Fig. 8a shows that an increase in garnet + cpx with increasing opx indeed is observed for the Lesotho garnet peridotites. However, the correlation improves if only garnet is plotted against opx (Fig. 8b), something previously found by Canil (1992) who compared harzburgites and lherzolites from Kaapvaal. Interestingly, the correlation entirely disappears if samples from other locations within the Kaapvaal (Kimberley) are included (gray dots in Fig. 8), indicating that the co-variations found for Lesotho might be fortuitous. In accordance with Canil (1992), and consistent with textural observations and results from thermobarometry, we therefore infer that only a fraction of the observed cpx was exsolved from opx and a different origin for the majority of the cpx is likely.

This conclusion is supported by mass balance calculations for selected Lesotho xenoliths. We compared residua from melting experiments of mantle peridotite at 1 (Falloon et al., 1999), 4, 6 and 7 GPa (Walter, 1998) with the bulk major oxide compositions of xenoliths M13 and LET38. All experimental compositions are residua from 35% to 50% melting and have Mg# in olivine between 92 and 93, hence in the range for Kaapvaal low-T peridotites. M13 is a garnet-spinel lherzolite with very Cr-rich garnet, and its bulk composition is close to the average for our Lesotho sample suite. LET38 is relatively Si and Al rich, has low Fe and Mg contents and, therefore, has high opx, garnet and cpx contents. The direct comparison of experimental residua and xenolith bulk composition shows significantly lower SiO_2 in the experiments (Table 4). Lower Cr and Ca and higher Mg contents in the experimental residua compared to M13 and LET38 are also evident. The contrast between the natural and experimental compositions becomes more apparent if modal compositions of

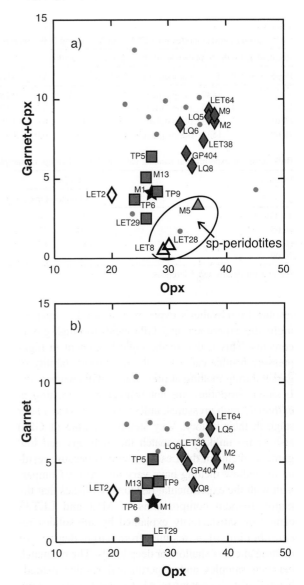

Fig. 8. (a) Correlation between modal garnet + cpx vs. opx content in Lesotho xenoliths. (b) Modal garnet vs. opx. Symbols as in Fig. 2.

minerals with compositions as in M13 and LET38 are calculated from the bulk compositions corresponding to the experimental residua. As expected, the ol/opx ratio is much higher because of the lower Si in the experimental residua than xenoliths. Cpx and garnet contents are also much lower in the experimental residua. The closest similarity between natural and experimental samples is obtained by using the

Table 4
Re-calculated mineral modes for LET38 and M13 using bulk compositions of residua from melting experiments

Recalculated bulk composition of the experimental residua

Experiment	P	T	Paragenesis	Melt %	SiO_2	TiO_2	Al_2O_3	Cr_2O_3	FeO	MnO	MgO	CaO	Na_2O	Total	Mg# (ol)
T-4330	1	1440	ol,opx,melt	65	42.3	0	0.18	0.63	7.34	0.13	49.0	0.38	0	100.0	0.92
40.05	4	1660	ol,opx,melt	61	43.8	0.01	1.17	0.31	6.90	0.11	46.6	0.55	0.02	99.4	0.93
60.03	6	1770	ol,opx,melt	50	43.8	0	0.90	0.25	6.33	0.09	48.8	0.39	0.02	100.5	0.93
70.09	7	1835	ol,opx,gt,melt	53	43.1	0.01	2.44	0.31	5.94	0.09	46.7	0.75	0.03	99.4	0.92

Bulk comp. of experimental residue: calculated modal comp. of samples LET38 and M13 using measured mineral comp.

Experiment	LET38					M13					
	Olivine	Opx	Cpx	Garnet	RMS	Olivine	Opx	Cpx	Garnet	Spinel	RMS
T-4330	93	4	1	2	0.119	92	5.5	1.7	0.6	0.1	0.004
40.05	80	13	1	6	0.543	80	14	0.7	5.1	0	0.021
60.03	87	10	1	2	0.193	87	11	1.1	1.4	0	0.429
70.09	82	7	1	9	0.02	83	8.4	0.7	8.2	0	0.897

See text for details and references.

residue from Walter's experiment at 4 GPa. Interestingly, the experiment at 7 GPa yields too high garnet contents. This is due to the high Al-content in high-pressure residua caused by the increasing stability of garnet during melting at pressures >6 GPa. Hence, the Lesotho peridotites are not residua from melting at extremely high pressures, unless they lost Al at a later stage in their history. Very shallow melting (1 GPa) also seems unlikely to match the high opx and low garnet contents of the Lesotho samples because residua of shallow melting have very low Al/Cr. Comparison with the experimental residua establishes that the major element compositions of M13 and LET38 cannot be satisfactorily explained by sub-solidus re-equilibration of a residue from large degrees of melting at either shallow or deep levels. The mismatch between samples and experimental residua extends beyond just Si content to Al, Cr and Ca. Therefore, these results are another indication that a significant proportion of the garnet and cpx in the Lesotho rocks was probably not exsolved from opx following initial melt extraction and subsequent cooling, but might have been introduced at a later stage by metasomatic processes.

5.2.2. Trace element equilibrium

In a further attempt to assess the degree of chemical equilibrium within the Lesotho peridotite xenoliths, a detailed evaluation of trace element partitioning was undertaken. Partitioning of elements between phases is most easily understood in terms of partition coefficients (Ds). Garnet/cpx Ds ($D^{gt/cpx}$) were calculated and compared to equilibrium Ds from the literature. We selected here the values obtained by Zack et al. (1997) on natural garnet pyroxenites from Kakanui since these rocks are similar in composition and are equilibrated at the same pressure and temperature conditions as our samples. There is no significant difference, however, between the $D^{gt/cpx}$ of Zack and other experimental and natural partitioning data from the literature (e.g., Hauri et al., 1994; Harte et al., 1996; Eggins et al., 1998; Glaser et al., 1999; Green et al., 2000).

$D^{gt/cpx}$ markedly differ between many mineral cores and rims. For most mineral cores, $D^{gt/cpx}_{LREE}$ are higher than the expected equilibrium values and $D^{gt/cpx}_{HREE}$ are lower (Fig. 9). In contrast, garnet–cpx partitioning relationships for the rims are consistent with equilibrium. $D^{gt/cpx}_{Sr}$ is identical for cores and rims and lies within the equilibrium range. The difference in behaviour of Sr and REE is consistent with faster equilibration of divalent cations compared to trivalent ions due to much higher diffusivities of the former (Ganguly et al., 1998a,b; Van Orman et al., 2001, 2002). As a whole, the sample suite appears to preserve different stages of re-equilibration between garnet and cpx. Taking textural observations into account, it seems that equilibration is far more advanced where garnets are significantly fractured and altered. "Clean" garnets usually preserve an LREE-

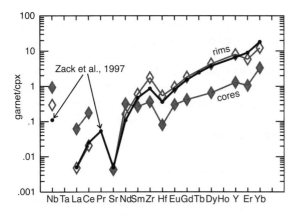

Fig. 9. Partitioning of trace elements between garnet and cpx cores and rims for sample LQ6, compared to equilibrium values of Zack et al. (1997).

enriched sigmoidal pattern. Consequently, zonation in REE in Liqhobong samples is opposite to the trend observed in zoned garnets in the other samples (Fig. 4c), since kelyphitization starts in the cores of the cracked garnets. Thus, garnet cores in LQ8 and LQ6 display "normal" REE patterns with chondrite normalized concentrations increasing from light to heavy REE, whereas the rims show sigmoidal patterns with a maximum at Nd. Thus, REE disequilibrium with cpx is preserved in the relatively fresh rims of the garnets while REE in the fractured cores equilibrated with cpx.

Equilibration of elements seems to have been facilitated by cracks, which were most likely caused by percolating fluids. These fluids enhanced chemical exchange and transport of elements. We surmise that the fracturing of the garnets occurred during rapid decompression, i.e., during entrapment of the xenoliths in the kimberlite and transport to the surface. This interpretation differs from that of Shimizu et al. (1999) who ascribed trace element zoning in Siberian garnets to crystallization and ring-like growth of the minerals in a disequilibrium process because the zoning pattern is not consistent with equilibrium diffusional exchange. We agree that the zonation is a short-lived feature and similar disequilibrium mechanisms might also be applicable in the case of the Lesotho garnets. This conclusion is consistent with the rapid nature of the kimberlite interaction with the lithosphere envisaged here as the probable cause for garnet trace element zonation.

The sigmoidal REE pattern of the original garnet was not introduced by the kimberlites or its precursors because La is inversely correlated with Ti. One would expect Ti to increase with increasing LREE if the enrichment was late stage and associated with a kimberlitic melt or kimberlite precursor. Therefore, the trace element enrichment recorded by garnet is likely to have happened earlier and reflects interaction with a highly fractionated fluid or melt with low Hf, Zr, Ti and Y contents. It is clear that clinopyroxene was probably not present at the stage of the original LREE enrichment of garnet because garnet REE have not equilibrated with cpx.

To emphasize the difference between the highly LREE enriched garnets in the Lesotho samples and a garnet that has exsolved from opx, we measured trace element contents in opx and garnet in a well-characterized garnet-spinel lherzolite from Jagersfontein (Simon, 1999: sample J8; $P–T$ and major and trace element compositions of minerals are given in Tables 1 and 3, respectively). Garnet and spinel in this rock form exsolution lamellae in opx and aggregate at opx rims (Fig. 1a). The REE pattern of opx from J8 is flat and REE contents are very low (< 0.1 times chondrite; Fig. 10a). The garnet that exsolved from this opx is also very low in REE and has a pattern that rises from $LREE_N \sim 0.1$ to $HREE_N \sim 1$, with a slight "hump" in MREE that can also be seen in the opx (Fig. 10a). Trace elements are preferentially partitioned into garnet: $D^{gt/opx}$ is close to unity for LREE and increases markedly towards HREE (Fig. 10b). $D^{gt/opx}$ for non-REE trace elements increases from ~ 1 to 21 in the order $Hf < Sr < Ti < Ba < Zr < Rb < Sc < Nb < Cr < Y$. We compared these measured partition coefficients with data from Glaser et al. (1999) who measured trace elements for garnet and garnet–spinel peridotite xenoliths from Vitim. $D^{gt/opx}$ for highly incompatible elements in J8 are comparable to Vitim xenoliths, but are considerably lower for more compatible elements (Fig. 10b). This could be due to: (i) much higher equilibration temperatures of the Vitim samples (1000–1100 °C, Glaser et al., 1999) compared to J8 (745 °C, Table 1), (ii) differing mineral and whole rock composition or (iii) incomplete equilibration between garnet and opx in J8. The important conclusion, however, is that even if garnet in J8 had not reached complete equilibrium with opx, garnet that exsolved from opx would not be as extremely LREE

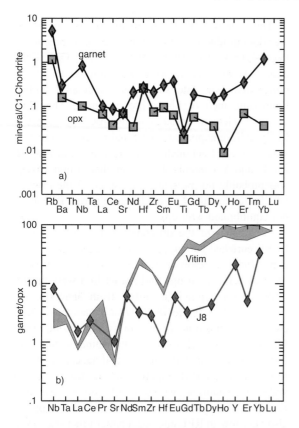

Fig. 10. (a) Trace element contents in opx, and garnet that exsolved from opx, in a low-T garnet–spinel lherzolite from Jagersfontein (J8). (b) Partitioning of trace elements between garnet and opx in J8. The grey field represents partition coefficients from Glaser et al. (1999) for garnet and garnet–spinel peridotites from Vitim.

enriched as is observed for the garnets from the Lesotho xenoliths.

5.2.3. The nature of the metasomatic agent

The calculated REE pattern of hypothetical melts in equilibrium with garnet, cpx and opx (mineral/melt partition coefficients from: Fujimaki et al., 1984; Hart and Dunn, 1993; Dunn and Sen, 1994; Zack et al., 1997; Green et al., 2000) places further constraints on the origin of these minerals (Fig. 11). The hypothetical melt parental to cpx ("cpx melt") has a REE pattern that is similar to a kimberlitic or lamproitic melt. This is consistent with the findings of Stachel and Harris (1997; their Figs. 9 and 11) for garnet and cpx inclusions in diamonds from Akwatia (Ghana).

A connection between cpx formation and kimberlite magmatism is also suggested by the observed close spatial relationship of cpx and phlogopite in many low-T peridotites. Boyd and Mertzman (1987) proposed that fine-grained, secondary diopside crystallized along with mica in Jagersfontein Low-T peridotites, possibly from a supercritical fluid phase via the reaction:

$$gt + opx + fluid = phlogopite + diopside.$$

Modal abundance evidence indicates a relationship between coarse phlogopite and diopside in other peridotite suites that may imply formation of coarse diopside via similar reactions. Strong correlation between the presence of diopside and coarse mica is observed for lherzolites (both garnet and spinel bearing) from the Premier kimberlite (Boyd, unpublished data) whereas harzburgites from the same pipe are almost mica-free. A similar relationship is evident for peridotites from Kimberley. Grégoire et al. (2002) demonstrated, on the basis of trace element and isotope analyses, that cpx in micaceous peridotites from the Kimberley area is linked to either Group I or Group II kimberlite. A similar conclusion was drawn by van Achterbergh et al. (2001) for metasomatized peridotites from Letlhakane. These authors showed that the successive increase in phlogopite and cpx in their rocks can be explained if the xenoliths represent wall rock adjacent to a major conduit for alkaline

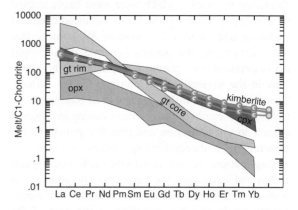

Fig. 11. Chondrite normalized trace element compositions (McDonough and Sun, 1995) of calculated hypothetical equilibrium melts for sample LQ6. See text for references. Kimberlite data are selected analyses of ~90 Ma Type I kimberlites from the Kaapvaal craton (Pearson, unpublished).

basic silicate melt, which percolated into the wall rock and caused the observed metasomatism. Our data also strongly suggests that cpx formation is somehow linked to kimberlite-like magmatism. The frequent occurrence of cpx attached to garnet therefore might not be caused by a common exsolution origin of both phases, but by preferential crystallization of melts on the energetically favourable surface of near spherical garnets.

The calculated equilibrium melt for garnet cores is very distinct from the "cpx melt". It shows a highly fractionated REE pattern and is very similar to the "megacryst melt" proposed by Burgess and Harte (submitted for publication). These authors studied garnets with sigmoidal REE patterns from a suite of garnet harzburgites from Jagersfontein and proposed that they might have crystallized from a melt that was fractionated by precipitation of Cr-poor garnet megacrysts. The inferred parental melt also closely resembles the REE pattern calculated for the melt in equilibrium with harzburgitic garnet inclusions in diamonds from Akwatia (Stachel and Harris, 1997). Following the classification of Griffin et al. (1999), the Lesotho samples (apart from TP9 and M13 that have relatively high Y in garnet) were affected by different extents of "phlogopite metasomatism", even though most samples do not contain phlogopite (only as a phase in garnet kelyphite rims). Griffin et al. (1999) state that the low-T metasomatism does not affect Mg# in the minerals. This is consistent with our data in so far that there is no significant zonation in Fe or Mg in any of the phases. The constant Mg# can also be explained, however, by fast diffusivity of Fe and Mg (Chakraborty and Rubie, 1996; Ganguly et al., 1998a). Therefore, if the metasomatism that led to the LREE enrichment in garnet was ancient, we would expect the Fe and Mg content to be homogenized in a relatively short time at mantle temperatures. There is also no correlation between Ti and LREE in garnet in the Lesotho samples.

Calculated "opx melts" have much lower REE contents (Fig. 9c), which is probably mainly due to the fact that the literature partitioning data is not appropriate for the $P-T$ conditions and compositions of opx in the Lesotho samples. The data used here are from experiments performed at a maximum of 2 GPa (Green et al., 2000), and Eggins et al. (1998) argued that partitioning of trace elements into opx is strongly dependent on pressure, temperature and probably also modal composition. Unfortunately, to date, there is no internally consistent data set for trace element distribution between all the phases in garnet peridotite and melt available in the literature. The slope of the "opx melt" REE pattern, however, is slightly steeper than that of the "cpx melt". Although these data might provide a further indication for incomplete REE equilibrium, at this time, we are unable to draw firm conclusions due to uncertainty in the opx Ds.

5.3. Implications for models of lithosphere formation

Kelemen et al. (1992, 1998) and Kelemen and Hart (1996) developed their lithosphere re-enrichment model based on a positive correlation between modal opx and Ni in olivine observed for garnet and spinel peridotites from Kaapvaal, Siberia, Greenland and Lac de Gras, Canada. In the Lesotho peridotites, opx shows no systematic co-variation with Ni in olivine. Kelemen et al. (1998) used high precision EPMA data for Ni contents in olivine, which is precise to within ± 50 ppm. Our Ni in olivine data is obtained by conventional EPMA measurements and therefore relative uncertainties of $\pm 10\%$ have to be assumed. In addition, the database shown here is much smaller, and consists of xenoliths from different kimberlite pipes. Kelemen et al. (1998) noted that not all kimberlite xenoliths show correlation between Ni in olivine and modal opx, and they obtained the best correlation for samples from Premier. Boyd (1997) re-analyzed many peridotites from the Kaapvaal craton with long microprobe counting times and found that only Premier xenoliths display a positive Ni in olivine vs. modal opx correlation. All other samples show a horizontal distribution with relatively constant Ni contents in olivine and variable opx concentrations. The lithospheric mantle at Premier was severely modified during the Bushveld event at 2 Ga (Carlson et al., 1999). This raises the possibility that the good correlation of Ni in olivine vs. modal opx seen in Premier samples might be caused by interaction with Bushveld magmas. Bushveld magmatism is unlikely to have modified the N. Lesotho mantle, and hence, this process is unlikely to be responsible for the craton-wide Si-enrichment typical of Kaapvaal peridotites.

For some relatively depleted samples from the Lesotho xenolith suite (e.g. LET8, LET28), opx and olivine are intimately intergrown and can also be found as inclusions within each other. This points to either a primary (non-metasomatic) origin of opx, in contradiction to the Kelemen model, or requires complete re-crystallization of the minerals.

However, we do not want to rule out the possibility of interaction of the depleted lithosphere with subduction-related fluids or melts. Eclogites are common components in the Kaapvaal lithosphere and a large proportion of them are believed to represent subducted oceanic crust (e.g., Barth et al., 2001, and references therein). Thus, subducted material is available as a source for metasomatic agents like hydrous fluids or silica-rich melts. Subducted oceanic crust converts to eclogite and is rich in garnet. Melting would, therefore, produce liquids with highly fractionated REE patterns similar to the calculated hypothetical melts in equilibrium with garnet in our samples. Yaxley (2000) carried out sandwich experiments with basalt and peridotite and showed that low-degree melts of oceanic crust would react with the surrounding lithosphere and enrich harzburgite in opx and magmaphile components. Complete melting of the eclogite leads to re-homogenization and re-fertilization of previously depleted mantle. Lherzolites formed by this process closely resemble "fertile" peridotite thought to comprise the asthenosphere. Re-melting of this material provides another source for metasomatic liquids with very complex trace element and isotope signatures. Recent Re–Os isotope work on sulfide inclusions in diamonds from the Kaapvaal craton (Pearson et al., 1998; Richardson et al., 2001, and references therein) shows that subduction was active from the late Archaean (2.9 Ga) until the Proterozoic (1 Ga). Thus, recycling of subducted components appears to provide an elegant way to explain several of the observed features in Kaapvaal low-*T* peridotites. In contrast, the formation of opx-rich cumulate is contemporaneous with melting (Herzberg, 1999) and therefore restricted to the Archaean. Such a process is inconsistent with the complex time-integrated isotope signature recorded by xenoliths (e.g., Pearson, 1999).

Cryptic metasomatism (modification of incompatible element content without precipitation of new phases) of a previously highly depleted residue cannot alone explain the observed high modal proportion of opx and garnet in Kaapvaal peridotites. A model is favoured that involves introduction of a silica (and incompatible element)-rich melt that precipitated opx and metasomatized or crystallized garnet. A similar model was proposed by Burgess and Harte (1998) and Burgess and Harte (submitted for publication) to explain the origin of zoned, LREE-enriched G9 and G10 garnets from Jagersfontein xenoliths. Even though it is likely that cpx formed during several events (exsolution from opx, crystallization from early metasomatic melts), our data indicate that the bulk of the cpx in Lesotho peridotites was introduced during a later event probably associated with either infiltration of the host kimberlite or by magmatism occurring just prior to kimberlite formation.

Acknowledgements

First, we would like to thank all the organizers and participants at the Slave-Kaapvaal workshop for generating stimulating discussion. Specific thanks go to Herman Grutter for his editorial handling of this manuscript. The manuscript also benefited from the constructive reviews of Joe Boyd and an anonymous reviewer. Invaluable assistance was provided by Jianhua Wang and Eric Hauri during SIMS and Paul Mason during LA-ICP-MS measurements. We would also like to thank Wim Lustenhouwer and Saskia Kars for EPMA and SEM analyses, respectively.

This research was carried out as part of the first author's PhD project funded by NWO; 809-31.001. Additional support for work at DTM was provided by NSF Grant EAR-9526840. This is NSG publication no. 20021002.

References

Arai, S., 1984. Pressure–temperature-dependent compositional variations of phlogopite micas in upper mantle peridotites. Contributions to Mineralogy and Petrology 87, 260–264.

Ballhaus, C., Berry, R.F., Green, D.H., 1991a. High-pressure experimental olivine-orthopyroxene-spinel-oxygen geobarometer—implications for the oxidation-state of the upper mantle. Contributions to Mineralogy and Petrology 107 (1), 27–40.

Ballhaus, C., Berry, R.F., Green, D.H., 1991b. Erratum, High-Pressure Experimental Calibration of the Olivine-Orthopyroxene-Spinel-Oxygen Geobarometer—Implications for the Oxidation-State of the Upper Mantle (vol. 107, p. 27, 1991). Contributions to Mineralogy and Petrology, 108, 384.

Barth, M.G., Rudnick, R.L., Horn, I., McDonough, W.F., Spicuzza, M.J., Valley, J.W., Haggerty, S.E., 2001. Geochemistry of xenolithic eclogites from West Africa. Part I: a link between low MgO eclogites and Archean crust formation. Geochimica et Cosmochimica Acta 65 (9), 1499–1527.

Bertrand, P., Mercier, J.-C.C., 1985. The mutual solubility of coexisting ortho- and clinopyroxene: toward an absolute geothermometer for the natural system. Earth and Planetary Science Letters 76, 109–122.

Boyd, F.R., 1989. Compositional distinction between oceanic and cratonic lithosphere. Earth and Planetary Science Letters 96, 15–26.

Boyd, F.R., 1997. Correlation of orthopyroxene abundance with the Ni content of coexisting olivine in cratonic peridotites. EOS Transactions of the American Geophysical Union 78, F746.

Boyd, F.R., England, J.L., 1964. The system enstatite-pyrope. Yearbook of the Carnegie Institution Washington 63, 157–161.

Boyd, F.R., Mertzman, S.A., 1987. Composition and structure of the Kaapvaal lithosphere, southern Africa. In: Mysen, B.O. (Ed.), Magmatic Processes: Physicochemical Principles. The Geochemical Society Special Publications, University Park, PA, pp. 13–24.

Boyd, F.R., Pokhilenko, N.P., Pearson, D.G., Mertzman, S.A., Sobolev, N.V., Finger, L.W., 1997. Composition of the Siberian cratonic mantle: evidence from Udachnaya peridotite xenoliths. Contributions to Mineralogy and Petrology 128 (2–3), 228–246.

Boyd, F.R., Pearson, D.G., Mertzman, S.A., 1999. Spinel-facies peridotites from the Kaapvaal root. In: Gurney, J.J., Gurney, J.L., Pascoe, M.D., Richardson, S.H. (Eds.), Proceedings of the 7th International Kimberlite Conference, Cape Town, 1998. Red Roof Design, Cape Town, South Africa, pp. 40–48.

Brey, G.P., 1989. Geothermobarometry for lherzolites: experiments from 10 to 60 kb, new thermobarometers and application to natural rocks. Habilitation Thesis, Technische Hochschule, Darmstadt. 227 pp.

Brey, G.P., Köhler, T., 1990. Geothermobarometry in four-phase lherzolites II. New thermobarometers, and practical assessment of existing thermobarometers. Journal of Petrology 31, 1353–1378.

Burgess, S.R., Harte, B., 1998. Tracing lithosphere evolution through the analysis of heterogeneous G9/G10 garnets in peridotite xenoliths. I: major element chemistry. In: Gurney, J.J., Gurney, J.L., Pascoe, M.D., Richardson, S.H. (Eds.), Proceedings of the 7th International Kimberlite Conference, Cape Town, 1998. Red Roof Design, Cape Town, pp. 66–80.

Burgess, S.R., Harte, B., 2002. Tracing lithosphere evolution trough the analysis of heterogeneous G9/G10 garnets in perdiotite xenoliths. II: REE chemistry. Journal of Petrology (submitted for publication).

Canil, D., 1991. Experimental-evidence for the exsolution of cratonic peridotite from high-temperature harzburgite. Earth and Planetary Science Letters 106 (1–4), 64–72.

Canil, D., 1992. Orthopyroxene stability along the peridotite solidus

and the origin of cratonic lithosphere beneath southern Africa. Earth and Planetary Science Letters 111 (1), 83–95.

Canil, D., Fedortchouk, Y., 1999. Garnet dissolution and the emplacement of kimberlites. Earth and Planetary Science Letters 167 (3–4), 227–237.

Carlson, R.W., Pearson, D.G., Boyd, F.R., Shirey, S.B., Irvine, G., Menzies, A.H., Gurney, J.J., 1999. Re–Os systematics of lithospheric peridotites: implications for lithosphere formation and preservation. Proceedings of the 7th International Kimberlite Conference, Cape Town, 1998. Red Roof Design, Cape Town, South Africa, pp. 99–108.

Carswell, D.A., 1975. Primary and secondary phlogopites and clinopyroxenes in garnet lherzolite xenoliths. In: Ahrens, L.H., Dawson, J.B., Duncan, A.R., Erlank, A.J. (Eds.), Proceeding of the First International Conference on Kimberlites, Cape Town, 1973. Physics and Chemistry of the Earth Pergamon, Oxford, pp. 417–429.

Chakraborty, S., Rubie, D.C., 1996. Mg tracer diffusion in aluminosilicate garnets at 750–850 degrees C, 1 atm and 1300 degrees C, 8.5 GPa. Contributions to Mineralogy and Petrology 122 (4), 406–414.

Cox, K.G., Smith, M.R., Beswetherick, S., 1987. Textural studies of garnet lherzolites: evidence of exsolution origin from high-temperature harzburgites. In: Nixon, P.H. (Ed.), Mantle Xenoliths. Wiley, Chichester, pp. 537–550.

Davis, C.L., 1977. The ages and uranium contents of zircons from kimberlites and associated rocks. Carnegie Institution of Washington Yearbook 76, 631–654.

de Hoog, J.C.M., Mason, P.R.D., van Bergen, M.J., 2001. Chalcophile elements in olivine-hosted melt inclusions from Galunggung, Indonesia: implications for the sulfur cycle in subduction zones. Geochimica et Cosmochimica Acta 65, 3147–3164.

Dunn, T., Sen, C., 1994. Mineral/matrix partition coefficients for orthopyroxene, plagioclase, and olivine in basaltic to andesitic systems; a combined analytical and experimental study. Geochimica et Cosmochimica Acta 58 (2), 717–733.

Eggins, S.M., Rudnick, R.L., McDonough, W.F., 1998. The composition of peridotites and their minerals: a laser-ablation ICP-MS study. Earth and Planetary Science Letters 154, 53–71.

Falloon, T.J., Green, D.H., 1987. Anhydrous partial melting of MORB pyrolite and other peridotite compositions at 10 kbar: implications for the origin of primitive MORB glasses. Mineralogy and Petrology 37, 181–219.

Falloon, T., Green, D., 1988. Anhydrous partial melting of perdiotite from 8 to 35 kbar and the petrogenesis of MORB. Journal of Petrology, 379–414.

Falloon, T.J., Green, D.H., Hatton, C.J., Harris, K.L., 1988. Anhydrous partial melting of a fertile and depleted peridotite from 2 to 30 kb and application to basalt petrogenesis. Journal of Petrology 29, 1257–1282.

Falloon, T.J., Green, D.H., Danyushevsky, L.V., Faul, U.H., 1999. Peridotite melting at 1.0 and 1.5 GPa: an experimental evaluation of techniques using diamond aggregates and mineral mixes for determination of near-solidus melts. Journal of Petrology 40 (9), 1343–1375.

Finnerty, A.A., Boyd, F.R., 1987. Thermobarometry for garnet peridotites: basis for the determination of thermal and compositional structure of the upper mantle. In: Nixon, P.H. (Ed.), Mantle Xenoliths. Wiley, Chichester, UK, pp. 381–402.

Fujimaki, H., Tatsumoto, M., Aoki, K., 1984. Partition coefficients of Hf, Zr, and REE between phenocrysts and groundmasses. Journal of Geophysical Research 89, B662–B672 (Supplement).

Ganguly, J., Cheng, W., Chakraborty, S., 1998a. Cation diffusion in aluminosilicate garnets: experimental determination in pyrope-almandine diffusion couples. Contribution to Mineralogy and Petrology 131, 171–180.

Ganguly, J., Tirone, M., Hervig, R.L., 1998b. Diffusion kinetics of samarium and neodymium in garnet, and a method for determining cooling rates of rocks. Science 281, 805–807.

Glaser, S.M., Foley, S.F., Günther, D., 1999. Trace element compositions of minerals in garnet and spinel peridotite xenoliths from the Vitim volcanic field, Transbaikalia, eastern Siberia. Lithos 48, 263–285.

Green, T.H., Blundy, J.D., Adam, J., Yaxley, G.M., 2000. SIMS determination of trace element partition coefficients between garnet, clinopyroxene and hydrous basaltic liquids at 2–7.5 GPa and 1080–1200 °C. Lithos 53, 165–187.

Grégoire, M., Bell, D.R., Le Roex, A.P., 2002. Trace element geochemistry of phlogopite-rich mafic mantle xenoliths: their classification and their relationship to phlogopite-bearing peridotites and kimberlites revisited. Contributions to Mineralogy and Petrology 142, 603–625.

Griffin, W.L., Shee, S.R., Ryan, C.G., Win, T.T., Wyatt, B.A., 1999. Harzburgite to lherzolite and back again: metasomatic processes in ultramafic xenoliths from the Wesselton kimberlite, Kimberley, South Africa. Contributions to Mineralogy and Petrology 134 (2–3), 232–250.

Günther, M., Jagoutz, E., 1994. Isotopic disequilibria (Sm/Nd, Rb/Sr) between minerals of coarse grained, low temperature garnet peridotites from Kimberley floors, Southern Africa. In: Meyer, H.O.A., Leonardos, O.H. (Eds.), Kimberlites, Related Rocks and Mantle Xenoliths. CPRM Spec. Publ., vol. 1A. Companhia de Pesquisa de Recursos Minerais, Rio de Janeiro, Brazil, pp. 359–365.

Harley, S.L., 1984. An experimental study of the partitioning of iron and magnesium between garnet and orthopyroxene. Contributions to Mineralogy and Petrology 86, 359–373.

Hart, S.R., Dunn, T., 1993. Experimental cpx/melt partitioning of 24 trace elements. Contributions to Mineralogy and Petrology 113, 1–8.

Harte, B., 1977. Rock nomenclature with particular relation to deformation and recrystallisation textures in olivine bearing xenoliths. Journal of Geology 85, 279–288.

Harte, B., Winterburn, P.A., Gurney, J.J., 1987. Metasomatic and enrichment phenomena in garnet peridotite facies mantle xenoliths from the Matsoku kimberlite pipe, Lesotho. In: Menzies, M. (Ed.), Mantle metasomatism. Academic Press, London, pp. 145–220.

Harte, B., Fitzsimons, I.C.W., Kinny, P.D., 1996. Clinopyroxene-garnet trace element partition coefficients for mantle peridotite and melt assemblages. V.M. Goldschmidt Conference Abstracts. Cambridge Publications, Heidelberg, Germany, p. 235.

Hauri, E.H., Wagner, T.P., Grove, T.L., 1994. Experimental and natural partitioning of Th, U, Pb and other trace elements between garnet, clinopyroxene and basaltic melts. Chemical Geology 117 (1–4), 149–166.

Herzberg, C., 1999. Phase equilibrium constraints on the formation of cratonic mantle. In: Fei, Y., Bertka, C.M., Mysen, B.O. (Eds.), Mantle Petrology: Field Observations and High Pressure Experimentation: A Tribute to Francis, R. (Joe) Boyd. The Geochemical Society Special Publications, Houston, USA, pp. 241–257.

Herzberg, C., Zhang, J.Z., 1996. Melting experiments on anhydrous peridotite KLB-1: compositions of magmas in the upper mantle and transition zone. Journal of Geophysical Research-Solid Earth 101 (B4), 8271–8295.

Hirose, K., Kushiro, I., 1993. Partial melting of dry peridotites at high-pressures-determination of compositions of melts segregated from peridotite using aggregates of diamond. Earth and Planetary Science Letters 114 (4), 477–489.

Irvine, G.J., 2002. Time constraints on the formation of lithospheric mantle beneath cratons: a Re–Os isotope and platinum group element study of peridotite xenoliths from Northern Canada and Lesotho. PhD Thesis, Durham University, Durham, UK.

Irvine, G.J., Pearson, D.G., Carlson, R.W., 2001. Lithospheric mantle evolution of the Kaapvaal craton: a Re–Os isotope study of peridotite xenoliths from Lesotho kimberlites. Geophysical Research Letters 28 (13), 2505–2508.

Jochum, K.P., Stolz, A.J., McOrist, G., 2000. Niobium and tantalum in carbonaceous chondrites; constraints on the solar system and primitive mantle niobium/tantalum, zirconium/niobium, and niobium/uranium ratios. Meteoritics and Planetary Science 35 (2), 229–235.

Kelemen, P.B., Hart, S.R., 1996. Silica enrichment in the continental lithosphere via melt/rock interaction. Goldschmidt Conference 1996, Heidelberg (Germany). Journal of Conference Abstracts, vol. 1-1. Cambridge Publications, Heidelberg, p. 308.

Kelemen, P.B., Dick, H.J.B., Quick, J.E., 1992. Formation of harzburgite by pervasive melt/rock reaction in the upper mantle. Nature 358, 635–641.

Kelemen, P.B., Hart, S.R., Bernstein, S., 1998. Silica enrichment in the continental upper mantle via melt/rock reaction. Earth and Planetary Science Letters 164 (1–2), 387–406.

Kesson, S.E., Ringwood, A.E., 1989. Slab-mantle interactions 2. The formation of diamonds. Chemical Geology 78, 97–118.

Kinzler, R.J., 1997. Melting of mantle peridotite at pressures approaching the spinel to garnet transition: application to mid-ocean ridge basalt petrogenesis. Journal of Geophysical Research-Solid Earth 102 (B1), 853–874.

Krogh, E.J., 1988. The garnet-clinopyroxene iron-magnesium geothermometer—a reinterpretation of existing experimental data. Contributions to Mineralogy and Petrology 99, 44–48.

Longerich, H.P., Jackson, S.E., Günther, D., 1996. Laser ablation inductively coupled plasma mass spectrometric transient signal data acquisition and analyte concentration calculation. Journal of Analytical Atomic Spectrometry 11, 899–904.

Luth, R.W., Virgo, D., Boyd, F.R., Wood, B.J., 1990. Ferric iron in mantle-derived garnets: implications for thermobarometry and for the oxidation state of the mantle. Contributions to Mineralogy and Petrology 104, 56–72.

McDonough, W.F., Sun, S.-S., 1995. The composition of the earth. Chemical Geology 120, 223–253.

Menzies, M.A., Rogers, N., Tindle, A., Hawkesworth, C.J., 1987. Metasomatic and enrichment processes in lithospheric peridotites, an effect of asthenosphere-lithosphere interaction. In: Menzies, M.A., Hawkesworth, C.J. (Eds.), Mantle Metasomatism. Academic Press, London, pp. 313–364.

Nixon, P.H., 1973. Preface. In: Nixon, P.H. (Ed.), Lesotho Kimberlites. Cape and Transvaal Printers, Cape Town, pp. v–vii.

Norman, M., 1998. Melting and metasomatism in the continental lithosphere: laser ablation ICP-MS analysis of minerals in spinel lherzolites from eastern Australia. Contributions to Mineralogy and Petrology 130, 240–255.

Norman, M.D., Pearson, N.J., Sharma, A., Griffin, W.L., 1996. Quantitative analysis of trace elements in geologic materials by laser ablation ICP-MS: instrumental operating conditions and calibration values of NIST glasses. Geostandards Newsletters 20, 247–261.

O'Neill, H.S.C., 1980. An experimental study of the iron-magnesium partitioning between garnet and olivine and its calibration as a geothermometer: corrections. Contributions to Mineralogy and Petrology 72, 337.

O'Neill, H.S.C., Wood, B.J., 1979. An experimental study of the iron-magnesium partitioning between garnet and olivine and its calibration as a geothermometer. Contributions to Mineralogy and Petrology 70, 59–70.

Pearce, N.J.G., Perkins, W.T., Westgate, J.A., Gorton, M.P., Jackson, S.E., Neal, C.R., Chenery, S.P., 1997. A compilation of new and published major and trace element data for NIST SRM 610 and NIST SRM 612 glass reference materials. Geostandards Newsletter 21, 115–144.

Pearson, D.G., 1999. Evolution of cratonic lithospheric mantle: an isotopic perspective. In: Fei, Y., Bertka, C.M., Mysen, B.O. (Eds.), Mantle Petrology: Field Observations and High-Pressure Experimentation. A Tribute to Francis, R. (Joe) Boyd. The Geochemical Society Special Publications, Houston, TX, pp. 57–78.

Pearson, D.G., Carlson, R.W., Shirey, S.B., Boyd, F.R., Nixon, P.H., 1995. Stabilization of archean lithospheric mantle: a Re–Os isotope study of peridotite xenoliths from the Kaapvaal craton. Earth and Planetary Science Letters 134 (3–4), 341–357.

Pearson, D.G., Shirey, S.B., Harris, J.W., Carlson, R.W., 1998. Sulphide inclusions in diamonds from the Koffiefontein kimberlite, S Africa: constraints on diamond ages and mantle Re–Os systematics. Earth and Planetary Science Letters 160 (3–4), 311–326.

Richardson, S.H., Shirey, S.B., Harris, J.W., Carlson, R.W., 2001. Archean subduction recorded by Re–Os isotopes in eclogitic sulfide inclusions in Kimberley diamonds. Earth and Planetary Science Letters 191 (3–4), 257–266.

Ross, K., Elthon, D., 1997. Extreme incompatible trace-element depletion of diopside in residual mantle from south of the Kane fracture zone. Proceedings of ODP. Scientific Results, vol. 153. Texas A&M University, College Station, TX, pp. 277–284.

Rudnick, R.L., Donough, W.L., Orpin, A., 1994. Northern Tanzanian peridotite xenoliths: a comparison with Kaapvaal peridotites and inferences on metasomatic interactions. In: Meyer, H.O.A., Leonardos, O.H. (Eds.), Kimberlites, Related Rocks and Mantle Xenoliths. CPRM Spec. Publ., Brasilia, pp. 336–353. Jan/94.

Rudnick, R.L., Nyblade, A.A., 1999. The thickness and heat production of Archaean lithosphere: constraints from xenolith thermobarometry and surface heat flow. In: Fei, Y., Bertka, C.M., Mysen, B.O. (Eds.), Mantle Petrology: Field Observations and High Pressure Experimentation: A Tribute to Francis, R. (Joe) Boyd. The Geochemical Society Special Publications, Houston, TX, pp. 3–12.

Saltzer, R.L., Chatterjee, N., Grove, T.L., 2001. The spatial distribution of garnets and pyroxenes in mantle peridotites: pressure–temperature history of peridotites from the Kaapvaal craton. Journal of Petrology 42 (12), 2215–2229.

Schmitz, M.D., Bowring, S.A., 2001. Constraints on southern African lithospheric thermal evolution from U–Pb rutile and titanite thermochronology of lower crustal xenoliths. In: Jones, A., Carlson, R.W., Grutter, H.S. (Eds.), Slave-Kaapvaal Workshop Ext. Abstr. Merrickville, ON, Canada.

Shimizu, N., 1975. Rare earth elements in garnets and clinopyroxenes from garnet lherzolite nodules in kimberlites. Earth and Planetary Science Letters 25, 26–32.

Shimizu, N., 1999. Young geochemical features in cratonic peridotites from Southern Africa and Siberia. In: Fei, Y., Bertka, C.M., Mysen, B.O. (Eds.), Mantle Petrology: Field Observations and High-Pressure Experimentation. A Tribute to Francis, R. (Joe) Boyd. The Geochemical Society Special Publications, Houston, TX, pp. 47–55.

Shimizu, N., Hart, S.R., 1982. Applications of the ion microprobe to geochemistry and cosmochemistry. Annual Reviews in Earth and Planetary Science 10, 483–526.

Shimizu, N., Richardson, S.H., 1987. Trace-element abundance patterns of garnet inclusions in peridotite-suite diamonds. Geochimica et Cosmochimica Acta 51 (3), 755–758.

Shimizu, N., Pokhilenko, N.P., Boyd, F.R., Pearson, D.G., 1999. Trace element characteristics of garnet dunites/harzburgites, host rocks for Siberian peridotitic diamonds. In: Gurney, J.J., Gurney, J.L., Pascoe, M.D., Richardson, S.H. (Eds.), Proceedings of the 7th International Kimberlite Conference, Cape Town, 1998. Red Roof Design, Cape Town, South Africa, pp. 773–782.

Simon, N.S.C., 1999. Geothermobarometrie an Harzburgiten: experimente von 1000 bis 1500 °C bei 4 und 5 GPa im natürlichen System. Diplom/Masters Thesis, Johann Wolfgang Goethe-Universität, Frankfurt a. M. 145 pp.

Simon, N.S.C., Davies, G.R., Pearson, D.G., Mason, P.R.D., Irvine, G.J., 2000. Multistage metasomatism and mineral growth of cratonic mantle recorded by a glass-bearing garnet lherzolite xenolith from Letseng-la-Terae, Lesotho. International V.M. Goldschmidt Conference Abstr., Oxford, UK.

Smith, D., 1999. Temperatures and pressures of mineral equilibration in peridotite xenoliths: review, discussion, and implications. In: Fei, Y., Bertka, C.M., Mysen, B.O. (Eds.), Mantle petrology: Field Observations and High Pressure Experimentation. A Tribute to Francis, R. (Joe) Boyd. The Geochemical Society Special Publications, Houston, TX, pp. 171–188.

Sobolev, N.V., Lavrent'ev, Y.u.G., Pokhilenko, N.P., Usova, L.V., 1973. Chrome-rich garnets from the kimberlites of Yakutia and their paragenesis. Contributions to Mineralogy and Petrology 40, 39–52.

Stachel, T., Harris, J.W., 1997. Diamond precipitation and mantle metasomatism—evidence from the trace element chemistry of silicate inclusions in diamonds from Akwatia, Ghana. Contributions to Mineralogy and Petrology 129 (2–3), 143–154.

Takahashi, E., 1986. Melting of dry peridotite KLB-1 up to 14 GPa: implications on the origin of peridotitic upper mantle. Journal of Geophysical Research 91, 9367–9382.

Van Achterbergh, E., Griffin, W.L., Stiefenhofer, J., 2001. Metasomatism in mantle xenoliths from the Letlhakane kimberlites: estimation of element fluxes. Contribution to Mineralogy and Petrology 141, 397–414.

Van Orman, J.A., Grove, T.L., Shimizu, N., 2001. Rare earth element diffusion in diopside: influence of temperature, pressure, and ionic radius, and an elastic model for diffusion in silicates. Contributions to Mineralogy and Petrology 141 (6), 687–703.

Van Orman, J.A., Grove, T.L., Shimizu, N., Layne, G.D., 2002. Rare Earth element diffusion in a natural pyrop single crystal at 2.8 GPa. Contributions to Mineralogy and Petrology 142 (4), 416–424.

Walker, R.J., Carlson, R.W., Shirey, S.B., Boyd, F.R., 1989. Os, Sr, Nd, and Pb isotope systematics of southern African peridotite xenoliths: implications for the chemical evolution of subcontinental mantle. Geochimica et Cosmochimica Acta 53, 1583–1595.

Walter, M.J., 1998. Melting of garnet peridotite and the origin of komatiite and depleted lithosphere. Journal of Petrology 39 (1), 29–60.

Walter, M.J., 1999. Melting residues of fertile peridotite and the origin of cratonic lithosphere. In: Fei, Y., Bertka, C., Mysen, B. (Eds.), Mantle Petrology: Field Observations and High Pressure Experimentation: A Tribute to Francis, R. (Joe) Boyd. The Geochemical Society Special Publications, Houston, TX, pp. 225–239.

Wells, P.R.A., 1977. Pyroxene thermometry in simple and complex systems. Contributions to Mineralogy and Petrology 62, 129–139.

Witt-Eickschen, G., Seck, H.A., 1991. Solubility of Ca and Al in orthopyroxene from spinel peridotite: an improved version of an empirical geothermometer. Contributions to Mineralogy and Petrology 106, 431–439.

Wood, B.J., Banno, S., 1973. Garnet–orthopyroxene and orthopyroxene–clinopyroxene relationships in simple and complex systems. Contributions to Mineralogy and Petrology 42, 109–124.

Yaxley, G.M., 2000. Experimental study of the phase and melting relations of homogenous basalt + peridotite mixtures and implications for the petrogenesis of flood basalts. Contributions to Mineralogy and Petrology 139, 326–338.

Zack, T., Foley, S.F., Jenner, G.A., 1997. A consistent partition coefficient set for clinopyroxene, amphibole and garnet from laser ablation microprobe analysis of garnet pyroxenites from Kakanui, New Zealand. Neues Jahrbuch Fur Mineralogie. Abhandlungen 172 (1), 23–41.

Available online at www.sciencedirect.com

SCIENCE @ DIRECT°

Lithos 71 (2003) 323–336

LITHOS

www.elsevier.com/locate/lithos

Re–Os systematics of diamond-bearing eclogites from the Newlands kimberlite

A.H. Menzies [a,1], R.W. Carlson [b,*], S.B. Shirey [b], J.J. Gurney [a]

[a] *Department of Geological Sciences, University of Cape Town, Rondebosch 7700, South Africa*
[b] *Department of Terrestrial Magnetism, Carnegie Institution of Washington, 5241 Broad Branch Road N.W., Washington, DC 20015, USA*

Abstract

A suite of 14 diamond-bearing and 3 diamond-free eclogite xenoliths from the Newlands kimberlite, South Africa, have been studied using the Re–Os isotopic system to provide constraints on the age and possible protoliths of eclogites and diamonds. Re concentrations in diamond-bearing eclogites are variable (0.03–1.34 ppb), while Os concentrations show a much more limited range (0.26–0.59 ppb). The three diamond-free eclogites have Re and Os concentrations that are at the extremes of the range of their diamond-bearing counterparts. $^{187}Os/^{188}Os$ ranges from 0.1579 to 1.4877, while $^{187}Re/^{188}Os$ varies from 0.54 to 26.2 in the diamond-bearing eclogites. The highly radiogenic Os in the diamond-bearing eclogites ($\gamma_{Os} = 23–1056$) is consistent with their high $^{187}Re/^{188}Os$ and requires long-term isolation from the convecting mantle. Re–Os model ages for 9 out of 14 diamond-bearing samples lie between 3.08 and 4.54 Ga, in agreement with FTIR spectra of Newlands diamonds that show nitrogen aggregation states consistent with diamond formation in the Archean. Re–Os isochron systematics for the Newlands samples do not define a precise isochron relationship, but lines drawn between subsets of the data provide ages ranging from 2.9 to 4.1 Ga, all of which are suggestive of formation in the Archean. The Re–Os systematics combined with mineral chemistry and stable isotopic composition of the diamond-bearing eclogites are consistent with a protolith that has interacted with surficial environments. Therefore, the favored model for the origin of the Newlands diamond-bearing eclogites is via subduction. The most likely precursors for the Kaapvaal eclogites include komatiitic ocean ridge products or primitive portions of oceanic plateaus or ocean islands.
© 2003 Elsevier B.V. All rights reserved.

Keywords: Eclogite; Re–Os; Archean; Lithosphere; Subduction

1. Introduction

Over 100 years ago, Bonney (1899) and Beck (1898) almost simultaneously recorded the first occur-

rence of diamond-bearing eclogite in the world in specimens from Newlands. Today, eclogite xenoliths found in kimberlites are known to be derived from a diverse range of temperatures and pressures and are of varying compositions that, in general, are broadly basaltic. This variety has led to a range of theories regarding the origin of eclogite: remnants of an early (>4 Ga) magma ocean (McCulloch, 1989; Snyder et al., 1993), metamorphic transformation of subducted ancient oceanic crust (Helmstaedt and Doig, 1975;

---

* Corresponding author. Tel.: +1-202-478-8474; fax: +1-202-478-8821.
E-mail addresses: Andrew.Menzies@minserv.co.za (A.H. Menzies), carlson@dtm.ciw.edu (R.W. Carlson).
[1] Current address: Mineral Services, P.O. Box 38668, Pinelands, Cape Town, South Africa.

0024-4937/$ - see front matter © 2003 Elsevier B.V. All rights reserved.
doi:10.1016/S0024-4937(03)00119-1

Jagoutz et al., 1984; Taylor and Neal, 1989; Neal et al., 1990; Jerde et al., 1993; Ireland et al., 1994; Jacob et al., 1994), metamorphic transformation of under-plated gabbroic rocks (Green and Ringwood, 1967; Griffin et al., 1990; El Fadili and Demaiffe, 1999), or primary mantle material derived from deep melts that fractionate in the upper mantle to form cumulates (O'Hara and Yoder, 1967; MacGregor and Carter, 1970; Smyth et al., 1989; Caporuscio and Smyth, 1990).

A major difficulty in determining the petrogenetic history of eclogite xenoliths is the mineralogical diversity of some suites and their complex history that sometimes includes multiple metamorphic events, partial melting, metasomatism, interaction with the host kimberlite, and extensive alteration. For example, Ireland et al. (1994) and Barth et al. (2001) conjectured that the protoliths of the Siberian and West African eclogites, respectively, were partially melted to form tonalites and other granitoids, leaving residues that are

the eclogitic xenoliths observed today. Moreover, Spetsius (1995) stated that all eclogites show evidence of having undergone partial melting. Therefore, constraining the origin of eclogites is difficult, as the effects of partial melting and mantle metasomatism may have significantly modified the protolith. A further impediment to constraining eclogite petrogenesis is that metamorphic recrystallization has led to substantial mineralogical heterogeneity and compositional layering, which can be sampled by the kimberlite in an uneven way.

Re and Os are strongly fractionated from one another during the production of mafic melts, a possible protolith to eclogite, by the partial melting of mantle peridotite. Compared to the Rb–Sr or Sm–Nd isotopic systems, the degree of parent–daughter fractionation in the Re–Os system during melt production can be several orders of magnitude greater because Re is moderately incompatible during melting whereas Os can be strongly compatible. The large change in the Re/

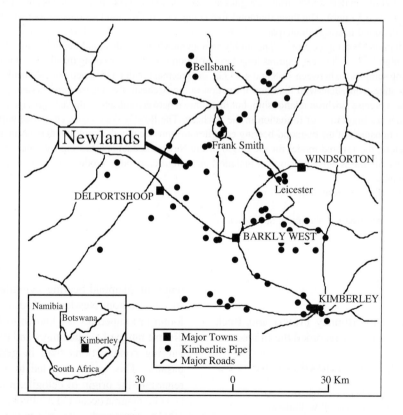

Fig. 1. Map of the Northern Cape Province, South Africa, showing the location of the Newlands kimberlites and other well-known kimberlite occurrences.

Os ratio that occurs during melting causes the Os isotopic composition of the melt to rapidly diverge from that of its mantle source. Because of this large shift in parent–daughter ratio in the Re–Os system, later chemical modification of the rock by metamorphism, metasomatism, or partial melting will potentially result in less obfuscation of the initial protolith formation age than would occur in either the Rb–Sr or Sm–Nd systems. The Sm–Nd system in eclogites, in particular, is very sensitive to modification accompanying garnet growth since garnet so strongly fractionates Sm from Nd. In this paper, we report the Re–Os isotopic systematics for a suite of eclogitic xenoliths from the Newlands kimberlite in order to constrain the formation age and subsequent events that may have occurred in their history and, in particular, their relationship with diamond paragenesis.

2. Sample description

The Newlands kimberlite cluster is a member of the Barkly West group located ~ 60 km NW of Kimberley, South Africa (Fig. 1). The kimberlite is Cretaceous in age (~ 114 Ma) and is of Group II affinity (Smith et al., 1985). Newlands consists of a series of en-echelon kimberlitic dykes of which there are at least five kimberlitic blows to the surface. All the xenoliths were recovered from blow 2 and are described in detail by Menzies (2001). Newlands kimberlite possesses a large range of mantle xenoliths (Menzies, 2001), including both diamond-bearing peridotite (Menzies et al., 1999) and diamond-bearing eclogite. All the specimens analysed in this study were obtained from the coarse concentrate produced during preliminary mining activities. Most samples are fragments of larger

Table 1
Garnet major (wt.%) and trace element (ppm) concentrations

Sample	AHM K1	AHM K3	AHM K4	AHM K5	AHM K6	AHM K7	AHM K8	AHM K9	AHM K10	AHM K12	AHM K13	AHM K14	AHM K15
SiO_2	39.10	39.48	39.38	39.03	39.91	39.14	40.16	39.80	39.35	39.45	39.81	39.49	38.99
TiO_2	0.23	0.28	0.29	0.26	0.28	0.29	0.25	0.25	0.29	0.27	0.26	0.29	0.28
Al_2O_3	22.22	22.31	22.49	22.57	22.77	22.66	23.16	22.89	22.70	22.29	22.74	22.39	22.32
Cr_2O_3	0.12	0.10	0.12	0.14	0.11	0.00	0.00	0.13	0.14	0.09	0.07	0.11	0.07
FeO	21.09	21.20	21.29	20.93	20.91	21.10	18.78	20.42	20.97	21.21	19.83	21.03	20.05
MnO	0.44	0.48	0.45	0.43	0.47	0.46	0.42	0.47	0.52	0.43	0.40	0.45	0.38
MgO	11.71	11.54	11.99	11.66	11.62	11.58	12.87	11.61	11.80	11.97	11.80	11.73	11.98
CaO	4.65	4.58	4.61	4.43	4.51	4.69	5.21	4.57	4.62	4.58	5.97	4.61	6.10
Na_2O	0.10	0.11	0.11	0.11	0.10	0.09	0.12	0.09	0.10	0.10	0.11	0.11	0.11
Ba	0.56	0.39		0.50	0.36	1.11	0.40	0.45	0.43		0.07		0.00
Nb	0.18	0.21		0.11	0.08	0.11	0.08	0.08	0.08				
La	0.05	0.05		0.04	0.04	1.02	0.04	0.04	0.04		0.02		0.02
Ce	0.24	0.20		0.19	0.21	1.48	0.23	0.21	0.20		0.15		0.16
Nd	0.65	0.74		0.68	0.68	0.92	0.86	0.70	0.66		0.52		0.51
Sr	1.02	1.19		1.03	1.01	5.70	1.17	1.09	1.09		0.56		0.53
Sm	0.46	0.48		0.44	0.39	0.45	0.55	0.38	0.35		0.41		0.50
Hf	0.57	0.92		1.03	0.67	0.81	0.86	0.65	0.74		0.53		0.30
Zr	13.1	14.6		13.9	12.4	10.9	18.1	11.2	11.1		11.1		11.5
Eu	0.25	0.28		0.35	0.26	0.26	0.35	0.24	0.24		0.31		0.38
Gd	0.96	0.95		1.39	0.99	0.97	1.22	0.96	0.96		1.60		1.77
Dy	2.84	2.82		3.78	2.93	2.79	3.31	2.86	2.98		3.67		4.10
Y	21.2	21.4		26.1	21.9	21.0	23.7	21.3	22.0		21.2		21.8
Er	2.74	2.76		3.29	2.86	2.97	3.16	2.91	3.02		2.57		3.09
Yb	3.67	3.74		3.92	3.71	3.86	3.82	3.73	3.80		3.29		3.36
Sc	62.4	60.9		48.4	40.6	39.9	37.0	38.7	38.0				
Cr	813	866		913	815	795	411	817	843				
T-EG79[a]	1064	1054	1055	1029	1045	1040	780	1058	1056	1052	1043	1113	1041
T-KR88[a]	998	985	986	954	974	972	695	991	988	981	1000	1092	999

[a] EG79 (Ellis and Green, 1979) and KR88 (Krogh, 1988) are calculated for an assumed equilibration pressure of 50 kbar and quoted in degrees Celsius.

xenoliths that were reduced to 2–5 cm in their longest dimension because of crushing during mining activity. Seventeen diamond-bearing eclogites and numerous diamond-free eclogites were obtained from Newlands kimberlite. For this study, 14 diamond-bearing eclogites and 3 diamond-free eclogites were analysed for Re and Os concentrations and isotopic ratios.

The diamond-bearing eclogites are modally dominated by garnet and clinopyroxene, but contain trace amounts of diamond, sulphides, and rutile. All samples are coarse grained with modal proportions of garnet to clinopyroxene ranging from 60:40 to 30:70. These modes are imprecise due the small sample size and large grain size (up to 1 cm). All the eclogites show variable degrees of fracture development. The garnets are pale orange, subequant, and free from alteration. In contrast, the dark to pale green interstitial clinopyroxenes display various degrees of alteration. Modally, the eclogites contained trace amounts of diamond, although most samples yielded numerous individual crystals or aggregates. FTIR studies indicate that the majority of run-of-mine diamonds from Newlands are eclogitic in origin (Menzies, 2001; Menzies et al., 1998a,b).

2.1. Mineral chemistry

Both garnet and clinopyroxene from the diamond-bearing eclogites display a lack of zoning and, with one exception (sample AHM K8), have a remarkably similar major element composition (Tables 1 and 2). The garnets are enriched in FeO ($\sim$ 21 wt.%) and depleted in MgO ($\sim$ 11.5 wt.%), while clinopyroxene is poor in MgO ($\sim$ 11 wt.%) and rich in Na_2O ($\sim$ 4.6 wt.%), relative to the various other diamond-free eclogitic suites studied from Newlands. The garnets have relatively high Na_2O concentrations (0.08–0.14 wt.%) and the clinopyroxenes (with one exception) have high K_2O concentrations (0.13–0.17 wt.%), both similar to the relatively high concentrations observed in eclogitic diamond inclusions from around the world. The mineral chemistry is equivalent to Group I eclogites (Fig. 2a) as defined by McCandless and Gurney (1989), or Group B eclogites (Fig. 2b) as defined by Taylor and Neal (1989). The sole exception (AHM K8) has clinopyroxene geochemistry that belongs to Group A of Taylor and Neal (1989).

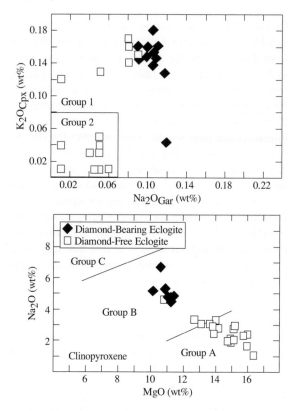

Fig. 2. Mineral chemistry classification of Newlands eclogites: (a) distribution of Na_2O in garnet and K_2O in clinopyroxene compositions. Note that all the diamond-bearing eclogites are classified as Group I eclogites, with Na_2O in garnet concentrations >0.07 wt.% and K_2O in clinopyroxene concentrations >0.08 wt.% (after McCandless and Gurney, 1989; Gurney et al., 1993). (b) Distribution of Na_2O–MgO compositions in clinopyroxenes from Newlands diamond-bearing eclogites in relation to the Newlands diamond-free eclogites. Note that all the clinopyroxenes from the diamond-bearing eclogites (with the exception of AHM K8) plot in the Group B field (after Taylor and Neal, 1989).

The garnets and clinopyroxenes display only a small range of trace element compositions (Tables 1 and 2, Fig. 3). Garnets are HREE enriched while clinopyroxenes are LREE enriched, resulting in a flat whole rock REE pattern at approximately five times chondritic (Fig. 3b; Menzies et al., 1998a,b). The clinopyroxene from sample AHM K8 is the solitary exception to the above compositional descriptions. Clinopyroxene in K8 is more akin to many of the other diamond-free eclogites from Newlands. In general, K8 clinopyroxene is relatively rich in Mg and Ca (MgO = 15 wt.% and CaO = 21.2 wt.%) and poor in Na (Na_2O = 2 wt.%).

Table 2
Clinopyroxene major (wt.%) and trace element (ppm) concentrations

Sample	AHM K1	AHM K3	AHM K4	AHM K5	AHM K6	AHM K7	AHM K8	AHM K9	AHM K10	AHM K12	AHM K13	AHM K14	AHM K15
SiO$_2$	54.88	55.02	55.24	55.05	55.11	54.74	54.74	55.25	55.24	55.10	53.84	54.91	55.21
TiO$_2$	0.43	0.42	0.42	0.44	0.40	0.43	0.16	0.43	0.43	0.42	0.41	0.44	0.40
Al$_2$O$_3$	7.80	7.78	7.70	7.86	7.83	7.57	3.83	7.41	7.92	7.99	8.75	7.75	9.04
Cr$_2$O$_3$	0.13	0.10	0.12	0.15	0.11	0.13	0.05	0.12	0.13	0.14	0.08	0.13	0.09
FeO	6.88	6.91	6.82	6.45	6.65	6.71	3.30	6.81	6.69	6.82	5.42	6.91	5.60
MnO	0.12	0.05	0.08	0.01	0.01	0.01	0.01	0.01	0.01	0.08	0.08	0.03	0.06
MgO	10.99	10.99	11.29	11.07	11.03	11.23	15.04	11.31	11.01	11.34	10.63	11.13	10.89
CaO	13.47	13.60	13.71	14.09	13.97	14.24	21.26	14.40	13.84	13.11	14.05	13.80	13.53
Na$_2$O	4.92	4.79	4.87	4.60	4.67	4.48	1.97	4.51	4.65	4.88	6.73	4.48	5.27
K$_2$O	0.15	0.15	0.15	0.16	0.15	0.16	0.04	0.15	0.16	0.14	0.18	0.15	0.16
Ba	3.67	1.73		1.65	1.47	1.49	1.12	1.21	1.33		0.32		0.33
Nb	0.24	0.23		0.20	0.20	0.20	0.06	0.19	0.17				
La	2.28	2.90		2.83	2.99	2.80	5.21	3.02	2.90		1.39		1.45
Ce	6.62	8.44		8.30	8.60	8.19	8.50	8.71	8.40		4.43		4.27
Nd	3.64	4.66		4.29	4.67	4.35	3.67	4.70	4.45		2.80		2.40
Sr	212	238		235	241	236	39	244	232		147		146
Sm	0.60	0.91		0.80	0.87	0.75	1.23	1.04	0.81		0.64		0.68
Hf	0.62	0.71		0.64	0.68	0.74	0.17	1.07	0.67		1.05		0.83
Zr	18.6	20.0		19.6	19.6	19.1	3.3	28.3	19.9		16.8		16.7
Eu	0.26	0.30		0.28	0.31	0.28	0.36	0.40	0.29		0.26		0.25
Gd	0.76	0.88		0.85	0.86	0.89	0.61	1.06	0.88		0.82		0.81
Dy	0.51	0.54		0.54	0.57	0.51	0.33	0.68	0.54		0.51		0.57
Y	1.99	2.18		2.22	2.18	2.13	1.02	2.58	2.14		1.60		1.61
Er	0.18	0.25		0.16	0.21	0.25	0.06	0.25	0.17		0.22		0.19
Yb	0.10	0.15		0.14	0.16	0.13	0.01	0.25	0.13		0.10		0.14
Sc	36.4	28.4		28.2	29.2	28.2	13.1	25.3	24.3				
Cr	732	855		859	847	879	539	871	874				

Furthermore, K8 clinopyroxene has a K$_2$O concentration of 0.04 wt.%, thus classifying as a Group II or Group A eclogite. The difference in K8 clinopyroxene compared to other samples is also seen in its lower Sr, Hf, Zr, and HREE abundances (Fig. 3b). Calculated whole rock REE patterns for the Newlands eclogites are similar to the low-Ca group of Beard et al. (1996) from Yakutia (Fig. 3). The diamond-bearing eclogites from Newlands, however, do not display the range of major or trace element variation observed in diamond-bearing eclogites from the Yakutian kimberlite fields, in particular Mir and Udachnaya (Jerde et al., 1993; Ireland et al., 1994; Taylor et al., 1996; Beard et al., 1996; Snyder et al., 1997).

2.2. Geothermometry

Using the geothermometers of Ellis and Green (1979) and Krogh (1988), the diamond-bearing eclogites yield a temperature range of 920–1080 °C at geotherm-dependent pressures of 42–58 kbar, assuming that the diamond-bearing eclogites were lying on a steady-state conductive geotherm that is determined for Newlands to be 37–38 mW/m^2 (Table 1; Menzies, 2001). This equates to a depth of approximately 130–170 km well within the diamond stability field at Newlands. The maximum temperature for the diamond-bearing eclogites is within error of the maximum temperature recorded at Newlands for any xenolith. Therefore, it is not possible to determine whether this temperature is a maximum for the diamond-bearing eclogites at Newlands or whether this represents the maximum depth of xenolith entrainment by the kimberlite magma.

3. Analytical procedures

All Re–Os analyses were determined on whole rock samples. Each sample was broken into small

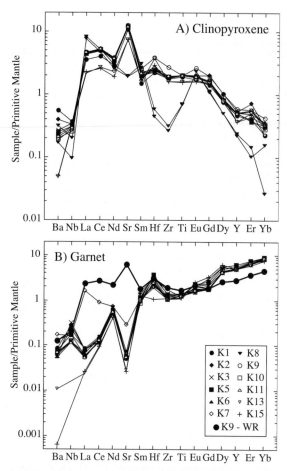

Fig. 3. Trace element abundances in Newlands eclogite minerals normalized to primitive mantle compositions (McDonough and Sun, 1995). The pattern for sample K9 whole rock is calculated assuming a mode of 50% garnet plus 50% pyroxene.

fragments and then powdered using an alumina jaw crusher and an alumina puck-mill. Any diamond was extracted before the powdering stage. The alumina equipment was cleaned between samples using deionized water, acetone, and compressed air. Quartz was used in a pre-crushing cleaning cycle as the sample size was limited.

All Re and Os isotope analyses were performed at the Department of Terrestrial Magnetism (DTM), Carnegie Institution of Washington in 1996, following procedures in use at that time. Detailed descriptions of the operating conditions used in this study are given in Menzies (2001) and are similar to those described in Shirey (1997) and Pearson et al. (1995a). Between 1 and 3 g of the sample were spiked with ^{190}Os and ^{185}Re, digested in a mixture of 4 ml concentrated HCl and 2 ml concentrated HNO_3 in a sealed Pyrex Carius tube following the method of Shirey and Walker (1995). After 2 days stored at 220 °C, the samples were allowed to cool and then distilled to separate the Re and Os. The digested sample was added to 6N H_2SO_4 and Os was separated from Re by two-stage distillation at ~ 110 °C; firstly into 6 M NaOH and then into concentrated HBr. The Os cut was further purified by microdistillation using the method of Roy-Barman and Allègre (1994). Re was extracted from the sample remaining after distillation by two sets of pre-cleaned AG1-X8 ion exchange columns using washes of 5N H_2SO_4 and 1N HCl followed by elution of Re in 4 N HNO_3. The average Re blank, Os blank, and Os yield are < 12 pg, < 5 pg, and 50–80%, respectively. Re and Os separates were then analysed by negative thermal ionization mass spectrometry (NTIMS) generally following the procedures outlined in Creaser et al. (1991).

4. Re and Os concentration systematics

4.1. Rhenium

Re concentrations of Newlands eclogites vary by nearly two orders of magnitude (Table 3). The 14 diamond-bearing eclogites range from 0.03 to 1.34 ppb Re, with the majority having concentrations greater than 0.2 ppb (Fig. 4). The solitary diamond-free eclogite (AHM C5) measured for Re has one of the lowest concentrations at 0.046 ppb. Replicate analyses were made on separate splits of sample powder from four diamond-bearing samples (Table 3). Three of the replicate analyses are within 0.01 ppb, while the fourth (sample AHM K8) has the highest and most variable Re concentration (1.18 and 1.34 ppb). The overall Re concentration ranges for the Newlands eclogites are remarkably similar to eclogites from Roberts Victor, South Africa (Shirey et al., 1999), but extend to significantly lower concentrations than diamond-bearing eclogites from Udachnaya, Siberia (Fig. 4; Pearson et al., 1995c).

Table 3
Re–Os isotopic data for Newlands eclogite xenoliths

Sample	Re (ppb)	Os (ppb)	$^{187}Re/^{188}Os$	$^{187}Os/^{188}Os$	Error	γOs_i	T_{MA} (Ga)
Diamond free							
AHM C5	0.0464	0.635	0.353	0.1427	0.0005	10.9	−13.10
AHM C6		0.269		0.3181	0.0009	147.2	
AHM C7		0.252		1.2824	0.0016	896.4	
Diamond bearing							
AHM K1/1	0.580	0.539	5.531	0.6524	0.0019	406.9	5.86
AHM K1/2		0.586		0.6054	0.0114	370.4	
AHM K3	0.916	0.348	14.06	0.9449	0.0024	634.2	3.49
AHM K4/1	0.620						
AHM K4/2	0.624	0.356	9.127	0.7659	0.0031	495.1	4.24
AHM K5/1	1.107						
AHM K5/2	1.097	0.554	10.53	0.9096	0.0099	606.8	4.47
AHM K5/3		0.356		1.3744	0.0034	967.9	
AHM K6/1	0.341	0.381	4.509	0.4926	0.0003	282.8	5.12
AHM K6/2	0.331	0.348	4.818	0.5219	0.0007	305.5	5.14
AHM K7	0.647	0.319	10.82	0.9461	0.0007	635.1	4.54
AHM K8/1	1.175	0.348	18.31	1.0854	0.0017	743.3	3.13
AHM K8/2	1.345	0.257	26.21	1.4877	0.0053	1056.0	3.08
AHM K9	0.518	0.348	7.650	0.6413	0.0010	398.3	4.11
AHM K10	1.039	0.421	13.12	0.9150	0.0011	610.9	3.61
AHM K11	0.319	0.363	4.396	0.4321	0.0008	235.8	4.42
AHM K12		0.297		0.6800	0.0005	428.3	
AHM K13	0.0347	0.300	0.565	0.1795	0.0031	39.5	18.50
AHM K14	0.718	0.432	8.619	0.7064	0.0058	448.9	4.09
AHM K15	0.0401	0.360	0.538	0.1579	0.0007	22.7	13.80

γOs is the percent deviation of the measured $^{187}Os/^{188}Os$ from that expected for primitive mantle (Meisel et al., 1996). T_{MA} is the Re–Os model age (e.g. Walker et al., 1989) calculated relative to these same primitive mantle Re–Os values.

4.2. Osmium

Osmium concentrations of Newlands eclogites show a limited range, varying by only a factor of 2 (Table 3). Diamond-bearing eclogites range from 0.26 to 0.59 ppb, while the three diamond-free eclogites yield the lowest and highest Os concentrations, respectively (Fig. 4). Replicate analyses of four diamond-bearing eclogites display considerable scatter in Os concentration. The lack of reproducibility is not unexpected as Os is largely concentrated in trace PGE-rich phases (Hart and Ravizza, 1995; Alard et al., 2000; Burton et al., 2000), particularly sulfide, which is a common trace phase in eclogites. Therefore, the powder used for whole rock analyses may not adequately sample the heterogeneous distribution of these trace Os-rich phases.

The Re and Os concentration spreads of eclogites for both Newlands and Roberts Victor are strikingly similar but significantly different from Udachnaya

(Fig. 5). The Newlands diamond-bearing eclogites (and Roberts Victor eclogites) overlap a variety of fields, including Archean basalts and komatiites (Walker et al., 1988), picrites from continental flood basalts (Ellam et al., 1992; Shirey, 1997), and primitive ocean island basalt (OIB) (Hauri et al., 1996). In contrast, the Udachnaya diamond-bearing eclogites overlap the low Os concentrations observed in mid-ocean ridge basalts (Roy-Barman and Allègre, 1994) and ocean island basalts (Martin, 1991; Hauri and Hart, 1993; Reisberg et al., 1993; Roy-Barman and Allègre, 1995; Hauri et al., 1996), and (with one exception) are distinctly lower in concentration to their Newlands counterparts analysed in this study.

5. Re–Os isotope systematics

The $^{187}Re/^{188}Os$ ratios are high in the Newlands eclogites, with diamond-bearing eclogites ranging

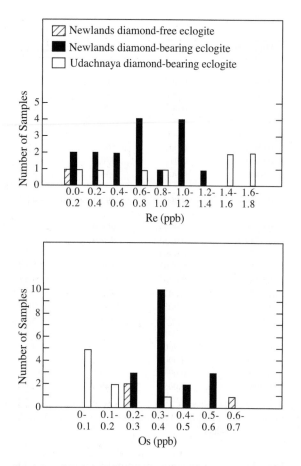

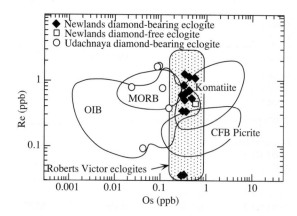

Fig. 5. Comparison of Re and Os concentrations in the Newlands eclogites with the concentration ranges for a variety of other mantle-derived melts including komatiites (Walker et al., 1988), picrites from continental flood basalts (Ellam et al., 1992; Shirey, 1997), ocean island basalts (Martin, 1991; Hauri and Hart, 1993; Reisberg et al., 1993; Roy-Barman and Allègre, 1995; Hauri et al., 1996), and mid-ocean ridge basalts (Roy-Barman and Allègre, 1994). The field for eclogites from Roberts Victor is from Shirey et al. (1999) and the Udachnaya datum is from Pearson et al. (1995c).

Fig. 4. Bar diagrams of whole rock (a) Re and (b) Os concentrations (in parts per billion (ppb)) of Newlands eclogites compared to data from Udachnaya eclogite xenoliths (Pearson et al., 1995c).

0.130; Roy-Barman and Allègre, 1994, 1995; Hauri and Hart, 1993; Reisberg et al., 1993; Hauri et al., 1996) and instead approach values typical of the present-day Os isotopic range of Archean komatiites (Fig. 6). The majority of Newlands diamond-bearing eclogites have lower $^{187}Os/^{188}Os$ than their diamond-bearing counter-parts from Udachnaya (Pearson et al., 1995c).

The highly radiogenic Os in the diamond-bearing eclogites ($\gamma_{Os} = 23$ to 1056) is consistent with their high

from 0.54 to 26.21 (Table 3). Even though this range in parent–daughter ratio covers nearly two orders of magnitude, it is significantly lower in absolute range when compared to the diamond-bearing eclogites from Udachnaya (Pearson et al., 1995c), but is similar to that observed in eclogites from the Roberts Victor kimberlite (Shirey et al., 2001). Only one diamond-free eclogite from Newlands (AHM C5) has been measured for Re and it has the lowest $^{187}Re/^{188}Os$ ratio (0.353) for this locality.

The diamond-bearing eclogites have measured $^{187}Os/^{188}Os$ ratios ranging from 0.1579 to 1.4877. The three diamond-free eclogites display a similar range of $^{187}Os/^{188}Os$ (Table 3). All of these ratios are considerably higher than any estimate of the present day Os isotopic composition of the mantle (0.124–

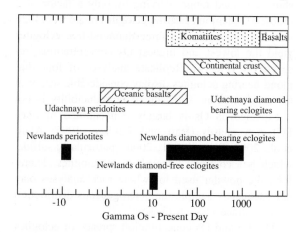

Fig. 6. Gamma Os diagram for a variety of mafic magmas (information sources as for Fig. 5).

^{187}Re/^{188}Os ratios and requires long-term isolation from the convecting mantle. Osmium isotope evolution trajectories for Newlands eclogites display a general convergence between 2.5 and 3.5 Ga (Fig. 7), with a minimum variance in calculated initial ^{187}Os/^{188}Os of ± 0.082 (1σ) at 2.9 Ga for the diamond-bearing samples. This point of convergence, however, is at ^{187}Os/^{188}Os = 0.264, well above the value expected for the mantle at this time (0.108). Thus, the Re–Os model ages, when calculated relative to a mantle undergoing chondritic Os isotopic evolution, are considerably older, varying from 3.1 to 18.5 Ga (see Table 3 for calculation parameters). Model ages older than the age of the Earth are a clear indication that at least some of the samples have not experienced the simple single-stage Re–Os evolution required by the model age calculation. The unrealistically old Re–Os model ages reflect Re/Os ratios too low to account for the high measured ^{187}Os/^{188}Os. Given the Re–Os characteristics of kimberlites (e.g. Walker et al., 1989; Carlson et al., 1996), contamination of the Newlands eclogites with the host Group II kimberlite can increase their model ages by about 200 My (for 1% kimberlite in the xenolith) to 1 Ga (for 5% kimberlite) with the offset depending on the exact Re–Os characteristics of the particular xenolith and the host kimberlite. Another likely process that can result in artificially old Re–Os model ages is Re loss either during metamorphism or partial melting long after the formation of the protoliths of the samples.

The complex history of Newlands eclogites is reflected by the considerable scatter they show about

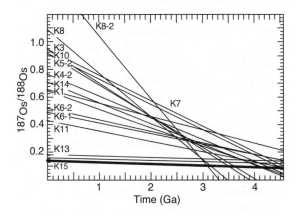

Fig. 7. Os isotopic evolution lines for the Newlands eclogites in comparison to a chondritic estimate for the Os isotopic growth in a mantle with chondritic Re/Os (Meisel et al., 1996).

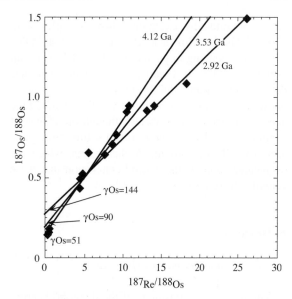

Fig. 8. Re–Os isochron plot for Newlands eclogite xenoliths. Errors for ^{187}Os/^{188}Os are less than the size of the symbol. Various eclogite combinations are possible to define "isochrons": (a) all diamond-bearing eclogites define an age of 2.92 ± 0.43 Ga, initial γ_{Os} = 144 ± 92 and MSWD = 291. Eliminating the most radiogenic sample (K8) increases the slope of the best fit line of the remaining samples to 3.53 ± 0.60 Ga, γ_{Os} = 90 ± 91, MSWD = 198. Eliminating the samples with ^{187}Re/^{188}Os > 11 (K3, K8, K10) provides the steepest slope corresponding to an age of 4.12 ± 0.57 Ga, γ_{Os} = 51 ± 93, and MSWD = 134. Uncertainties in age and initial Os isotopic compositions are at the 95% confidence limit. All ages are calculated using a Re decay constant of λ = 1.666 × 10^{-11}/year.

any best-fit line on a Re–Os isochron diagram. All line fitting reported here is carried out with the ISOPLOT program (Ludwig, 1991) with errors reported at the 95% confidence level. The data for all diamond-bearing samples scatter (MSWD = 291) about a best-fit line that would correspond to an age of 2.92 ± 0.43 Ga with a very elevated initial γ_{Os} = 144 ± 79, but this line is strongly controlled by replicate analyses of sample AHM K8 that has the highest Re/Os of this sample set. The two replicates of AHM K8 have sufficient spread in Re/Os and ^{187}Os/^{188}Os to define a chord with slope equivalent to an age of 2.98 ± 0.23 Ga with initial ^{187}Os/^{188}Os of 0.153 ± 0.087 (γ_{Os} = 43 ± 81). If sample AHM K8 and the diamond-free samples are excluded, a best-fit line (MSWD = 198) to the remaining 13 points (including one replicate analysis) corresponds to an age of 3.53 ± 0.60 Ga with an initial ^{187}Os/^{188}Os of 0.196 ± 0.088 (γ_{Os} = 90 ± 85; Fig. 8). The steepest (oldest) slope found for the data set is obtained by excluding those samples with $^{187}Re/^{188}Os > 11$. This data set gives a line (MSWD = 134) corresponding to an age of 4.12 ± 0.57 Ga with initial $\gamma_{Os} = 51 \pm 70$. Clearly, the Newlands eclogites do not define meaningful Re–Os isochrons, and we do not wish to strongly argue for the validity of any of the ages noted above. Nevertheless, though the choice of samples included in these various line fittings is arbitrary, all lines provide Archean ages, which we do believe supports the general conclusion that these eclogites were formed in the Archean.

6. Discussion

6.1. Formation and metamorphic ages

The mineral compositions of the diamond-bearing eclogites are remarkably constant for both major and trace elements and record similar ambient temperatures just prior to capture in the kimberlite (the exception is sample AHM K8, see below). Therefore, it is likely that the diamond-bearing eclogites are cogenetic, derived from the same protolith, and have experienced similar mantle conditions over the geological aeons. The similarity of most of the eclogites is supported by the relatively small range in Os concentrations in comparison to modern basalt suites where variations in Os content are often significantly larger than in Re content. Exceptions are the two diamond-bearing specimens and one diamond-free sample that have unusually low Re concentrations, approximately an order of magnitude lower, than other Newlands eclogites. As seen in Fig. 7, these samples have shallow Os isotope evolution lines, paralleling the chondritic growth curve for the mantle. To reach their level of excess ^{187}Os compared to the mantle would require only tens to a couple hundred million years of radiogenic ingrowth if they initially had the high Re/Os ratios characterizing the majority of the Newlands eclogites. Consequently, if these samples formed at roughly the same time as the other Newlands samples, they must have experienced a Re-loss event within less than a few hundred million years after their formation in the Archean. Alternatively, if these samples originally had Re contents similar to even the low end of the Re concentration range of the remainder of the samples (circa 0.3 ppb), their rela-

tively unradiogenic Os would result in relatively young Re–Os model ages of between 430 (AHM K15) and 680 Ma (AHM K13).

Diamond-bearing sample AHM K8 displays a major and trace element composition and temperature significantly different from the other 16 diamond-bearing eclogites from Newlands. This specimen also displays Re concentrations and Re–Os isotopic characteristics at the extremes for the Newlands diamond-bearing eclogitic suite, i.e. the highest $^{187}Re/^{188}Os$ and $^{187}Os/^{188}Os$ ratios. Nevertheless, the sample yields a model age consistent with the picture outlined previously—the replicates provide an "age" of 2.98 Ga.

A mid-Archean age of the Newlands eclogites would overlap the age of the diamond-bearing eclogites from Roberts Victor (Shirey et al., 2001), Udachnaya (2.90 ± 0.38 Ga; Pearson et al., 1995c), the 2.89 ± 0.06 Ga Re–Os age for eclogitic paragenesis sulfide inclusions in diamonds from Kimberley (Richardson et al., 2001), circa 2.8 Ga Re–Os model ages for two sulfide inclusions in diamonds from Koffiefontein (Pearson et al., 1998), and similar ages for eclogitic sulfide inclusions from Orapa and Jwaneng (Shirey et al., 2001). These data suggest that eclogite formation occurred widely across the deep lithosphere of the western Kaapvaal craton in the mid to late Archean (Shirey et al., 2002).

6.2. Protolith characteristics and origin

Preliminary stable isotope studies on garnet, clinopyroxene, and diamonds from the Newlands eclogites suggest that the protolith underwent chemical exchange in a low-temperature surface environment and thus entered the deep Kaapvaal lithosphere through subduction (Menzies, 2001). Oxygen isotopic compositions of Newlands garnet and clinopyroxene range from 5.5‰ to 8‰ (Menzies, 2001), overlapping but extending to values higher than the mantle (~ 5.5‰; Kyser, 1990; Mattey et al., 1994). Three carbon isotopic analyses indicate that Newlands eclogitic diamonds range down to approximately − 10‰ (Menzies, 2001), compared to mantle carbon which is approximately − 4 to − 5‰ (Kirkley et al., 1991). These oxygen isotope signatures are indicative of low-temperature fractionation typical of the interaction of eclogite protoliths with surficial environments. The carbon isotopic compositions could be interpreted similarly, though a high-temperature fractionation origin for C and N isotopic variability in diamonds has also been proposed (e.g. Cartigny et al., 1999, 2001). The heavy O isotope compositions are not consistent with formation of the protoliths at magmatic temperatures, as would be required if the eclogites were related to deep-seated melts emplaced within the lithospheric mantle.

The Archean Re–Os ages obtained for the Kaapvaal (Newlands and Roberts Victor (Shirey et al., 2001)) and Siberian (Udachnaya) eclogites overlap the Re-depletion model ages for peridotitic samples from both cratons (Walker et al., 1989; Pearson et al., 1995a,b; Carlson et al., 1999; Menzies et al., 1999). Moreover, all these ages overlap, to slightly post-date, major crust building and craton stabilisation periods in these cratons (DeWit et al., 1992). The association of continental crust formation, basalt depletion of mantle peridotite, and the presence of subducted oceanic crust, all of overlapping age, suggest that craton lithospheres were created in settings similar to modern convergent margins. Once formed, these assemblages remain coupled in the continental lithosphere and can survive the destructive forces of plate tectonics over billion year time scales.

The Newlands diamond-bearing eclogites have Re and Os concentrations that are similar to those in eclogites from Roberts Victor (Shirey et al., 1999) (both located on the Kaapvaal craton) but are significantly different from those from Udachnaya (Pearson et al., 1995c) (located on the Siberian craton). These data sets are too small, however, to determine if this difference between the cratons is significant. The majority of the Newlands samples not only overlap the komatiite field in Re–Os concentration space (Walker et al., 1988), as also noted for the Roberts Victor eclogites (Shirey et al., 2001), but also extend into the fields for primitive ocean island basalt (OIB) (Roy-Barman and Allègre, 1995) and continental flood basalt (CFB) picrites (Ellam et al., 1992; Shirey, 1997) (Fig. 5). In contrast, the samples from Udachnaya primarily plot in the evolved mid-ocean ridge basalt field (Roy-Barman and Allègre, 1994). If the Newlands eclogitic protolith was formed in an Archean ocean floor environment and accreted to the lithospheric mantle through subduction, the Re–Os concentrations suggest a more magnesian, i.e. komatiitic to picritic, oceanic crust composition compared to present day MORB. Alternatively, the Newlands eclogites could reflect the subduction of primitive portions of oceanic plateaus or islands.

7. Conclusions

Whole rock Re concentrations of diamond-bearing eclogites from Newlands are variable while Os concentrations are relatively constant. The Re–Os concentrations of Newlands eclogites overlap the ranges seen in komatiites, continental flood basalt picrites, and ocean island basalts, but the Os contents are higher than typical of modern mid-ocean ridge basalts.

The scatter in Re–Os systematics reflects a complex history for these eclogites that makes it impossible to define a precise age. Nevertheless, the elevated $^{187}Os/^{188}Os$ of the samples coupled with only moderately high Re/Os requires that these eclogites be old, most likely Archean. An Archean age is similar to that determined for diamond-bearing eclogites from Udachnaya, Siberia (Pearson et al., 1995c), and sulfide inclusions in diamonds from Koffiefontein (Pearson et al., 1998), Kimberley (Richardson et al., 2001), Orapa, and Jwaneng (Shirey et al., 2002).

The Re–Os systematics combined with mineral chemistry and stable isotopic composition of the diamond-bearing eclogites are consistent with a protolith that has interacted with surficial environments. Therefore, the subduction hypothesis is the favored model for the origin of the Newlands eclogites. The Re–Os concentrations and isotope systematics are strikingly similar to eclogites from Roberts Victor (Shirey et al., 1999, 2001) and distinctly different to those from Udachnaya (Pearson et al., 1995c). This suggests that the eclogitic protoliths accreted to the lithospheric mantle beneath the Kaapvaal and Siberian cratons may be fundamentally different. The most likely precursors for the Newlands eclogites include komatiitic/picritic ocean ridge products or primitive portions of oceanic plateaus or ocean islands.

Acknowledgements

Detailed reviews by Bill Griffin and Roberta Rudnick along with the editorial comments of Herman

Grutter are much appreciated. This work was supported by NSF Grant EAR-9526840.

References

Alard, A., Griffin, W.L., Lorand, J.-P., Jackson, S., O'Reilly, S.Y., 2000. Non-chondritic distribution of the highly siderophile elements in mantle sulphides. Nature 407, 891–894.

Barth, M.G., Rudnick, R.L., Horn, I., McDonough, W.F., Spicuzza, M.J., Valley, J.W., Haggerty, S.E., 2001. Geochemistry of xenolithic eclogites from West Africa, part I: a link between low MgO eclogites and Archean crust formation. Geochim. Cosmochim. Acta 65, 1499–1527.

Beard, B.L., Fraracci, K.N., Taylor, L.A., Snyder, G.A., Clayton, R.A., Mayeda, T.K., Sobolev, N.V., 1996. Petrography and geochemistry of eclogites from the Mir kimberlite, Yakutia, Russia. Contrib. Mineral. Petrol. 125, 293–310.

Beck, R., 1898. Die diamantenlagerstatte von Newland in Griqua Land West: Zeitschrift fur praktische Geologie. Mai, 163–164.

Bonney, T.G., 1899. The parent-rock of the diamond in South Africa. Geol. Mag. 6, 309–321.

Burton, K.W., Schiano, P., Birck, J.L., Allegre, C.J., Rehkaemper, M., Halliday, A.N., Dawson, J.B., 2000. The distribution and behaviour of rhenium and osmium amongst mantle minerals and the age of the lithospheric mantle beneath Tanzania. Earth Planet. Sci. Lett. 183, 93–106.

Caporuscio, F.A., Smyth, J.R.S., 1990. Trace element crystal chemistry of mantle eclogites. Contrib. Mineral. Petrol. 105, 550–561.

Carlson, R.W., Esperanca, S., Svisero, D.P., 1996. Chemical and Os isotopic study of Cretaceous potassic rocks from southern Brazil. Contrib. Mineral. Petrol. 125, 393–405.

Carlson, R.W., Pearson, D.G., Boyd, F.R., Shirey, S.B., Irvine, G., Menzies, A.H., Gurney, J.J., 1999. Re–Os systematics of lithospheric peridotites: implications for lithosphere formation and preservation. In: Gurney, J.J., Gurney, J.L., Pascoe, M.D., Richardson, S.H. (Eds.), Proc. 7th Int. Kimberlite Conf., Red Roof Design, Cape Town, pp. 99–108.

Cartigny, P., Harris, J.W., Javoy, M., 1999. Eclogitic, peridotitic and metamorphic diamonds and the problems of carbon recycling— the case for Orapa (Botswana). In: Gurney, J.J., Gurney, J.L., Pascoe, M.D., Richardson, S.H. (Eds.), Proc. 7th Int. Kimberlite Conf., Red Roof Design, Cape Town, pp. 117–124.

Cartigny, P., Harris, J.W., Javoy, M., 2001. Diamond genesis, mantle fractionations and mantle nitrogen content: a study of δ13C and N concentrations in diamonds. Earth Planet. Sci. Lett. 185, 85–98.

Creaser, R.A., Papanastassiou, D.A., Wasserburg, G.J., 1991. Negative thermal ion mass spectrometry of osmium, rhenium, and iridium. Geochim. Cosmochim. Acta 55, 397–401.

DeWit, M.J., Roering, C., Hart, R.J., Armstrong, R.A., Ronde, C.E.J.D., Green, R.W.E., Tredoux, M., Peberdy, E., Hart, R.A., 1992. Formation of an Archaean continent. Nature 357, 553–562.

El Fadili, S., Demaiffe, D., 1999. Petrology of eclogite and granulite nodules from the Mbuji Mayi kimberlites (Kasai, Congo): significance of kyanite–omphacite intergrowths. In: Gurney,

J.J., Gurney, J.L., Pascoe, M.D., Richardson, S.H. (Eds.), Proc. 7th Int. Kimberlite Conf., Red Roof Design, Cape Town, pp. 205–213.

Ellam, R.M., Carlson, R.W., Shirey, S.B., 1992. Evidence from Re–Os isotopes for plume–lithosphere mixing in Karoo flood basalt genesis. Nature 359, 718–721.

Ellis, D.G., Green, D.H., 1979. An experimental study of the effect of Ca upon garnet–clinopyroxene Fe–Mg exchange equilibria. Contrib. Mineral. Petrol. 71, 13–22.

Green, D.H., Ringwood, A.E., 1967. An experimental investigation of the gabbro to eclogite transformation and its petrological applications. Geochim. Cosmochim. Acta 31, 767–833.

Griffin, W.L., O'Reilly, S.Y., Pearson, N.J., 1990. Eclogitic stability near the crust–mantle boundary. Eclogite Facies Rocks. Blackie, Glasgow, pp. 291–314.

Gurney, J.J., Helmstaedt, H.H., Moore, R.O., 1993. A review of the use and application of mantle mineral geochemistry in diamond exploration. Pure Appl. Chem. 65, 2423–2442.

Hart, S.R., Ravizza, G.E., 1995. Os partitioning between phases in lherzolite and basalt. American Geophysical Union Monograph, vol. 95. Am. Geophys. Union, Washington, pp. 123–144.

Hauri, E.H., Hart, S.R., 1993. Re–Os isotope systematics of HIMU and EMII oceanic island basalts from the south Pacific Ocean. Earth Planet. Sci. Lett. 114, 353–371.

Hauri, E.H., Lassiter, J.C., DePaolo, D.J., 1996. Osmium isotope systematics of drilled lavas from Mauna Loa, Hawaii. J. Geophys. Res. 101, 11793–11806.

Helmstaedt, H.H., Doig, R., 1975. Eclogite nodules from kimberlite pipes of the Colorado Plateau—samples of subducted Franciscan type oceanic lithosphere. Phys. Chem. Earth 9, 95–111.

Ireland, T.R., Rudnick, R.L., Spetsius, Z., 1994. Trace elements in diamond inclusions from eclogites reveal link to Archean granites. Earth Planet. Sci. Lett. 128, 199–213.

Jacob, D.E., Jagoutz, E., Lowry, D., Mattey, D., Kudrjavtseva, G., 1994. Diamondiferous eclogites from Siberia: remnants of Archean oceanic crust. Geochim. Cosmochim. Acta 58, 5191–5207.

Jagoutz, E., Dawson, J.B., Hoernes, S., Spettel, B., Wanke, H., 1984. Anorthositic oceanic crust in the Archean Earth. Lunar Planet. Sci. XV, 395–396.

Jerde, E.A., Taylor, L.A., Crozaz, G., Sobolev, N.V., Sobolev, V.N., 1993. Diamondiferous eclogites from Yakutia, Siberia: evidence for a diversity of protoliths. Contrib. Mineral. Petrol. 114, 189–202.

Kirkley, M.B., Gurney, J.J., Otter, M.L., Hill, J.S., Daniels, L.R.M., 1991. The application of C isotope measurements to the identification of the sources of C in diamonds; a review. Appl. Geochem. 6, 477–494.

Krogh, E.J., 1988. The garnet–clinopyroxene Fe–Mg geothermometer—a reinterpretation of existing experimental data. Contrib. Mineral. Petrol. 99, 8–44.

Kyser, T.K., 1990. Stable isotopes in the continental lithospheric mantle. In: Menzies, M.A. (Ed.), Continental Mantle. Oxford University Press, Oxford, pp. 127–156.

Ludwig, K.R., 1991. ISOPLOT: a plotting and regression program for radiogenic–isotope data. U.S. Geol. Survey Open-File Report, 91-445. 39 pp.

MacGregor, I.D., Carter, J.L., 1970. The chemistry of clinopyroxene and garnets of eclogite and peridotite xenoliths from the Roberts Victor mine, South Africa. Phys. Earth Planet. Inter. 3, 391–397.

Martin, C.E., 1991. Os isotopic characteristics of mantle derived rocks. Geochim. Cosmochim. Acta 55, 1421–1434.

Mattey, D., Lowry, D., Macpherson, C., 1994. Oxygen isotope composition of mantle peridotite. Earth Planet. Sci. Lett. 128, 231–241.

McCandless, T.E., Gurney, J.J., 1989. Sodium in garnet and potassium in clinopyroxene: criteria for classifying mantle eclogites. Proc. 4th International Kimberlite Conf. Geol. Soc. Australia Spec. Pub., vol. 14, pp. 827–833.

McCulloch, M.T., 1989. Sm–Nd systematics in eclogite and garnet peridotite nodules from kimberlites: implications for the early differentiation of the Earth. Kimberlites and Related Rocks. Geol. Soc. Australia Spec. Pub., vol. 14, pp. 864–876.

McDonough, W.F., Sun, S.-S., 1995. The composition of the Earth. Chem. Geol. 120, 223–253.

Meisel, T., Walker, R.J., Morgan, J.W., 1996. The osmium isotopic composition of the Earth's primitive upper mantle. Nature 383, 517–520.

Menzies, A.H., 2001. A detailed investigation into diamond-bearing xenoliths from Newlands kimberlite, South Africa, unpublished PhD Thesis, The University of Cape Town.

Menzies, A.H., Gurney, J.J., Harte, B., Hauri, E., 1998a. REE patterns in diamond-bearing eclogite and peridotite from Newlands kimberlite, South Africa. Extended Abstracts, 7th Int., Kimberlite Conf.

Menzies, A.H., Milledge, H.J.M., Gurney, J.J., 1998b. Fourier transform infra-red (FTIR) spectroscopy of Newlands diamonds. Extended Abstracts, 7th Int. Kimberlite Conf., Cape Town, South Africa.

Menzies, A.H., Carlson, R.W., Shirey, S.B., Gurney, J.J., 1999. Re–Os systematics of Newlands peridotite xenoliths: implications for diamond and lithosphere formation. In: Gurney, J.J., Gurney, J.L., Pascoe, M.D., Richardson, S.H. (Eds.), Proc. 7th Int. Kimberlite Conf., Red Roof Design, Cape Town, pp. 566–573.

Neal, C.R., Taylor, L.A., Davidson, J.P., Holden, P., Halliday, A.N., Nixon, P.H., Paces, J.B., Clayton, R.N., Mayeda, T.K., 1990. Eclogites with oceanic crustal and mantle signatures from the Bellsbank kimberlite, South Africa, part 2: Sr, Nd, and O isotope geochemistry. Earth Planet. Sci. Lett. 99, 362–379.

O'Hara, M.J., Yoder, H.S., 1967. Formation and fractionation of basic magmas at high pressures. Geology 3, 67–117.

Pearson, D.G., Carlson, R.W., Shirey, S.B., Boyd, F.R., Nixon, P.H., 1995a. Stabilization of Archean lithospheric mantle: a Re–Os isotope study of peridotite xenoliths from the Kaapvaal craton. Earth Planet. Sci. Lett. 134, 341–357.

Pearson, D.G., Shirey, S.B., Carlson, R.W., Boyd, F.R., Pokhilenko, N.P., Shimizu, N., 1995b. Re–Os, Sm–Nd and Rb–Sr isotope evidence for thick Archaean lithospheric mantle beneath the Siberian craton modified by multistage metasomatism. Geochim. Cosmochim. Acta 59, 959–977.

Pearson, D.G., Snyder, G.A., Shirey, S.B., Taylor, L.A., Carlson, R.W., Sobolev, N.V., 1995c. Archaean Re–Os age for Siberian eclogites and constraints on Archaean tectonics. Nature 374, 711–713.

Pearson, D.G., Shirey, S.B., Harris, J.W., Carlson, R.W., 1998. Sulphide inclusions in diamonds from the Koffiefontein kimberlite, S. Africa: constraints on diamond ages and mantle Re–Os systematics. Earth Planet. Sci. Lett. 160, 311–326.

Reisberg, L., Zindler, A., Marcantonio, F., White, W., Wyman, D., Weaver, B., 1993. Os isotope systematics in ocean island basalts. Earth Planet. Sci. Lett. 120, 149–167.

Richardson, S.H., Shirey, S.B., Harris, J.W., Carlson, R.W., 2001. Archean subduction recorded by Re–Os isotopes in eclogitic sulfide inclusions in Kimberley diamonds. Earth Planet. Sci. Lett. 191, 257–266.

Roy-Barman, M., Allègre, C.J., 1994. ^{187}Os/^{186}Os ratios of mid-ocean ridge basalts and abyssal peridotites. Geochim. Cosmochim. Acta 58, 5043–5054.

Roy-Barman, M., Allègre, C.J., 1995. ^{187}Os/^{186}Os in oceanic island basalts: tracing oceanic crust recycling in the mantle. Earth Planet. Sci. Lett. 129, 145–161.

Shirey, S.B., 1997. Re–Os isotopic compositions of midcontinent rift system picrites and tholeiites: implications for plume–lithosphere interaction and enriched mantle sources. Can. J. Earth Sci. 34, 489–503.

Shirey, S.B., Walker, R.J., 1995. Carius tube digestions for low-blank rhenium–osmium analysis. Anal. Chem. 67, 2136–2141.

Shirey, S.B., Wiechert, U., Carlson, R.W., Gurney, J.J., Heerden, L.V., 1999. Re–Os and oxygen isotopic systematics of diamondiferous and non-diamondiferous eclogites from the Roberts Victor kimberlite, South Africa. Ninth Annual V.M. Goldschmidt Conference. Lunar and Planetary Institute, Houston, TX, pp. 273–274.

Shirey, S.B., Carlson, R.W., Richardson, S.H., Menzies, A., Gurney, J.J., Pearson, D.G., Harris, J.W., Wiechert, U., 2001. Archean emplacement of eclogitic components into the lithospheric mantle during formation of the Kaapvaal craton. Geophys. Res. Lett. 28, 2509–2512.

Shirey, S.B., Harris, J.W., Richardson, S.H., Fouch, M.J., James, D.E., Cartigny, P., Deines, P., Viljoen, F., 2002. Diamond genesis, seismic structure and evolution of the Kaapvaal–Zimbabwe craton. Science 297, 1683–1686.

Smith, C.B., Gurney, J.J., Skinner, E.M.W., Clement, C.R., Ebrahim, N., 1985. Geochemical character of southern African kimberlites: a new approach based on isotopic constraints. Trans. Geol. Soc. South Africa 88, 267–280.

Smyth, J.R., Caporuscio, F.A., McCormick, T.C., 1989. Mantle eclogites: evidence of igneous fractionation in the mantle. Earth Planet. Sci. Lett. 93, 133–141.

Snyder, G.A., Jerde, E.A., Taylor, L.A., Halliday, A.N., Sobolev, V.N., Sobolev, N.V., 1993. Nd and Sr isotopes from diamondiferous eclogites, Udachnaya kimberlite pipe, Yakutia, Siberia: evidence of differentiation in the early Earth? Earth Planet. Sci. Lett. 118, 91–100.

Snyder, G.A., Taylor, L.A., Crozaz, G., Halliday, A.N., Beard, B.L., Sobolev, V.N., Sobolev, N.V., 1997. The origin of Yakutian eclogite xenoliths. J. Petrol. 38, 85–113.

Spetsius, Z.V., 1995. Occurrence of diamond in the mantle: a case study from the Siberian platform. J. Geochem. Explor., 25–39.

Taylor, L.A., Neal, C.R., 1989. Eclogites with oceanic crustal and mantle signatures from the Bellsbank kimberlite, South Africa: Part I. Mineralogy, petrography, and whole rock chemistry. J. Geol. 97, 551–567.

Taylor, L.A., Snyder, G.A., Crozaz, G., Sobolev, V.N., Yefimova, E.S., Sobolev, N.V., 1996. Eclogitic inclusions in diamonds: evidence of complex mantle process over time. Earth Planet. Sci. Lett. 142, 535–551.

Walker, R.J., Shirey, S.B., Stecher, O., 1988. Comparative Re–Os,

Sm–Nd and Rb–Sr isotope and trace element systematics for Archean komatiite flows from Munro Township, Abitibi Belt, Ontario. Earth Planet. Sci. Lett. 87, 1–12.

Walker, R.J., Carlson, R.W., Shirey, S.B., Boyd, F.R., 1989. Os, Sr, Nd, and Pb isotope systematics of southern African peridotite xenoliths: implications for the chemical evolution of subcontinental mantle. Geochim. Cosmochim. Acta 53, 1583–1595.

Available online at www.sciencedirect.com

SCIENCE @ DIRECT°

Lithos 71 (2003) 337–351

ELSEVIER

LITHOS

www.elsevier.com/locate/lithos

Trace element geochemistry of coesite-bearing eclogites from the Roberts Victor kimberlite, Kaapvaal craton

D.E. Jacob[a,*,1], B. Schmickler[a,2], D.J. Schulze[b]

[a] Institut für Geologische Wissenschaften, Universität Greifswald, F.L. Jahnstr. 17a, D-17487 Greifswald, Germany
[b] Department of Geology, University of Toronto, Erindale College, Mississauga, Ontario, Canada L5L 1C6

Abstract

Trace element characteristics of seven coesite-bearing eclogitic xenoliths from the Roberts Victor kimberlite demonstrate that this suite of eclogites originated as gabbroic cumulates in oceanic crust that was subsequently subducted. All but one of the garnets show positive Eu anomalies, accompanied by a flat heavy rare earth pattern, which is atypical of garnet, but characteristic of plagioclase, arguing for a considerable amount of plagioclase in the protoliths. Forward modelling of the accumulation of liquidus minerals from primitive komatiitic, picritic, and basaltic liquids suggests that at least some of the eclogite protoliths were not derived from basaltic parental liquids, whereas derivation from either komatiitic or picritic liquids is possible. The reconstructed eclogite bulk rocks compare favourably with oceanic gabbros from ODP hole 735B (SW Indian Ridge), even to the extent that oxygen isotopic systematics show signs of low-temperature seawater alteration. However, the oxygen isotope trends are the reverse of what is expected for cumulates in the lower section of the oceanic crust. These new findings show that $\delta^{18}O$ values in eclogitic xenoliths, despite being sound indicators for their interaction with hydrothermal fluids at low pressure, do not necessarily bear a simple relationship with the inferred oceanic crustal stratigraphy of the protoliths.
© 2003 Elsevier B.V. All rights reserved.

Keywords: Eclogite xenoliths; Trace elements; Roberts Victor; Coesite; Cumulates; Archean ocean crust

1. Introduction

Eclogitic xenoliths from the Roberts Victor kimberlite in South Africa were among the first for which an origin as subducted oceanic crust was proposed (e.g., Helmstaedt and Doig, 1975; Jagoutz et al., 1984). They belong to one of the best studied suites

* Corresponding author. Fax: +49-3834-864572.
 E-mail address: djacob@uni-greifswald.de (D.E. Jacob).
 [1] Present address: Max-Planck Institut für Chemie, Abt. Geochemie, Postfach 3060, D-55020 Mainz, Germany.
 [2] Present address: Bayerisches Geoinstitut, Universität Bayreuth, Universitätsstr. 30, D-95447 Bayreuth, Germany.

worldwide, and while the origin of other eclogite xenoliths as high-pressure melts/cumulates or as subducted oceanic basalts is still debated (e.g., Jacob et al., 1998; Snyder et al., 1998), the following evidence for a low-pressure origin of the protoliths for the majority of eclogites at Roberts Victor is convincing:

(1) Oxygen isotopic values of the eclogites deviate from those of mantle peridotites ($5.5 \pm 0.4\permil$, Mattey et al., 1994), and their range compares favourably with those of ophiolites and the oceanic crust (e.g., Gregory and Taylor, 1981). The Roberts Victor eclogites display the widest

0024-4937/$ - see front matter © 2003 Elsevier B.V. All rights reserved.
doi:10.1016/S0024-4937(03)00120-8

spectrum of $\delta^{18}O$ values known from any eclogite xenolith suite, namely, $+2\%_o$ to $+8\%_o$.

(2) The occasional occurrence of sanidine (MacGregor and Carter, 1970; Smyth and Hatton, 1977) and especially of coesite or quartz pseudomorphs after coesite in eclogite xenoliths with a large range of Mg-numbers (Schulze and Helmstaedt, 1988; Schulze et al., 2000; Smyth, 1977) excludes a direct origin of the eclogites as melts from the Earth's mantle. Partial melts of peridotite are not quartz-normative at any pressure higher than 8 kbar, but are olivine + hypersthene-normative and may be nepheline-normative at higher pressures (Green and Falloon, 1998). Although free silica may occur in olivine tholeiites or olivine basalt compositions in eclogite facies conditions, these melts could not have originated by melting of peridotite at pressures greater than 20 kbar (Green and Falloon, 1998).

(3) Many garnets from Roberts Victor eclogites, including those presented in this study, show rare earth element (REE) patterns with positive Eu anomalies and flat heavy rare earth elements (HREE) that are incompatible with the trace element partitioning characteristics of garnet, but resemble those of plagioclase (Green, 1994), implying a metamorphic origin of the garnets from protoliths derived from lower pressures.

The Roberts Victor eclogite suite was determined to be of late Archean age (ca. 2.7 Ga; Jagoutz et al., 1984; Shirey et al., 2001) and therefore offers a rare and interesting possibility to study late Archean oceanic crust. It has been shown that it is possible to reconstruct some original characteristics of the Archean oceanic crust using trace elements (Jacob and Foley, 1999), despite the rather complicated history of these rocks. An additional uncertainty arises from the fact that some eclogite suites show signs of a loss of a tonalitic melt during subduction (Barth et al., 2001; Jacob and Foley, 1999) which needs to be restored to arrive at the original whole rock composition.

In this contribution, we present trace element data for Roberts Victor eclogite xenoliths containing several percent of coesite or its quartz pseudomorphs, as well as sanidine. Based on experimental studies of the suprasolidus stability of coesite and quartz (Yaxley

and Green, 1998), the presence of these minerals limits possible loss of a partial melt to a maximum of about 13%, so that the trace elements of this suite may give a more direct impression of the characteristics of the subducted Archean oceanic crust.

2. Samples and analytical techniques

Seven eclogite xenoliths containing coesite or quartz pseudomorphs after coesite, mostly from a new collection (Schulze et al., 2000) were included in this study. Major element mineral chemistry of a larger suite including these seven samples was presented and discussed by Schulze et al. (2000). One sample (R-71) has been described previously (MacGregor and Carter, 1970; Manton and Tatsumoto, 1971; Schulze and Helmstaedt, 1988). The samples are relatively small, measuring between 1 and 4 cm in the longest dimension. All samples contain coesite or quartz, garnet and cpx, and some contain the additional phases rutile (13-64-100, 13-64-122), sanidine (R-71), and kyanite (13-64-100). Cpx and garnet modal abundances are variable. The xenoliths were sectioned to deliberately expose the coesite crystals that make up between approximately 5 and 20 vol.% on the exposed surface (Schulze et al., 2000).

In situ trace element analyses were obtained by Laser Ablation Microprobe (LAM) at the Department of Earth Sciences, Memorial University of Newfoundland, equipped with a frequency-quadrupled 266-nm wavelength Nd:YAG laser integrated with an enhanced sensitivity Fisons PQII+ "S" ICPMS (Günther et al., 1996). Material was ablated from a polished thin section of sample R-71 and from polished surfaces of cut blocks of the other six quartz-bearing eclogites. Calcium concentrations in garnet and clinopyroxene determined by microprobe (Schulze et al., 2000) were used as internal standards for the ICPMS measurements, titanium was used as internal standard for rutiles from two samples which were assumed to contain 100 wt.% TiO_2. TiO_2-contents in rutiles from eclogitic xenoliths have been found to be as low as 92 wt.% (Rudnick et al., 2000). An assumption of 100 wt.% TiO_2 therefore introduces an error of 5–10% on the trace element concentrations in rutile, but does not affect trace element ratios in rutiles. NIST SRM 612 glass, doped with approximately 40 ppm of a large range of trace elements, was used as the external standard. The USGS glass standard BCR-2g was measured as an unknown to monitor accuracy. Further details of the analytical procedures, data reduction, and calculation of detection limits are given in Longerich et al. (1996).

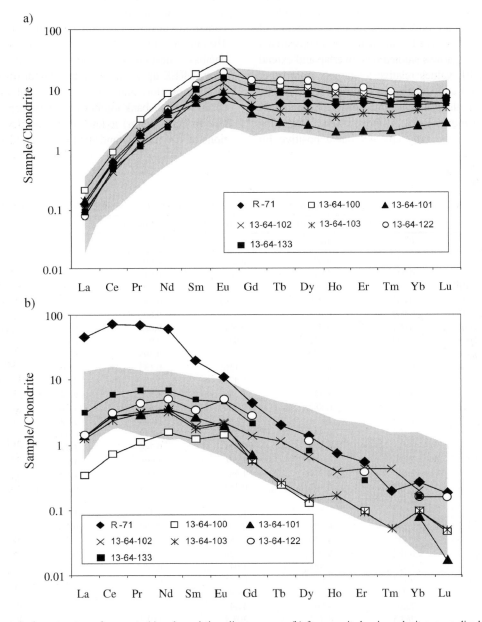

Fig. 1. Rare earth element patterns for garnets (a) and coexisting clinopyroxenes (b) from coesite-bearing eclogites, normalized to chondritic values (Sun and McDonough, 1989). Garnets with these characteristic REE patterns are not restricted to the Roberts Victor kimberlite pipe, but are observed in several other localities in Siberia and South Africa (shaded area in (a); data from our worldwide database). Coexisting clinopyroxenes are shown by the shaded area in (b).

3. Results

All garnets have chondrite normalized rare earth element patterns (Fig. 1a) with flat HREE between three and nine times chondritic and positive Eu anomalies with chondrite-normalized Eu/Eu* (Eu*=(Sm + Gd)/2) up to 2.1 (Table 1). Roberts Victor and Udachnaya are both known for the occurrence of eclogites with positive Eu anomalies, and comparison shows that the garnets studied here overlap and extend to higher REE values relative to garnets from other Roberts Victor and Udachnaya samples (Fig. 1a). The only garnet without a detectable positive Eu anomaly is that in sample R-71. Clinopyroxene from this sample is also the only cpx without a positive Eu

anomaly and is strongly enriched in LREE with Ce >70 times chondritic as well as in other incompatible elements (Fig. 1b). Clinopyroxenes from all other studied samples have positive Eu anomalies with Eu/Eu* up to 1.8 and LREE abundances up to approximately six times chondritic. They overlap with the range observed for cpx coexisting with garnets containing positive Eu anomalies (shaded area in Fig. 1b) and are comparable to those from our worldwide database: most cpx from eclogite xenoliths worldwide show LREE up to ca. 30 times chondritic. Cpx from R-71 is therefore unusually enriched; only one sample (from the Bellsbank kimberlite, a Group 2 kimberlite, like Roberts Victor) is known to have La concentrations of 100 times chondritic (Taylor and Neal, 1989).

Table 1
Trace element composition of garnets measured in situ by Laser-ICP-MS, n.d. = not detectable, n = number of spot analyses. Mn in wt%

	R-71	13-64-100	13-64-101	13-64-102	13-64-103	13-64-122	13-64-133
n	7	9	10	9	6	8	8
Ba	n.d.	0.039	0.099	n.d.	0.005	0.013	n.d.
Rb	n.d.	0.02	0.02	<0.01	0.01	<0.05	<0.07
Th	0.010	0.008	0.006	<0.03	0.003	0.001	0.001
U	0.012	0.005	0.009	<0.01	0.005	0.002	0.002
Nb	0.013	0.009	0.026	0.019	0.005	0.015	0.011
Ta	0.001	0.001	n.d.	0.001	0.001	0.001	0.004
La	0.028	0.049	0.032	0.023	0.032	0.018	0.021
Ce	0.370	0.547	0.370	0.253	0.398	0.299	0.283
Pr	0.164	0.295	0.161	0.116	0.183	0.165	0.108
Sr	0.835	9.26	2.30	1.47	4.53	2.58	4.34
Nd	1.93	3.88	1.73	1.24	1.92	2.12	1.09
Sm	1.07	2.75	0.882	0.944	1.036	1.762	1.503
Zr	3.53	22.55	16.7	19.6	20.9	28.8	58.0
Hf	0.042	0.439	0.343	0.382	0.419	0.552	1.228
Eu	0.388	1.81	0.515	0.511	0.746	1.096	0.907
Ti	420	2638	1199	1559	1439	1858	1799
Gd	0.987	2.66	0.777	1.59	1.09	2.89	2.17
Tb	0.214	0.424	0.102	0.353	0.159	0.513	0.328
Dy	1.41	2.59	0.593	2.41	1.03	3.35	2.00
Y	8.51	12.3	2.75	12.3	5.37	15.4	9.14
Ho	0.311	0.500	0.109	0.475	0.192	0.610	0.337
Er	0.958	1.41	0.319	1.28	0.646	1.744	1.055
Tm	0.153	0.187	0.051	0.156	0.096	0.231	0.138
Yb	1.16	1.23	0.404	1.04	0.748	1.46	0.930
Lu	0.179	0.173	0.0689	0.145	0.124	0.217	0.146
Sc	n.d.	n.d.	n.d.	39.8	32.4	n.d.	n.d.
V	48.1	72.0	42.6	47.4	35.6	54.3	46.7
Cr	568	263.5	330.2	301.4	180.2	184.2	164.8
Mn	0.173	0.109	0.105	0.137	0.153	0.322	0.275
Co	63.0	43.6	49.5	50.7	56.6	46.6	51.6
Ni	23.5	52.6	55.7	49.2	57.4	23.1	30.0

Table 2
Trace element composition of clinopyroxenes, n.d. = not detectable, *n* = number of spot analyses. Mn in wt%

	R-71	13-64-100	13-64-101	13-64-102	13-64-103	13-64-122	13-64-133
n	6	6	3	8	8	10	3
Ba	2.08	<0.24	n.d.	n.d.	n.d.	n.d.	n.d.
Rb	3.38	n.d.	n.d.	0.070	0.160	n.d.	n.d.
Th	0.088	n.d.	n.d.	0.016	0.009	n.d.	n.d.
U	0.013	n.d.	n.d.	0.012	n.d.	n.d.	n.d.
Nb	0.0056	<0.01	<0.03	0.038	0.022	0.016	0.016
Ta	0.0004	n.d.	n.d.	0.003	0.003	n.d.	n.d.
La	11.0	0.082	0.325	0.320	0.302	0.330	0.748
Ce	44.7	0.452	1.76	1.74	1.48	1.91	3.64
Pr	6.78	0.107	0.282	0.311	0.293	0.419	0.663
Sr	1074	162	135	136	197	196	276
Nd	28.2	0.746	1.70	1.69	1.51	2.40	3.20
Sm	3.05	0.189	0.408	0.300	0.278	0.530	0.768
Zr	8.31	7.90	6.42	9.86	4.13	13.7	11.56
Hf	0.231	0.489	0.343	0.555	0.207	1.41	0.688
Eu	0.637	0.082	0.112	0.131	0.124	0.290	0.267
Ti	959	2338	1439	2038	1259	2518	2398
Gd	0.917	0.121	0.145	0.286	0.114	0.576	0.438
Tb	0.077	0.009	n.d.	0.042	0.010	n.d.	n.d.
Dy	0.339	0.031	<0.06	0.162	0.037	0.293	0.197
Y	0.808	0.103	0.131	0.392	0.147	0.506	0.551
Ho	0.042	n.d.	n.d.	0.022	0.010	n.d.	n.d.
Er	0.090	0.016	<0.08	0.071	0.015	0.065	0.047
Tm	0.005	n.d.	n.d.	0.011	0.001	n.d.	n.d.
Yb	0.046	0.016	0.013	0.034	0.064	0.027	0.027
Lu	0.005	0.001	0.0004	<0.01	0.001	0.004	0.008
V	224	n.d.	n.d.	167	129	n.d.	n.d.
Cr	1013	n.d.	n.d.	454	304	n.d.	n.d.
Mn	0.036	n.d.	n.d.	0.016	0.017	n.d.	n.d.
Co	21.9	n.d.	n.d.	17.8	19.0	n.d.	n.d.
Ni	250	n.d.	n.d.	284	248	n.d.	n.d.

Typically, Sr concentrations in cpx from eclogite xenoliths worldwide are in the range of 100–300 ppm. The concentration of Sr in R-71 cpx greatly exceeds this range (Table 2) and, together with this sample's elevated LREE-contents, suggests a metasomatic overprint. The occurrence of sanidine is not necessarily connected to metasomatism, because the only other coesite–sanidine eclogite known from this locality is not enriched in elements thought to be indicative of metasomatism (Harte and Kirkley, 1997; Smyth and Hatton, 1977). Rutiles from two samples were analysed for selected trace elements (Table 3) and show chondritic and slightly subchondritic Nb/Ta ratios of 17.3 and 11.3, respectively (chondritic Nb/Ta = 17.4, Sun and McDonough, 1989). Nb and Ta concentrations as well as Nb/Ta ratios in eclogitic rutiles show large variations.

Rudnick et al. (2000) reported Nb concentrations between 14 and 16,301 ppm and Nb/Ta ratios between 4 and 343 for eclogites from Africa and Siberia, whereas rutiles with very low (<1 ppm) Nb and Ta abundances are known from Roberts Victor (Jacob et al., 2003). When interpreting these rutile

Table 3
Trace element composition of rutiles, *n* = number of spot analyses

	13-64-100	13-64-122
n	5	3
Nb	126	90.80
Ta	7.31	8.01
Zr	1026	759
Hf	27.8	24.3
V	590	n.d.
Cr	521	1294

data, it is crucial to bear in mind that rutile is often not primary in eclogite xenoliths, but instead, the product of metamorphic reactions on entering the eclogite facies, or of exsolution from garnet during cooling (Green and Ringwood, 1967a). Furthermore, since the xenoliths may reside in the Earth's mantle over a long timespan, textural evidence for exsolution of rutile is often eradicated. Foley et al. (2002) demonstrated that those rutiles with superchondritic Nb/Ta must be the products of metamorphism of rocks that had been partially melted as amphibolites.

4. Reconstructed whole rocks

Since all eclogite xenoliths suffer from infiltration of kimberlitic material, whole rock compositions have to be reconstructed based on mineral data and modal estimations. These estimations necessarily bear a large uncertainty because of the large grain size and often small sample size, and because they are normally carried out two dimensionally on a cut surface of the nodule. However, this method is accurate enough to constrain general trends or geochemical character-

Table 4
Reconstructed bulk major element composition and CIPW norm assuming 10 vol.% (8.6 wt.%) coesite

Measured Mode (vol.%)	R-71	13-64-100*	13-64-101	13-64-102	13-64-103	13-64-122*	13-64-133
	50gt, 1san, 39 cpx	44gt, 7ky, 38.5 cpx	66gt, 24 cpx	62gt, 28 cpx	79gt, 11 cpx	50gt, 39.5 cpx	50gt, 40cpx
SiO_2	52.13	50.71	49.85	49.89	47.28	51.06	51.14
	(49.90–56.70)	(48.47–55.29)	(47.68–54.28)	(48.04–54.52)	(45.11–51.70)	(48.77–55.62)	(48.87–55.20)
TiO_2	0.10	0.99	0.19	0.26	0.22	0.96	0.31
Al_2O_3	15.87	21.53	19.39	18.56	20.57	18.00	17.13
	(14.42–16.57)	(19.82–22.36)	(17.75–20.20)	(16.97–19.49)	(18.89–21.39)	(16.28–18.83)	(15.45–17.92)
Cr_2O_3	0.14	0.04	0.06	0.05	0.01	0.05	0.03
FeO	7.48	5.59	7.19	8.37	9.31	11.35	11.07
	(6.81–7.81)	(5.03–5.87)	(6.69–7.43)	(7.76–8.74)	(8.78–9.57)	(10.34–11.85)	(10.06–11.55)
MnO	0.15	0.07	0.13	0.16	0.17	0.32	0.28
MgO	14.37	5.54	12.72	11.91	11.32	7.77	9.16
	(13.0–15.0)	(5.0–5.8)	(11.6–13.2)	(10.90–12.50)	(10.40–11.80)	(7.00–8.10)	(8.20–9.60)
CaO	8.06	13.22	9.14	8.83	10.29	7.49	7.82
	(7.19–8.48)	(11.83–13.90)	(8.18–9.61)	(7.90–9.36)	(9.30–10.78)	(6.75–7.86)	(6.98–8.20)
Na_2O	1.78	2.44	1.47	1.65	0.77	3.02	2.67
	(1.58–1.88)	(2.16–2.58)	(1.19–1.60)	(1.38–1.79)	(0.45–0.93)	(2.68–3.19)	(2.34–2.81)
K_2O	0.15	0.07	0.03	0.03	0.01	0.03	0.04
Total	100.2	100.2	100.2	99.7	100.0	100.0	99.7
Mg-no.	77.39	63.86	75.93	71.73	68.44	54.95	59.59
Q	–	–	–	–	–	–	–
or	0.89	0.41	0.18	0.18	0.06	0.18	0.24
ab	15.1	20.7	12.4	14.0	6.52	25.6	25.6
an	35.0	47.5	45.3	43.3	51.1	35.5	34.6
C		–	0.33	–	0.61	–	–
di	4.05	14.7	32.8	0.44	–	1.40	3.49
hy	38.0	13.8	–	32.6	30.8	27.2	29.0
ol	6.97	1.14	8.70	8.77	10.5	8.32	9.13
il	0.19	1.88	0.36	0.49	0.42	1.82	0.59
sum	100.12	100.14	100.12	99.68	99.99	99.96	99.59
Σ (or + an + ab)	51.0	68.6	57.9	57.5	57.7	61.3	60.4

The ranges given in parentheses are for bulk compositions reconstructed assuming 5 and 20 vol.% (4.3 and 17.5 wt.%) coesite, respectively. For samples 13-64-100 and 13-64-122 (both with asterisk), in which rutile was observed 0.5% rutile was assumed for the whole rock reconstruction. All modes in vol.% were converted to wt.% contents of minerals prior to whole rock reconstruction using densities for the respective minerals from the literature. Mineral compositions are taken from Schulze et al. (2000). Gt = garnet, cpx = clinopyroxene, ky = kyanite, san = sanidine, Mg-no. = 100*Mg/(Mg + Fe).

istics (e.g., Jacob and Foley, 1999). A conservative estimate implies that a maximum error of 20% for a modal estimation on small xenoliths is realistic, and error bars are shown for reconstructed whole rock compositions in diagrams where modal proportions appear to be crucial. Modal proportions in this study (Table 4) were estimated using a Zeiss digital imaging system at the University of Greifswald and by element mapping using a JEOL JXA 8900 RL microprobe at University of Göttingen, Germany.

Exact amounts of coesite and rutile are difficult to constrain. Coesite varies between 5 and 20 vol.% (Schulze et al., 2000) in thin section, but is not evenly distributed throughout the xenoliths. Assuming 20% free modal coesite in a bulk rock reconstruction is considered an overestimate as it yields unrealistically high SiO_2-contents between 54 and 60 wt.%, whereas oceanic gabbros typically contain 47–53 wt.% SiO_2 (Bach et al., 2001; Niu et al., 2002). Furthermore, the majority of oceanic gabbros or basalts are olivine- and/or hypersthene-normative which would translate to eclogite compositions with ca. 5–15 vol.% coesite. Therefore, a more realistic, uniform amount of 10 vol.% (8.6 wt.%) coesite was assumed for each

sample in Table 4 and ranges for bulk compositions reconstructed with 5 and 20 vol.% (4.3 and 17.5 wt.%) coesite are given in parentheses. The uncertainty introduced in this way is smaller than that caused by the estimation of modes and mainly has a diluting effect. Although rutile was observed only in three samples, it is inferred that phases in all studied samples were equilibrated with accessory rutile because of the unusually low Nb and Ta concentrations in the reconstructed silicate whole rocks (cf. Jacob and Foley, 1999). Barth et al. (2001) determined rutile abundances in eclogitic xenoliths from West Africa by Ti mass balance to lie between 0.1 and 0.9 wt.%. Therefore, an average amount of 0.5 wt.% rutile was assumed for the whole rock reconstruction of the two samples in which rutiles were analysed for trace elements. Normalized to primitive mantle (Sun and McDonough, 1989), this amount of rutile yields Ti abundances of up to six times primitive mantle (Fig. 2), compared with about 10 times for 1 wt.% rutile and ca. 2 times for 0.1 wt.% rutile.

For comparison, rutile-free eclogites from the Udachnaya kimberlite in Siberia have Ti abundances of two to three times primitive mantle (Jacob and

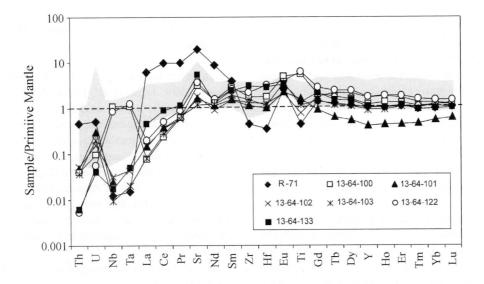

Fig. 2. Trace element patterns for reconstructed coesite–eclogite bulk compositions using mineral modal abundances given in Table 4, normalized to primitive mantle (Sun and McDonough, 1989) and compared with oceanic gabbros from ODP Hole 735B (shaded area, Bach et al., 2001). Note that for those samples where rutiles could not be analysed for trace elements, only silicate phases were used for the reconstruction of Ti, Nb, Ta, Zr, and Hf-contents. Abundances of the trace HFSE in samples other than 13-64-100 and 13-64-122 are therefore artificially low.

Foley, 1999) and eclogites from Koidu, West Africa have between about two and seven times primitive mantle (Barth et al., 2001).

Fresh oceanic gabbros from the SW Indian Ridge show a range between 0.2 and 0.8 wt.% TiO_2 and average 0.44 wt.% TiO_2 (Bach et al., 2001), resulting in Ti abundances between one and five times that of primitive mantle.

Reconstructed bulk eclogites have Mg-numbers between 55 and 77 (Table 4). Two samples have reconstructed MgO-contents >12 wt.% and are thus picritic, whereas all others are of basaltic/gabbroic composition. Al_2O_3 ranges between 15.9 and 21.5 wt.%. The most aluminous sample is 13-64-100, which is kyanite bearing. This sample also has the most calcic garnet of the suite (0.52 mol% grossular content, Schulze et al., 2000) and can thus be classified as a grospydite. The average Na_2O-content of this suite is 2.0 wt.% which is well within the range observed in Roberts Victor eclogites (Hatton, 1978) and in low MgO eclogites from Koidu (Barth et al., 2001), but higher than that of high MgO eclogites from Koidu and eclogites from the Zero kimberlite (Schmickler, 2002), suites that are both interpreted as cumulates.

5. Discussion

Several characteristics of this coesite-bearing eclogite suite argue in favour of an origin at low pressure, probably as part of the oceanic crust. The occurrence of coesite in itself is strong evidence against an origin as a high-pressure melt (>20 kbar), since partial melts from peridotite are hypersthene-normative or, at higher pressures, nepheline-normative (Green and Falloon, 1998). Only at pressures less than 8 kbar are melts quartz-normative (Falloon and Green, 1988). It has been argued that eclogites with coesite (and sanidine) could be the more evolved members of a suite of high-pressure igneous fractionates (e.g., Hatton and Gurney, 1987; O'Hara and Yoder, 1967). However, Schulze and Helmstaedt (1988) and Schulze et al. (2000) showed for eclogite suites from several kimberlites, including that of the Roberts Victor pipe, that Mg-numbers of coesite and sanidine eclogites may be as high as 80 and completely overlap the range of Mg-numbers interpreted by some authors as an igneous fractionation trend. During prograde metamorphism of

plagioclase-bearing basic rocks, however, quartz is routinely formed following the reactions albite = jadeite + quartz and anorthite = grossular + kyanite + quartz. Furthermore, Green and Ringwood (1967a,b) demonstrated experimentally that up to approximately 17% free silica can be stable in eclogites formed by high-pressure metamorphism of tholeiites. A low-pressure origin is further supported by REE patterns of the eclogitic garnets (Fig. 1a) that show pronounced positive Eu anomalies and flat heavy REE patterns, typical for plagioclase-bearing magmatic rocks. Similar patterns can be found in eclogite xenoliths from other localities worldwide (Fig. 1a), but they are very unlike equilibrium REE patterns for garnet from high-pressure rocks that are expected to have high abundances of heavy REE due to its partitioning characteristics.

Taking this evidence together with oxygen isotopic values deviating from those of the unchanged Earth's mantle (Schulze et al., 2000) and the occurrence of these samples within a suite of eclogitic xenoliths for which an origin as Archean oceanic crust has been proposed, it is convincing that these coesite eclogites represent samples from the same Archean oceanic igneous succession brought to mantle depths before being sampled by the Roberts Victor kimberlite.

6. Pre-metamorphic composition of the coesite eclogites

In earlier studies (e.g., Jacob et al., 1994; Jacob and Foley, 1999; Jacob et al., 1998), it has been shown that despite the complicated history of eclogite xenoliths, many petrogenetically crucial original characteristics survived, and it is possible to reconstruct the composition of Archean oceanic crust. The study of eclogite xenoliths may therefore place important constraints on the nature of Archean oceanic crust, since they are possibly the only direct relict samples. Evidence for Archean oceanic crustal remnants in greenstone belts is equivocal, because in the only known non-tectonic contacts with basement, greenstone belts lie on continental crust (Bickle, 1978; Bickle et al., 1994).

Seawater alteration, dehydration during subduction and partial melting are the main processes which may have affected the major and trace element budgets of unmetasomatized eclogite xenoliths. For the suite under investigaton, however, partial melting effects can be assumed to be minimal (≤ 13%), since quartz and sanidine are preserved in the paragenesis of some of the eclogites. Dehydration and loss of a small melt component in the amphibolite or eclogite facies would mainly affect the abundances of the LREE and LILE (Brenan et al., 1995; Stalder et al., 1998; Rapp et al., 1991; Foley et al., 2002). Element fluxes during seawater alteration of modern oceanic crust are difficult to quantify, but general trends include enrichment in alkalis and large ion lithophile elements (LILE) and depletion in Si, Mg, and Ca (for a very general summary of effects on eclogites, see Jacob et al., 1994). In modern altered oceanic crust, as well as in their subducted eclogitic equivalents, these changes can be monitored using stable oxygen isotopes (e.g.,

Bach et al., 2001; Jacob et al., 1994; Jacob and Foley, 1999). As a rather simplistic generalization, it could be stated that the relatively immobile elements are those that remain compatible in minerals during prograde metamorphism and eventually in the eclogite phases (cpx, gt, rutile). Although this still needs to be studied in detail, it seems to apply in general for a number of trace elements including the HFSE, Cr, Ni, middle and heavy REE in eclogite xenoliths from the mantle (Jacob and Foley, 1999).

This suite of coesite eclogites has reconstructed bulk compositions of picritic to basaltic/gabbroic composition with MgO-contents ranging from 14.4 to 5.5 wt.%, Mg-numbers between 77 and 55 (Table 4) and Ni-contents as expected for basaltic/gabbroic rocks within this range of Mg-numbers (Table 5). Low Cr_2O_3-

Table 5
Reconstructed bulk trace element compositions using mineral data from Tables 1 and 2 and modal mineral amounts from Table 4. Mn in wt%

	R-71	13-64-100*	13-64-101	13-64-102	13-64-103	13-64-122*	13-64-133
Ba	0.814	n.d.	0.067	n.d.	0.004	0.007	–
Th	0.039	0.003	0.004	0.005	0.003	–	–
U	0.011	0.002	0.006	–	0.004	–	–
Nb	0.009	0.801	0.018	0.023	0.007	0.614	0.012
Ta	0.001	0.047	n.d.	0.002	n.d.	0.051	0.002
La	4.31	0.053	0.099	0.104	0.058	0.140	0.310
Ce	17.7	0.420	0.671	0.645	0.482	0.909	1.60
Pr	2.73	0.174	0.176	0.161	0.179	0.250	0.320
Sr	420	66.2	33.8	38.8	25.1	78.5	112
Nd	12.0	2.04	1.57	1.26	1.71	2.03	1.84
Sm	1.74	1.31	0.693	0.684	0.865	1.11	1.08
Zr	5.07	19.7	12.8	15.2	17.3	25.0	34.5
Hf	0.112	0.561	0.314	0.397	0.360	0.992	0.907
Eu	0.449	0.847	0.374	0.361	0.615	0.677	0.573
Ti	590	7390	2126	1030	1511	8273	1771
Gd	0.868	1.25	0.560	1.09	0.892	1.71	1.29
Tb	0.140	0.195	0.069	0.236	0.130	0.264	0.169
Dy	0.860	1.18	0.401	1.58	0.833	1.84	1.11
Y	4.71	5.61	1.89	7.90	4.35	8.10	4.92
Ho	0.177	0.226	0.073	0.308	0.156	0.313	0.173
Er	0.530	0.644	0.216	0.833	0.522	0.921	0.561
Tm	0.081	0.085	0.034	0.102	0.077	0.119	0.071
Yb	0.618	0.562	0.276	0.670	0.610	0.761	0.489
Lu	0.094	0.078	0.047	0.092	0.100	0.113	0.078
V	112	n.m.	n.m.	81.6	46.6	n.m.	n.m.
Cr	689	n.m.	n.m.	320	181	n.m.	n.m.
Mn	0.093	n.m.	n.m.	0.083	0.116	n.m.	n.m.
Co	37.60	n.m.	n.m.	34.61	44.75	n.m.	n.m.
Ni	121	n.m.	n.m.	121	81.5	n.m.	n.m.

Asterisks designate samples in which rutiles were measured (Table 3) and whose analyses were considered for the reconstructions. All other samples, although rutile-bearing, were reconstructed without rutile except for Ti and are therefore unrealistically low in Nb, Ta, Zr, and Hf. n.d. = not detected, n.m. = not measured.

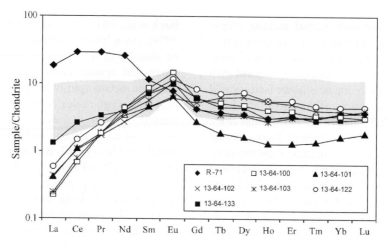

Fig. 3. Chondrite-normalized rare earth element patterns for reconstructed bulk coesite eclogites using mineral modes in Table 4, compared with measured bulk rock REE patterns from gabbros from the SW Indian Ridge (ODP Hole 735B, Bach et al., 2001).

contents argue against a high percentage of Cr-spinel in the eclogite protoliths, whereas relatively high Na_2O-contents in most samples and reconstructed bulk rock trace element patterns with distinct positive Eu and Sr anomalies combined with flat HREE patterns (Figs. 2, 3) are strong evidence for plagioclase as a major phase in the protoliths, accompanied by clinopyroxene. These REE patterns are not typical of primitive basaltic melt compositions, nor of melt compositions produced by fractionation of basalts, but they may correspond to accumulations of phenocryst assemblages that would crystallize from basaltic melts. The reconstructed HREE pattern is dominated by cpx, because the partition coefficients for plagioclase liquid pairs are much lower (Bindeman et al., 1998).

It is therefore likely that within an oceanic crustal environment, the protoliths could be plagioclase-bearing cumulates, such as gabbros or troctolites, consisting mainly of plagioclase, clinopyroxene, and/or olivine + orthopyroxene. Calculated CIPW-Norms (Table 4) are in fact very similar to those of oceanic gabbros, yielding between 51% and 68% normative calcic feldspar. In addition, the reconstructed REE patterns are, in general, very similar to whole rock REE patterns of comparably evolved fresh gabbros from ODP hole 735B on the SW Indian Ridge (Bach et al., 2001, Fig. 3). Depletions in the LREE (La, Ce, Pr) in the eclogites compared to modern gabbros are attributed to the combined effects of dehydration during subduction (Brenan et al., 1995; Stalder et al.,

1998) and small degrees of partial melting. However, the presence of coesite/quartz and sanidine shows that partial melting must have played a restricted role, as loss of more than ca. 13% melt would have resulted in loss of these phases. Rare earth element patterns of samples R-71 and 13-64-101 deviate significantly from those expected for subducted gabbro equivalents. As outlined above, sample R-71 contains unusually high amounts of trace elements that are typically enriched during cryptic mantle metasomatism (e.g., Dawson, 1984; Frey and Green, 1974), and it is therefore argued that its original composition was overprinted by metasomatism. The REE pattern of sample 13-64-101 shows distinctly lower abundances of HREE that cannot be explained by simple variation in modal abundances of garnet and cpx in this sample. Garnet, the main host of HREE, has an approximate modal abundance of 66% (Table 4); only the unrealistic assumption of 100% garnet would bring the REE pattern close to the range observed in modern gabbros. A possible explanation could be orthopyroxene or spinel in the protolith. Since these minerals incorporate only insignificant amounts of REE, they effectively "dilute" the REE abundances of the whole rock.

7. Oxygen isotopes

It was mainly on the basis of the oxygen isotopic systematics of eclogitic xenoliths that they were initially proposed to be remnants of subducted oceanic crust (Jagoutz et al., 1984), since the variation of $\delta^{18}O$ values observed in them cannot be generated by fractionation at mantle depths and temperatures (Clayton et al., 1975), whereas it resembles the range of $\delta^{18}O$ values in ophiolites caused by interaction with seawater (e.g., Gregory and Taylor, 1981). In ophiolites and the oceanic crust, metamorphism at low temperatures involving seawater causes high $\delta^{18}O$ values, whereas metamorphism at higher temperatures (greenschist facies) generates low $\delta^{18}O$ values. In general, as for example in the Semail ophiolite (Gregory and Taylor, 1981), low-temperature metamorphism affects the upper part of the oceanic crust, whereas deeper parts of the oceanic crust experience greenschist facies metamorphism.

This principle proved to be transferable to eclogite xenoliths from the Udachnaya kimberlite in Siberia, where those rocks with low $\delta^{18}O$ values show geochemical characteristics of gabbroic protoliths, and those with elevated $\delta^{18}O$ values are postulated to have picritic to basaltic lava protoliths (Jacob and Foley, 1999). However, new data for the eclogite suite from Roberts Victor (McDade, 1999, D. Lowry, unpublished data; Jacob et al., 2003) show an exactly opposite relationship between reconstructed protolith mineralogy and $\delta^{18}O$ values. Here, the rocks postulated to have gabbroic protoliths on the basis of their trace element geochemistry show elevated $\delta^{18}O$ values and were altered at low temperature, whereas those thought to represent lavas have low $\delta^{18}O$ values, indicating alteration at higher temperatures. This observation is further supported by results from this study and those of Schulze et al. (2000). Alteration of rocks by fluids is not homogeneous, as fluid percolation is mainly controlled by permeability (Norton and Taylor, 1979) and plastic and brittle deformation play an important role. Contrasting trends of oxygen isotopic values with degree of differentiation (although caused by late-stage hydrothermal alteration and genetically unrelated to fractional crystallization) are known from the Skaergaard and the Muskox layered intrusions (Epstein and Taylor, 1967; Norton and Taylor, 1979) and new ODP results report en echelon trends of $\delta^{18}O$ with depth (Hart et al., 1999). Gabbros from ODP hole 735B used for comparison in this study show a reverse oxygen isotope profile with $\delta^{18}O$ values increasing with depth (Bach et al., 2001),

much like that inferred for the protoliths of the Roberts Victor eclogites.

In view of these new data on eclogite xenoliths and on modern oceanic crust, it appears that although $\delta^{18}O$ values in eclogites are a reliable indicator of seawater alteration of the protoliths, they may not necessarily in each case indicate their position within the stratigraphy of the oceanic crust.

8. Forward modelling of cumulates

The parental magmas of Archean oceanic crust may have been picritic or komatiitic rather than basaltic (e.g., Arndt, 1983; Jacob and Foley, 1999; Nisbet and Fowler, 1983), and so, given the 2.7-Ga age of the Roberts Victor eclogites, it is of interest to investigate the nature of the parental liquids of their gabbroic protoliths. We have therefore used a forward approach, modelling cumulates with the petrological program *pmelts* (Ghiorso and Hirschmann, 1998). This program calculates minerals and coexisting liquids from a given bulk chemical composition using mineral thermodynamic data to minimize the Gibbs free energy of the system. The modelling was carried out for three different melt compositions produced in high-pressure experiments on MORB-Pyrolite: komatiitic, picritic, and basaltic (taken from Falloon and Green (1988), their melts 11, 8, and 6, respectively) with 19.99, 17.80, and 10.89 wt.% MgO, corresponding to Mg-numbers of 79.8, 79.7, and 68.25. The bulk composition fed into the program was modelled as completely melted and cumulate minerals were fractionated from the melt by isobaric cooling from supraliquidus conditions at a constant oxygen fugacity of IW + 2 at 1 GPa. Compositions and modal abundances of the crystallizing minerals were then taken to calculate the bulk cumulate compositions at a given temperature. Mantle potential temperatures may have been about 200 °C higher during the Archean than today, in which case the oceanic crust may have had a maximal thickness of around 40 km (McKenzie and Bickle, 1988). Therefore, a pressure of 1 GPa, corresponding to a depth of about 35 km, was chosen to represent the maximum pressure for the occurrence of cumulates in thick Archean oceanic crust.

The results of the modelling are shown in Fig. 4, and selected compositions of cumulates and coexisting

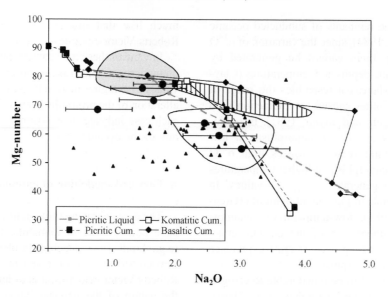

Fig. 4. Results of forward modelling of fractional accumulation using the program *pmelts*. Three lines are shown for the changing composition of cumulates with degree of fractionation from a primitive picritic ("Picritic Cum.," 17.8 wt.% MgO), komatiitic ("Komatiitic Cum.," 19.99 wt.% MgO), and basaltic liquids ("Basaltic Cum.," 10.89 wt.% MgO) from a starting point of 100% melt. Composition of the coexisting liquid for the picrite (grey line: "Picritic Liquid") is shown for comparison. Data field for MORB (white), oceanic gabbros (hatched, Bach et al., 2001), and high-MgO eclogites from Koidu (grey, Barth et al., 2002), as well as data points for low-MgO Koidu eclogites (triangles, Barth et al., 2001) are shown for comparison. Reconstructed coesite–eclogite compositions (solid circles) have error bars showing maximum uncertainties (20% in garnet and cpx abundances) due to estimation of modal mineralogy. See text for description of modelling procedure and further explanation.

liquids are listed in Table 6. Plagioclase-bearing cumulates start to form from all three modelled compositions when the liquids have Mg-numbers of approximately 57 (Table 6). At this point, the resulting cumulates have Mg-numbers between 78 and 80, coinciding with the highest Mg-numbers observed in this eclogite suite (77). Compositions of cumulates from primary komatiitic and picritic liquids give very similar trends, because their parental liquids have similar Mg-numbers, whereas basaltic cumulates are higher in Na_2O at Mg-numbers below 80 than cumulates from picritic and komatiitic liquids. This model can therefore clearly distinguish between cumulates from different parental liquids only at Mg-numbers below ca. 78 and/or Na_2O-contents of less than ca. 2.5 wt.%. Above this threshold, cumulates from picritic, komatiitic, as well as basaltic liquids contain small amounts of feldspar and have indistinguishable major element compositions. The protoliths of those eclogites with lower Mg-number (samples 13-64-103, 13-64-122, 13-64-133) cannot be cumulates from

basaltic liquids, because Na_2O contents in the modelled basaltic cumulates are generally approximately 50% higher at a given Mg-number than those in the reconstructed eclogite whole rocks. Even if a small percentage of partial melting ($\leq 13\%$) changed the bulk composition of the rocks, it could not result in loss of half of the Na_2O because Na_2O is not fractionated appreciably during melting of eclogites or amphibolites (Rapp et al., 1991).

For those eclogite samples with high Mg-numbers, a basaltic affinity cannot be excluded on the basis of this model, but a pricritic/komatiitic protolith is likely under the assumption that all samples belong to the same suite. As the eclogite compositions overlap with the cumulate compositions from either komatiitic or picritic liquid over a range of Mg-numbers, this method cannot distinguish between these possibilities.

Eclogite bulk compositions from the Koidu kimberlite, West Africa, plot to generally higher Mg-numbers in Fig. 4 in the area of cumulates consisting mainly of spinel and cpx, but lacking plagioclase.

Table 6
Results for forward modelling of cumulates and liquids from different starting compositions using *pmelts*

Basaltic system

Cumulate			Basaltic liquid	
Paragenesis	Na$_2$O	Mg-no.	Na$_2$O	Mg-no.
cpx + sp	0.62	82.31	4.35	62.77
cpx + fs + sp	1.55	80.46	4.84	58.96
fs + cpx + sp	2.77	78.09	5.2	56.04
fs + cpx + sp	3	76.3	5.61	52.61
fs + cpx + sp	3.54	71.1	6.72	43.69
fs + cpx + ol + sp	4.76	68.5	7.38	38.27
fs + sp + ol	4.41	43.2	8.77	32.46
fs + ol + sp	4.77	39.1	11.13	22.21

Picritic system

Cumulate			Picritic liquid	
Paragenesis	Na$_2$O	Mg-no.	Na$_2$O	Mg-no.
cpx + sp	0.45	82.75	2.86	63.2
fs + sp	1.98	77.5	3.56	54.68
fs + cpx + sp	2.78	68.64	4.64	40.24
fs + ol + sp + cpx	3.84	34.54	7.76	20.3

Komatiitic system

Cumulate			Komatiitic liquid	
Paragenesis	Na$_2$O	Mg-no.	Na$_2$O	Mg-no.
cpx + sp	0.49	80.65	2.81	57.7
fs + cpx + sp	2.16	78.48	3.28	51.49
cpx + fs	2.83	65.63	4.23	35.53
fs + ol + sp + cpx	3.8	31.85	7.64	18.07

Shown are selected compositions for cumulates and coexisting liquids (in wt.% Na$_2$O) in the range of parageneses (i.e., plagioclase-bearing) and Mg-numbers of relevance for the eclogite samples studied. fs = feldspar, cpx = clinopyroxene, sp = spinel, ol = olivine, Mg-no. = Mg-number.

This is consistent with their proposed derivation from cpx-spinel cumulate protoliths based on an interpretation using trace elements (e.g., elevated Cr-contents: Barth et al., 2002).

9. Conclusions

The major and trace element composition of coesite-bearing eclogites from the Roberts Victor kimberlite are consistent with an origin as Archean oceanic gabbroic cumulates. In many characteristics, the suite compares well with drilled modern oceanic gabbros from ODP hole 735B on the SW Indian Ridge (e.g., flat HREE pattern, positive Sr and Eu anomalies in reconstructed whole rocks), even to the extent that oxygen isotope ratios imply low-temperature seawater alteration with trends opposite to those expected from the protolith's inferred stratigraphic position within oceanic crust (e.g., Gregory and Taylor, 1981). An important result of this investigation, therefore, is that δ^{18}O values in eclogitic xenoliths, despite being sound indicators for their interaction with hydrothermal fluids at low pressure, do not necessarily bear a simple relationship with the inferred oceanic crustal stratigraphy of the protoliths.

The reconstructed bulk eclogites are depleted in LILE and LREE relative to oceanic gabbro, postulated to result from the combined effects of fluid loss during subduction and small degrees of partial melting. However, the amounts of partial melt lost must have been small (< 13%) as non-refractory minerals (coesite/quartz and sanidine) are preserved.

Using a forward model for crystal accumulation, it can be shown that the protoliths of the eclogites are unlikely to have been cumulates from primitive basaltic liquids, whereas derivation from more magnesian liquids such as picrite or komatiite agrees with the major element characteristics of the bulk eclogites.

Acknowledgements

Reviews by Chris Hatton, Hermann Grutter, and an anonymous reviewer, comments from Vincent Salters on an earlier version, as well as discussions with Steve Foley helped to improve the manuscript. Thomas Zack contributed to the data collection at the LAM in St. John's, Newfoundland, which is kept in excellent running condition by Mike Tubrett. Financial support by the Deutsche Forschungsgemeinschaft to D.J. and B.S. is gratefully acknowledged.

References

Arndt, N., 1983. Role of a thin, komatiite-rich oceanic crust in the Archean plate-tectonic process. Geology 11, 372–375.

Bach, W., Alt, J.C., Niu, Y., Humphris, S., Erzinger, J., Dick, H.J.B., 2001. The geochemical consequences of late-stage low-grade alteration of lower oceanic crust at the SW Indian Ridge: results from ODP Hole 735B (Leg 176). Geochim. Cosmochim. Acta 65 (19), 3267–3288.

Barth, M., Rudnick, R.L., Horn, I., McDonough, W.F., Spicuzza, M., Valley, J.W., Haggerty, S.E., 2001. Geochemistry of xenolithic eclogites from West Africa: part I. A link between low

MgO eclogites and Archean crust formation. Geochim. Cosmochim. Acta 65, 1499–1527.

Barth, M., Rudnick, R.L., Horn, I., McDonough, W.F., Spicuzza, M., Valley, J.W., Haggerty, S.E., 2002. Geochemistry of xenolithic eclogites from West Africa: part II. Origins of the high MgO eclogites. Geochim. Cosmochim. Acta 66, 4325–4345.

Bickle, M.J., 1978. Heat loss from the Earth: a constraint on Archaean tectonics from the relation between geothermal gradients and the rate of plate production. Earth Planet. Sci. Lett. 40, 301–315.

Bickle, M.J., Nisbet, E.G., Martin, A., 1994. Archean greenstone belts are not oceanic crust. J. Geol. 102, 121–138.

Bindeman, I.N., Davis, A.M., Drake, M.J., 1998. Ion microprobe study of plagioclase-basalt partition experiments at natural concentration levels of trace elements. Geochim. Cosmochim. Acta 62, 1175–1194.

Brenan, J.M., Shaw, H.F., Ryerson, F.J., Phinney, D.L., 1995. Mineral-aqueous fluid partitioning of trace elements at 900C and 2.0 GPa: constraints on the trace element chemistry of mantle and deep crustal fluids. Geochim. Cosmochim. Acta 59, 3331–3350.

Clayton, R.N., Goldsmith, J.R., Karel, V.J., Mayeda, T.K., Newton, R.C., 1975. Limits on the effect of pressure on isotopic fractionation. Geochim. Cosmochim. Acta 39, 1197–1201.

Dawson, J.B., 1984. Contrasting types of upper-mantle metasomatism? In: Kornprobst, J. (Ed.), Kimberlites II: The Mantle and Crust–Mantle Relationships. Elsevier, Amsterdam, pp. 289–294.

Epstein, S., Taylor, H.P., 1967. Variation of O^{18}/O^{16} in minerals and rocks. In: Abelson, P.H. (Ed.), Researches in Geochemistry. Wiley, New York, pp. 29–62.

Falloon, T., Green, D., 1988. Anhydrous partial melting of peridotite from 8 to 35 kb and the petrogenesis of MORB. J. Petrol., 379–414 (special lithospheric issue).

Foley, S.F., Tiepolo, M., Vannucci, R., 2002. Growth of continental crust controlled by melting of amphibolite, not eclogite. Nature 417, 837–840.

Frey, F.A., Green, D.H., 1974. The mineralogy, geochemistry and origin of lherzolite inclusions in Victorian basanites. Geochim. Cosmochim. Acta 38, 1023–1059.

Ghiorso, M.S., Hirschmann, M.M., 1998. pMELTS: A Revised Calibration of MELTS for Modeling Peridotite Melting at High Pressure. EOS Trans. Am. Geophys. Union Suppl. 79 (45), F1005.

Green, T., 1994. Experimental studies of trace-element partitioning applicable to igneous petrogenesis—Sedona 16 years later. Chem. Geol. 117, 1–36.

Green, D.H., Falloon, T.J., 1998. Pyrolite: a Ringwood concept and its current expression. In: Jackson, I. (Ed.), The Earth's mantle. Cambridge Univ. Press, Melbourne, pp. 311–380.

Green, D.H., Ringwood, A.E., 1967a. An experimental investigation of the gabbro to eclogite transformation and its petrological applications. Geochim. Cosmochim. Acta 31, 767–833.

Green, D.H., Ringwood, A.E., 1967b. The genesis of basaltic magmas. Contrib. Mineral. Petrol. 15, 103–190.

Gregory, R.T., Taylor, H.P., 1981. An oxygen isotope profile in a section of cretaceous oceanic crust, Samail ophiolite, Oman: evidence for $\delta^{18}O$ buffering of the oceans by deep (>5 km) seawater-hydrothermal circulation at mid-ocean ridges. J. Geophys. Res. 86 (B4), 2737–2755.

Günther, D., Longerich, H.P., Jackson, S.E., 1996. A new enhanced sensitivity quadruple inductively coupled plasma-mass spectrometer (ICPMS). Can. J. Appl. Spectrosc. 40, 111–116.

Hart, S.R., Blusztajn, J., Dick, H.J.B., Meyer, P.S., Muehlenbachs, K., 1999. The fingerprint of seawater circulation in a 500-meter section of ocean crust gabbros. Geochim. Cosmochim. Acta 63, 4059–4080.

Harte, B., Kirkley, M.B., 1997. Partitioning of trace elements between clinopyroxene and garnet: data from mantle eclogites. Chem. Geol. 136, 1–24.

Hatton, C.J., The geochemistry and origin of xenoliths from the Roberts Victor mine. PhD thesis, University of Cape Town.

Hatton, C.J., Gurney, J.J., 1987. Roberts Victor eclogites and their relation to the mantle. In: Nixon, P.H. (Ed.), Mantle Xenoliths. Wiley, London, pp. 453–463.

Helmstaedt, H., Doig, R., 1975. Eclogite nodules from the Colorado plateau—samples of subducted Franciscan type oceanic lithosphere. Phys. Chem. Earth 9, 95–111.

Jacob, D.E., Foley, S.F., 1999. Evidence for Archean ocean crust with low high field strength element signature from diamondiferous eclogite xenoliths. Lithos 48, 317–336.

Jacob, D., Jagoutz, E., Lowry, D., Mattey, D., Kudrjavtseva, G., 1994. Diamondiferous eclogites from Siberia: remnants of Archean oceanic crust. Geochim. Cosmochim. Acta 58, 5191–5207.

Jacob, D.E., Jagoutz, E., Lowry, D., Zinngrebe, E., 1998. Comment on the origin of Yakutian eclogite xenoliths. J. Petrol. 39, 1527–1533.

Jacob, D.E., Bizimis, M., Salters, V.J.M., 2003. Lu–Hf isotopic systematics of subducted ancient oceanic crust: Roberts Victor eclogites. Contrib. Mineral. Petrol. (submitted for publication).

Jagoutz, E., Dawson, J.B., Hoernes, S., Spettel, B., Wänke, H., 1984. Anorthositic oceanic crust in the Archean Earth. 15th Lunar Planet. Sci. Conf., Houston, TX, 395–396 (abstract).

Longerich, H.P., Jackson, S.E., Günther, D., 1996. Laser ablation inductively coupled plasma mass spectrometric transient signal data acquisition and analyte concentration calculation. J. Anal. Atom. Spectr. 11, 899–904.

MacGregor, I.D., Carter, J.L., 1970. The chemistry of clinopyroxenes and garnets of eclogite and peridotite xenoliths from the Roberts Victor mine, South Africa. Phys. Earth Planet. Int. 3, 391–397.

Manton, W.I., Tatsumoto, M., 1971. Some Pb and Sr isotopic measurements on eclogites from the Roberts Victor mine, South Africa. Earth Planet. Sci. Lett. 10, 217–226.

Mattey, D., Lowry, D., MacPherson, C., 1994. Oxygen isotope composition of mantle peridotite. Earth Planet. Sci. Lett. 128, 231–241.

McDade, P., 1999. An experimental and geochemical study of the products of extreme metamorphic processes: ultrahigh-temperature granulites and the Roberts Victor eclogites. PhD thesis, University of Edinburgh. 390 pp.

McKenzie, D., Bickle, M.J., 1988. The volume and composition of melt generated by extension of the lithosphere. J. Petrol. 29, 625–679.

Nisbet, E., Fowler, C., 1983. Model for Archean plate tectonics. Geology 11, 376–379.

Niu, Y., Gilmore, T., Mackie, S., Greig, A., Bach, W., 2002. Mineral Chemistry, whole-rock compositions, and petrogenesis of Leg 176 gabbros: data and discussion. In: Natland, J.H., Dick, H.J.B., Miller, D.J., et al. (Eds.), Proc. ODP, Scientific Results, vol. 176. Online available from world wide web: http://www-odp.tamu.edu/publications/176_SR/176TOC.HTM (Cited 2002-05-30).

Norton, D., Taylor, H.P., 1979. Quantitative simulation of the hydrothermal systems of crystallizing magmas on the basis of transport theory and oxygen isotope data: an analysis of the Skaergaard Intrusion. J. Petrol. 20, 421–486.

O'Hara, M.J., Yoder, H.S., 1967. Formation and fractionation of basic magmas at high pressures. Scott. J. Geol. 3, 67–117.

Rapp, R., Watson, E., Miller, C., 1991. Partial melting of amphibolite/eclogite and the origin of Archean trondhjemites and tonalites. Precambrian Res. 51, 1–25.

Rudnick, R.L., Barth, M., Horn, I., McDonough, W.F., 2000. Rutile-bearing refractory eclogites: missing link between continents and depleted mantle. Science 287, 278–281.

Schmickler, B., 2002. Petrogenesis of eclogite xenoliths from the Zero kimberlite, Kuruman Province South Africa. PhD thesis, University of Göttingen.

Schulze, D.J., Helmstaedt, H., 1988. Coesite–sanidine eclogites from kimberlite: products of mantle fractionation or subduction? J. Geol. 96, 435–443.

Schulze, D.J., Valley, J.W., Spicuzza, M., 2000. Coesite eclogites from the Roberts Victor kimberlite, South Africa. Lithos 54, 23–32.

Shirey, S.B., Carlson, R.W., Richardson, S.H., Menzies, A., Gurney, J.J., Pearson, D.G., Harris, J.W., Wiechert, U., 2001. Archean emplacement of eclogitic components into the lithospheric mantle during formation of the Kaapvaal Craton. Geophys. Res. Lett. 28, 2509–2512.

Smyth, J.R., 1977. Quartz pseudomorphs after coesite. Am. Mineral. 62, 828–830.

Smyth, J.R., Hatton, C.J., 1977. A coesite–sanidine grospydite from the Roberts Victor kimberlite. Earth Planet. Sci. Lett. 34, 284–290.

Snyder, G.A., Taylor, L.A., Beard, B.L., Crozaz, G., Halliday, A.L., Sobolev, V.L., Sobolev, N.V., 1998. Reply to a comment by D. Jacob et al. on the origins of Yakutian eclogite xenoliths. J. Petrol. 39, 1535–1543.

Stalder, R., Foley, S.F., Brey, G.P., Horn, I., 1998. Mineral–aqueous fluid partitioning of trace elements at 900 °C–1200 °C and 3.0 GPa to 5.7 GPa: new experimental data for garnet, clinopyroxene and rutile and implications for mantle metasomatism. Geochim. Cosmochim. Acta 62, 1781–1801.

Sun, S.-S., McDonough, W.F., 1989. Chemical and isotopic systematics of oceanic basalts: implications for mantle composition and processes. In: Saunders, A.D., Norry, M.J. (Eds.), Magmatism in the Ocean Basins. Geological Society Special Publication, London, pp. 313–345.

Taylor, L.A., Neal, C.R., 1989. Eclogites with oceanic crustal and mantle signatures from the Bellsbank kimberlite, South Africa: part I. Mineralogy, petrography and whole rock chemistry. J. Geol. 97 (5), 551–567.

Yaxley, G.M., Green, D.H., 1998. Reactions between eclogite and peridotite: mantle refertilisation by subduction of oceanic crust. Schweiz. Mineral. Petrogr. Mitt. 78, 243–255.

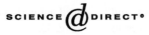

Available online at www.sciencedirect.com

SCIENCE @ DIRECT°

LITHOS

Lithos 71 (2003) 353–367

www.elsevier.com/locate/lithos

Mantle structure and composition to 800-km depth beneath southern Africa and surrounding oceans from broadband body waves

R.E. Simon[a,b], C. Wright[a,*], M.T.O. Kwadiba[a,c], E.M. Kgaswane[a,1]

[a]Bernard Price Institute of Geophysical Research, The University of the Witwatersrand, Johannesburg, Private Bag 3, Wits 2050, South Africa
[b]Department of Physics, University of Botswana, Private Bag UB704, Gaborone, Botswana
[c]Department of Geological Survey, Private Bag 14, Lobatse, Botswana

Abstract

Average one-dimensional P and S wavespeed models from the surface to depths of 800 km were derived for the southern African region using travel times and waveforms from earthquakes recorded at stations of the Kaapvaal and South African seismic networks. The Herglotz–Wiechert method combined with ray tracing was used to derive a preliminary P wavespeed model, followed by refinements using phase-weighted stacking and synthetic seismograms to yield the final model. Travel times combined with ray tracing were used to derive the S wavespeed model, which was also refined using phase-weighted stacking and synthetic seismograms. The presence of a high wavespeed upper mantle lid in the S model overlying a low wavespeed zone (LWZ) around 210- to ~ 345-km depth that is not observed in the P wavespeed model was inferred.

The 410-km discontinuity shows similar characteristics to that in other continental regions, but occurs slightly deeper at 420 km. Depletion of iron and/or enrichment in aluminium relative to other regions are the preferred explanation, since the P wavespeeds throughout the transition zone are slightly higher than average. The average S wavespeed structure beneath southern Africa within and below the transition zone is similar to that of the IASP91 model. There is no evidence for discontinuity at 520-km depth. The 660-km discontinuity also appears to be slightly deeper than average (668 km), although the estimated thickness of the transition zone is 248 km, similar to the global average of 241 km. The small size of the 660-km discontinuity for P waves, compared with many other regions, suggests that interpretation of the discontinuity as the transformation of spinel to perovskite and magnesiowüstite may require modification. Alternative explanations include the presence of garnetite-rich material or ilmenite-forming phase transformations above the 660-km discontinuity, and the garnet–perovskite transformation as the discontinuity.
© 2003 Elsevier B.V. All rights reserved.

Keywords: Body waves; Discontinuities; Kaapvaal craton; Low wavespeed zone; Transition zone

1. Introduction

In the present study, we take advantage of the high density of seismic stations and the high quality and great quantity of the database of the Kaapvaal craton seismological project to define the average body

* Corresponding author. Fax: +27-11-717-6579.
E-mail address: wrightc@geosciences.wits.ac.za (C. Wright).
[1] Present address: Council for Geoscience, Private Bag X112, Pretoria 0001, South Africa.

0024-4937/$ - see front matter © 2003 Elsevier B.V. All rights reserved.
doi:10.1016/S0024-4937(03)00121-X

wavespeed structure beneath southern Africa and surrounding oceans from broadband body waves to depths of about 800 km. This work complements the previous studies (e.g. Qiu et al., 1996; Zhao et al., 1999) by improving the resolution of deep-seated subtle features. It is also an extension of the work of Simon et al. (2002) and Wright et al. (2002). We summarize the P wavespeed results reported by Simon et al. (2002) and present an average S wavespeed model for this region. We compare the P wavespeed results with the preferred model (SATZ) of Zhao et al. (1999) because it was derived using a similar kind of data for a region just to the north of the present study area. The recent Generalised Northern Eurasia Model (GNEM) of Ryberg et al. (1998) for northern Eurasia allows a comparison with the upper mantle and transition zone below a region in a different continent. The S wavespeed results are compared with those of previous shear wave models obtained for southern Africa (Cichowicz and Green, 1992; Qiu et al., 1996; Zhao et al., 1999). A comparison with the IASP91 model of Kennett (1991a) is also provided since it is a fairly recent reference wavespeed model in which the upper mantle component is weighted by data from both oceanic and continental regions.

2. Data acquisition and selection

Data used in this study are from local and regional earthquakes and mining-induced tremors recorded by the Kaapvaal craton broadband seismic network supplemented by data from the South African network at short distances (Simon et al., 2002; Wright et al., 2002). The Kaapvaal craton network consisted of approximately 80 broadband stations that formed part of the international Kaapvaal craton programme (Carlson et al., 1996, 2000). The seismometers were deployed at locations across southern Africa from April 1997 through April 1999, with about 50 stations operating at any one time. The project involved universities in Botswana, South Africa, Zimbabwe, the mining industry in southern Africa and two American research institutions: the Carnegie Institution of Washington and the Massachusetts Institute of Technology. The digital data consist of continuous three-component recordings of seismic waves recorded at 20 samples/s. The broadband seismometers formed a

grid network with an average spacing of about 100 km (Fig. 1).

In this work, the mantle structure to depths of 800 km mainly beneath southern Africa, but with some weighting from surrounding oceans at depths below 200 km, is investigated using travel times and waveforms from body waves. We select broadband seismograms from the Kaapvaal craton experiment, concentrating on data from regional and teleseismic earthquakes at epicentral distances of less than 34° (Fig. 1). This experiment recorded many tectonic earthquakes occurring on the African continent and below the surrounding oceans, and yielded waveforms that include relevant arrivals from upper mantle triplications of the travel-time curves that are caused by both the 410- and 660-km discontinuities. Hypocentral data were obtained from the Preliminary Determination of Epicentre bulletins issued by the United States Geological Survey.

3. P wave models

The travel times used by Simon et al. (2002) are in the range 0–34°, which provides information from the surface to depths of about 800 km. These travel times (Fig. 2) comprised 15 tectonic earthquakes and 14 mining-induced tremors that occurred on the periphery of the Witwatersrand basin recorded by the Kaapvaal network, and 120 events recorded by the South African network. A detailed description of the method used to estimate the baseline correction for each event and relative origin time errors was provided by Wright et al. (2002). For events beyond the 15° distance, resolution of the 410- and 660-km discontinuities has been improved by using the phase-weighted stacking technique (Schimmel and Paulssen, 1997) to identify later P wave arrivals of different slownesses associated with triplications in the travel times. For P wave work, all seismograms were filtered between 0.4 and 4.0 Hz. The seismograms were stacked at different slownesses to enhance coherent signals and diminish the amount of random and signal-generated noise.

Because the data used by Simon et al. (2002) differed slightly at short distances from those of Wright et al. (2002), due to the addition of two extra events at distances less than 20°, the P wavespeeds in

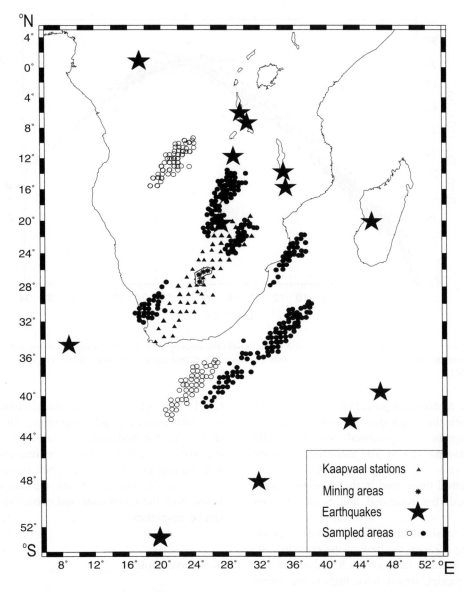

Fig. 1. Map showing locations of mine tremors and earthquakes recorded by broadband stations of the Kaapvaal craton network used in the present study. The four enclosed small stars show the main areas of mining-induced seismicity: from bottom to top—Welkom, Klerksdorp, Far West and West Rand, Central and East Rand. The map shows big stars, triangles, closed circles and open circles denoting earthquake source locations, the Kaapvaal craton broadband station locations, sampled areas above and within the transition zone and sampled areas below the transition zone, respectively. The circles define the midpoints between sources and stations.

the crust and upper mantle show small differences from model BPI1 derived by Wright et al. (2002) which extends to 320-km depth. Therefore, the revised model was denoted BPI1A. The deeper part of the average P wavespeed model was constructed by evaluating the Herglotz–Wiechert integral (Aki and

Richards, 1980, pp. 643–651) to provide a preliminary smooth model. The method of constructing the preliminary P wavespeed model and the subsequent refinements were described in Simon et al. (2002). The refinements involve ray tracing, and use of later arrival times obtained from the phase-weighted stack

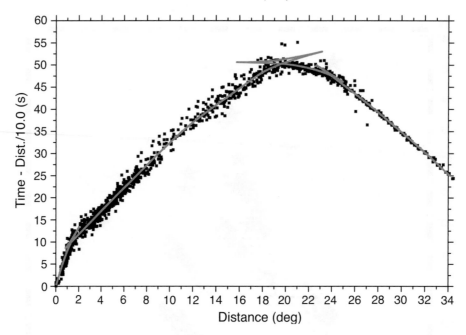

Fig. 2. Plot showing all P wave first-arrival times used in the travel-time analysis from both the Kaapvaal craton and South African networks, with baseline corrections applied, and overlain with the reduced travel-time curve from model BPI1A with $V_{red} = 10.0$ km/s (after Simon et al., 2002).

and synthetic seismograms to define the triplications associated with the major discontinuities. The synthetic seismograms were generated using the WKBJ technique (Chapman, 1978). The resulting travel-time curve is shown in Fig. 2, superimposed on all corrected first arrival times used in the travel-time analysis from both the Kaapvaal and South African networks.

According to Simon et al. (2002), the wavespeeds of BPI1A (Fig. 3) from the base of the crust to 270-km depth lie between those of the SATZ model and the IASP91 model, which have higher and lower wavespeeds, respectively. Between depths of 270 km and the 410-km discontinuity, models BPI1A, IASP91 and SATZ have similar wavespeeds. Model GNEM for Eurasia has lower wavespeeds than the other three models above the 410-km discontinuity. Within the transition zone, models BPI1A and SATZ converge as the depth increases, with wavespeeds exceeding those of IASP91 below 500-km depth. These models and model GNEM all have similar wavespeeds below 750-km depth. Both major discontinuities in BPI1A show similar patterns to those in North America reported recently by Neves et al.

(2001) from waveform inversion of broadband seismic data using a genetic algorithm. Model BPI1A shows that the transition zone average thickness is approximately 248 km. The travel times from events with turning points shown as open circles in Fig. 1 reveal that the mantle beneath the transition zone, below both the continental and oceanic regions, has similar properties.

4. S wave models

S wave first arrivals from events with body wave magnitudes between 4.0 and 5.8 have been used to investigate the S wavespeed structure of the mantle beneath the southern African region. The horizontal-component broadband recordings were rotated to radial and transverse components and then band-pass filtered from 0.2 to 2.0 Hz before picking the travel times. Fig. 4 shows all the SV travel times used from the tectonic events. Although the data are generally scattered, most of the stations have reasonably clear S wave arrivals. This scatter appears to be predominant for stations of epicentral distances between 18° and

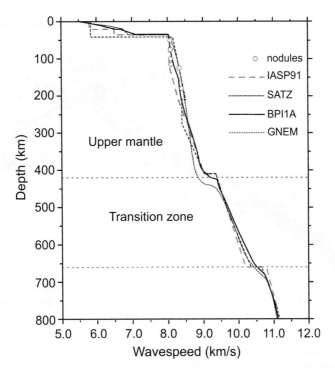

Fig. 3. Comparison of P wavespeed models: IASP91 (Kennett, 1991a), SATZ (Zhao et al., 1999), BPI1A (Simon et al., 2002) and GNEM (Ryberg et al., 1998). The gray circles denote values of P wavespeed for mantle xenoliths estimated by Qiu et al. (1996).

23.5° (Fig. 4), and may be due to complexity in the S wave structure of the mantle above the transition zone below the continental and oceanic lithospheres, or to errors in picking (i.e. phase misidentification) of the S wave arrivals due to the complexity of the P coda resulting from scattering (Kaiho and Kennett, 2000). It appears that at these distances, travel times from continental events are slightly longer than those from oceanic events. SH times show similar variations to those of Fig. 4, and fit a model not significantly different from that derived from SV arrivals. Thus, we are not able to isolate any effect due to anisotropy, and the following discussion is confined to results from SV arrivals.

Because of the large scatter in the S wave data, the Herglotz–Wiechert inversion was not performed, even though the travel times are reasonably uniformly distributed to epicentral distances of 34°. Instead, we used ray tracing to fit a model to the S wave travel times. The starting model was calculated from the P wavespeed model, assuming a constant Poisson's ratio of 0.25. Iterative ray tracing was then used to

fit a model to the travel times. The results were then compared with the travel times predicted by the IASP91 shear wave model (Fig. 4). All S wave data were adjusted for origin time errors using corrections computed for the P wave data from the same events used by Simon et al. (2002). The wavespeed model, which gave the best fit to the observed data, was taken as a smoothed approximation to the actual average S wavespeed distribution for this region. The phase-weighted stacking technique (Schimmel and Paulssen, 1997) was then used to identify later shear wave arrivals of different slownesses associated with the upper mantle triplications for both the 410- and 660-km discontinuities, thereby enabling the adjustment of the wavespeeds close to the discontinuities. In addition, synthetic seismograms were generated to assist in using the results from the phase-weighted stacking to refine the model in exactly the same manner as the P wave data. The depths of the 410- and 660-km discontinuities were constrained to lie within 10 km of the corresponding discontinuities for P waves.

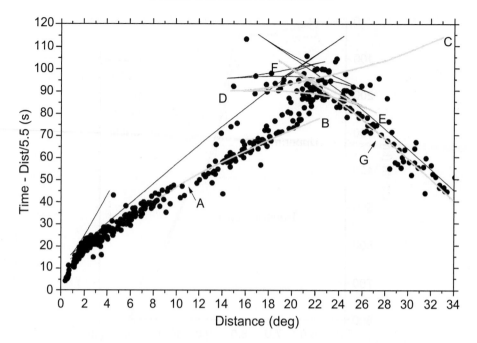

Fig. 4. Plot showing all SV first arrival times used in the travel-time analysis from the Kaapvaal craton network, with baseline corrections applied, and overlain with the reduced travel time curves from shear wave models IASP91 (dark) and BPISM (gray) with $V_{red} = 5.5$ km/s. For BPISM, travel time branches are as follows: AB—arrivals from the mantle above the LWZ; CD—refracted energy below the LWZ and wide-angle reflections from the top of the 410-km discontinuity; DE—refracted arrivals from within the transition zone; EF—wide angle reflections from the top of the 660-km discontinuity; FG—refracted arrivals from the lower mantle.

The final reduced travel-time curve for S waves is overlain on times measured from tectonic- and mining-induced events recorded by the Kaapvaal network (Fig. 4). Fig. 5 shows seismograms from six earthquakes to illustrate the rapid decrease in first arrival amplitudes of S beyond distances of 19°, which provides evidence for the presence of the low wavespeed zone (LWZ) shown in Fig. 6. The retrograde branches associated with the LWZ and 410-km discontinuity and with the 660-km discontinuity are well defined, and the width of triplication associated with the 660-km discontinuity is much greater than for P waves. The travel-time curve was computed from the average S wave model (BPISM) derived from this work, which is shown in Fig. 6. Other recently published S wave models for this region extend only to depths of 500 km, and include those from Cichowicz and Green (1992), Qiu et al. (1996) and Zhao et al. (1999). Because the models of Zhao et al. (1999), Cichowicz and Green (1992) and Qiu et al. (1996) [hereafter referred to as Vp_Poisson, CGSM92 and QSM96, respectively] are all deficient in some way,

we chose not to provide their travel times here. The problems are evident from Fig. 8 of Zhao et al. (1999), which shows that beyond the triplication associated with the 410-km discontinuity, none of the models presented in their work is consistent with their observed data. However, our shear wavespeed model shows travel times that are consistent with observed data up to 34°. Although the observed data show that there is energy arriving where the shadow zone is expected after 22.4° (Fig. 5), this is probably diffracted energy that is not explained by ray theory. The observed shear wave data in Zhao et al. (1999) and our observed data both show a consistent pattern, although in the former, the data are rather sparse beyond distances of 13°.

BPISM provides a better agreement than QSM96 between the seismic wavespeed structure and the wavespeeds estimated by Qiu et al. (1996) from peridotite xenoliths found within the Kaapvaal craton, shown as circles in Fig. 6. This agreement determined from seismology and petrology suggests that the nodules are representative of the average upper mantle

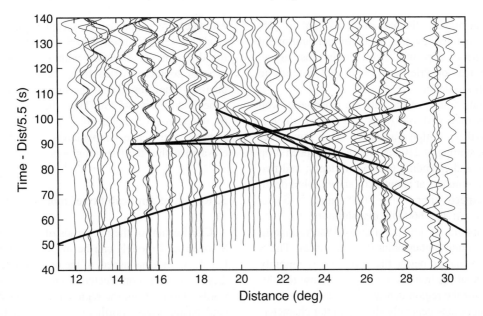

Fig. 5. Composite record section of radial component SV waveforms from tectonic events. The waveforms have been bandpass filtered from 0.2 to 2.0 Hz and overlain with the reduced travel-time curve from model BPISM (V_{red} = 5.5 km/s). The travel-time curves calculated from BPISM and also plotted in Fig. 4 have been superimposed.

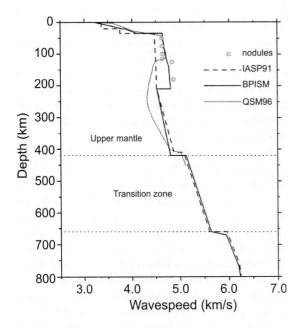

Fig. 6. Comparison of shear wavespeed models IASP91 (Kennett, 1991a), BPISM and QSM96 (Qiu et al., 1996). The gray circles denote values of shear wavespeed for mantle xenoliths estimated by Qiu et al. (1996).

composition beneath this region (Kaapvaal craton with more weighting from the adjacent mobile belts as the depth increases) to depths of 180 km. This is a departure from the QSM96 model, which limits the agreement between seismology and petrology to depths shallower than 120 km. However, our average shear wavespeed model has a LWZ from 210- to ~ 345-km depths while the LWZ extends from 100 to ~ 250, 120 to ~ 400 and 160 to ~ 350 km in CGSM92, QSM96 and Priestley (1999), respectively. Priestley's (1999) S wave model is a revised version of QSM96 to satisfy both the regional seismic waveform data and the fundamental mode Rayleigh wave data for the region.

The trough in S wavespeed defining the LWZ runs from 4.52 to 4.70 km/s in model BPISM (Fig. 6), while model Vp_Poisson has no LWZ. The LWZ in model BPISM shows a similar pattern to the LWZ in the shear wave model IASP89 (Kennett, 1991b). BPISM generates a shadow zone in the travel-time curve starting at around 22.36° and extends to at least 33.06°, where the reflected energy from below the LWZ starts (points B and C in Fig. 4). Our results from ray tracing place the base of the LWZ at ~ 345-km depth, defined by the depth of penetration of refracted energy at point C, which is in close agreement with that of Priestley (1999) that extends to around 350-km depth.

The main features of the data are reasonably well matched by model BPISM (Figs 4, 5 and 6) which is characterized by: a high wavespeed lid, a low wavespeed zone starting at 210 km with a wavespeed drop of ~ 5.8% and extending to 345-km depth; and two major discontinuities at about 420- and 665-km depth with wavespeed jumps of ~ 6.7% and ~ 5.6%, respectively. At the 660-km discontinuity, the data are not capable of unambiguously discriminating between first and higher order discontinuities. The S wavespeed structure below the 410-km discontinuity and throughout the transition zone and below to 800-km depths for the region beneath southern Africa and surrounding oceans generally shows similar characteristics to the P wavespeed distribution (Figs. 3 and 6). The major difference is the evidence for the LWZ around 210-km depth overlain by the high wavespeed upper mantle lid in the S wave model that is absent for P waves.

Above the 210-km depth, the IASP91 S wave model shows wavespeeds that are too slow to be representative of the average wavespeed structure beneath our region of study (Fig. 6). The S wave models BPISM and IASP91 have very minor differences in their wavespeeds within and below the transition zone (Fig. 6). This is evident from the 5 s measured difference (i.e. constant difference over a considerable distance range) in travel-time curves and the similarity in the travel-time slopes of the two models both within and below the transition zone. The systematic difference in travel-time curves is largely due to the upper mantle wavespeed lid in BPISM. This implies that the derived average S wavespeed model of the structure beneath southern Africa within and below the transition zone is similar to that of the global average.

Furthermore, the wavespeeds from model BPISM reveal that the transition zone is approximately 245 km in thickness (248 km from P waves), and these results are consistent with the global average thickness estimates of 241 and 243 km obtained from recent SS precursor studies by Flanagan and Shearer (1998) and Gu et al. (1998), respectively. These

estimates are supported by recent results from Gao et al. (2002) who used P-to-S conversions to image the discontinuity structure within the top 1000 km beneath southern Africa from the Kaapvaal seismic data set. They found that the mean mantle transition zone thickness is 245 km. Thus, there is little thinning of the transition zone compared with the global average, and hence temperatures are unlikely to be elevated within the transition zone.

5. Region of applicability of models and errors

Clarification of the properties of the derived P and S wavespeed models can be made with reference to Fig. 1. An unusual aspect of the present models is that they have been constructed from the surface downwards, starting from the region around the Witwatersrand basin, whose northern and western boundaries are defined approximately by the areas of mining-induced seismicity. The top 100 km of the models is broadly representative of the Kaapvaal craton, but become increasingly weighted by information from the surrounding mobile belts as the depth increases. Below depths of about 200 km, information from the oceanic regions and the more northerly regions of the African continent starts to influence the results, until, at the maximum depth of 800 km, the models are weighted averages over a broad region of southern Africa and the oceanic regions to the south. Regional differences in the deep structure are relatively small, detectable, but not accurately resolvable without more data (Simon et al., 2002).

Detailed interpretations of the derived compressional model BPI1A (Fig. 3) and its uncertainties have been described by Simon et al. (2002). The determination of the uncertainties in model BPI1A model has been achieved by using both errors in summary gradients in the fitting of a slowness curve (Bolt, 1978) and cumulative errors in baseline corrections to obtain 95% confidence limits on the wavespeed model through the Herglotz–Wiechert integral (Fig. 7). The intervals just below the Moho depth (Fig. 7), where the dark gray limits spread away from model BPI1A (black line), indicate larger bounds (± 0.12 km/s) compared with the rest of the model that shows average confidence limits of ± 0.03 km/s (Simon et al., 2002). The wider bounds below the Moho are due

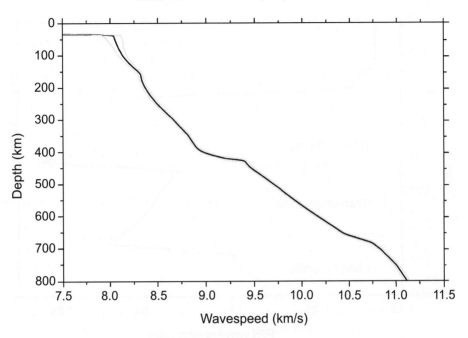

Fig. 7. 95% confidence limits for model BPI1A (dark gray lines) based on the assumption that the assumed sharpness of the 410- and 660-km discontinuities is correct in each case. The error bounds were found assuming fixed crustal wavespeeds to Moho depth (34 km), and have an average deviation of ± 0.03 km/s (after Simon et al., 2002).

to the 'hump' observed in the data and from fixing crustal wavespeeds and Moho depth. The error bounds on BPI1A are smaller than those on earlier published models (e.g. Wiggins et al., 1973; Walck, 1984), since they were computed using a probabilistic rather than an extremal method (Simon et al., 2002).

While it is relatively easy to constrain confidence limits on the P wavespeed model, the presence of a LWZ makes this very difficult for S. Because no seismic energy can reach its maximum depth within the LWZ, there is no way of reliably constraining the wavespeeds within the zone from the data alone, and the problem of non-uniqueness in this instance was discussed by Gerver and Markushevitch (1966). The actual wavespeeds within the zone have to be assumed using 'a priori' information, which is the range of petrologically acceptable models. In the present instance, the IASP91 S model (Kennett, 1991a) has been used to constrain the S wavespeeds in the low wavespeed zone.

Reliable statistical estimates of the errors in depths of the 410- and 660-km discontinuities are not easily derived. Perhaps the simplest way of approaching the

problem is to stress that a change in depth of either discontinuity of 5 or 10 km for P or S, respectively, results in an observable misfit of the times to those estimated by ray tracing. 5 and 10 km for P and S are therefore approximate values for uncertainties in the depths, subject to the constraint that the errors in the overlying wavespeeds are small.

6. P/S wavespeed ratios

Direct calculations of the P/S wavespeed ratio as a function of depth are particularly valuable, since the main features of the P and S wavespeed distributions were established from the same events (Fig. 8). There are few results for this ratio, which provides insight into temperature variations, partial melting and phase transformations. The reason is that in most parts of the world, P and S models have been constructed by different methods, and using different distributions of events and stations. We estimate the errors in the wavespeed ratio to be about ±0.01 for the upper mantle above the LWZ and the uppermost lower

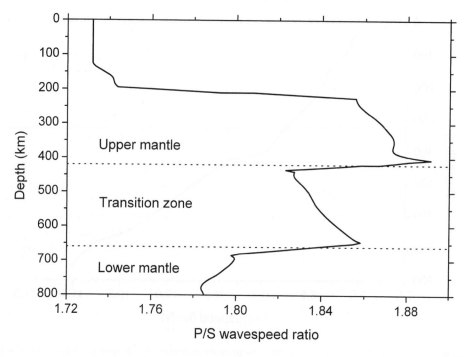

Fig. 8. The wavespeed ratio between P and S waves (P/S) as a function of depth calculated from models BPI1A and BPISM with a moderate amount of smoothing.

mantle, ±0.03 for the LWZ and ±0.02 for the transition zone. From the top of the mantle to depths of 200 km, the wavespeed ratio increases slightly from 1.73 to 1.75, indicating very similar gradients for both P and S waves. This ratio then increases within and below the LWZ from 1.86 at 210 km to 1.92 at 420 km, although the S wavespeeds are poorly constrained from depths of 210 to 420 km. The LWZ is not detected for P waves, and, in such a situation, the LWZ for S is due to temperature gradients alone without partial melting.

At 420-km depth, where the first major discontinuity occurs, the P/S ratio sharply drops from 1.92 to 1.80. However, within the transition zone, the ratio gradually increases from 1.83 to 1.87, indicating a minor reduction in shear wave gradients relative to P. At the base of the transition zone, the ratio sharply drops from 1.87 to 1.79. The sharp drop in ratio upon passing through both boundaries marking the major discontinuities is probably caused by changes in crystal structure, since the increase in wavespeed at these boundaries is due to phase changes inferred from experimental petrology. The P/S ratio stabilizes

to values of around 1.79 for the uppermost lower mantle, which is consistent with results from Kennett et al. (1994) who obtained a similar value from their broadband observations for the structure beneath northern Australia. The main features in the P/S wavespeed ratios deduced from our work (Fig. 8) are very similar to those in Kennett et al. (1994) for depths below the 410-km discontinuity through the transition zone down to the lower mantle.

7. The low wavespeed zone in the upper mantle

Our results reflect evidence of a LWZ in S waves in agreement with other studies across southern Africa (e.g. Bloch et al., 1969; Cichowicz and Green, 1992; Qiu et al., 1996; Priestley, 1999; Stankiewicz et al., 2002). However, our model shows greater thickness of the seismic lithosphere (~ 210 km thick) than other studies if the top of the LWZ defines the base of the lithosphere. But these results are inconsistent with those reported by other investigators (Frey-bourger et al., 2001; James et al., 2001; Gao et al.,

2002) whose results show no evidence of a LWZ for S waves within the Kaapvaal craton. However, we observe from the regions where the rays turn or bottom (Fig. 1) that the LWZ in model BPISM appears to be located outside the craton (i.e. on the cratonic margins and adjacent mobile belts). This probably explains the inconsistency of the results, since sub-cratonic and adjacent continental regions may be different. Our average models reveal that the use of the IASP91 model as a reference model for southern Africa is inadequate in the uppermost mantle, since our models show high wavespeeds compared with the global average. This has important consequences for teleseismic travel-time tomography based on a modified IASP91 model (e.g. results from James et al., 2001). Our results confirm suggestions by Kearey and Vine (1996) that a LWZ is present at a variety of depths between about 80 km and below 300 km, and that it appears to be generally present for S waves beneath stable continental regions, except, possibly, below some Archaean cratons. It may, however, be absent in certain regions for P waves, especially beneath stable continental regions where temperature gradients are lower.

The low seismic wavespeeds could arise from a number of different mechanisms, including an anomalously high temperature, a phase change, a compositional change, the presence of open fluid-filled cracks or fissures and partial melting (Kearey and Vine, 1996). Liebermann and Schreiber (1969) found that critical thermal gradients based upon ultrasonic laboratory data for a great variety of minerals confirm that it is possible to have a LWZ for shear waves without requiring one for compressional waves. Birch (1969) suggested that the LWZ in the upper mantle coincides with a region of increasing iron content. However, Liebermann and Schreiber (1969) argued that there is no compelling evidence that an increase in iron content occurs at depths of less than 400 km. They further point out that in order for the LWZ to be a consequence of partial melting, the LWZ for P and S waves must occur at the same depth. If the S wavespeed decrease is caused by a decrease in the shear modulus, the P wavespeed could decrease at a lesser rate or perhaps not at all, provided that the bulk modulus is increasing fast enough with depth (Dowling and Nuttli, 1964). Anderson and Sammis (1970) suggested that the

LWZ should begin and end gradually if it is due to temperature effects. Hence, as a result of these ambiguities, it would be unwise to rule out any of the plausible explanations of the LWZ because they can also occur as a combination.

We interpret the high wavespeeds that persist to depths of ~ 210 km in the shear wave model to indicate that the minimum lithosphere thickness beneath southern Africa is ~ 210 km. However, we note that the LWZ corresponds to the lithosphere beneath the cratonic margins and adjacent mobile belts. Therefore, the cratonic lithosphere could be greater than 210 km thick. This result is consistent with the findings of Ritsema et al. (1998) that the cratonic lithosphere beneath the Tanzania craton is at least 200 km thick. They further argue that the cratonic lithosphere is tectonically "stable" to the extent that it has not been completely destroyed by extensional tectonics that have altered the lithosphere of the younger surrounding mobile belts. A relatively thick (>200 km) cratonic lithosphere is also consistent with the low heat flow from the craton (e.g. Nyblade et al., 1990). Furthermore, our estimate of the lithospheric thickness is comparable to that observed in other regions where ultra-deep diamond inclusions are found, such as the Slave craton in Canada. From their geochemical results, O'Reilly et al. (2001) found that the lithosphere–asthenosphere boundary beneath the Slave craton lies at 200–220-km depth, consistent with the magnetotelluric studies of Jones et al. (2001). The analysis of available Rayleigh wave dispersion data indicates that Precambrian shield areas (such as the Kaapvaal craton) have a deep, relatively rigid, high wavespeed root extending to depths of at least 200 km (Calcagnile, 1991).

The high wavespeeds observed in the upper mantle lid probably indicate a characteristic feature of this region (depleted peridotite), since similar results have been reported from different data sets. These high wavespeeds do not result from neglecting anisotropy because low-frequency surface wave observations (Nataf et al., 1984; Montagner, 1994) show weak anisotropy in the upper mantle beneath southern Africa. The LWZ is of major importance to plate tectonics as it represents a low viscosity layer along which relative movements of the lithosphere and asthenosphere can be accommodated (Kearey and Vine, 1996).

8. Phase transformations

The possible phase transformations in the mantle are shown in Fig. 9. Because the depth of the 410-km discontinuity in model BPI1A closely matches that of the global average, this discontinuity can probably be explained by the transformation of olivine to the β-phase (e.g. Jeanloz and Thompson, 1983). However, the depth and form of the discontinuity are influenced by the pyroxene–garnet transformation, because the two transformations interact through iron partitioning (Weidner and Wang, 2000). The phases that are stable at any depth are influenced by temperature and by variations in aluminium and iron content. As the temperature increases in a mantle of pyrolite composition, the depth of the discontinuity increases, while increases in aluminium and decreases in iron content have the same effect. The preferred depth of the discontinuity (defined by the point of inflection in the wavespeed–depth profile) is 420 km, with an even greater depth suggested by Zhao et al. (1999) for a region of southern Africa to the north of the present study. The slightly greater depth of the discontinuity compared with other regions suggests higher aluminium content and/or iron depletion as the major con-

tributors, since the P wavespeeds are also marginally higher throughout the transition zone (Simon et al., 2002).

Around depths of 550 km, the β-phase transforms to the spinel structure and this transformation is accompanied by a slight increase in seismic wavespeed gradients in some regions, but is not resolved in the present data (Fig. 9). The most distinctive feature of the 660-km discontinuity compared with other regions is the relatively small size of the jump in wavespeed for P, defined by the small lateral extent of the triplication for P, in which the differences in slowness between the two refracted branches are also small compared with the triplication produced by the 410-km discontinuity (Fig. 2). A similar small jump in the P wavespeed at the discontinuity is also present in the results of Zhao et al. (1999) for southern Africa. However, the triplication produced by the 660-km discontinuity appears more pronounced for S (Figs. 4 and 5). Whether this is an effect produced by the 660-km discontinuity itself or is due to differences in frequency content between the high-frequency P arrivals and low-frequency S arrivals remains unclear.

Below the 660-km discontinuity, the dominant phase is perovskite. There are several possible phase

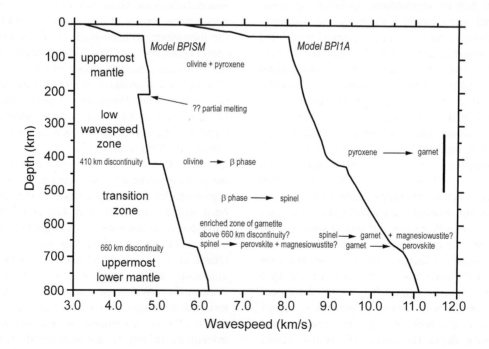

Fig. 9. Possible phase transformations occurring in the mantle superimposed on models BPISM and BPI1A for southern Africa.

transformations which can lead to this (Weidner and Wang, 2000). Garnet may transform partially to ilmenite and then to perovskite, while the spinel (γ) phase may also transform completely to garnet and magnesiowüstite, followed by the garnet to perovskite transformation. Ringwood (1991) interpreted the 660-km discontinuity as a transformation from spinel to perovskite and magnesiowüstite. However, Irifune et al. (1998) reported new data on the breakdown of spinel to perovskite in synthetic forsterite, from which Weidner and Wang (2000) inferred that the spinel form of olivine breaks down first to garnet and magnesiowüstite in a pyrolite transition zone. The garnet would then transform to perovskite at a greater pressure to form the 660-km discontinuity. The magnitude of the 660-km discontinuity is also influenced by the aluminium content (Weidner and Wang, 2000).

Ringwood (1994) discussed the possible accumulation of subducted former oceanic crust to form a gravitationally stable layer of garnetite some 50 km thick on top of the 660-km discontinuity: a process that may have been most effective during the Archaean when mantle temperatures were higher. This would result in an increase in P and S wavespeed gradients towards the base of the transition zone with a consequent reduction of the wavespeed jump at the 660-km discontinuity. The present P and S wavespeed models provide some support for such a process having occurred beneath southern Africa. However, Weidner and Wang (2000) have emphasized that ilmenite-forming transformations could be mistaken for a subducted slab resting on top of the 660-km discontinuity.

9. Conclusions

Observations from broadband body waves confirm the presence of resolvable differences between P and S wavespeed variations above the transition zone. We find evidence of a high wavespeed upper mantle lid in the S wavespeed model overlying a LWZ around 210- to ~ 345-km depth that is not resolved in the P wavespeed model. The LWZ is representative of the structure beneath the mobile belts adjacent to the Kaapvaal craton, and is most probably a temperature effect. Recent results reported by Stankiewicz et al. (2002) from P-to-S converted waves suggest that a reduced wavespeed gradient zone does exist beneath the Kaapvaal craton, although its depth has remained enigmatic. We estimate the minimum thickness of the lithosphere beneath the periphery of the Kaapvaal craton to be ~ 210 km. Hence, the lithosphere could be thicker than 210 km beneath the craton, which is consistent with other Precambrian areas. The base of the LWZ beneath southern Africa, which occurs at ~ 345 km from our S wave model, is consistent with that at ~ 350 km from Priestley's (1999) results that satisfy both the regional seismic waveform data and the fundamental mode Rayleigh wave data for southern Africa. In the upper mantle, both P and S wave models are in good agreement with wavespeeds estimated from peridotite xenoliths found within the Kaapvaal craton. The sampled area at lithospheric depths (Fig. 1) is dominated by the continental mantle, so that the upper parts of the models are largely representative of the continental lithosphere and similar to models derived for other continental regions.

The 410-km discontinuity shows similar characteristics to that in other continental regions, but appears to be depressed slightly to depths of 420 km. Depletion of iron and/or enrichment in aluminium relative to other regions rather than anomalously high temperatures are the preferred explanation, since the P wavespeeds throughout the transition zone appear to be slightly higher than average. The average shear wavespeed structure beneath southern Africa within and below the transition zone is similar to that of the IASP91 model. There is no evidence for a discontinuity at 520-km depth.

The 660-km discontinuity also appears to be slightly deeper than average (668 km), although the estimated thickness of the transition zone is 248 km for P, which is only marginally greater than the global average of 241 km (Flanagan and Shearer, 1998) or 243 km (Gu et al., 1998). The most interesting feature of the results is the small size of the 660-km discontinuity for P waves compared with many other regions, which suggests that the interpretation of the discontinuity as the transformation of spinel to perovskite and magnesiowüstite may require modification. Possibilities are the presence of a garnetite layer sitting on top of the 660-km discontinuity, ilmenite-forming phase transformations and explanation of the discontinuity as the garnet–perovskite transformation.

Acknowledgements

This research was supported by funding from the National Research Foundation (South Africa), the National Science Foundation (USA), the University Research Council of the University of the Witwatersrand and by mining companies operating in southern Africa. The University of Botswana sponsored R.E. Simon. M.T.O. Kwadiba received financial support from the Government of Botswana and from the Kellogg Foundation (USA), and E.M. Kgaswane from the National Research Foundation and the Council for Geoscience. We thank those who made the Kaapvaal project successful. We thank Alan Jones (Geological Survey of Canada) and two anonymous reviewers for their helpful suggestions towards improving this manuscript.

References

Aki, K., Richards, P.G., 1980. Quantitative Seismology: Theory and Methods, vol. 2. Freeman, San Francisco, pp. 559–932.

Anderson, D.L., Sammis, C., 1970. Partial melting in the upper mantle. Phys. Earth Planet. Inter. 3, 41–50.

Birch, F., 1969. Density and composition of the upper mantle: first approximation as an olivine layer. In: Hart, P.J. (Ed.), The Earth's Crust and Upper Mantle. AGU Monograph, vol. 13, pp. 18–36.

Bloch, S., Hales, A.L., Landisman, M., 1969. Velocities in the crust and upper mantle of southern Africa from multi-mode surface wave dispersion. Bull. Seismol. Soc. Am. 59, 1599–1629.

Bolt, B.A., 1978. Summary value smoothing of physical time series with unequal intervals. J. Comput. Phys. 29, 357–369.

Calcagnile, G., 1991. Deep structure of Fennoscandia from fundamental and higher mode dispersion of Rayleigh waves. Tectonophysics 195, 139–149.

Carlson, R.W., Grove, T.L., de Wit, M.J., Gurney, J.J., 1996. Program to study crust and mantle of the Archean craton in southern Africa. EOS, Transactions, American Geophys. Union 77, 273 and 277.

Carlson, R.W., Boyd, F.R., Shirey, S.B., Janney, P.E., Grove, T.L., Bowring, S.A., Schmitz, M.D., Dann, J.C., Bell, D.R., Gurney, J.J., Richardson, S.H., Tredoux, M., Menzies, A.H., Pearson, D.G., Hart, R.J., Wilson, A.H., Moser, D., 2000. Continental growth, preservation and modification in southern Africa. GSA Today 10, 1–7.

Chapman, C.H., 1978. A new method for computing synthetic seismograms. Geophys. J. R. Astron. Soc. 54, 481–518.

Cichowicz, A., Green, R.W.E., 1992. Tomographic study of upper mantle structure of the South African continent, using waveform inversion. Phys. Earth Planet. Inter. 72, 276–285.

Dowling, J., Nuttli, O., 1964. Travel-time curves for a low veloc-ity channel in the upper mantle. Bull. Seismol. Soc. Am. 54, 1981–1996.

Flanagan, M.P., Shearer, P.M., 1998. Global mapping of topography on transition zone velocity discontinuities by stacking SS precursors. J. Geophys. Res. 103, 2673–2692.

Freybourger, M., Gaherty, J.B., Jordan, T.H., Kaapvaal Seismic Group, 2001. Structure of the Kaapvaal craton from surface waves. Geophys. Res. Lett. 28, 2489–2492.

Gao, S.S., Silver, P.G., Liu, K.H., Kaapvaal Seismic Group, 2002. Mantle discontinuities beneath southern Africa. Unpublished manuscript.

Gerver, M., Markushevitch, V., 1966. Determination of a seismic wave velocity from the travel time curve. Geophys. J. R. Astron. Soc. 11, 165–173.

Gu, Y., Dziewonski, A.M., Agee, C.B., 1998. Global de-correlation of the topography of transition zone discontinuities. Earth Planet. Sci. Lett. 157, 57–67.

Irifune, T., Nishiyama, N., Kuroda, K., Inoue, T., Isshiki, M., Utsumi, W., Funakoshi, K., Urakawa, S., Uchida, T., Katsura, T., Ohtaka, O., 1998. The postspinel phase boundary in Mg_2SiO_4 determined by in situ x-ray diffraction. Science 279, 1698–1700.

James, D.E., Fouch, M.J., VanDecar, J.C., van der Lee, S., Kaapvaal Seismic Group, 2001. Tectospheric structure beneath southern Africa. Geophys. Res. Lett. 28, 2485–2488.

Jeanloz, R., Thompson, A.B., 1983. Phase transitions and mantle discontinuities. Rev. Geophys. 21, 51–74.

Jones, A.G., Ferguson, I.J., Chave, A.D., Evans, R.L., McNeice, G.W., 2001. The electric lithosphere in the Canadian shield. J. Geophys. Res. 103, 15269–15286.

Kaiho, Y., Kennett, B.L.N., 2000. Three-dimensional seismic structure beneath the Australasian region from refracted wave observation. Geophys. J. Int. 142, 651–668.

Kearey, P., Vine, F.J., 1996. Global Tectonics, 2nd ed. Blackwell, Oxford. 333 pp.

Kennett, B.L.N. (Ed.), 1991a. IASPEI 1991 Seismological Tables. Research School of Earth Sciences. Australian National University, Canberra, Australia. 167 pp.

Kennett, B.L.N., 1991b. Seismic velocity gradients in the upper mantle. Geophys. Res. Lett. 18, 1115–1118.

Kennett, B.L.N., Gudmundsson, O., Tong, C., 1994. The upper mantle S and P velocity structure beneath northern Australia from broadband observations. Phys. Earth Planet. Inter. 86, 85–98.

Liebermann, R.C., Schreiber, E., 1969. Critical thermal gradients in the mantle. Earth Planet. Sci. Lett. 7, 77–81.

Montagner, J.P., 1994. Can seismology tell us anything about convection in the mantle? Rev. Geophys. 32, 115–137.

Nataf, H.C., Nakanishi, I., Anderson, D.L., 1984. Anisotropy and shear-velocity heterogeneties in the upper mantle. Geophys. Res. Lett. 11, 109–112.

Neves, F.A., Singh, S.C., Priestley, K., 2001. Velocity structure of the upper mantle discontinuities beneath North America from waveform inversion of broadband seismic data using a genetic algorithm. J. Geophys. Res. 106, 21883–21895.

Nyblade, A.A., Pollack, H.N., Jones, D.L., Podmore, F., Mushayandebvu, M., 1990. Terrestrial heat flow in east and southern Africa. J. Geophys. Res. 95, 17371–17384.

O'Reilly, S.Y., Griffin, W.L., Djomani, Y.P., Natapov, L.M., Pearson, N.J., Davies, R.M., Doyle, B.J., Kivi, K., 2001. The mantle beneath the Slave craton (Canada): composition and architecture. Extended Abstract. The Slave–Kaapvaal Workshop, September 5–9, 2001, Merrickville, Ontario, Canada. Geological Survey of Canada, Ottawa. 5 pp.

Priestley, K., 1999. Velocity structure of the continental upper mantle: evidence from southern Africa. Lithos 48, 45–56.

Qiu, X., Priestley, K., McKenzie, D., 1996. Average lithospheric structure of southern Africa. Geophys. J. Int. 127, 563–587.

Ringwood, A.E., 1991. Phase transformations and their bearing on the constitution and dynamics of the mantle. Geochem. Cosmochim. Acta 55, 2083–2110.

Ringwood, A.E., 1994. Role of the transition zone and 660 km discontinuity in mantle dynamics. Phys. Earth Planet. Inter. 86, 5–24.

Ritsema, J., Nyblade, A.A., Owens, T.J., Langston, C.A., VanDecar, J.C., 1998. Upper mantle seismic velocity structure beneath Tanzania, east Africa: implications for the stability of cratonic lithosphere. J. Geophys. Res. 103, 21201–21213.

Ryberg, T., Wenzel, F., Egorkin, A.V., Solodilov, L., 1998. Properties of the mantle transition zone in northern Eurasia. J. Geophys. Res. 103, 811–822.

Schimmel, M., Paulssen, H., 1997. Noise reduction and detection of weak, coherent signals through phase-weighted stacks. Geophys. J. Int. 130, 497–505.

Simon, R.E., Wright, C., Kgaswane, E.M., Kwadiba, M.T.O., 2002. The P wavespeed structure below and around the Kaapvaal craton to depths of 800 km, from traveltimes and waveforms of local and regional earthquakes and mining-induced tremors. Geophys. J. Int. 151, 132–145.

Stankiewicz, J., Chevrot, S., van der Hilst, R.D., de Wit, M.J., 2002. Crustal thickness, discontinuity depth, and upper mantle structure beneath southern Africa: constraints from body wave conversions. Phys. Earth Planet. Inter. 130, 235–251.

Walck, M.C., 1984. The P wave upper mantle structure beneath an active spreading center: the Gulf of California. Geophys. J. R. Astron. Soc. 76, 697–723.

Weidner, D.J., Wang, Y., 2000. Phase transformations: implications for mantle structure. In: Karato, S., Forte, A.M., Liebermann, R.C., Masters, G., Stixrude, L. (Eds.), Earth's Deep Interior. Mineral Physics and Tomography from the Atomic to the Global Scale. Geophysical Monograph, vol. 117. American Geophysical Union, Washington, DC, pp. 215–235.

Wiggins, R.A., McMechan, G.A., Toksöz, M.N., 1973. Range of earth structure nonuniqueness implied by body wave observations. Rev. Geophys. Space Phys. 11, 87–113.

Wright, C., Kwadiba, M.T.O., Kgaswane, E.M., Simon, R.E., 2002. The structure of the crust and upper mantle to depths of 320 km beneath the Kaapvaal craton, from P wave arrivals generated by regional earthquakes and mining-induced tremors. J. Afr. Earth Sci. 35, 477–488.

Zhao, M., Langston, C.A., Nyblade, A.A., 1999. Upper mantle velocity structure beneath southern Africa from modeling regional seismic data. J. Geophys. Res. 104, 4783–4794.

Available online at www.sciencedirect.com

SCIENCE  DIRECT°

ELSEVIER

Lithos 71 (2003) 369–392

LITHOS

www.elsevier.com/locate/lithos

South African seismicity, April 1997 to April 1999, and regional variations in the crust and uppermost mantle of the Kaapvaal craton

C. Wright[a,*], E.M. Kgaswane[a,1], M.T.O. Kwadiba[a,b], R.E. Simon[a,c], T.K. Nguuri[a,2], R. McRae-Samuel[a]

[a] Bernard Price Institute of Geophysical Research, University of the Witwatersrand, Johannesburg, Private Bag 3, Wits 2050, South Africa
[b] Department of Geological Survey, Private Bag 14, Lobatse, Botswana
[c] Department of Physics, University of Botswana, Private Bag UB704, Gaborone, Botswana

Abstract

Events induced by deep gold-mining activity on the edge of the Witwatersrand basin dominate the seismicity of South Africa. The deployment of 54 broad-band seismic stations at 84 separate locations across southern Africa between April 1997 and April 1999 (Kaapvaal network) enabled the seismicity of South Africa to be better defined over a 2-year period. Seismic events located by the South African national network, and by localized seismic networks deployed in mines or across gold-mining areas, were used to evaluate earthquake location procedures and to show that the Kaapvaal network locates mining-induced tremors with an average error of 1.56 ± 0.10 km compared with 9.50 ± 0.36 km for the South African network. Travel times of seismic events from the mines recorded at the Kaapvaal network indicate regional variations in the thickness of the crust but no clearly resolved variations in seismic wavespeeds in the uppermost mantle. Greater average crustal thicknesses (48–50 km compared with 41–43 km) are observed in the northern parts of the Kaapvaal craton that were affected by the Bushveld magmatism at 2.05 Ga. Estimates of average crustal thickness for the southern part of the Kaapvaal craton from receiver functions (38 km) agree well with those from refracted arrivals from mining-induced earthquakes if the crustal thicknesses below the sources are assumed to be 40–43 km. In contrast, the average crustal thickness inferred from refracted arrivals for the northern part of the Kaapvaal craton is larger by about 7 km (51 km) than that inferred from receiver functions (44 km), suggesting a thick mafic lower crust of variable seismic properties due to variations in composition and metamorphic grade. Pn wavespeeds are high (8.3–8.4 km/s), indicating the presence of highly depleted magnesium-rich peridotite throughout the uppermost mantle of the craton. Seismic Pg and Sg phases indicate that the upper crust around the Witwatersrand basin is comparatively uniform in composition when averaged over several kilometres.
© 2003 Elsevier B.V. All rights reserved.

Keywords: Crustal thickness; Kaapvaal craton; Mining-induced earthquakes; Composition of crust; Composition of uppermost mantle

* Corresponding author. Fax: +27-11-717-6579.
E-mail address: wrightc@geosciences.wits.ac.za (C. Wright).
[1] Present address: Council for Geoscience, Private Bag X112, Pretoria 0001, South Africa.
[2] Present address: PREPCOMM CTBTO, Vienna International Centre, PO Box 1250, A-1400, Vienna, Austria.

1. Introduction

The Archaean Kaapvaal craton and adjacent Proterozoic mobile belts cover much of southern Africa (Fig. 1). Natural seismicity of southern Africa (Fig. 2)

0024-4937/$ - see front matter © 2003 Elsevier B.V. All rights reserved.
doi:10.1016/S0024-4937(03)00122-1

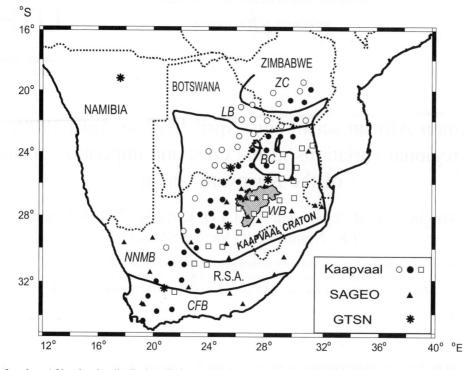

Fig. 1. Map of southern Africa showing distribution of seismic stations and main tectonic elements. BC = Outer boundary of surface outcrops of Bushveld complex; CFB = Cape Fold belt; LB = Limpopo belt; NNMB = Namaqua–Natal mobile belt; WB = Witwatersrand basin; ZC = Zimbabwe craton. Open and closed circles and squares denote stations of Kaapvaal broad-band network that operated from April 1997 to April 1998, April 1997 to April 1999 and April 1998 to April 1999, respectively. Triangles denote stations of South African network, and asterisks indicate broad-band stations of the Global Telemetered Seismic Network, or long-period stations.

is fairly typical of stable continental regions, with some areas of pronounced neotectonic activity (Andreoli et al., 1996), and with a largest instrumentally determined earthquake magnitude within South Africa of 6.3 (Green and Bloch, 1971). Almost a century ago, however, Wood (1913) reported that there was considerable seismicity in the Johannesburg area resulting from gold mining on the margin of the Witwatersrand basin. Such seismicity has persisted until the present day and, as mining operations move to greater depths, continues to pose a significant safety hazard to workers; it is monitored by networks of seismic instruments deployed in individual mines or distributed across particular goldfields (Mendecki, 1997).

A feasibility study on the use of mine tremors in South Africa for determining crustal structure was pioneered more than 50 years ago by Gane et al. (1946). P and S waves from these mining-induced events were subsequently used to determine the wave-

speed structure of the crust and uppermost mantle around the Witwatersrand basin (Willmore et al., 1952; Gane et al., 1956; Hales and Sacks, 1959; Durrheim and Green, 1992; Durrheim, 1998a). However, only recently have such events been used in conjunction with local and regional tectonic events to determine earth structure from the surface to depths of 320 km (Wright et al., 2002).

There have been few seismological studies of earth structure within southern Africa compared with many other areas of the world such as Australia, North America and western Europe, so that there is a need to define regional travel time curves and the structure of the crust and upper mantle to better understand cratonic evolution and to enable earthquakes to be more accurately located.

Between April 1997 and April 1999, a network of 54 broad-band seismic instruments was deployed at 84 different locations across southern Africa (Carlson et al., 1996) as part of the international Kaapvaal

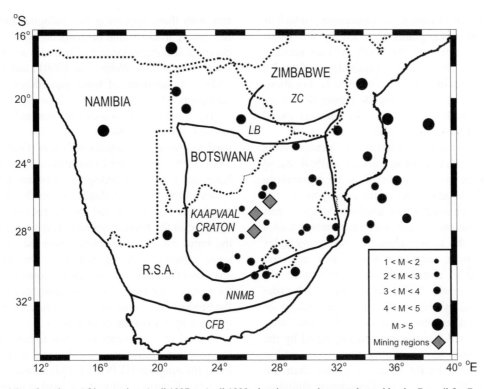

Fig. 2. Seismicity of southern African region, April 1997 to April 1999, showing tectonic events located by the Council for Geoscience. CFB, LB, NNMB and ZC are described in the caption to Fig. 1. The most seismically active areas of gold mining are, from north to south: Far West and West Rand, Klerksdorp and Welkom.

craton program (Fig. 1). The main objective of this seismic experiment was to derive tomographic images of the crust and mantle to depths of about 700 km using recordings of teleseismic earthquakes (i.e., earthquakes at distances greater than 3300 km). However, the network also recorded local (distances less than 300 km) and regional earthquakes (distances between 300 and 3300 km), thus enabling more detailed information on the seismicity of southern Africa to be obtained over a 25-month period.

The purpose of the present work is to show how a temporary network of three-component, broad-band seismometers deployed across southern Africa supplements the South African National seismic network in defining seismicity and in improving the accuracy of the determination of catalogued seismic events. The temporary network is then used to define average P and S travel time curves and wavespeed models for the southern African region to improve earthquake locations. Locations of mining-induced

earthquakes from mine seismic networks provide a large number of seismic events distributed over five distinct mining regions that are sufficiently accurate to be regarded as equivalent to explosive sources used in crustal refraction studies in other parts of the world. Regional differences in the structure of the crust and uppermost mantle of the Kaapvaal craton are then defined through analysis of the seismograms of these mining-induced events. These differences are interpreted with information from receiver functions to provide constraints on deep-seated petrological variations.

2. Sources of information on earthquakes

Information on the locations and magnitudes of earthquakes within South Africa, neighbouring countries and adjacent oceanic regions, for the period April 1997 to April 1999, comes from bulletins published by the Council for Geoscience, which is derived from seismograms recorded by the short-period seismometers of the South African network (Graham, 1997, 1998, 1999). Magnitudes are from 1.0 for some mining-induced events upwards. Locations of most events of local magnitude greater than 4 are also listed in preliminary earthquake bulletins issued by the United States Geological Survey.

The South African seismic network locates events using a limited number of stations (typically 4–8), so that epicentres are not expected to be particularly accurate. The large number of three-component seismic stations deployed across southern Africa as part of the international Kaapvaal experiment enables relocation of the larger seismic events (magnitudes >2.5) with more data, and with a new regional reference model and travel times (Simon et al., 2002; Wright et al., 2002). To compare earthquake locations made by the South African and Kaapvaal networks, accurate locations of events by some other means is required. Seismic networks operated by the mining industry across mining regions or within individual mines locate far more events than are published in the Council for Geoscience Bulletins (Graham, 1997, 1998, 1999), and for which the location errors in most cases are less than 400 m (Webb et al., 2001). Catalogues of earthquakes prepared by the mining industry can therefore be used to evaluate the relative location accuracies of mining-induced events by the South African and Kaapvaal networks, for which the location errors are expected to be greater than a kilometre. The resulting estimates of location errors will then give insight into the errors in locating tectonic earthquakes away from the mining areas, which will be used in future for studying earth structure.

3. Correlations and database

The production of a comprehensive catalogue of seismic events for South Africa of magnitude greater than 2 for the period April 1997 to April 1999, and the development of a database of waveforms of events listed in the Council for Geoscience bulletins recorded by the Kaapvaal network, was required as the first stage in this work. The procedures used for correlating events reported in the Council for Geoscience bulle-

tins with those appearing in catalogues prepared by the mining industry, and for reducing all locations to a common coordinate system of latitudes and longitudes were described by Webb et al. (2001) and Kgaswane (2002). A summary of these procedures, and some comments on the results are as follows.

The Council for Geoscience operated 29 seismic stations throughout South Africa (Fig. 1) that were used to produce earthquake bulletins (Graham, 1997, 1998, 1999) during the 25 months of operation of the Kaapvaal network. Catalogues of events from seven mines in the Far West Rand, two in the West Rand, one in the East Rand, and for the regional networks operated in the Klerksdorp and Welkom areas formed the input data (not complete) from the gold-mining areas, which are shown in Fig. 3. The steps in the analysis of the data, the search for common events in the Council for Geoscience and mine catalogues and the production of a database for future research were outlined by Kgaswane et al. (2002).

A total of 429 events common to both mine and Council for Geoscience catalogues was relocated using the program HYPOELLIPSE (Lahr, 1989), data from the Kaapvaal network, a local reference travel time curve for P that was used to derive earth model BPI1 of Wright et al. (2002) and a similar travel time curve for S whose derivation is discussed in Section 4. Assuming that errors in locations from mine catalogues could be neglected, the average errors in locations from the South African network and from the Kaapvaal network were 9.50 and 1.56 km, respectively (Kgaswane, 2002; Kgaswane et al., 2002).

Table 1 summarises the most important results from the catalogue correlations and relocations. Data from the mine catalogues were cut off below a moment magnitude of 1.7. The magnitude range of the 1578 correlated events is 1.7–4.5 from the mine catalogues and 1.4–5.1 for the local magnitudes of the Council bulletins (Graham, 1997, 1998, 1999). A combined total of 4414 events in the gold-mining regions was located with a minimum magnitude of 1.1 in the Council bulletins, and 1.7 in the mine catalogues. The largest event in the mine catalogues had a magnitude of 4.5, but the same event in the Council bulletins was given a magnitude of 5.1. The magnitudes from the mine catalogues are systematically lower at high magnitudes and similar or slightly higher at low magnitudes than those in the Council

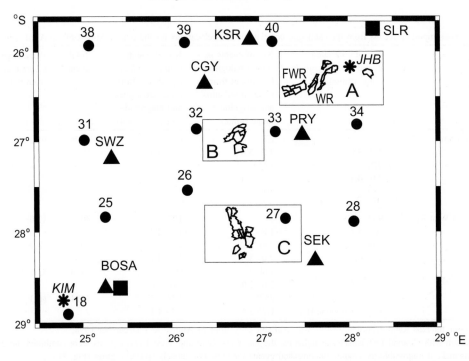

Fig. 3. Map of gold-mining areas on the margin of the Witwatersrand basin. (A) Far West Rand, West Rand and Central and East Rand from west to east. (B) Klerksdorp. (C) Welkom. KIM: Kimberley. FWR: Far West Rand. JHB: Johannesburg. WR: West Rand. Numbers and circles denote stations of the Kaapvaal network. Three- or four-letter codes and triangles denote stations of the South African network. Permanent broad-band or long-period stations are shown as squares.

bulletins, and the reasons for these differences were discussed by Kgaswane et al. (2002).

The catalogue of events summarised in Table 1 shows that over 4400 seismic events with mine magnitudes greater than 1.7 occurred during the period of the Kaapvaal experiment, of which over 4000 have been located with accuracies better than 400 m. Many of the smaller events of magnitudes less than 2.5 are located with errors less than 100 m. These events are all in the central region of the Kaapvaal craton. The merged catalogue is clearly incomplete, since separate catalogues were not available for many seismically active mines in the Far West Rand, West Rand and Central and East Rand regions. Furthermore, 16.5% of the events in mining regions that appeared in the Council catalogues were absent from the mine catalogues, including the regional networks for Klerksdorp and Welkom (8.9% and 33.5%, respectively). Thus, periods during which both the regional seismic networks and the networks in individual mines were not functioning also contribute to

the lack of completeness of the merged catalogue. Only 42 tectonic events were located by the South African network during the period April 1997 to April 1999 (Fig. 2), of which 18 were outside South Africa (Webb et al., 2001), though many more tectonic events within southern Africa were recorded by the Kaapvaal network. These events are undergoing further analysis, so that they can be used for tomographic imaging of the crust and uppermost mantle of the region.

4. Reference travel time curves

Fig. 4(a) shows part of a reference P wave travel time curve plotted as reduced time to distances of 1200 km, derived from travel times of earthquakes within South Africa, adjacent countries and surrounding oceans. The time data from both the South African and Kaapvaal networks are superimposed and, at short distances, consist predominantly of times from events

Table 1
Summary of the most important results from the catalogue correlations and relocations for the period April 1997 to April 1999, inclusive

Region	No. of correlated events and magnitude range (mine)	No. of events in mine catalogues, but absent from Council bulletins, and magnitude range (mine)	No. of events in Council bulletins, but absent from mine catalogues, and magnitude range (Council)	Total events	Percentage of events in Council bulletins absent from mine catalogues
East and Central Rand[a]	6	27	63	96	91.3
	1.7–3.0	1.7–3.0	1.8–3.8		
Far West Rand	773	1849			
	2.0–4.1	1.8–3.4			
West Rand	159	105			
	2.0–3.4	2.0–3.1			
Far West Rand and West Rand[b]	932	1954	127	3013	12.0
	2.0–4.1	1.8–3.4	1.3–3.2		
Klerksdorp	503	433	49	985	8.9
	2.0–4.0	2.0–3.3	1.1–4.1		
Welkom	137	114	69	320	33.5
	2.0–4.5	2.0–2.9	2.0–3.8		
All regions	1578[c]	2528	308	4414	16.3
	1.4–5.1 (Council)	1.7–3.4	1.1–4.1		

[a] Mine catalogue for East Rand Proprietary Mine was available only from 24 August to 1 December 1998.

[b] For comparison with Council for Geoscience bulletins, the Far West Rand and West Rand regions were combined, because errors in Council locations make it impossible to separate uncorrelated events for these two closely spaced regions (Fig. 3).

[c] One event in Council for Geoscience bulletins corresponds to two events at the Elandsrand mine closely separated in space and time, so that there were only 1577 correlated events in the Council for Geoscience catalogues.

in gold-mining areas recorded at stations in and around the Witwatersrand basin. These data were derived before accurate locations of mining-induced events were available, so that the epicentres of all events within South Africa and surrounding countries and oceans were obtained from the Council for Geoscience earthquake bulletins (Graham, 1997, 1998, 1999), while epicentres of more distant events were taken from the United States Geological Survey Preliminary Determinations of Epicenters. The numerical and statistical methods used to derive the reference travel time curve were described by Wright et al. (2002). The most important features of the data are the 'hump' in the reduced travel time curve with a peak at a distance of 230 km, and the relatively large scatter of the data beyond distances of 200 km, where the expected first arrivals have paths through rocks with upper mantle wavespeeds.

Fig. 5 shows two average P wavespeed models to depths of 180 km that fit the data according to different criteria (Wright et al., 2002). The first model BPI1 was chosen as the simplest, using the criterion

that the wavespeed gradients must be positive throughout the mantle. The peak in the reduced travel time curve at a distance of 230 km (Fig. 4(a)) was assumed to result from failure to identify the Pn arrivals at distances between about 170 and 240 km. Beyond 240 km distance, the Pn signals were believed to become larger, so that the Pn arrival was picked more frequently, thus pulling the smooth curve down towards the time at which it should be observed, merging with the expected times at distances near 400 km. The difficulty with model BPI1 is that it gives an average crustal thickness of 34 km, which is systematically lower by about 6 km than the average over the same region estimated from the receiver functions of Nguuri et al. (2001), but is in satisfactory agreement with crustal thicknesses estimated by earlier workers using mining-induced events (Willmore et al., 1952; Gane et al., 1956; Hales and Sacks, 1959; Durrheim and Green, 1992). Model BPI2 (Fig. 5) was constructed in an attempt to reconcile the differences in the average thickness of the crust estimated from refracted arrivals and from receiver functions, and to

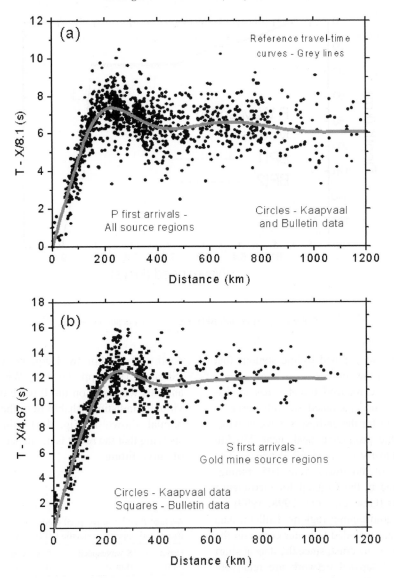

Fig. 4. (a) Reference P wave travel time curve for the southern African region plotted as reduced times to a distance of 1200 km, together with the data from which it was derived. (b) Corresponding S wave travel time curve for the southern African region plotted as reduced time, together with the data from which it was derived.

explain some of the anomalous features of the travel times (Wright et al., 2002). It has a region of seismic wavespeeds of about 8.0 km/s between depths of 36 and 47 km, whose petrological nature (crust or mantle?) remains uncertain, with a relatively high wavespeed of 8.23 km/s at depths between 47 and 65 km. A low wavespeed zone between depths of about 65 and 125 km with an average wavespeed of 8.1 km/s is then required to prevent the travel times computed from the model from becoming too early at distances greater than 1000 km. Since the most of the seismic sources are concentrated in the gold-mining areas of the central part of the Kaapvaal craton, it is not clear if the requirement for a low wavespeed zone in the one-dimensional model BPI2 (Fig. 5) arises from lateral variations in structure as the margins of

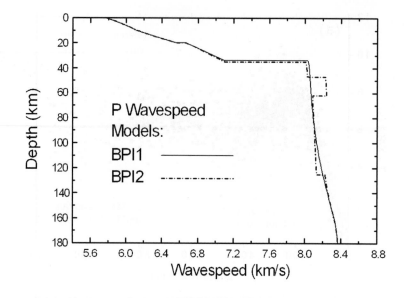

Fig. 5. P wavespeed models BPI1 and BPI2 to a depth of 180 km.

the cratonic nucleus are sampled by the stations that are more distant from the sources.

Fig. 4(b) shows the reduced travel times to distances of 1200 km and calculated smooth curve to distances of 1040 km for the earliest S wave arrivals, which also show the prominent peak near 230 km distance that was clearly resolved for P waves. The input data consist of 506 times from 105 mining-induced events listed in the Council for Geoscience earthquake bulletins (Graham, 1997, 1998, 1999) for earthquakes in the gold-mining regions. Bulletin data for both P and S were included to better constrain the seismic wavespeeds in the crust, since the data at short distances from the Kaapvaal network are relatively sparse. The remaining S wave data consist of times from eight accurately located mining-induced events (see Section 5). The times were corrected for origin time errors in a similar manner to the P wave data (Wright et al., 2002).

A smooth curve was then fitted through the combined first arrival times (Sg and Sn) times to distances of 1000 km (Fig. 4(b)) by the method of summary values (Jeffreys, 1937; Bolt, 1978). A cubic smoothing spline was then fitted through the summary points using the algorithm of Reinsch (1967) to give both a reference S wave travel time curve and a slowness curve that was used to derive a wavespeed model to

depths of 40 km by Herglotz–Wiechert inversion (Shearer, 1999, pp. 66–67). When this method of analysis was used on the P wave data, the result was virtually indistinguishable from the 14th degree polynomial shown in Fig. 4(a) (Simon et al., 2002), showing that the result is not dependent on the method of curve fitting.

Table 2
Average P and S wavespeed models of the crust for the Kaapvaal region and uppermost mantle (BPI1) for southern Africa[a]

Depth (km)	S wavespeed (km/s)	P wavespeed (km/s) (range)	Poisson's ratio
0	3.462	5.594–5.927	0.190–0.241
5.0	3.553	5.903–6.133	0.216–0.247
10.0	3.659	6.133–6.286	0.224–0.244
15.0	3.752	6.385–6.454	0.236–0.245
20.0	3.806	6.591	0.250
20.0	3.926	6.800	0.250
27.0	3.954	6.950	0.261
34.0	3.982	7.100	0.271
34.0	4.639	8.035	0.250
40.0	4.647	8.049	0.250
50.0	4.653	8.059	0.250
60.0	4.658	8.067	0.250

[a] The P wavespeeds listed are a small modification of model BPI1 called BPI1A, due to the addition of two events and construction of a new model (Simon et al., 2002).

The smooth model was then modified by comparison with the equivalently derived smooth P wavespeed model, to give the shear-wave equivalent to model BPI1 of Wright et al. (2002) with discontinuities at depths of 20 and 34 km (Table 2). This model has an increase in Poisson's ratio from 0.19 at the surface to 0.27 at the crust–mantle boundary. The fitting of an average S model for the upper mantle is subject to the same problems encountered for P due an increase in slowness with increasing distance (Fig. 4). The average model for S (BPI1) is not shown here, but it is incorporated into the work of Simon et al. (2003). The average models for P and S at mantle depths are for the southern African region, and therefore represent a weighted average for the Kaapvaal craton and surrounding mobile belts. The next step is to focus on resolving regional differences in seismic structure with emphasis on the Kaapvaal craton.

5. Regional P and S wave times across southern Africa

Earliest P arrival times from 20 accurately located mining-induced events for each of regions A, B and C of Fig. 3 were picked from seismograms bandpass-filtered between 0.4 and 4.0 Hz, and the data for region A are shown in Fig. 6. The data for each source region were divided into two groups according to source–receiver paths: those corresponding to paths predominantly through the undisturbed southern part of the Kaapvaal craton and the Namaqua–Natal and Cape Fold belts (Fig. 7), and those with paths through the northern regions in which the original Archaean lower crust and possibly the uppermost mantle may have been altered by the Bushveld event at 2.05 Ga (Nguuri et al., 2001). Times are significantly longer for the northern stations to distances of about 650 km but become closer to those for southern stations at larger distances. The data to the south show an offset or upward curvature at distances greater than 650 km. Wright et al. (2002) suggested that this effect could be produced by a low wavespeed layer in the upper mantle at depths between 65 and 125 km. An alternative explanation is a significant decrease in seismic wavespeeds in the upper mantle across the southern margin of the Kaapvaal craton.

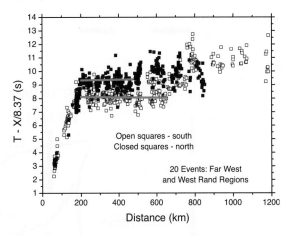

Fig. 6. Reduced travel times for P wave first arrivals for 20 accurately located mining-induced earthquakes in the Far West and West Rand regions (Fig. 3). Regression lines through the Pn arrivals at stations in the northern and southern areas of the craton (Fig. 7) are the upper and lower lines, respectively.

Regression lines have been fitted through the Pn times from the northern and southern regions for distances beyond 200 km for stations that lie within the Kaapvaal craton, and the results are shown in Table 3. From the slopes of these lines, the estimated average seismic wavespeeds in the uppermost mantle for the three separate source regions are in the range 8.33–8.43 and 8.30–8.32 km/s for the northern and southern regions, respectively, suggesting only minor differences in mantle wavespeeds between the southern region and the northern region that includes the Bushveld complex.

Average P wavespeed models for the northern and southern parts of the craton were derived from the data of Fig. 6 by ray tracing for a spherical earth. No attempt was made to revise the P wavespeeds in the crust, the values from BPI1A (Table 3; Simon et al., 2002) being used. There is at present no clear evidence for differences in wavespeeds in the upper crust for different parts of the craton, and wavespeeds in the lower crust are not well resolved. The thickness of the crust was varied between 34 and 52 km, keeping the wavespeeds at 20 km depth and at the crust–mantle boundary constant at 6.80 and 7.10 km/s with a constant wavespeed gradient between these depths. Upper mantle wavespeeds with low wavespeed gradients below the crust–mantle transition, but with average values close to 8.33 and 8.31 km/s were used

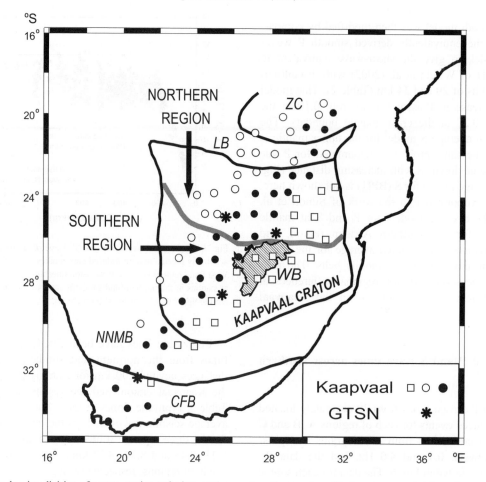

Fig. 7. Map showing division of source–station paths into northern and southern regions. Circles, squares, asterisks and two- to four-letter abbreviations have the same meaning as in Fig. 1.

for the northern and southern regions, respectively. The best model fits to the data are shown in Fig. 8 for events for the Far West and West Rand regions and were chosen to be those that produced a good fit to the regression lines of Table 3. Reduced travel time curves are shown for the best fitting models and for

Table 3
Regression lines of form $T = a + bx$ for Pn and Sn for southern and northern regions of the Kaapvaal craton

Description	a (s)	b^{-1} (km/s)	Distance range (km)	S.D. on residuals (s)	No. of observations	V (km/s, corrected for curvature of Earth)
P North FWR	9.293 ± 0.15	8.386 ± 0.032	205–474	0.48	205	8.333 ± 0.032
P North KLE	9.182 ± 0.12	8.468 ± 0.027	228–553	0.47	204	8.415 ± 0.022
P North WEL	9.498 ± 0.15	8.485 ± 0.022	340–648	0.47	208	8.432 ± 0.022
P South FWR	8.143 ± 0.12	8.360 ± 0.020	228–652	0.46	249	8.308 ± 0.020
P South KLE	7.659 ± 0.16	8.357 ± 0.029	200–555	0.65	209	8.305 ± 0.029
P South WEL	7.997 ± 0.16	8.365 ± 0.032	201–488	0.65	262	8.313 ± 0.032
S North (all)	16.148 ± 0.48	4.826 ± 0.030	211–648	1.12	50	4.796 ± 0.030
S South (all)	14.343 ± 0.57	4.857 ± 0.035	200–646	1.41	58	4.827 ± 0.035

FWR, KLW and WEL denote the Far West and West Rand, Klerksdorp and Welkom mining regions, respectively.

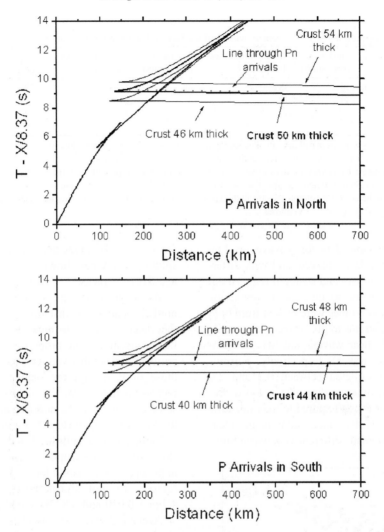

Fig. 8. Average P wavespeed models for the northern (top) and southern (bottom) regions of the Kaapvaal craton, derived from 20 mine tremors in the Far West and West Rand regions.

models with crustal thicknesses 4 km thicker and thinner. The average crustal thicknesses in the northern and southern regions are 50 and 44 km, respectively, with an estimated precision of about 1 km in each case.

Table 4 shows the separate results for the three main source regions, indicating a mean thickness for paths from the Witwatersrand basin to the northern and southern parts of the craton (Fig. 7) of about 50 and 42–43 km, respectively. These thicknesses are strongly weighted by the thicknesses in the source regions, and they tend to be larger than those estimat-

ed from receiver functions, especially for the northern part of the craton (Kwadiba et al., 2003). Using estimates of crustal thicknesses at each station, assuming the same thickness at each source region, there is a systematic difference of 2.5 ± 0.30 km (34 stations) and 2.2 ± 0.35 km (31 stations), respectively, between results from the West Rand and Klerksdorp and from the West Rand and Welkom source regions. This establishes that the crustal thicknesses are 2.5 and 2.2 km less below the Klerksdorp and Welkom regions, respectively, than beneath the Far West and West Rand (Kwadiba et al., 2003).

Table 4
Average crustal thicknesses in the northern and southern regions of the Kaapvaal craton estimated from refracted P and S arrivals

Source region and wave type	Thickness (km, North)	Minimum thickness (km, North)	Thickness (km, South)	Minimum thickness (km, South)	No. of events used	Pg–Pn cross-over distance (km)[a]
Far West and West Rand, Pn	50	48	44	42	20	207 N, 181 S
Klerksdorp, Pn	49	47	41	39	20	198 N, 169 S
Welkom, Pn	51	49	42	40	20	205 N, 176 S
All regions, Sn	52	–	44	–	8	216 N, 188 S

[a] The cross-over distances were estimated from the Pg, Sg, Pn and Sn times used in this paper. However, the cross-over distances in Figs. 8 and 10 are at larger distances. The reason is that the travel time curves in these figures correspond to the models of Table 2, which were derived from events for which accurate locations were not available, and for which it was not possible to reliably separate Pg and Pn or Sg and Sn arrivals at distances between about 170 and 220 km. The estimates of crustal thickness are not seriously biased by this effect, since it is the average wavespeed within the crust that controls the crustal thicknesses. Nevertheless, the need to refine the models of Table 2 using only data from accurately located mining tremors is recognized.

Fig. 9 shows the same data for S waves as Fig. 4(b), with the bulletin data removed, and with different symbols for the northern and southern regions. All seismograms were rotated to give radial and transverse components, and S times were picked from both sets of rotated seismograms to give travel times of SV (vertically polarised shear waves) and SH (horizontally polarised shear waves), respectively. Seismograms were bandpass-filtered between 0.2 and 2.0 Hz prior to picking times, noting that a lower frequency band than for P was required to maximize the clarity of the S arrivals. Since there is no clear evidence for any systematic differences between times

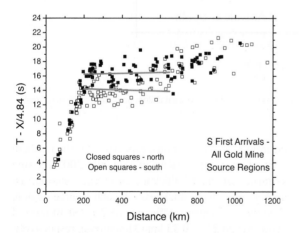

Fig. 9. Reduced travel times for S wave first arrivals for accurately located mining-induced earthquakes divided into two regions: stations in the southern part of the Kaapvaal craton, and in the northern part of the Kaapvaal craton overlying deep crust and uppermost mantle influenced by the Bushveld magmatism (Fig. 7). The upper and lower regression lines are fitted through the northern and southern Sn arrivals, respectively.

for SV and SH waves that might indicate shear-wave splitting, the times shown are the average of the SV and SH time picks.

As for P, times are significantly longer for the northern stations to distances of about 650 km, but they become closer to those for southern stations at longer distances. The properties of the S data for the northern and southern regions are generally similar to those of P, though the scatter is greater. Separate regression lines were fitted through the times measured at stations in the northern and southern parts of the Kaapvaal craton (Fig. 7) for distances greater than 200 km, and the results are listed in Table 3. From the slopes of these lines, approximate seismic wavespeeds in the uppermost mantle for the northern and southern regions are not significantly different, and are 4.796 ± 0.030 and 4.827 ± 0.035 km/s, respectively.

Average S wavespeed models for the northern and southern parts of the craton (Fig. 10) were derived from the data of Fig. 9 in exactly the same way as the P wave models of Fig. 6. Reduced travel time curves are shown for the best fitting models and for models with crustal thicknesses 4 km thicker and thinner. The estimated average crustal thicknesses in the northern and southern regions are 52 and 44 km, respectively, with an estimated precision of about 2 km in each case and are slightly higher than the values derived from P waves (Table 4).

6. Accuracy of origin times

Since the average crustal thicknesses imply values at individual stations greater than those from receiver

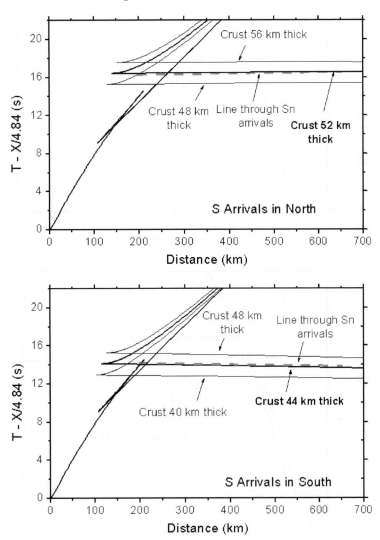

Fig. 10. Average S wavespeed models for the northern (top) and southern (bottom) regions of the Kaapvaal craton, derived from mine tremors in the Far West and West Rand (3), Klerksdorp (3) and Welkom (2) regions.

functions (see Section 8), we must determine if there is any bias in origin times that might affect the estimates of crustal thicknesses from refracted arrivals. The origin times provided in earthquake bulletins were corrected by reference to a travel time curve derived from times of P arrivals (Pg phase) at stations of both the South African and Kaapvaal networks (Simon et al., 2002; Wright et al., 2002). The model derived from the travel time curve (Table 2) does have relatively low near-surface wavespeeds, which could result from inaccurate and biased epicentre locations

at short distances. The average wavespeeds of near-surface rocks in the Witwatersrand basin area could be as high as 6.0 km/s if crack porosity is low (e.g., Green and Chetty, 1990). The estimates of crustal thicknesses have been computed again using the higher P wavespeeds of Table 3, which provide a 'maximum wavespeed' model (Fig. 11); it yields a P travel time curve that produces an average forward shift in origin times of 0.46 s. The result is faster times for Pg and Pn than before, but the effect on Pg is greater than for Pn. The consequence is that all

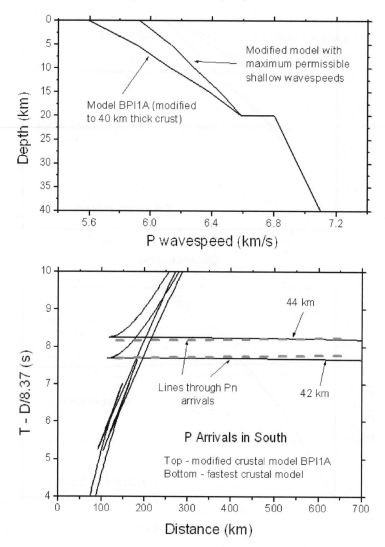

Fig. 11. Top: P wavespeed model BPI1A and modified model with 'maximum possible' wavespeeds in the upper crust. Moho is at 40 km depth. Bottom: Reduced travel times for 'best fit' models through data, using both wavespeed models shown above.

estimates of crustal thickness are reduced by 2 km (Fig. 11 and Table 4), which can be regarded as 'minimum thicknesses'.

A check on the suitability of the P reference travel time curve was obtained using the S travel times, since they were all corrected using data from P arrivals. The S times of Figs. 4b and 9 provide travel time curves that yield acceptable wavespeeds at shallow depths (Table 5). Furthermore, reducing the travel times at short distances by 0.46 s, results in a slight increase in slope with increasing distance to get zero time at zero offset, suggesting that a revised correction of 0.46 s to origin times is too large. The S times thus confirm that use of times computed from the higher wavespeed model of Table 2 will yield minimum estimates of crustal thicknesses.

7. Regional seismic phases

At distances greater than about 200 km, the most commonly observed later phases (i.e., signals that

Table 5
Regression lines of form $T = a + bx$ for Pg and Sg for the Kaapvaal craton (distance ranges 0–200 and 200–670 km for P, and 0–200 and 200–1170 km for S)

Description	a (s)	b^{-1} (km/s, average V)	S.D. on residuals (s)	No. of observations
Pg (0–200 km)	0.683 ± 0.20	6.214 ± 0.054	0.46	52
Pg (200–670 km)	0.290 ± 0.53	6.178 ± 0.050	1.15	56
Sg (0–200 km)	-0.037 ± 0.45	3.544 ± 0.039	1.03	50
Sg (200–1170 km)	0.192 ± 0.30	3.634 ± 0.0071	1.67	236

arrive later than the earliest arrivals that traverse the mantle, Pn and Sn) are denoted Pg and Sg, which are waves that travel entirely within the upper and middle crust. They are observed in the southern African region to distances of about 700 and 1200 km, respectively (Fig. 12). Figs. 13 and 14 show specimen seismograms for P and S arrivals, respectively, for a mine tremor in the Klerksdorp area on 13 September 1997. Fig. 13 shows a clear Pn arrival followed by a large, clear Pg phase about 7 s later at a distance of 364.6 km recorded by a station south of the Limpopo belt in eastern Botswana. Later Pg phases of this clarity are relatively uncommon. Fig. 14 shows a weak Sn arrival at a distance of 661.3 km, identifiable only on the transverse component, followed by a strong Sg arrival some 26 s afterwards. The station is on the Zimbabwe craton, near the southwestern margin. It is easier to identify the Sg arrivals at distances greater than 200 km, which tend to have clearer onsets than the Pg phase, resulting in the larger number of Sg observations in Fig. 12.

The slopes of the Pg and Sg curves can be used to constrain P and S wavespeed models for the upper and middle crust, provided there are sufficient data. Table 5 shows the results of fitting two regression lines through the Pg and Sg times for the eight events used to determine the Sn wavespeeds: (i) at distances less than 200 km, where they are the earliest P or S arrivals, and (ii) at greater distances where they are later phases and the onsets less clearly defined. The scatter of the data at distances beyond 200 km is greater, as indicated by the standard deviations on the residuals to the linear fits. The results show that there is no clearly resolvable increase in wavespeed with increasing distance for Pg or Sg either as first arrivals or as later arrivals, though for Sg the average value of 3.634 ± 0.0071 km/s for the later arrivals between distances of 200 and 1170 km is significantly greater

than the value of 3.544 ± 0.039 km/s for short distances. The data of Table 5 yield values of Poisson's ratios of 0.262 and 0.235 for short distances (< 200 km) and long-distance observations of Pg and Sg, respectively. Unfortunately, the difference between these two estimates of Poisson's ratio for the upper crust is too large to enable much to be inferred about

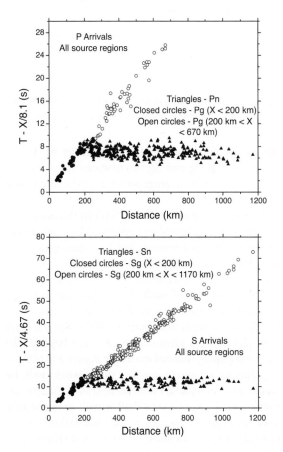

Fig. 12. Observations of the crustal phases, Pg and Sg, plotted as reduced travel times with Pn and Sn, for the eight events used in Fig. 9.

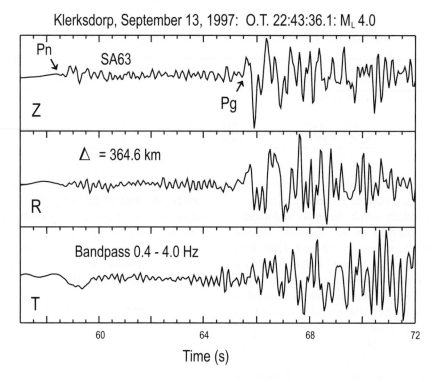

Fig. 13. Seismograms from a mine tremor at Klerksdorp on 13 September 1997 to illustrate observed P phases. Z, R and T denote vertical, radial and transverse components of motion, respectively. Zero of time scale is from the start of the file, and not from the origin time of the event.

the average composition. However, further work on a larger suite of Pg and Sg observations may assist in better defining the average structure and composition of the upper crust.

The mantle phases PmP and SmS (reflections from the crust–mantle boundary) generated by earthquakes are often large in amplitude but have emergent arrivals. Although, in principle, they can be used to estimate crustal thicknesses in continental regions, reliable estimates of such thicknesses are rare. An exception is the study of PmP and crustal thicknesses by Richards-Dinger and Shearer (1997). Several clear PmP phases were identified at distances beyond 110 km from mining-induced events by Kgaswane (2002), but due to emergent onsets, more work will be required to get reliable estimates of crustal thicknesses from them. A clear suite of PmP arrivals at one station from an array of sources in the Klerksdorp region yielded a crustal thickness of 40–42 km (Kgaswane, 2002), in good agreement with the results from receiver functions in the vicinity of the Witwatersrand basin.

8. Regional variations in the deep crust and uppermost mantle

Variations in the thickness of the crust for the Kaapvaal craton and surrounding regions estimated from receiver functions were described by Nguuri et al. (2001), who found much greater thicknesses for the northern region (Fig. 7) compared with those of the southern region of the Kaapvaal craton and the Zimbabwe craton. Useful inferences on the nature of the crust can be made by comparing estimates of crustal thicknesses using Pn or Sn times at individual stations with the results from receiver functions (Kwadiba et al., 2003).

The general features of the Archaean crust and uppermost mantle inferred in the present work are slightly different from those of Durrheim and Green (1992) (35–40 km deep Moho overlain by a crust–mantle transition zone of thickness 1–3 km) below the Witwatersrand basin and those of earlier workers (Willmore et al., 1952; Gane et al., 1956; Hales and Sacks, 1959). There are several reasons for this. The

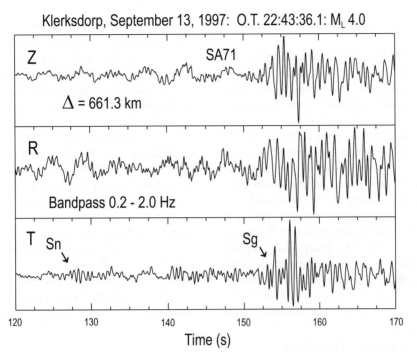

Fig. 14. Seismograms from a mine tremor at Klerksdorp on 13 September 1997 to illustrate observed S phases. Z, R and T denote vertical, radial and tranverse components of motion, respectively. Zero of time scale is from the start of the file, and not from the origin time of the event.

average cross-over distances at which the Pn phase takes over from Pg as the first arrival are now placed at 198–207 and 169–181 km, respectively, for the northern and southern regions, with the lowest and highest values for the Klerksdorp and West Rand regions where the crust below the sources is thinnest and thickest, respectively, (Table 4). Equivalent values, averaged for all three source regions, for the distances at which Sn takes over from Sg are 216 and 188 km, respectively. Previous work by Durrheim and Green (1992) placed the cross-over between Pg and Pn at about 160–170 km for the Witwatersrand basin area (southern part of craton) and suggested an upper mantle wavespeed of 8.1 km/s. The explanation offered for this discrepancy is that the older work did not use arrivals beyond distances of about 300 km, making it difficult to accurately resolve the Pn wavespeeds. An underestimate of the upper mantle wavespeed will also result in crustal thickness estimates that are too low (about 1 km for an error of 0.1 km/s). The present work supports a crustal thickness of 40–43 km and an upper mantle P wavespeed of 8.3–8.4

km/s in the Witwatersrand basin area (Kwadiba et al., 2003).

In the Witwatersrand basin region, seismic reflection profiling shows that the deep crust is no more reflective than the upper crust and there is no identifiable reflection Moho (Durrheim, 1998b), while receiver functions indicate a strong impedance contrast at an average depth of about 38 km. The wavelengths of reflected seismic signals from seismic reflection surveys are around 200–250 m, indicating that large juxtaposed bodies of different compositions due to extensive crustal mixing give rise to relatively weak reflections in the undisturbed regions of the Kaapvaal craton. On the other hand, the wavelengths of the signals that show strong P-to-S conversions (receiver functions) are around 12–15 km. This would require a major change in impedance (bulk composition and/or metamorphic grade) with depth over a depth range of less than about 3 km. This suggests a change from rocks of intermediate composition in the lower crust to highly depleted peridotite over this depth range, with little or no mafic material sandwiched in between, as proposed by Durrheim and Mooney (1994).

The use of the Pn times to estimate thicknesses of the crust at individual stations is described by Kwadiba et al. (2003), using the assumption of a small bias in origin times to yield minimum crustal thicknesses. If the thickness of the crust in the Witwatersrand basin region is greater than 40 km on the average, the Pn and Sn arrivals imply average crustal thicknesses for 19 stations in the southern region of the craton of 38.07 ± 0.85 km, in agreement with the results from receiver functions, which yield an average of 37.58 ± 0.70 km for the same 19 stations. This is in a region where the P-to-SV conversions in the receiver functions are clear (Fig. 15). These results support the interpretation of the Moho in the southern part of the craton as a relatively sharp transition from rocks of intermediate composition in the lower crust to highly depleted peridotite. However, the average crustal

thicknesses inferred from Pn paths to 19 selected northern stations is 50.52 ± 0.88 km, compared with 43.58 ± 0.57 km: some 7 km greater than the average inferred from receiver functions. The greater average crustal thickness determined from refracted arrivals is attributed to the presence of mafic material introduced by underplating. In this northern region, however, receiver functions vary in clarity with estimated thicknesses varying from 38 to 50 km (Fig. 15). How can such differences between two methods of estimating crustal thicknesses be explained by a petrological model of the lower crust?

Hurich et al. (2001) concluded that seismic wavespeed and density in the upper crust, neglecting the effects of open fractures, are dominated by composition. In the middle and lower crust, however, composition and metamorphic grade are equally important. We make use of their conceptual model of continental crust to explain the differences between crustal prop-

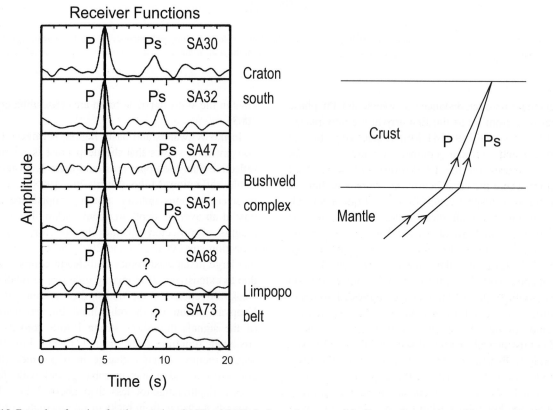

Fig. 15. Examples of receiver functions: stations SA30 and SA32, in the southern part of the Kaapvaal craton, have clear converted S arrivals (Ps); SA47 and SA51, in the Bushveld complex, have complicated receiver functions in which the Ps arrival from the crust–mantle boundary is less clearly defined; SA68 and SA73 lie in the Limpopo belt, where the receiver functions give no clear indication of crustal thickness.

erties for the northern parts of the Kaapvaal craton inferred by receiver functions and refracted arrivals.

Seismic wavespeeds for P and S in the uppermost mantle appear to be similar on the average below the northern and southern parts of the craton. There is also the suggestion that the P wavespeeds are slightly higher in the northern region by about 0.08 km/s (Table 3). The average values for P are in the range 8.3–8.4 km/s, and there is no clear evidence for significant differences in more localised regions (Kwadiba et al., 2003). Such high wavespeeds for Pn indicate that the uppermost mantle consists of a highly depleted, magnesium-rich peridotite across most of the Kaapvaal craton, including the areas affected by the Bushveld magmatism. These values are similar to those obtained for other cratons, particularly the recent values of Langston et al. (2002) for the Tanzanian craton. Most importantly, the processes that resulted in the thick crust beneath the Bushveld complex and much of the remaining northern regions of the Kaapvaal craton clearly had little effect on the peridotites of the uppermost mantle.

Both Hynes and Snyder (1995) and Hurich et al. (2001) have indicated that it is possible for significant portions of the lower continental crust to have wavespeeds indistinguishable from mantle peridotites, provided that the lower crust is almost entirely of mafic composition. This is less likely to occur in Archaean cratons where the wavespeeds of mantle peridotites are very high. Nevertheless, the impedance contrasts between mafic lower crust and peridotitic upper mantle may be so low that appreciable P-to-SV

conversion cannot occur. A larger impedance contrast will then occur across the boundary between the mafic lower crust and intermediate middle crust (Table 6; Fig. 16). We suggest that this occurs in parts of the northern and western parts of the Kaapvaal craton, where receiver functions show complexity or variability between nearby stations.

Table 6 shows P wavespeeds, densities and impedances for rock types likely to be found in the middle and lower crust, based on data from Rudnick and Fountain (1995) and Hurich et al. (2001). Because shear-wave data to match those of Hurich et al. are not available, shear wavespeeds were calculated from the P wavespeeds assuming Poisson's ratios of 0.26 and 0.25 for felsic-intermediate and ultramafic rocks, respectively, 0.28 for mafic rocks in the lower crust and 0.27 for eclogites. The values of Poisson's ratio are based on the data compiled by Rudnick and Fountain (1995). All laboratory measurements correspond to room temperature and confining pressures of 400 MPa (Hurich et al., 2001) and 600 MPa (Rudnick and Fountain, 1995). The calculations of impedances are approximate, and no corrections for temperature or pressure have been applied. Wavespeeds of mantle peridotites are those estimated from Pn and Sn data (Table 3).

The amplitudes of P-to-SV conversions will depend most strongly on changes in S wave impedance or S wavespeed, and the effects of changes in P wavespeed are negligible (Appendix A). Fig. 16 shows a simplified, conceptual model for the lower crust of the Kaapvaal craton. The top half of the figure

Table 6

P and S wavespeeds, densities and impedances of rocks that might be found in the deeper parts of the Kaapvaal craton

Rock type	Source of information	P and S wavespeeds (km/s)	Density (kg/m^3)	P wave impedance (kg/(m^2 s))	S wave impedance (kg/(m^2 s))
Granulite facies, intermediate composition	Rudnick and Fountain, 1995; Hurich et al., 2001	6.70, 3.82	2900	19.4×10^6	11.1×10^6
Mafic rocks, upper amphibolite facies	Hurich et al., 2001	7.00, 3.87	3070	22.8×10^6	11.9×10^6
Mafic granulites, plagioclase-rich	Hurich et al., 2001	7.27, 4.02	3190	23.2×10^6	12.8×10^6
Mafic granulites, plagioclase-poor	Hurich et al., 2001	7.60, 4.20	3280	24.9×10^6	13.8×10^6
Mafic rocks in eclogite facies	Rudnick and Fountain, 1995; Hurich et al., 2001	8.00, 4.63	3390	27.1×10^6	15.7×10^6
Kaapvaal peridotites	Durrheim and Mooney, 1994; Table 3	8.20–8.40,[a] 4.84	3200–3300	27.0×10^6	15.7×10^6

[a] P wavespeed estimates are based on data from Table 3, noting that a higher wavespeed is more likely to be associated with iron depletion and therefore a lower density.

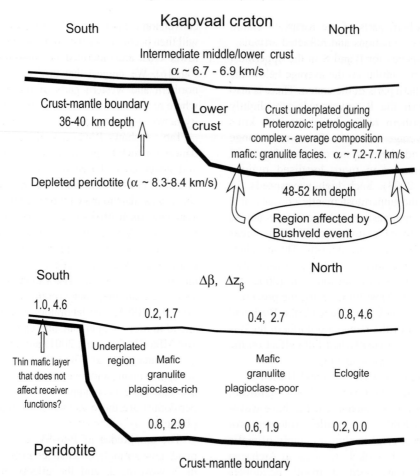

Fig. 16. Schematic model of the lower crust and upper mantle of the Kaapvaal craton (not to scale). $\Delta\beta$ and Δz_β represent changes in shear wavespeed (in km/s) and shear wave impedance with the factor of 10^6 kg/(m^2 s) omitted. The numerical values are from Table 6. The lower section is an expanded representation of the northern region in which the effect of changes in composition and metamorphic grade at the top and bottom of the mafic lower crust are shown. P and S wavespeed are represented by α and β, respectively.

shows the crust–mantle boundary (thick line) in the southern part of the craton between intermediate granulite and depleted peridotite at depths between 36 and 40 km (40–44 km below the Witwatersrand basin). To the north of the boundary marked in Fig. 7, mafic rocks were introduced below the old Archaean Moho, mainly during the Bushveld event, accompanied by intrusion of mafic material (Bushveld complex) into the lower crust in some localities. This mafic material results in a crust–mantle boundary at 48–52 km depth over much of the northern part of the Kaapvaal craton. Because of variations in composition and metamorphic grade across the region, the seismic properties of the lower crust are highly variable, so

that receiver functions give no clear indication of crustal thicknesses. This idea is illustrated in the lower half of Fig. 16. In the southern part of the craton, the impedance contrast for S is 4.6×10^6 kg/m^2 s, and the change in S wavespeed is more than 1.0 km/s, giving rise to strong P-to-SV conversions. For the north, three possible situations are shown. If the lower crust is predominantly plagioclase-rich mafic granulite, the impedance and wavespeed contrasts are higher at the crust–mantle boundary than at the top of the lower crust, giving rise to a stronger P-to-SV conversion at the Moho. If, however, the lower crust consists of plagioclase-poor mafic granulites, the stronger conversion will be at the top of the lower crust. In the extreme case of an eclogitic lower crust, there will be very little P-to-SV conversion at the petrological Moho, and a relatively strong conversion at the boundary between mafic lower crust and intermediate upper crust. The real situation is most likely one of extreme complexity, in which the lower crust is petrologically complex, and contains mafic rocks of variable composition and metamorphic grade. This would give rise to situations when the receiver functions are sometimes dominated by P-to-SV conversions at the Moho, and at other locations by conversions at the top of the mafic lower crust. This phenomenon is proposed as an explanation for the greater crustal thicknesses inferred from Pn and Sn arrivals due to the variable seismic properties of the Moho. The crust–mantle model of Fig. 16, particularly the thicknesses of the underplated and undisturbed regions of the Kaapvaal craton, is also geometrically similar to that of the Archaean Kapukasing structure (Boland and Ellis, 1989). The estimates of crustal thicknesses for the northern part of the craton will be underestimated, because of the fitting of data to crustal models with a wavespeed of 7.1 km/s at the base of the crust (Table 2); actual values for a mafic lower crust will be significantly higher.

Information from mafic granulite xenoliths that might assist in better understanding the properties of the lower crust is restricted largely to the southern part of the craton (Huang et al., 1995; Dawson et al., 1997). Mafic granulites are rare in kimberlite pipes in the southern part of the craton, but more common in kimberlite pipes of the adjacent mobile belts (Griffin et al., 1979), supporting the model of the Archaean crust/mantle transition of Durrheim and Mooney (1994) in which there is little mafic material in the lower crust.

9. Conclusions

Accurate locations of more than 4400 earthquakes associated with gold mining on the margin of the Witwatersrand basin have been obtained over a 25-month period from specialised seismic networks deployed by the gold-mining industry to monitor seismic activity around deep-mining operations. Most of these events are located with errors of less than 400 m, making them comparable with large explosive sources for the determination of the structure of the crust and uppermost mantle. Errors in locations of events in the mining areas made by the South African network and the Kaapvaal network average about 9.5 and 1.6 km, respectively. Many larger events that were not present in the mine catalogues can therefore be relocated using the Kaapvaal network and used in future studies of earth structure.

The average thickness of the crust below the Kaapvaal craton determined from P and S refracted arrivals from accurately located mining-induced earthquakes is about 47–49 km for paths to the northern and 40–42 km to southern regions of the Kaapvaal craton from sources on the edge of the Witwatersrand basin. If the thickness of the crust is assumed to lie in the range 40–43 km below and around the Witwatersrand basin, as suggested by the results from receiver functions, the average crustal thicknesses determined by the two methods are in good agreement (38 km) for the southern regions of the craton. In these southern regions, where receiver functions imply a sharp crust–mantle boundary, the crust–mantle transition is adequately explained as a boundary between granulites of intermediate composition and a highly depleted perodite with little or no mafic material in the lowermost crust. The estimated crustal thicknesses around the Witwatersrand basin are also greater by 4–8 km than in previous studies in South Africa that have made use of recordings of mining-induced earthquakes.

The refracted arrivals give an average crustal thickness for the northern parts of the Kaapvaal craton of 51 km, compared with 44 km from receiver functions. The estimated difference in crustal thickness between the two methods for the region that was disturbed by post-Archaean tectonism is attributed to the complexity of the lowermost crust resulting from underplating, particularly during the Proterozoic Bushveld event at 2.05 Ga. Mafic rocks within the lowermost crust can have a wide range of seismic properties that depend on both composition and metamorphic grade. Thus, whenever the lower crust is predominantly mafic, S wavespeed contrasts at the crust–mantle boundary and at the upper boundary of the mafic region, can vary widely from one region to another, causing wide variability in the appearance of receiver functions. In contrast, the Pn or Sn energy travels within the high wavespeed uppermost mantle, even when the seismic properties of the lower crust approach those of the upper mantle, thereby providing more reliable definition of the petrological crust–mantle boundary.

The uppermost mantle beneath the Kaapvaal craton is remarkably uniform in properties, having relatively high seismic wavespeeds that imply a highly depleted magnesium-rich perodotite. However, the southern boundary of the craton is clearly defined seismically by an increase in crustal thickness and a decrease in upper mantle wavespeed. Changes in seismic character at the northern margin of the craton are less clearly defined.

Acknowledgements

Financial support from the National Research Foundation (South Africa), the National Science Foundation (USA), the University of the Witwatersrand Research Council and from several South African and foreign mining companies who contributed to the running costs and establishment of the Kaapvaal broad-band network is gratefully acknowledged. We also thank members of the Kaapvaal working group from many different organisations in southern Africa and the USA for their assistance. Staff of AngloGold, the Council for Geoscience, East Rand Proprietary Mines, Gold Fields South Africa and ISS International provided valuable sources of information on earth tremors located by the South African network and mine networks, for which we are very grateful. E.M. Kgaswane received financial support from the National Research Foundation and the Council for Geoscience, M.T.O. Kwadiba from the Government of Botswana and the Kellogg Foundation, R.E. Simon from the University of Botswana, and T.K. Nguuri from the Mellon Foundation. The contributions of S.J. Webb to the studies of seismicity are also acknowledged.

Appendix A

At seismic wavespeeds and densities likely to be found in the lowermost crust and uppermost mantle (Table 6), the following results have been derived to illustrate the dependence of P-to-SV conversion amplitudes on changes in impedance, P and S wavespeed, and density. If the average impedances for P and S (products of wavespeed and density), the P and S wavespeeds and densities at the crust–mantle boundary are denoted Z_α, Z_β, α, β and ρ, respectively, and the corresponding changes in these parameters across the crust–mantle boundary ΔZ_α, ΔZ_β, $\Delta\alpha$, $\Delta\beta$ and $\Delta\rho$,

$$\Delta Z_\alpha / Z_\alpha = \Delta\rho/\rho + \Delta\alpha/\alpha \tag{A1}$$

and

$$\Delta Z_\beta / Z_\beta = \Delta\rho/\rho + \Delta\beta/\beta \tag{A2}$$

where

$$\alpha = (\alpha_1 + \alpha_2)/2, \quad \beta = (\beta_1 + \beta_2)/2 \quad \text{and}$$
$$\rho = (\rho_1 + \rho_2)/2 \tag{A3}$$

$$\Delta\alpha = \alpha_1 - \alpha_2, \quad \Delta\beta = \beta_1 - \beta_2 \quad \text{and}$$
$$\Delta\rho = \rho_1 - \rho_2 \tag{A4}$$

The subscripts 1 and 2 denote parameter values in the uppermost mantle and lowermost crust, respectively, and Eqs. (A1) and (A2) hold provided the changes in parameters are small.

Suppose α and β are kept constant and the P-to-SV transmission coefficient C is plotted as a function of $\Delta\rho/\rho$, for a boundary between ultramafic and mafic rocks, and for a P wave arrival of apparent wavespeed of 16.0 km/s, corresponding to a shallow earthquake at a distance of about 59°. The slope of the graph is almost constant at 0.10 over a range of $\Delta\rho/\rho$ values of about 0.1. Similarly, if α and ρ are kept constant, the slope of the plot of C against $\Delta\beta/\beta$ is almost constant at 0.51. Finally, when β and ρ are kept constant, the slope of the plot of C against $\Delta\alpha/\alpha$ is 0.039.

Thus, C depends most strongly on the change in S wavespeed, with the change in density providing a smaller but significant dependance. The change in P wavespeed has only a slight effect on C. C therefore depends on ΔZ_α through the density change only, whereas the dependence on ΔZ_β involves both changes in S wavespeed and density, but with a fractional change in β having about five times the effect of a similar change in ρ (Eqs. (A1) and (A2)).

All calculations have been undertaken using the Zoeppritz equations listed on page 150 of Aki and Richards (1980). It is also noteworthy that over the teleseismic distance range (30–95°), C decreases by a factor of about 2 due to the decrease in angle of incidence of the P wave at the crust–mantle boundary.

References

Aki, K., Richards, P.G., 1980. Quantitative Seismology, Theory and Methods, vol. 1. Freeman, San Francisco. 557 pp.

Andreoli, M.A.G., Doucouré, M., Van Bever Donker, J., Brandt, D., Andersen, N.J.B., 1996. Neotectonics of southern Africa—a review. Afr. Geosci. Rev. 3, 1–16.

Boland, A.V., Ellis, R.M., 1989. Velocity structure of the Kapuskasing Uplift, Northern Ontario, from seismic refraction studies. J. Geophys. Res. 94, 7189–7204.

Bolt, B.A., 1978. Summary value smoothing of physical time series with unequal intervals. J. Comput. Phys. 29, 357–369.

Carlson, R.W., Grove, T.L., de Wit, M.J., Gurney, J.J., 1996. Program to study crust and mantle of the Archean craton in southern Africa. Eos Trans. Am. Geophys. Union 77 (29), 273 and 277.

Dawson, J.B., Harley, S.C., Rudnick, R.L., Ireland, T.L., 1997. Equilibration and reaction in Archaean quartz–sapphirine granulite xenoliths from the Lace kimberlite pipe, South Africa. J. Metamorph. Petrol. 15, 253–266.

Durrheim, R.J., 1998a. Seismic refraction investigations of the Kaapvaal craton. S. Afr. Geophys. Rev. 2, 29–35.

Durrheim, R.J., 1998b. A deep seismic reflection profile across the Witwatersrand basin. S. Afr. Geophys. Rev. 2, 69–73.

Durrheim, R.J., Green, R.W.E., 1992. A seismic refraction investigation of the Archaean Kaapvaal Craton, South Africa, using mine tremors as the energy source. Geophys. J. Int. 108, 812–832.

Durrheim, R.J., Mooney, W.D., 1994. Evolution of the Precambrian lithosphere: seismological and geochemical constraints. J. Geophys. Res. 99, 15359–15374.

Gane, P.G., Hales, A.L., Oliver, H.O., 1946. A seismic investigation of the Witwatersrand earth tremors. Bull. Seismol. Soc. Am. 36, 49–80.

Gane, P.G., Atkins, A.R., Sellschop, J.P.F., Seligman, P., 1956. Crustal structure in the Transvaal. Bull. Seismol. Soc. Am. 46, 293–316.

Graham, G. (Ed.), 1997. Seismological Bulletins (produced monthly). Council for Geoscience, Pretoria, South Africa, April–December issues.

Graham, G. (Ed.), 1998. Seismological Bulletins (produced quarterly). Council for Geoscience, Pretoria, South Africa, January–March, April–June, July–September and October–December issues.

Graham, G. (Ed.), 1999. Seismological Bulletins (produced quarterly). Council for Geoscience, Pretoria, South Africa, January–March and April–June issues.

Green, R.W.E., Bloch, S., 1971. The Ceres, South Africa, earthquake of September 29, 1969. Bull. Seismol. Soc. Am. 61, 851–859.

Green, R.W.E., Chetty, P., 1990. Seismic studies in the basement of the Vredefort structure. Tectonophysics 171, 105–113.

Griffin, W.L., Carswell, D.A., Nixon, P.H., 1979. Lower crust granulites and eclogites from Lesotho, southern Africa. In: Boyd, F.R., Meyer, H.O.R. (Eds.), The Mantle Sample: Inclusions from Kimberlites and Other Volcanics. American Geophysical Union, Washington, pp. 59–86.

Hales, A.L., Sacks, I.S., 1959. Evidence for an intermediate layer from crustal structure studies in the Eastern Transvaal. Geophys. J. R. Astron. Soc. 2, 15–33.

Huang, Y.-M., Van Calsteren, P., Hawkesworth, C.J., 1995. The evolution of the lithosphere in southern Africa: a perspective on the basic granulite xenoliths from kimberlites in South Africa. Geochim. Cosmochim. Acta 59, 4905–4920.

Hurich, C.A., Deemer, S.J., Indares, A., 2001. Compositional and metamorphic controls on velocity and reflectivity in the continental crust: an example from the Grenville Province of eastern Québec. J. Geophys. Res. 106, 665–682.

Hynes, A., Snyder, D.B., 1995. Deep crustal mineral assemblages and potential for crustal rocks below the Moho in the Scottish Caledonides. Geophys. J. Int., 323–339.

Jeffreys, H., 1937. On the smoothing of observed data. Proc. Camb. Philos. Soc. 33, 444–450.

Kgaswane, E.M., 2002. The Characterisation of Natural and Human-Related Seismicity in Southern Africa: April 1997–April 1999. MSc dissertation, University of the Witwatersrand, Johannesburg, 110 pp.

Kgaswane, E.M., Wright, C., Kwadiba, M.T.O., Webb, S.J., McRae-Samuel, R., 2002. A new look at South African seismicity using a temporary network of seismometers. S. Afr. J. Sci. 98, 377–384.

Kwadiba, M.T.O., Wright, C., Kgaswane, E.M., Simon, R.E., Nguuri, T.K., 2003. *Pn* arrivals and lateral variations of Moho geometry beneath the Kaapvaal craton. Lithos 71, 393–411 (this issue).

Lahr, J.C., 1989. HYPOELLIPSE/Version 2.0: a computer program for determining local earthquake hypocentral parameters, magnitudes and first motion patterns. U.S. Geological Survey Open-File Report 89-116, 92 pp.

Langston, C.A., Nyblade, A.A., Owens, T.J., 2002. Regional wave propagation in Tanzania, East Africa. J. Geophys. Res. 107 (B1), 1–18 (ESE1).

Mendecki, A.J. (Ed.), 1997. Seismic Monitoring in Mines. Chapman & Hall, London. 262 pp.

Nguuri, T.K., Gore, J., James, D.E., Webb, S.J., Wright, C., Zengeni, T.G., Gwavava, O., Snoke, J.A.Kaapvaal Seismic Group, 2001. Crustal structure beneath southern Africa and its implications for the formation and evolution of the Kaapvaal and Zimbabwe cratons. Geophys. Res. Lett. 28, 2501–2504.

Reinsch, C.H., 1967. Smoothing by spline functions. Numer. Math. 19, 177–183.

Richards-Dinger, K.B., Shearer, P.M., 1997. Estimating crustal thickness in southern California by stacking PmP arrivals. J. Geophys. Res. 102, 15211–15224.

Rudnick, R.L., Fountain, D.W., 1995. Nature and composition of the continental crust: a lower crustal perspective. Rev. Geophys., 267–309.

Shearer, P.M., 1999. Introduction to Seismology. Cambridge Univ. Press, Cambridge, UK. 260 pp.

Simon, R.E., Wright, C., Kgaswane, E.M., Kwadiba, M.T.O., 2002. The P wavespeed structure below and around the Kaapvaal craton to depths of 800 km, from travel times and waveforms of local and regional earthquakes and mining-induced tremors. Geophys. J. Int. 151, 132–145.

Simon, R.E., Wright, C., Kwadiba, M.T.O., Kgaswane, E.M., 2003. Mantle structure and composition to 800-km depth beneath southern Africa and surrounding oceans from broadband body waves. Lithos 71, 353–367 (this issue).

Webb, S.J., Wright, C., Kgaswane, E.M., 2001. Characterization of natural and human-related seismicity in South Africa: April 1997 to April 1999. Report (with database) sponsored by U.S. Defense Special Weapons Agency (principal investigator, T.H. Jordan, University of Southern California).

Willmore, P.L., Hales, A.L., Gane, P.G., 1952. A seismic investigation of crustal structure in the western Transvaal. Bull. Seismol. Soc. Am. 42, 53–80.

Wood, H.E., 1913. On the occurrence of earthquakes in South Africa. Bull. Seismol. Soc. Am. 3, 113–120.

Wright, C., Kwadiba, M.T.O., Kgaswane, E.M., Simon, R.E., 2002. The structure of the crust and upper mantle to depths of 320 km beneath the Kaapvaal craton, from P wave arrivals generated by regional earthquakes and mining-induced tremors. J. Afr. Earth Sci. 35, 477–488.

Available online at www.sciencedirect.com

ELSEVIER

LITHOS

Lithos 71 (2003) 393–411

www.elsevier.com/locate/lithos

Pn arrivals and lateral variations of Moho geometry beneath the Kaapvaal craton

M.T.O.G. Kwadiba[a,b], C. Wright[a,*], E.M. Kgaswane[a,1],
R.E. Simon[a,c], T.K. Nguuri[a,2]

[a] *Bernard Price Institute of Geophysical Research, School of Earth Sciences, University of the Witwatersrand, Johannesburg, Private Bag 3, Wits 2050, South Africa*
[b] *Geophysics Division, Department of Geological Survey, Private Bag 14, Lobatse, Botswana*
[c] *Department of Physics, University of Botswana, Private Bag UB707, Gaborone, Botswana*

Abstract

Pn arrivals from mining-induced earthquakes on the edge of the Witwatersrand basin show that the P wavespeeds in the uppermost mantle are almost constant throughout most of the Kaapvaal craton. The presence of only small wavespeed variations allows the use of a simple method of estimating crustal thicknesses below the stations of the Kaapvaal broad-band network using *Pn* times that has been compared with results from receiver functions. One thousand three hundred thirty-seven *Pn* arrivals were used to derive crustal thicknesses at 46 stations on the Kaapvaal craton. The average crustal thicknesses for 19 centrally located stations on each of the northern and southern regions of the craton that yielded well-constrained thicknesses were 50.52 ± 0.88 km and 38.07 ± 0.85 km, respectively. In contrast, the corresponding average thicknesses determined from receiver functions were 43.58 ± 0.57 km and 37.58 ± 0.70 km, respectively. The systematically lower values for receiver functions in the northern part of the Kaapvaal craton that was affected by the Bushveld magmatism at 2.05 Ga, suggest that the receiver functions do not enable the petrological crust mantle boundary to be reliably resolved due to variations in composition and metamorphic grade in a mafic lower crust. The *Pn* times also suggest pervasive azimuthal anisotropy with maximum wavespeeds of about 8.40 km/s at azimuths of about 15° and 217° in the northern and southern regions of the craton, respectively, and minimum wavespeeds of about 8.25 km/s.
© 2003 Elsevier B.V. All rights reserved.

Keywords: Crustal thickness; Kaapvaal craton; Mining-induced earthquakes; *Pn* arrivals; Properties of lower crust

1. Introduction

Southern Africa comprises a large continental craton of Precambrian age with moderate seismic activity (Kulhanek and Meyer, 1979). Most of the earthquakes in the southern part of the continent are associated with the East African Rift, which is characterized by extensional tectonics (Kulhanek and Meyer, 1979). South Africa lies in this intraplate region, which is situated thousands of kilometres away from the seismically active mid-oceanic ridges in the south Atlantic and in the southwest Indian ocean (Andreoli et al., 1996). Few tectonic events occur within South Africa, and most seismic events

---

* Corresponding author. Fax: +27-11-717-6579.
  E-mail address: wrightc@geosciences.wits.ac.za (C. Wright).
[1] Present address: Council for Geoscience, Private Bag X112, Pretoria 0001, South Africa.
[2] Present address: PREPCOM CTBTO, Vienna International Centre, P.O. Box 1250, A-1400, Vienna, Austria.

0024-4937/$ - see front matter © 2003 Elsevier B.V. All rights reserved.
doi:10.1016/j.lithos.2003.07.008

located within the country are related to gold mining in the Witwatersrand basin. These mining-induced events have been used to determine the structure of the crust and upper mantle in the central regions of the Kaapvaal craton, and earlier work was summarised by Wright and Fernandez (2003) and Wright et al. (2002, 2003). The Southern African Broadband Seismic experiment (SABSE) provided the opportunity to better define South African seismicity over a 2-year period. A database of waveforms recorded during the SABSE, which consists mainly of mining-induced events, was assembled (Kgaswane et al., 2002; Wright et al., 2003), and provided the data used in this paper.

Receiver functions provide another approach to measuring crustal thicknesses that was used to show that the crust beneath the southern region of the Kaapvaal craton that has remained undisturbed since the Archaean has a simple crust–mantle transition at depths typically between 35 and 40 km. In contrast, the adjacent Proterozoic mobile belts exhibit complex Moho signatures and thicknesses of 45 to 50 km (Nguuri et al., 2001). The northern parts of the Kaapvaal craton around the Bushveld complex, however, show receiver functions that indicate more complicated crust–mantle transitions that exhibit similar character to those in the adjacent mobile belts.

The objective of the present study is to define variations in crustal thicknesses beneath the Kaapvaal craton using *Pn* arrivals recorded at those SABSE stations located within the craton or very close to the craton margin. Comparison of our results with those from receiver functions (Nguuri et al., 2001) has been made to provide complementary information on the structure and properties of the lower crust and uppermost mantle. According to Baumont et al. (2001), *Pn* travel times are sensitive to Moho geometry, the crustal V_p, the upper mantle V_p to a lesser extent, and not sensitive to crustal V_s. V_p and V_s are the P and S wavespeeds, respectively. The time delays between the P and P-to-S converted arrivals in receiver functions are sensitive to Moho depth, and both the average crustal V_p and V_s. The effectiveness of the method also depends primarily on the S wavespeed contrast at the Moho. Therefore, comparison of crustal thicknesses estimated using receiver functions and refracted arrivals

will assist in better understanding the seismic properties and petrology of the lower crust (Wright et al., 2003). Furthermore, our results will be used to define starting models for tomographic inversion for the structure of the crust and uppermost mantle of the Kaapvaal craton and surrounding regions. The present work provides the basis for more comprehensive 3-D studies of seismological structure, in which *Pn* times are used to define crustal thicknesses below individual stations of the SABSE. Comments on identification of the reflected phase *PmP* to provide further constraints on crustal thicknesses, and discussion of 'record sections' are also provided.

2. General geological setting

The geology of the mine source regions was described by Pretorius (1986). On a larger scale, de Wit et al. (1992) gave an account of the Archaean history of South Africa. Southern Africa comprises a well-defined Archaean continental nucleus surrounded by Proterozoic mobile belts (de Wit et al., 1992; Carlson et al., 1996). The Archaean region (Fig. 1) is made up of the Kaapvaal and Zimbabwe cratons, consisting of low-grade metamorphosed granite–greenstone terrains, welded together by a 250-km wide ENE-trending high-grade tectono-metamorphic terrain of the Limpopo belt (Van Zijl, 1978; de Wit et al., 1992). The high-grade gneisses and granulites constituting the Limpopo belt have been considered to be the ensialic reworking of cratonic material (Mason, 1973). The cratonic regions are bordered to the south by the Namaqua-Natal mobile belt, to the west by the Kheis belt and to the east by the Lebombo monocline (de Wit et al., 1992).

The Kaapvaal craton formed and stabilised between 3.7 and 2.7 Ga (Carlson et al., 1996) and covers an area of 1.2×10^6 km^2 (de Wit et al., 1992); it has been extensively studied because of its great concentration of mineral resources (Boyd et al., 1985; Boyd and Gurney, 1986), but relatively little geophysical work has been done.

The composition of the continental crust is established to be andesitic, which cannot have been produced by the basaltic magmatism that presently

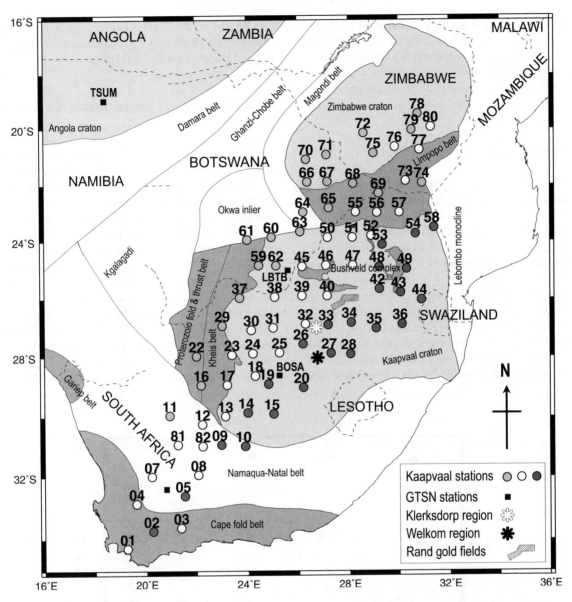

Fig. 1. Map of southern Africa showing distribution of stations of the April 1997–April 1999 Southern African Broadband Seismic Experiment superimposed on the major crustal blocks of southern Africa (After de Wit et al., 1992; Carlson et al., 1996; Nguuri et al., 2001). The grey circles, open circles and dark grey circles denote stations deployed during the period April 1997–April 1998, April 1997–April 1999 and April 1998–April 1999, respectively. The closed squares symbolise the Global Telemetered Seismological Network (GTSN). Locations of the Klerksdorp and Welkom mining regions are shown as open and closed asterisks, respectively.

operates at convergent margins and within plates (Rudnick, 1995). This makes it more difficult to accept that the continental crust is derived by partial melting of the underlying mantle, leaving a

highly depleted refractory peridotite (Hofmann, 1988). Furthermore, laboratory-melting experiments carried out on mantle peridotite produce basaltic and picritic magmas (Ireland et al., 1994; Rudnick,

1995). This poses a paradox: if the crust is formed by processes of partial melting at high pressures, why is the crust not basaltic? Seismological investigations provide important clues to the processes of formation that led to the present composition of the continents.

3. Sources of data and methodology

The data consist of seismograms from 60 mining-induced tremors recorded by stations of the SABSE: 20 in each of the Far West and West Rand, Klerksdorp and Welkom areas (Fig. 2). Events with local magnitudes greater than 3.0 were used to ensure reasonably strong arrivals to distances of at least 700 km and the locations are shown in Fig. 3. Epicentres were obtained from mine catalogues and converted to latitudes and longitudes relative to the World Geodetic Service, 1984 (WGS84) (Kgaswane et al., 2002). Origin times were derived from the Council for Geoscience earthquake bulletins (Graham, 1997, 1998, 1999) and corrected by the method described

by Wright et al. (2003). For time picking of first arrivals (*Pg* or *Pn*), all seismograms were bandpass filtered between 0.4 and 4.0 Hz.

4. Determination of crustal thicknesses from *Pn* arrivals

The present data consist of *Pn* arrivals distributed over a wide area (Fig. 1) and differ from those of Baumont et al. (2001) who used approximately linear profiles of broad-band stations to estimate crustal thicknesses. The method of determining crustal thickness is similar in principle to that of Baumont et al., and relies on the result that the upper mantle wavespeed is practically uniform across the entire Kaapvaal craton. Similar studies that use *Pn* arrivals to define Moho topography have been described by Vetter and Minster (1981), Shearer and Oppenheimer (1982) and Zervas and Crosson (1986). A fundamental difficulty with the present data is the concentration of sources in three restricted geographical areas, which does not allow the construction of

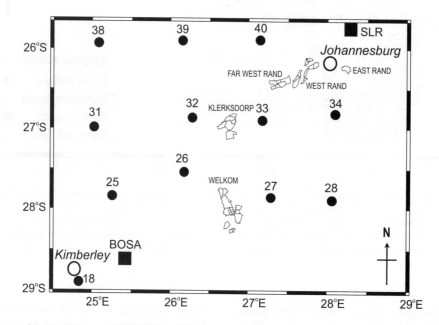

Fig. 2. Location map of the East Rand, West Rand, Far West Rand, Klerksdorp and Welkom source regions in the Witwatersrand basin. Also shown on the map are some of the seismic stations of the SABSE project in Fig. 1 (circles). Squares denote permanent broadband or long-period seismic stations.

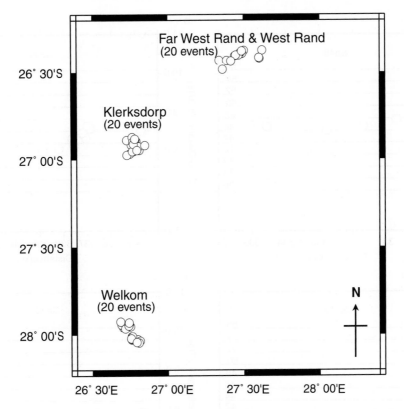

Fig. 3. Location map of the 60 events used as seismic sources in this study. The epicentres are denoted by open circles, which are distributed in clusters around the mining regions as shown on the map. These events generated 1337 *Pn* arrival time picks for stations on the Kaapvaal craton or very close to the craton margin. The names of the source regions are also indicated together with the number of events from each region.

reversed refraction profiles that are commonly used in crustal-scale seismic refraction work. There is also the need to estimate crustal thicknesses at the sources using other data.

Wavespeed models for the crust corresponding to those of Table 2 of Wright et al. (2003) were used in which depth variations were implemented by keeping the wavespeeds constant at 6.80 and 7.10 km/s at 20 km depth and at the base of the crust, respectively, and having a constant wavespeed gradient within that region. In the uppermost mantle, wavespeed models with low positive gradients that give an average ray parameter of 1/8.37 s/km were chosen. Ray tracing was then undertaken through spherically symmetric models with crustal thicknesses at 2 km separations, varying from 36 to 56 km, and the travel–time curves and *Pn* times were superimposed on re-

duced-time (T-D/8.37) graphs (Fig. 4). For each station, all time observations for the *Pn* phase for a particular source region were plotted interactively on the reduced-time plots, and an average crustal thickness for that source-to-station path estimated for each time value.

Each estimated thickness can be regarded as approximately the average of the crustal thickness beneath the source and the recording station. To obtain a crustal thickness below the station, an assumption has to be made concerning the thickness below the source. Since the average crustal thicknesses estimated from *Pn* arrivals are slightly greater than those estimated from receiver functions, and the thicknesses estimated from receiver functions are slightly larger below and around the Witwatersrand basin than below much of the south-

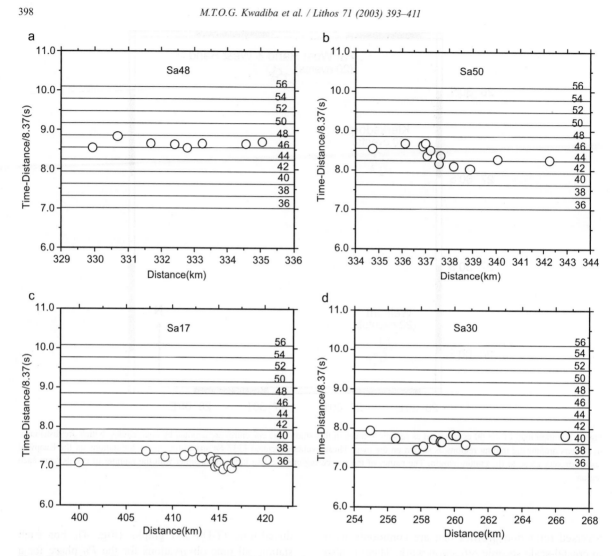

Fig. 4. Reduced travel-time curves from P wavespeed models used in the estimation of average crustal thicknesses. a, b, c and d are examples of plotted *Pn* times against the predicted thicknesses in the background for stations Sa48, Sa50, Sa17 and Sa30, respectively, for 20 events in the Klerksdorp region (Fig. 3).

ern part of the Kaapvaal craton, a starting thickness of 45 km was assumed for events from the Far West and West Rand regions. The average crustal

thickness from receiver functions for stations close to this source region was 42 km (Wright et al., 2003).

Notes to Table 1:

[a] Estimates of crustal thicknesses from *Pn* arrivals were determined from one or more source regions: KL—Klerksdorp; WE—Welkom; WR—Far West and West Rand; All—All three source regions.

[b] Stations marked [b] were used to compute averages for comparison of thicknesses from *Pn* arrivals with those from receiver functions. Stations yielding a bimodal distribution of thicknesses from *Pn* times have two entries marked (a) and (b). Stations SA139 and SA155 are stations less than 40 km from SA39 and SA55 respectively; they are not marked as separate stations in Figs. 1 and 7.

Table 1
List of crustal thicknesses based on *Pn* arrival times and receiver function analysis for the southern and northern regions of the Kaapvaal craton

Station	Mean thickness (km)	Standard error (km)	No. of estimates	Source regions[a]	Receiver function thickness (km)	Difference (km)
South						
BOSA[b]	38.58	0.89	36	KL, WR	35.0	3.6
SA12[b]	36.63	0.64	58	All	45.0	− 8.4
SA13[b]	34.29	0.54	53	All	35.0	− 0.7
SA14[b]	36.47	1.03	25	All	34.0	2.5
SA15[b]	35.20	1.57	14	All	38.0	− 2.8
SA16	47.01	0.94	32	All	36.0	11.0
SA17[b]	31.13	0.61	52	All	38.0	− 6.9
SA18[b]	35.49	0.88	42	All	36.0	− 0.5
SA19[b]	37.33	0.97	26	All	36.0	1.3
SA20[b]	38.21	1.33	17	KL, WR	38.0	0.2
SA22	56.49	0.77	21	All	35.0	21.5
SA23[b]	35.08	0.81	41	All	44.0	− 8.9
SA24[b]	34.16	0.85	35	All	38.0	− 3.8
SA25[b]	39.15	1.25	16	WR	38.0	1.2
SA29[b]	42.21	0.98	21	All	35.0	7.2
SA30[b]	39.27	1.09	38	All	35.0	4.3
SA31[b]	41.78	1.05	30	WE, WR	40.0	1.8
SA35[b]	37.25	1.19	10	KL, WE	40.0	− 2.8
SA36[b]	43.10	1.43	21	All	38.0	5.1
SA37[b]	46.46	0.96	31	All	34.0	12.5
SA38(a)[b]	41.50	0.89	27	KL, WR	37.0	4.5
SA38(b)	53.53	1.44	15	WE	37.0	16.5
North						
LBTB[b]	50.95	0.60	51	All	45.0	6.0
SA39[b]	51.97	4.96	6	WE	42.0	10.0
SA139[b]	51.10	2.83	6	WE	40.0	11.1
SA40[b]	40.84	0.67	20	WE	45.0	− 4.2
SA42[b]	41.42	1.05	13	KL, WE	42.0	− 0.6
SA44[b]	49.74	1.60	23	All	40.0	9.7
SA45[b]	56.74	0.92	53	All	45.0	11.7
SA46[b]	54.66	0.84	38	KL, WE	44.0	10.7
SA47[b]	51.03	0.75	29	KL, WE	45.0	6.0
SA48[b]	49.87	1.02	24	All	45.0	4.9
SA50[b]	48.94	0.72	36	All	43.0	5.9
SA51(a)	45.00	0.79	38	KL, WE	50.0	− 5.0
SA51(b)	53.63	1.14	20	WR	50.0	3.6
SA53[b]	51.06	1.25	13	All	43.0	8.1
SA54[b]	50.47	2.04	7	KL, WE	38.0	12.5
SA55[b]	50.68	1.19	33	All	44.0	6.7
SA155[b]	50.87	1.48	21	All	42.0	8.9
SA56(a)	42.77	0.82	37	KL, WE	44.0	− 1.2
SA56(b)	53.63	1.14	20	WR	44.0	9.6
SA57(a)	41.94	1.02	17	KL	42.0	− 0.1
SA57(b)	49.25	1.08	36	WE, WR	42.0	7.3
SA59[b]	50.56	1.12	19	All	45.0	5.6
SA60[b]	54.54	1.91	17	All	45.0	9.5
SA61	55.16	1.88	17	All	46.0	9.2
SA62[b]	50.94	0.59	17	All	45.0	5.9
SA63[b]	53.48	1.14	20	All	47.0	6.5
SA64	59.13	1.23	17	All	41.0	18.1
SA65	56.53	0.94	20	All	45.0	11.5

Table 2
List of *Pn* wavespeeds estimated for the azimuth ranges in Fig. 7

Region and azimuth range (Klerksdorp) (°)	Distance range (km) and source region	Intercept (s)	Reciprocal slope (km/s)	Standard deviation (s)	No. of times
R1 118-233	232–555 (KL)	7.735 ± 0.13	8.398 ± 0.023	0.38	94
	211–488 (WE)	7.703 ± 0.16	8.389 ± 0.029	0.65	67
	265–652 (WR)	8.272 ± 0.11	8.393 ± 0.017	0.35	119
R2 216-252	269–555 (KL)	7.438 ± 0.27	8.340 ± 0.044	0.68	122
	211–488 (WE)	6.943 ± 0.18	8.235 ± 0.033	0.49	133
	265–652 (WR)	7.918 ± 0.17	8.350 ± 0.024	0.47	134
R3 233-303	200–510 (KL)	7.367 ± 0.31	8.271 ± 0.058	0.80	104
	200–471 (WE)	7.272 ± 0.27	8.203 ± 0.056	0.69	132
	228–602 (WR)	7.712 ± 0.20	8.275 ± 0.034	0.65	132
R4 252-338	200–492 (KL)	7.225 ± 0.25	8.093 ± 0.052	0.57	80
	200–532 (WE)	8.179 ± 0.22	8.272 ± 0.045	0.55	106
	205–579 (WR)	8.729 ± 0.19	8.361 ± 0.038	0.68	152
R5 303-10	227–463 (KL)	8.623 ± 0.19	8.304 ± 0.043	0.47	95
	232–580 (WE)	8.894 ± 0.19	8.358 ± 0.032	0.44	108
	205–473 (WR)	9.211 ± 0.17	8.353 ± 0.036	0.40	110
R6 338-35	227–499 (KL)	9.382 ± 0.14	8.501 ± 0.026	0.40	125
	228–603 (WE)	8.620 ± 0.18	8.364 ± 0.027	0.57	155
	279–424 (WR)	8.973 ± 0.33	8.313 ± 0.064	0.49	99
R7 10-83	264–553 (KL)	8.922 ± 0.19	8.445 ± 0.031	0.41	109
	228–648 (WE)	8.119 ± 0.15	8.323 ± 0.021	0.48	140
	230–474 (WR)	9.621 ± 0.28	8.456 ± 0.052	0.49	93
R8 35-118	226–553 (KL)	7.697 ± 0.28	8.282 ± 0.046	0.56	54
	251–648 (WE)	7.411 ± 0.22	8.234 ± 0.030	0.47	69
	230–474 (WR)	9.467 ± 0.33	8.430 ± 0.061	0.50	39
R9 83-216	226–372 (KL)	7.908 ± 0.41	8.445 ± 0.11	0.47	43
	251–468 (WE)	6.465 ± 0.35	8.091 ± 0.060	0.43	30
	255–617 (WR)	8.297 ± 0.24	8.359 ± 0.044	0.42	40
North	228–553 (KL)	9.182 ± 0.12	8.350 ± 0.024	0.47	204
	340–648 (WE)	9.498 ± 0.15	8.485 ± 0.022	0.47	208
	205–474 (WR)	9.293 ± 0.15	8.386 ± 0.032	0.48	205
South	200–555 (KL)	7.659 ± 0.16	8.357 ± 0.029	0.65	209
	201–488 (WE)	7.998 ± 0.16	8.375 ± 0.032	0.65	262
	228–652 (WR)	8.143 ± 0.12	8.360 ± 0.020	0.46	249

Approximate mean crustal thicknesses at each station for each source region were estimated from the average crustal thicknesses computed by ray tracing, assuming a crustal thickness of 45 km below each source region. The crustal thickness estimates were then used to estimate systematic differences between the thicknesses of the crust in the three source regions. This was achieved in the following

way. If the systematic difference between source regions 1 and 2 is c_{12},

$$c_{12} = \sum_{i=1}^{N} w_{i12}(T_{i1} - T_{i2}) / \sum_{i=1}^{N} w_{i12} \qquad (1)$$

where w_{i12} is the number of separate estimates of crustal thickness at station *i* for source regions 1 and

Fig. 5. Grey-toned contour maps of the estimated crustal thicknesses listed in Table 1. The contour map (a) is for crustal thicknesses inferred from the Pn arrival times and (b) for crustal thicknesses estimated from receiver functions (Nguuri, 2003; Nguuri et al., 2001). Dots represent stations of the SABSE. The hatched area on the insert map indicates the region for which crustal thicknesses were derived, based on *Pn* arrival times. Field of view: 21°; rotation: 0°; tilt: 90°; projection: orthographic.

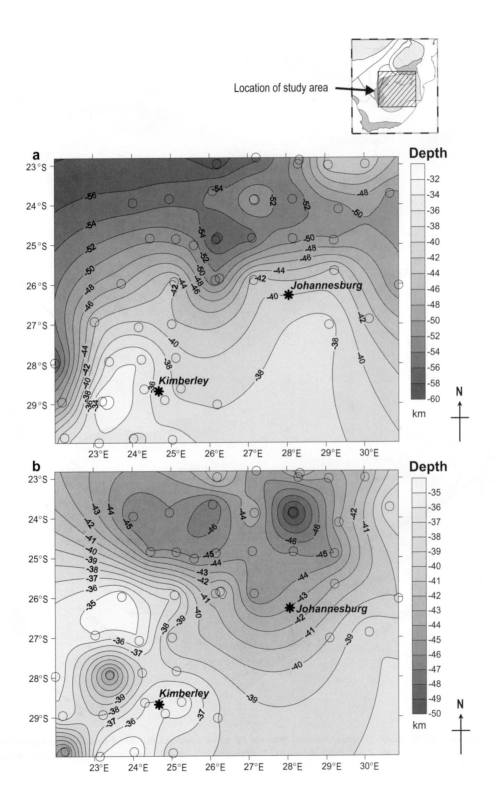

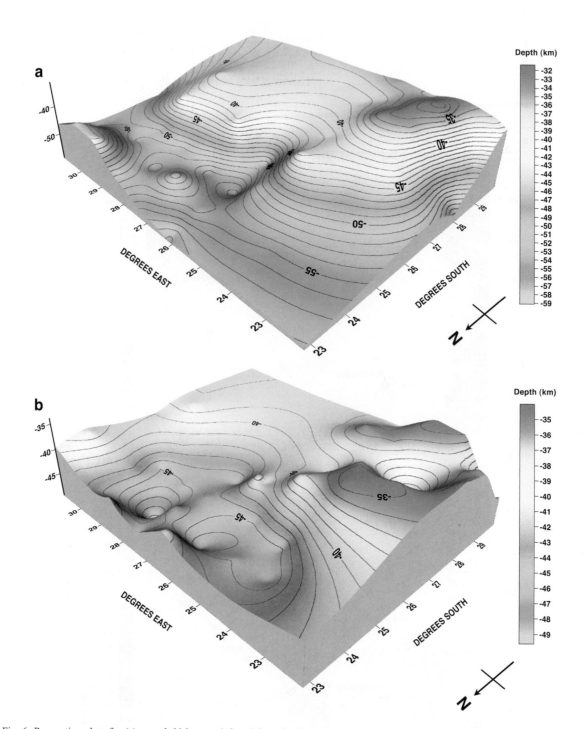

Fig. 6. Perspective plots for (a) crustal thicknesses inferred from the *Pn* arrival times and (b) crustal thicknesses estimated from receiver functions (Nguuri, 2003; Nguuri et al., 2001). Field of view: 45°; rotation: 135°; tilt: 30°; projection: perspective; light position angles: horizontal: 135° and vertical: 45°.

2, and T_{i1} and T_{i2} are the average thicknesses for station i estimated for source regions 1 and 2, respectively. When region 1 is the Far West and West Rand, and 2 is firstly Klerksdorp and secondly Welkom, c_{12} has values of 2.5 ± 0.30 km ($N=34$ stations) and 2.2 ± 0.35 km ($N=31$ stations), respectively.

Wright et al. (2003) investigated the possibility of bias in event origin times due to errors in the reference P travel–time curve used to refine the origin times. They found a maximum possible bias of 0.46 s (later origin times). When the resulting faster wavespeeds are used to compute crustal thicknesses, there is a shift of 2.0 km to lower thicknesses. In computing the crustal thicknesses listed in Table 1, this maximum

correction has been used to make the thicknesses as low as possible to minimize disagreement with earlier seismic studies. To compute crustal thicknesses below each station i, the following formulae were used.

$$\text{Far West and West Rand}: \quad H_{ij1} = 2T_{ij1} - H - 4.0 \tag{2a}$$

$$\text{Klerksdorp}: \quad H_{ij2} = 2T_{ij2} - H - 1.5 \tag{2b}$$

$$\text{Welkom}: \quad H_{ij3} = 2T_{ij3} - H - 1.8 \tag{2c}$$

where H_{ijk} and T_{ijk} are the estimated crustal thickness below station i and the average crustal thickness (without corrections for origin time bias) for station i

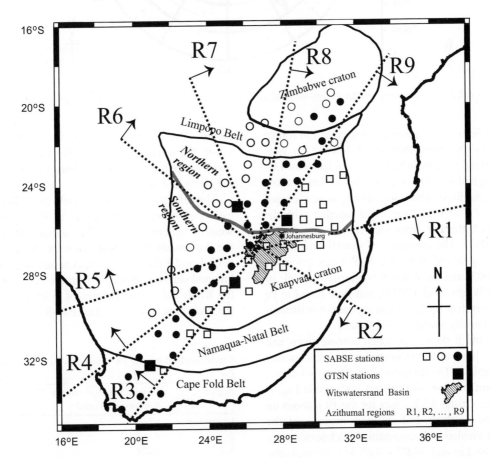

Fig. 7. Map showing subdivision of the Kaapvaal craton region into nine overlapping azimuth ranges for the Klerksdorp region, and division of craton into northern and southern regions. Each azimuth range comprises the segment clockwise from the numbered arrow through the start of the next range to the start of following range. Thus, range 1 covers the region from the line defining the start of range 1 to the line defining the start of range 3, and so on. The subdivisions for the West Rand and Welkom source regions were made as close as possible to those of the Klerksdorp region. Open circles, closed circles and open squares denote SABSE stations that operated from April 1997–April 1998, April 1997–April 1999 and April 1998–April 1999, respectively.

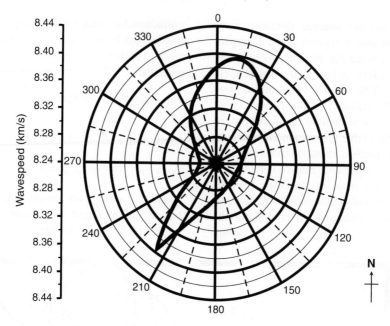

Fig. 8. Polar plot of average wavespeeds for *Pn* arrivals listed in Table 3, using cubic spline smoothing to generate a continuous curve.

and source region *k*, respectively, for event *j* in source region *k*. *H* is the crustal thickness below the Far West and West Rand regions, which has been taken as 43.0 km (the original 45.0 km corrected by 2.0 km to allow for origin time bias). The values of H_{ijk} were averaged for each value of *i*, and are listed together with the thicknesses estimated from receiver functions from Nguuri (2003). The estimates of crustal thicknesses were also calculated from the Ps-P times given by Nguuri (2003) and the P and S wavespeed models of Table 2 of Wright et al. (2003), and were found to be slightly lower than her listed values, but not sufficiently lower to justify using revised values.

Figs. 5 and 6 show grey-toned contour maps and perspective plots of the estimated crustal thicknesses listed in Table 1. The division of the Kaapvaal craton into northern and southern regions is shown in Fig. 7. The average crustal thicknesses for the 19 stations on each of the northern and southern regions of the craton that yielded well-constrained thicknesses were 50.52 ± 0.88 km and 38.07 ± 0.85 km, respectively. In contrast, the corresponding average thicknesses determined from receiver functions were 43.58 ± 0.57 km and 37.58 ± 0.70 km, respectively. The contour maps for thicknesses estimated by both methods agree satisfactorily in the southern parts of the

craton. The differences are in the northern regions where the refracted arrivals yield greater thicknesses and a steeper, more clearly defined transition between the north and south.

The refracted arrivals also yield anomalously thick estimates of crustal thickness in the southern and western regions. In particular, stations 16 and 22 in the Proterozoic thrust belt just west of the Kheis belt (Fig. 1 and Table 1) yield thicknesses in excess of 45 km, and stations 29 and 37 near the eastern margin of the Kheis belt have thicknesses in excess of 40 km. Receiver functions give clear P-to-S conversions that

Table 3
List of weighted average *Pn* wavespeeds for the azimuth ranges in Fig. 7

Region	Weighted average seismic P wavespeed (km/s)	Azimuth (°)
1	8.393	215
2	8.315	233
3	8.259	255
4	8.269	290
5	8.344	330
6	8.425	10
7	8.371	40
8	8.275	80
9	8.282	120

imply thicknesses of just 34–36 km for these four stations. The reason for the much larger thicknesses from refracted arrivals is probably a gradual decrease in upper mantle wavespeed to the west, which cannot be resolved without seismic sources west of the Kaapvaal craton. A similar situation occurs with stations 60–61 and 63–65, which lie in the Okwa inlier and the Magondi belt, where the refracted

arrivals yield crustal thicknesses 6–18 km greater than those from receiver functions; a decrease in upper mantle wavespeed again seems to be the most plausible explanation. The results thus suggest that the highly depleted uppermost mantle of P wavespeed 8.3–8.4 km extends only as far west as the Kheis belt, and also terminates to the northwest below the Okwa inlier and the Magondi belt.

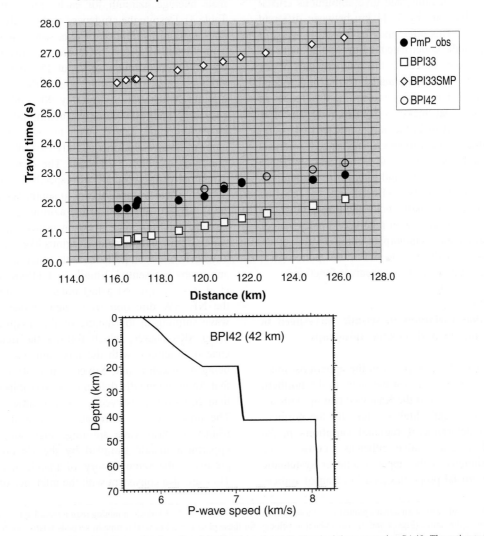

Fig. 9. Travel-time diagram showing observed and predicted *PmP* and predicted *SmP* arrival times at station SA40. The estimated *PmP* arrival times were derived from a source array of ten events in the Klerksdorp mining region. Open diamond symbols indicate theoretical *SmP* times for model BPI33 (P and S, as in Table 2 of Wright et al., 2003), open squares predicted *PmP* times for model BPI33, and open circles predicted *PmP* times for model BPI42. Solid circles denote observed *PmP* arrivals, which indicate a crustal thickness of about 42 km. BPI33 and BPI42 denote models with crusts 33 and 42 km thick, respectively. The insert is a graphical presentation of P model BPI42.

In Table 1, *Pn* arrivals at station 38 to the north and west of the seismic sources suggest a crustal thickness some 12 km greater for paths from Welkom than for the other two source regions, and similar results are found for *Sn* arrivals. This suggests that paths from the Far West and West Rand, and Klerksdorp regions miss the underplated mafic lower crust whose boundary is defined approximately in Fig. 7 (Wright et al., 2003). Even more intriguing are the results for stations 51, 56 and 57, which lie in similar azimuths from the source regions, and give ambiguous crustal thicknesses that vary by 7–11 km between different source regions. This effect may be caused by gaps or 'thin zones' in the underplated region, so that there are 'fast paths' from sources in the southern part of the craton to some stations situated in the underplated region. Wright et al. (2003) explain the differences in estimates of crustal thickness from receiver functions and refracted arrivals in the northern Kaapvaal craton as being due to the variable S wavespeed contrasts between mantle peridotites and mafic granulites (base of lower crust) and mafic and intermediate granulites (top of lower crust). This effect arises because the seismic properties of mafic rocks in the deep crust can vary quite significantly with relatively small changes in composition and metamorphic grade (Hurich et al., 2001), making it difficult to define the petrological crust–mantle boundary using receiver functions.

5. Azimuthal variations in seismic wavespeed in the upper mantle and possible anisotropy

The average P wavespeeds in the uppermost mantle were found to be almost the same in the northern and southern regions of the Kaapvaal craton, with the suggestion of slightly higher values for the northern region. To determine if regional variations in *Pn* wavespeed were detectable, either as a consequence of real variations in the upper mantle or systematic changes in crustal properties as a function of source-to-station distance, the time data were divided into nine overlapping azimuth ranges, as shown in Fig. 7 for Klerksdorp. The subdivisions for the Far West and West Rand and Welkom regions were made as close as possible to those of Fig. 7, and the results of the regression line fits are shown in Table 2.

To test for evidence of pervasive azimuthal anisotropy across the entire craton, the weighted average reciprocal slopes of Table 1 were calculated for each of the nine regions, and plotted against an approximate average azimuth for each region (Fig. 8 and Table 3). Despite the crudeness of the analysis and the non-uniform distribution of data with respect to both azimuth and distance, there are maxima in seismic wavespeed at azimuths of about 15° and 217°, and, although data to the east are sparse, the suggestion of minima at azimuths of roughly 90° and 270°. Fig. 8 clearly suggests coherent azimuthal anisotropy across the entire craton, with the wavespeed maximum of about 8.40 km/s slightly east of north, and a minimum of about 8.25 km/s. The distortion of the azimuthal distribution of wavespeeds is due to very poor coverage from northeast to south. We cannot be sure that we are measuring azimuthal anisotropy because the wavespeeds are measured radially outwards from the three source regions like the spokes of a wheel, and comparison with other evidence for anisotropy is essential. Vinnik et al. (1996) and Silver et al. (2001) inferred a fast polarization direction in the NE–SW direction from their studies of shear-wave splitting in the mantle of the Kaapvaal craton using SKS phases, which follows the trend of Archaean structures, with the most anisotropic regions being Archaean in age. Silver et al. (2001) concluded that Archaean mantle deformation to depths of greater than 200 km is preserved as fossil mantle anisotropy. The present results correlate satisfactorily with the results of shear-wave splitting, suggesting that the uppermost mantle sampled by the *Pn* arrivals has preserved the same history of mantle deformation. They are also consistent with the inference of Schulte-

---

Fig. 10. Vertical-component seismograms plotted as reduced times for two events from the Klerksdorp mining region recorded by stations to the north and west of the source (Figs. 1 and 7). The reduction velocity for these plots is 8.2 km/s, with time in seconds relative to the respective origin times of the earthquakes. The original seismograms were bandpass-filtered between 2.0 and 7.0 Hz. Amplitudes were automatically scaled by the SAC program (Tapley and Tull, 1992). The solid horizontal bars at the base of the traces indicate the times of the first P arrivals. Arrows indicate a region of anticipated change of structure across the Limpopo belt (Fig. 1). Epicentral parameters of events in a and b, respectively, are: July 21, 1997, 26.8962°S and 26.7793°E, depth 0.72 km, origin time 08:45:50.2 UT, $M_L = 4.0$; January 17, 1998, 26.8824°S and 26.7798°E, depth 0.62 km, origin time 13:06:45.1 UT, $M_L = 3.8$.

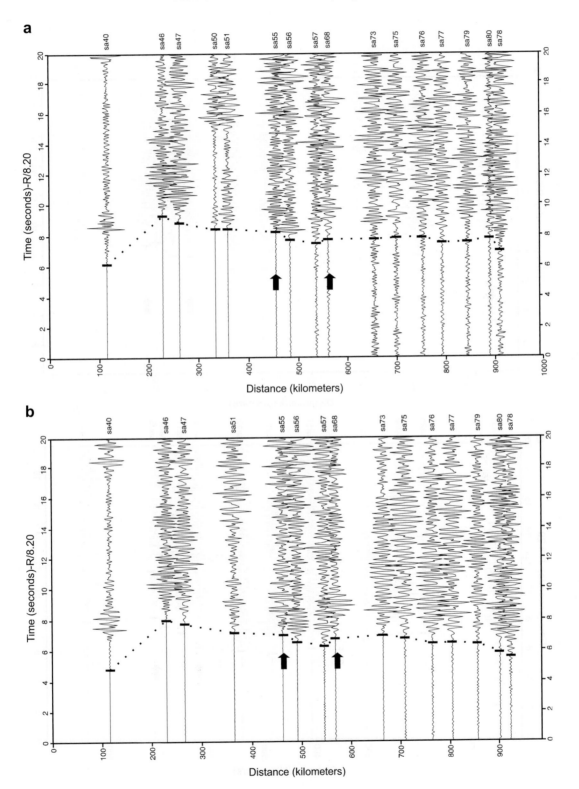

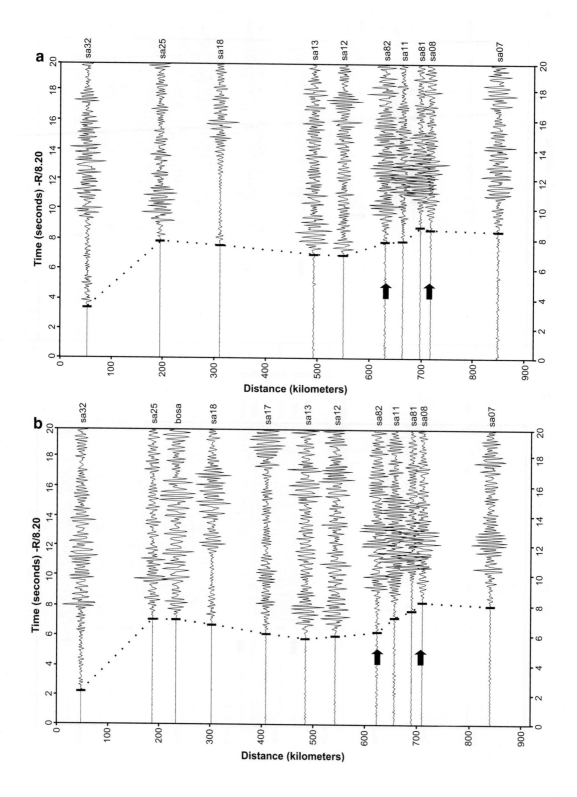

Pelkum et al. (2001) that the fast directions on cratons inferred by P wave polarization, *Pn* travel times and SKS splitting are aligned with absolute plate motions.

6. *PmP* phases and seismic records

The *PmP* phase, which is a wide-angle reflection from the Moho, can have large amplitudes, but the onsets are often emergent, making its use to estimate reliable crustal thicknesses difficult (Wright et al., 2003). A few examples of *PmP* phases have been obtained at distances less than 200 km using source arrays (accurately located earthquakes clustered within an area of a few hundred square kilometres). Fig. 9 shows *PmP* arrival times at station 40 (Fig. 1) from a source array of 10 events in the Klerksdorp mining region at distances between 116 and 126 km, together with computed *PmP* and *SmP* travel times for different crustal thicknesses. A crustal thickness of 42 km provides a good fit to the data for a location half way between Klerksdorp and station 40. This value appears reasonable when compared with the estimates of crustal thickness at station 40 (41 km from refracted arrivals and 45 km from receiver functions (Table 1)).

Figs. 10 and 11 are vertical-component records for two different events from the Klerksdorp region, recorded at stations to the northern (Fig. 10) and southern (Fig. 11) regions of Fig. 7. In Fig 10, arrows at stations 55 and 68 (Fig. 1) mark locations at the southern and northern margins of the Limpopo belt, respectively, beyond which there is an offset from the linear trend of reduced times to significantly later times at distances of 560 km onwards, where all stations lie within the northern marginal zone of the Limpopo belt or within the Zimbabwe craton. Fig. 10 shows how reproducible the travel time anomalies are for nearby seismic events, and illustrates the likelihood of both changes in crustal thickness and mantle wavespeeds that make the method of estimating

crustal thicknesses applicable only to the northern Kaapvaal craton and the southern marginal zone of the Limpopo belt. Fig. 11 shows much larger time delays beyond station 12, which is just off the southern margin of the Kaapvaal craton. The late arrival times at stations within the Namaqua-Natal belt suggest both lateral reductions in mantle wavespeed and the increases in crustal thickness that are confirmed by the receiver functions (Nguuri et al., 2001). However, the presence of a low wavespeed zone within the upper mantle as well as an increase in crustal thickness (Wright et al., 2003) is an alternative explanation.

7. Conclusions

Pn arrivals from mining-induced tremors recorded at stations of the SABSE indicate that the seismic wavespeeds in the uppermost mantle of the Kaapvaal craton are relatively uniform and generally lie between 8.3 and 8.4 km/s, indicating the presence of highly depleted Mg-rich peridotites. This comparative uniformity of wavespeed allows use of a simple method of estimating crustal thicknesses for seismic stations on the craton. The estimated crustal thicknesses within the southern part of the craton vary between 31 and 56 km, with an average of 38.07 ± 0.85 km for the 19 most centrally located stations. This average compares well with the average of 37.58 ± 0.70 km estimated from receiver functions for the same stations. Anomalously high estimates of crustal thicknesses from refracted arrivals compared with corresponding results for receiver functions occur in the western part of the craton, and are attributed to a decrease in upper mantle P wavespeed towards the west. The *Pn* arrivals indicate crustal thicknesses in the northern part of the craton that lie between 41 and 59 km, and an average crustal thickness of 50.52 ± 0.88 km for the 19 most centrally located stations: some 7 km greater than the

Fig. 11. Vertical-component seismograms plotted as reduced times for two events from the Klerksdorp mining region recorded by stations to the south and west of the source (Figs. 1 and 7). The reduction velocity for these plots is 8.2 km/s, with time in seconds relative to the respective origin times of the earthquakes. The original seismograms were bandpass-filtered between 2.0 and 7.0 Hz. Amplitudes were automatically scaled by the SAC program (Tapley and Tull, 1992). The solid horizontal bars at the base of the traces indicate the times of the first P arrivals. Arrows indicate a region of rapid change of structure in the northern part of the Namaqua-Natal belt (Fig. 1). Epicentral parameters of events in a and b, respectively, are: July 21, 1997, 26.8962°S and 26.7793°E, depth 0.72 km, origin time 08:45:50.2 UT, $M_L = 4.0$; December 12, 1997, 26.6640°S and 26.7481°E, depth 0.63 km, origin time 16:42:47.3 UT, $M_L = 4.3$.

corresponding average thicknesses determined from receiver functions (43.58 ± 0.57 km). This difference is attributed to the variable seismic properties of mafic granulites in the underplated northern part of the craton, which results in lack of a clearly defined Moho in the receiver functions. Furthermore, the transition between the northern and southern regions is much more clearly defined by the Pn arrivals than by receiver functions. For some directions from sources in the Witwatersrand basin to stations in the northern region, estimates of crustal thickness are bimodal and are attributed to the availability of paths through the mantle that narrowly miss the incompletely underplated northern region. In the northern region of the craton, crustal thicknesses are probably slightly underestimated from both receiver functions and Pn arrivals because the P and S wavespeeds in the lower crust used in the calculations are too low for mafic granulites.

There is a detectable dependence of Pn wavespeed on source-to-station azimuth which may indicate a small but pervasive azimuthal anisotropy throughout the Kaapvaal craton, in which the maximum wavespeeds are at azimuths of about 15° and 217° in the northern and southern regions of the craton, respectively. Such anisotropy would result in slight underestimates and overestimates, respectively, of crustal thicknesses at stations situated on fast and slow source-to-station azimuths, respectively.

Acknowledgements

Financial support for the running of the Southern African Broadband Seismic Experiment (SABSE) was provided by the National Research Foundation (South Africa), the National Science Foundation (USA), the University of the Witwatersrand Research Council, the Department of Geological Survey (Botswana), the University of Botswana and several southern African mining companies. M.T.O.G. Kwadiba received financial support from the Government of Botswana, the Kellogg Foundation and the University of the Witwatersrand, E.M. Kgaswane from the National Research Foundation and the Council for Geoscience, R.E. Simon from the University of Botswana, and T.K. Nguuri from the Mellon Foundation. Assistance provided by staff of the Carnegie Institution of Washington (USA) and members of the Kaapvaal Research Group is gratefully acknowledged. David E. James offered helpful suggestions and valuable advice, for which we are grateful. We are also most grateful to Fenglin Niu and Derek Schutt for providing numerous UNIX scripts that were used in the processing of data.

References

Andreoli, M.A.G., Doucouré, M., Van Bever Donker, J., Brandt, D., Andersen, N.J.B., 1996. Neotectonics of southern Africa—a review. Afr. Geo. Rev. 3, 1–16.

Baumont, D., Paul, A., Zandt, G., Beck, S.L., 2001. Inversion of Pn travel times for lateral variations of Moho geometry beneath the Central Andes and comparison with the receiver functions. Geophys. Res. Lett. 28, 1663–1666.

Boyd, F.R., Gurney, J.J., 1986. Diamonds and the African lithosphere. Science 232, 472–477.

Boyd, F.R., Gurney, J.J., Richardson, S.H., 1985. Evidence for a 150–200 km thick Archean lithosphere from diamond inclusion thermobarometry. Nature 315, 387–389.

Carlson, R.W., Grove, T.L., de Wit, M.J., Gurney, J.J., 1996. Program to study crust and mantle of the Archean craton in southern Africa. EOS 77 (29), 273 and 277.

De Wit, M.J., Roering, C., Hart, R.J., Armstrong, R.A., Ronde, C.E.J., Green, R.W.E., Tredoux, M., Peberdy, E., Hart, R.A., 1992. Formation of an Archaean continent. Nature 357, 553–562.

Graham, G. (Ed.), 1997. Seismological Bulletins (produced monthly). Council for Geoscience, Pretoria, South Africa, April–December issues.

Graham, G. (Ed.), 1998. Seismological Bulletins (produced quarterly). Council for Geoscience, Pretoria, South Africa, January–March, April–June, July–September and October–December issues.

Graham, G. (Ed.), 1999. Seismological Bulletins (produced quarterly). Council for Geoscience, Pretoria, South Africa, January–March and April–June issues.

Hofmann, A.W., 1988. Chemical differentiation of the earth: the relationship between mantle, continental crust, and oceanic crust. Earth Planet. Sci. Lett. 90, 297–314.

Hurich, C.A., Deemer, S.J., Indares, A., 2001. Compositional and metamorphic controls on velocity and reflectivity in the continental crust: an example from the Grenville Province of eastern Québec. J. Geophys. Res. 106, 665–682.

Ireland, T.R., Rudnick, R.L., Spetsius, Z., 1994. Trace elements in diamond inclusions from eclogites reveal link to Archaean granites. Earth Planet. Sci. Lett. 128, 199–213.

Kgaswane, E.M., Wright, C., Kwadiba, M.T.O., Webb, S.J., McRae-Samuel, R., 2002. A new look at South African seismicity using a temporary network of seismometers. S. Afr. J. Sci. 98, 377–384.

Kulhanek, O., Meyer, K., 1979. A proposal for a seismograph station network in Ngamiland, Botswana. Seismological Institute, Uppsala, Sweden.

Mason, R., 1973. The Limpopo mobile belt-southern Africa. Phil. Trans. R. Soc. London, Series A 273, 463–485.

Nguuri, T.K., 2003. Crustal structure of the Kaapvaal craton and surrounding mobile belts: analysis of teleseismic P waveforms and surface wave inversions. PhD thesis, the University of the Witwatersrand, Johannesburg.

Nguuri, T.K., Gore, J., James, D.E., Webb, S.J., Wright, C., Zengeni, T.G., Gwavava, O., Snoke, J.A., The Kaapvaal Seismic Group, 2001. Crustal structure beneath southern Africa and its implications for the formation and evolution of the Kaapvaal and Zimbabwe cratons. Geophys. Res. Lett. 28, 2501–2504.

Pretorius, D.A., 1986. The Witwatersrand Basin: Surface and Subsurface Geology and Structure. Economic Geology Research Unit, University of the Witwatersrand, Johannesburg, South Africa.

Rudnick, R.L., 1995. Making continental crust. Nature 378, 571–577.

Schulte-Pelkum, V., Masters, G., Shearer, P.M., 2001. Upper mantle anisotropy from long-period P polarization. J. Geophys. Res. 106, 21917–21934.

Shearer, P.M., Oppenheimer, D.H., 1982. A dipping Moho and crustal-velocity zone from *Pn* arrivals at the Geysers-Clear Lake, California. Bull. Seismol. Soc. Am. 72, 1551–1566.

Silver, P.G., Gao, S.S., Liu, K.H., the Kaapvaal Seismic Group, 2001. Mantle deformation beneath southern Africa. Geophys. Res. Lett. 28, 2493–2496.

Tapley, W.C., Tull, J.E., 1992. SAC-Seismic Analysis Code, Users Manual, Revision 4. Lawrence Livermore National Laboratory, Livermore, CA. 278 pp.

Van Zijl, J.S.V., 1978. The relationship between the deep electrical resistivity structure and tectonic provinces in southern Africa. Part I. Results obtained by Schlumberger soundings. Trans. Geol. Soc. Afr. 81, 129–142.

Vetter, U., Minster, J.B., 1981. *Pn* velocity anisotropy in southern California. Bull. Seismol. Soc. Am. 71, 1511–1530.

Vinnik, L.P., Green, R.W.E., Nicolaysen, L.O., 1996. Seismic constraints on dynamics of the mantle of the Kaapvaal craton. Phys. Earth Planet. Inter. 95, 139–151.

Wright, C., Fernandez, L.M., 2003. Earthquakes, seismic hazard and earth structure in South Africa (national report for South Africa). In: Lee, W.H.K., Kanamori, H., Jennings, P.C., Kisslinger, C. (Eds.), International Handbook of Earthquake and Engineering Seismology, Part B, International Association of Seismology and the Physics of the Earth's Interior (I.A.S. P.E.I.). Elsevier, Amsterdam.

Wright, C., Kwadiba, M.T.O., Kgaswane, E.M., Simon, R.E., 2002. The structure of the crust and uppermantle to depths of 320 km beneath the Kaapvaal craton, from P wave arrivals generated by regional earthquakes and mining-induced tremors. J. Afr. Earth Sci. 35, 477–488.

Wright, C., Kgaswane, E.M., Kwadiba, M.T.O., Simon, R.E., Nguuri, T.K., McRae-Samuel, R., 2003. South African seismicity, April 1997–April 1999, and regional variations in the composition of the crust and uppermost mantle. Lithos 71, 369–392 (this issue).

Zervas, C., Crosson, R.S., 1986. *Pn* observation and interpretation in Washington. Bull. Seismol. Soc. Am. 76, 521–546.

Available online at www.sciencedirect.com

SCIENCE  DIRECT°

ELSEVIER

Lithos 71 (2003) 413–429

LITHOS

www.elsevier.com/locate/lithos

Crustal structure of the Kaapvaal craton and its significance for early crustal evolution

David E. James[a,*], Fenglin Niu[b], Juliana Rokosky[a,c]

[a]*Department of Terrestrial Magnetism, Carnegie Institution of Washington, 5241 Broad Branch Road, N.W., Washington, DC 20015, USA*
[b]*Department of Earth Science, Rice University, Houston, TX 77251, USA*
[c]*Macalester College, St. Paul, MN, USA*

Abstract

High-quality seismic data obtained from a dense broadband array near Kimberley, South Africa, exhibit crustal reverberations of remarkable clarity that provide well-resolved constraints on the structure of the lowermost crust and Moho. Receiver function analysis of Moho conversions and crustal multiples beneath the Kimberley array shows that the crust is 35 km thick with an average Poisson's ratio of 0.25. The density contrast across the Moho is ~ 15%, indicating a crustal density about 2.86 gm/cc just above the Moho, appropriate for felsic to intermediate rock compositions. Analysis of waveform broadening of the crustal reverberation phases suggests that the Moho transition can be no more than 0.5 km thick and the total variation in crustal thickness over the 2400 km^2 footprint of the array no more than 1 km. Waveform and travel time analysis of a large earthquake triggered by deep gold mining operations (the Welkom mine event) some 200 km away from the array yield an average crustal thickness of 35 km along the propagation path between the Kimberley array and the event. P- and S-wave velocities for the lowermost crust are modeled to be 6.75 and 3.90 km/s, respectively, with uppermost mantle velocities of 8.2 and 4.79 km/s, respectively. Seismograms from the Welkom event exhibit theoretically predicted but rarely observed crustal reverberation phases that involve reflection or conversion at the Moho. Correlation between observed and synthetic waveforms and phase amplitudes of the Moho reverberations suggests that the crust along the propagation path between source and receiver is highly uniform in both thickness and average seismic velocity and that the Moho transition zone is everywhere less than about 2 km thick. While the extremely flat Moho, sharp transition zone and low crustal densities beneath the region of study may date from the time of crustal formation, a more geologically plausible interpretation involves extensive crustal melting and ductile flow during the major craton-wide Ventersdorp tectonomagmatic event near the end of Archean time.
© 2003 Elsevier B.V. All rights reserved.

Keywords: Seismic crustal structure; Archean crustal formation; Southern Africa; Lower crustal composition; Kaapvaal craton; Moho structure

1. Introduction

Despite a voluminous literature on both the composition and structure of Archean crust, many contentious issues remain as to how continental crust formed and evolved in Archean time. Of particular interest is evidence suggesting that in southern Africa, both the structure and composition of Archean crust differ in fundamental ways from that of crust formed in post-Archean time (e.g. (Durrheim and Mooney, 1994; Griffin and O'Reilly, 1987; Nguuri et al., 2001; Niu and James, 2002). That there are differences in the

* Corresponding author. Tel.: +1-202-478-8838; fax: +1-202-478-8821.
E-mail address: james@dtm.ciw.edu (D.E. James).

0024-4937/$ - see front matter © 2003 Elsevier B.V. All rights reserved.
doi:10.1016/j.lithos.2003.07.009

processes of crustal formation between Archean and post-Archean time is not unexpected, but the nonintuitive nature of these differences is significant. Previous studies in southern Africa suggest that the crust beneath the undisturbed regions of Kaapvaal and Zimbabwe cratons is not only thin relative to that of adjacent Proterozoic terranes, but it lacks both the lower crustal reflectors and the basal high-velocity (7 ×) layers that typify many post-Archean crustal sections (Durrheim and Mooney, 1991, 1994; Nguuri et al., 2001; Niu and James, 2002). Thus, while there is evidence that the lower continental crust in general is dominated by mafic compositions (Christensen and Mooney, 1995; Rudnick, 1992, 1995; Rudnick and Fountain, 1995), the lower crust beneath the cratons of southern Africa may be intermediate or even felsic in composition[1] (Griffin and O'Reilly, 1987).

In this paper, we analyze data from a dense 31 station broadband seismic array in the vicinity of Kimberley, South Africa. The unique dataset from the array and its unusually high quality make it possible to image the crust and Moho beneath the undisturbed Kaapvaal craton in unprecedented detail. We will show that in the vicinity of Kimberley, the crust is relatively thin, that it is of average felsic to intermediate composition in its lowermost section and that it is characterized by a sharp Moho that is almost perfectly flat over an extensive area. The remarkable nature of the Moho structure and the unambiguous composition of the lowermost crust place first-order constraints on possible mechanisms of crustal formation and stabilization of the Kaapvaal craton in the late Archean.

1.1. Kaapvaal Project

The data for this study were collected as part of the Southern Africa Seismic Experiment. The seismic experiment was a major component of the Kaapvaal Project, a multidisciplinary program to study cratonic formation and evolution. The locations of broadband seismic stations of both the large aperture regional

array and the dense Kimberley array are shown in Fig. 1 in relation to major geologic provinces. Technical details of the large aperture seismic experiment have been described in previous papers (James et al., 2001b; Nguuri et al., 2001) and for the Kimberley array in Niu and James (2002).

1.1.1. Geologic setting

The Archean Kaapvaal and Zimbabwe cratons form the continental nucleus of southern Africa. The Kaapvaal craton itself is comprised of an eastern domain with age ca. 3.5 Ga and younger and a western domain (in which is included the principal area of study for this paper) mostly post 3.2 Ga in age. While stabilization of the Kaapvaal craton was essentially complete by about 2.6 Ga, not all of the craton has been stable since that time. The north-central Kaapvaal craton was disrupted ca. 2.05 Ga by the major intracratonic Bushveld event. The Bushveld Complex, the outcrop limbs of which are shown in Fig. 1, is the largest known layered mafic intrusion in the world ($0.5–1.0 \times 10^6$ km³) (Von Gruenewaldt et al., 1985). The Bushveld appears to be a major geologic province, extending westward as far as Botswana (Kampunzu, personal communication, 2000) and well into the mantle (James et al., 2001b). The Limpopo belt sandwiched between the cratons is a collisional terrane generally considered to have formed when the Kaapvaal and Zimbabwe cratons were consolidated in late Archean time (Van Reenen et al., 1992). The Limpopo belt is comprised of a Northern Marginal Zone, a Central Zone and a Southern Marginal Zone. The two marginal zones have been interpreted to be overthrust belts atop cratonic crust (e.g. Van Reenen et al., 1992; Fedo et al., 1995). The Central Zone appears to be a deep zone of tectonic deformation, subsequently uplifted and exhumed to expose high-grade metamorphic rocks (Treloar et al., 1992). The Kaapvaal craton is bounded on the south and east by the Namaqua–Natal Proterozoic mobile belt (ca. 1.1–1.9 Ga) and on the west by the Kheis overthrust belt (ca. 2.0 Ga) (De Wit et al., 1992). The regional seismic array extends across the Phanerozoic the Cape Fold belt at the southernmost extent of the continent.

1.1.2. Crustal structure beneath the regional array

Recent seismic studies based on single station receiver function analysis of P- to S-wave conversions

---

[1] Here, we loosely follow the International Union of Geological Sciences nomenclature for broad categories of igneous rocks: mafic rocks have <52% SiO_2, intermediate rocks have 52–63% SiO_2 and felsic rocks have >63% SiO_2. These compositions are sufficiently distinct as to have relevance in terms of quantifiable seismic wave velocities.

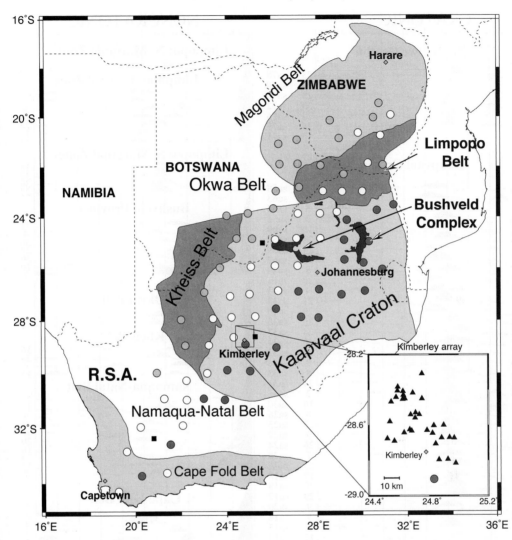

Fig. 1. Map showing the principal geologic provinces in southern Africa. The 81 stations of the Southern Africa Seismic Experiment are shown as circles, with shading indicating different periods of deployment. Black squares denote global digital seismic stations. Station locations of the Kimberley array are shown in the map inset and are indicated by solid triangles. Open squares in the inset denote Moho conversion points, 0P1S, for the analyzed teleseismic receiver function event (see text). All seismometers are sited on bedrock.

(Ps) at the Moho have been carried out to obtain crustal thickness beneath each station of the large aperture array (see Fig. 1) (Nguuri et al., 2001). The receiver function depth images obtained by Nguuri et al. are shown in Fig. 2, arranged by geological province as in Fig. 1. As seen from Fig. 2, only one consistent Ps signal occurs, and it is readily associated with the M-discontinuity. A composite summary of these receiver function results is presented as a color-coded map of

crustal thickness reproduced in Fig. 3 (from Nguuri et al., 2001).

Stations located within regions of the Kaapvaal or Zimbabwe cratons that have remained largely undisturbed since Archean time typically exhibit sharp, large-amplitude depth images for the Moho with arrival times that indicate thin crust. There is no evidence for mid-crustal reflectors anywhere in the craton. Moho depths for the Zimbabwe craton

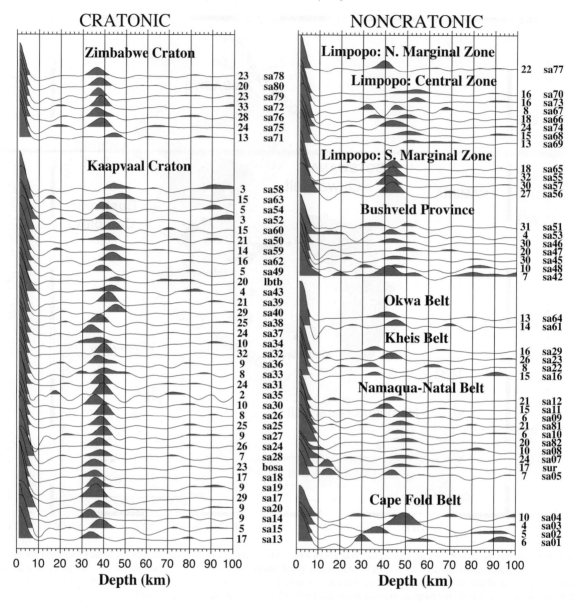

Fig. 2. One-dimensional receiver function stacks for southern Africa organized by geologic province. The number of events included in the stack and the station name is shown to the right of each trace. Individual receiver functions have been corrected for move-out. The Ps conversion from the Moho is the dominant signal on most of the depth images shown. Relatively thin crust and sharp, well-defined Ps Moho conversions are associated with undisturbed craton. Ps arrivals associated with modified regions of the craton and post-Archean terranes tend to be more diffuse and of smaller amplitude (from Nguuri et al., 2001; © 2001 American Geophysical Union).

are with one exception less than 40 km; those for the Kaapvaal craton average about 38 km and range from about 34 to 47 km. Uncertainty in Moho depth is difficult to determine, but we estimate it to be less than about ± 3 km based on tests of

varying velocity models. Post-Archean and modified cratonic regions characteristically exhibit smaller amplitude Ps converted phases, and the crust is commonly thicker than that beneath undisturbed craton. Among the most significant findings pre-

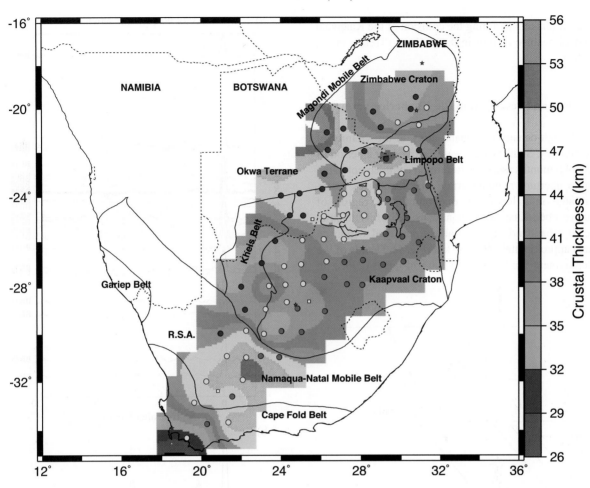

Fig. 3. Color-coded contour map of depth to Moho beneath the southern Africa array based on depth images in Fig. 2 (from Nguuri et al., 2001; © 2001 American Geophysical Union). Crustal thickness color scale is shown on right. Thin crust tends to be associated with undisturbed areas of craton, particularly in the southern and eastern parts of the Kaapvaal craton and in the Zimbabwe craton north of the Limpopo belt. Greater crustal thickness is associated with the Bushveld region and its westward extension into the Okwa and Magondi belts, and with the central zone of the Limpopo belt and the Proterozoic Namaqua–Natal mobile belt.

sented by Nguuri et al. (2001) is evidence for pervasive Proterozoic (ca. ~ 2 Ga) modification of Archean crust across a broad east–west zone bounded by the Bushveld on the east and by the Okwa/ Magondi terranes on the west (see also (Carney et al., 1994; Shirey et al., 2001). Moho Ps conversions for stations in this region of disturbed craton tend to be low in amplitude and, in some cases, ambiguous, suggesting that the Moho may a weak and/or broad transitional (e.g. >3–5 km) boundary. Both crustal thickness and the Moho signature observed in the region of modified Archean crust are similar to those

observed at stations in the Proterozoic Namaqua–Natal belt.

Depth images for the northern and southern marginal zones of the inter-cratonic Limpopo Belt exhibit a characteristic cratonic signature in both Ps and in crustal thickness (Nguuri et al., 2001). The seismic results are consistent with geologic interpretations of the marginal zones as overthrust belts atop cratonic crust (Van Reenen et al., 1992). The Central Zone displays thickened crust (up to 50 km or more) and poorly developed Moho Ps conversions, consistent with geologic models of pervasive shortening and crustal thickening during the collision of the Kaapvaal and Zimbabwe cratons in the Archean (Treloar et al., 1992).

2. Data and results from the Kimberley array

A high-density broadband array in the vicinity of Kimberley, South Africa, was deployed for a period of about 6 months as a high-resolution complement to the large aperture regional array. The Kimberley array, situated in a legendary diamondiferous region of undisturbed Kaapvaal craton, consisted of 31 telemetered broadband seismic stations deployed within an area approximately 40 × 60 km (Fig. 1). Details of the deployment can be found in Niu and James (2002). In

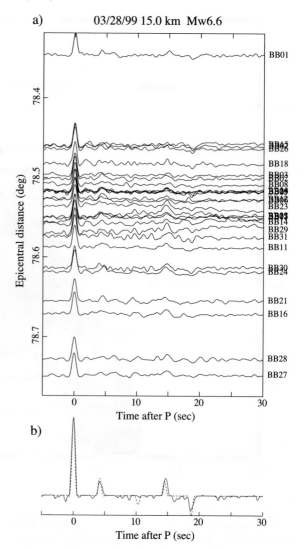

Fig. 5. (a) Deconvolved radial components of seismograms (receiver functions) recorded by the Kimberley array for the event 03/28/99 that occurred in the Tibet–India border region. (b) Fourth-root-stacked observed waveform (solid trace) with overlaid fourth-root stack of synthetic seismograms (dashed trace) calculated for a one-layer crustal model with a P and S wave velocity and density of 6.46 km/s, 3.71 km/s and 2.86 gm/cc, respectively (from Niu and James, 2002).

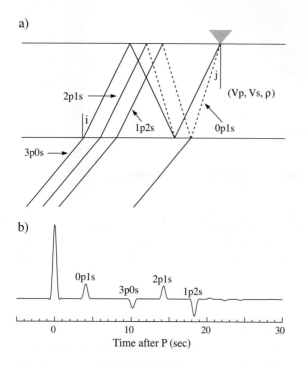

Fig. 4. (a) Ray paths of the direct P-wave, P to S converted wave and the reverberation phases for a single-layered crustal model. We use the general notation, nPmS, where n and m connote the respective number of P- and S-wave legs within the crust. For example, the P to S converted wave at the Moho is termed 0P1S, which means that it has no P-wave rays and 1 S-wave ray within the crust. (b) The radial component of a synthetic seismogram calculated by the Thomson–Haskell method for a one-layer crustal model modified from the IASP91 global model (from Niu and James, 2002).

this section, we elaborate on previous high-resolution receiver function analyses (Niu and James, 2002) and present new results from the analysis of a remarkable set of seismic records clean from a magnitude 5.7 earthquake triggered by mining activity at the Welkom deep gold mine.

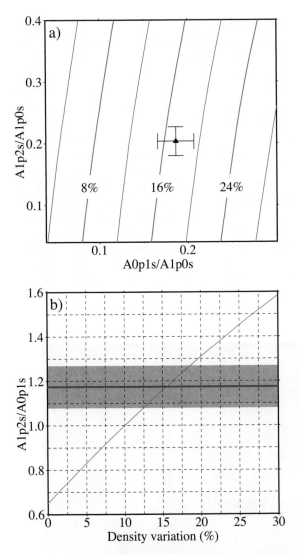

Fig. 6. (a) Contour diagram of S-wave velocity jump across the Moho, shown as a function of amplitude ratios of 0P1S/1P0S and 1P2S/1P0S. The triangle and bars represent the observed amplitude ratios and their 1σ error. (b) Relative amplitude of 1P2S to 0P1S is shown as a function of density contrast across the Moho (thin solid line). The observed amplitude ratio of 1P2S/0P1S (thick solid horizontal line) is shown with the its calculated 1σ error (gray shadowed region) (from Niu and James, 2002).

2.1. Teleseismic receiver functions

In typical receiver function analysis, the P- to S-wave conversion from the Moho is the dominant and commonly the only reliably identifiable signal on the

record. While this phase alone is sufficient to obtain reasonably accurate estimates of crustal thickness, the later arriving crustal reverberation phases (see Fig. 4 for raypath descriptions and notation) provide much additional information that is rarely available simply because of poor signal-to-noise ratio. When present, however, these reverberations provide detailed information about the velocity and density structure of the lowermost crust and the fine structure of the Moho beneath the array.

Niu and James (2002) analyzed the seismograms from the Kimberley array for a large event that

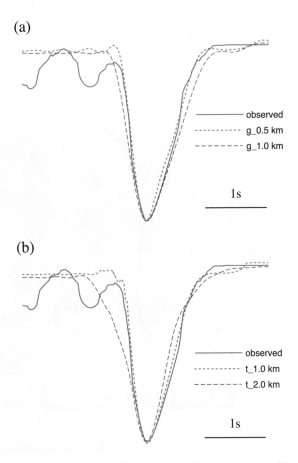

Fig. 7. (a) Observed 1P2S waveform (solid line) is shown with stacked synthetics calculated for Moho models with a transition thickness of 0.5 km (dotted line) and 1.0 km (dashed line). (b) Random variation in crustal thickness of 1.0 km (dotted) and 2.0 km (dashed). Synthetic seismograms are calculated by the Thomson–Haskell method and stacked as for the observed data. See text for discussion (from Niu and James, 2002).

occurred in the Tibet–India border region on March 28, 1999. Individual receiver functions (Fig. 5) were summed using a fourth-root stacking procedure (Kanasewich, 1973; Muirhead, 1968) to produce the receiver function shown as the bottom trace in Fig. 5. Here, both the P to S converted phase (0P1S) and the reverberation phases (2P1S, 1P2S) are clearly observed. The average V_P/V_S ratio for the crust beneath the array derives directly from the lapse time between 0P1S and 2P1S to yield a Poisson's ratio of 0.254. The lapse time between direct P and 0P1S gives an estimated crustal thickness is 35.4 ± 0.2 km, based on an assumed an average V_P in the crust of 6.46 km/s, estimated from waveform analysis and travel time studies of the Welkom event records across the Kimberley array (discussed below).

The P to S conversion coefficient for a near-vertically incident P wave is most sensitive to the

S-wave velocity contrast across the Moho, whereas the reflection coefficient of the near-vertically incident S wave at the Moho depends equally on the S-wave velocity and density variation (Kato and Kawakatsu, 2001). While the minimum S-wave contrast across the Moho can be shown to be approximately 17% (Fig. 6a), the more definitive measure is that of density. The density contrast across the Moho is derived from the relative amplitude of 1P2S to 0P1S and is calculated to be $15.4 \pm 2.3\%$ (Fig. 6b). Thus, for an assumed density of 3.30 gm/cc in the uppermost mantle beneath Kimberley (Boyd and McCallister, 1976), the calculated density of the lowermost crustal layer is 2.86 ± 0.06 gm/cc. This density is consistent with a lowermost crust of intermediate, not mafic, composition: gabbros and mafic granulites of the lower crust typically have densities in excess of 3.0 gm/cc (Christensen, 1982). These seismic results are consistent

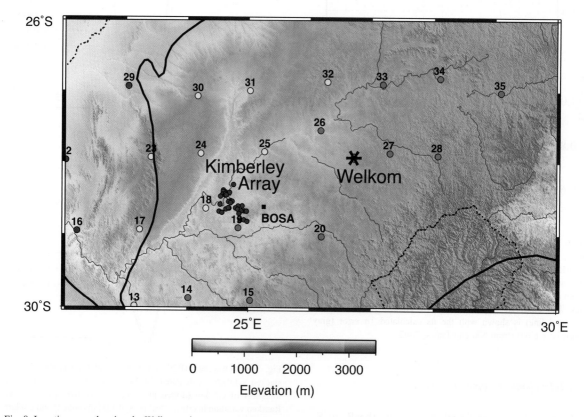

Fig. 8. Location map showing the Welkom mine event used in this study in relationship to the stations of the Kimberley array and the regional topography. Numbered stations are those of the large aperture regional array. The epicenter of the Welkom mine event: latitude = 27.9338°S, longitude = 26.7128°E; depth = 2.1 km.

with xenolith evidence suggesting a notable paucity of mafic material in the lowermost cratonic crust (Niu and James, 2002; Schmitz and Bowring, 2003; Van Calsteren et al., 1986).

Analysis of waveform broadening of reverberation phase 1P2S places two additional and critically important constraints on the structure and morphology of the Moho beneath the Kimberley array as shown in Fig. 7. Niu and James showed that if the observed broadening is caused entirely by a velocity gradient across the Moho (Fig. 7a), then the width of the waveform of 1P2S constrains the allowable extent of the transition zone to <0.5 km. If the broadening is produced entirely by lateral variation in depth of the Moho and/or by velocity heterogeneity within the crust, the maximum allowable depth variation must be <1 km (Fig. 7b). While these results are strictly applicable only to the crust beneath Kimberley, both the extraordinary sharpness of the Moho and flatness of its topography over an area of ~ 2400 km^2 are

first-order constraints on models for crustal evolution of the Kaapvaal craton.

2.2. Welkom mine event

A large earthquake (Mb 5.7) occurred on April 22, 1999 in the Welkom gold mine approximately 190–230 km from the Kimberley array (Fig. 8). The earthquake source was simple and short duration and unusually clean signals were recorded on-scale (24 bit) at all stations of the array. The location (see Fig. 8 caption) of the event is very well constrained by a local mine network. The data contain a wealth of rarely observed crustal reverberation phases that place tight constraints on the nature of the crust and the Moho in the region between earthquake source and seismic stations.

2.2.1. Travel time analysis

Record sections for the vertical and transverse components are shown in Fig. 9, on which are super-

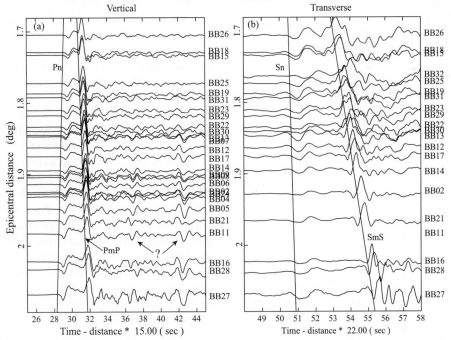

Fig. 9. Welkom mine event record sections for P and S waves. Left panel: vertical component travel time plot with Pn and PmP phases indicated by solid lines for reference. The two phases shown by arrows with a question mark are PmPPmP and 3P1S crustal reverberations as discussed in the text. Right panel: transverse component travel time plot with Sn and SmS indicated for reference. Predicted travel times are based on the velocity–depth model for a 35-km crust given in Table 1.

imposed predicted travel time curves for the major arriving phases based on the velocity–depth model given in Table 1. The most prominent phases on the record are the mantle refractions Pn and Sn and the wide-angle Moho reflections PmP and SmS. The crustal velocity–depth model in Table 1 is constructed to match the arrival times and apparent velocities of these major phases. Together, these primary phases are used to construct a crustal velocity–depth model. While details of the crustal model are not well constrained by our data, the average crustal velocity, crustal thickness, uppermost mantle velocity and average Poisson's ratio in both crust and mantle are well determined. The velocity model also produces accurate travel times for the crustal reverberations discussed below. Of particular interest is the uniformly low Poisson's ratio both in the crust (about 0.25) and in the mantle (about 0.24). The low Poisson's ratio in the uppermost mantle may well be a reflection of both the

low cratonic geotherm and a relatively high-proportion orthopyroxene in the mantle rocks (James et al., 2001a).

2.2.2. Crustal reverberations

A series of remarkably clear crustal reverberation phases, indicated by arrows with a question mark on Fig. 9a, are observed in the time interval following the direct-arriving Pn phase. Identifiable Moho interacting P-wave phases along with their particle motions and ray paths are shown in Fig. 10. While the identification of PmP and PmPPmp phases was unambiguous, the identification of the phase labeled 3P1S was considerably more problematic despite the fact that the same phase was clearly observed on synthetic seismograms as discussed below. Several characteristics of the phase provided the essential clues as to its identity. First, as shown in Fig. 9, the phase exhibits a systematic increase in amplitude with distance from the event. The particle motion diagram shown in Fig. 10 confirms that all of the marked phases arrive as incident P waves. Predicted arrival times shown in Fig. 11 for the 3P1S phase indicate that the times are correct for a reverberation phase in which three legs are P waves and one leg is an S wave. Further analysis based on modeling with reflectivity synthetic seismograms proved conclusive. Fig. 12 shows a series of computed broadband reflectivity seismograms with frequency passband 0.01–2.0 Hz to mimic the observed signals from the Welkom mine event. In Fig. 12, the top two traces are vertical component records and the lower two traces are radial component records. For the purposes of this particular reflectivity diagnostic, the source was constrained to be a point explosion (isotropic source) with moment tensors Mrr = Mtt = Mff = 1.0 and Mrt = Mrf = Mtf = 0.0. For a point explosion, only P-wave energy will be produced. The crustal model used in the computation is as in Table 1, with the Moho assumed to be a first-order discontinuity. The first and third traces shown in Fig. 12 are vertical and radial component synthetic seismograms computed to include only refractions and reflections. The second and fourth traces (again, vertical and radial) were computed to include all possible phases, including P/S conversions (P-to-S or S-to-P). The presence of the 3P1S phase only on those seismograms for which converted phases were included in the computation demonstrates conclusively that the phase must

Table 1
Velocity–depth model for travel time calculations and computation of synthetic seismograms

Layer thickness (km)	V_P (km/s)	V_S (km/s)	Density (gm/cc)	Q_P	Q_S
1.0	5.0	2.91	2.65	100.0	40.0
2.0	6.03	3.51	2.70	200.0	80.0
2.0	6.09	3.54	2.71	400.0	175.0
2.0	6.15	3.57	2.72	600.0	270.0
2.0	6.21	3.60	2.73	1000.0	450.0
2.0	6.26	3.64	2.74	1000.0	450.0
2.0	6.31	3.68	2.75	1000.0	450.0
2.0	6.37	3.72	2.76	1000.0	450.0
2.0	6.43	3.75	2.77	1000.0	450.0
2.0	6.48	3.77	2.78	1000.0	450.0
2.0	6.51	3.78	2.79	1000.0	450.0
2.0	6.54	3.79	2.80	1000.0	450.0
2.0	6.58	3.81	2.81	1000.0	450.0
2.0	6.62	3.82	2.82	1000.0	450.0
2.0	6.64	3.83	2.83	1000.0	450.0
2.0	6.67	3.85	2.83	1000.0	450.0
2.0	6.70	3.87	2.84	1000.0	450.0
2.0	6.73	3.89	2.86	1000.0	450.0
5.00	8.20	4.79	3.31	1000.0	450.0
5.00	8.24	4.82	3.33	800.0	350.0
5.00	8.28	4.84	3.34	800.0	350.0
5.00	8.32	4.86	3.35	800.0	350.0
40.00	8.45	4.91	3.41	800.0	350.0
00.00	8.60	5.03	3.50	700.0	300.0

All parameters are constant within layers.

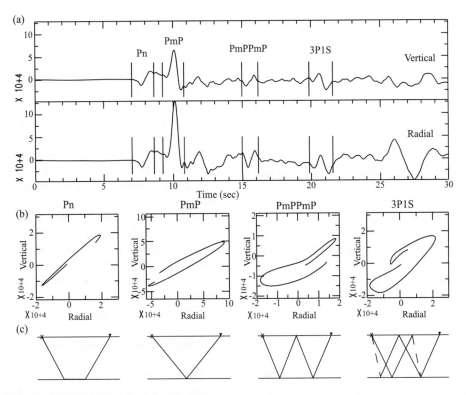

Fig. 10. (a) Vertical and radial component waveforms with phases Pn, PmP, PmPPmP and 3P1S indicated by lined intervals. (b) Plots showing particle motion for phases indicated in (a). (c) Ray paths for the four phases analyzed. Note that 3P1S is a multiplet, with three kinematically equivalent ray paths including SPPP, PSPP and PPSP.

involve at least one converted reflection. In fact, since the source in this case contains no S-wave energy and the wave arrives at the station as a P wave, then two converted reflections must be involved. Thus, there must be a P-to-S conversion either at the Moho or the free surface followed by an S-to-P conversion either at the free surface or the Moho (see ray diagram in Fig. 10). For the Welkom event, where S-wave energy is also generated, the first conversion may be an S-to-P at the Moho. A final test of the phase identification and waveform matching is shown in Fig. 13, where pairs of observed and synthetic seismograms are shown at selected distances over the range of observations. A diagnostic characteristic of 3P1S in both observed and synthetic records is increasing amplitude with distance.

The fact that 3P1S is a multiplet and yet produces a clean and well-correlated signal over a considerable range of distances and azimuths on observed seismograms implies a simple Moho structure and consistent

crustal thickness between the earthquake source and the seismic stations. Waveform and amplitude correlation between observed and synthetic seismograms provide a basis to estimate the maximum allowable thickness of the Moho transition zone as shown in Fig. 14. The top trace in Fig. 14 is the vertical component seismogram for station BB05. The lower traces (b–e) are synthetic seismograms computed for successively broader Moho transition zones, in 1-km increments, from zero thickness in frame (b) to 3 km in frame (e). While this test is not highly quantitative, it reveals that the observed 3P1S signal is similar in amplitude to that computed for a zero thickness Moho transition. Models with a Moho transition thickness in excess of about 2 km produce synthetic signals that are clearly too small in amplitude to match the observed signal. We note that a velocity gradient across the Moho is but one factor that will act to degrade the 3P1S reverberation signal. In particular, the effects of even modest variations in Moho topog-

04/22/99, 2.1 km, M5.7, vertical

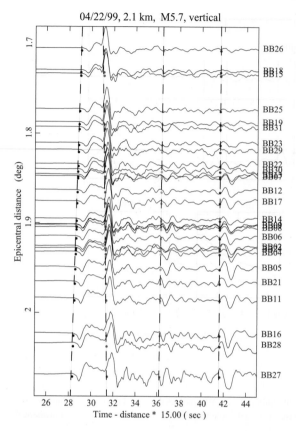

Fig. 11. Vertical component record section for the Welkom event with predicted times of arrival for Pn and reverberation phases indicated by the dashed lines. Seismograms have been aligned on the Pn phase.

raphy or average velocity in the crust will produce phase shifts between the kinematically equivalent arrivals that go to make up the 3P1S reverberation. All such phase shifts will result in interference that acts to degrade the 3P1S signal. We conclude, therefore, that the presence of an uncomplicated signal is further evidence for the uniformity of crust and Moho and the flatness of the Moho not only beneath the Kimberley array but in the extensive region between the Welkom mine and the seismic array.

3. Discussion and conclusion

The analysis of both teleseismic and local events recorded on the Kimberley broadband array reveals a remarkably uniform crustal structure both beneath the array (Fig. 15) and in the region between array and the Welkom mine. The results presented in this paper represent the most precise quantitative seismic measurements of crustal and Moho properties to date for southern Africa.

Three first-order conclusions about the nature of crust and Moho in the Kimberley region emerge from the analysis of the broadband array data. (1) The crust is thin, about 35 km thick; it is of average felsic to intermediate composition even in its lowermost section, and it is uniform over thousands of square kilometers. (2) The Moho is a sharp discontinuity, with the crust–mantle transition less than 0.5 km beneath the array and less than 2 km along the Welkom transect outside the array. (3) The Moho is almost perfectly flat, with less than 1 km of relief beneath the array. Beneath the sampled region of the Welkom transect, there is virtually no discernable difference in crustal thickness with that beneath the array.

3.1. Implications of crustal composition

The low Poisson's ratio of 0.25 for the crust as a whole and a calculated density of 2.86 gm/cc for lowermost crustal rocks provide compelling evidence that the lower crust beneath the Kaapvaal craton is dominated by rocks of intermediate to felsic rather than mafic composition (Christensen and Fountain, 1975). The absence of a mafic signature in the seismic results is consistent with the fact that mafic granulite xenoliths from the lower Kaapvaal crust are rare (Schmitz and Bowring, 2003; Van Calsteren et al., 1986). These observations highlight one of the great dilemmas of Archean crustal composition: if the continental crust was derived from a hotter mantle in the Archean, why is the crust not basaltic or even picritic in composition (Rudnick, 1995)? Instead, not even the lowermost crust beneath the Kimberley area is significantly mafic in composition. In this respect, the crust beneath the Kaapvaal appears to be less mafic than that beneath adjacent Proterozoic mobile belts, where mafic granulite xenoliths from the lower crust are common (Schmitz and Bowring, 2003).

We propose that the present composition of the Kaapvaal crust is most plausibly explained as the result of extensive remelting of the lower crust. This conclusion is buttressed by a recent study showing that the lower Kaapvaal crust was subjected to ultrahigh-tem-

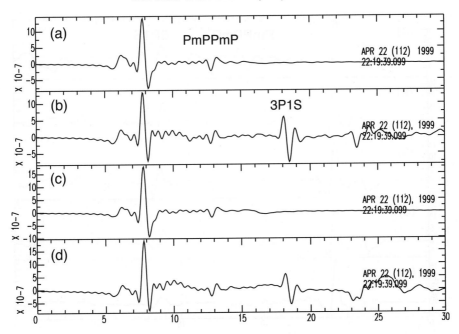

Fig. 12. Frames (a) and (b) are vertical component reflectivity synthetic seismograms for an explosive source (moment tensor Mrr = Mtt = Mff = 1.0 and Mrt = Mrf = Mtf = 0.0) at the Welkom event location. Only P-wave energy is produced by the source. Frames (c) and (d) are radial component synthetic seismograms for the same moment tensor source. Traces (a) and (c) show synthetic seismograms for which refracted and reflected phases have been computed but for which no conversions have been allowed. Traces (b) and (d) are synthetic seismograms for which all phases (including conversions) have been calculated. PmP and PmPPmP are seen on traces (a) and (c) as predicted, but the later phase labeled 3P1S is found only on synthetic seismograms where conversions are included in the computation. This result, along with the travel time correlation shown in Fig. 10, confirms that 3P1S is correctly identified. See text for further discussion.

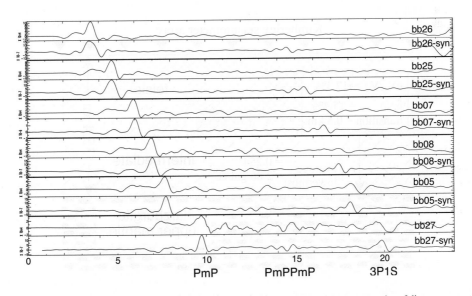

Fig. 13. Vertical component observed and synthetic seismogram pairs for the Welkom mine event over a series of distance ranges from a nearby station (top) to a distant station (bottom). Synthetic seismogram is the bottom trace in each pair of panels. The figure confirms the increasing amplitude of 3P1S with distance on both the synthetic seismograms and the observed records.

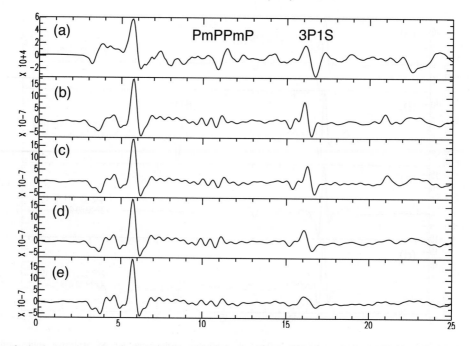

Fig. 14. Series of seismic traces showing the impact of an increasingly gradational Moho on the 3P1S signal strength for the Welkom mine event. The top panel (a) shows the observed seismogram for station bb05. Subsequent panels are reflectivity synthetic seismograms for (b) zero thickness transition zone from crust to mantle at the Moho; (c) 1-km Moho transition; (d) 2-km Moho transition; and (e) 3-km Moho transition. The amplitude of PmP/3P1S of the observed vs. synthetic signals suggests that the Moho transition can be no more than about 2 km in width and may be much sharper.

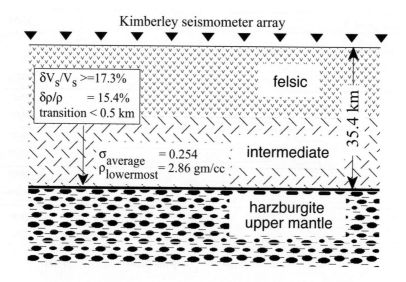

Fig. 15. Interpretive cross-section of crust and Moho structure of the Kaapvaal craton beneath the Kimberley array. σ = Poisson's ratio averaged for the entire crust. ρ = crustal density at the crust–mantle boundary. Compositional terms "felsic" and "intermediate" are meant as averages. Stations are shown schematically as solid inverted triangles (from Niu and James, 2002).

perature metamorphism during the Ventersdorp tectonomagmatic event around 2.7 Ga (Schmitz and Bowring, 2003). The Ventersdorp event, which produced temperatures in excess of 1000 °C in the lower crust of the craton, affected the whole of the western Kaapvaal (including the Kimberley area). While the effects of very high temperatures on the lower crust beneath Kimberley associated with the Ventersdorp event are difficult to quantify, a case study of a similar tectonomagmatic event exposed in the Vredefort meteorite impact structure provides considerable insight into the effect of high temperatures on the lower crust. A major thermal event similar to that of the Ventersdorp, but predating the docking of the Kimberlite block to the nuclear Kaapvaal craton, was analyzed as part of a geochronological study of crustal rocks exposed in the Vredefort dome located some 300 km NW of Kimberley (Moser et al., 2001). The Vredefort complex is an extraordinary structure comprising an almost complete section of cratonic crust that has been turned on-edge by isostatic rebound following a massive meteorite impact around 2.0 Ga (Tredoux et al., 1999). Moser et al. (2001) carried out a detailed study of middle and lower crustal rocks (which are dominantly felsic to intermediate in composition with subordinate mafic interlayering) exposed in the eroded remains of the crater. They presented evidence that the rocks exposed in the Vredefort dome record a complex history of crustal formation around 3.5 Ga followed by large-scale remelting during a craton-wide thermal event at 3.11 Ga. Based on the distribution and ages of rocks in the Vredefort crustal section, they concluded that as much as 40% of the crust, chiefly the lower crust, was remelted during a craton-wide thermal event at 3.11 Ga. They suggested, moreover, that lithospheric strain associated with the tectonomagmatic event was not propagated into the upper crust, but was instead accommodated by "deep crustal ductile flow in a hotter Archean crust" (Moser et al., 2001, p. 467). We hypothesize that temperatures in excess of 1000 °C in the lowermost crust, coupled with ample evidence for widespread anatectic melting during the Ventersdorp event, could well imply similar or even greater degrees of partial melting and ductile flow in the crust beneath Kimberley. Certainly, the very high degree of partial melting proposed by Moser et al. for the middle and lower crust is sufficient to produce widespread flow and magmatic differentiation in the lowermost

crust. Layering associated with the crystallization of melt in the lower crust is one possible means for producing both a flat and sharp Moho as well as the evolved rock compositions that we infer from the seismic data.

3.2. Implications of Moho structure

Both within the area of the array and in the region between the Welkom mine and the array, the Moho appears to be remarkably sharp and flat. While it may be possible under special conditions to form a primary crustal section such as that observed beneath the Kimberley array by accretionary processes, the Moho structures observed are more plausibly produced by reworking during subsequent tectonomagmatic events. Models of the assembly history of the Kaapvaal craton typically involve extensive collisional accretion of island arcs and microcontinental blocks to form nuclear continental masses (De Wit et al., 1992). Such accretionary processes may be expected to produce a complicated mosaic of varying Moho structures and diverse crustal lithology. We observe just such structures in the Limpopo Belt, a late Archean collisional zone sandwiched between the Zimbabwe and Kaapvaal cratons that exhibits many traits of tectonic deformation seen in modern continent–continent collision, including extensive crustal thickening and a highly complex Moho (Nguuri et al., 2001). Such structures are notably absent beneath undisturbed Kaapvaal craton.

Lack of Moho relief in regions where topography on the Moho might be expected is surprisingly common. That fact is widely attributed to post-formation modification of the lower crust through magmatic reworking and ductile flow (e.g. Mooney and Meissner, 1992), and the role of thermal and deformational processes in modifying the lower crust has been discussed at length in the literature (e.g. Murrell, 1986). The relatively flat Moho in extensional provinces such as the Basin and Range in the western US, for example, is commonly interpreted to be due to reforming of the Moho by igneous and ductile deformation in the deep crust (e.g. Klemperer et al., 1986). As noted by Mooney and Meissner (1992), such igneous reworking implies that the lowermost crust and the crust–mantle transition is effectively the "youngest" layer within the continental crust. Thus, the interpretation of the seismic evidence from southern Africa must be considered in the context of the tectonothermal history of the Kaapvaal craton. It therefore follows that extensive remelting and ductile flow in the lower Kaapvaal crust are probable consequences of the profound thermal perturbation resulting from the Ventersdorp intracratonic event, which involved flood basalt magmatism, crustal extension and widespread anatexis (Schmitz and Bowring, 2003). Indeed, Schmitz and Bowring (2003) speculate that "during the Ventersdorp event, middle to lower crustal flow spatially coherent with simultaneous brittle extension and crustal magmatism operated across much of the northern and western [Kaapvaal] craton, erasing primary variations in Moho topography".

In terms of geologically plausible scenarios, we view thermal reactivation and large volume melting of the cratonic crust in late Archean time as the most viable hypothesis for flattening and sharpening of the Moho and for the absence of mafic layering in the lowermost crust. Alternative explanations for the observed Moho structures do remain, however, and one is particularly relevant to the discussion here. Moho and lower crustal structure imaged as part of Lithoprobe seismic reflection profiling across the Phanerozoic accreted terranes of southwestern Canada suggest that a relatively flat Moho is not necessarily inconsistent with accretionary tectonics or with large-scale crustal structures that are geometrically discordant with a flat Moho (Cook, 1995). As evidenced from the seismic reflection data, upper crustal structures appear to flatten into the deep crust and Moho, tending to converge into what Cook interprets to be zones of regional detachment. This mechanism for producing a flat and relatively sharp Moho may also, of course, imply some degree of ductile flow, given the clear geometrical relationship between Moho and deep crustal reflections with higher angle fault structures in the shallower crust (Cook, 1995). While the Canadian model is not necessarily applicable to the Kaapvaal craton, regional detachment at the Moho may be a viable means by which to produce primary Moho structures that are relatively flat during the processes of crustal aggregation. In that hypothetical circumstance, the Ventersdorp thermal remobilization of the lower crust and the ductile stretching resulting from crustal rifting would act to enhance flattening of the Moho and to sharpen the crust–mantle transition zone even further. Such a combination of primary and late stage tectonothermal processes may suffice to explain both lowermost crust-al compositions and the remarkable Moho structures observed beneath the Kaapvaal craton.

Acknowledgements

The Kaapvaal Project involved the efforts of more than 100 people affiliated with about 30 institutions. Details of participants and a project summary can be found on the Kaapvaal website, http://www.ciw.edu/kaapvaal. We owe a special debt of thanks to the PASSCAL team that manages the telemetered array, notably Frank Vernon and Jim Fowler, with the able technical support of Glen Offield. Rod Green, with support from Jock Robey, Josh Harvey, Lindsey Kennedy and others of the de Beers geology group in Kimberley, sited and constructed the stations of the array. As always, supertech Randy Kuehnel was indispensable. The data were processed through Frank Vernon's laboratory at UCSD (IGPP) and are archived at the IRIS Data Management Center from which were obtained the selected data used in this study. The Kaapvaal Project is funded by the National Science Foundation Continental Dynamics Program (EAR-9526840) and by several public and private sources in southern Africa. F.N. thanks the Carnegie Institution for support under its postdoctoral fellowship program. Map figures were produced with GMT (Wessel and Smith, 1991). Comments from Mark Schmitz, Roberta Rudnick, Andy Nyblade, Alan Linde, Selwyn Sacks, Liangxing Wen, Alan Jones and two anonymous reviewers for Lithos were very helpful in improving earlier versions of this work.

References

Boyd, F.R., McCallister, R.H., 1976. Densities of fertile and sterile garnet peridotites. Geophys. Res. Lett. 3 (9), 509–512.

Carney, J.N., Aldiss, D.T., Lock, N.P., 1994. The Geology of Botswana. Bulletin, vol. 37. Geological Survey Department, Garorone.

Christensen, N.I., 1982. Seismic velocities. In: Carmichael, R.S. (Ed.), Handbook of Physical Properties of Rocks. CRC Press, Boca Raton, FL, USA, p. 228.

Christensen, N.I., Fountain, D.M., 1975. Constitution of the lower continental crust based on experimental studies of seismic velocities in granulite. Geol. Soc. Am. Bull. 86, 227–236.

Christensen, N.I., Mooney, W.D., 1995. Seismic velocity structure and composition of the continental crust; a global view. J. Geophys. Res. 100 (6), 9761–9788.

Cook, F.A., 1995. The reflection Moho beneath the southern Canadian Cordillera. Can. J. Earth Sci. 32, 1520–1530.

De Wit, M.J., et al., 1992. Formation of an Archaean continent. Nature 357 (6379), 553–562.

Durrheim, R.J., Mooney, W.D., 1991. Archean and Proterozoic crustal evolution; evidence from crustal seismology. Geology (Boulder) 19 (6), 606–609.

Durrheim, R.J., Mooney, W.D., 1994. Evolution of the Precambrian lithosphere; seismological and geochemical constraints. J. Geophys. Res. 99, 15359–15374.

Fedo, C.M., Eriksson, K.A., Blenkinsop, T.G., 1995. Geologic history of the Archean Bukwa Greenstone Belt and surrounding granite-gneiss terrane, Zimbabwe, with implications for the evolution of the Limpopo Belt. Can. J. Earth Sci. 32, 1997–1990.

Griffin, W.L., O'Reilly, S.Y., 1987. The composition of the lower crust and the nature of the continental Moho—xenolith evidence. In: Nixon, P.H. (Ed.), Mantle Xenoliths. Wiley, Chichester, UK, pp. 413–432.

James, D.E., Carlson, R.W., Boyd, F.B., Janney, P.E., 2001a. Petrologic constraints on seismic velocity variations in the upper mantle beneath southern Africa. EOS, Trans. - Am. Geophys. Union 82 (20 (Abstract Suppl.)), S247.

James, D.E., Fouch, M.J., VanDecar, J.C., van der Lee, S., and Kaapvaal Seismic Group, 2001b. Tectospheric structure beneath southern Africa. Geophys. Res. Lett. 28 (13), 2485–2488.

Kanasewich, E.R., 1973. Time sequence analysis in geophysics. University of Alberta Press, Alberta, Canada. 352 pp.

Kato, M., Kawakatsu, H., 2001. Seismological in situ estimation of density jump across transition zone discontinuities beneath Japan. Geophys. Res. Lett. 28, 2541–2544.

Klemperer, S.L., Hauge, T.A., C, H.E., Oliver, J.E., Potter, C.J., 1986. The Moho in the northern Basin and Range province, Nevada, along the COCORP 40 degree N seismicreflection transect. Geol. Soc. Am. Bull. 97, 603–618.

Mooney, W.D., Meissner, R., 1992. Multi-genetic origin of crustal reflectivity: continental lower crust and Moho. In: Fountain, D.M., Arculus, R., Kay, R.W. (Eds.), Continental Lower Crust. Elsevier, Amsterdam, pp. 45–79.

Moser, D.E., Flowers, R.M., Hart, R.J., 2001. Birth of the Kaapvaal tectosphere 3.08 billion years ago. Science 291, 465–468.

Muirhead, K.J., 1968. Eliminating false alarms when detecting seismic events automatically. Nature 217, 533–534.

Murrell, S.A.F., 1986. The role of deformation, heat, and thermal processes in the formation of the lower continental crust. In: Dawson, J.B., Hall, J., Wedepohl, K.H. (Eds.), The Nature of the Lower Continental Crust. Geological Society Special Publication, vol. 24. Blackwell, Oxford, pp. 107–117.

Nguuri, T., et al., 2001. Crustal structure beneath southern Africa and its implications for the formation and evolution of the Kaapvaal and Zimbabwe cratons. Geophys. Res. Lett. 28 (13), 2501–2504.

Niu, F., James, D.E., 2002. Constraints on the formation and composition of crust beneath the Kaapvaal craton from Moho reverberations. Earth Planet. Sci. Lett. 200, 121–130.

Rudnick, R.L., 1992. Xenoliths—samples of the lower continental crust. In: Fountain, D.M., Kay, R.W. (Eds.), The Continental Lower Crust. Elsevier, Amsterdam, pp. 269–315.

Rudnick, R.L., 1995. Making continental crust. Nature 378, 571–578.

Rudnick, R.L., Fountain, D.M., 1995. Nature and composition of the continental crust: a lower crustal perspective. Rev. Geophys. 33 (3), 267–309.

Schmitz, M.D., Bowring, S.A., 2000. The temporal diversity of lower crustal granulites throughout the Kaapvaal craton, Southern Africa. EOS, Trans. - Am. Geophys. Union 81, 1249 (Fall Meet. Suppl.).

Schmitz, M.D., Bowring, S.A., 2003. Ultrahigh-temperature metamorphism in the lower crust during Neoarchean Ventersdorp rifting and magmatism, Kaapvaal craton, southern Africa. Geol. Soc. Am. Bull. 115, 533–548.

Shirey, S.B., et al., 2001. Emplacement of eclogite components into the lithospheric mantle during craton formation. Geophys. Res. Lett. 28 (13), 2509–2512.

Tredoux, M., Hart, R.J., Carlson, R.W., Shirey, S.B., 1999. Ultramafic rocks at the center of the Vredefort structure: further evidence for the crust on edge model. Geology (Boulder) 27, 923–926.

Treloar, P.J., Coward, M.P., Harris, N.B.W., 1992. Himalayan–Tibetan analogies for the evolution of the Zimbabwean Craton and Limpopo belt. Precambrian Res. 55, 571–587.

Van Calsteren, P.W.C., Harris, N.B.W., Hawkesworth, C.J., Menzies, M.A., Rogers, N.W., 1986. Xenoliths from southern Africa: a perspective on the lower crust. In: Dawson, J.B., Carswell, D.A., Hall, J., Wedepohl, K.H. (Eds.), The Nature of the Lower Continental Crust. Oxford University Press, London, pp. 351–362.

Van Reenen, D.D., Roering, C., Ashwal, L.D., de Wit, M.J., 1992. Regional geological setting of the Limpopo belt. Precambrian Res. 55, 1–5.

Von Gruenewaldt, G., Sharpe, M.R., Hatton, C.J., 1985. The Bushveld Complex; introduction and review. Econ. Geol. 80, 803–812.

Wessel, P., Smith, W.H.F., 1991. Free software helps map and display data. EOS, Trans. - Am. Geophys. Union 72, 445–446.

Available online at www.sciencedirect.com

Lithos 71 (2003) 431–460

ELSEVIER

LITHOS

www.elsevier.com/locate/lithos

Major and trace element geochemistry of plutonic rocks from Francistown, NE Botswana: evidence for a Neoarchaean continental active margin in the Zimbabwe craton

A.B. Kampunzu[a,*], A.R. Tombale[b], M. Zhai[a], Z. Bagai[a], T. Majaule[c], M.P. Modisi[a]

[a] *Department of Geology, University of Botswana, Private Bag 0022, Gaborone, Botswana*
[b] *Ministry of Mineral, Energy and Water Affairs, Private Bag 0018, Gaborone, Botswana*
[c] *Department of Geological Survey, Private Bag 14, Lobatse, Botswana*

Abstract

The Neoarchaean Tati granite–greenstone terrane occurs within the southwestern part of the Zimbabwe craton in NE Botswana. It comprises 10 intrusive bodies forming part of three distinct plutonic suites: (1) an earlier TTG suite dominated by tonalites, trondhjemites, Na-granites distributed into high-Al (Group 1) and low-Al (Group 2) TTG sub-suite rocks; (2) a Sanukitoid suite including gabbros and Mg-diorites; and (3) a younger high-K granite suite displaying I-type, calc-alkaline affinities.

The Group 1 TTG sub-suite rocks are marked by high Sr/Y values and strongly fractionated chondrite-normalized rare earth element (REE) patterns, with no Eu anomaly. The Group 2 TTG sub-suite displays higher LREE contents, negative Eu anomaly and small to no fractionation of HREE. The primordial mantle-normalized patterns of the Francistown TTGs are marked by negative Nb–Ti anomalies. The geochemical characteristics of the TTG rocks are consistent with features of silicate melts from partial melting of flat subducting slabs for the Group 1 sub-suite and partial melting of arc mafic magmas underplated in the lower crust for the Group 2 sub-suite. The gabbros and high-Mg diorites of the Sanukitoid suite are marked by $Mg\# > 0.5$, high Al_2O_3 ($>>16\%$), low TiO_2 ($<0.6\%$) and variable enrichment of HFSE and LILE. Their chondrite-normalized REE patterns are flat in gabbros and mildly to substantially fractionated in high-Mg diorites, with minor negative or positive Eu anomalies. The primordial mantle-normalized diagrams display negative Nb–Ti (and Zr in gabbros) anomalies. Variable but high Sr/Y, Sr/Ce, La/Nb, Th/Ta and Cs/La and low Ce/Pb ratios mark the Sanukitoid suite rocks. These geochemical features are consistent with melting of a sub-arc heterogeneously metasomatised mantle wedge source predominantly enriched by earlier TTG melts and fluids from dehydration of a subducting slab. Melting of the mantle wedge is consistent with a steeper subduction system. The late to post-kinematic high-K granite suite includes I-type calc-alkaline rocks generated through crustal partial melting of earlier TTG material. The Neoarchaean tectonic evolution of the Zimbabwe craton is shown to mark a broad continental magmatic arc (and related accretionary thrusts and sedimentary basins) linked to a subduction zone, which operated within the Limpopo–Shashe belt at ~ 2.8–2.65 Ga. The detachment of the subducting slab led to the uprise of a hotter mantle section as the source of heat inducing crustal partial melting of juvenile TTG material to produce the high-K granite suite.
© 2003 Elsevier B.V. All rights reserved.

Keywords: Petrogenesis; Continental arc; TTG; Sanukitoids; High-K granites; Zimbabwe craton; Botswana

---

* Corresponding author.
E-mail address: kampunzu@mopipi.ub.bw (A.B. Kampunzu).

0024-4937/$ - see front matter © 2003 Elsevier B.V. All rights reserved.
doi:10.1016/S0024-4937(03)00125-7

1. Introduction

The Zimbabwe craton has a complicated tectonic and magmatic history, which is the subject of much controversy. The oldest crustal rocks preserved in this craton (Fig. 1) are ~ 3.57–3.45 Ga in age and occur within the Tokwe segment (Wilson et al., 1995; Horstwood et al., 1999; Dodson et al., 2001), although Re–Os isotopic analyses of chromites from Archaean ultramafic rocks indicated that the Zimbabwean subcontinental lithospheric mantle began to be separated from the asthenospheric mantle before 3.8 Ga (Nägler et al., 1997). Kusky and Kidd (1992) suggested that the Belingwe greenstone belt is a remnant of an oceanic plateau allochthon. However, the existence of an old (Mesoarchaean) continental crust allowed several workers to model the Neoarchaean evolution of the Zimbabwe craton in terms of continental rifting above mantle plumes (e.g. Bickle et al., 1994; Jelsma et al., 1996; Hunter et al., 1998). Blenkinsop et al. (1993) and Bickle et al. (1994) contend that the Belingwe greenstone belt was emplaced in an ensialic setting. The 2.75–2.65 Ga igneous rocks from the Zimbabwe craton are tonalite–trondhjemite–granite and tholeiitic and calc-alkaline volcanic suites emplaced in arc and back-arc settings (e.g. Condie and Harrison, 1976; Jelsma et al., 2001). The greenstone belts (e.g. Tati) aligned along the southern part of the Zimbabwe craton (Fig. 1) and the Matsitama greenstone belt in northeast Botswana were taken for allochtonous sheets along the suture of an ocean closed at ca. 2.7 Ga (Kusky, 1998). The Matsitama greenstone belt is part of the Limpopo–Shashe orogenic belt (Ranganai et al., 2002) and this raises the question of the geotectonic relations between the evolution of the Zimbabwe craton and the Limpopo belt. A number of workers (Coward et al., 1976; McCourt and Wilson, 1992) indicated that the Zimbabwe craton was affected by shortening in response to Neoarchaean northerly thrusting of the Northern Marginal Zone (NMZ) of the Limpopo belt. The shear zones in the Zimbabwe craton are inferred to originate during the 2.6 Ga collision of the craton with the central zone of the Limpopo belt (Treloar et al., 1992; Rollinson, 1993; Treloar and Blenkinsop, 1995), although there is no agreement on this interpretation. Structural studies showed that several Neoarchaean greenstone belts of the Zimbabwe craton were affected by accretionary layer-parallel shear zones between ~ 2.7 and 2.6 Ga (Kusky and Kidd, 1992; Jelsma and Dirks, 2000). Dirks et al. (2002) reported an early episode of thin-skinned thrusting and a later steep west-directed thrusting and interpret these structures as indicating progressive accretion and crustal thickening marking a low-angle subduction or underplating. Steep geothermal gradients (~ 50 °C/km) characterise this tectonic crustal thickening (e.g. Jelsma and Dirks, 2000).

During the last decade, published geotectonic interpretations of the Zimbabwe craton focused on geochronological and structural data, with little emphasis on the geochemistry of igneous rocks (e.g. Condie and Harrison, 1976; Luais and Hawkesworth, 1994; Majaule et al., 1997; Jelsma et al., 2001). Geochemical studies of Archaean igneous rocks show a diversity of composition that matches that documented in modern igneous provinces emplaced in various tectonic settings (e.g. Dostal and Mueller, 1992; Hollings et al., 1999). Some workers (e.g. Hamilton, 1998) claim that Archaean granite–greenstone terranes mark igneous and tectonic processes different from Proterozoic and Phanerozoic plate tectonics. However, most igneous rock types recorded in Archaean terranes are known in Phanerozoic igneous provinces (e.g. Drummond and Defant, 1990; Martin, 1993) and therefore these modern analogues can be used for petrogenetic and geotectonic reconstruction of Archaean terranes.

The objectives of this paper are: (1) to present new chemical data of Neoarchaean plutonic rocks from the Zimbabwe craton exposed in the Tati granite–greenstone terrane in northeast Botswana; (2) to discuss the petrogenesis of the plutonic rocks and the tectonic setting during their emplacement; (3) to re-evaluate the relationship between the Zimbabwe craton and the Limpopo–Shashe belt during the emplacement of the Francistown granitoids and coeval igneous rocks in the Zimbabwe craton. The data in this paper are important for the understanding of Archaean accretionary processes in southern Africa and the formation, growth and preservation of the Archaean crust in the Zimbabwe craton.

2. Geological setting

The Zimbabwe craton is composed of granite–greenstone terranes emplaced between ca. 3.5 and 2.5

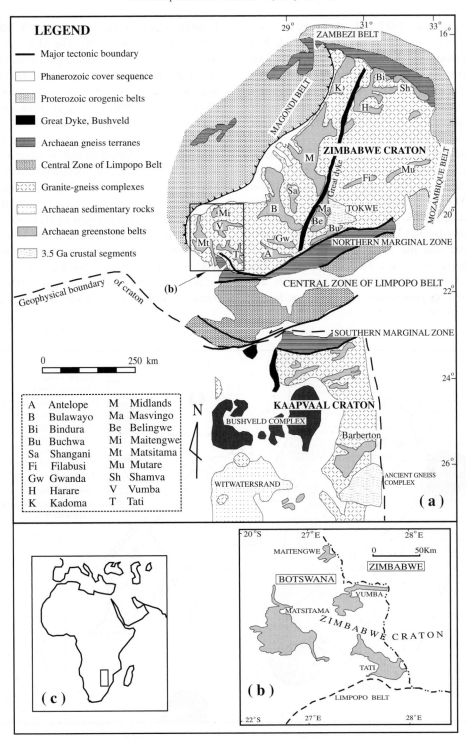

Fig. 1. (a) Map showing the main geological units of southern Africa cratons and adjacent Proterozoic belts (Bagai et al., 2002 and references therein). The rectangle locates the map of (b). (b) Distribution of the four main granite–greenstone terranes of the Zimbabwe craton in NE Botswana (slightly modified from Key et al., 1976). (c) Location of the southern Africa cratons in Africa.

Ga (Blenkinsop et al., 1997 and references therein). The southwestern margin of this craton extends into northeastern Botswana (Key et al., 1976) where four granite–greenstone terranes (Tati, Vumba, Maitengwe and Matsitama; Fig. 1) occur. The Matsitama granite–greenstone complex is dominated by shallow water clastic and chemical sedimentary rocks and minor mafic–ultramafic rocks (Aldiss, 1991; Majaule et al., 1997). Gravity data show that this granite–greenstone terrane is located within the Limpopo–Shashe belt (Ranganai et al., 2002) and the crustal thickness beneath the Matsitama area is ~ 50–55 km (Nguuri et al., 2001), in contrast to the cratonic areas where the crust is thinner (~ 35–37 km). U–Pb single zircon dating showed that the Matsitama granitoids were emplaced between 2710 ± 19 and 2646 ± 3 Ma (Majaule and Davis, 1998). These granitoids are time-equivalent to voluminous felsic plutons in the

Limpopo belt (e.g. Berger et al., 1995; Mkweli et al., 1995; Kröner et al., 1999). The Matsitama granitoids are also coeval to the Vumba granitoids (Bagai et al., 2002) and to Upper Bulawayan greenstone belts and Sesombi granitoid suites in the Zimbabwe craton (Wilson et al., 1995; Luais and Hawkesworth, 1994; Jelsma et al., 1996).

The Tati granite–greenstone terrane (Fig. 2) includes the Tati greenstone belt, voluminous granitoids and gabbros, and minor metasedimentary rocks (e.g. Key et al., 1976; Tombale, 1992). Preliminary U–Pb zircon geochronological data (Kampunzu, unpublished) and the continuation of granitoid bodies from Francistown (Tati terrane) up to the Vumba granite–greenstone terrane suggest similar emplacement ages (e.g. Key et al., 1976; Tombale, 1992), i.e. between ~ 2.73 and 2.65 Ga (Bagai et al., 2002). Ten main intrusive complexes were identified in the Tati gran-

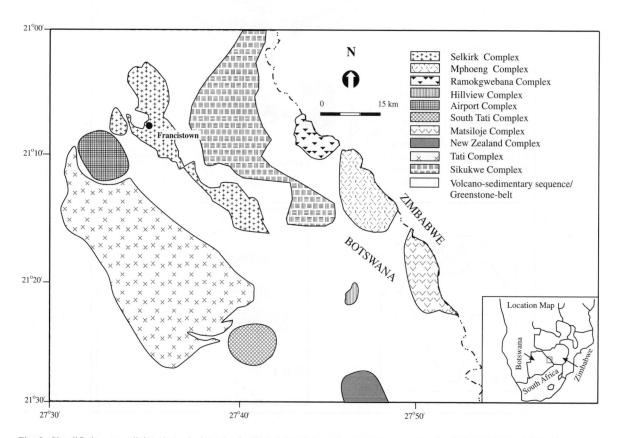

Fig. 2. Simplified map outlining the main intrusive bodies of the Tati granite–greenstone terrane (modified from Tombale, 1992). Insert: Location of the main map in southern Africa.

ite–greenstone terrane (Fig. 2). Four (Mphoeng, Ramokgwebana, Matsiloje and Sekukwe) are located northeast of the Tati greenstone belt and extends into Zimbabwe to the east and northeast. This group is referred to as the northeastern igneous complexes. The Ramokgwebana granitoids include coarse to porphyritic granites made of K-feldspar (megacrystic), quartz, biotite, amphibole, zircon and titanite. The Matsiloje batholith is made of coarse tonalites and trondhjemites. The Mphoeng complex consists of coarse to porphyritic gabbros, tonalites and trondhjemites. The gabbros consist of plagioclase in a dark coarse-grained groundmass of pyroxene, plagioclase, minor hornblende, biotite and opaque minerals. TTGs exposed in the northeastern igneous complexes are large bodies preserving igneous fabrics, except along discrete shear zones. These rocks are equigranular and contain plagioclase, quartz, amphibole, biotite, zircon, titanite, apatite and magnetite. The southwestern igneous complexes, located southwest of the Tati greenstone belt, include the Tati complex (Fig. 2) and its satellites called Airport, South Tati and New Zealand complexes which intrude the Tati greenstone belt lithologies. The most important lithologies in the southwestern igneous complexes are coarse grey tonalites, trondhjemites and granites converted into orthogneisses (Tombale, 1992). Tonalite and trondhjemite of the southwestern complex are foliated to banded grey gneisses (Martin, 1994) containing ortho-amphibolitic enclaves. The main fabric in these rocks trend NW–SE. The most frequent minerals in these gneisses are plagioclase, quartz, biotite, amphibole, titanite, epidote, zircon, magnetite and apatite. Microcline occurs in some samples. In the southwestern igneous complexes, an intense tectonic foliation occurs close to the Limpopo–Shashe belt. Foliated pink K-feldspar megacrystic granites and granite dykes intruding the southwestern igneous complex close to the Limpopo–Shashe belt could be linked to the evolution of that belt. These granites were excluded from this study that is focused on the evolution of the Zimbabwe craton. The central igneous complexes include Selkirk and Hillview complexes. The Selkirk complex is elongated NW–SE, parallel to the trend of the Tati greenstone belt. The absence of a pervasive NW–SE solid-state fabric in this complex suggests that it post-dates the emplacement of the southwestern igneous complexes. It includes troctolites, gabbros

and diorites intruding the Tati greenstone belt lithologies. The primary minerals in these rocks include plagioclase, olivine, pyroxene and opaque minerals. The Ni–Cu sulphide deposits of Selkirk and Phoenix mines are hosted in this complex.

3. Sampling and analytical techniques

Hundred rock samples selected for geochemical analyses were collected from mine drillcores and from large grey gneiss complex outcrops where the least altered and deformed rocks are exposed. We also sampled plutonic bodies preserving igneous mineralogy and/or igneous foliation in the northern and central complexes. The locations of the samples used in this study are compiled on the new 1/125,000 map of the Botswana Geological Survey Department (Francistown Quarter Degree sheet, to be published in 2003).

Samples with initials AR were analysed at Memorial University of Newfoundland (Canada). Major element compositions were determined using a Perkin Elmer (except for P_2O_5: colourimetry) and trace elements Rb, Sr, Ga, Zr, Y, Nb, Th, U, Pb, Zn, V, Cr and Ni were analysed using a Philips 1450 XRF. The other trace elements, including REE, were determined using an ICP-MS (Jenner et al., 1990). The remaining samples were analysed at Chemex laboratories (Canada) using ICP-AES (major elements) and ICP-MS (trace elements), except Li, Cr, Pb and Ni, which were determined using AAS. The precision for all the chemical analyses is better than 1% for major elements and better than 5% for trace elements. Representative analyses of the main intrusive rock-types are listed in Table 1. Several samples yielded relatively high LOI values, reflecting the alteration of the rocks. However, the variation trends and concentrations of major elements, high-field-strength elements (HFSE), REE and transition metals are similar to modern plutonic rocks and are thought to reflect primary magmatic distributions. Alkalies show some scatter and were probably slightly modified by secondary processes in some samples and REE contents of some samples (e.g. NN6 and NN8) show abnormal low values correlating with petrographic features (e.g. sericitisation) suggesting that these samples were affected by hydrothermal alteration. These rocks were

Table 1

| | Sanukitoid suite | | | | | | | | | TTG suite | | | | | |
| | Gabbros | | | | High-Mg diorites | | | | | Enclaves | | Tonalites – Trondjhemites – Na-granites | | | |
	MK115	AR227	TS86	TS165	AR128	AR234	AR71	AR245	MK32	AR77	AR78	TS169	MK85	AR63	MKQ3
SiO_2	44.00	48.80	49.84	49.83	52.40	53.10	53.40	56.50	59.00	48.90	51.10	60.59	61.77	62.22	67.33
TiO_2	0.09	0.50	0.21	0.21	0.52	0.36	0.28	0.36	0.56	1.63	1.24	0.75	0.73	0.91	0.58
Al_2O_3	26.00	21.70	18.66	16.59	14.30	17.30	16.20	17.20	10.76	14.00	14.01	14.80	14.27	15.71	15.06
Fe_2O_3	4.33	0.71	7.23	7.78	0.98	0.59	0.65	0.58	8.32	1.34	1.59	4.38	6.08	0.71	4.31
FeO		5.71			7.90	4.78	5.28	4.66		10.81	12.87			5.77	
MnO	0.06	0.11	0.13	0.14	0.17	0.08	0.11	0.08	0.13	0.17	0.19	0.06	0.09	0.12	0.06
MgO	8.98	4.88	7.82	8.31	7.39	7.74	7.63	7.74	8.02	4.49	4.18	3.15	3.14	3.38	1.41
CaO	13.69	13.08	13.80	13.24	7.76	6.66	12.94	6.66	8.84	10.60	8.10	7.41	5.45	5.23	4.03
Na_2O	1.10	1.85	1.60	1.43	3.82	2.42	1.80	2.42	2.33	2.90	2.93	6.17	3.63	3.74	4.24
K_2O	0.07	0.14	0.14	0.20	0.20	2.31	0.13	2.31	0.78	0.36	0.38	0.19	1.96	2.02	2.01
P_2O_5	0.01	0.03	0.01	0.01	0.08	0.14	0.04	0.14	0.08	2.15	1.93	0.18	0.26	0.19	0.16
LOI	1.75	1.94	1.41	1.52	1.76	2.27	1.06	2.33	0.88	0.47	0.76	2.88	0.88	0.88	0.58
Total	100.08	99.46	100.85	99.26	97.28	97.75	99.52	100.98	99.70	97.82	99.28	100.56	98.26	100.88	99.77
Mg#	0.80	0.57	0.68	0.68	0.60	0.72	0.69	0.72	0.65	0.40	0.34	0.59	0.51	0.48	0.39
Cr	694		16.0	24.0	234	188		194	118		74.0	29.0	52.0		12.0
Ni	275		26.0	32.0	81.0	340		339	48.0		2.0	6.0	43.0		10.0
Co	41.5		45.5	50.0					43.0			13.5	27.5		21.5
V			125	115	248	76.0		73.0	165		146	95.0	105		55.0
Cu	5.0		80.0	75.0	16.0	53.0		53.0	65.0		6.0	5.0	20.0		5.0
Pb	1.0	0.5	1.0	1.0		16.0	1.6	16.0	2.0		1.0	1.0	3.0		3.0
Zn	15.0		25.0	30.0	51.0	56.0		56.0	55.0		80.0	35.0	75.0		60.0
Rb	2.20	3.11	2.80	4.80	2.00	115	2.00	116	19.40	10.00	12.00	2.60	97		70.6
Cs	0.60	0.31	0.20	0.30	0.05	5.50	0.10	5.66	0.30	0.26		0.40	11.60	5.92	2.50
Ba	16.0	35.0	69.5	72.0	27.0	490	61.0	504	139	136		465	458	350	708
Sr	100	143	214	184	151	251	177	252	136	229	218	244	352	298	296
Li	1.0	12.0	3.0	4.0	6.1	29.0	7.6	32.2	6.0	8.7		3.0	31.0	191	25.0
Ta	1.00	1.11	2.00	1.50	0.10	0.76	0.08	0.75	3.00	0.26		2.50	4.00	0.55	5.50
Nb	1.00	0.97	1.00	1.00	3.00	4.00	1.00	5.00	4.00	3.29	5.00	7.00	8.00	6.28	11.00
Hf	1.00		1.00	1.00					3.00			5.00	4.00	0.00	5.00
Zr	1.00	8.00	10.00	7.50	45.00	108	16.00	111	99.5	33.5	81.0	169	195	65	207
Y	2.00	8.00	5.00	6.00	19.00	7.00	7.00	6.00	14.00	26.80	29.00	23.50	18.50	21.29	22.50
Th	3.00	0.20	1.00	1.00	1.00	20.00	0.80	18.00	6.00	2.20	1.00	3.00	7.00	3.20	6.00
U	0.50		0.50	0.50	0.30	6.00	0.20	5.00	2.00	0.70		0.50	3.00	2.07	3.00
La	1.00	1.61	2.50	3.00	3.30	24.13	4.00	23.80	12.00	21.10		16.50	35.00	26.19	45.5
Ce	1.50	4.00	4.50	5.00	7.89	46.66	8.10	44.77	25.50	45.81		36.5	65.0	41.6	85.5
Pr	0.20	0.60	0.70	0.70	1.09	5.28	1.00	5.07	3.10	6.13		4.60	8.00	5.86	8.90
Nd	1.00	2.87	3.00	3.50	5.04	19.03	4.14	18.48	12.00	27.57		18.50	28.50	23.12	31.00
Sm	0.20	1.00	0.70	0.90	1.56	3.41	1.00	3.37	2.90	6.16		4.00	4.90	4.60	5.70
Eu	0.10	0.47	0.40	0.60	0.50	0.74	0.41	0.77	0.90	2.10		1.30	1.50	1.49	1.50
Gd	0.30	1.28	1.10	1.10	2.14	2.91	1.19	3.13	2.80	6.94		5.00	4.90	4.40	6.50
Tb	0.10	0.22	0.10	0.10	0.38	0.33	0.18	0.30	0.50	0.89		0.70	0.80	0.65	0.80
Dy	0.30	1.47	1.00	1.10	2.59	1.67	1.16	1.78	2.80	5.23		4.60	4.30	3.87	4.50
Ho	0.10	0.31	0.10	0.20	0.59	0.32	0.25	0.31	0.50	1.00		1.00	0.60	0.78	0.80
Er	0.20	0.95	0.60	0.90	1.69	0.91	0.78	0.90	1.60	2.83		2.50	1.70	2.15	2.30
Tm	0.10	0.13	0.10	0.10	0.25	0.13	0.11	0.13	0.20	0.37		0.30	0.30	0.31	0.30
Yb	0.10	0.88	0.50	0.50	1.68	0.75	0.77	0.75	1.30	2.25		2.20	2.10	1.99	2.00
Lu	0.10	0.13	0.10	0.10	0.24	0.12	0.11	0.12	0.20	0.35		0.30	0.20	0.29	0.30

Except in samples with initial AR, Fe_2O_3 represents total iron.

excluded from the geochemical diagrams and the discussion of results. In the text, the compositions used when describing individual samples were recalculated to 100% anhydrous to minimise the effect of alteration on the samples.

4. Geochemical classification

The chemical analyses illustrate that the rocks exposed in the Francistown plutonic complex range from gabbros to granites (Table 1). The chemical

MK80	TS2	MKQ4	MK104	MK68	AR150	AR60	NN82B	AR188	TS168A	TS168B	MK36	AR181	MKQ2	TS167	MK34B
67.57	68.10	68.13	68.40	68.50	69.30	69.62	70.00	71.10	71.21	71.61	72.50	72.90	73.50	73.82	73.86
0.57	0.51	0.53	0.40	0.66	0.40	0.73	0.19	0.24	0.16	0.19	0.21	0.16	0.23	0.28	0.16
13.77	15.86	15.03	15.58	13.17	16.00	14.25	16.68	15.30	15.07	15.30	13.46	13.00	13.65	11.49	13.84
4.19	3.34	3.18	3.12	4.43	0.22	0.46	1.06	0.12	1.58	1.43	2.04	0.14	2.15	4.68	1.49
					1.79	3.70		0.99				1.12			
0.06	0.05	0.03	0.04	0.05	0.03	0.06	0.02	0.03	0.02	0.01	0.03	0.04	0.03	0.06	0.03
1.08	1.06	1.11	1.33	0.76	0.74	1.68	0.43	0.56	0.47	0.52	0.47	0.42	0.49	0.15	0.45
2.69	3.49	4.16	3.56	2.54	2.20	3.72	3.35	2.32	1.72	1.34	2.56	1.24	2.52	1.56	2.28
4.81	4.87	4.46	4.62	4.23	4.86	4.12	5.11	4.84	5.62	5.37	4.49	4.04	4.49	4.04	3.86
1.85	1.59	1.46	2.14	2.56	3.56	1.51	2.47	2.28	1.73	1.81	2.10	2.86	1.64	1.63	2.60
0.16	0.18	0.18	0.09	0.20	0.19	0.17	0.05	0.13	0.04	0.05	0.05	0.02	0.04	0.03	0.01
1.86	0.63	0.61	0.69	1.80	0.46	0.53	0.97	1.02	1.42	1.42	0.64	2.93	0.36	1.02	1.82
98.61	99.68	98.88	99.97	98.90	99.75	100.53	100.33	98.93	99.04	99.05	98.55	98.87	99.10	98.76	100.40
0.34	0.39	0.41	0.46	0.25	0.40	0.42	0.45	0.48	0.37	0.42	0.31	0.38	0.31	0.60	0.37
14.0	9.0	9.0	22.0	7.0	21.0	62.6	8.0	18.0	6.0	7.0	7.0	14.0	6.0	6.0	20.0
6.0	10.0	10.0	12.0	1.0	8.0	17.2	4.0	11.0	2.0	4.0	1.0	6.0	3.0	1.0	2.0
15.5	13.5	16.5	15.0	19.5		0.0	16.0		10.5	11.5	22.5		16.0	14.0	18.5
45.0	25.0	35.0	45.0	40.0	30.0	70.7	15.0	13.0	10.0	15.0	15.0	8.0	5.0	5.0	10.0
5.0	5.0	5.0	20.0	5.0	19.0	11.1	5.0	15.0	5.0	15.0	5.0	34.0	5.0	5.0	20.0
7.0	2.0	1.0	4.0	10.0	22.0	8.1	18.0	19.0	4.0	3.0	7.0	18.0	3.0	3.0	1.0
110	85.0	80.0	50.0	105	63.0	60.6	25.0	49.0	35.0	30.0	90.0	39.0	60.0	40.0	25.0
177	104	67.8	62.2	119.0	102.0	87.0	63.0	75.0	51.0	55.0	83.0	144	76.4	40.2	101
4.60	3.50	3.20	2.40	1.50	1.52	4.72	2.00	3.49	1.10	1.00	2.40	1.73	2.90	0.30	1.70
342	212	421	628	1210	1007	449	789	1232	756	459	358	543	247	905	943
179	228	325	397	280	493	260	501	430	402	223	222	101	162	89	177
60.0	35.0	31.0	24.0	21.0	24.4	108	19.0	54.4	4.0	5.0	21.0	3.6	25.0	3.0	2.0
5.50	4.00	4.50	2.50	4.00	0.33	0.66	5.00	0.31	2.50	3.50	8.00	1.23	6.00	5.50	5.50
29.00	12.00	9.00	3.00	12.00	4.00	9.09	1.00	3.00	2.00	2.00	10.00	9.00	9.00	13.00	6.00
13.00	5.00	5.00	3.00	11.00			1.00		2.00	2.00	4.00	3.00		4.00	9.00
445	215	235	108	458	163	202	79	116	72	83	123	91	150	304	108
75.5	11.00	7.00	6.00	17.50	2.00	17.18	2.50	1.00	3.50	6.50	9.50	18.00	7.00	39.0	12.00
38.00	5.00	4.00	5.00	21.00		12.12	2.00	2.00	1.00	1.00	4.00	11.00	4.00	7.00	17.00
4.50	5.00	3.00	1.50	1.50	4.00	1.01	1.50	0.90	1.00	0.50	5.00	3.00	3.50	3.00	3.50
167	31.5	30.00	18.50	143	26.57	51.5	7.50	10.66	9.00	18.50	20.50	14.24	39.0	36.5	26.50
373	56.5	54.5	35.5	263	49.4	78.3	14.0	20.0	14.5	17.5	33.5	27.2	62.5	72.5	47.5
42.10	6.50	5.60	4.00	27.7	5.30	9.08	1.70	2.42	2.00	3.50	3.70	2.94	6.50	8.50	4.60
160	21.50	20.50	15.50	89.5	18.50	30.9	7.00	8.40	5.50	12.00	12.50	1.68	19.50	32.00	15.50
32.30	3.60	3.20	2.90	11.60	2.95	5.09	1.20	1.57	1.30	2.70	2.20	2.46	3.20	6.30	3.00
2.20	0.80	0.90	0.80	1.80	0.81	1.36	0.40	0.41	0.40	0.70	0.60	0.47	0.50	1.00	0.60
30.50	3.70	4.40	2.00	10.40	2.48	4.48	1.10	1.27	1.10	2.10	2.30	2.86	2.70	6.40	3.10
4.20	0.60	0.30	0.30	1.10	0.21	0.60	0.10	0.11	0.10	0.30	0.30	0.43	0.30	1.00	0.40
20.50	2.60	2.00	1.30	5.10	0.97	3.37	0.50	0.60	0.60	1.40	2.10	2.68	1.40	7.40	2.20
3.40	0.30	0.20	0.30	0.60	0.15	0.66	0.10	0.10	0.10	0.30	0.30	0.54	0.20	1.40	0.40
7.20	1.30	0.50	0.70	1.60	0.41	1.77	0.20	0.21	0.30	0.60	1.20	1.70	1.10	4.30	1.10
0.90	0.10	0.10	0.10	0.30	0.06	0.24	0.10	0.04	0.10	0.10	0.10	0.26	0.10	0.70	0.10
5.00	0.80	0.60	0.50	1.20	0.33	1.59	0.20	0.21	0.40	0.50	0.80	1.80	0.80	4.20	1.10
0.70	0.10	0.10	0.10	0.10	0.05	0.25	0.10	0.03	0.10	0.10	0.20	0.27	0.10	0.60	0.10

(continued on next page)

compositions define three distinct petrological suites. The first suite includes rocks defining a tonalite–trondhjemite–granite association, hereafter called TTG suite (Fig. 3a). The MgO content in the TTG suite is < 5 wt.%, and Mg# ($Mg/(Mg + Fe^{2+})$) calculated with $Fe^{3+}/Fe^{2+} = 0.2$ is ≤ 0.5, including in ortho-amphibolitic enclaves hosted in the trondhjemites and marked by $SiO_2 < 53\%$ (e.g. sample AR77, Table 1). The Al_2O_3 content of TTG rocks containing SiO_2 in the range 67–73% defines two groups. The

Table 1 (*continued*)

	TTG suite							High-K granite suite								
	Tonalites–Trondhjemites–Na-granites							Potassic granites								
	MKQ1	TS82	TS70	BR24A	MK43A	MK43B	TS168C	MK54	MK108	MKQDK1	MK112A	MK49	NN8	MKQ5	NN6	MK 19
SiO_2	74.00	74.39	74.46	74.50	75.22	76.00	77.01	70.73	71.86	72.20	73.50	74.27	75.82	76.50	76.50	77.58
TiO_2	0.13	0.08	0.26	0.06	0.03	0.06	0.04	0.28	0.22	0.28	0.21	0.01	0.04	0.06	0.03	0.04
Al_2O_3	14.09	13.53	10.94	13.86	13.53	14.20	12.94	13.57	13.11	13.47	12.65	12.60	13.63	12.70	13.27	12.41
Fe_2O_3	1.37	1.36	3.96	0.85	0.70	0.75	0.86	2.25	1.84	2.42	1.61	0.45	0.46	0.76	0.44	0.63
FeO																
MnO	0.02	0.04	0.06	0.01	0.01	0.01	0.01	0.04	0.02	0.02	0.02	0.01	0.02	0.01	0.01	0.01
MgO	0.27	0.20	0.12	0.33	0.07	0.16	0.10	0.55	0.47	0.39	0.29	0.05	0.17	0.11	0.13	0.21
CaO	2.01	1.28	1.11	1.91	1.29	1.65	0.47	1.66	1.71	1.33	1.03	0.44	0.99	1.15	0.95	1.36
Na_2O	4.32	4.40	4.14	5.17	4.55	4.56	3.65	3.65	4.07	3.03	3.21	2.53	3.46	3.27	3.58	3.22
K_2O	3.09	3.04	2.22	1.80	3.10	2.86	3.51	4.51	4.17	5.74	5.07	7.70	4.52	5.60	4.82	4.47
P_2O_5	0.01	0.03	0.11	0.02	0.01	0.01	0.01	0.08	0.07	0.06	0.04	0.01	0.01	0.01	0.01	0.01
LOI	0.53	1.67	1.00	0.68	0.41	0.38	0.82	1.39	0.64	0.87	0.73	0.36	0.47	0.31	0.73	0.90
Total	99.84	100.02	98.38	99.19	98.92	100.64	99.42	98.71	98.18	99.81	98.36	98.43	99.59	100.48	100.47	100.84
Mg#	0.28	0.23	0.70	0.43	0.17	0.30	0.19	0.33	0.34	0.24	0.26	0.18	0.42	0.22	0.37	0.40
Cr	5.0	4.0	5.0	26.0	26.0	46.0	4.0	8.0	7.0	5.0	6.0	6.0	12.0	3.0	10.0	14.0
Ni	3.0	7.0	3.0	15.0	5.0	13.0	1.0	2.0	2.0	3.0	2.0	1.0	3.0	2.0	1.0	14.0
Co	19.0	9.0	15.5	12.5	12.5	16.0	16.0	13.5	14.0	15.0	9.0	26.5	23.0	16.0	24.5	9.5
V	5.0	5.0	5.0	5.0	5.0	5.0	5.0	20.0	15.0	5.0	15.0	5.0	5.0	5.0	5.0	5.0
Cu	5.0	5.0	5.0	10.0	5.0	5.0	35.0	5.0	5.0	5.0	5.0	15.0	5.0	35.0	5.0	10.0
Pb	5.0	15.0	4.0	16.0	21.0	15.0	4.0	18.0	15.0	19.0	12.0	6.0	11.0	7.0	17.0	4.0
Zn	45.0	40.0	40.0	15.0	5.0	10.0	5.0	60.0	55.0	55.0	35.0	5.0	5.0	5.0	5.0	10.0
Rb	99.2	86.0	48.0	24.2	93.4	62.2	61.8	149	172	198	297	202	115	145	117	114
Cs	2.40	2.00	0.30	0.40	0.50	0.40	0.20	0.80	1.80	1.40	4.70	1.20	1.20	1.50	1.00	1.60
Ba	570	751	984	679	525	693	1010	966	838	1085	706	1835	1115	1265	1040	1055
Sr	168	128	75	196	120	184	82	180	189	124	101	280	184	154	178	151
Li	16.0	7.0	3.0	1.0	1.0	1.0	1.0	14.0	15.0	26.0	13.0	2.0	3.0	7.0	3.0	7.0
Ta	8.50	3.00	6.00	5.00	4.50	6.00	7.00	4.00	5.50	5.50	5.50	8.50	7.00	6.50	8.00	4.00
Nb	9.00	5.00	14.00	6.00	1.00	6.00	9.00	10.00	12.00	11.00	18.00	2.00	3.00	4.00	2.00	3.00
Hf	3.00	2.00	9.00	1.00	4.00	1.00	4.00	7.00	6.00	8.00	6.00	1.00	3.00	1.00	1.00	1.00
Zr	94	43	292	57	117	58	81	222	182	276	188	25	93	65	66	29
Y	9.00	6.00	35.5	8.00	11.00	11.50	18.50	16.50	9.50	19.00	17.00	3.50	1.50	5.00	1.50	4.00
Th	4.00	3.00	6.00	11.00	49.0	13.00	9.00	25.00	22.00	37.00	22.00	1.00	1.00	4.00	2.00	10.00
U	7.00	2.50	3.00	2.50	5.00	2.50	5.50	7.50	8.00	6.50	4.00	0.50	3.00	19.00	3.00	4.50
La	21.50	13.00	32.00	18.50	48.5	22.00	9.50	76.5	69.5	113	9.00	9.00	2.50	16.00	2.50	6.50
Ce	35.0	24.0	64.0	32.0	87.5	38.0	28.5	142	122	223	80.0	15.0	5.0	27.5	3.5	10.5
Pr	3.50	2.70	7.20	3.00	8.80	3.70	2.40	15.10	12.20	23.80	2.10	2.10	0.40	2.90	0.40	1.20
Nd	12.50	11.00	27.50	9.50	30.00	12.00	8.00	47.00	42.00	79.00	7.00	5.50	1.50	10.00	1.50	4.50
Sm	2.30	2.10	5.90	1.80	5.00	2.00	1.80	7.90	6.10	10.60	1.50	1.80	0.60	2.00	0.30	0.80
Eu	0.80	0.50	1.10	0.60	0.70	0.60	0.30	0.80	0.80	1.20	0.30	0.60	0.50	0.50	0.40	0.50
Gd	2.10	1.50	6.40	1.60	3.80	2.00	2.50	8.40	5.60	11.10	2.50	1.20	0.40	1.90	0.30	1.00
Tb	0.30	0.20	1.00	0.30	0.60	0.30	0.50	1.00	0.60	1.20	0.40	0.20	0.10	0.20	0.10	0.10
Dy	2.20	1.10	6.50	1.20	2.10	1.80	3.20	4.10	2.50	4.50	2.40	0.50	0.20	1.40	0.20	0.80
Ho	0.40	0.30	1.30	0.30	0.40	0.40	0.60	0.60	0.30	0.70	0.80	0.10	0.10	0.10	0.10	0.10
Er	1.10	0.90	4.10	0.70	1.00	1.10	1.80	1.60	0.90	2.20	2.10	0.50	0.10	0.70	0.20	0.50
Tm	0.10	0.10	0.70	0.10	0.10	0.10	0.40	0.30	0.10	0.30	0.30	0.10	0.10	0.10	0.10	0.10
Yb	0.80	0.40	4.00	1.00	1.10	1.20	2.00	1.40	0.70	1.70	2.20	0.40	0.30	0.80	0.20	0.50
Lu	0.10	0.10	0.70	0.10	0.10	0.20	0.40	0.30	0.10	0.30	0.40	0.10	0.10	0.10	0.10	0.10

first group (hereafter Group 1 TTG sub-suite) contains >15% Al_2O_3 indicating a high-Al trondhjemitic sub-suite, whereas the second group (hereafter Group 2 TTG sub-suite) has <15% Al_2O_3 and corresponds to a low-Al trondhjemitic sub-suite (Barker, 1979; Martin, 1993, 1994). The high-Al trondhjemitic sub-suite occurs in the southwestern igneous complexes whereas the low-Al trondhjemitic sub-suite forms the northeastern igneous complexes. In the diagram K_2O–Na_2O–CaO (Fig. 3b), the majority of the Francistown

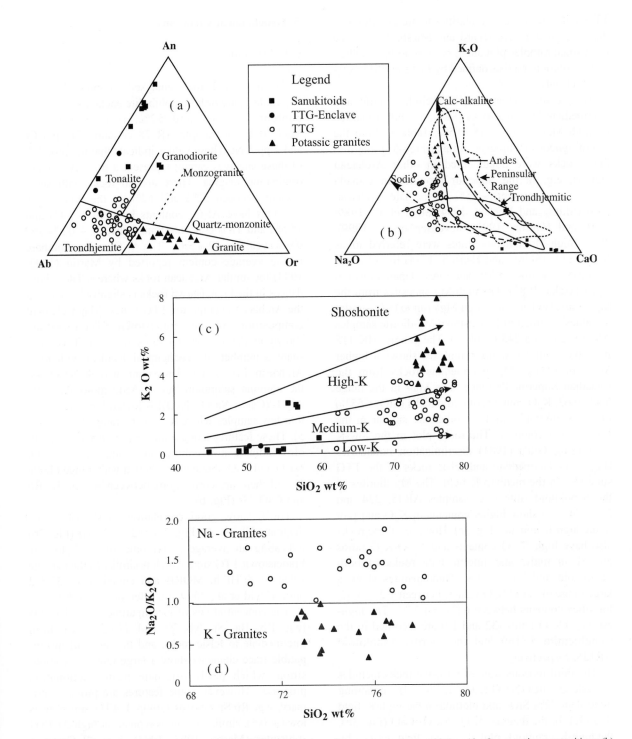

Fig. 3. Geochemical classification of the intrusive rocks from the Tati granite–greenstone terrane. (a) An–Ab–Or normative compositions; (b) K_2O–Na_2O–CaO diagram with the various fields and trends drawn according to Luais and Hawkesworth (1994) and references therein; (c) K_2O–SiO_2 diagram with the fields drawn based on Gill (1981); (d) Na_2O/K_2O vs. SiO_2.

TTG suite rocks show similarities to arc granitoids in the Peninsular Range (Baird and Miesh, 1984). The most sodic samples plot along or close to the "sodic" TTG evolution trend as defined by Luais and Hawkesworth (1994).

The second igneous suite includes mafic and intermediate intrusive rocks (SiO_2: 45–60%) marked by high MgO contents (5–9.5%; Table 1) from the central igneous complexes. Similar high-Mg intermediate rocks were documented in other Archaean cratons, e.g. in the Superior Province in Canada where they represent ~ 15% of the late Archaean crust (Shirey and Hanson, 1984; Stern et al., 1989; Stern and Hanson, 1991; Stevenson et al., 1999). These high-Mg igneous suites were referred to as Sanukitoids (Shirey and Hanson, 1984) by reference to the Setouchi high-Mg andesites, Japan (Tatsumi and Ishizaka, 1982). The high-Mg andesites from the Japanese island arc have high Mg# (up to 0.7), similar to values recorded in Francistown Mg-diorite samples AR234, 237 and 245 (Table 1). The gabbro MK 115 from the central igneous complexes shows a higher Mg# (0.8). High-Mg intermediate rocks from the Canadian Superior Province have Mg# in the range 0.43–0.62. K_2O contents of mafic rocks ($SiO_2 < 55\%$) of the Sanukitoid and TTG suites are similar in the Francistown complex. The rocks plot in the low-K field using Gill's (1981) discrimination boundaries (Fig. 3c). Intermediate and felsic rocks of the TTG suite plot in the medium-K field. The Mg-diorites of the Sanukitoid suite (e.g. samples AR45, 234 and 237; Table 1) show higher contents of K_2O and plot in the high-K domain (Fig. 3c). However, these rocks also have high Na_2O contents and Na_2O/K_2O ratios are >1 in mafic and intermediate rocks of both Sanukitoid and TTG suites. Trondhjemites show a large range of Na_2O/K_2O values, between 1.2 and 32, for silica contents between 62% and 77%. The highest Na_2O/K_2O ratios (32 and 19) are recorded in the trondhjemite TS169 and the dioritic Sanukitoid AR128, respectively.

The third igneous suite of intrusive rocks includes potassic granites ($Na_2O/K_2O < 1$; Fig. 3d), containing more than 72% SiO_2 and plotting in the high-K field (Fig. 3c). In the diagram $K_2O–Na_2O–CaO$ (Fig. 3b), the high-K granitoids plot also in the field defined by Peninsular Range arc granitoids (Baird and Miesh, 1984).

5. Geochemical variations

5.1. TTG suite

The most mafic rocks analysed (samples AR77 and 78; Table 1) are ortho-amphibolite enclaves with SiO_2 contents of 50% and 52%, respectively, and Mg# < 0.5. The sample AR 78 contains 74 ppm Cr and 2 ppm Ni. These values indicate that the protolith of these enclaves cannot represent a primary mantle-originating magma. There are no rocks with silica content between 52% and 62% in the TTG suite. Tonalites have Al_2O_3 contents between ~ 14% and 15.5% and TiO_2 concentrations between ~ 0.5% and 0.9% at ca. 63–69% SiO_2. The Al_2O_3 values compare to the average content reported by Martin (1993, 1994) for similar Archaean rocks whereas TiO_2 abundances in the Francistown rocks is slightly higher than the Archaean average for TTG suites. Major element compositions of the Francistown TTG rocks are shown in the Harker diagrams (Fig. 4). The rocks share a number of geochemical features with most Archaean TTG suites. They contain high Na_2O and the alumina saturation index (ASI = molecular ratio $Al_2O_3/(CaO + Na_2O + K_2O)$) is lower than one except in a few samples where it is in the range 1–1.1 (Fig. 5). These features along with the presence of amphiboles indicate a metaluminous composition. CaO, Na_2O and K_2O abundances show a wide range (Table 1) and there are correlations between K–Ba, K–Rb and CaO–Sr (Fig. 6).

HFSE show good correlation as shown in the diagrams Nb vs. Zr (Fig. 7a) and Zr vs. Hf (Fig. 7b) indicating an average Zr/Hf ratio of ~ 37 for the Francistown TTG suite which is similar to the average value of Zr/Hf in MORB and chondrite (~ 37 ± 2 ppm: David et al., 2000 and references therein).

The concentrations of incompatible trace elements (e.g. Rb, Ba, Sr, Nb, Zr and Th) increase from intermediate to felsic rocks, and the ratios of incompatible trace elements show a large range of composition, which partly originate from fractionation processes. However, some features are probably primary, e.g. Rb/Sr ratio of Group 1 TTG sub-suite is low (~ 0.5), similar to values typical for high-Al TTG sub-suites (Martin, 1993, 1994). Low-Al Group 2 TTG sub-suite displays variable Rb/Sr ratios in the range 0.4–1. Th/Ta ratio is ~ 8 in the enclave sample

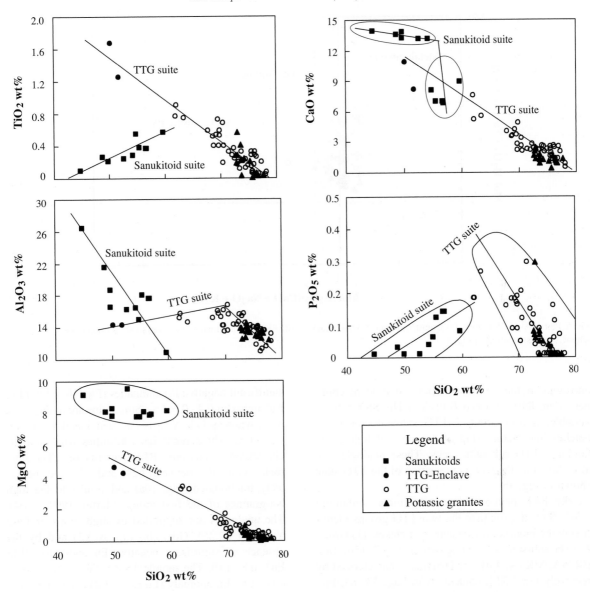

Fig. 4. Harker-type diagrams for intrusive rocks from the Tati granite–greenstone terrane.

AR 77 of mafic rocks contained in the Group 1 TTG rocks ($SiO_2 \sim 50\%$). This enclave is also marked by a high value for La/Nb (~ 6) and a low Nb/Ta ratio (~ 13). These features are similar to those documented in arc mafic rocks (e.g. Sun and McDonough, 1989). Low Nb/Ta ratio (~ 13) in this sample of metamorphosed igneous mafic rock is similar to values reported in mafic rocks affected by retrograde

hydration with fluids expulsed from a subducting slab (Kamber and Collerson, 2000).

Rb, Y, Nb, Th, Ta, Zr and Yb relationships in tonalites, trondhjemites and granites indicate also that the Francistown intermediate and felsic rocks are similar to equivalent igneous rocks in Phanerozoic arcs (Fig. 8a–c). High Th/Yb (>5) ratios correlated to high values (>10, up to ca. 100) for La/Yb show the

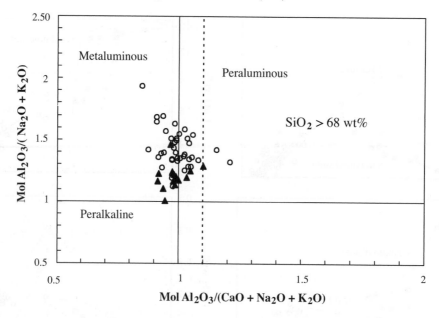

Fig. 5. Plot of $Al_2O_3/(Na_2O + K_2O)$ against ASI {molar $Al_2O_3/(CaO + Na_2O + K_2O)$} for intrusive felsic rocks from the Tati granite–greenstone terrane. Same symbols as in Fig. 3.

analogies of these intrusive rocks to modern continental arc felsic magmas (Fig. 8c). The Sr/Y ratio is variable, low in Group 2 TTG sub-suite rocks but reaches high values (Fig. 9a) typical of high-Al in Group 1 TTG sub-suite rocks. These analogies are also shown by the REE composition of the TTG suite shown in Fig. 9b.

The REE patterns of trondhjemites containing >70% SiO_2 discriminate the two TTG groups identified in the Francistown igneous complexes: (1) Group 1 TTG sub-suite rocks (e.g. samples AR 150, 188, BR24A, MK 34, 43B, TS 168B) are characterized by relatively low LREE contents (e.g. La_N: 33–83; Fig. 10c), low Yb_N: 1–5, La_N/Yb_N in the range 12–54 and the absence of significant Eu negative anomaly (Eu/Eu* = 0.9–1.1). Both LREE and HREE are fractionated as shown by La_N/Sm_N (4–7) and Gd_N/Yb_N (1.3–6; average = 3) ratios. These rocks share similarities with Archaean TTG which are characterised by $5 < La_N/Yb_N < 150$, $0.3 < Yb_N < 9$ and the lack of negative Eu anomaly (Martin, 1994); (2) Group 2 TTG sub-suite rocks (e.g. samples TS 70, 167, 168c, AR65, 81, 181) are marked by higher LREE contents (e.g. La_N up to 114), Yb_N up to ~ 20, lower La_N/Yb_N ratios in the range 3–10, and the REE patterns display

significant negative Eu anomalies (Fig. 10e), with Eu/Eu* between 0.4–0.6. La_N/Sm_N ratios are in the range 3–6, whereas Gd_N/Yb_N is between 1 and 1.5 (average = 1.3). These rocks have affinities with post-Archaean arc granitoids. REE patterns of TTG suite rocks with SiO_2 between 62% and 70% are moderately fractionated (Fig. 10a) and similar to the high Na-granites of the first group (Martin, 1993, 1994). The analogies are supported by high ratios of La_N/Yb_N (up to 25), Gd_N/Yb_N (up to 3.2) and by the absence of significant negative Eu anomalies (Eu/Eu*: 0.8–1.0). The sample TS 80 (Fig. 10a) displays a positive Eu anomaly (Eu/Eu* = 1.4). Light REE are more fractionated (La_N/Sm_N = 3–6) than heavy REE (Gd_N/Yb_N = 1–3). The mafic enclave sample AR 77 from the TTG suite (Fig. 10g) is also enriched in REE (La_N~ 66, Yb_N~ 11 and La_N/Yb_N~ 6) and shows fractionation for both LREE and HREE (La_N/Sm_N and Gd_N/Yb_N~ 2). The absence of Eu anomaly is also characteristic for this sample (Eu/Eu* = 1). The relationships between La_N/Yb_N vs. Yb_N (Fig. 9b) suggest that Group 1 TTG sub-suite rocks have affinities with Archaean TTGs, whereas Group 2 TTG sub-suite rocks are similar to post-Archaean arc granitoids (Martin, 1993, 1994).

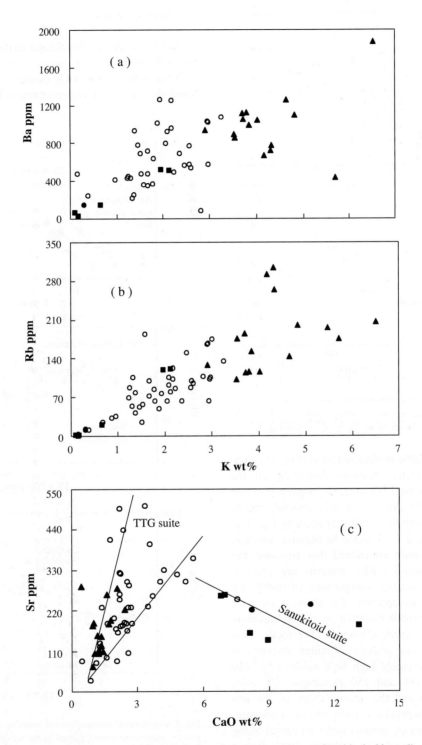

Fig. 6. Geochemical compositions of intrusive rocks from the Tati granite–greenstone terrane plotted in the binary diagrams: (a) Ba–K, (b) Rb–K, (c) Sr–CaO. Symbols as in Fig. 3.

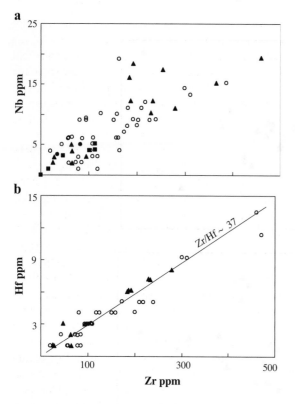

Fig. 7. High-field-strength-element compositions of intrusive rocks from the Tati granite–greenstone terrane plotted in the binary diagrams: (a) Nb vs. Zr, (b) Hf vs. Zr. Symbols as in Fig. 3.

Primordial mantle normalized plots (Fig. 11) summarize the distinctive features shared by all the igneous rocks of the TTG suite, i.e. negative anomalies for Ti and Nb (Ta). A negative anomaly for Sr and a positive anomaly for Hf (Zr) occur in Fig. 11e, within the Group 2 TTG rocks. Sr negative anomaly in primordial mantle normalised diagrams and flat chondrite-normalised HREE patterns are features indicating a non-adakitic composition of this group of TTG rocks. In contrast, the absence of a Sr anomaly and a small negative to no Hf anomaly (Fig. 11a and c) mark Group 1 TTG rocks. This group of rocks has adakitic affinities and this is supported by extremely high Sr/Y values (e.g. 430 in sample AR 188 and 250 in sample AR 150). These anomalies and the general shape of the primordial-mantle normalised spider diagrams are similar to the features of igneous rocks emplaced along convergent plate margins (e.g. Pearce, 1982; Pearce et al., 1984).

5.2. Sanukitoid suite

Al_2O_3 content in Sanukitoid mafic rocks ($<53\%$ SiO_2) is high, in the range $\sim 16–26\%$ (average $\sim 20\%$), whereas the TiO_2 content is low ($<0.3\%$). Similarly, high Al_2O_3 concentrations (16–18%, except

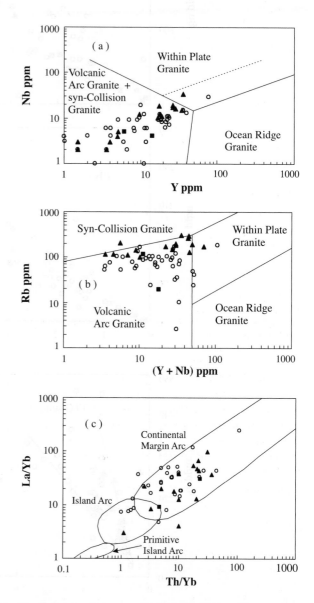

Fig. 8. Geochemical compositions of intrusive rocks from the Tati granite–greenstone terrane plotted in the tectonic setting discrimination diagrams: (a) Y vs. Nb, (b) Rb vs. Y + Nb (Pearce et al., 1984), (c) La/Yb vs. Th/Yb (Condie, 1989). Symbols as in Fig. 3.

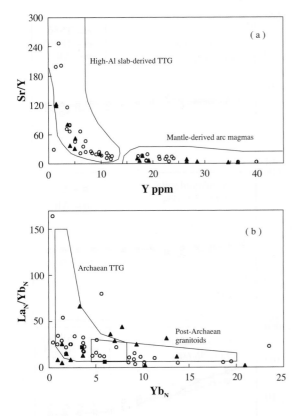

Fig. 9. Geochemical compositions of intrusive rocks from the Tati granite–greenstone terrane plotted in the diagrams: (a) Sr/Y vs. Y, (b) La_N/Yb_N vs. Yb_N. The fields are according to Martin (1993, 1994). Symbols as in Fig. 3.

in samples MK32 and AR128) and low TiO_2 abundances (<0.6%) occur in intermediate rocks (SiO_2: 54–60%). These features are similar to those marking high-Al mafic and intermediate magmas in modern arc settings (e.g. Gill, 1981; Thorpe, 1982). MgO contents are high in both mafic (5–9.5%) and intermediate (7.7–8.1%) rocks. The Mg# values are in the range 0.58–0.80. The highest Mg# (0.8) occurs in the gabbro MK115, which contains ~ 45% SiO_2. Diorites from this group have also high Mg# between 0.6 and 0.7. Within the Sanukitoid suite, the sample MK 115 contains the lowest SiO_2 content (~ 45%) correlated to the lowest abundances of most incompatible elements, e.g. Ba, Sr, Zr, Y and REE (Table 1). In contrast, it shows the highest content in Ni (280 ppm) and Cr (706 ppm) suggesting that it is representative of the most primitive magmas in this igneous suite. However, except for the most primitive gabbro

MK 115, the transition metal abundances are generally higher in the Mg-diorites than in the gabbros of the Sanukitoid suite. The Mg-diorites of the Sanukitoid suite display higher contents of most incompatible trace elements (e.g. HFSE and REE) when compared to the gabbros of the same suite. Incompatible element ratios show a large range of values in the Sanukitoid suite. Th/Ta ratios are low (0.2–3, average ~ 1) in the mafic rocks, similar to values reported in mafic rocks emplaced in extensional tectonic settings such as MORB and OIB (e.g. Wood et al., 1979). High-Mg diorites show a very large range for Th/Ta values (2–26, average ~ 14). The high Th/Ta values compare to ratios in arc magmas (Wood et al., 1979). La/Nb values display a similar trend with low ratios (1–3; average ~ 2) similar to values recorded in average MORB and OIB worldwide (e.g. Sun and McDonough, 1989). A broader La/Nb range (1–6) and a higher average (~ 4) similar to La/Nb ratios in arc magmas characterise high-Mg diorites. Nb/Ta ratio is extremely variable in the Francistown Sanukitoid suite, but low values (<13) predominate, except in the high-Mg diorite sample AR 128 (Nb/Ta~ 30). Nb/Ta ratios in MORB and OIB and chondrites are ~ 17 ± 2 (e.g. Kamber and Collerson, 2000 and references therein). Sedimentary rocks and continental crust are marked by lower Nb/Ta values of ca. 12 ± 2. Blueschist- and eclogite-facies mafic rocks from obducted oceanic crust show a large Nb/Ta range, between 7 and 77, with an average of ~ 34 ± 4 (Kamber and Collerson, 2000).

Cs/La (0.08–0.6; average 0.24), Ba/Th (70–174, except in sample MK 115) and Ba/La (16–28; average 23) ratios are generally higher in mafic rocks than in high-Mg diorites (Cs/La: 0.02–0.24, average 0.11; Ba/Th: 23–76 and Ba/La 8–21, average 15). MORB and OIB are usually characterized by Cs/La<0.04, Ba/Th<100 and Ba/La<10 (e.g. Sun and McDonough, 1989). Higher ratios commonly occur in arc magmas (e.g. Gill, 1981; Ryan et al., 1995), especially at the volcanic arc front. For example, in the Central American arc, Cs/La>0.1 occur only at and close (<30 km) to the volcanic arc front, whereas this ratio is lower (<1, with most data <0.05) in igneous rocks emplaced behind the volcanic arc front (Walker et al., 2000). Ce/Pb ratio is low in both mafic (2–8) and high-Mg diorites (3–13) of the Francistown Sanukitoid suite. The Ce/Pb ratios of these rocks are similar to values marking arc magmas (e.g. Chauvel et al.,

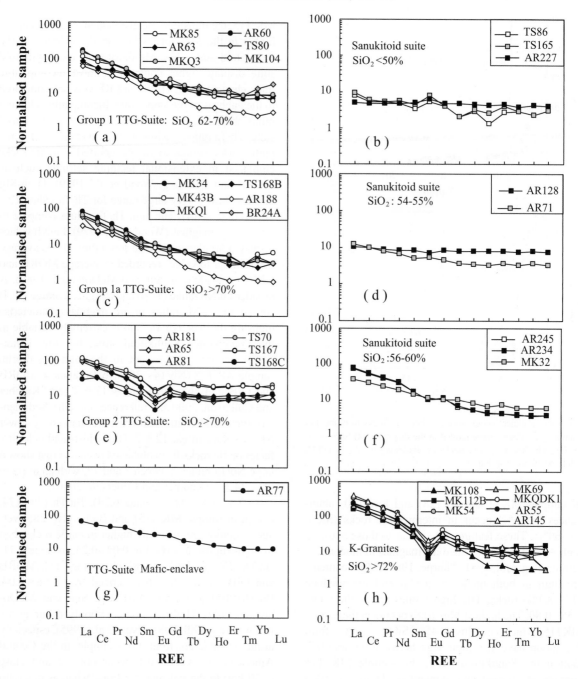

Fig. 10. Representative chondrite-normalised REE patterns for intrusive rocks from the Tati granite–greenstone terrane. Normalising values of Sun and McDonough (1989).

1995). Sr/Y values are variable in the Sanukitoid suite, ranging between ~ 18 and 50 in mafic rocks and 7 and 42 in high-Mg diorites. Sr/Ce ratios are also variable but high in gabbros (36–67) and low to high in Mg diorites (~ 5–22). For comparison, Sr/Ce is ~ 10 in MORB and OIB. Lower values (average ~ 6)

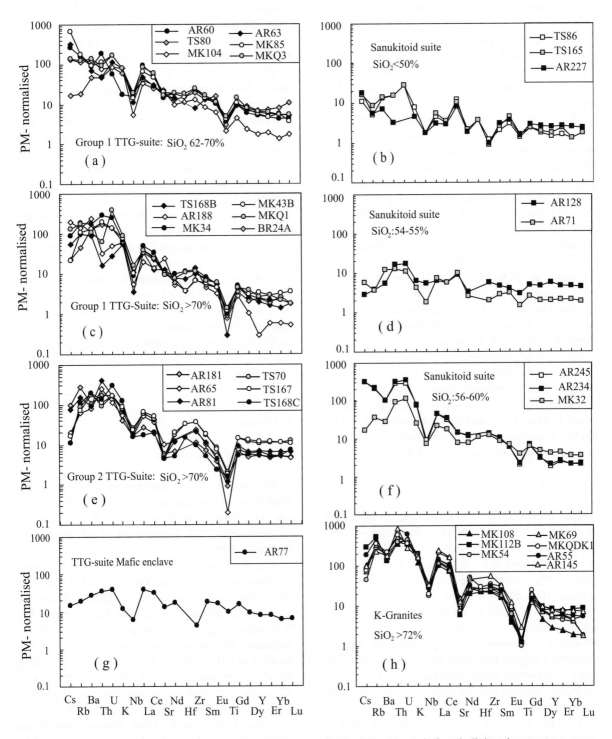

Fig. 11. Representative primordial mantle (PM)-normalised multi-element diagrams for intrusive rocks from the Tati granite–greenstone terrane. Normalising values of Sun and McDonough (1989).

mark sedimentary rocks (e.g. Plank and Langmuir, 1998).

The REE patterns of igneous rocks of the Sanukitoid suite are displayed in Fig. 10b, d and f. Rocks with $SiO_2 < 50\%$ (Fig. 10b) are characterized by flat REE patterns, with $La_N/Yb_N \sim 1–3$, $La_N/Sm_N \sim 1–2$ and $Gd_N/Yb_N \sim 1–2$. Small to moderate Eu positive anomalies $(Eu/Eu^* = 1.3–1.9)$ occur in these mafic rocks (Fig. 10b). Intermediate rocks with SiO_2 between 54% and 55% (Fig. 10d) show flat to slightly fractionated REE patterns $(La_N/Yb_N \sim 1–4$, $La_N/Sm_N = 1–3$ and $Gd_N/Yb_N \sim 1)$ and very small negative and positive anomalies of Eu $(Eu/Eu^* = 0.8–1.2)$. High-Mg diorites with higher SiO_2 content (56–60%) show more fractionated REE patterns (Fig. 10f) marked by higher $La_N/Yb_N = 6–22$, $La_N/Sm_N = 3–4$, $Gd_N/Yb_N = 2–3$ and a moderate Eu negative anomaly in several samples $(Eu/Eu^* \sim 0.7)$. The REE pattern features change with the increase of SiO_2 content in the Sanukitoid suite. This change correlates with the increase of incompatible element abundances reported above but does not correlate with a decrease of transition metal abundance.

Primordial mantle normalized spidergrams of Sanukitoid suite rocks are characterized by negative anomalies for Ti, Nb (Ta) and positive anomalies for U and Th (Fig. 11b, d and f). Rocks with $SiO_2 \leq 55\%$ display in addition a Sr positive anomaly and a small Zr negative anomaly (Fig. 11b and d).

5.3. High-K granites

The silica-rich high-K granites (Fig. 3b and c) are characterized by Al_2O_3 contents in the range $\sim 12.5–14.5\%$, Na_2O of $\sim 2.5–4.5\%$, K_2O in the range $\sim 3.4–7.9\%$ and $0.9 < A/CNK < 1.1$ (average A/CNK = 0.99). These granites are predominantly metaluminous (Fig. 5), with a few slightly peraluminous rocks (A/CNK>1, up to 1.09). These features along with the absence of mineral phases marking strong peraluminous compositions (e.g. cordierite, primary muscovite) suggest that these are I-type granites. The range of transition metal concentrations (Table 1) in the high-K granites and Na-granites of the TTG suite are similar.

The abundances of other trace elements such as alkali, alkali-earth, HFSE and REE are variable, from values similar to those reported in TTG suite rocks of

similar SiO_2 content to lower concentrations. A negative anomaly of Eu in chondrite-normalised patterns (Fig. 10h) and negative anomalies for Sr and Ba in primordial-mantle normalised spidergrams (Fig. 11h) mark high-K granites. The incompatible element content of high-K granites and TTG suite rocks show a number of similarities. The high-K granites are marked by a large variation of LREE content (e.g. $La_N \sim 8–397$) but, like TTG suite rocks, show variably fractionated REE patterns $(La_N/Sm_N = 2–7$, $Gd_N/Yb_N = 0.9–6$ and $La_N/Yb_N = 2–66)$ and a significantly variable Eu anomaly $(Eu/Eu^* = 0.3–4)$. The samples with positive Eu anomalies $(Eu/Eu^*$ up to 4) correspond to granite samples affected by hydrothermal alteration (not plotted in Fig. 10 and 11).

The primordial mantle-normalized diagrams (Fig. 11h) show that, at similar SiO_2 content, the high-K granites have higher contents of most incompatible elements (except HREE) than TTG suite rocks. However, the general shape of the spidergrams is similar and both high-K granites (Fig. 11h) and TTG suite rocks (Fig. 11e) are characterized by negative anomalies for Ti, Sr, Nb and Ba in primordial mantle-normalized diagrams.

6. Discussion

6.1. General observations

TTG suite rocks are abundant in the Kaapvaal and Zimbabwe cratons and in the Limpopo belt in southern Africa (e.g. de Wit et al., 1992; Luais and Hawkesworth, 1994; Berger et al., 1995; Berger and Rollinson, 1997).

The Francistown TTG suite defines two distinct petrological groups. The Group 2 TTG sub-suite shows flat chrondrite-normalized HREE patterns and negative Eu anomalies (Fig. 10e) and relatively low Sr/Y ratios (Fig. 9a) indicating its affinities with arc magmas. The geochemical diagrams, e.g. K_2O–Na_2O–CaO (Fig. 3b) and La_N/Yb_N vs. Yb_N (Fig. 9b) show the analogies of the Francistown Group 2 TTG sub-suite rocks with the Cretaceous plutonic rocks from the Peninsular Range and Idaho in USA (e.g. Baird and Miesh, 1984; Gromet and Silver, 1987). These analogies were previously reported by Luais and Hawkesworth (1994) for similar TTG suites exposed in the centre of the Zimbabwe craton, ~ 300 km NE of the Francistown igneous complex. Group 1 TTG sub-suite rocks display fractionation of both LREE and HREE and no significant negative Eu anomaly in the chondrite-normalized diagrams (Fig. 10c) and are characterised by high Sr/Y ratios (Fig. 9a) marking strongly sodic Archaean TTG suites (Martin, 1993, 1994).

In this paper, Sanukitoid suite rocks are documented for the first time in the Archaean of northeastern Botswana. Jelsma et al. (2001) (and references therein) were the first to describe Neoarchaean high-Mg intermediate rocks from Zimbabwe as Sanukitoids, although the first analyses of high-Mg andesites exposed in the Zimbabwe craton were published by Condie and Harrison (1976). The Sanukitoids provide a strong indication of the overall similarity of the Francistown Neoarchaean plutonic rocks from the Zimbabwe craton to modern volcanic arc magmatism (Tatsumi and Ishizaka, 1982; Shirey and Hanson, 1984; Stern et al., 1989; Stern and Hanson, 1991). The Francistown high-K granitoids display also several petrological analogies with calc-alkaline potassic granites from Phanerozoic orogenic provinces (e.g. Gromet and Silver, 1987; Pitcher, 1993). There are strong chemical similarities between the various granitoids exposed in the Francistown area (Fig. 11), despite some differences marking distinct petrogenetic processes. Petrogenetic models for these rocks must account for their systematic arc-like signature shown by high Al_2O_3, La/Nb, Th/Ta, low TiO_2 and Ce/Pb (in mafic-intermediate rocks) and negative anomalies of Nb–Ti in the primordial mantle normalized diagrams. These features indicate that these igneous rocks formed along a destructive plate margin or were derived largely from a crustal source that in itself was formed within an arc setting.

In the following section, we discuss in more details our data for the Francistown Neoarchaean plutonic suites in order to constrain the petrogenesis of the rocks and the tectonic setting evolution.

6.2. TTG suite

There is an agreement to consider that the geochemical features of TTG magmas originate from partial melting of a mafic precursor. However, there is a divergence of opinion on the processes involved

during the petrogenesis of TTGs. Drummond and Defant (1990) attribute their genesis to partial melting of mafic rocks within a subducting oceanic slab. Alternatively, Petford and Atherton (1996) suggested that TTG magmas form during partial melting of mafic rocks underplated within the crust. Geochemical data do not allow direct discrimination of these two processes. However, melting at the base of (or within) the crust requires the conversion of underplated mafic rocks into garnet amphibolite/eclogite prior to partial melting. The first interpretation requires a tectonically overthickened crust or a steep geothermal gradient beneath the Zimbabwe craton. Tomographic data indicate that the crustal thickness of the Zimbabwe craton is ~ 34–37 km (Nguuri et al., 2001) and exposed rocks are mainly low-grade (greenschist facies) metamorphic rocks. Therefore this crust has never been overthickened. The second interpretation requires a source of heat to create the steep geothermal gradient. In arc settings, a mechanism that can supply heat for crustal melting is underplating of mafic/ultramafic arc magmas into the lower crust (e.g. Hyndman and Foster, 1988). However, there is not yet evidence to support or reject a linkage between the emplacement of mafic/ultramafic magmas in the crust (magma underplating) and the generation and ascent of the Francistown TTG rocks. The low concentrations of Y, the strong depletion of HREE in the chondrite-normalized diagram (Fig. 10a and c), the absence of Sr negative anomaly in the primordial mantle normalized diagram (Fig. 11c), the overall high concentration of Sr in Group 1 TTG sub-suite rocks and the strong Nb negative anomalies require the presence of garnet and amphibole and no plagioclase in the melt residue. Rushmer (1991) pointed out that amphibole is a stable phase up to 18 kb at 950 °C and therefore it could be among residual minerals during dehydration melting of the oceanic crust. Experimental studies (e.g. Rapp et al., 1991) showed that, during dehydration melting of mafic rocks at ~ 700–1000 °C, the trondhjemitic melt generated has Al-rich composition and the melt residue is plagioclace-free and rich in amphibole and garnet only at high pressure ($\geq$ 1.5 GPa). At lower pressure, the generated melt has calc-alkaline affinity and its composition evolves from Al-poor trondhjemites to diorites/granodiorites, i.e. a magma similar to Group 2 TTG sub-suite rocks in the Tati granite–

greenstone terrane. Plagioclase is stable and therefore its presence in the melt residue could explain the Eu negative anomaly in the Group 2 TTG sub-suite (Fig. 10e). However, it is unlikely that the Group 2 TTG sub-suite was generated under low pressure (< 8 kb) since melts generated by both water-undersaturated and water-saturated experiments at such pressures are granodioritic to tonalitic (Helz, 1976) and/or strongly peraluminous (e.g. Holloway and Burnham, 1972; Beard and Lofgren, 1991). In contrast, the Group 2 TTG sub-suite rocks are predominantly metaluminous to slightly peraluminous. Their chemical composition is similar to that of experimental melts produced from garnet amphibolite at pressures between 8 and 15 kb (Rapp, 1997; Wyllie et al., 1997). These data indicate that partial melting of mafic igneous rocks under-plated in the lower crust is the most likely source for Group 2 TTG sub-suite rocks in the Tati granite–greenstone terrane.

Numerical models (Peacock et al., 1994) indicated that water-undersaturated (~ 5% H_2O) partial melting of mafic rocks produces a plagioclase-free residual assemblage containing garnet + amphibole. A water-absent melting produces a plagioclase-bearing residue containing two pyroxenes. Thus, the compositional differences between Francistown Groups 1 and 2 TTG rocks could reflect partial melting under different water pressure, i.e. water-absent conditions for Group 2 and water-undersaturated conditions for Group 1. The geochemical characteristics require a subduction process during or before the generation of both TTG rocks. In a subduction setting, Group 1 high-Al TTG magmas could originate from partial melting of young, hot, flat subducting oceanic slab with amphibole and garnet as residual phases (e.g. McCulloh, 1993). Sajona et al. (1993) pointed out that melting of old oceanic crust is also possible during the early stages of subduction and/or during fast or oblique subduction. At the regional scale, the Group 1 TTG sub-suite (Grey gneisses) was emplaced early as shown by similar U–Pb SHRIMP ages for TTG rocks and greenstone belt in the Vumba area (Bagai et al., 2002). Preliminary data related to Zimbabwe show a similar relation between ~ 2.7 Ga Neoarchaean TTG and greenstone belts (Jelsma, personal communication, 2000). Therefore, the Neoarchaean high-Al TTG sub-suite in the Zimbabwe craton most probably indicate partial melting of a flat subducting slab as

documented in modern settings (e.g. Gutscher et al., 2000).

6.3. Sanukitoid suite

Sanukitoids and Sanukitoid-like igneous suites occur in both Archaean and post-Archaean igneous provinces. The high Mg# and relatively high transition metal contents (Table 1) suggest an ultramafic mantle source for the primary magmas of the Francistown Sanukitoid suite. The high-Mg diorites display a number of geochemical features indicating their affinity with modern arc igneous rocks. These include high values for Th/Ta, La/Nb, Cs/La, Ba/Th, Ba/La and negative Ti–Nb (Ta) anomalies in primordial mantle normalized diagrams (Fig. 11). The gabbros show most of the above features but their Th/Ta and La/Nb ratios are close to values marking extensional mafic igneous rocks. Both Sanukitoid gabbros and Mg-diorites are enriched (Figs. 10 and 11), with the strongest REE enrichment occurring in the Mg-diorites containing >55% SiO_2 and marked by the highest Ni concentrations (Table 1). As stated before, the Mg-diorites were not formed by mafic mineral fractionation from gabbroic magmas since they usually have similar or higher Mg# and transition metal concentrations. Although small degree of partial melting of a sub-arc mantle could produce an enrichment in incompatible trace elements (e.g. LILE and HFSE), it cannot explain the extremely high and variable ratios between incompatible elements such as the LILE/HFSE ratios (e.g. Th/Ta and La/Nb) reported above. The geochemical characteristics suggest that the Sanukitoid gabbros and Mg-diorites originate from an heterogenous mantle source.

The elemental compositions and inter-element ratios reported above are similar to values documented in arc igneous provinces and linked to three main processes: (1) mantle wedge enrichment by sediments recycled in the mantle along subducting slabs (e.g. Plank and Langmuir, 1993); (2) mantle wedge enrichment by fluids from the dehydration of a subducting slab (e.g. Keppler, 1996; Tatsumi and Kogiso, 1997); (3) enrichment of the sub-arc mantle by slab-derived melts (e.g. Kelemen, 1995; Yogodzinski et al., 1995; Kepezhinskas et al., 1996; Sajona et al., 2000). The studies related to the mobility of elements in the aqueous fluids escaping from the subducting slab (e.g. Stolper and Newman, 1994; You et al., 1996) showed that more than 95% by mass are made of LILE and other highly incompatible elements. LREE have low to moderate mobility whereas HREE are the least mobile. Therefore, arc magmas derived from partial melting of a mantle wedge enriched by fluids escaping from a subducting slab will be characterized by a strong enrichment in LILE (e.g. Sr, Pb and Hf) versus a moderate enrichment in LREE, leading to high LILE/LREE in the melt. Mafic magmas generated in a sub-arc mantle source enriched by slab-derived fluids are generally marked by high Sr/Ce values ($\gg 20$), low Sr/Y ratios (Fig. 9a) and low Ce/Pb ratios ($\ll 20$) (e.g. Chauvel et al., 1995; Kapenda et al., 1998). However, Sajona et al. (2000) stressed that the same enrichment style characterizes mantle wedge sections metasomatized by slab melts. In contrast, mafic magmas originating from a sub-arc mantle enriched by subducting sediment melts are marked by low Sr/Ce, Ce/Pb and Sr/Y values. Shimoda et al. (1998) suggested that subducting sediment melts represent the main mantle wedge enrichment agent at the source of Setouchi Mg-andesites. However, this interpretation is unlikely for the Francistown Sanukitoid gabbros and high-Mg diorites marked by high to very high Sr/Ce ratios and high Sr/Y values (Fig. 9a). Mg-diorites with low values of Sr/Y and Sr/Ce (e.g. down to 5) could, however, originate from a mantle source enriched by sediments melts. The increase of silica correlated to an increase of LREE and fractionation of HREE in the Francistown Sanukitoid suite indicates a mantle source enrichment predominantly controlled by a silicate melt. The strong fractionation of LREE in high-Mg diorites suggest melting under high pressure, within the stability field of garnet, whereas the strong fractionation of HREE indicates partial melting of an enriched mantle wedge. Both features could originate from partial melting of a mantle wedge that was enriched by silicate melts formed within the stability field of garnet. The negative Nb anomaly suggests the presence of amphibole in the melt residue. The high K_2O content of high-Mg diorites requires a K-rich mineral phase(s) at the source, most probably phlogopite and/or K-amphibole. However, K_2O contents are not high in all the rocks of this suite. This indicates that the distribution of the K-rich mineral phase(s) in the melt source was heterogeneous. The heterogeneity of the

mantle source is also indicated by the large variation of ratios between incompatible trace elements, e.g. Th/Ta, Nb/Ta, La/Nb, Sr/Ce, etc. The coexistence of enriched rocks with both flat (Fig. 10b and d) and fractionated (Fig. 10f) REE patterns support this interpretation. In addition to SiO_2, Na_2O/K_2O and Sr/Y ratios are high in the Sanukitoid rocks, including in K_2O-rich high-Mg diorites. These features indicate that the mantle source metasomatizing agent was a silica- and Na-rich melt, pointing towards a sub-arc mantle enrichment controlled by Al-rich (Group 1) TTG melts, which were emplaced before the Sanukitoids in the Francistown region. Schiano et al. (1995) documented mantle xenoliths from arc settings containing hydrous silica-rich melt inclusions of high-Mg andesitic composition in olivine grains. This observation supports the mantle origin of Mg-dioritic magmas. Experimental studies (e.g. Sen and Dunn, 1995; Rapp et al., 1999) showed that the interaction between silica-rich melts and peridotites produces metasomatic ultramafic rocks containing garnet, amphibole, phlogopite and two pyroxenes. Partial melting of a metasomatized mantle with such a paragenesis would be an appropriate source for the Francistown Sanukitoid rock suite.

Yogodzinski et al. (1995) suggested that Archaean Sanukitoids could originate from direct partial melting of subducted mafic oceanic crust. This could explain several geochemical features of the Francistown Sanukitoids such as Ce/Pb, Sr/Y, Th/Ta, La/Nb, Cs/La, Ba/Th and Ba/La ratios and Ce negative anomaly documented in a few rock samples (e.g. TS 168B; Fig. 10c). Cerium negative anomaly is known to occur in oceanic mafic rocks affected by seawater alteration (e.g. Hole et al., 1984). It has also been documented in fresh arc mafic rocks where it is inherited from the mantle source enriched by fluids or melts from an oceanic crust affected by seawater alteration (e.g. Hole et al., 1984; Elliot et al., 1997). However, the Francistown Sanukitoid rocks are not the product of direct partial melting of a subducting slab because they are characterized by high Mg# and high concentration of transition metals in high-Mg diorites.

A number of workers (e.g. Stern and Hanson, 1991; Stevenson et al., 1999) stressed the geochemical similarities between Sanukitoids and calc-alkaline magmas and suggested the derivation of Sanukitoids from partial melting of a sub-arc peridotitic mantle wedge enriched by fluids expelled from a subducting slab. Although this interpretation can explain a number of geochemical characteristics of the Francistown Sanukitoids, it cannot explain the high SiO_2, Ni and Cr contents, low Sr/Ce ratios of some high-Mg diorites and high Sr/Y values of mafic rocks of the Francistown Sanukitoid suite. In addition, it is not consistent with the low Mg# and transition metal abundances marking experimental melts from amphibolite or eclogite (Rapp et al., 1991; Sen and Dunn, 1994). Kelemen et al. (1993) pointed out that high Mg# and transition metal contents of high-Mg andesites are acquired during interaction of silica-rich slab-derived melt with the overlying mantle and this is consistent with the following geochemical features of the Francistown Sanukitoid suite: (1) an increase of SiO_2 from gabbros to Mg-diorites and a correlative increase of LREE shown by an increase of La_N/Sm_N and La_N/Yb_N; (2) a substantial LREE and HREE fractionation in the high-Mg diorites (Fig. 10f) and no REE fractionation in the mafic rocks (Fig. 10b); (3) high Ni and Cr abundances requiring an ultramafic source. Altogether, the geochemical data indicate that the source of the Sanukitoid suite is a sub-arc mantle wedge variably metasomatized by both TTG melts and fluids escaping from the subducting slab. The high-Mg diorites originate from a mantle section strongly metasomatized during the ascent of earlier TTG melts. High pressure experimental investigations by Rapp et al. (1999) showed that slab-derived silica-rich melt interacting with mantle peridotite during ascent becomes Mg-rich but preserves its main geochemical features, e.g. high SiO_2 content, high LREE enrichment, fractionation of REE and high Sr/Y ratios. In addition, the ascending silicate melt is used in reactions converting the mantle peridotite assemblages into a metasomatic assemblage made of garnet-phlogopite-amphibole-two pyroxenes, a paragenesis inferred above for the source of the Francistown Sanukitoids and documented in mantle xenoliths from arc settings (e.g. Vidal et al., 1989).

6.4. High-K granites

Two main models can account for the genesis of granites: (1) fractional crystallisation of a mantle-originating mafic magma, combined or not with crustal assimilation; (2) partial melting of crustal rocks. The Francistown high-K granites are part of a large late to post-orogenic potassic granite province known as the Chilimanzi granite suite in the Zimbabwe craton. There is no evidence suggesting a linkage of this large potassic granitic province to mantle-originating mafic magmas, although these granitoids are partly coeval to the emplacement of the Great Dyke in Zimbabwe (e.g. Armstrong and Wilson, 2000). The high SiO_2 contents of these granites preclude a direct mantle origin.

The Francistown high-K granites are metaluminous to slightly peraluminous (Fig. 5) and show affinities to I-type potassic calc-alkaline granites. Their chondrite normalized REE patterns are similar to those of calc-alkaline felsic rocks from modern continental active margins, e.g. rhyolites and granites from the Cascades in California, USA (e.g. Tepper et al., 1993; Borg and Clynne, 1998). However, similar rocks are also emplaced during late-to post-orogenic evolution of orogenic belts (e.g. Pitcher, 1993) and they mark late to post-orogenic evolution of most Archaean cratons (e.g. Tchameni et al., 2000).

In California, high-K felsic rocks chemically similar to the Francistown high-K granites were attributed to large proportions ($\sim 35-45\%$) of partial melting of a mafic lower crust. Dehydration melting experiments of mafic rocks between ~ 800 and 1100 °C yielded felsic melts with calc-alkaline composition (e.g. Beard and Lofgren, 1991; Rushmer, 1991; Wolf and Wyllie, 1994; Patiño Douce and Beard, 1995). The negative Eu anomaly in the REE patterns (Fig. 10) suggests substantial plagioclase in the residual assemblage and, according to Tepper et al. (1993), this would indicate melting under low $a_{H_2}O$. The flat HREE patterns of some high-K granites (e.g. $0.9 < Gd_N/Yb_N < 1.9$ in samples MK112A, B, MKQ 5 and MK 31B) and their relatively high Y and Yb contents indicate a garnet-free source. Although host–rock composition dependent, garnet is generally absent in dehydration melting experiments conducted at <8 kb (Rushmer, 1991; Rapp et al., 1991), whereas it is present in similar experiments at ~ 10 kb (Wolf and Wyllie, 1994). Therefore, the Francistown high-K granites have most probably been generated at pressures <8 kb, i.e. within the middle or the lower crust as already proposed above. To reach the critical melt fraction of $\sim 30-40\%$ required for felsic melt to separate from its source and define discrete magma bodies (Wickham, 1987), we infer that temperatures in excess of 900 °C were required.

The composition of the source rocks can be inferred from the chemical composition of the granites. Roberts and Clemens (1993) pointed out that, because of their low K_2O contents, metabasaltic rocks are unsuitable sources for high-K, I-type granitoids. According to these authors, these granites are derived from partial melting of metamorphosed hydrous intermediate calc-alkaline rocks. Experimental data of Carrol and Wyllie (1989) indicate that partial melting of tonalites could produce high-K granite melts and thus we infer that TTG material represents the potential source rocks for the high-K granites in the Zimbabwe craton. The Francistown high-K granites show a large variation of composition in terms of LILE and REE (Figs. 10 and 11). However, this suite includes rocks having the highest incompatible element contents in the Francistown granitoids. The overall shape of the Francistown high-K granites in primordial mantle normalized diagrams (Fig. 11) are identical to the patterns of the TTG suite rocks (Fig. 11c and e). Thus, the high-K granites most probably formed by partial melting of TTG rocks. Particularly, the granites with flat HREE patterns (e.g. $La_N/Yb_N \sim 0.94$ in sample MK 31B) most probably were formed by partial melting of Group 2-like TTG rocks. The high-K granite MK 108 shows a fractionated HREE pattern (Fig. 11h), with $La_N/Yb_N \sim 6.4$ suggesting that it could originate from Group 1-like TTG sub-suite rocks. Melting of a source including both Groups 1 and 2 TTG sub-suite rocks could produce melts with geochemical features intermediate between these two end-member source rocks. Our interpretation is in line with the results of previous workers (e.g. Luais and Hawkesworth, 1994) who showed that, in the Zimbabwe craton, the youngest granitoids originate from partial melting, within the crust, of earlier TTG material.

6.5. Tectonic implications

The geochemical data indicate that the Neoarchaean Francistown granitoids are marked by: (1) a high-Al (Group 1) TTG sub-suite originating from partial melting of a subducting slab; (2) a low-Al (Group 2) TTG sub-suite originating from partial melting of arc mafic igneous rocks underplated and metamorphosed in the lower crust; (3) a Sanukitoid suite ultimately derived from a sub-arc mantle wedge enriched by silica-rich TTG melts and fluids escaping from a subducting slab; and (4) younger crustally derived high-K granites resulting from partial melting of TTG material. Treloar et al. (1992) suggested the genesis of these high-K granites could indicate a major crustal thickening following continental collision. However, there are no S-type granites in the Francistown region and in the rest of the Zimbabwe craton despite the presence of supracrustal sedimentary rocks and low-grade metasedimentary assemblages. Thus, there was no major tectonically induced overthickening of the crust beneath the Zimbabwe craton and this is compatible with geophysical data indicating a crustal thickness between 34 and 37 km (Nguuri et al., 2001). The layer-parallel shear zones and thrusts active between ~ 2.68 and 2.6 Ga in the Zimbabwe craton (e.g. Kusky and Kidd, 1992; Dirks and Van der Merwe, 1997; Jelsma and Dirks, 2000; Bagai et al., 2002; Dirks et al., 2002) induced a limited thickening of the crust, probably similar to tectonic thickening in modern accretionary orogens (e.g. Andes). Dirks et al. (2002) suggested that these layer-parallel shear zones and thrusts formed during a shallow subduction process or underplating. Geochemical data in this paper support a flat slab model.

High-K granites are widespread in the Zimbabwe craton whereas coeval mafic rocks are only developed along the Great Dyke (e.g. Armstrong and Wilson, 2000). The absence of large volume of mafic rocks spatially closely related to the high-K granites implies that mafic magma underplating is probably not the source of heat responsible for the generation of the granites. The alternative process allowing to heat the lower crust is the juxtaposition of the asthenospheric mantle against the base of the crust. This could happen during delamination (detachment) of a subducting slab (e.g. Houseman et al., 1981) and does not require Himalayan-type continental collision.

The magmatic suites identified in the Francistown igneous province define a consistent tectono-magmatic evolution pattern marked by the following:

(1) A shallow-dipping (flat) subduction during the earliest stage leading to partial melting of the subducting slab to produce the high-Al TTG sub-suite (Group 1). The earliest ductile fabric reported in the

Francistown TTG rocks and unknown in the adjacent (younger) Sanukitoid rocks mark a shortening event during this earliest igneous event and this is compatible with a flat subduction. During flat subduction, there is a higher interplate coupling and the cold, strong rheology of the overriding lithosphere enables stress and deformation to be transmitted far inboard into the upper plate (e.g. Pubellier and Cobbold, 1996). Shallow-dipping subduction systems are marked by wide arc systems, extending much further (≥ 400 km) from the trench (Gutscher et al., 2000). In the case of central Chile, Kay and Abbruzzi (1996) indicated that the arc above the central Andean flat slab extends between 250 and 800 km from the trench. This could explain the large area covered by Neoarchaean TTG rocks and affected by ca. 2.7–2.6 Ga layer-parallel shear zones and thrusts in the Zimbabwe craton.

(2) Generation of the Sanukitoid suite by partial melting of a sub-arc mantle wedge enriched during the ascent of the earlier TTG melts. The absence of TTG igneous rocks emplaced at the same time or mixed with the Sanukitoid suite rocks suggest that the partial melting of the subducting slab had ceased when the Sanukitoid magmas were produced. We speculate that the transition from TTG to Sanukitoids could correspond to a change from shallow to a steep subduction. It is known that flat subduction alters the thermal structure of an active margin because of the insertion of cold oceanic lithosphere beneath the upper lithosphere in the area where normally hot asthenosphere occurs in the case of steep subduction (e.g. Davies, 1999). Therefore, prolonged flat subduction cools both plates and increases the strength of the upper plate (Vlaar, 1983; Spencer, 1994). We infer that a substantial cooling of the downgoing slab beneath the Zimbabwe craton shut its melting and favoured the conversion of basaltic rocks into eclogites, leading to a steeper subduction regime and related partial melting of the sub-arc mantle wedge (cf. Sanukitoid suite). An important observation from our geochemical data is that the ratios Cs/La, Ba/Th, Ba/La point to the emplacement of the Francistown Sanukitoids at or close (<50 km) to the volcanic arc front. This requires a trench, which was relatively close to Francistown during the Neoarchaean and, as there is no evidence of extreme crustal overthickening in the region, presumably the trench sediment infill should

still be preserved in the vicinity. In our interpretation, the supracrustal (meta) sedimentary rocks exposed in the Shashe belt represent remnants of the Neoarchaean accretionary sedimentary packages. The Matsitama greenstone belt, which is close to this inferred trench sedimentary package, contains mafic rocks with geochemical affinities with oceanic arc tholeiites (Majaule et al., 1997). Geophysical studies have shown that the Shashe belt is the western extension of the Neoarchaean Limpopo belt (Ranganai et al., 2002). Therefore, the subduction along the Limpopo–Shashe belt presumably controlled the genesis and emplacement of the 2.7–2.6 TTG and Sanukitoid suites in the Zimbabwe craton. This implies a northerly-dipping subduction zone, and this is in agreement with the north–south Neoarchaean shortening documented in the southern part of the Zimbabwe craton by Treloar and Blenkinsop (1995). Granitoids exposed in the Limpopo–Shashe belt are coeval to the TTG suite in the Zimbabwe craton (Majaule and Davis, 1998; Kröner et al., 1999). The convex-shape of the Limpopo–Shashe belt compares to the shape of the Pacific subduction system along the Aleutian trench in Alaska (North America) or the Makran plate convergence zone between the Arabian and Iranian microplates, and is most probably a primary feature. A convex subduction along the Limpopo–Shashe belt implies, in present day coordinates, a west to north-west-dipping subduction zone to the east of the Zimbabwe craton and a northeast-dipping subduction zone to the west of the craton. In such a continental active margin, the layer-parallel shear zones and thrust developing in the overriding plate will be marked by cratonward transport direction, i.e. towards the northeast in the west of the craton and westward/northwestward in the east. This is compatible with available structural data (Dirks and Van der Merwe, 1997; Jelsma and Dirks, 2000; Dirks et al., 2002; Paya and Kampunzu, unpublished data).

Several authors (e.g. Cloos, 1993; Gutscher et al., 2000 and references therein) indicated that flat subduction are common where ridge or oceanic plateaus are subducting. The Central Zone of the Limpopo belt includes mafic–ultramafic rocks chemically similar to mafic rocks along the Chile-type oceanic ridge and emplaced when this ridge was subducting beneath southern America (Kampunzu et al., 2002, this volume). The subduction and partial melting of this type of oceanic basalts would provide an adequate source for the high-Al TTG sub-suite in the Zimbabwe craton. Seismic tomography investigations showed that slab melting is closely linked to flat subduction of thick oceanic crust (e.g. Gutscher et al., 2000). Therefore, the Francistown TTG probably reflects a flat subduction of a thick oceanic crust, i.e. an oceanic plateau or an arc. Presumably, this process was common during the Archaean, leading to frequent slab melting (e.g. Martin, 1993, 1994). It is known that both the overriding plate and the subducting slab are cooled by flat subduction (e.g. Vlaar, 1983; Spencer, 1994). Cooling the subducting lithosphere delays the basalt to eclogite transition but once this happens, the average density of the downgoing slab would increase and this should induce a change from flat to steeper subduction (e.g. Abbott et al., 1994). The production of TTG magmas will cease with that change. A steep subduction favours dehydration of fluids from the subducting-slab which enrich the mantle wedge leading to the generation of calc-alkaline mafic magmas (e.g. Pearce, 1982). The Francistown Sanukitoid suite rocks have geochemical affinities with calc-alkaline magmas and this probably indicate a fluid input at the mantle source but bear also the geochemical imprint of a sub-arc mantle metasomatism induced by the ascent of earlier TTG melts.

Condie and Harrison (1976) reported "tholeiitic" mafic rocks in the Zimbabwe craton (between ~ 18–20°S and 29–30°E). These rocks show chemical similarities with the Francistown Sanukitoid suite mafic rocks. In addition, the andesites reported by these authors in the same area contain on average ~ 6% MgO at ~ 57% SiO_2 and corresponds to Mg-andesites similar to the Sanukitoid Mg-diorites documented in this paper. According to Jelsma (personal communication, 2002), the Nd and Pb isotopic compositions published by Jelsma et al. (1996) relate to the Sanukitoid rocks in Zimbabwe. They show positive (+2 to +3) epsilon Nd_T and identical U–Pb zircon and T_{DM} model ages at ca. ~ 2.64 Ga. These data point to mantle origin for the Sanukitoids as proposed in this study. The high-μ value reported by Jelsma et al. (1996) for the source of these rocks support the subduction model proposed during the genesis of the Francistown Sanukitoids. The areal coverage of the Sanukitoid volcanism in the Zimbabwe craton is yet not known but conservative

estimate using the data from this study and from Condie and Harrison (1976) is >1000 km^2. It indicates an important continental active margin and probably a steeper subduction process in the evolution of the Limpopo–Shashe accretionary system. Our interpretation is close to that of Berger and Rollinson (1997) who pointed out that Neoarchaean crustal growth within the Northern Marginal Zone of the Limpopo belt in Southern Zimbabwe compares with modern continental convergent margins such as the Andes (see also Berger et al., 1995). We further contend that the Neoarchaean magmatism at ~ 2.8–2.6 Ga in the Zimbabwe craton and the Limpopo–Shashe belt are linked to various evolutionary stages of a single Andes-type long-lived active continental margin, with an earlier flat subduction stage (generation of high-Al TTGs) relayed by a steeper subduction (emplacement of Sanukitoids). The seismic tomography map of the Zimbabwe craton (James et al., 2001) shows a fast mantle and a keel beneath the Zimbabwe craton. We infer that this mantle root represents the residual mantle wedge after extraction of the arc magmas.

Previous geodynamic models assuming that the ~ 2.8–2.5 Ga magmatism in the Zimbabwe craton was emplaced in a classical continental rift setting (e.g. Nisbet et al., 1981; Blenkinsop et al., 1993; Jelsma et al., 1996) are not supported by our data. Hunter et al. (1998) linked the ~ 2.7 Ga Belingwe greenstone belt (Zimbabwe) to an extensional (rift) setting based on sedimentological data (e.g. proximal thin sedimentary deposits, heterogeneity of the sedimentary rocks indicating very local sources) and the presence of komatiites inferred to mark a plume. However, the above sedimentological features could as well be accommodated in a steep subduction model known to be characterised by the development of extensional basins in the upper plate. In addition, komatiites can form in arc settings (T. Grove, personal communication, 2000; see also Parman et al., 1997), although there is no agreement on this question (Herzberg and O'Hara, 1998 and references therein).

Kusky (1998) proposed that the Tati and Vumba greenstone–granite terranes represent an oceanic plateau accreted onto the Zimbabwe craton and subsequently affected by arc magmatism. Geochemical data (Bagai, 2000; this paper and our unpublished data) do not support this interpretation. The

Vumba and Tati greenstone belts and related granitoids are part of arc magmatism; there is no evidence for an oceanic plateau in these two greenstone
belts.

The Francistown high-K granites and correlative
granitoids of the Chilimanzi suite in Zimbabwe have
structural features (e.g. Mkweli et al., 1995; Frei et al.,
1999) and geochemical characteristics (this paper)
which are compatible with their genesis in a late to
post-orogenic environment. There is no evidence of
magmatism due to overthickening of continental crust
in the Zimbabwe craton. S-type plutons, which would
be significant in such a setting, are absent and the I-
type composition of the voluminous late-to post-
orogenic high-K granites in the Zimbabwe craton
indicates partial melting of TTG bodies underplated
in the lower crust. A potential tectonic model for the
high-K granites, in line with the geotectonic interpretation of the Francistown Sanukitoid and TTG suites,
would be the detachment (delamination) of the subducting slab (e.g. Houseman et al., 1981; Kampunzu
et al., 1998). Steepening of the subducting slab most
probably led to its detachment, allowing upwelling of
a hotter and deeper mantle section supplying heat for
crustal melting of earlier TTG material. This interpretation is supported by metamorphic studies indicating
a ~ 2.6–2.5 LP-HT granulite facies metamorphism
(Berger et al., 1995; Blenkinsop and Frei, 1996;
Kamber et al., 1996) showing an anticlockwise $P–
T–t$ path in the NMZ (Kamber and Biino, 1995). On
this reasoning the main region of delamination lay
within and north of the NMZ, consistent with the
occurrence of large volumes of Chilimanzi-type granitoid plutons in Zimbabwe.

7. Conclusion

The Neoarchaean (~ 2.7–2.6 Ga) intrusive bodies
exposed at the southwestern margin of the Zimbabwe
craton in NE Botswana represent three distinct magmatic suites: (1) TTG suite made of tonalites–trondhjemites and Na-granites; (2) Sanukitoid suite including gabbros and Mg-diorites; and (3) high-K granites
which are part of the Zimbabwe craton-wide Chilimanzi late- to post-orogenic granitoid suite. These
three magmatic suites occur also in the centre of the
Zimbabwe craton.

Major and trace element compositions of these
three magmatic suites indicate the following features:

(1) High-Na, Sr and Sr/Y, enrichment in LILE and
REE, fractionation of both LREE and HREE and
no negative Eu anomaly in the chondrite-
normalized patterns of Al-rich (Group 1) TTG
sub-suite rocks. These features indicate the
genesis of these magmas by partial melting of a
subducting slab. The Al-poor (Group 2) TTG
sub-suite rocks originate from partial melting (8
kb < P < 15 kb) of garnet amphibolites representing underplated arc mafic rocks metamorphosed
at the base or within the lower crust;
(2) High-Mg diorites which are rich in transition
metals, LILE and LREE and possess steeply
fractionated REE profiles. These geochemical
features are typical for Archaean Sanukitoids and
modern Mg-andesites in arc settings. The transition metal and incompatible trace element
contents are high in these silica-rich magmas.
The enrichment style (e.g. Nb–Ti negative
anomalies, low Ce/Pb, high SiO_2 and transition
metal contents, high La/Nb, Th/Ta, Ba/La, Cs/La,
Ba/Th) are consistent with a sub-arc enriched
mantle source for the Sanukitoid suite. TTG melts
from melting of the subducting slab and fluids
escaping from the downgoing slab represent the
most important metasomatising agents of this
mantle source;
(3) High-K granites originate from crustal melting of
earlier TTG material. There is no evidence for or
against partial melting triggered by underplating
of arc mafic magmas to produce these granites.

However, the evolution from TTG to Sanukitoids
and to high-K granites is here taken to indicate a
progressive change from an earlier flat subduction of
a hot lithosphere to a steeper subduction induced by
the cooling of the slab during the subduction process
and finally a break-off and detachment of this steep
slab. This detachment allowed the uprise of hotter
mantle material, which supplied the required heat for
crustal melting of earlier TTG to produce high-K
granites. We believe that the Francistown igneous
suites and coeval igneous rocks in the Zimbabwe
craton and Limpopo–Shashe belt represent a single
accretionary system including: (1) a continental mag-

matic arc within the Zimbabwe craton and (2) an accretionary sedimentary and volcanic/plutonic assemblage within the southward-convex Limpopo–Shashe belt.

Acknowledgements

This is a contribution to the Kaapvaal Craton Project. A.B.K., M.M., M.Z. and Z.B. acknowledge the logistic and financial support of University of Botswana (UB) to the Kaapvaal Craton Project (RPC Grant R#442) and the logistic support of Geological Survey of Botswana (DGS) to the UB Kaapvaal Craton Team. We acknowledge the contribution to this research of Year 4 students from University of Botswana (Segwabe and Keeletsang: academic year 1997/1998; Keitumetse, Ntere and Rachere: academic year 1999/2000). A.R.T. and T.M. acknowledge the DGS for supporting the involvement of their staff into the Kaapvaal Project. This paper is published with the authorisation of the Minister of Mineral, Energy and Water Affairs and of the Director of Geological Survey Department, Botswana.

References

Abbott, D., Drury, R., Smith, W.H.F., 1994. Flat to steep transition in subduction style. Geology 22, 937–940.

Aldiss, D.T., 1991. The Motloutse Complex and the Zimbabwe craton/Limpopo belt transition in Botswana. Precambrian Res. 50, 89–109.

Armstrong, R., Wilson, A.H., 2000. A SHRIMP U–Pb study of zircons from the layered sequence of the Great Dyke, Zimbabwe, a granitoid anatectic dyke. Earth Planet. Sci. Lett. 180, 1–12.

Bagai, Z., 2000. Geochemical and geochronological investigations of the Vumba granite–greenstone terrain of NE Botswana. MPhil Thesis, Univ. Durham, UK. 174 pp.

Bagai, Z., Armstrong, R., Kampunzu, A.B., 2002. U–Pb single zircon geochronology of granitoids in the Vumba granite–greenstone terrain (NE Botswana): implication for the Archaean Zimbabwe craton. Precambrian Res. 118, 149–168.

Baird, A.K., Miesh, A.T., 1984. Batholithic rocks of southern California—a model for the petrological nature of their source material. U.S. Geol. Surv. Prof. Paper, vol. 1284. 42 pp.

Barker, F., 1979. Trondhjemite: definition, environment, and hypotheses of origin. In: Barker, F. (Ed.), Trondhjemites, Dacites and Related Rocks. Elsevier, Amsterdam, pp. 1–12.

Beard, J.S., Lofgren, G.E., 1991. Dehydration melting and water-saturated melting of basaltic and andesitic greenstones and amphibolites at 1, 3 and 6.9 kb. J. Petrol. 32, 465–501.

Berger, M., Rollinson, H., 1997. Isotopic and geochemical evidence for crust–mantle interaction during late Archaean crustal growth. Geochim. Cosmochim. Acta 61, 4809–4829.

Berger, M., Kramers, J.D., Nagler, Th.F., 1995. Geochemistry and geochronology of charnockites and enderbites in the Northern Marginal Zone of the Limpopo belt, Southern Africa, and genetic models. Schweiz. Mineral. Petrogr. Mitt. 75, 17–42.

Bickle, M.J., Nisbet, E.G., Martin, A., 1994. Archaean greenstone belts are not oceanic crust. J. Geol. 102, 121–138.

Blenkinsop, T.G., Frei, R., 1996. Archaean and Proterozoic mineralisation and tectonics at Renco mine (Northern Marginal Zone, Limpopo belt, Zimbabwe). Econ. Geol. 91, 1225–1238.

Blenkinsop, T.G., Fedo, C.M., Bickle, M.J., Eriksson, K.A., Martin, A., Nisbet, E.G., Wilson, J.F., 1993. Ensialic origin for the Ngezi Group, Belingwe greenstone belt, Zimbabwe. Geology 21, 1135–1138.

Blenkinsop, T.G., Martin, A., Jelsma, H.A., Vinyu, M.L., 1997. The Zimbabwe craton. In: de Wit, M.J., Ashwal, L.D. (Eds.), Greenstone Belts. Clarendon Press, New York, pp. 567–580.

Borg, L.E., Clynne, M.A., 1998. The petrogenesis of felsic calc-alkaline magmas from the southernmost Cascades, California: origin by partial melting of basaltic lower crust. J. Petrol. 39, 1197–1222.

Carroll, M.J., Wyllie, P.J., 1989. Experimental phase relations in the system tonalite–peridotite–H_2O at 15 kbar, implications for assimilation and differentiation processes near the crust–mantle boundary. J. Petrol. 30, 1351–1382.

Chauvel, C., Goldstein, S.L., Hofman, A.W., 1995. Hydration and dehydration of oceanic crust controls Pb evolution in the mantle. Chem. Geol. 126, 65–75.

Cloos, M., 1993. Lithospheric buoyancy and collisional orogenesis: subduction of oceanic plateaus, continental margins, island arcs, spreading ridges, and seamounts. Geol. Soc. Am. Bull. 105, 715–737.

Condie, K.C., 1989. Geochemical changes in basalts and andesites across the Archean–Proterozoic boundary: identification and significance. Lithos 23, 1–18.

Condie, K.C., Harrison, N.M., 1976. Geochemistry of the Archean Bulawayan Group, Midlands greenstone belt, Rhodesia. Precambrian Res. 3, 253–271.

Coward, M.P., James, P.R., Wright, L., 1976. Northern margin of the Limpopo orogenic belt, southern Africa. Geol. Soc. Am. Bull. 87, 601–611.

David, K., Schiano, P., Allègre, C.J., 2000. Assessment of the Zr/Hf fractionation in oceanic basalts and continental materials during petrogenetic processes. Earth Planet. Sci. Lett. 178, 285–301.

Davies, J.H., 1999. Simple analytic model for subduction zone thermal structure. Geophys. J. Int. 139, 823–828.

de Wit, J.M., Roering, C., Hart, R.J., Armstrong, R.A., De Ronde, C.E.J., Green, R.W.E., Tredoux, M., Peberdy, E., Hart, R.A., 1992. Formation of an Archaean continent. Nature 357, 553–562.

Dirks, P.H.G.M., Van der Merwe, J., 1997. Early duplexing in an Archaean greenstone sequence and its control on gold mineralization. J. Afr. Earth Sci. 24, 603–620.

Dirks, P.H.G.M., Jelsma, H.A., Hofmann, A., 2002. Thrust-related accretion of an Archaean greenstone belt in the Midlands of Zimbabwe. J. Struct. Geol. 24, 1707–1727.

Dodson, M.H., Williams, I.S., Kramers, J.D., 2001. The Mushandike granite: further evidence for 3.4 Ga magmatism in the Zimbabwe craton. Geol. Mag. 138, 31–38.

Dostal, J., Mueller, W., 1992. Archean shoshonites from the Abitibi greenstone belt, Chibougamau (Quebec, Canada): geochemistry and tectonic setting. J. Volcanol. Geotherm. Res. 53, 145–165.

Drummond, M.S., Defant, M.J., 1990. A model for trondhjemite–tonalite–dacite genesis and crustal growth via slab melting: Archaean to modern comparisons. J. Geophys. Res. 95, 21503–21521.

Elliot, T., Plank, T., Zindler, A., White, W., Bourdon, B., 1997. Element transport from slab to volcanic front at the Mariana arc. J. Geophys. Res. 102, 14991–15019.

Frei, R., Blenkinsop, T.G., Schönberg, R., 1999. Geochronology of the late Archaean Razi and Chilimanzi suites of granites in Zimbabwe: implications for the late Archaean tectonics of the Limpopo belt and Zimbabwe craton. S. Afr. J. Geol. 102, 55–63.

Gill, J.B., 1981. Orogenic Andesites and Plate Tectonics Springer-Verlag, Berlin. 389 pp.

Gromet, L.P., Silver, L.T., 1987. REE variations across the Peninsular Ranges batholith: implications for batholithic petrogenesis and crustal growth in magmatic arcs. J. Petrol. 28, 75–125.

Gutscher, M.A., Maury, R., Eissen, J.P., Bourdon, E., 2000. Can slab melting be caused by flat subduction? Geology 28, 535–538.

Hamilton, W.B., 1998. Archean magmatism and deformation were not products of plate tectonics. Precambrian Res. 91, 143–179.

Helz, R., 1976. Phase relations of basalts in their melting ranges at $P_{H_2O} = 5$ kb: Part 2. Melt compositions. J. Petrol. 17, 139–193.

Herzberg, C., O'Hara, M.J., 1998. Phase equilibrium constraints on the origin of basalts, picrites, and komatiites. Earth Sci. Rev. 44, 39–79.

Hole, M.J., Saunders, A.D., Marriner, G.F., Tarney, J., 1984. Subduction of pelagic sediments: implications for the origin of Ce-anomalous basalts from the Mariana islands. J. Geol. Soc. Lond. 141, 453–472.

Hollings, P., Wyman, D., Kerrich, R., 1999. Komatiite–basalt–rhyolite volcanic associations in Northern Superior Province greenstone belts: significance of plume–arc interaction in the generation of the proto continental Superior Province. Lithos 46, 137–161.

Holloway, J.R., Burnham, C.W., 1972. Melting relations of basalt with equilibrium water pressure less than total pressure. J. Petrol. 13, 1–29.

Horstwood, M.S.A., Nesbitt, R.W., Noble, S.R., Wilson, J.F., 1999. U–Pb zircon evidence for an extensive early Archean craton in Zimbabwe: a reassessment of the timing of craton formation, stabilization and growth. Geology 27, 707–710.

Houseman, G.A., McKenzie, D.P., Molnar, P., 1981. Convective instability of a thickened boundary layer and its relevance for the thermal evoluion of continental convergence belts. J. Geophys. Res. 86, 6115–6132.

Hunter, M.A., Bickle, M.J., Nisbet, E.G., Martin, A., Chapman, H.J., 1998. Continental extensional setting for the Archean Belingwe greenstone belt, Zimbabwe. Geology 26, 883–886.

Hyndman, D.W., Foster, D.A., 1988. The role of tonalites and mafic dykes in the generation of Idaho batholith. J. Petrol. 25, 894–929.

James, D.E., Fouch, M.J., VanDecar, J.C., van der Lee, S., the Kaapvaal Seismic Group, 2001. Tectospheric structure beneath southern Africa. Geophys. Res. Lett. 28, 2485–2488.

Jelsma, H.A., Dirks, P.H.G.M., 2000. Tectonic evolution of a greenstone sequence in northern Zimbabwe: sequential early stacking and pluton diapirism. Tectonics 19, 135–152.

Jelsma, H.A., Vinyu, M.L., Valbracht, P.J., Davies, G.R., Wijbrans, J.R., Verdurmen, E.A.T., 1996. Constraints on Archaean crustal evolution of the Zimbabwe craton: a U–Pb zircon, Sm–Nd and Pb–Pb whole-rock isotope study. Contrib. Mineral. Petrol. 124, 55–70.

Jelsma, H.A., Becker, J.K., Siegesmund, S., 2001. Geochemical characteristics and tectonomagmatic evolution of the Chinamora batholith, Zimbabwe. Z. Dtsch. Geol. Ges. 152, 199–225.

Jenner, G.A., Longerich, H.P., Jackson, S.E., Fryer, B.J., 1990. ICP-MS—a powerful tool for high-precision trace-element analysis in Earth Sciences: evidence from analysis of selected U.S.G.S. reference samples. Chem. Geol. 83, 133–148.

Kamber, B.S., Biino, G.G., 1995. The evolution of high T–low P granulites in the Northern Marginal Zone sensu stricto, Limpopo belt, Zimbabwe—the case for petrography. Schweiz. Mineral. Petrogr. Mitt. 75, 427–454.

Kamber, B.S., Collerson, K.D., 2000. Role of "hidden" deeply subducted slabs in mantle depletion. Chem. Geol. 166, 241–254.

Kamber, B.S., Biino, G.G., Wijbrans, J.R., Davies, G.R., Villa, I.M., 1996. Archaean granulites of the Limpopo belt, Zimbabwe: one slow exhumation or two rapid events? Tectonics 15, 1414–1430.

Kampunzu, A.B., Akanyang, P., Mapeo, R.B.M., Modie, B.N., Wendorff, M., 1998. Geochemistry and tectonic significance of Mesoproterozoic Kgwebe metavolcanic rocks in northwest Botswana: implications for the evolution of the Kibaran Namaqua-Natal belt. Geol. Mag. 133, 669–683.

Kampunzu, A.B., Ramakgala, M., Paya, B.K., McCourt, S., Mitchell, A., Hoffmann, D., Witley, J., 2002. Archaean podiform chromitite from the Phikwe Ni–Cu sulphide deposit (Limpopo belt, NE Botswana): chrome spinel and host-rock chemistry, implications for geotectonic setting. Lithos (this volume, submitted).

Kapenda, D., Kampunzu, A.B., Cabanis, B., Namegabe, M., Tshimanga, K., 1998. Petrology and geochemistry of post-kinematic mafic rocks from the Paleoproterozoic Ubendian belt, NE Katanga (Democratic Republic of Congo). Geol. Rundsch. 87, 345–362.

Kay, S.M., Abbruzzi, J.M., 1996. Magmatic evidence for Neogene lithospheric evolution of the central Andean "flat-slab" between 30°S and 32°S. Tectonophysics 259, 15–28.

Kelemen, P.B., 1995. Genesis of the high Mg andesites and the continental crust. Contrib. Mineral. Petrol. 120, 1–19.

Kelemen, P.B., Shimizu, N., Dunn, T., 1993. Relative depletion of niobium in some arc magmas and the continental crust: partitioning of K, Nb, La and Ce during melt/ rock reaction in the upper mantle. Earth Planet. Sci. Lett. 120, 111–134.

Kepezhinskas, P.K., Defant, M.J., Drummond, M.S., 1996. Progressive enrichment of island arc mantle by melt–peridotite interaction inferred from Kamchatka xenoliths. Geochim. Cosmochim. Acta 60, 1217–1229.

Keppler, H., 1996. Constraints from partitioning experiments on the composition of subduction-zone fluids. Nature 380, 237–240.

Key, R.M., Litherland, M., Hepworth, J.V., 1976. The evolution of the Archaean crust of northeast Botswana. Precambrian Res. 3, 375–413.

Kröner, A., Jaeckel, P., Brandle, G., Nemchin, A.A., Pidgeon, R.T., 1999. Single zircon ages for granitoid gneisses in the Central Zone of the Limpopo belt, southern Africa and geodynamic significance. Precambrian Res. 93, 299–337.

Kusky, T.M., 1998. Tectonic setting and terrane accretion of the Archean craton. Geology 26, 163–166.

Kusky, T.M., Kidd, W.S.F., 1992. Remnants of an Archaean oceanic plateau, Belingwe greenstone belt, Zimbabwe. Geology 20, 43–46.

Luais, B., Hawkesworth, C.J., 1994. The generation of continental crust: an integrated study of crust-forming processes in the Archaean of Zimbabwe. J. Petrol. 35, 43–93.

Majaule, T., Davis, D.W., 1998. U–Pb zircon dating and geochemistry of granitoids in the Mosetse area, NE Botswana, and tectonic implications. Geol. Surv. Botswana, 50th Anniversary Internat. Conf. Abstr. Vol, pp. 46–48.

Majaule, T., Hall, P., Hughes, D., 1997. Geochemistry of mafic and ultramafic igneous rocks of the Matsitama supracrustal belt, northeastern Botswana—provenance implications. S. Afr. J. Geol. 100, 169–179.

Martin, H., 1993. The mechanism of petrogenesis of the Archean continental crust—comparison with modern processes. Lithos 30, 373–388.

Martin, H., 1994. The Archaean grey gneisses and the genesis of the continental crust. In: Condie, K.C. (Ed.), The Archaean Crustal Evolution. Elsevier, Amsterdam, pp. 205–259.

McCourt, S., Wilson, J.F., 1992. Late Archaean and early Proterozoic tectonics, Limpopo and Zimbabwe. In: Glover, J.E., Ho, S.E. (Eds.), The Archaean: Terrains, Processes and Metallogeny. Proceedings Third Internat. Archaean Symposium, Perth, University Western Australia Public, Nr. 22, 237–246.

McCulloh, M.T., 1993. The role of subducted slabs in an evolving earth. Earth Planet. Sci. Lett. 115, 89–100.

Mkweli, S., Kamber, B., Berger, M., 1995. Westward continuation of the Craton–Limpopo belt tectonic break in Zimbabwe and new age constraints on the timing of the thrusting. J. Geol. Soc. Lond. 152, 77–83.

Nägler, Th.F., Kramers, J.D., Kamber, B.S., Frei, R., Predergast, M.D.A., 1997. Growth of subcontinental lithospheric mantle beneath Zimbabwe started at or before 3.8 Ga: Re–Os study on chromites. Geology 25, 983–986.

Nguuri, T.K., Gore, J., James, D.E., Wright, C., Zengeni, T.D., Gwavava, O., Webb, S.J., Snoke, J.A., the Kaapvaal Seismic Group, 2001. Crustal structure beneath southern Africa and its implications for the formation and evolution of the Kaapvaal and Zimbabwe cratons. Geophys. Res. Lett. 28, 2501–2504.

Nisbet, E.G., Wilson, J.F., Bickle, M.J., 1981. The evolution of the Rhodesian craton and adjacent Archaean terrain: tectonic models. In: Kröner, A. (Ed.), Precambrian Plate Tectonics. Elsevier, Amsterdam, pp. 161–183.

Parman, S.W., Dann, J.C., Grove, T.L., de Wit, M.J., 1997. Emplacement conditions of komatiites magmas from the 3.49 Ga Komati Formation, Barberton Greenstone Belt, South Africa. Earth Planet. Sci. Lett. 150, 303–323.

Patiño Douce, A.E., Beard, J.S., 1995. Dehydration melting of biotite gneiss and quartz amphibolite from 3 to 15 kbar. J. Petrol. 36, 707–738.

Peacock, S.M., Rushmer, T., Thompson, A.B., 1994. Partial melting of subducting oceanic crust. Earth Planet. Sci. Lett. 121, 224–244.

Pearce, J.A., 1982. Trace element characteristics of lavas from destructive plate boundaries. In: Thorpe, R.S. (Ed.), Andesites. Wiley, New York, pp. 525–548.

Pearce, J.A., Harris, N.B.W., Tindle, A.G., 1984. Trace element discrimination diagrams for the tectonic interpretation of granitic rocks. J. Petrol. 25, 956–983.

Petford, N., Atherton, M., 1996. Na-rich partial melts from newly underplated basaltic crust: the Cordillera Blanca Batholith, Peru. J. Petrol. 37, 1491–1521.

Pitcher, W.S., 1993. The Nature and Origin of Granites. Blackie, Glasgow. 316 pp.

Plank, T., Langmuir, C.H., 1993. Tracing trace elements from sediment input to volcanic output at subduction zones. Nature 362, 739–742.

Plank, T., Langmuir, C.H., 1998. The chemical composition of subducting sediment and its consequences for the crust and mantle. Chem. Geol. 145, 325–394.

Pubellier, M., Cobbold, P., 1996. Analogue models for the transpressional docking of volcanic arcs in the western Pacific. Tectonophysics 253, 33–52.

Ranganai, R.T., Kampunzu, A.B., Atekwana, E.A., Paya, B.K., King, J.G., Koosimile, D.I., Stettler, E.H., 2002. Gravity evidence for a larger Limpopo belt in southern Africa and geodynamic implications. Geophys. J. Int. 149, F9–F14.

Rapp, 1997. Heterogeneous source for Archaean granitoids: experimental and geochemical evidence. In: de Wit, M.J., Ashwal, L.D. (Eds.), 1997. Greenstone Belts. Clarendon Press, New York, pp. 267–279.

Rapp, R.P., Watson, E.B., Miller, C.F., 1991. Partial melting of amphibolite/eclogite and the origin of Archean trondhjemites and tonalites. Precambrian Res. 94, 4619–4633.

Rapp, R.P., Shimizu, N., Norman, M.D., Applegate, G.S., 1999. Reaction between slab melts and peridotite in the mantle wedge: experimental constraints at 3.8 GPa. Chem. Geol. 160, 335–356.

Roberts, M.P., Clemens, J.D., 1993. Origin of high-potassium, calc-alkaline, I-type granitoids. Geology 21, 825–828.

Rollinson, H.R., 1993. A terrane interpretation of the Archaean Limpopo belt. Geol. Mag. 130, 755–765.

Rushmer, T., 1991. Partial melting of two amphibolites: contrasting experimental results under fluid-absent conditions. Contrib. Mineral. Petrol. 107, 41–59.

Ryan, J.G., Morris, J., Tera, F., Leeman, W.P., Tsvetkov, A., 1995. Cross-arc geochemical variations in the Kurile arc as a function of slab depth. Science 270, 625–627.

Sajona, F.G., Maury, R.C., Bellon, H., Cotten, J., Defant, M.J., Pubellier, M., Rangin, C., 1993. Initiation of subduction and the generation of slab melts in western and eastern Mindanao, Phillippines. Geology 21, 1007–1010.

Sajona, F.G., Maury, R.C., Pubellier, M., Leterrier, J., Bellon, H., Cotton, J., 2000. Magmatic source enrichment by slab-derived melts in a young post-collision setting, central Mindanao (Philippines). Lithos 54, 173–206.

Schiano, P., Clochiatti, R., Shimizu, N., Maury, R.C., Jochum, K.P., Hofmann, A.W., 1995. Hydrous silica-rich melts in the sub-arc mantle and their relationship with erupted arc lavas. Nature 377, 595–600.

Sen, C., Dunn, T., 1994. Dehydration melting of a basaltic composition amphibolite at 1.5 and 2.0 GPa: implications for the origin of adakites. Contrib. Mineral. Petrol. 117, 394–409.

Sen, C., Dunn, T., 1995. Experimental modal metasomatism of a spinel lherzolite and the production of amphibole bearing peridotite. Contrib. Mineral. Petrol. 119, 394–409.

Shimoda, G., Tatsumi, Y., Nohda, S., Ishizaka, K., Jahn, B.M., 1998. Setouchi high-Mg andesites revisited: geochemical evidence for melting of subducting sediments. Earth Planet. Sci. Lett. 160, 479–492.

Shirey, S.B., Hanson, G.N., 1984. Mantle-derived Archean monzodiorites and trachyandesites. Nature 310, 222–224.

Spencer, J.E., 1994. A numerical assessment of slab strength during high- and low-angle subduction and implications for Laramide orogenesis. J. Geophys. Res. 99, 9227–9236.

Stern, R.A., Hanson, G.N., 1991. Archean high-Mg granodiorite: a derivative of light rare earth element-enriched monzodiorite of mantle origin. J. Petrol. 32, 201–238.

Stern, R.A., Hanson, G.N., Shirey, S.B., 1989. Petrogenesis of mantle-derived, LILE-enriched Archean monzodiorite and trachyandesites (Sanukitoids) in Southwestern Superior Province. Can. J. Earth Sci. 26, 1688–1712.

Stevenson, R., Henry, P., Gariepy, C., 1999. Assimilation-fractional crystallization origin of Archean Sanukitoid suites: Western Superior Province, Canada. Precambrian Res. 96, 83–99.

Stolper, E., Newman, S., 1994. The role of water in the petrogenesis of Mariana trough magmas. Earth Planet. Sci. Lett. 121, 293–325.

Sun, S.S., McDonough, W.F., 1989. Chemical and isotopic systematics of oceanic basalts: implications for mantle composition and processes. In: Saunders, A.D., Norry, M.J. (Eds.), Magmatism in the Ocean Basins. Geol. Soc. London Spec. Publ., vol. 42, pp. 313–345.

Tatsumi, Y., Ishizaka, K., 1982. Origin of high-magnesian andesites in the Setouchi volcanic belt, southwest Japan: I. Petrographical and chemical characteristics. Earth Planet. Sci. Lett. 60, 293–304.

Tatsumi, Y., Kogiso, T., 1997. Trace element transport during dehydration processes in the subducted oceanic crust: 2. Origin of chemical and physical characteristics in arc magmatism. Earth Planet. Sci. Lett. 148, 207–221.

Tchameni, R., Mezger, K., Nsifa, N.E., Pouclet, A., 2000. Neoarchaean crustal evolution of the Congo craton: evidence from K-rich granitoids of the Ntem Complex, southern Cameroon. J. Afr. Earth Sci. 30, 133–147.

Tepper, J.H., Nelson, B.K., Bergantz, G.W., Irving, A.J., 1993. Petrology of the Chilliwack batholith, North Cascades, Washington: generation of calc-alkaline granitoids by melting of mafic lower crust with variable water fugacity. Contrib. Mineral. Petrol. 113, 333–351.

Thorpe, R.S. (Ed.), 1982. Andesites: Orogenic Andesites and Related Rocks. Wiley, Chichester. 724 pp.

Tombale, A.R., 1992. The geology, geochemistry and metallogeny of the Tati Greenstone belt, northeastern Botswana. PhD Thesis, Memorial University of Newfoundland, Canada. 383 pp.

Treloar, P.J., Blenkinsop, T.G., 1995. Archaean deformation patterns in Zimbabwe: true indicators of Tibetan-style crustal extrusion or not? Geol. Soc. London Spec. Publ. 95, 87–108.

Treloar, P.J., Coward, M.P., Harris, N.B.W., 1992. Himalayan–Tibetan analogies for the evolution of the Zimbabwe craton and Limpopo belt. Precambrian Res. 55, 571–587.

Vidal, P., Dupuy, C., Maury, R., Richard, M., 1989. Mantle metasomatism above subduction zones: trace element and radiogenic isotope in xenoliths from Bataan island (Philippines). Geology 17, 1115–1118.

Vlaar, N.J., 1983. Thermal anomalies and magmatism due to lithospheric doubling and shifting. Earth Planet. Sci. Lett. 65, 322–330.

Walker, J.A., Patino, L.C., Cameron, B.I., Carr, M.J., 2000. Petrogenetic insights provided by compositional transects across the Central American arc: southeastern Guatemala and Honduras. J. Geophys. Res. 105 (B8), 18949–18963.

Wickham, S.M., 1987. The generation and emplacement of granitic magmas. J. Geol. Soc. Lond. 144, 281–297.

Wilson, J.F., Nesbitt, R.W., Fanning, C.M., 1995. Zircon geochronology of Archaean felsic sequences in the Zimbabwe craton: a revision of greenstone stratigraphy and a model for crustal growth. In: Coward, M.P., Ries, A.C. (Eds.), Early Precambrian Processes. Geol. Soc. London Spec. Publ., vol. 95, pp. 109–126.

Wolf, M.B., Wyllie, P.J., 1994. Dehydration melting of solid amphibolite at 10 kbar: the effects of temperature and time. Contrib. Mineral. Petrol. 115, 369–383.

Wood, D.A., Joron, J.L., Treuil, M., 1979. A re-appraisal of the use of trace elements to classify and discriminate between magma series erupted in different tectonic settings. Earth Planet. Sci. Lett. 45, 326–336.

Wyllie, P.J., Wolf, M.B., van der Laan, S.R., 1997. Conditions for formation of tonalites and trondhjemites: magmatic sources and products. In: de Wit, M.J., Ashwal, L.D. (Eds.), Greenstone Belts. Clarendon Press, New York, pp. 256–266.

Yogodzinski, G.M., Kay, R.W., Volynets, O.N., Koloskov, A.V., Kay, S.M., 1995. Magnesian andesite in the western Aleutian Komandorsky region: implications for slab melting and processes in the mantle wedge. Geol. Soc. Am. Bull. 107, 505–519.

You, C.F., Castillo, P.R., Gieskes, J.M., Chan, L.H., Spivack, A.J., 1996. Trace element behavior in hydrothermal experiments: implications for fluid processes at shallow depths in subduction zones. Earth Planet. Sci. Lett. 140, 41–52.

Available online at www.sciencedirect.com

Lithos 71 (2003) 461–488

ELSEVIER

LITHOS

www.elsevier.com/locate/lithos

A Re–Os isotope and PGE study of kimberlite-derived peridotite xenoliths from Somerset Island and a comparison to the Slave and Kaapvaal cratons

Gordon J. Irvine[a], D. Graham Pearson[a],*, B.A. Kjarsgaard[b], R.W. Carlson[c],
M.G. Kopylova[d], G. Dreibus[e]

[a] *Department of Geological Sciences, Durham University, South Road, Durham DH1 3LE, UK*
[b] *Geological Survey of Canada, 601 Booth Street, Ottawa, Canada K1A 0E8*
[c] *Department of Terrestrial Magnetism, Carnegie Institution of Washington, 5241 Broad Branch Road N.W., Washington, DC 20015, USA*
[d] *Department of Earth and Ocean Sciences, University of British Columbia, 6339 Stores Road, Vancouver, BC, Canada V6T 1Z4*
[e] *Max-Planck-Institut für Chemie, Abt. Kosmochemie, Postfach 3060, 55020 Mainz, Germany*

Abstract

The concentrations of platinum-group elements (PGE; Os, Ir, Ru, Pd and Pt) and Re, and the Os isotopic compositions were determined for 33 lithospheric mantle peridotite xenoliths from the Somerset Island kimberlite field. The Os isotopic compositions are exclusively less radiogenic than estimates of bulk-earth ($^{187}Os/^{188}Os$ as low as 0.1084) and require a long-term evolution in a low Re–Os environment. Re depletion model ages (T_{RD}) indicate that the cratonic lithosphere of Somerset Island stabilised by at least 2.8 Ga, i.e. in the Neoarchean and survived into the Mesozoic to be sampled by Cretaceous kimberlite magmatism. An Archean origin also is supported by thermobarometry (Archean lithospheric keels are characterised by >150 km thick lithosphere), modal mineralogy and mineral chemistry observations. The oldest ages recorded in the lithospheric mantle beneath Somerset Island are younger than the Mesoarchean (>3 Ga) ages recorded in the Slave craton lithospheric mantle to the southwest [Irvine, G.J., et al., 1999. Age of the lithospheric mantle beneath and around the Slave craton: a Rhenium–Osmium isotopic study of peridotite xenoliths from the Jericho and Somerset Island kimberlites. Ninth Annual V.M. Goldschmidt Conf., LPI Cont., 971: 134–135; Irvine, G.J., et al., 2001. The age of two cratons: a PGE and Os-Isotopic study of peridotite xenoliths from the Jericho kimberlite (Slave craton) and the Somerset Island kimberlite field (Churchill Province). The Slave–Kaapvaal Workshop, Merrickville, Ontario, Canada]. Younger, Paleoproterozoic, T_{RD} model ages for Somerset Island samples are generally interpreted as the result of open system behaviour during metasomatic and/or magmatic processes, with possibly the addition of new lithospheric material during tectono-thermal events related to the Taltson–Thelon orogen. PGE patterns highly depleted in Pt and Pd generally correspond to older Archean T_{RD} model ages indicating closed system behaviour since the time of initial melt extraction. Younger Proterozoic T_{RD} model ages generally correspond to more complex PGE patterns, indicating open system behaviour with possible sulfide or melt addition. There is no correlation between the age of the lithosphere and depth, at Somerset Island.
© 2003 Elsevier B.V. All rights reserved.

Keywords: Canadian craton; Mantle xenolith; Peridotite; Somerset Island; Re–Os; Platinum-group element (PGE)

---

* Corresponding author.
E-mail address: d.g.pearson@durham.ac.uk (D.G. Pearson).

0024-4937/$ - see front matter © 2003 Elsevier B.V. All rights reserved.
doi:10.1016/S0024-4937(03)00126-9

1. Introduction

Until recently, samples of sub-cratonic mantle, erupted by kimberlitic magmatism, were dominated by xenoliths from the Kaapvaal and Siberian cratons (Boyd and Mertzman, 1987; Boyd, 1989; Griffin et al., 1999). Studies of xenoliths from the cratonic lithospheric mantle highlight the refractory nature of mantle roots beneath Archean cratons, i.e. residues depleted in incompatible major elements from melt extraction (Nixon, 1987; Herzberg, 1993; Boyd et al., 1997).

The presence of kimberlites on Somerset Island (Mitchell and Fritz, 1973; Pell, 1993) has provided an opportunity to assess the areal and depth extent of the Archean Canadian Shield via entrained mantle xenoliths. To understand the mechanism and rate of formation of cratonic lithosphere, it is essential to understand the processes involved in the formation and evolution of both crustal and mantle components, and their interrelationship. The evolution of continental crust is, in general, well constrained for most Archean cratons. However, the timing and processes involved in the stabilisation of the underlying lithospheric mantle remains poorly understood.

In this study, we have applied the Re–Os isotopic system to a suite of previously well-characterised peridotite xenoliths from the Batty Bay and Nord kimberlites (Kjarsgaard and Peterson, 1992), and xenoliths from the JP kimberlite (collected by G.J. Irvine, D.G. Pearson and B.A. Kjarsgaard in 1999), in order to constrain the age and evolution of the lithosphere underlying the northern Churchill Province. We use full platinum-group element (PGE) patterns to systematically evaluate the significance of the Os isotope model ages in whole-rock peridotite analyses, and the open-closed nature of the Os isotopic system. Similarities have been noted between events recorded in the lithospheric mantle section and those recorded in the overlying crust. This study has also enabled us to compare age systematics with the extensive data set available for the Kaapvaal craton, and with data we have previously produced for Slave craton lithospheric mantle (Irvine et al., 1999, 2001a).

2. Geologic setting

Somerset Island is considered to be part of the Churchill structural province, on the northern margin of the Laurentian Shield. The Churchill Province differs from most Archean cratonic areas because of its variable and widespread Paleoproterozoic reactivation (Percival, 1996). The northern part of the Churchill Province consists of rare gneisses (3.3–2.6 Ga; Percival, 1996), ca. 2.74–2.70 Ga greenstone belts (e.g. Mary River, Prince Albert and Woodburn Groups dominated by komatiite and tholeittic rocks; Jenner et al., 2002), and voluminous granite magmatism at 2.6 Ga (Le Cheminant and Roddick, 1991). Proterozoic basic magmatism, interpreted to be related to large igneous province style magmatism, is manifested as MacQuoid (2.19 Ga; Tella, 2001) and Tulamela dykes, and mafic volcanic rocks of the Ketyet Group (>2.0 Ga, <2.5 Ga; Pehrsson et al., 2002). The western margin of the Churchill Province incorporates elements of a major Paleoproterozoic orogen (Fig. 1a), the Taltson–Thelon orogen, between 2.02 and 1.9 Ga (Hoffman, 1988). This orogen formed during the eastward subduction of the Slave craton beneath the Churchill Province, followed by continent–continent collision (van Breeman and Loveridge, 1987; Hoffman, 1988; Hanmer et al., 1992). UPb zircon ages from a granodiorite gneiss in Somerset Island have produced an upper intercept age of $2776 + 67/ - 56$ Ma with a lower intercept ages of 2200 ± 68 Ma. Two populations of zircon suggest that the gneiss represented a Proterozoic protolith with inherited late Archean zircons (Frisch and Hunt, 1993). A U–Pb zircon age of 1.93 Ga was determined for the Cape Bird syenite, which was interpreted to be the age of granulite facies metamorphism (Frisch and Sandeman, 1991). Kimberlite derived xenocrystic zircons (assumed to be derived from the lower crust) have U–Pb ages of 1976 ± 5 Ma and 2500 Ma (Frisch and Sandeman, 1991).

Somerset Island kimberlites form a northeast–southwest trending belt (Fig. 1b) controlled by fracture sets in the country rocks (Mitchell, 1975). These fractures can be seen as lineaments in the Precambrian Boothia terrane exposed on the west of Somerset Island and are continuous, extending from the basement into the Paleozoic cover (Blackadar, 1967). U–Pb perovskite dating produced Cretaceous ages for the Ham kimberlite (88 Ma; Heaman, 1989) and the Georgia melnoite pipe (105 Ma). Additional perovskite dating has produced ages of 98–94 Ma for six other Somerset Island kimberlites (Kjarsgaard and Heaman, unpublished data).

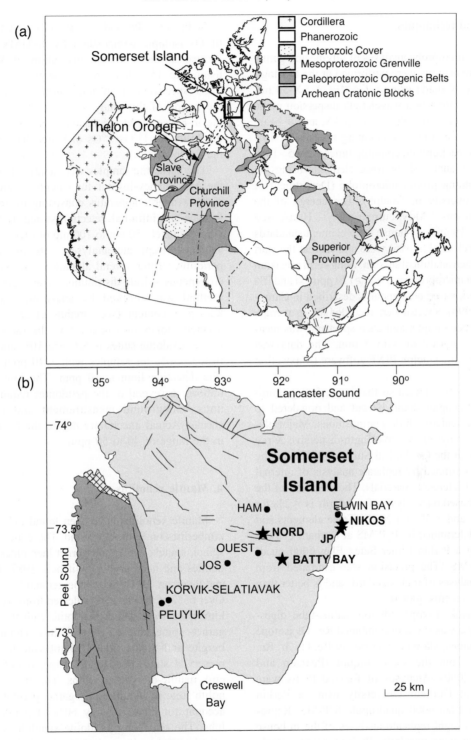

Fig. 1. (a) Map of Canada showing tectonic elements (constructed from Wheeler et al., 1996). (b) Geological map of Somerset Island (after Stewart, 1987; Kjarsgaard, 1996) showing location of kimberlites. Peridotite xenoliths used in this study are from the kimberlites highlighted with stars and bold. Legend: light grey, Paleozoic cover; striped, late Proterozoic cover; dark grey, Precambrian basement; lines, normal faults.

3. Analytical techniques

Electron-microprobe analyses of olivine and garnet mineral compositions were performed at the Geological Survey of Canada using a CAMECA SX-50 microprobe, operating in a wavelength dispersion mode with an acceleration voltage of 15 kV and a beam current of 25 nA. On-peak counting times were 10–20 s, and the background counting times were 5–10 s. Garnets were analysed for trace elements using the Guelph scanning proton microprobe (Department of Physics, University of Guelph). The energy of the proton beam was 3 MeV, current was ~ 10 nA, and spot size ~ 15 μm^2. In-house trace element standards (e.g. Campbell et al., 1996) were analysed at the start and end of each analytical session. Long-term precision of the proton microprobe is 3% at >120 ppm Ni and 7% at 40 ppm Ni, based on standards described in Campbell et al. (1996). Spectra were recorded for an average of 10 spots per sample, and each spectrum was measured over a period of 3 to 5 min. The data was processed via the Guelph PIXE software (Maxwell et al., 1995).

Samples were selected so that large, 500 g (minimum) bulk samples could be cut and powdered to provide representative bulk compositions. Major elements were analysed by wavelength-dispersive X-ray fluorescence at the Geological Survey of Canada, the precision determined by replicate analysis of internal standard and reference materials. The precision of the GSC XRF, based on analysis of standards is <2% for all elements and <3% for Na_2O. Trace elements and REE were determined by ICP-MS at Durham University utilising a Perkin Elmer Sciex Elan 6000 quadrupole ICP-MS. The precision was estimated from duplicate analyses of rock standards and is better than 3% for all elements quoted.

A low-blank, isotope dilution, carius-tube digestion technique was used for combined Re–Os isotope and PGE studies, allowing analysis of Re, Os, Ir, Ru, Pt and Pd from the same aliquot (Pearson and Woodland, 2000). Analyses of Re and PGEs were performed at Durham University using a Perkin Elmer Sciex Elan 6000 quadrupole ICP-MS. Reproducibility for eight replicate analyses of the in-house Durham University peridotite PGE standard, sample GP13, a spinel lherzolite from the Beni Bousera peridotite massif were 9% Os, 18% Ir, 7% Ru,

15% Pt, 13% Pd and 3% Re (2σ RSD). Analyses of Os isotopic composition by N-TIMS were performed at the Carnegie Institution of Washington using the 15" magnetic sector thermal ionisation mass spectrometer in ion-counting mode. During the period of this study, replicate determinations of the in-house DTM Os standard gave external reproducibility for $^{187}Os/^{188}Os$ of better than ± 2.4‰ at the 2σ level.

The peridotite samples were analysed for sulfur at the Max-Planck-Institut für Chemie using commercially available apparatus consisting of an induction furnace and infra-red detection system (CSA, 2002). Approximately 50 mg to 100 mg of rock sample were analysed. Replicate analyses were made for each peridotite, using several aliquots to account for heterogeneous sulfide distribution. The accuracy of this method was checked by analysing standards of known S content (see Dreibus et al., 1995). The accuracy for S measurement in the range observed in the peridotite suites is between 10% and 15%, but rises to 20% for samples with <40 ppm S content. The detection limit is 10 ppm. The low S concentrations in several of the peridotites required continuous background measurement and calibration checks. Actual uncertainty due to the blank value is in the range of 40 to 50 ppm.

4. Mantle xenoliths

Mantle xenoliths have been found in a number of kimberlites on Somerset Island. These are dominantly spinel, spinel–garnet and garnet lherzolites, with rare dunites and harzburgites (Mitchell, 1987; Kjarsgaard and Peterson, 1992; Schmidberger and Francis, 1999). Over 65% of the samples collected from the Batty Bay kimberlite were garnet peridotite with the remainder garnet–spinel and spinel peridotites, with minor harzburgite and dunite. The JP kimberlite hosts a well-preserved suite of generally large, ovoid (15 to 40 cm), mantle-derived xenoliths. The JP peridotite xenolith suite is dominated by garnet peridotites, which account for approximately 60% of peridotite xenoliths. The remaining peridotite xenoliths are garnet–spinel (30%) and spinel (10%) peridotites. The xenoliths are mainly lherzolites and harzburgites, with minor dunite.

A representative selection of 33 peridotite xenoliths were chosen for this study from the various kimberlite bodies on Somerset Island. Using the textural nomenclature of Harte (1977), most xenoliths are classified as coarse equant to coarse tabular. Large subhedral olivine (up to 12 mm) and orthopyroxene (2–7 mm) crystals dominate the mineral assemblage. Clinopyroxene occurs as emerald-green, anhedral equant crystals (up to 5 mm), with smooth boundaries. Lilac coloured garnets (1–3 mm) are subhedral to anhedral, and variably surrounded by kelyphitic rims containing secondary phlogopite and spinel. A subordinate number of samples exhibit porphyroclastic, mosaic porphyroclastic and disrupted porphyroclastic textures.

5. Analytical data

5.1. Mineral chemistry

5.1.1. Olivine

Olivine in the JP peridotites is magnesium rich and ranges in magnesium number ($Mg^\# = Mg/\{Mg + Fe\}$; Table 1; Fig. 2) from 0.917 to 0.926 (average 0.921). This is comparable to the reported range of 0.913 to 0.927, for the Batty Bay peridotite xenolith suite (Kjarsgaard and Peterson, 1992). Mineral data for Batty Bay peridotite xenoliths are available from the authors on request. The most magnesian olivine (Fo92–93) occurs in harzburgites and ranges in $Mg^\#$ from 0.922 to 0.926 (average 0.924), with the $Mg^\#$ for lherzolites ranging from 0.917 to 0.923 (average 0.920). The average Somerset Island olivine $Mg^\#$ for this study at 0.921 ± 0.003 (1σ) is within error of that reported by Schmidberger and Francis (1999) at 0.923. The average Somerset Island olivine $Mg^\#$ in this study is higher than that reported for peridotite xenoliths from the Jericho kimberlite (average 0.913, Kopylova et al., 1999), and more comparable with peridotite xenoliths from the Torrie kimberlite (average 0.920, MacKenzie and Canil, 1999), both in the Slave craton. These magnesium numbers are generally lower than the reported range of 0.92 to 0.93 (average 0.926) of Kaapvaal peridotite xenoliths (Boyd, 1989).

Concentrations of NiO in olivine for the JP and Nikos xenoliths range from 0.33 to 0.43 wt.% (average 0.38 wt.% NiO), comparable to previous studies from Somerset Island kimberlite-derived xenoliths

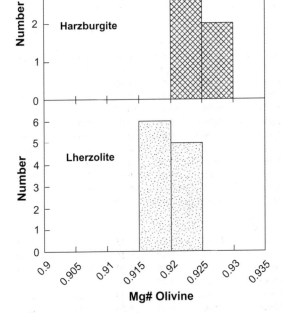

Fig. 2. Histogram of olivine $Mg^\#$ $\{Mg/(Mg + Fe)\}$ for Somerset Island peridotite xenoliths.

(Batty Bay, Kjarsgaard and Peterson, 1992; and Nikos/JP, Schmidberger and Francis, 1999). This is comparable to those reported for the Slave craton (Jericho average 0.39 wt.%, Kopylova et al., 1999; and Torrie average 0.36 wt.%, MacKenzie and Canil, 1999).

5.1.2. Garnet

The garnet from Somerset Island peridotites is a chromium pyrope with high MgO, low CaO and $Mg^\#$ ranging from 0.81 to 0.86 (Table 1). The garnets are homogeneous and exhibit no chemical zoning on a thin-section scale, and very little variation is observed between garnet from peridotites of different facies or textures. The Cr_2O_3 content within the garnets ranges from 2.5 to 7.7 wt.%. On average, garnet from harzburgites is more Cr_2O_3 rich (6.6 ± 0.8 wt.%) than garnet from lherzolites (4.7 ± 1.6 wt.%). The garnets can generally be classified as group 9 (G9) garnets in the statistical classification scheme of Dawson and

Table 1
Representative olivine and garnet analyses for Somerset Island peridotites

Sample	XO4	XO5	XO6	JP1-X2	JP3-X	JPS-1	JPS-6A	JPS-6B	JPN-3A	JPN-9	JPN-11
Type	lherz	lherz	lherz	lherz	harz	lherz	lherz	lherz	harz	lherz	lherz
Facies	Gnt	Gnt	Gnt–Sp	Gnt–Sp	Gnt	Gnt	Gnt	Gnt	Gnt	Gnt	Gnt
Kimberlite	Nikos	Nikos	Nikos	JP (North)	JP (South)	JP (South)	JP (South)	JP (South)	JP (North)	JP (North)	JP (North)
Olivine											
SiO_2	39.99	39.67	40.30	39.81	39.66	40.05	39.73	39.36	39.96	39.83	39.66
TiO_2	0.00	0.02	0.01	0.00	0.04	0.03	0.04	0.02	0.01	0.03	0.04
Al_2O_3	0.03	0.02	0.00	0.00	0.04	0.03	0.06	0.01	0.05	0.03	0.01
Cr_2O_3	0.00	0.04	0.02	0.00	0.08	0.02	0.08	0.00	0.04	0.10	0.10
FeO	8.09	7.85	8.00	7.69	7.56	7.45	8.13	7.78	7.54	7.64	8.31
MnO	0.11	0.11	0.11	0.07	0.10	0.09	0.12	0.11	0.13	0.15	0.11
MgO	50.94	50.68	50.37	50.75	50.41	50.83	50.55	50.25	50.90	50.64	50.24
CaO	0.04	0.04	0.02	0.01	0.04	0.00	0.05	0.04	0.00	0.06	0.03
Na_2O	0.00	0.00	0.00	0.00	0.00	0.00	0.00	0.00	0.00	0.00	0.00
K_2O	0.02	0.02	0.01	0.00	0.01	0.00	0.01	0.02	0.01	0.01	0.02
NiO	0.37	0.36	0.35	0.35	0.37	0.42	0.41	0.39	0.35	0.40	0.41
V_2O_3	0.00	0.01	0.00	0.00	0.00	0.00	0.01	0.01	0.06	0.02	0.00
Total	99.60	98.80	99.20	98.69	98.30	98.93	99.19	98.01	99.05	98.92	98.93
$Mg^{\#}$	0.918	0.920	0.918	0.922	0.922	0.924	0.917	0.920	0.923	0.922	0.923
Garnet											
SiO_2	41.09	40.70	40.45	40.52	40.32	40.54	40.80	41.00	40.87	37.00	39.99
TiO_2	0.18	0.40	0.19	0.09	0.18	0.10	0.23	0.30	0.26	0.12	0.29
Al_2O_3	21.59	20.06	21.03	20.52	18.26	18.31	20.68	20.86	19.33	20.86	17.69
Cr_2O_3	2.49	4.29	3.49	4.13	7.21	6.97	3.53	3.68	5.39	4.53	7.69
FeO	7.21	6.31	8.33	7.23	5.99	6.40	6.73	6.92	6.57	7.97	6.62
MnO	0.40	0.30	0.53	0.41	0.30	0.32	0.33	0.36	0.30	0.38	0.31
MgO	21.51	21.46	19.96	20.94	20.65	20.84	21.53	21.66	21.31	23.80	19.84
CaO	4.45	5.14	5.10	5.02	6.18	5.78	4.57	4.72	5.32	3.63	6.29
Na_2O	0.00	0.01	0.00	0.00	0.00	0.00	0.02	0.01	0.00	0.00	0.00
K_2O	0.01	0.02	0.01	0.00	0.00	0.03	0.02	0.03	0.00	0.04	0.02
NiO	0.02	0.00	0.02	0.00	0.05	0.02	0.04	0.06	0.02	0.00	0.03
V_2O_3	0.03	0.10	0.03	0.05	0.09	0.08	0.01	0.04	0.02	0.04	0.07
Total	99.00	98.80	99.15	98.91	99.23	99.39	98.48	99.65	99.39	98.36	98.83
Trace elements (ppm)											
Zn	11.8	7.6	8.1	7.3	12.2	11.6	13.7	14.2	10.6	10.6	13
Ga	10.3	5.6	6.3	9.2	8	0	10.2	8.8	2.5	3.9	1.2
Y	19.4	19.3	9.1	16.5	9.1	5.7	19.5	13.3	12.9	14.6	21.6
Zr	20.7	50.6	7.5	15.1	28.4	128.2	60.3	59.2	34.3	53	101
Ni	32.5	71.4	22.4	24.2	68.4	69.4	53	80.8	74.8	23.9	74.9
Ni temperatures											
T (°C)	951	1226	865	884	1200	1152	1066	1242	1262	850	1190
P (GPa)	3.36	5.1	2.94	3.06	4.89	4.57	4.03	5.3	5.8	2.88	4.56

Trace element composition of peridotitic garnet (ppm) and estimated Ni-in-garnet temperatures. Pressures estimated from the steady state geotherm of Somerset Island lithospheric mantle (44 mW/m^2, Kjarsgaard and Peterson, 1992) Gnt: Garnet, Sp: Spinel, lherz: lherzolite and harz: harzburgite.

Stephens (1975), in agreement with previous studies of Somerset Island peridotite xenoliths (Mitchell, 1978, 1987; Jago and Mitchell, 1987; Kjarsgaard and Peterson, 1992; Schmidberger and Francis, 1999).

Trace elements were measured in garnet and these contain 22–81 ppm Ni, 0–10.3 ppm Ga, 6–22 ppm Y, and 8–128 ppm Zr (Table 1). More than half of the Somerset Island peridotitic garnets have Y concentrations exceeding 10 ppm, and Zr exceeding 30 ppm. This is similar to levels reported for peridotite xenoliths from the Jericho kimberlite (Kopylova et al., 1999).

5.2. Temperature and pressure

For the majority of the Batty Bay and Nord samples, P and T was determined using conventional (i.e. Brey et al., 1990) thermobarometric methods. For the remaining samples (Nikos, JP, Batty Bay), temperature of equilibration was calculated using the partitioning of Ni between chrome-pyrope garnet and olivine (Griffin et al., 1989; Kjarsgaard, 1992; Ryan et al., 1996). This was done so that a stratigraphy could be constructed for the lithospheric mantle beneath Somerset Island based on Re–Os model age determinations.

Temperatures were calculated for the Somerset Island xenoliths using measured Ni concentrations in olivine (Table 1), which were commonly higher than the fixed value of 2900 ppm used by Ryan et al.

(1996). Temperatures were calculated using the following parameters and equation:

$$T = \left(\frac{1000}{-0.428 \times \log_{10}(100 \times K_d^{gt/ol}) + 0.84} \right) - 273.3$$

Previous studies of Somerset Island peridotites (Kjarsgaard and Peterson, 1992; Schmidberger and Francis, 1999), using standard three- and four-phase geothermobarometers, defined a "steady-state" geotherm for the lithospheric mantle beneath Somerset Island of 44 mW/m² (Fig. 3). The pressure of formation for an individual xenolith was determined by the intersection of calculated Ni temperature with the known geotherm. Equilibration temperatures within the Somerset Island peridotite suite range from 850 to 1260 °C, with a corresponding pressure range from 2.9 to 5.8 GPa. These pressures indicate lithospheric depths that clearly extend into the diamond stability

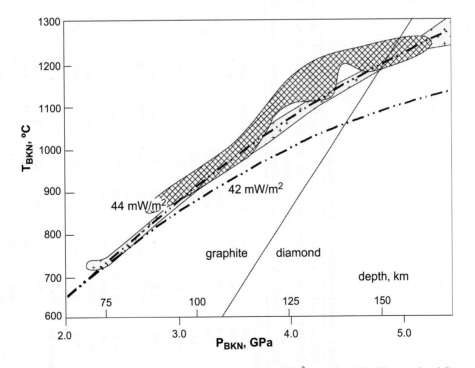

Fig. 3. Steady state geotherm for Somerset Island peridotite xenoliths (44 mW/m²), calculated by Kjarsgaard and Peterson (1992) and Schmidberger and Francis (1999) using the experimentally calibrated thermobarometers of Brey et al. (1990). Data from Kjarsgaard and Peterson (1992) represented by the hatched field (available on request from authors), Schmidberger and Francis (1999) by the stippled field.

Table 2
Major element compositions for Somerset Island peridotites

Sample	K11A14	K11A15	K11A16	K11A17	K11A18	K12A1	K13A1	K13A3	K13A4	K13A5	K13B4	K15A4	N1C	N2B	XO4
Type	Gnt lherz	Gnt-Sp lherz	Gnt lherz	Gnt-Sp lherz	Gnt-Sp lherz	Sp lherz	Sp lherz	Sp lherz	Sp lherz	Sp lherz	Gnt period	Gnt-Sp lherz	Gnt-Sp lherz	Gnt-Sp lherz	Gnt lherz
Texture	Coarse	Coarse	Coarse	Coarse	Coarse	Porphyr	Coarse	Coarse	Coarse	Coarse	Coarse	Coarse	Coarse	Coarse	Coarse
Major elements in wt.%															
SiO_2	42.40	41.30	41.60	41.20	41.10	42.00	38.60	43.50	38.80	41.40	44.10	41.70	41.50	42.40	44.30
TiO_2	0.04	0.08	0.03	0.06	0.04	0.01	0.02	0.04	0.04	0.23	0.08	0.02	0.11	0.09	0.09
Al_2O_3	1.10	1.00	1.30	0.90	0.80	0.90	0.90	0.40	1.40	1.10	0.80	0.90	0.80	0.70	2.30
FeO	7.20	7.02	6.57	6.93	6.93	7.02	7.02	6.75	7.02	6.93	6.93	6.84	7.56	7.29	7.02
MnO	0.12	0.11	0.10	0.11	0.11	0.11	0.10	0.11	0.09	0.19	0.10	0.09	0.26	0.11	0.12
MgO	44.72	43.47	43.27	43.27	45.44	44.85	38.84	44.96	34.41	39.50	43.96	43.11	44.32	45.93	39.90
CaO	0.73	1.10	1.22	0.93	0.47	0.60	1.37	0.50	2.90	0.82	0.63	0.30	0.69	0.72	2.28
Na_2O	0.00	0.00	0.10	0.00	0.00	0.00	0.00	0.00	0.00	0.00	0.00	0.00	0.00	0.00	0.10
K_2O	0.22	0.22	0.26	0.21	0.10	0.06	0.06	0.05	0.03	0.08	0.09	0.05	0.14	0.24	0.10
P_2O_5	0.03	0.03	0.04	0.04	0.02	0.01	0.03	0.02	0.03	0.04	0.02	0.04	0.03	0.02	0.02
Cr_2O_3	0.37	0.41	0.55	0.46	0.68	0.46	0.54	0.22	0.42	0.38	0.37	0.39	0.46	0.27	0.41
LOI	3.80	5.20	5.30	5.80	4.90	4.20	12.30	3.60	14.60	9.70	3.20	6.60	4.20	2.60	3.00
Total	100.73	99.94	100.34	99.91	100.59	100.22	99.78	100.15	99.74	100.37	100.28	100.04	100.07	100.37	99.64
Mg#	0.92	0.92	0.92	0.92	0.92	0.92	0.91	0.92	0.90	0.91	0.92	0.92	0.91	0.92	0.91
Calculated modes in wt.%															
Ol	75.24	77.43	74.43	76.73	83.18	76.18	67.88	73.07	52.6	59.19	67.25	73.13	77.53	80.83	56.49
OPX	17.14	14.32	15.06	16.19	12.89	19.01	21.86	24.15	27.81	33.75	27.53	23	15.67	14.04	23.91
CPX	1.28	4.01	3.89	3.37	1.47	2.4	6.28	1.9	13.9	3.36	1.09	0.27	2.1	2.5	7.38
Sp		0.06		0.11	0.01	2.4	3.98	0.88	5.63	3.7		0.01	1.05	0.01	
Gnt	6.34	4.19	6.63	3.6	2.45						4.14	3.6	3.65	2.62	12.22
Trace elements in ppm															
Ni	2380	2441	2435	2188	2408	2350	2407	2468	2128	2170	2519	2441	2495	2550	2034

Sample	XO5	XO6	XO7	JP1-X2	JP2-X2	JP3-X1	JP3-X	JPS-1	JPS-6A	JPS-6B	JPN-2	JPN-3A	JPN-3B	JPN-4	JPN-9	JPN-11
Type	Gnt lherz	Gnt–Sp lherz	Gnt–Sp harz	Gnt–Sp lherz	Sp lherz	Sp harz	Gnt harz	Gnt–Sp harz	Gnt lherz	Gnt lherz	Gnt–Sp lherz	Gnt harz	Gnt perid	Gnt lherz	Gnt lherz	Gnt lherz
Texture	Coarse	Coarse	Coarse	Coarse	Coarse	Coarse	Coarse	Coarse	Coarse	Coarse	Coarse	Coarse	Coarse	Coarse	Coarse	Coarse
Major Elements in wt.%																
SiO_2	42.70	41.10	44.40	43.00	42.20	41.50	41.50	42.30	41.50	41.30	41.40	44.00	43.70	43.40	42.30	44.60
TiO_2	0.05	0.07	0.03	0.04	0.08	0.04	0.06	0.05	0.06	0.05	0.13	0.08	0.07	0.10	0.02	0.03
Al_2O_3	1.60	3.10	0.20	1.30	0.60	0.10	0.80	0.60	1.40	1.60	0.70	1.10	1.10	1.70	0.70	0.70
FeO	7.47	7.20	6.84	7.29	6.75	7.29	7.02	6.75	7.29	7.29	8.01	6.84	6.75	7.02	7.38	7.11
MnO	0.11	0.13	0.10	0.21	0.10	0.10	0.08	0.09	0.10	0.10	0.10	0.15	0.10	0.11	0.10	0.11
MgO	42.46	39.16	44.80	43.58	43.22	45.72	44.20	43.20	43.17	42.68	44.59	42.74	42.13	42.84	44.44	43.05
CaO	1.23	1.68	0.24	1.32	0.95	0.40	0.47	0.98	1.28	1.65	0.52	0.76	1.18	0.98	0.74	0.85
Na_2O	0.00	0.00	0.00	0.00	0.00	0.00	0.00	0.00	0.00	0.00	0.00	0.00	0.00	0.00	0.00	0.00
K_2O	0.07	0.20	0.04	0.04	0.37	0.03	0.04	0.10	0.06	0.06	0.35	0.14	0.13	0.22	0.04	0.07
P_2O_5	0.01	0.03	0.01	0.03	0.02	0.01	0.01	0.02	0.02	0.02	0.02	0.02	0.02	0.02	0.02	0.02
Cr_2O_3	0.43	0.56	0.25	0.40	0.23	0.28	0.37	0.38	0.34	0.40	0.32	0.36	0.38	0.42	0.26	0.38
LOI	3.80	7.00	3.40	3.40	5.20	4.60	5.40	5.20	4.90	5.20	3.90	4.70	4.80	4.00	4.50	3.50
Total	99.93	100.23	100.31	100.61	99.72	100.07	99.95	99.67	100.12	100.35	100.04	100.89	100.36	100.81	100.50	100.42
$Mg^{\#}$	0.91	0.91	0.92	0.91	0.92	0.92	0.92	0.92	0.91	0.91	0.91	0.92	0.92	0.92	0.91	0.92
Calculated modes in wt.%																
Ol	67.8	64.2	69.75	72.06	71.93	81.84	76.25	73.11	74.45	74.32	78.1	63.67	63.78	66.02	75.1	63.33
OPX	19.65	16.39	29.38	17.4	22.49	15.37	18.56	20.97	13.46	10.79	15.73	29.26	27.01	23.23	18.47	30.49
CPX	2.9	0.1	0.85	4.41	3.99	1.55	0.4	4	3.76	5.46	1.29	1.26	3.6	1.57	1.84	2.21
Sp		trace	0.01	0.37	1.6	1.24		0.01			1.31					
Gnt	9.65	15.43	0.01	5.75			4.78	1.91	8.33	9.42	3.58	5.81	5.62	9.17	4.58	3.98
Trace elements in ppm																
Ni	2303	1793	2722	2351	2367	2566	2439	2648	2498	2516	2455	2267	2309	2107	2619	2529

LOI: loss on ignition. Total Fe given as FeO. Lherz: lherzolite, Harz: harzburgite, Gnt: garnet, Sp: spinel, Ol: olivine, PX: orthopyroxene, CPX: clinopyroxene.

field. It is important to note that equilibration pressures calculated for circumcratonic xenolith suites from around the Kaapvaal craton do not extend into the diamond stability field (Finnerty and Boyd, 1987). The pressure extent observed for the Somerset samples thus is more consistent with that observed for other Archean cratons rather than the shallower maximum depths generally found in xenoliths from post-Archean terranes globally.

5.3. Whole-rock chemistry

Whole-rock analyses for Somerset Island peridotites (Table 2) indicate that they are strongly depleted in incompatible elements such as Ti, Ca, Al and Na compared to estimates of fertile mantle composition (McDonough, 1990), and generally depleted in Fe, in keeping with the magnesian nature of their olivines. Good negative correlations exist between Ca and Al, and Mg (Fig. 4), and there is a positive correlation that exists between Ni and Mg for the Somerset Island peridotite xenoliths, as previously noted by Schmidberger and Francis (1999). MgO contents of the peridotite xenoliths range from 40.4 to 47.9 wt.%. Whole-rock $Mg^{#}$ for lherzolites (average 0.914) and harzburgites (average 0.918) indicate a shift towards higher values in the more refractory peridotites (Table 2). The average whole-rock $Mg^{#}$ for the Somerset Island peridotite suite in this study is 0.915. This is lower than the average olivine $Mg^{#}$ (0.921) for the peridotite suite and suggests whole-rock enrichment in Fe, probably by a melt phase, lowering bulk $Mg^{#}$.

5.4. Mineral modes

The modal mineral proportions of Somerset Island xenoliths (Table 2) were calculated using whole-rock compositions and converting to an equivalent modal mineralogy using mass balance relationships related to the observed mineralogy in Somerset Island peridotites (after the method described by Kopylova and Russell, 2000).

Somerset Island peridotites from this study contain higher modal abundances of olivine (average 71%) than peridotites from Kaapvaal (61%, Boyd, 1989), and are similar to spinel peridotites from the Jericho kimberlite (72%, Kopylova and Russell, 2000). The

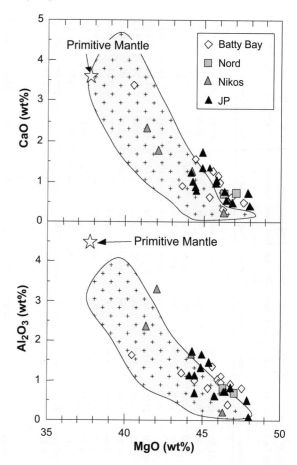

Fig. 4. Variation of major (CaO and Al_2O_3 wt.%) elements versus MgO (wt.%) for Somerset Island peridotite xenoliths. Primitive mantle composition after McDonough (1990). Stippled field represents Somerset Island peridotite data from Schmidberger and Francis (1999).

high modal olivine content is indicative of the highly refractory nature of the Somerset Island lithospheric mantle and suggests these are residues from high degrees of partial melting. The chemical/modal mineralogy characteristics of the Somerset Island peridotites are similar to compositions reported for low-temperature peridotites from the Archean Kaapvaal and Siberian cratons but clearly distinct from oceanic domains (Fig. 5). The average modal orthopyroxene content of Somerset Island peridotites is 20% (range of 11% to 34%). This is similar to that reported for peridotite xenoliths from kimberlites in the Siberian craton (Boyd et al., 1997) and the Jericho kimberlite in the Slave craton (Kopylova and Russell, 2000).

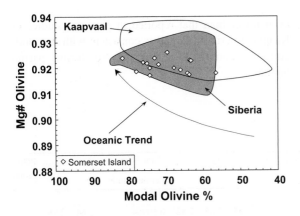

Fig. 5. Olivine Mg$^{\#}$ vs. modal olivine content (%) for Somerset Island peridotite xenoliths. Fields for Low-T peridotite xenoliths from the Kaapvaal (white) and Siberian cratons (light grey) after Boyd (1989) and Boyd et al. (1997). Trend defined by oceanic peridotites after Boyd (1989).

The Kaapvaal craton in contrast has an average orthopyroxene mode of 31% (Boyd, 1989). The average modal clinopyroxene content of Somerset Island peridotites (3%) is slightly higher than that observed in Kaapvaal cratonic peridotites (1.8%, Boyd, 1989), consistent with the slightly less refractory character of the Somerset Island peridotites, or possibly later cpx re-enrichment (Simon et al., 2003).

5.5. Re–Os isotope and PGE chemistry

5.5.1. Re variation

Re contents in the Somerset Island peridotite xenoliths (Table 3) show a considerable scatter (0.012 to 2.58 ppb), extending over a larger range than that reported from other studies of Archean (Walker et al., 1989; Carlson and Irving, 1994; Pearson et al., 1995a,b; Carlson et al., 1999; Chesley et al., 1999) and post-Archean (Meisel et al., 1996; Handler et al., 1997; Peslier et al., 2000) cratonic peridotite xenoliths.

The mean Re concentration for all Somerset Island peridotite xenoliths in this study is 0.343 ppb (S.D. = 0.637) with a median of 0.066 ppb ($n = 33$), compared to an estimated 0.26 ppb for fertile mantle (Morgan, 1986). This illustrates that many peridotite xenoliths from Somerset Island are, in general, highly depleted in Re relative to fertile mantle compositions but also that a few samples are anomalously enriched

in Re. If we consider only those samples with Re abundances less than fertile mantle, the mean Re content is 0.072 ppb (S.D. = 0.063) with a median of 0.047 ppb ($n = 26$). This is comparable to the mean Re concentration of 0.084 ppb (S.D. = 0.078) and median of 0.072 ppb for Kaapvaal peridotites (Carlson et al., 1999).

5.5.2. Os variation

Measured Os concentrations for the Somerset Island peridotites (Table 3) show considerable scatter (0.55 to 6.7 ppb) and span a similar range to that reported for Kaapvaal and Siberian peridotites (Walker et al., 1989; Pearson et al., 1995a,b; Carlson et al., 1999). On average, the Os content of Somerset Island peridotites is slightly higher (mean = 3.54 ppb, median = 3.41 ppb) than estimates for fertile mantle (3.4 ppb, Morgan, 1986), as would be expected for compatible element behaviour of Os in melt residues. Os concentrations reported for Kaapvaal peridotite xenoliths (mean = 4.61 ppb, median = 4.05 ppb, Carlson et al., 1999) and Alpine peridotites (mean = 3.74 ppb, median = 3.84 ppb, Reisberg and Lorand, 1995) are also higher than fertile mantle. The Os content of Somerset Island xenolith peridotites, although spanning a similar range to both these groups, tends to be lower on average and tends more towards that of fertile mantle. In contrast, rift-related spinel peridotite xenoliths (McBride et al., 1996; Meisel et al., 1996; Handler et al., 1997; Peslier et al., 2000), and arc-related spinel peridotite xenoliths (Brandon et al., 1996) have Os contents lower than fertile mantle estimates.

5.5.3. Re–Os isotopic variation

The range in Re and Os concentrations of the Somerset Island peridotite xenolith suite (Table 3) results in a large range in $^{187}Re/^{188}Os$ ratios, from 0.018 to 6.23 (subchondritic to suprachondritic), with a mean of 0.548 (S.D. = 1.22) and a median of 0.104. This is higher than $^{187}Re/^{188}Os$ ratios reported for cratonic peridotites from the Kaapvaal craton (mean $^{187}Re/^{188}Os = 0.109$, Carlson et al., 1999) and Alpine peridotites (mean $^{187}Re/^{188}Os = 0.207$, Reisberg and Lorand, 1995), but similar to that seen in spinel peridotite xenoliths from the Northern Canadian Cordillera (mean $^{187}Re/^{188}Os = 0.537$, Peslier et al., 2000). Of the 33 Somerset Island samples analysed

(Table 3), seven have ^{187}Re/^{188}Os ratios greater than fertile mantle. If we exclude these seven samples, the average ^{187}Re/^{188}Os ratio is reduced considerably to 0.072 (S.D. = 0.063) with a median of 0.047. This is even lower than that reported in previous studies of Kaapvaal peridotites, and is consistent with the moderately incompatible nature of Re during partial melting.

Table 3
Re–Os Isotopic data for Somerset Island peridotites

Samples	Re (ppb)	Os (ppb)	^{187}Re/^{188}Os	^{187}Os/^{188}Os (m)	±	^{187}Os /^{188}Os (i) Chondrite	γOs	T_{RD}	T_{MA}	^{187}Os/^{188}Os (i) Fertile mantle	T_{RD}	T_{MA}	PGE group
Batty Bay													
K11A6	0.023	4.57	0.0245	0.10968	8	0.10964	−13.65	2.65	2.82	0.10964	2.64	2.79	
K11A14	0.026	3.61	0.0351	0.11291	9	0.11285	−11.11	2.18	2.38	0.11285	2.20	2.39	C
K11A15	0.019	1.74	0.0519	0.11431	14	0.11422	−10.04	1.98	2.26	0.11422	2.01	2.28	C
K11A16	0.048	4.57	0.0503	0.11227	10	0.11218	−11.64	2.28	2.59	0.11218	2.29	2.58	A
K11A17	0.019	2.38	0.0375	0.11157	16	0.11151	−12.17	2.38	2.61	0.11151	2.38	2.60	C
K11A18	0.012	0.55	0.1096	0.11493	15	0.11475	−9.62	1.91	2.58	0.11475	1.94	2.57	C
K12A1	0.052	3.42	0.0732	0.11460	15	0.11448	−9.84	1.95	2.36	0.11448	1.98	2.36	C
K13A1	0.129	4.32	0.1440	0.11113	12	0.11089	−12.66	2.47	3.78	0.11089	2.47	3.65	B
K13A3	0.025	3.80	0.0314	0.11100	13	0.11095	−12.61	2.46	2.66	0.11095	2.46	2.64	A
K13A4	0.034	1.89	0.0875	0.11955	18	0.11941	−5.95	1.22	1.53	0.11941	1.30	1.61	C
K13A5	0.197	3.15	0.3004	0.11279	10	0.11229	−11.56	2.27	8.53	0.11229	2.28		D
K13B4	0.012	3.21	0.0182	0.10940	12	0.10937	−13.86	2.69	2.81	0.10937	2.67	2.79	A
K15A4	1.402	1.08	6.2295	0.11871	10	0.10836	−14.65	2.84		0.10836	2.81		B
Nord													
N1C	0.194	3.15	0.2957	0.11448	11	0.11399	−10.22	2.02		0.11399	2.05		D
N2B	0.034	1.42	0.1140	0.11348	16	0.11329	−10.77	2.12	2.91	0.11329	2.14	2.87	C
Nikos													
X04	0.229	3.31	0.3335	0.11644	16	0.11589	−8.72	1.74		0.11589	1.79		D
X05	0.386	3.23	0.5750	0.11474	16	0.11378	−10.38	2.05		0.11378	2.07		D
X06	0.126	2.76	0.2190	0.11393	27	0.11356	−10.55	2.08	4.43	0.11356	2.10	4.17	D
X07	0.066	5.95	0.0534	0.10973	12	0.10964	−13.64	2.65	3.04	0.10964	2.64	2.99	A
JP													
JP1-X2	0.111	3.41	0.1565	0.11339	12	0.11313	−10.89	2.14	3.44	0.11313	2.16	3.34	C
JP2-X2	0.029	4.07	0.0342	0.11150	11	0.11145	−12.22	2.39	2.60	0.11145	2.39	2.59	B
JP3-X	0.046	2.60	0.0857	0.11388	10	0.11374	−10.42	2.06	2.58	0.11374	2.08	2.57	C
JP3-X1	0.974	3.30	1.4201	0.11135	27	0.10899	−14.16	2.74		0.10899	2.73		B
JPN2	0.058	4.91	0.0563	0.11086	12	0.11077	−12.76	2.49	2.87	0.11077	2.48	2.84	C
JPN3A	0.079	6.70	0.0564	0.11274	10	0.11265	−11.27	2.21	2.56	0.11265	2.23	2.55	C
JPN3B	0.159	5.81	0.1319	0.11445	11	0.11423	−10.03	1.98	2.90	0.11423	2.01	2.86	C
JPN4	0.027	2.91	0.0439	0.11593	10	0.11586	−8.75	1.74	1.95	0.11586	1.79	1.98	C
JPN9	0.092	4.25	0.1040	0.11256	12	0.11238	−11.48	2.25	3.00	0.11238	2.27	2.95	C
JPN11	0.033	2.77	0.0572	0.11487	14	0.11477	−9.60	1.90	2.20	0.11477	1.94	2.22	B
JPS1	1.646	6.07	1.3031	0.11092	9	0.10875	−14.34	2.78		0.10875	2.76		B
JPS4	0.449	3.48	0.6196	0.11036	10	0.10933	−13.89	2.69		0.10933	2.68		D
JPS6A	2.585	4.43	2.8043	0.11410	11	0.10945	−13.80	2.68		0.10945	2.66		D
JPS6B	1.989	3.97	2.4060	0.11390	10	0.10990	−13.44	2.61		0.10990	2.60		D

γOs, T_{RD} and T_{MA} calculated with respect to chondrite (^{187}Os/^{188}Os = 0.3972 and ^{187}Re/^{188}Os = 0.1276, Walker et al., 1989) and fertile mantle (^{187}Os/^{188}Os = 0.4243 and ^{187}Re/^{188}Os = 0.1287, Meisel et al., 1996) with ^{187}Re decay constant of 1.666×10^{-11} year^{-1}. γOs is the difference (in %) in ^{187}Os/^{188}Os between the sample and chondrite at the eruption age of the kimberlite. Errors are 2σ within run. T_{RD} and T_{MA} expressed in Ga.

5.5.4. Os isotopic variation

The Os isotopic composition of Somerset Island peridotite xenoliths ranges from $^{187}Os/^{188}Os$ of 0.1084 to 0.1194 (Table 3), comparable to the range observed for Kaapvaal and Siberian peridotite xenolith suites (Pearson et al., 1995a,b; Carlson et al., 1999), but not showing the tail to more radiogenic isotopic compositions found in other cratons (Fig. 6). Somerset Island peridotites are highly unradiogenic relative to chondritic values (negative γOs, Fig. 6) and fertile mantle estimates at the time of eruption. This is a general characteristic of old lithospheric mantle. The unradiogenic initial Os isotopic compositions of the Somerset Island peridotites are comparable to other Archean cratonic peridotites (Fig. 6). They are clearly distinct from oceanic mantle and are indicative of long-term Re depletion and isolation from the convecting mantle. The lack of a more radiogenic tail to the Os isotope distribution of the Somerset Island samples suggests that the Somerset lithospheric mantle has experienced less Re addition through metasomatism than is found for other cratons. For those Somerset samples with high Re contents, their unradiogenic Os isotopic compositions indicate that the Re addition must have occurred recently, perhaps associated with capture and transport by the host kimberlite. Somerset Island peridotite xenoliths are also typically less radiogenic than circum-cratonic mantle xenoliths, e.g. from East Griqualand and Namibia (Pearson et al., 2002), suggesting that they are cratonic.

5.5.5. Whole-rock PGE abundances

The Somerset Island peridotite suite shows a large variation in PGE and Re abundances (Table 4). The overall range is larger than that observed for Pyrenean orogenic lherzolites (Lorand et al., 1999), or alkali-basalt hosted spinel peridotites (Morgan et al., 1981; Handler and Bennett, 1999). On average, IPGE (Ir, Os, Ru) abundances for Somerset Island peridotites (Table 4) are similar to primitive mantle estimates

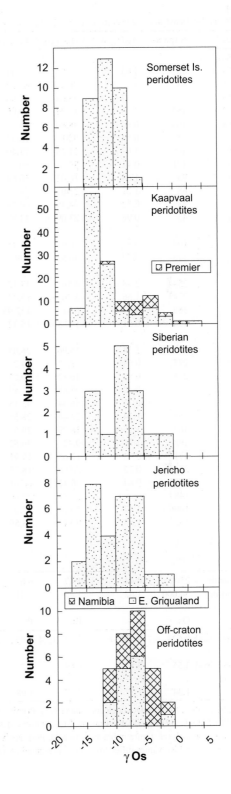

Fig. 6. Histograms of Os isotopic compositions of Somerset Island xenoliths expressed as initial γOs values (at the time of kimberlite eruption) compared to cratonic and circumcratonic peridotite xenoliths from other locales. (Data sources: Walker et al., 1989; Pearson et al., 1995a,b; Carlson et al., 1999; Menzies et al., 1999; Irvine et al., 2001b; Irvine, 2002.)

Table 4

PGE, Re and selected major and trace element abundances for Somerset Island peridotites

Samples	S (ppm)	Al_2O_3 (wt.%)	$(La/Yb)_n$	Ba (ppm)	Os (ppb)	Ir (ppb)	Ru (ppb)	Pt (ppb)	Pd (ppb)	Re (ppb)	$(Pd/Ir)_n$
Batty Bay											
K11A14	101	1.03	12.95	73.41	2.39	2.20	4.43	1.98	1.42	0.03	0.53
K11A15	68	1.06	26.06	68.57	2.35	2.50	5.40	1.11	6.89	0.02	2.28
K11A16	126	1.37	33.63	97.90	3.25	3.66	6.64	4.21	1.40	0.05	0.32
K11A17	157	0.96	16.74	69.06	2.58	2.75	6.02	1.25	0.96	0.02	0.29
K11A18	108	0.84	38.80	25.64	0.62	0.64	1.23	0.09	0.41	0.01	0.53
K12A1	78	0.94	9.31	16.22	3.60	2.56	4.53	2.82	1.44	0.05	0.47
K13A1	40	1.03	21.38	73.40	4.49	4.42	9.81	7.54	1.77	0.13	0.33
K13A3	171	0.41	59.77	50.05	4.20	3.97	7.19	2.08	0.58	0.03	0.12
K13A4	93	1.64	16.02	56.64	2.25	3.32	6.19	5.24	3.95	0.04	0.98
K13A5	68	1.21	2.90	58.76	3.73	4.66	8.67	2.81	1.72	0.22	0.31
K13B4	73	0.82	74.48	60.79	4.18	4.25	6.14	8.32	0.46	0.01	0.09
K15A4	43	0.96	23.09	43.88	1.80	2.20	7.52	2.92	0.69	1.50	0.26
Nord											
N1C	194	0.83	36.65	63.45	3.38	3.31	6.31	2.19	1.22	0.21	0.30
N2B	123	0.72	40.52	109.27	1.74	1.87	3.85	1.59	0.91	0.03	0.40
Nikos											
X04	2544	2.38	4.86	34.12	3.31	3.10	5.28	3.86	3.38	0.23	0.90
X05	2741	1.66	3.26	15.33	3.23	2.97	6.78	1.93	1.34	0.39	0.37
X06	369	3.33	8.81	132.93	2.87	2.66	6.12	2.40	1.64	0.14	0.51
X07	190	0.21	140.38	15.32	5.95	5.93	10.81	5.35	1.50	0.07	0.21
JP											
JP1-X2	260	1.34	15.96	26.05	3.41	3.21	5.61	3.07	2.43	0.11	0.63
JP2-X2	277	0.63	38.83	24.25	3.58	3.18	5.57	2.90	0.16	0.03	0.04
JP3-X	275	0.85	16.12	10.25	2.89	2.82	5.13	2.27	1.37	0.05	0.40
JP3-X1	177	0.10	78.07	11.42	2.65	2.65	3.42	1.12	0.27	1.09	0.08
JPN2	339	0.73	56.34	43.18	4.91	4.61	9.12	4.01	2.48	0.06	0.45
JPN3A	268	1.14	23.44	28.55	6.70	6.52	12.38	7.52	4.20	0.08	0.53
JPN3B	407	1.15	18.79	29.53	5.81	5.78	9.77	5.78	3.82	0.16	0.55
JPN4	298	1.76	9.42	36.52	2.91	2.59	4.77	1.93	1.43	0.03	0.46
JPN9	471	0.73	13.63	19.91	4.25	3.79	6.87	2.85	1.89	0.09	0.41
JPN11	236	0.72	20.60	18.75	2.77	2.93	5.17	1.47	0.08	0.03	0.02
JPS1	1702	0.64	36.42	66.61	6.07	5.63	10.88	4.43	1.18	1.65	0.17
JPS4	1484		33.81	75.86	3.48	4.04	7.40	2.42	1.47	0.45	0.30
JPS6A	1006	1.47	10.70	47.63	4.43	4.15	7.80	3.71	2.28	2.58	0.45
JPS6B	1028	1.68	9.84	38.59	3.97	3.90	7.42	4.17	2.59	1.99	0.55
GP13	274				3.87	3.56	6.97	7.00	5.64	0.33	
	(12)				(0.17)	(0.33)	(0.23)	(0.52)	(0.35)	(0.01)	
Prim. mantle	250	4.44	1.47	6.6	3.4	3.2	5	7.1	3.9	0.28	
CI chondrite	54,000	1.62	1.47	2.41	490	455	710	1010	550	40	

	Os	Ir	Ru	Pt	Pd	Re
Somerset island range	0.623–6.7	0.636–6.52	1.23–12.4	0.094–8.32	0.080–6.89	0.013–2.59
Somerset island average	3.55	3.52	6.69	3.29	1.79	0.36
S.D.	1.34	1.27	2.37	1.98	1.42	0.40

CI chondrite values from McDonough and Sun (1995). Primitive mantle values taken as $0.007 \times$ CI chondrite after method of McDonough and Sun (1995). Sample GP13 (peridotite from the Beni Bousera massif) was used as a standard for sulfur and PGE analyses. Sulfur data for GP13 is the mean for 14 analyses, figure in brackets 1σ S.D. PGE analyses for GP13 are the mean of eight separate dissolutions, figure in brackets 1σ S.D. Where Os and Re values differ from those presented in Table 3, these represent separate dissolutions of the sample.

(McDonough and Sun, 1995), whilst PPGE (Pt, Pd) abundances are generally depleted. In comparison to peridotites from other tectonic settings, IPGEs within the Somerset Island peridotites have similar abundances to those observed in orogenic massifs (Pattou et al., 1996; Lorand et al., 1999), and higher than those observed for alkali-basalt hosted spinel peridotites (Morgan et al., 1981; Handler and Bennett, 1999).

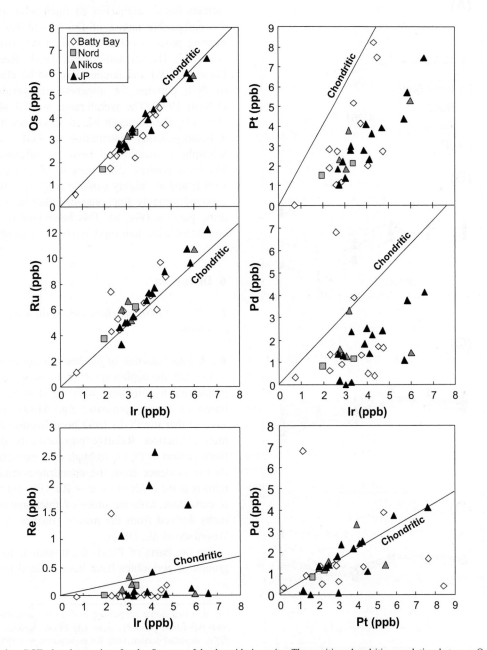

Fig. 7. Bivariate PGE abundance plots for the Somerset Island peridotite suite. The positive chondritic correlation between Os, Ru and Ir (r = 0.95 and 0.93, respectively) may suggest these elements are controlled by the same mineral phase(s) in the xenoliths. None of the remaining PGEs or Re correlate with Ir or each other. Chondritic ratio after McDonough and Sun (1995).

In contrast, PPGEs within the Somerset Island peridotites are depleted relative to both orogenic massifs and alkali basalt hosted spinel peridotites.

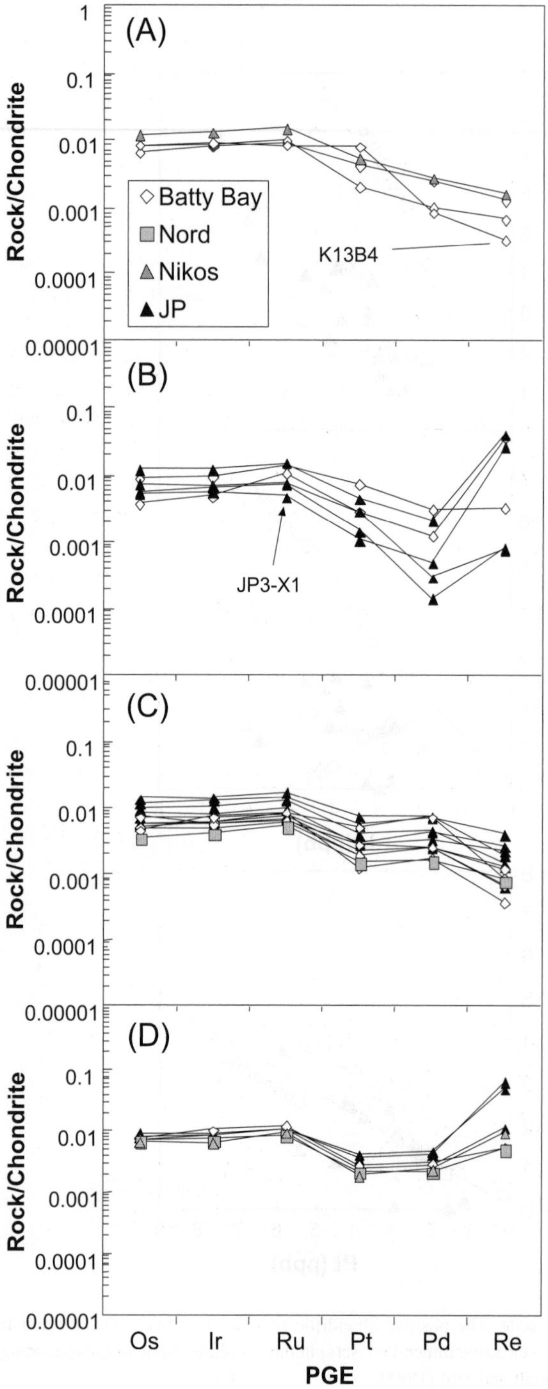

Osmium and Ru show good positive correlations with Ir, with close to chondritic ratios (Fig. 7). IPGEs range between 0.001 and $0.017 \times CI$ chondrite (mean $= 0.008 \times CI$ chondrite) and generally show a flat chondrite normalised pattern with very little intra-element fractionation (Fig. 8). Such behaviour reflects the compatible nature of Os, Ir and Ru in mantle melting processes (Barnes et al., 1985; Brügmann et al., 1987; Barnes and Picard, 1993; Reisberg and Lorand, 1995) and is considered to be characteristic of PGE patterns for upper-mantle peridotites (cf. O'Neill, 1991). The overall range in IPGE abundances shown by the Somerset Island peridotites may reflect a heterogeneous distribution of PGEs within the lithospheric mantle and between different samples. PPGEs in contrast show poor systematic correlations with Ir and are highly variable (Fig. 7), with samples showing strongly fractionated, rather than flat chondritic patterns (Fig. 8). This behaviour is essentially consistent with their moderately incompatible nature.

6. Discussion

6.1. Re and PGE geochemistry of the somerset island peridotites

6.1.1. Fractionation of platinum group elements

Overall, the depletion of PPGEs in Somerset Island peridotites is consistent with the degree of melting based on major elements, e.g. Al_2O_3, and would suggest that the PGEs have been fractionated during melt extraction. Relative compatibility during the fractionation of PGEs, through melt extraction, based on the evidence from the chondrite-normalised patterns is in the order $Os \sim Ir \sim Ru < Pt < Pd \leq Re$. This is consistent with the order of fractionation seen in melts derived from the mantle (Barnes et al., 1985; Woodland et al., 2002).

On the basis of PPGE fractionation, four distinct groups of peridotites have been defined in the Som-

Fig. 8. CI-chondrite normalised PGE (+Re) abundances for the Somerset Island peridotite suite. (A) PPGE depleted pattern, (B) PPGE depleted pattern with Re enrichment, (C) PPGE depleted pattern with Pd and minor Re enrichment, and (D) PPGE depleted pattern with Pd and Re enrichment. CI-chondrite normalising values after McDonough and Sun (1995).

erset Island peridotite suite. Four samples from Somerset Island have highly fractionated PPGEs (Fig. 8A), with marked depletions in Pd and Re, and intermediate depletion in Pt (Group A, PPGE depleted peridotites). This is the expected pattern for peridotites that have undergone extensive degrees of melt extraction. In the Somerset Island suite, the degree of this depletion is comparable to that observed in peridotite xenoliths from other Archean cratons (Irvine, 2002), and more evident than that observed in harzburgitic residues from Pyrenean orogenic massifs (cf. Lorand et al., 1999). Chondrite normalised Pd/Ir ratios for these samples range from $(Pd/Ir)_n$ of 0.09 to 0.32. The depleted PGE pattern of these samples suggests that the system has remained closed since the time of melt extraction, i.e. they have not experienced significant addition of Re or PPGE.

A further six samples have depletions in PPGEs (Fig. 8B), but with variable Re contents and marked upward inflections in the extended PGE pattern at Re (PPGE depleted peridotites with Re enrichment, Group B). Low $(Pd/Ir)_n$ of 0.02 to 0.33 support the notion that these samples represent variably but extensively depleted melt residues. The elevated Re abundances suggests that the samples have experienced Re addition after the initial melt extraction event(s). The remaining Somerset Island peridotite xenoliths in this study show varying degrees of disturbance to the PGE systematics. Eleven peridotites show a noticeable depletion in Pt similar to that observed in groups A and B (Fig. 8C), but these are considerably more enriched in Pd (PPGE depleted peridotites with Pd enrichment, Group C). This suggests that Pd has been re-introduced to the peridotites after an initial melt extraction event. The wide variations in Re abundance and Re/Pt within this group indicate that some samples may also have experienced some degree of Re enrichment. The final group of six samples (Fig. 8D) again show a noticeable depletion in Pt, as observed for all Somerset Island peridotites, but both Re and Pd have clearly been enriched (PPGE depleted peridotites with Pd and Re enrichment, Group D).

6.1.2. PGE variation with whole-rock sulfur chemistry

The control of sulfides on PGEs within the Somerset Island peridotites can be considered through comparison of PGE and S abundances. Sulfur is a moderately incompatible element in the mantle and behaves coherently during mantle melting processes (Lorand, 1989; Lorand, 1991). PGEs, however, have high partition coefficients into sulfides and unless all sulfide within the mantle is consumed (~ 25% melting) and the melt nears saturation (Keays, 1995), no correlation of S with PGEs should necessarily exist. Estimates for the S content of fertile mantle vary but are generally less than 300 ppm (Peach and Mathez, 1996). The S contents within Somerset Island peridotites range from 40 to 2741 ppm (Fig. 9; Table 4), i.e. from depleted to highly enriched relative to fertile mantle source. Little overall correlation exists between PGEs and S within the Somerset Island peridotites. In detail, systematic correlations between PPGE variations and S exist in the Somerset Island peridotites if considered in terms of their chondrite normalised PGE patterns (Table 5), with Pd enrichment (Groups C and D) associated with enrichment in S. The Somerset Island kimberlites are emplaced in platform sedimentary sequences, which contain locally, thick gypsum deposits. It is thus possible that the kimberlites and entrained peridotite xenoliths may also have experienced some S enrichment from S-rich groundwater interaction, post-emplacement (i.e. in the last 100 Ma).

6.1.3. Behaviour of Re, PGEs during partial melt extraction

If the Re characteristics of the Somerset Island xenoliths reflect partial melting processes operating

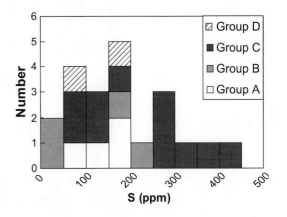

Fig. 9. Histogram of S concentrations in Somerset Island xenoliths expressed in terms of PGE pattern groupings.

Table 5
Summary of PGE behaviour within the Somerset Island peridotites, on the basis of the shape of the PPGE part of the PGE patterns

Group	PGE Pattern	Pd (ppb)	Re (ppb)	S (ppm)	Al_2O_3 (wt%)	Ba (ppm)	Nb (ppm)	$(Pd/Ir)_n$	$(Re/Ir)_n$	$(Re/Pt)_n$	TRD model ages Range	TRD model ages Mean	TRD model ages +/-	Interpretation
A		0.98	0.04	140	0.70	56.02	3.32	0.18	0.10	0.25	2.28 – 2.69	2.5	0.2	Melt residue
B		0.69	0.74	155	0.69	39.72	2.10	0.15	2.72	8.04	1.90 – 2.84	2.5	0.4	Melt residue + Re addition
C		2.27	0.06	236	1.10	46.21	2.50	0.52	0.19	0.47	1.22 – 2.49	2.1	0.3	Melt addition + Pd (PGE) + Re addition
D		1.77	0.97	1087	1.37	49.94	2.73	0.38	2.81	7.30	1.74 – 2.69	2.3	0.4	Melt addition + Pd (PGE) + Re addition

Average whole-rock elemental abundances and PGE ratios on the basis of PPGE determined groups.

on a homogeneous mantle source, Re concentrations should correlate with indices of melt extraction. Studies on Alpine peridotites have suggested that Re partitions into melts with a distribution coefficient similar to that of Al (Reisberg and Lorand, 1995). Similar correlations have been reported for rift-related spinel peridotites erupted in alkali basalts (Meisel et al., 1996; Handler et al., 1997). In contrast, arc-related spinel peridotites show little correlation between Re and Al (Brandon et al., 1996). Peridotite xenoliths from Somerset Island show no overall correlation between Re and indices of melt depletion, such as Al_2O_3, suggesting that Re variation is not exclusively controlled by partial melt extraction (Fig. 10A). Studies of cratonic peridotite xenoliths from the Kaapvaal craton (Pearson et al., 1995a; Carlson et al., 1999) also found no correlation between Re and melt-extraction indices and these studies suggested that the large variation in Re could be attributed to "metasomatism". The generally low Re–Os of Somerset Island peridotites, however, is consistent with the melt-depleted nature of both bulk-rock and mineral compositions that indicate depletion in magmaphile elements.

No overall systematic variation is observed between PGE abundance and major elements for the Somerset Island peridotites, despite the large range observed in Al_2O_3. Os and Ir are compatible during mantle melting and therefore plots of Os and Ir against melt-indices such as Al_2O_3 should produce sub-horizontal trends. This is the case for peridotites from most tectonic settings (Fig. 10B). Peridotites from Somerset Island together with cratonic peridotites from the Kaapvaal craton (Pearson et al., 1995a; Carlson et al., 1999) show a wide, non-systematic variation in Os concentrations with a near vertical trend for a low Al_2O_3 (wt.%) content. The reason for this observed trend is not known and even extreme cases of melt extraction could not produce this range in Os content. The scatter in Os concentration may reflect the heterogeneous distribution of sulfides or ultra-trace Os-alloys within the Somerset Island lithospheric mantle. It is unclear why this feature appears to be exclusive to cratonic lithospheric peridotites such as seen at Somerset Island and the Kaapvaal craton, and not peridotites from other tectonic settings. It is possible that the variations observed in the IPGE contents for cratonic peridotites are produced

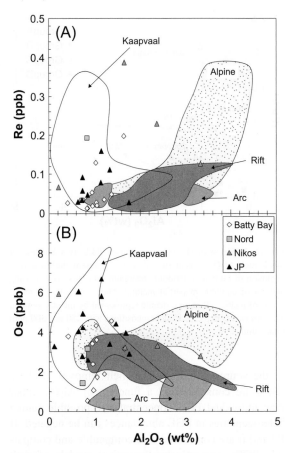

Fig. 10. Re and Os concentrations (in ppb) against Al_2O_3 content (in wt.%). Cratonic peridotites from the Kaapvaal craton (Carlson et al., 1999) represented by the white field. Alpine peridotite data (Reisberg et al., 1991; Reisberg and Lorand, 1995) are shown by the stippled field and rift-related spinel peridotites (Meisel et al., 1996; Handler et al., 1997) by the dark grey field. Spinel peridotites from "arc" settings are shown by the light grey fields, one each for samples from the Simcoe volcano in the Cascades and the Itchinomegata volcano in the back arc of the Japanese convergent margin (Brandon et al., 1996).

from processes other than melt removal. The PPGEs, which are moderately incompatible and depleted within the Somerset Island suite, show no distinct correlations either.

Considerable vertical variation in Pd concentrations on plots of PGE vs. melt indices (e.g. Al_2O_3) can result simply from the observed PGE heterogeneity in mantle peridotites (Pearson et al., in press). This will obviously create scatter on any melt trends. Assuming chondritic PGE ratios were present initially

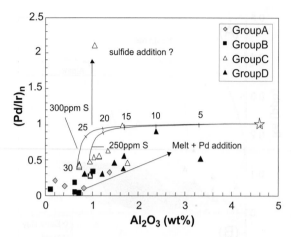

Fig. 11. Variation of chondrite-normalised Pd/Ir as a function of whole-rock Al_2O_3 content (wt.%). Lines represent the theoretical evolution of Pd/Ir in a peridotite residuum submitted to increasing degrees of equilibrium partial melting. Marks indicate 5% increments of melting. Primitive mantle represented by the star symbol. The two curves depict starting compositions with 250 and 300 ppm total sulfur.

in the source mantle before melt removal, if these ratios are considered as a function of Al_2O_3 rather than individual PGE concentrations, possible source heterogeneities in PGE abundances can be negated. If Pd and Ir are considered as incompatible and compatible PGEs, respectively, theoretical models of melt extraction should constrain the behaviour and fractionation of these groups (Fig. 11). There is limited correlation between measured Pd/Ir fractionation and the theoretical behaviour of Pd/Ir and Al_2O_3 for melt extraction models in Somerset Island peridotites, but some systematics do emerge. Correlations between the shape of PGE patterns and bulk-rock Al_2O_3 content are evident (Table 5). Samples enriched in Pd (Groups C, D) also show higher Al_2O_3 (Fig. 11). For several of the Somerset Island peridotite xenoliths, depleted bulk-rock Al_2O_3 content correlates with PGE abundances indicating melt depletion, i.e. low $(Pd/Ir)_n$ (Group A and B; Table 5). For these samples, a case can be made for PGEs having remained closed since the time of initial melt extraction. In contrast, displacement of other samples from the theoretical model trends on a Pd/Ir vs. Al_2O_3 plot is consistent with their disturbed PGE patterns, often showing Re and/or Pd enrichment. These samples generally plot on vectors that may be attributed to melt addition (Fig.

11), but not to the addition of an isolated sulfide component.

6.2. Timing of lithospheric mantle differentiation

6.2.1. Re–Os isotope systematics

The observed range in Os isotopic compositions for the Somerset Island peridotites (Table 3) equate to T_{RD} model ages of 2.8 to 1.3 Ga (Fig. 12A). Nine samples have T_{RD} model ages between 2.5 and 3.0 Ga, and 24 samples have T_{RD} ages <2.5 Ga. Three features are prominent in the frequency distribution of T_{RD} model ages shown in Fig. 12A; (i) there are no samples with T_{RD}>3.0 Ga, (ii) there is a prominent mode in the model ages between 2.0 and 2.25 Ga, and (iii) the broad distribution of T_{RD} model ages from

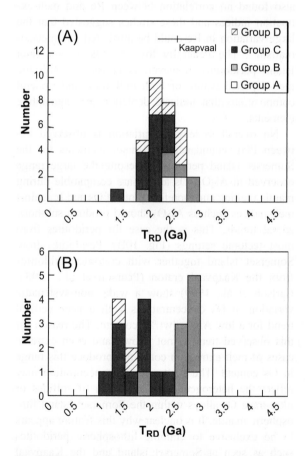

Fig. 12. Histograms of Re-depletion (T_{RD}) model ages for Somerset Island (A) and Jericho (B) peridotites expressed in terms of PGE pattern groupings. Also shown is the mean T_{RD} value for Kaapvaal from Carlson et al. (1999) at 2.5 Ga.

2.25 to 3.0 Ga. The mean T_{RD} model age for all Somerset Island peridotite samples is 2.3 ± 0.3 Ga (1 S.D.) with a median of 2.3 Ga. This is lower than the mean T_{RD} model age reported by Carlson et al. (1999) for cratonic peridotites from the Kaapvaal craton (mean = 2.5 ± 0.4 Ga), but higher than the mean T_{RD} model age reported by Irvine et al. (1999, 2001a) for cratonic peridotites from the Jericho kimberlite on the Slave craton (mean = 2.1 ± 0.7 Ga). On average, peridotite xenoliths from Somerset Island with group A (mean $2.5 = \pm 0.2$ Ga) and B (mean = 2.5 ± 0.4 Ga) PGE patterns have older T_{RD} model ages than those from either group C (mean $2.1 = \pm 0.3$ Ga) or D (mean $2.3 = \pm 0.4$ Ga) (Table 5).

There is no clear overall correlation between T_{RD} model ages and depth of origin for the Somerset Island peridotites (Fig. 13). The maximum T_{RD} model age for spinel peridotites at Somerset Island is 2.7 Ga, for garnet–spinel peridotites it is 2.8 Ga, and for garnet peridotites 2.7 Ga. The similarity of the oldest T_{RD} model ages at different depths in the mantle beneath Somerset Island indicates that ancient, Archean lithospheric mantle persists to depth at Somerset Island. The apparent young T_{RD} model ages at shallower depths may therefore indicate that Archean lithospheric mantle has been modified by reworking at a later stage, and/or younger material has been added to the lithospheric keel. The overall range in T_{RD} ages for the Somerset Island lithospheric mantle indicates that parts have persisted since at least the Late Archean/Early Proterozoic.

T_{MA} model ages for the Somerset Island xenoliths range from 4.2 to 1.6 Ga (Table 3), excluding those that produce future ages or give ages that are greater than the age of the Earth. The significance of T_{MA} vs. T_{RD} is discussed below.

6.2.2. Constraints from PGE systematics on the significance of model ages for the lithospheric mantle beneath somerset island

Recent studies of mineral separates and sulfides from peridotites (Burton et al., 1999) and in-situ laser-ablation analysis of sulfide Re–Os isotope systematics (Alard et al., 2000; Pearson et al., 2000; Aulbach et al., 2001) have indicated the presence of multiple generations of sulfide in some mantle peridotites. In fertile peridotite, sulfides are considered to be the host for the majority of Re and Os (Mitchell

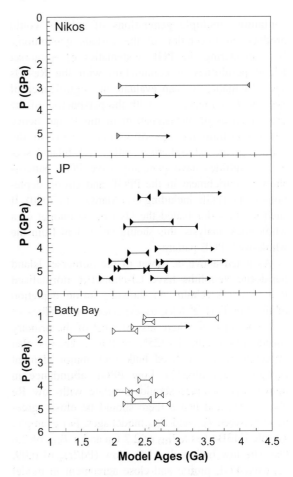

Fig. 13. T_{RD} (right facing triangles) and T_{MA} (left facing triangles) model ages for peridotite xenoliths from individual kimberlite pipes plotted against the pressure of equilibration as determined by mineral geothermobarometry. All samples plotting at pressures less than 2 GPa are spinel peridotites (pressure arbitrarily assigned in 0.2 GPa increments up to 2 GPa). Samples with arrows have T_{MA} ages in excess of the age of the Earth.

and Keays, 1981; Hart and Ravizza, 1996; Pearson et al., 1998; Burton et al., 1999; Alard et al., 2000), and secondary sulfides may be introduced during melt/fluid influxes. Re–Os model ages of primary sulfide inclusions are generally older than interstitial sulfides (Alard et al., 2000; Pearson et al., 2000). The introduction of melt/fluid fluxes also seems to affect the major element chemistry of the host peridotite, as evident from the low-angle correlation between $(Pd/Ir)_n$ and Al_2O_3 (Fig. 11) and hence seem likely to be due to the movement of silicate melt rather than pure sulfide melt. A whole-rock analysis of a sample containing multiple generations of sulfide would produce an integration of the various generations. By considering the PGE systematics of Somerset Island peridotites in conjunction with the Re–Os age systematics we can evaluate the significance of T_{RD} vs. T_{MA} model ages with the perspective of the consequences of metasomatism in mind, and hence better constrain the age of various events in the history of the lithospheric mantle. Melt-related secondary sulfides have generally have PGE patterns showing enrichment in the PPGE and strong depletions in the IPGE, including Os (Alard et al., 2000). If such sulfides dominated the Os isotope budget of a whole-rock analysis, this should be evident from the whole-rock PGE pattern.

As noted above, a subset of the Somerset Island peridotite xenoliths have CI-chondrite normalised PGE patterns (Group A) showing strong depletion of the PPGE (Fig. 8A), as expected for residues of extensive melting where loss of most of the primary sulfide has occurred (>25% melting). This is also reflected in the depleted bulk rock major element composition (Table 2). Low PPGE abundance in undisturbed samples should correlate with low Re abundances and hence there should be close agreement between T_{RD} and T_{MA} model ages. For example, sample K13B4 has a T_{RD} of 2.7 Ga and a T_{MA} of 2.8 Ga. The low bulk-rock Al_2O_3, low $(Pd/Ir)_n$ of 0.09, depleted PGE profile and close agreement in model ages for this sample suggest that the Re–Os isotope systematics for at least some of the Somerset Island peridotites have remained closed since initial melt extraction and that their Re–Os model ages reflect the age of this depletion event. The T_{MA} model age of sample K13B4 therefore is a likely approximation of the age of melting, and provides a maximum age for the lithospheric mantle as sampled by the Somerset Island kimberlites from the present sample suite. There is a significant melt extraction event and formation of peridotitic komatiite, high Mg-tholeiite and tholeiite at this time (2.74–2.7 Ga) in the Mary River and Prince Albert Groups (Jenner et al., 2002), which are geographically close to Somerset Island (<500 km). The small difference between T_{RD} and T_{MA} model ages shows that in this case, although T_{RD} represents the minimum age of the sample, it would appear to be a good approximation of the actual age for PPGE depleted samples. This provides robust

evidence for Archean lithosphere below Somerset Island.

The Somerset Island peridotites that display highly fractionated PGE patterns but with marked inflection in the Re abundance (Group B; Fig. 8B) show more complex model age systematics. The variation in Re concentrations do not correlate with indices of melt depletion, such as Al_2O_3, suggesting that Re variation is not controlled by partial melt extraction alone. Previous studies have attributed the Re variation in cratonic peridotite xenoliths to "metasomatism" (Pearson et al., 1995a; Carlson et al., 1999) and this is supported by the observed disturbance to the PGE systematics of many Somerset Island peridotites. $(Pd/Ir)_n$ values for these samples (Table 5) suggest they have undergone melt extraction but give little indication of the disturbed nature of the PGE profiles. The extended PGE pattern for samples such as JP3-X1 shows a Re inflection and this leads to a considerable difference between T_{RD} (2.7 Ga) and T_{MA} (future) model ages. Re enrichment in this sample appears to be accompanied by additions of S and other LIL elements (this sample has among the highest La/Yb and Ba content of the Somerset peridotites), but not Al_2O_3 (Table 4). This metasomatic pattern suggests that the agent that added Re was also LIL-rich, but was not an aluminous silicate melt. The host kimberlite has suitable enrichment in Re, S and LIL's as well as low Al_2O_3 content and thus could be responsible for enhancing the Re content of this sample during transport to the surface. Sample JP3-X1 has a similar T_{RD} model age to sample K13B4, consistent with a major melt extraction event at 2.7 Ga, and may suggest that Os was not disturbed by the metasomatism, and that the metasomatism causing enrichment of Re was quite recent. For example, back extrapolating the measured Os isotopic composition for sample JP3-X1 using its measured Re–Os causes its Os isotopic composition to go below that expected for 3.5 Ga old Re-free mantle by 393 Ma, and below that of the solar system initial Os isotopic composition by 677 Ma. Thus, the enrichment in Re experienced by this sample must have occurred well less than 400–700 Ma, and could have occurred during entrainment in the host kimberlite, since the T_{RD} of this sample is the same as that of the Re-depleted sample K13B4.

In contrast, samples with PGE patterns that also show evidence of Pd addition (Groups C and D) tend to have younger T_{RD} model ages (Table 5), suggesting that the $^{187}Os/^{188}Os$ isotopic signatures have also been disturbed during this enrichment event.

The generally unradiogenic $^{187}Os/^{188}Os$ of many samples together with the clustering of T_{RD}, but diverse T_{MA} model ages suggests to us that the elevated Re contents in this suite are often a result of recent processes, possibly associated with the kimberlite. For this reason we focus our discussion of lithospheric evolution around the geological significance of the xenolith T_{RD} ages and PGE patterns.

The discovery of Late Archean ages in the lithospheric mantle beneath Somerset Island indicates that cratonic mantle exists beneath this region. The age of this mantle is not quite as old as that beneath the Slave craton, further to the southwest, where Re-depletion ages as old as 3.2 Ga have been reported from xenolith suites (Irvine et al., 1999, 2001a). The Somerset Island lithospheric mantle may be mantle on the edge of an Archean craton that has been reworked and chemically disturbed by subsequent Proterozoic tectonomagmatic events.

6.3. Evolution of the lithospheric mantle beneath the Churchill Province

The Re–Os age systematics of the Somerset Island peridotites suggest that the oldest remnant lithospheric mantle is ca. 2.8 Ga. Isotopic work on the Precambrian Churchill basement of Somerset Island have produced Sm–Nd T_{DM} crustal separation ages of 3.0 to 2.2 Ga (Frisch and Sandeman, 1991; Frisch and Hunt, 1993), although meso-Archean ages have been reported in the Queen Maud block to the southwest (W. Davis, pers. comm.). We recognise, however, that T_{DM} Nd ages probably represent a minimum estimate for crustal stabilisation. No peridotites in this study were found to be as old as 3.0 Ga, although the error on the Re–Os model age calculation is ~ 200 Ma. Given the poor age constraints for crustal stabilisation in this area, we cannot realistically resolve the relationship between crust and lithospheric mantle evolution at present.

The frequency distribution of T_{RD} ages shows data points in the range of 1.8 to 2.8 Ga and it is difficult to interpret these ages. Here we simply note that numerous crustal ages have been reported in this time interval. The peridotite ages could simply represent disturbance at these times. The upper constraints of this age range are consistent with melt-extraction of ultrabasic and basic magmas at 2.74–2.70 Ga, the formation of granite magma at 2.6 Ga, and with an upper intercept U–Pb zircon age recorded from a granodiorite gneiss at 2.77 Ga as previously discussed. Re–Os T_{RD} model ages of 2.2 to 1.8 Ga coincide with a period of intense tectonic activity

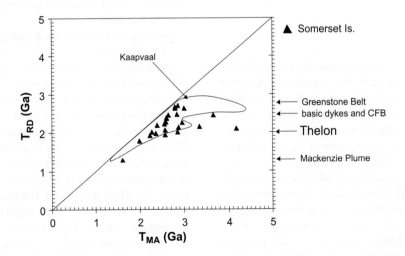

Fig. 14. Plot of T_{RD} vs. T_{MA} model ages for individual Somerset Island peridotite suites. The line in the figure represents the $T_{RD} = T_{MA}$ line. Data for peridotite xenoliths from the Kaapvaal craton is from Carlson et al. (1999). Greenstone belt formation at 2.74–2.70 Ga (Jenner et al., 2002). Proterozoic basic magmatism ca. 2.2 Ga (Tella, 2001; Pehrsson et al., 2002). The Taltson–Thelon orogen, between 2.02 and 1.9 Ga (Hoffman, 1988). Mackenzie plume event ca. 1.27 Ga (Le Cheminant and Heaman, 1989).

within the Churchill Province related to collision with the Slave, Superior and Nain cratons (Hoffman, 1990; Fig. 14). Evidence for Paleoproterozoic rifting in the northern Churchill is manifested as continental flood basalts of the Ketyet group and the MacQuoid (2.19 Ga) and Tulamela diabase dykes. Furthermore, the Thelon Orogen (2.02–1.9 Ga) is of particular importance to lithosphere development and modification of the northern Churchill Province (Fig. 1a). The 2.2 to 1.8 Ga Re–Os T_{RD} model ages of the Somerset Island peridotite xenoliths may therefore represent either (i) the addition of new, subduction-related material to the Churchill lithosphere during the Thelon orogeny, or (ii) that there has been a substantial disturbance of PGE and Os isotope systematics due to tectonomagmatism related to this event. Option (ii) allows the possibility that a meso-Archean keel once existed beneath the area and has been completely modified. However, the fact that undisturbed PGE patterns (Group A) do not extend to meso-Archean ages would argue against that.

A single, young T_{RD} model age of 1.3 Ga may relate to disturbance of the lithospheric mantle by Proterozoic (1400 and 1000 Ma) rifting on nearby northern Baffin Island (Okulitch et al., 1986). Alternatively, this young Re–Os model age may relate to a disturbance by the Mackenzie plume event ca. 1.27 Ga (Le Cheminant and Heaman, 1989).

7. Conclusion

Using combined PGE characteristics with Re–Os isotope systematics and the correlation between PGEs and magmaphile elements in Somerset Island peridotite xenoliths we gain better insights into the age systematics of the Somerset Island lithospheric mantle. This allows us to improve our understanding of initial differentiation and subsequent modification of the Somerset Island lithosphere. From the extended CI-chondrite normalised PGE patterns it is evident that the majority of samples have undergone Re enrichment, most likely by interaction with the host kimberlite during eruption. As all Somerset Island peridotite samples show vestiges of major-element and PPGE melt depletion, if the observed Re addition did occur at the time of kimberlite eruption, the T_{RD} model ages of the samples provide

a reliable indicator of the age of a particular sample. For those samples that show systematic behaviour between indicators of melt depletion, such as high IPGE/PPGE ratios and low Al_2O_3 contents (PGE group A), the T_{MA} ages, all of which are younger than 2.8 Ga, may provide the best indication of the time of melt depletion and lithosphere formation in this area.

Consideration of Re–Os model age constraints, thermobarometry, modal mineralogy and mineral chemistry indicate that Archean age lithospheric mantle clearly exists beneath the Churchill Province at Somerset Island. As in the crustal section, the Archean lithospheric mantle in the Churchill has been substantially modified or disturbed, possibly associated with the Proterozoic rifting (ca. 2.2 Ga). Somerset Island is also located on the NW margin of the Churchill Province, close to the Thelon orogeny (ca. 2.0 Ga; Fig. 1a). This event, reflected in granulite metamorphism in the Somerset Island crust (Frisch and Sandeman, 1991), may also have significantly modified the Archean lithospheric mantle. The Archean mantle identified beneath Somerset Island does not appear to be quite as old as the mid-Archean lithospheric mantle found beneath the Slave craton to the southwest. The younger age of the lithospheric mantle below Somerset Island, combined with evidence for extensive mid-Proterozoic re-working, may explain the paucity of diamonds in this area, as compared to the Slave craton.

New age constraints for peridotite xenoliths from the Churchill Province cratonic lithosphere aid our understanding of cratonic root formation. In addition, Re–Os and PGE data from this study allow us to track the subsequent evolution, as indicated by modification of the lithospheric mantle beneath this region associated with orogenesis-related tectono-magmatic events. This is comparable to the reported modification or addition of new material to the lithospheric mantle beneath the Premier kimberlite in South Africa, related to the Bushveld event (Pearson et al., 1995a; Carlson et al., 1999). This suggests that, although lithospheric mantle keels persist beneath Archean cratons, they are locally subject to modification by major thermal events, which may overprint the primary PGE and Re–Os systematics and modify the thermal profile of the lithosphere sufficiently to affect diamond occurrence.

Acknowledgements

We thank C.J. Ottley for assistance with PGE measurements, and M.F. Horan with laboratory help. J. Armstrong of DIAND, Yellowknife provided additional Nikos and JP xenolith samples. The Polar Continental Shelf Project is thanked for logistical and helicopter support for the Somerset Island field-work. Olivier Alard and Bill Davis provided much appreciated reviews of this paper. This study forms part of a NERC PhD studentship for G.J. Irvine Ref. GT04/97/73/ES. Analyses at DTM and field-work in Somerset Island were supported by Carnegie Canada. This is Geological Survey of Canada contribution number 2002-134.

References

Alard, O., Griffin, W.L., Lorand, J.P., Jackson, S.E., O'Reilly, S.Y., 2000. Nonchondritic distribution of the highly siderophile elements in mantle sulphides. Nature 407, 891–894.

Aulbach, S., et al., 2001. Re–Os Isotope Evidence for Meso-Archaean Mantle beneath 2.7 Ga Contwoyto Terrane, Slave Craton, Canada. Implications for the Tectonic History of the Slave Craton, The Slave–Kaapvaal Workshop, Merrickville, Ontario, Canada.

Barnes, S.J., Picard, C.P., 1993. The behaviour of platinum-group elements during partial melting, crystal fractionation and sulfide segregation: an example from the Cape Smith Fold Belt, Northern Quebec. Geochimica et Cosmochimica Acta 57, 79–87.

Barnes, S.J., Naldrett, A.J., Gorton, M.P., 1985. The origin and fractionation of platinum-group elements in terrestrial magmas. Chemical Geology 53, 303–323.

Blackadar, R.G., 1967. Precambrian geology of the Boothia Peninsula, Somerset Island, and the Prince of Wales Island, District of Franklin. Geological Survey of Canada Bulletin 151, 62 pp.

Boyd, F.R., 1989. Compositional distinction between oceanic and cratonic lithosphere. Earth and Planetary Science Letters 96, 15–26.

Boyd, F.R., Mertzman, S.A., 1987. Composition and structure of the Kaapvaal lithosphere, Southern Africa. In: Mysen, B.O. (Ed.), Magmatic Processes: Physicochemical Principles. The Geochemical Society, Penn State, pp. 13–24.

Boyd, F.R., et al., 1997. Composition of the Siberian cratonic mantle: evidence from Udachnaya peridotite xenoliths. Contributions to Mineralogy and Petrology 128, 228–246.

Brandon, A.D., Creaser, R.A., Shirey, S.B., Carlson, R.W., 1996. Os recycling in subduction zones. Science 272, 825–861.

Brey, G.P., Kohler, T., Nickel, K.G., 1990. Geothermobarometry in four-phase lherzolites: I. Experimental results from 10 to 60 kb. J. Petrol. 31, 1313–1352.

Brügmann, G.E., Arndt, N.T., Hofmann, A.W., Tobschall, H.J., 1987. Noble metal abundances in komatiite suites from Alexo, Ontario, and Gorgona Island, Columbia. Geochimica et Cosmochimica Acta 51, 2159–2169.

Burton, K.W., Schiano, P., Birk, J.L., Allegre, C.J., 1999. Osmium isotope disequilibrium between mantle minerals in a spinel-lherzolite. Earth and Planetary Science Letters 172, 311–322.

Campbell, J.L., Teesdale, W.J., Kjarsgaard, B.A., Cabri, L.J., 1996. Micro-PIXE Analysis of Silicate Reference Standards for Trace Ni, Cu, Zn, Ga, Ge, As, Rb, Sr, Y, Zr, Nb, Mo and Pb, with emphasis on Ni for application of the Ni-garnet geothermometer. The Canadian Mineralogist 34, 37–48.

Carlson, R.W., Irving, A.J., 1994. Depletion and enrichment history of subcontinental lithospheric mantle: Os, Sr, Nd and Pb evidence for xenoliths from the Wyoming Craton. Earth and Planetary Science Letters 126, 457–472.

Carlson, R.W., et al., 1999. Re–Os systematics of lithospheric peridotites: implications for lithosphere formation and preservation. In: Gurney, J.J., Gurney, J.L., Pascoe, M.D., Richardson, S.H. (Eds.), Proceedings of the VIIth International Kimberlite Conference. Red Roof Design, Cape Town, pp. 99–108.

Chesley, J.T., Rudnick, R.L., Lee, C.T., 1999. Re–Os systematics of mantle xenoliths from the east African rift: age, structure and history of the Tanzanian craton. Geochimica et Cosmochimica Acta 63 (7/8), 1203–1217.

Dawson, J.B., Stephens, W.E., 1975. Statistical classification of garnets from kimberlites and associated xenoliths. Journal of Geology 83, 589–607.

Dreibus, G., Palme, H., Spettel, B., Zipfel, J., Wänke, H., 1995. Sulfur and selenium in chondritic meteorites. Meteoritics 30, 439–445.

Finnerty, A., Boyd, F., 1987. Thermobarometry for garnet peridotites: basis for the determination of thermal and compositional structure of the upper mantle. In: Nixon, P. (Ed.), Mantle Xenoliths. Wiley, Chichester, UK, pp. 381–402.

Frisch, T., Hunt, P.A., 1993. Reconnaissance U–Pb geochronology of the crystalline core of the Boothia Uplift, District of Franklin, Northwest Territories. Paper 93-2, Geological Survey of Canada.

Frisch, T., Sandeman, H.A.I., 1991. Reconnaissance geology of the Precambrian Shield of the Boothia Uplift, northwestern Somerset Island and eastern Prince of Wales Island, District of Franklin. Current Research Part C, Paper 91-1C, Geological Survey of Canada.

Griffin, W.L., Cousens, D.R., Ryan, C.G., Sie, S.H., Suter, G.F., 1989. Ni in chromepyrope garnets: a new geothermobarometer. Contributions to Mineralogy and Petrology 103, 199–202.

Griffin, W.L., et al., 1999. Layered mantle lithosphere in the lac de gras area, Slave craton: composition, structure and origin. Journal of Petrology 40 (5), 705–727.

Handler, M.R., Bennett, V.C., 1999. Behaviour of Platinum-group elements in the subcontinental mantle of Eastern Australia during variable metasomatism and melt depletion. Geochimica et Cosmochimica Acta 63 (21), 3597–3618.

Handler, M.R., Bennett, V.C., Esat, T.Z., 1997. The persistence of off-cratonic lithospheric mantle: Os isotopic systematics of variably metasomatised southeast Australian xenoliths. Earth and Planetary Science Letters 151, 61–75.

Hanmer, S., Bowring, S., vanBreeman, O., Parrish, R., 1992. Great Slave Lake shear zone, NW Canada: mylonitic record of early Proterozoic continental convergence, collision and indentation. Journal of Structural Geology 14, 757–773.

Hart, S.R., Ravizza, G.E., 1996. Os partitioning between phases in lherzolite and basalt. In: Basu, A., Hart, S.R. (Eds.), Earth Processes: Reading the Isotopic Code. American Geophysical Union, Washington, pp. 123–134.

Harte, B., 1977. Rock nomenclature with particular relation to deformation and recrystallisation textures in olivine-bearing xenoliths. Journal of Geology 85, 279–288.

Heaman, L.M., 1989. The nature of the subcontinental mantle from Sr–Nd–Pb isotopic studies on kimberlitic perovskite. Earth and Planetary Science Letters 92, 323–334.

Herzberg, C.T., 1993. Lithosphere peridotites of the Kaapvaal craton. Earth and Planetary Science Letters 120, 13–29.

Hoffman, P.F., 1988. United plates of America, the birth of a continent: early Proterozoic assembly and growth of Laurentia. Annual Reviews Earth and Planetary Science 16, 543–603.

Hoffman, P.F., 1990. Geological constraints on the origin of the mantle root beneath the Canadian shield. Philosophical Transactions of the Royal Society of London. A 331, 523–532.

Irvine, G.J., 2002. Time Constraints on the Formation of Lithospheric Mantle beneath Cratons: A Re–Os Isotope and Platinum Group Element Study of Peridotite Xenoliths from Northern Canada and Lesotho. PhD Thesis, University of Durham, Durham, 384 pp.

Irvine, G.J., et al., 1999. Age of the lithospheric mantle beneath and around the Slave craton: a Rhenium–Osmium isotopic study of peridotite xenoliths from the Jericho and Somerset Island kimberlites. Ninth Annual V.M. Goldschmidt Conf., LPI Cont., vol. 971, pp. 134–135.

Irvine, G.J., et al., 2001a. The age of Two cratons: A PGE and Os-Isotopic study of peridotite xenoliths from the Jericho kimberlite (Slave Craton) and the Somerset Island kimberlite field (Churchill Province). The Slave–Kaapvaal Workshop, Merrickville, Ontario, Canada.

Irvine, G.J., Pearson, D.G., Carlson, R.W., 2001b. Lithospheric mantle evolution of the Kaapvaal Craton: a Re–Os isotope study of peridotite xenoliths from Lesotho kimberlites. Geophysical Research Letters 28 (13), 2505–2508.

Jago, B.C., Mitchell, R.H., 1987. Ultrabasic xenoliths from the Ham kimberlite, Somerset Island, Northwest Territories. Canadian Mineralogist 25, 515–525.

Jenner, G.A., Shirey, S.B., Hyde, D., Elkins, L., Kjarsgaard, B.A., Kerswill, J.A., 2002. Late Archean komatiites and BIFS, Woodburn Lake Group, Churchill Province, Nunavut. Geochimica et Cosmochimica Acta 66 (15), 365.

Keays, R.R., 1995. The role of komatiitic and picritic magmatism and S-saturation in the formation of ore-deposits. Lithos 34, 1–18.

Kjarsgaard, B.A., 1992. Is Ni in pyrope garnet a valid diamond exploration tool? Paper 92-1E, Geological Survey of Canada.

Kjarsgaard, B.A., 1996. Somerset Island kimberlite field, District of Franklin, N.W.T. In: LeCheminant, A.N., Richardson, D.G., DiLabio, R.N.W., Richardson, K.A. (Eds.), Searching for Diamonds in Canada, pp. 61–66. Geological Survey of Canada, Open File.

Kjarsgaard, B.A., Peterson, T.D., 1992. Kimberlite-derived ultramafic xenoliths from the diamond stability field: a new Cretaceous geotherm for Somerset Island, Northwest Territories. Current Research, Part B; Geological Survey of Canada. Paper 92-1B: 1–6.

Kopylova, M.G., Russell, J.K., 2000. Chemical stratification of cratonic lithosphere: constraints from the Northern Slave craton, Canada. Earth and Planetary Science Letters 181, 71–87.

Kopylova, M.G., Russell, J.K., Cookenboo, H., 1999. Petrology of peridotite and pyroxenite xenoliths from the Jericho Kimberlite: implications for the thermal state of the mantle beneath the Slave Craton, Northern Canada. Journal of Petrology 40 (1), 79–104.

Le Cheminant, A.N., Heaman, L.M., 1989. Mackenzie igneous events, Canada: middle Proterozoic hotspot magmatism associated with ocean opening. Earth and Planetary Science Letters 96, 38–48.

Le Cheminant, A.N., Roddick, J.C., 1991. U–Pb zircon evidence for widespread 2.6 Ga felsic magmatism in the central district of Keewatin, N.W.T. Radiogenic Age and Isotopic Studies, Report 4, Geological Survey of Canada, Paper 90-2: 91–99.

Lorand, J.P., 1989. Abundances and distribution of Cu–Fe–Ni sulfides, sulfur, copper and platinum-group elements in orogenic-type spinel peridotites of Ariege (Northeastern Pyrenees, France). Earth and Planetary Science Letters 93, 50–64.

Lorand, J.P., 1991. Sulphide petrology and sulphur geochemistry of orogenic lherzolites: a comparative study of the Pyrenean bodies (France) and the Lanzo Massif (Italy). Journal of Petrology, Special Lherzolites Issue, 77–95.

Lorand, J.P., Pattou, L., Gros, M., 1999. Fractionation of platinum-group elements in the upper mantle: a detailed study in pyrenean orogenic lherzolites. Journal of Petrology 40 (6), 957–981.

MacKenzie, J.M., Canil, D., 1999. Composition and thermal evolution of the upper mantle beneath the central Archean Slave Province, NWT, Canada. Contributions to Mineralogy and Petrology 134, 313–324.

McBride, J.S., Lambert, D.D., Greig, A., Nicholls, I.A., 1996. Multi-stage evolution of Australian subcontinental mantle: Re–Os isotopic constraints from Victorian mantle xenoliths. Geology 24, 631–634.

McDonough, W.F., 1990. Constraints on the composition of the continental lithospheric mantle. Earth and Planetary Science Letters 101, 1–18.

McDonough, W.F., Sun, S.S., 1995. The composition of the Earth. Chemical Geology 120, 223–253.

Maxwell, J.A., Teesdale, W.J., Campbell, J.L., 1995. The Guelph PIXE software package II. Nuclear Instruments and Methods in Physics Research, B 95, 407–421.

Meisel, T., Walker, R.J., Morgan, J.W., 1996. The osmium isotopic composition of the Earth's primitive upper mantle. Nature 383, 517–520.

Menzies, A.H., Shirey, S.B., Carlson, R.W., Gurney, J.J., 1999. Re–Os isotope systematics of diamond-bearing eclogites and peridotites from Newlands kimberlite. In: Gurney, J.J., Gurney, J.L., Pascoe, M.D., Richardson, S.H. (Eds.), Proceedings of the VIIth International Kimberlite Conference. Red Roof Design, Cape Town, pp. 579–581.

Mitchell, R.H., 1975. Geology, magnetic expression, and structural control of the central Somerset Island kimberlites. Canadian Journal of Earth Science 12, 757–764.

Mitchell, R.H., 1978. Garnet lherzolites from Somerset Island, Canada and aspects of the nature of perturbed geotherms. Contributions to Mineralogy and Petrology 67, 341–347.

Mitchell, R.H., 1987. Mantle-derived xenoliths in Canada. In: Nixon, P.H. (Ed.), Mantle Xenoliths. Wiley, Chichester, pp. 33–40.

Mitchell, R.H., Fritz, P., 1973. Kimberlite from Somerset Island, District of Franklin, NWT. Canadian Journal of Earth Science 10, 384–393.

Mitchell, R.H., Keays, R.R., 1981. Abundance and distribution of gold, palladium and iridium in some spinel and garnet lherzolites: implications for the nature and origin of precious metal-rich intergranular components in the upper mantle. Geochimica et Cosmochimica Acta 45, 2425–2442.

Morgan, J.W., 1986. Ultramafic xenoliths: clues to earth's late accretionary history. Journal of Geophysical Research 91 (B12), 12375–12387.

Morgan, J.W., Wandless, G.A., Petrie, R.K., Irving, A.J., 1981. Composition of the Earth's upper mantle: I. Siderophile trace elements in ultramafic nodules. Tectonophysics 75, 47–67.

Nixon, P.H., 1987. Kimberlitic xenoliths and their cratonic setting. In: Nixon, P.H. (Ed.), Mantle Xenoliths. Wiley, Chichester, pp. 215–239.

Okulitch, A.V., Packard, J.J., Zolnai, A.I., 1986. Evolution of the Boothia Uplift, Arctic Canada. Canadian Journal of Earth Science 23, 350–358.

O'Neill, H.S.C., 1991. The origin and the early history of the Earth—a chemical model: Part 2. The earth. Geochimica et Cosmochimica Acta 55, 1159–1172.

Pattou, L., Lorand, J.P., Gros, M., 1996. Non-chondritic platinum-group element ratios in the Earth's mantle. Nature 379, 712–715.

Peach, C.L., Mathez, E.A., 1996. Constraints on the formation of PGE deposits in igneous rocks. Economic Geology, 439–450.

Pearson, D.G., Woodland, S.J., 2000. Solvent extraction/anion exchange separation and determination of PGEs (Os, Ir, Pt, Pd, Ru) and Re–Os isotopes in geological samples by isotope dilution ICP-MS. Chemical Geology 165, 87–107.

Pearson, D.G., Carlson, R.W., Shirey, S.B., Boyd, F.R., Nixon, P.H., 1995a. The stabilisation of Archaean lithospheric mantle: a Re–Os isotope study of peridotite xenoliths from the Kaapvaal craton. Earth and Planetary Science Letters 134, 341–357.

Pearson, D.G., et al., 1995b. Re–Os, Sm–Nd and Rb–Sr isotope evidence for thick Archaean lithospheric mantle beneath the Siberia craton modified by multi-stage metasomatism. Geochimica et Cosmochimica Acta 59, 959–977.

Pearson, D.G., Shirey, S.B., Harris, J.W., Carlson, R.W., 1998. A Re–Os isotope study of sulfide diamond inclusions from the Koffiefontein kimberlite, S.Africa: constraints on diamond crystallisation ages and mantle Re–Os systematics. Earth and Planetary Science Letters 160, 311–326.

Pearson, N.J., Allard, O., Griffin, W.L., Graham, S., Jackson, S.E., 2000. LAM-MCICPMS analysis of mantle derived sulfides: the key to Re–Os systematics of mantle peridotites. Journal of Conference Abstracts 5 (2), 777.

Pearson, D.G., Irvine, G.J., Carlson, R.W., Kopylova, M.G., Ionov, D.A., 2002. The development of lithospheric keels beneath the earliest continents: time constraints using PGE and Re–Os systematics. In: Fowler, C.M.R., Ebinger, C.J., Hawkesworth, C.J. (Eds.), The Early Earth: Physical, Chemical and Biological Development. Special Publications, vol. 199. Geological Society, London, pp. 65–90.

Pearson, D.G., Canil, D.C., Shirey, S.B., in press. Mantle xenoliths and diamonds. Treatise on Geochemistry, vol. 2: The Mantle and Core. Elsevier.

Pehrsson, S.J., Jenner, G.A., Kjarsgaard, B.A., 2002. The Ketyet River Group: correlation with Paeoproterozoic supracrustal sequences of northeastern Rae and implications for Proterozoic orogenesis in the Churchill Province. Geological Association of Canada, Saskatoon, p. 90. Program with Abstracts, 27.

Pell, J.P., 1993. New kimberlite discoveries on Somerset Island. In: Goff, S.P. (Ed.), Exploration Overview 1993. NWT Geology Division, Department of Indian and Northern Affairs, Yellowknife, p. 47.

Percival, J.A., 1996. Archean cratons. In: Le Cheminant, A.N., Richardson, D.G., DiLabio, R.N.W., Richardson, K.A. (Eds.), Searching for Diamonds in Canada, pp. 161–169. Geological Survey of Canada, Open File.

Peslier, A.H., Reisberg, L., Ludden, J., Francis, D., 2000. Re–Os constraints on harzburgite and lherzolite formation in the lithospheric mantle: a study of northern Cordillera xenoliths. Geochimica et Cosmochimica Acta 64 (17), 3061–3071.

Reisberg, L., Lorand, J.P., 1995. Longevity of sub-continental mantle lithosphere from osmium isotope systematics in orogenic peridotite massifs. Nature 376, 159–162.

Reisberg, L.C., Allegre, C.J., Luck, J.-M., 1991. The Re–Os systematics of the Ronda ultramafic complex in southern Spain. Earth and Planetary Science Letters 105, 196–213.

Ryan, C.G., Griffin, W.L., Pearson, N.L., 1996. Garnet geotherms: pressuretemperature data from Cr-pyrope garnet xenocrysts in volcanic rocks. Journal of Geophysical Research 101 (B3), 5611–5625.

Schmidberger, S.S., Francis, D., 1999. Nature of the mantle roots beneath the North American craton: mantle xenolith evidence from Somerset Island kimberlites. Lithos 48, 195–216.

Simon, N.S.C., Irvine, G.J., Davies, G.R., Pearson, D.G., Carlson, R.W., 2003. The origin of garnet and clinopyroxene in "depleted" Kaapvaal peridotites. Lithos 71, 289–322 (this issue).

Stewart, W.D., 1987. Late Proterozoic to Early Tertiary stratigraphy of Somerset Island and northern Boothia Peninsula, District of Franklin, Paper 83-26. Geological Survey of Canada.

Tella, S., 2001. MacQuoid Lake–Gibson lake–Akunak Bay area, Nunavut. Geological Survey of Canada, Map 2008A, scale 1:100,000.

van Breeman, O.J.B.H., Loveridge, W.D., 1987. U–Pb zircon and monazite geochronology and zircon morphology of granulites and granite from the Thelon Tectonic Zone, Healey Lake and Artillery Lake map areas. Current Research Part A, Paper 87-1A, Geological Survey of Canada.

Walker, R.J., Carlson, R.W., Shirey, S.B., Boyd, F.R., 1989. Os, Sr, Nd, and Pb isotope systematics of southern African peridotite xenoliths: Implications for the chemical evolution of subcontinental mantle. Geochimica et Cosmochimica Acta 53, 1583–1595.

Wheeler, et al., 1996. Geological Map of Canada. Geological Survey of Canada, Map 1860A.

Woodland, S.J., Pearson, D.G., Thirlwall, M.F., 2002. A platinum group element and Re–Os isotope investigation of siderophile element recycling in subduction zones: comparison of Grenada, Lesser Antilles Arc, and Izu–Bonin Arc. Journal of Petrology 43, 171–198.

Available online at www.sciencedirect.com

ELSEVIER

Lithos 71 (2003) 489–503

LITHOS

www.elsevier.com/locate/lithos

Peridotitic diamonds from the Slave and the Kaapvaal cratons—similarities and differences based on a preliminary data set

Thomas Stachel[a,*], Jeff W. Harris[b], Ralf Tappert[a,c], Gerhard P. Brey[c]

[a] *Department of Earth and Atmospheric Sciences, University of Alberta, Edmonton, Alberta, Canada T6G 2E3*
[b] *Division of Earth Sciences, University of Glasgow, Glasgow G12 8QQ, UK*
[c] *Institut für Mineralogie, Universität Frankfurt, 60054 Frankfurt, Germany*

Abstract

A comparison of the diamond productions from Panda (Ekati Mine) and Snap Lake with those from southern Africa shows significant differences: diamonds from the Slave typically are un-resorbed octahedrals or macles, often with opaque coats, and yellow colours are very rare. Diamonds from the Kaapvaal are dominated by resorbed, dodecahedral shapes, coats are absent and yellow colours are common. The first two features suggest exposure to oxidizing fluids/melts during mantle storage and/or transport to the Earth's surface, for the Kaapvaal diamond population.

Comparing peridotitic inclusions in diamonds from the central and southern Slave (Panda, DO27 and Snap Lake kimberlites) and the Kaapvaal indicates that the diamondiferous mantle lithosphere beneath the Slave is chemically less depleted. Most notable are the almost complete absence of garnet inclusions derived from low-Ca harzburgites and a generally lower Mg-number of Slave inclusions.

Geothermobarometric calculations suggest that Slave diamonds originally formed at very similar thermal conditions as observed beneath the Kaapvaal (geothermal gradients corresponding to 40–42 mW/m² surface heat flow), but the diamond source regions subsequently cooled by about 100–150 °C to fall on a 37–38 mW/m² (surface heat flow) conductive geotherm, as is evidenced from touching (re-equilibrated) inclusions in diamonds, and from xenocrysts and xenoliths. In the Kaapvaal, a similar thermal evolution has previously been recognized for diamonds from the De Beers Pool kimberlites. In part very low aggregation levels of nitrogen impurities in Slave diamonds imply that cooling occurred soon after diamond formation. This may relate elevated temperatures during diamond formation to short-lived magmatic perturbations.

Generally high Cr-contents of pyrope garnets (inside and outside of diamonds) indicate that the mantle lithosphere beneath the Slave originally formed as a residue of melt extraction at relatively low pressures (within the stability field of spinelperidotites), possibly during the extraction of oceanic crust. After emplacement of this depleted, oceanic mantle lithosphere into the Slave lithosphere during a subduction event, secondary metasomatic enrichment occurred leading to strong re-enrichment of the deeper (>140 km) lithosphere. Because of the extent of this event and the occurrence of lower mantle diamonds, this may be related to an upwelling plume, but it may equally just reflect a long term evolution with lower mantle diamonds being transported upwards in the course of "normal" mantle convection.
© 2003 Elsevier B.V. All rights reserved.

Keywords: Diamond; Inclusion; Peridotite; Craton; Kaapvaal; Slave; Geotherm

* Corresponding author.
E-mail address: tstachel@ualberta.ca (T. Stachel).

0024-4937/$ - see front matter © 2003 Elsevier B.V. All rights reserved.
doi:10.1016/S0024-4937(03)00127-0

1. Introduction

The predominantly peridotitic mantle roots of cratonic cores such as the Kaapvaal or the Slave originate in the early (>3 Ga) to late Archean (deWit, 1998; Carlson et al., 1999b; Griffin et al., 1999a; Pearson, 1999; Bleeker, 2003). Then, as they remained open systems, they are expected to reflect a complex history, involving stages of cooling and heating (plume events), partial melting and infiltration by melts and fluids. In consequence, xenoliths of cratonic peridotites are probably unreliable witnesses to the early stages of the formation of mantle lithosphere.

Diamonds, which formed in these cratonic garnet peridotites, are also assumed to be generally of Archean age (Richardson et al., 1984, 2001; Pearson and Shirey, 1999) some exceeding 3 Ga. In contrast to xenoliths, the syngenetic inclusions in such diamonds have been encapsulated in a closed system for that entire period. Therefore, they represent important samples with which to study the early evolution of the subcratonic mantle lithosphere through time. In addition, diamonds can record the PT-conditions at the time of their formation through non-touching (i.e. separated) inclusions, which have remained chemically isolated and thus unable to adjust chemical equilibria to changing physical conditions. Therefore, comparing diamonds and their inclusions from the Slave and the Kaapvaal cratons will supplement and clarify the information collected using mantle xenoliths and will aid to clearly identify Archean signatures as opposed to later overprint.

Compared to the Kaapvaal, where diamonds have been mined for more than a century, mining activity on the Slave only just commenced with the opening of the Ekati Mine (Fig. 1) in 1998. As a consequence, only few systematic studies on diamonds and their inclusions from the Slave craton have been carried out so far. Davies et al. (1999) studied inclusions in diamonds from the southeast side of Lac de Gras (pipe DO27 of the Tli Kwi Cho kimberlite complex, see Doyle et al., 1999) and Chinn et al. (1998) examined the production of the Misery, Sable and Jay diatremes, which are part of the Ekati Mine, located about 25 km north of Lac de Gras (see Carlson et al., 1999a). In addition to these published data for the central Slave, we present new analyses from the Panda diatreme of the Ekati mine. For the Southern Slave (Fig. 1), Pokhilenko et al. (2001) studied inclusions in diamonds from the Snap Lake kimberlite dike (near Camsell Lake, 220 km NE of Yellowknife); no inclusion data are currently available for the Northern Slave.

1.1. Database

We examined 132 inclusions from 88 Panda diamonds and observed inclusion abundances are given in Table 1. For the purpose of the present study, diamonds containing eclogitic, sub-lithospheric (lower mantle) or epigenetic inclusions are not considered, excluding less than 20% of the Panda diamonds (see Table 1). A preliminary "Slave" data base was established by addition of published analyses of peridotitic inclusions from DO27 (12 inclusions from 11 diamonds, Davies et al., 1999) and Snap Lake (20 inclusions from 14 diamonds, Pokhilenko et al., 2001).

The data set for the Kaapvaal comprises inclusion analyses from Bellsbank, Bultfontein, De Beers Pool, Finsch, Jagersfontein, Koffiefontein, Letlakane, Letseng-la-Terai, Monastery, Premier, Roberts Victor and Wesselton (Deines et al., 1984, 1987, 1989, 1991, 1997; Deines and Harris, 1995; Gurney et al., 1979, 1984, 1985; McDade and Harris, 1999; Moore and Gurney, 1989; Phillips and Harris, 1995; Rickard et al., 1989; Wilding et al., 1994).

1.2. Analytical techniques

Inclusions were released by breakage of the host diamond in a steel crusher. The inclusions were embedded in Araldite® and polished on a Pb–Sb plate using 0.25 μm diamond powder. Major and trace element compositions (Table 2) were obtained by electron microprobe analysis (Jeol JXA-8900 RL) at 20 kV gun potential and 20 nA beam current using silicate, oxide and metal standards. Count times ranged between 30 and 100 s, and three analytical points were averaged to ensure detection limits of 100 ppm or better for all oxides except Na_2O (200 ppm). A spectrometer with small Rowland circle (100 mm radius) and a LIF crystal with Johannson type geometry was used for Mn, Fe, Ni and Zn, resulting in ca. three times higher count rates compared to a standard

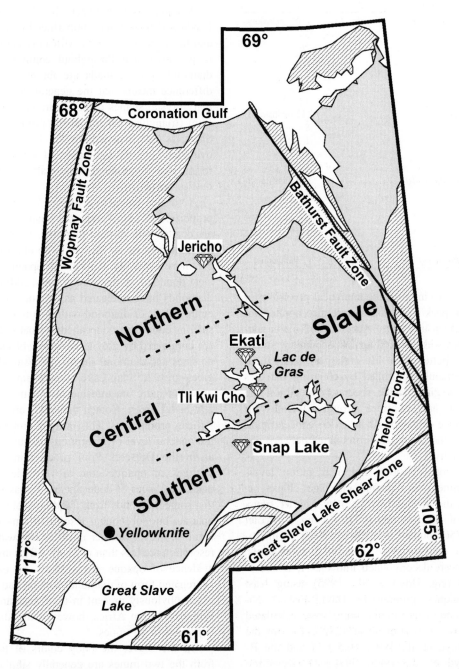

Fig. 1. Location of sample sources (Ekati, Tli Kwi Cho and Snap Lake) in the Slave Province. Subdivisions for the Slave are from Grütter et al. (1999) and are based on compositional differences of xenocrystic garnet derived from the lithospheric mantle. The diamondiferous Jericho kimberlite (Northern Slave) is shown with reference to the geothermobarometric work of Kopylova et al. (1999) discussed in the text. Boundaries between the Slave Craton (light grey) and Proterozoic mobile belts and Proterozoic and Phanerozoic cover sequences (hatched) are taken from Bleeker (2003).

Table 1
Inclusion abundance table for the 88 diamonds (132 inclusions) from Panda studied

Peridotitic (72 diamonds)	(103 inclusions)
Pyrope garnet	30
Olivine	26
Cr-diopside	8
Enstatite	2
Chromite	18
NiFe-sulfide	19
Eclogitic (7 diamonds)	(12 inclusions)
Pyr-alm garnet	8
Omphacitic cpx	1
Rutile	1
Fe-sulfide	2
Lower mantle (4 diamonds)	(10 inclusions)
Ferropericlase	4
CaSi-perovskite	1
Olivine	3
SiO_2	1
$MgAl_2O_4$	1
Epigenetic (5 diamonds)	(7 inclusions)

configuration. To improve the analytical precision for Zn in spinel, peak count times were also applied for background measurements. Accuracy of major element analyses was checked against secondary standards and is better than 1% (relative). Analytical precision is mainly controlled by counting statistics (which, for a given oxide species, depend on the concentration) and therefore strongly declines for minor and trace elements. At an oxide concentration of 0.10 wt.%, the two sigma errors range between 4% and 15% (relative).

Nitrogen concentrations and aggregation levels were determined on diamond cleavage chips (i.e. not through the diamond surface) by Fourier transform infrared spectroscopy (FTIR) using a Nicolet Magna IR 550 coupled to a Spectra-Tech IR-microscope. After conversion to absorption coefficients, the spectra were deconvoluted into the A, B and D components (e.g. Boyd et al., 1995) using least square techniques (program by David Fisher). Nitrogen concentrations (atomic ppm) were calculated using the absorption strength at 1282 cm^{-1} for the A-centre (Boyd et al., 1994, 16.5 ± 1) and the B-centre (Boyd et al., 1995, 79.4 ± 8). Detection limits and errors strongly depend on the quality of the cleavage chip, but typically range between 10–20 ppm and about 10–20% of the concentration respectively.

2. Diamond characteristics

A comparison between the diamonds from the Panda and Snap Lake kimberlites in the Slave craton and the general southern African (Kaapvaal) productions shows that throughout common sizes ranges diamonds from Canada are far less resorbed. This difference means that the general dominance of primary crystal shapes of octahedra and the spinel twin form (triangular macle), in the productions from the Slave, is not matched on the Kaapvaal, where an overwhelming predominance of secondary diamond forms like dodecahedra and flattened dodecahedral macles occurs.

A further distinctive feature is that a significant proportion of both Panda and Snap Lake productions are coated with impure diamond. In the larger sizes, the coat may be so thin as to be almost transparent, but more generally, it forms an opaque cover over a clear diamond. In the smaller sizes, additionally, some diamonds have nucleated as opaque cubes, interpenetrant cubes, or diamond with a more complex shape, such as cubo-octa-dodecahedra. Coat colors at Panda are dominated by grey to dark grey/black, throughout all sizes and diamond shapes. Occasionally, some are green-grey. At Snap Lake, some opaque coats are also grey, but more are translucent/opaque yellow, with a few being green. Coated diamonds are rare in South African productions. This rarity is probably related to the general level of resorption, because occasionally, as in the DeBeers Pool production for example, vestiges of opaque coat are recorded on residual octahedral faces of dodecahedra. It may be, therefore, that some diamonds from the Kaapvaal were similarly opaque, but subsequent resorption removed this material. However, further north in Botswana, levels of resorption are less than in South Africa and there is no evidence of opaque coats. Boart, an opaque, often aggregated form of diamond, with no particular shape, is not obviously present in the Canadian production. Within southern Africa, however, this diamond form is present in all mines.

On the Slave, the body colors of the diamonds from the two mines are generally similar, especially once the dark diamonds from these mines (consisting of colorless diamond full of graphite inclusions) have been removed. At Panda, brown is the dominant color throughout all size ranges, followed by colorless. The

Table 2
Electron microprobe analyses of peridotitic inclusions in diamonds from Panda

Sample mineral T_{OW79} (°C) paragen.	PA-1b (1) garnet 1043 gt, gt-ol	PA-1b (2) olivine	PA-3a garnet gt	PA-5a (1) garnet 1076	PA-5a (2) olivine gt-ol-opx	PA-5a (3) opx	PA-13a garnet gt, chr	PA-13b chromite	PA-19a (1) garnet 1169	PA-19a (2) opx gt-opx, ol	PA-19b olivine
P_2O_5	0.01	≤ 0.01	0.01	0.02	≤ 0.01	≤ 0.01	0.04	≤ 0.01	0.05	≤ 0.01	≤ 0.01
SiO_2	41.31	41.11	40.79	41.71	41.13	57.90	41.02	0.20	41.80	57.94	40.93
TiO_2	0.02	≤ 0.01	0.03	0.08	≤ 0.01	0.02	0.09	0.16	0.07	0.02	≤ 0.01
Al_2O_3	16.25	0.01	13.21	16.83	0.02	0.40	13.93	4.12	16.82	0.45	0.02
Cr_2O_3	10.45	0.12	13.66	9.63	0.21	0.34	12.69	65.72	9.08	0.38	0.05
FeO	6.37	6.87	5.59	6.45	7.08	4.13	6.71	16.68	6.04	3.79	7.19
MnO	0.31	0.09	0.28	0.32	0.10	0.10	0.37	0.28	0.29	0.09	0.11
NiO	≤ 0.01	0.39	0.01	0.01	0.40	0.11	0.02	0.10	0.01	0.14	0.35
MgO	20.79	50.75	21.67	21.45	50.40	35.57	19.36	12.84	20.50	35.95	50.65
CaO	4.67	0.03	3.71	3.93	0.04	0.34	6.14	≤ 0.01	5.96	0.54	0.04
Na_2O	0.31	0.23	0.02	≤ 0.02	≤ 0.02	≤ 0.02	≤ 0.02	≤ 0.02	≤ 0.02	≤ 0.02	≤ 0.02
K_2O	≤ 0.01	≤ 0.01	≤ 0.01	≤ 0.01	≤ 0.01	≤ 0.01	≤ 0.01	0.08	≤ 0.01	≤ 0.01	≤ 0.01
Total	100.50	99.60	99.00	100.42	99.37	98.89	100.37	100.15	100.60	99.30	99.33

Sample mineral T_{OW79} (°C) paragen.	PA-23a garnet 1214 gt, ol	PA-23b olivine	PA-24a chromite 2chr	PA-26a chromite chr	PA-29a chromite chr	PA-31b chromite 2chr	PA-40a olivine 1196 gt, ol	PA-40b garnet	PA-42a olivine 1118 gt, ol	PA-42b garnet
P_2O_5	0.01	≤ 0.01	≤ 0.01	≤ 0.01	≤ 0.01	≤ 0.01	≤ 0.01	0.03	≤ 0.01	0.01
SiO_2	41.84	40.87	0.12	0.10	0.09	0.10	40.85	42.40	40.87	42.12
TiO_2	0.16	≤ 0.01	0.04	0.10	0.22	0.10	≤ 0.01	0.13	≤ 0.01	0.03
Al_2O_3	16.94	0.03	5.74	7.25	7.20	7.16	0.02	16.30	0.02	15.02
Cr_2O_3	8.14	0.07	65.77	63.90	63.73	63.99	0.06	8.59	0.04	10.74
FeO	6.10	7.44	16.23	15.01	14.87	15.02	7.13	5.95	7.57	6.61
MnO	0.29	0.11	0.30	0.25	0.25	0.25	0.10	0.28	0.11	0.34
NiO	0.01	0.34	0.06	0.10	0.10	0.09	0.35	0.01	0.38	0.01
MgO	20.75	50.09	11.60	13.73	13.82	13.79	50.20	20.72	50.06	20.56
CaO	5.51	0.05	≤ 0.01	≤ 0.01	≤ 0.01	≤ 0.01	0.04	5.79	0.04	5.21
Na_2O	≤ 0.02	≤ 0.02	≤ 0.02	≤ 0.02	0.04	≤ 0.02	≤ 0.02	0.03	≤ 0.02	≤ 0.02
K_2O	≤ 0.01	≤ 0.01	≤ 0.01	≤ 0.01	≤ 0.01	≤ 0.01	≤ 0.01	≤ 0.01	≤ 0.01	≤ 0.01
Total	99.75	99.00	99.85	100.44	100.32	100.50	98.75	100.24	99.09	100.65

T_{OW79} refers to temperatures (at $P = 5$ GPa) calculated from the Mg–Fe exchange between garnet and olivine (O'Neill and Wood, 1979; O'Neill, 1980).

brown color is linked to high levels of plastic deformation (see Harris, 1992). Yellow diamonds are extremely rare in the production. At Snap Lake, colorless dominates over brown plastically deformed diamonds and yellow diamonds are extremely rare. Additionally, at Snap Lake, diamonds with transparent green coats are common. The green transparent coat is quite distinct from the opaque or translucent coats discussed above and results from radiation damage (see Vance et al., 1973). Its absence in the Panda production probably relates to the much younger age of kimberlite emplacement (Snap Lake: 523 Ma,

Geospec Consultants, 1999; Panda: 53.2 Ma, Carlson et al., 1999a).

In Kaapvaal productions principal diamond colors are colorless, yellow and brown. If mining is within the upper oxidized part of a kimberlite, then diamonds with transparent green coats are a common feature. For a more detailed assessment of color in diamonds from southern Africa, see Harris (1992) and references therein.

Yellow in diamond results from the impurity nitrogen, occurring in the so-called N_3 configuration (three nitrogens and a vacancy). Its virtual absence in Canadian diamonds means that the N_3 centre is not a significant by-product of the more common nitrogen aggregation from the A-centre (a nitrogen pair) to the B-centre (four nitrogens and a vacancy) (Evans, personal communication, but also see Collins, 2001, for a detailed discussion of color in diamonds). Thus, at present, there is a clear color distinction between diamonds from Canada and southern Africa.

3. Inclusion compositions

Comparing the composition of pyrope garnet inclusions from the southern and central Slave and the Kaapvaal in a Ca–Cr plot (Fig. 2) shows that a low-Ca harzburgitic garnet population, which is characteristic for the Kaapvaal, is absent for the Slave samples. On average, Slave garnet inclusions are higher in Ca and Cr and many harzburgitic garnets plot close to the lherzolitic trend, which suggests that their source rocks were near saturation

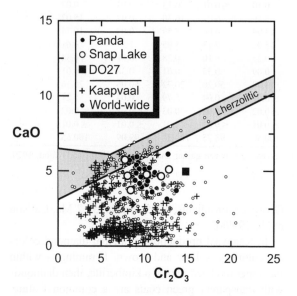

Fig. 2. Diagram of CaO versus Cr_2O_3 (wt.%) in garnet with the lherzolite field (shaded area) as defined by Sobolev et al. (1973). As in all following diagrams, data for DO27 and Snap Lake are taken from Davies et al. (1999) and Pokhilenko et al. (2001), respectively, references for Kaapvaal diamonds are given in the text and sources for the data base on inclusions from mines world-wide are given in Stachel and Harris (1997) and Stachel et al. (2000).

in clinopyroxene. A Ca–Cr plot for garnet inclusions from Misery, Sable and Jay in Chinn et al. (1998, their Fig. 3) suggests compositional similarity to Panda and Snap Lake but indicates lower Cr_2O_3 (4–6 wt.%) for the lherzolitic inclusion paragenesis observed at Sable and reveals the presence of a minor low-Ca harzburgitic suite (2 out of 18 garnets have $Cr_2O_3 < 2$ wt.%).

A comparison of the Mg-number of garnet inclusions from Panda and Snap Lake with the Kaapvaal and other diamond sources world-wide also indicates a higher fertility for the Slave (Fig. 3). However, the Mg-number of garnet is not only a function of bulk composition but is also strongly controlled by the temperature dependent Mg–Fe exchange with olivine, orthopyroxene, clinopyroxene and Mg-chromite. If the low Mg-number of Slave garnets only indicates formation at unusually low temperatures, then olivine and orthopyroxene should show elevated Mg-numbers. However, Fig. 4 shows that the Mg-number of Slave olivine inclusions perfectly overlaps with the world-wide data set and falls on the Mg-poor side of the Kaapvaal distribution. Including 43 inclusions from Misery, Jay and Sable (Chinn et al., 1998), olivine inclusions from the Slave show a sharp mode in Mg-number in class 92–93, as opposed to a mode in class 94–95 for olivines from the Kaapvaal. Only nine orthopyroxene inclusions have been recovered for Panda, DO27 and Snap Lake but they all have a Mg-number < 94.5, which is clearly distinct from the Kaapvaal (mode in Mg-number in class 95–96). An interesting feature in Fig. 4 is the clear chemical distinction between Panda and DO27 olivine, with Panda being high and DO27 being low in Ni. In contrast, Ni contents of DO27 orthopyroxenes are normal (not illustrated). This suggests that the low Ni contents in olivine inclusions probably reflect the modal ol/opx ratio rather than bulk Ni contents, with the source regions of DO27 being orthopyroxene poor compared to Panda.

The composition of chromite inclusions is shown in Fig. 5 and again Panda and Snap Lake (no chromites have been observed at DO27) on average fall on the Fe-rich side of the Kaapvaal data. In terms of ferric iron content the available chromite data appear to fall into two groups, one (five chromites from Panda, ferric iron number of 6.7–9.2) being

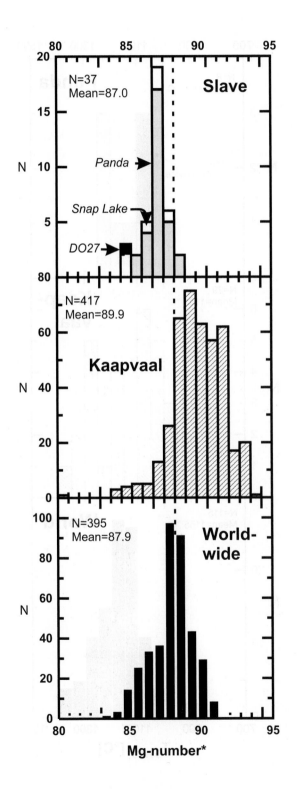

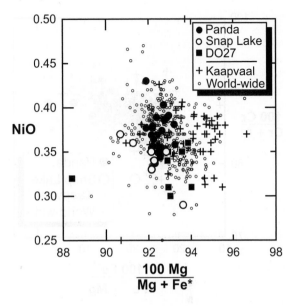

Fig. 4. NiO (wt.%) versus molar Mg-number for olivine inclusions in diamonds.

similar to the reduced Mg-chromites from the Kaapvaal, the other (15 inclusions from Panda, 5 inclusions from Snap Lake, ferric iron number of 16.0–24.5) being slightly more oxidized.

For Snap Lake, Pokhilenko et al. (2001) report two knorringite rich (Cr_2O_3 of 11.8 and 12.8 wt.%) garnet inclusions with a significant majorite component (6.23 and 6.33 cations Si at [O] = 24). From the experimental work of Irifune (1987), this may be interpreted to suggest a continuation of the subcratonic lithospheric mantle beneath the southern Slave to about 300–350 km depth, at least up to the emplacement of the Snap Lake dike in the Cambrian. Apart from this unique feature, the inclusion compositions show that diamonds from the Slave are derived from a chemical environment that is chemically less depleted than the

Fig. 3. Histograms showing the Mg-number (molar Mg/(Mg + Fe*)) of garnet inclusions in diamonds from the Slave, Kaapvaal and other cratonic areas. The effect of Ca on the Mg–Fe partitioning between garnet and olivine was eliminated by recalculating all garnets to a Ca-free composition. From the data of O'Neill and Wood (1979), it may be estimated that at constant temperature the Mg-number of garnet decreases by 2.0 per cation Ca (based on a formula with 24 oxygens) for the temperature range (1100–1200 °C, 5 GPa) and olivine composition (Mg-number around 93) relevant here.

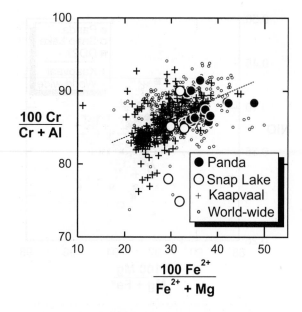

Fig. 5. Cr-number versus Fe-number (molar ratios) for Mg-chromite inclusions. Fe^{2+} is estimated from total iron contents (EPMA) using the equation of Droop (1987). For crystallo-chemical reasons, Fe/Mg and Cr/Al ratios in spinel are positively correlated. The fact that a number of inclusions from Panda and Snap Lake plot below the world-wide trend thus implies a source with elevated Fe/Mg (as unusually low temperatures may be excluded).

Kaapvaal but very similar to other diamond sources world-wide.

4. Geothermobarometry

Equilibration temperatures based on the Mg–Fe exchange between olivine and garnet (O'Neill and Wood, 1979; O'Neill, 1980) were obtained for eight Panda diamonds (see Table 2). Calculated for an assumed pressure of 5 GPa, temperatures range from 1060 to 1210 °C, with an average of 1140 °C (Fig. 6). Garnet–olivine temperatures of 1180 °C reported by Chinn et al. (1998) for two more diamonds from the Ekati mine (Misery, Sable or Jay) and of 1130 °C calculated for diamond SL$_3$-31/00 from Snap Lake

Fig. 6. Histograms showing calculated equilibration temperatures (at $P = 5$ GPa) for coexisting garnet and olivine (O'Neill and Wood, 1979; O'Neill, 1980) inclusions in diamonds from the Slave (only Panda), Kaapvaal and other cratonic areas. The two Panda diamonds, which contain touching pairs of garnet–olivine, fall on the low-temperature end and yield temperatures of about 1070 °C.

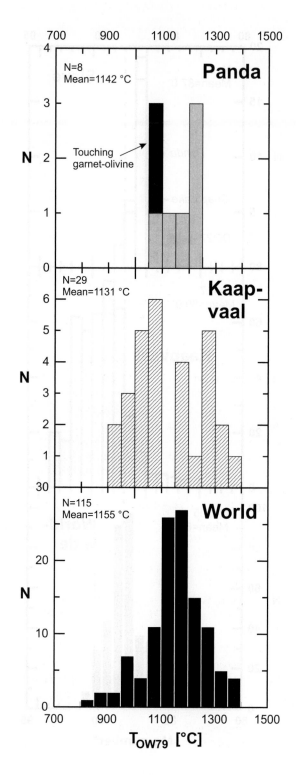

Table 3
Zn contents in Mg-chromite inclusions and corresponding temperatures (Ryan et al., 1996)

Sample	ZnO (ppm)	T-Zn (°C)
PA-13	433	1287
PA-20	429	1293
PA-24	821	962
PA-25	423	1303
PA-26	498	1202
PA-27	505	1194
PA-28	713	1021
PA-29	458	1252
PA-30	500	1200
PA-31	551	1146
PA-32	550	1147
PA-33	715	1020
PA-34	630	1079
PA-35	595	1107
Average		1158

(Pokhilenko et al., 2001) are in good agreement with the Panda data. These results for the Slave are similar to the Kaapvaal (29 diamonds with an average of 1130 °C) and also to data from world-wide sources ($N = 115$), which form a normal distribution with a mode and average of approximately 1150 °C (Fig. 6).

The Zn content of Mg-chromite inclusions indicates very similar temperatures (calculated after Ryan et al., 1996), resulting in averages of 1150 °C for both Panda (see Table 3) and world-wide sources. From all these estimates, it appears that diamond formation at the Slave and Kaapvaal cratons and in other cratonic areas world-wide took place in a very similar geothermal regime.

Applying the relationship between crustal heat production and heat transported from the mantle as established by Pollack and Chapman (1977), steady state palaeo-geotherms of about 40–42 mW/m² (surface heat flow) have been calculated for the formation of diamonds by a number of authors (e.g. Boyd and Gurney, 1986; Griffin et al., 1992; Stachel and Harris, 1997). Kopylova et al. (1999) and Russell and Kopylova (1999) demonstrated that crustal heat production in the Slave exceeds the values observed in other cratons (including the Kaapvaal), leading to significantly higher surface heat flow. However, to compare palaeo-geotherms in the lithospheric mantle, it is common practice to ignore differences in crustal heat production and to look at apparent surface heat flow values instead, with a fixed amount of 60% (Pollack and Chapman, 1977) of the surface heat flux coming from the mantle. The very good agreement of the geothermometers applied, indicates that diamond formation beneath the Slave took place at a similar geothermal gradient as in the Kaapvaal, corresponding to about 40–42 mW/m² surface heat flow. This palaeo-geotherm is hotter than the gradients of about 38 mW/m² derived from the study of xenoliths and xenocrysts for the deeper (>140 km) mantle lithosphere at Lac de Gras (Pearson et al., 1999; Griffin et al., 1999a; Cretaceous to Tertiary kimberlites activity) and beneath the Jurassic Jericho kimberlite pipe (Kopylova et al., 1999).

Four diamonds from Snap Lake and Panda contain inclusions of garnet and orthopyroxene, a paragenesis from which pressure and temperature can be calculat-

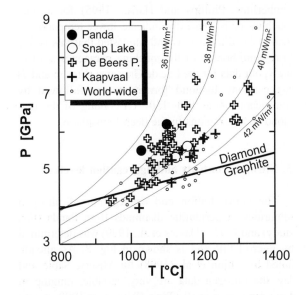

Fig. 7. Pressure and temperature estimates for coexisting inclusions of garnet and orthopyroxene in diamonds from the Slave (Panda and Snap Lake), Kaapvaal and other cratonic areas. Geothermal gradients are calculated after Pollack and Chapman (1977). The diamond graphite transition is from Kennedy and Kennedy (1976). Data points for the De Beers Pool kimberlites are shown separately from the rest of the Kaapvaal data, because they reveal a cooling history from a geotherm corresponding to 40 mW/m² surface heat flow (non-touching inclusion pairs) to about 38 mW/m² for touching pairs (Phillips and Harris, 1995). The two Panda diamonds containing touching pairs of garnet–orthopyroxene also indicate a lower geotherm than deduced from the two non-touching pairs from Snap Lake and garnet–olivine and Zn in spinel thermometry. Thus, re-equilibration of touching inclusions during cooling subsequent to diamond formation is also assumed for the two Panda diamonds.

ed simultaneously, based both on the Al (Brey and
Köhler, 1990) and Mg–Fe exchange (Harley, 1984)
between these two phases. For Snap Lake, the two
data points calculated from analyses SL$_3$-12/00 and
SL$_5$-8/00 (Pokhilenko et al., 2001, their Tables 2 and
3) plot on a 40 mW/m^2 surface heat flow geotherm
(Fig. 7), consistent with the results of garnet–olivine
thermometry. The PT-conditions derived for Panda
from two touching pairs of garnet–orthopyroxene
indicate a lower geothermal gradient corresponding
to about 37 mW/m^2 surface heat flow. This result is
within error of Jurassic to Tertiary palaeo-geotherms
determined from xenolith studies by Kopylova et al.
(1999) and Griffin et al. (1999a). Fig. 7 shows the
results obtained by Phillips and Harris (1995) for
touching and non-touching garnet and orthopyroxene
inclusions in diamonds from the De Beers Pool
kimberlites. Phillips and Harris (1995) found that
touching pairs generally indicate lower temperatures
than isolated inclusions and suggested that this is due
to re-equilibration of the touching parageneses. Similarly, our data for the Lac de Gras area (Figs. 6 and 7)
suggest that diamond formation was followed by
significant (about 100–150 °C) cooling of the diamond source regions in the deep lithosphere.

5. Nitrogen contents and aggregation levels

The concentration and aggregation of nitrogen
impurities in peridotitic diamonds from Panda (this
study) and DO27 (Davies et al., 1999) was determined
by micro-FTIR and is shown in Fig. 8. As already
noted by Chinn et al. (1998) for Misery, Sable and
Jay, the nitrogen data are very variable, ranging in
concentration from 0 (Type II) to over 1000 atomic
ppm and showing a range in nitrogen aggregation
from pure Type IaA to pure Type IaB.

The dependence of nitrogen aggregation on mantle
residence time and temperature was derived by Evans
and Harris (1989) and Taylor et al. (1990). Under
mantle conditions, the aggregation of nitrogen from
single substitutional atoms to pairs ("A-centre")
occurs so fast (Taylor et al., 1996) that it may be
ignored in a thermo-chronometric analysis. The second step, the aggregation from the A-centre to groups
of four nitrogens surrounding a vacancy ("B-centre"),
occurs at a much lower rate and thus, in principle, it is

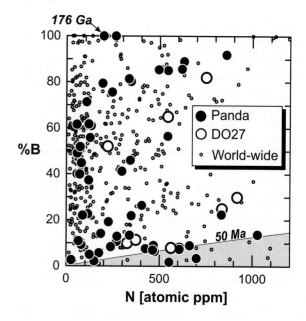

Fig. 8. Aggregation level (percentage of the higher aggregate
B-centre relative to the A-centre) versus concentration (atomic ppm)
of nitrogen impurities in peridotitic diamonds from the Slave (Panda
and DO27) and world-wide sources (including Kaapvaal data). The
grey sector indicates mantle residence times of 0–50 Ma assuming a
residence temperature of 1150 °C, which is consistent with the
results of garnet–olivine and Zn in spinel thermometry for Panda
inclusions. The 50 Ma envelope and the residence time of 176 Ga for
a Type IaB diamond (top of diagram) were calculated using the
activation energy for the conversion of the A- to the B-centre derived
by Taylor et al. (1990).

possible to constrain either the mantle residence time
(given the residence temperature) or the residence
temperature (given the residence time) of a diamond
from the concentration and aggregation of nitrogen
impurities. Estimates of mantle residence temperatures are fairly insensitive to the choice of residence
time (based on values generally >1 Ga) and normally
show satisfactory agreement with independent estimates based on inclusion geothermobarometry (e.g.
Taylor et al., 1995; Leahy and Taylor, 1997).

However, plastic deformation is common in diamonds and for other minerals has been shown to
increase diffusion rates significantly (see Cole and
Chakraborty, 2001 for a recent review). This renders
mantle residence temperatures and residence times
based on nitrogen data maximum estimates. Considering the very strong control of temperature on
nitrogen aggregation, introduced errors will probably

be within the accuracy of conventional geothermobar-ometry (as indicated by the overall agreement of both methods) but the application of nitrogen data as a chronometer may be seriously hampered. However, *maximum* residence times may be safely calculated using this method. Some of the Slave diamonds show medium to high nitrogen concentrations at very low aggregation levels (grey sector in Fig. 8). Maximum residence times for these diamonds are in the range of 10–50 Ma (Fig. 8), if the average temperature of 1150 °C derived from various geothermometers (see above) is used for the calculations. If, however, fairly rapid cooling after diamond formation by about 100 °C is assumed, residence times in excess of 1 Ga are calculated. This is in good agreement with the cooling of the diamond source regions by about 100–150 °C derived from garnet–orthopyroxene geothermobar-ometry and may indicate that diamond formation

was accompanied by transient thermal perturbations (magmatic intrusions). Spatially confined lithospheric heating through magmatic pulses may also be indi-cated by an unusually high paleo-geotherm (equiva-lent to 42 mW/m^2 surface heat flow) derived by MacKenzie and Canil (1999) for the late Creta-ceous–early Paleocene Torrie kimberlite, which is in stark contrast to the 38 mW/m^2 (at $T>900$ °C) observed by Griffin et al. (1999a) for xenocrysts and xenoliths from penecontemporaneous kimberlites at nearby Lac de Gras. Compared to the Lac de Gras area, the elevated paleo-geotherm at Torrie is also accompanied by a highly re-fertilized mantle litho-sphere (Griffin et al., 1999c) providing further evi-dence for a locally confined magmatic event.

Some Panda diamonds show very high aggregation levels which result in un-reasonably long residence times at 1150 °C (Fig. 8). These diamonds obviously

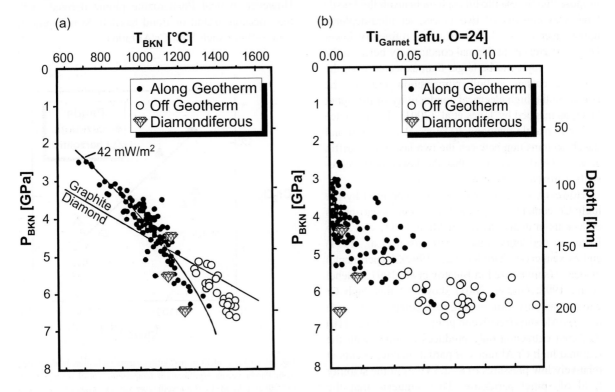

Fig. 9. (a) Pressure (gt-opx) and temperature (cpx-opx) estimates (calculated after Brey and Köhler, 1990) for garnet peridotites occurring as xenoliths in Kaapvaal kimberlites (for details, see Brey, 1989). The majority of xenoliths define a geotherm of about 42 mW/m^2 surface heat flow (Pollack and Chapman, 1977). Between about 5 and 6.5 GPa, a second array of xenoliths exists that indicates significantly hotter conditions. (b) Using Ti in garnet as a measure of chemical fertility, a positive correlation between pressure (i.e. depth) and chemical enrichment becomes apparent. This relationship is particularly prominent for garnets from the xenoliths defining the "hot" array in (a). As firstly suggested by Brey (1989), these hot enriched xenoliths are likely to reflect the infiltration of asthenospheric melts into the base of the subcratonic lithosphere.

were exposed to higher temperatures, possibly implying formation at greater depth or very close to magmatic intrusions.

6. Discussion

Stabilization of the Slave craton is presently dated to about 2.58 Ga (cessation of terminal "granite bloom", Bleeker, 2003) and possibly relates to the formation of a late Archean supercontinent ("Sclavia") including the Dharwar, Wyoming and Zimbabwe cratons (Bleeker, 2003). Re–Os model ages on sulphide inclusions in olivine (Aulbach et al., 2003) indicate that the diamondiferous mantle root beneath the central Slave may be significantly older (>3 Ga), thus implying tectonic emplacement of a pre-existing root at around 2.6 Ga. Griffin et al. (1999a) propose that the mantle lithosphere beneath the Lac de Gras area consists of two layers, an ultra-depleted upper layer (<145 km) and a less refractory lower layer. Differences in thermal conductivity between the two layers appear to be larger than expected from the observed change in bulk composition and may be related to the high electrical conductivity of the upper layer, as mapped by Jones et al. (2001). The presence of lower mantle diamonds (Davies et al., 1999) and the sharp transition between the two layers led Griffin et al. (1999a) to propose that the lower layer represents a frozen plume head.

However, this conclusion is not supported by the high Cr contents of peridotitic garnet occurring as inclusions in diamonds (Ekati Mine, DO27 and Snap Lake), in peridotitic xenoliths (Pearson et al., 1999) and as xenocrysts (Grütter et al., 1999). Experimental evidence (summarized in Bulatov et al., 1991; Stachel et al., 1998; Grütter, 2001) shows (i) that high Cr garnets can only form in bulk rock compositions with high Cr/Al ratios (relative to primitive mantle) and (ii) that melt extraction only produces residues with the required high Cr/Al ratios, if partial melting occurs at relatively low pressures (0.8–2.5 GPa) in the stability field of spinel peridotite. This suggests that the primary depletion event occurred at the Archean equivalent of mid-ocean ridges or island arc settings (see also Grütter and Anckar, this volume). These depleted protoliths were subsequently transferred to garnet–facies depths.

Based on this evidence, we propose that both the upper and the lower layer of the mantle lithosphere beneath Lac de Gras represent depleted former oceanic lithosphere, which became imbricated beneath the Slave during Archean subduction of relatively young and hot slabs (Helmstaedt and Schulze, 1989). The age mismatch between the stabilization of the Slave craton and its deep mantle lithosphere in the Lac de Gras area may imply a more complex tectonic evolution: Aulbach et al. (this volume) suggest thrusting of old mantle lithosphere attached to the central Slave Basement Complex beneath the Contwoyto terrane during collision of these terranes. An alternative explanation, which could reconcile the occurrence of high Cr garnets with the plume model of Griffin et al. (1999a), is derivation of the deep lithosphere from an ascending megalith consisting of accumulated old oceanic slabs (Ringwood, 1991). However, typical Phanerozoic plume derived melts (e.g. oceanic island or flood basalts) do not indicate sources with such high Cr/Al ratios.

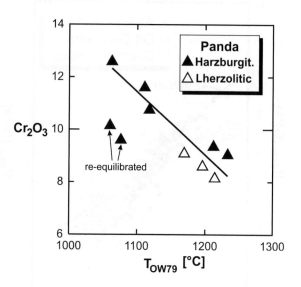

Fig. 10. Cr_2O_3 content in peridotitic garnet inclusions in diamonds from Panda versus equilibration temperature (gt-ol thermometry: O'Neill and Wood, 1979; O'Neill, 1980) as a measure of depth. The two samples with touching inclusions of garnet–olivine were excluded from the regression and are indicated as "re-equilibrated". As shown by Griffin et al. (1999b), Cr in garnet may be employed as a direct proxy for the bulk rock composition of lithospheric garnet peridotites and thus decreasing Cr contents indicate that the diamond sources become less depleted with depth.

Our preferred model, therefore, is a similar (although probably not contemporaneous) origin of the upper and lower layer of the mantle lithosphere beneath Lac de Gras, with the less depleted character of the lower layer being in part a secondary feature, due to metasomatic overprint. Increasing fertility of the sub-cratonic lithosphere with depth is also observed for the Kaapvaal. This is particularly prominent for xenoliths falling on the hot side of the local palaeo-geotherm which show a marked increase in Ti with increasing pressure (Fig. 9). Taking temperature as a measure of depth, similar co-variations are also determined for the Slave, e.g. a significant decrease in the Cr content of garnet (Fig. 10) towards the base of the lithosphere. Re-fertilization of cratonic roots from beneath, through melts or fluids ascending from the asthenosphere, is also shown by Griffin et al. (1999b) for a number of cratons, including the Kaapvaal. The strong metasomatic overprint observed for the deeper lithosphere in the Slave may relate to a plume event, thereby explaining the occurrence of lower mantle diamonds, but the evidence for such an event remains inconclusive.

Acknowledgements

David Fisher (DTC Research Laboratories, Maidenhead) is thanked for his help with the deconvolution of the infrared spectra. The manuscript was considerably improved through formal reviews by Dante Canil, Maya Kopylova and Nick Pokhilenko, and significant input from Herman Grütter as the responsible editor. Funding through the German Research Foundation (DFG), NSERC and the Canada Research Chairs program (CRC) is gratefully acknowledged. Additional support and co-operation was provided by The Diamond Trading, a De Beers Group Company, De Beers Consolidated Mines and BHP-Billiton.

References

Aulbach, S., et al., 2003. Mantle formation and evolution, Slave Craton: constraints from HSE abundances and Re–Os isotope systematics of sulfide inclusions in mantle xenocrysts. Chem. Geol. (in press).

Bleeker, W., 2003. The late Archean record: a puzzle in ca. 35 pieces. Lithos 71 (This issue).

Boyd, F.R., Gurney, J.J., 1986. Diamonds and the African lithosphere. Science 232, 472–477.

Boyd, S.R., Kiflawi, I., Woods, G.S., 1994. The relationship between infrared absorption and the A defect concentration in diamond. Philos. Mag. B 69, 1149–1153.

Boyd, S.R., Kiflawi, I., Woods, G.S., 1995. Infrared absorption by the B nitrogen aggregate in diamond. Philos. Mag. B 72, 351–361.

Brey, G.P., 1989. Geothermobarometry for lherzolites: experiments from 10 to 60 kb, new thermobarometers and application to natural rocks. Habilitation Thesis, TU Darmstadt. 227 pp.

Brey, G.P., Köhler, T., 1990. Geothermobarometry in four-phase lherzolites: II. New thermobarometers, and practical assessment of existing thermobarometers. J. Petrol. 31, 1353–1378.

Bulatov, V., Brey, G.P., Foley, S.F., 1991. Origin of low-Ca, high-Cr garnets by recrystallization of low-pressure harzburgites. Fifth International Kimberlite Conference, Extended Abstracts, CPRM Spec. Publ., vol. 2/91, pp. 29–31.

Carlson, J.A., Kirkley, M.B., Thomas, E.M., Hillier, W.D., 1999a. Recent Canadian kimberlite discoveries. In: Gurney, J.J., Gurney, J.L., Pascoe, M.D., Richardson, S.H. (Eds.), The J.B. Dawson Volume, Proceedings of the VIIth International Kimberlite Conference. Red Roof Design, Cape Town, pp. 81–89.

Carlson, R.W., et al., 1999b. Re–Os systematics of lithospheric peridotites: implications for lithosphere formation and preservation. In: Gurney, J.J., Gurney, J.L., Pascoe, M.D., Richardson, S.H. (Eds.), The J.B. Dawson Volume, Proceedings of the VIIth International Kimberlite Conference. Red Roof Design, Cape Town, pp. 99–108.

Chinn, I.L., Gurney, J.J., Kyser, K.T., 1998. Diamonds and mineral inclusions from the NWT, Canada. 7th International Kimberlite Conference, Cape Town. Addendum, not paginated.

Cole, D.R., Chakraborty, S., 2001. Rates and mechanisms of isotopic exchange. In: Stable Isotope Geochemistry. Reviews in Mineralogy and Geochemistry. Mineralogical Society of America, Washington, pp. 83–223.

Collins, A.T., 2001. The colour of diamond and how it may be changed. J. Gemmol. 27, 341–359.

Davies, R.M., et al., 1999. Diamonds from the deep: pipe DO-27, Slave craton, Canada. In: Gurney, J.J., Gurney, J.L., Pascoe, M.D., Richardson, S.H. (Eds.), The J.B. Dawson Volume, Proceedings of the VIIth International Kimberlite Conference. Red Roof Design, Cape Town, pp. 148–155.

Deines, P., Harris, J.W., 1995. Sulfide inclusion chemistry and carbon isotopes of African diamonds. Geochim. Cosmochim. Acta 59 (15), 3173–3188.

Deines, P., Gurney, J.J., Harris, J.W., 1984. Associated chemical and carbon isotopic composition variations in diamonds from Finsch and Premier kimberlite, South Africa. Geochim. Cosmochim. Acta 51, 1227–1243.

Deines, P., Harris, J.W., Gurney, J.J., 1987. Carbon isotopic composition, nitrogen content, and inclusion composition of diamonds from the Roberts Victor kimberlite, South Africa: evidence for ^{13}C depletion in the mantle. Geochim. Cosmochim. Acta 51, 1227–1243.

Deines, P., Harris, J.W., Spear, P.M., Gurney, J.J., 1989. Nitrogen and ^{13}C content of Finsch and Premier diamonds and their implications. Geochim. Cosmochim. Acta 53, 1367–1378.

Deines, P., Harris, J.W., Gurney, J.J., 1991. The carbon isotopic composition and nitrogen content of lithospheric and asthenospheric diamonds from the Jagersfontein and Koffiefontein kimberlite, South Africa. Geochim. Cosmochim. Acta 55, 2615–2625.

Deines, P., Harris, J.W., Gurney, J.J., 1997. Carbon isotope ratios, nitrogen content and aggregation state, and inclusion chemistry of diamonds from Jwaneng, Botswana. Geochim. Cosmochim. Acta 61 (18), 3993–4005.

deWit, M.J., 1998. On Archean granites, greenstones, cratons and tectonics: does the evidence demand a verdict? Precambrian Res. 91 (1–2), 181–226.

Doyle, B.J., Kivi, K., Scott Smith, B.H., 1999. The Tli Kwi Cho (DO27 and DO18) diamondiferous kimberlite complex, Northwest Territories, Canada. In: Gurney, J.J., Gurney, J.L., Pascoe, M.D., Richardson, S.H. (Eds.), The J.B. Dawson Volume, Proceedings of the VIIth International Kimberlite Conference. Red Roof Design, Cape Town, pp. 194–204.

Droop, G.T.R., 1987. A general equation for estimating Fe^{3+} concentrations in ferromagnesian silicates and oxides from microprobe analyses, using stoichiometric criteria. Geol. Mag. 51, 431–435.

Evans, T., Harris, J.W., 1989. Nitrogen aggregation, inclusion equilibration temperatures and the age of diamonds. In: Ross, J., et al. (Eds.), Kimberlites and Related Rocks. GSA Spec. Publ., vol. 14. Blackwell, Carlton, pp. 1001–1006.

Geospec Consultants, 1999. Rb–Sr Isotopic Analyses for Winspear Resources. Internal Report, De Beers Canada Mining, June 1999.

Griffin, W.L., Gurney, J.J., Ryan, C.G., 1992. Variations in trapping temperatures and trace-elements in peridotite-suite inclusions from African diamonds—evidence for 2 inclusion suites, and implications for lithosphere stratigraphy. Contrib. Mineral. Petrol. 110 (1), 1–15.

Griffin, W.L., et al., 1999a. Layered mantle lithosphere in the Lac de Gras area, Slave craton: composition, structure and origin. J. Petrol. 40 (5), 705–727.

Griffin, W.L., O'Reilly, S.Y., Ryan, C.G., 1999b. The composition and origin of subcontinental lithospheric mantle. In: Fei, Y., Bertka, C.M., Mysen, B.O. (Eds.), Mantle Petrology: Field Observations and High Pressure Experimentation: A tribute to Francis R. (Joe) Boyd. Special Publication. Geochemical Society, Houston, pp. 13–45.

Griffin, W.L., et al., 1999c. Lithospheric structure and mantle terranes: slave craton, Canada. In: Gurney, J.J., Gurney, J.L., Pascoe, M.D., Richardson, S.H. (Eds.), The J.B. Dawson Volume, Proceedings of the VIIth International Kimberlite Conference. Red Roof Design, Cape Town, pp. 299–306.

Grütter, H.S., 2001. The Genesis of High Cr/Al Garnet Peridotite, With Implications for Cratonic Crust–Mantle Architecture. The Slave-Kaapvaal Workshop, Merrickville, Ontario. 3 pp.

Grütter, H., Anckar, E., 2001. Cr–Ca and related characteristics of mantle pyropes from the central Slave craton—a comparison with the type areas on the central Kaapvaal craton, with implications for carbon in peridotite. Presented at the Slave-Kaapvaal Workshop, Merrickville, Ontario, Canada, September 2001.

Grütter, H.S., Apter, D.B., Kong, J., 1999. Crust–mantle coupling: evidence from mantle-derived xenocrystic garnets. In: Gurney, J.J., Gurney, J.L., Pascoe, M.D., Richardson, S.H. (Eds.), The J.B. Dawson Volume, Proceedings of the VIIth International Kimberlite Conference. Red Roof Design, Cape Town, pp. 307–313.

Gurney, J.J., Harris, J.W., Rickard, R.S., 1979. Silicate and oxide inclusions in diamonds from the Finsch kimberlite pipe. In: Boyd, F.R., Meyer, H.O.A. (Eds.), Kimberlites, Diatremes and Diamonds. AGU, Washington, pp. 1–15.

Gurney, J.J., Harris, J.W., Rickard, R.S., 1984. Minerals associated with diamonds from the Roberts Victor Mine. In: Kornprobst, J. (Ed.), Kimberlites II: The Mantle and Crust–Mantle Relationships. Elsevier, Amsterdam, pp. 25–32.

Gurney, J.J., Harris, J.W., Rickard, R.S., Moore, R.O., 1985. Premier Mine diamond inclusions. Trans. Geol. Soc. S. Afr. 88, 301–310.

Harley, S.L., 1984. An experimental study of the partitioning of iron and magnesium between garnet and orthopyroxene. Contrib. Mineral. Petrol. 86, 359–373.

Harris, J.W., 1992. Diamond geology. In: Field, J.E. (Ed.), The Properties of Natural and Synthetic Diamond. Academic Press, London.

Helmstaedt, H., Schulze, D.J., 1989. Southern African kimberlites and their mantle sample: implications for Archean tectonics and lithosphere evolution. In: Ross, J., et al. (Eds.), Kimberlites and Related Rocks. GSA Spec. Publ., vol. 14. Blackwell, Carlton, pp. 358–368.

Irifune, T., 1987. An experimental investigation of the pyroxene–garnet transformation in a pyrolite composition and its bearing on the constitution of the mantle. Earth. Planet. Sci. Lett. 45, 324–336.

Jones, A.G., Ferguson, I.J., Chave, A.D., Evans, R.L., McNeice, G.W., 2001. Electric lithosphere of the Slave craton. Geology 29 (5), 423–426.

Kennedy, C., Kennedy, G., 1976. The equilibrium boundary between graphite and diamond. J. Geophys. Res. 81, 2467–2470.

Kopylova, M.G., Russell, J.K., Cookenboo, H., 1999. Petrology of peridotite and pyroxenite xenoliths from the Jericho kimberlite: implications for the thermal state of the mantle beneath the Slave craton, Northern Canada. J. Petrol. 40 (1), 79–104.

Leahy, K., Taylor, W.R., 1997. The influence of the Glennie domain deep structure on the diamonds in Saskatchewan kimberlites. Geol. Geofiz. 38 (2), 451–460.

MacKenzie, J.M., Canil, D., 1999. Composition and thermal evolution of cratonic mantle beneath the central Archean Slave Province, NWT, Canada. Contrib. Mineral. Petrol. 134 (4), 313–324.

McDade, P., Harris, J.W., 1999. Syngenetic inclusion bearing diamonds from Letseng-la-Terai, Lesotho. In: Gurney, J.J., Gurney, J.L., Pascoe, M.D., Richardson, S.H. (Eds.), The P.H. Nixon Volume, Proceedings of the VIIth International Kimberlite Conference. Red Roof Design, Cape Town, pp. 557–565.

Moore, R.O., Gurney, J.J., 1989. Mineral inclusions in diamond from Monastery kimberlite, South Africa. In: Ross, J., et al. (Eds.), Kimberlites and Related Rocks. GSA Spec. Publ., vol. 14. Blackwell, Carlton, pp. 1029–1041.

O'Neill, H.S.C., 1980. An experimental study of the iron–magnesium partitioning between garnet and olivine and its calibration as a geothermometer: corrections. Contrib. Mineral. Petrol. 72, 337.

O'Neill, H.S.C., Wood, B.J., 1979. An experimental study of the iron–magnesium partitioning between garnet and olivine and its calibration as a geothermometer. Contrib. Mineral. Petrol. 70, 59–70.

Pearson, D.G., 1999. The age of continental roots. Lithos 48 (1–4), 171–194.

Pearson, D.G., Shirey, S.B., 1999. Isotopic dating of diamonds. In: Ruiz, J., Lambert, D.D. (Eds.), Applications of Radiogenic Isotopes to Ore Deposit Research. Economic Geology Special Publication: SEG Reviews in Economic Geology, pp. 143–171.

Pearson, N.J., et al., 1999. Xenoliths from kimberlite pipes of the Lac de Gras Area, Slave craton, Canada. In: Gurney, J.J., Gurney, J.L., Pascoe, M.D., Richardson, S.H. (Eds.), The P.H. Nixon Volume, Proceedings of the VIIth International Kimberlite Conference. Red Roof Design, Cape Town, pp. 644–658.

Phillips, D., Harris, J.W., 1995. Geothermobarometry of diamond inclusions from the De Beers Pool Mines, Kimberley, South Africa. Sixth International Kimberlite Conference, Novosibirsk, Extended Abstracts, 441–443.

Pokhilenko, N.P., et al., 2001. Crystalline inclusions in diamonds from kimberlites of the Snap Lake area (Slave craton, Canada): new evidences for the anomalous lithospheric structure. Dokl. Earth Sci. 380 (7), 806–811.

Pollack, H.N., Chapman, D.S., 1977. On the regional variation of heat flow, geotherms, and lithospheric thickness. Tectonophysics 38, 279–296.

Richardson, S.H., Gurney, J.J., Erlank, A.J., Harris, J.W., 1984. Origin of diamonds from old continental mantle. Nature 310, 198–202.

Richardson, S.H., Shirey, S.B., Harris, J.W., Carlson, R.W., 2001. Archean subduction recorded by Re–Os isotopes in eclogitic sulfide inclusions in Kimberley diamonds. Earth Planet. Sci. Lett. 191 (3–4), 257–266.

Rickard, R.S., Harris, J.W., Gurney, J.J., Cardoso, P., 1989. Mineral inclusions from Koffiefontein mine. In: Ross, J., et al. (Eds.), Kimberlites and Related Rocks. GSA Spec. Publ., vol. 14. Blackwell, Carlton, pp. 1054–1062.

Ringwood, A.E., 1991. Phase transformations and their bearing on the constitution and dynamics of the mantle. Geochim. Cosmochim. Acta 55 (8), 2083–2110.

Russell, J.K., Kopylova, M.G., 1999. A steady state conductive geotherm for the north central Slave, Canada: inversion of petrological data from the Jericho Kimberlite pipe. J. Geophys. Res. Solid Earth 104 (B4), 7089–7101.

Ryan, C.G., Griffin, W.L., Pearson, N.J., 1996. Garnet geotherms—pressure–temperature data from Cr-pyrope garnet xenocrysts in volcanic rocks. J. Geophys. Res. 101 (B3), 5611–5625.

Sobolev, N.V., Lavrent'ev, Y.G., Pokhilenko, N.P., Usova, L.V., 1973. Chrome-rich garnets from the kimberlites of Yakutia and their paragenesis. Contrib. Mineral. Petrol. 40, 39–52.

Stachel, T., Harris, J.W., 1997. Syngenetic inclusions in diamond from the Birim field (Ghana)—a deep peridotitic profile with a history of depletion and re-enrichment. Contrib. Mineral. Petrol. 127 (4), 336–352.

Stachel, T., Viljoen, K.S., Brey, G., Harris, J.W., 1998. Metasomatic processes in lherzolitic and harzburgitic domains of diamondiferous lithospheric mantle: REE in garnets from xenoliths and inclusions in diamonds. Earth Planet. Sci. Lett. 159 (1–2), 1–12.

Stachel, T., Brey, G.P., Harris, J.W., 2000. Kankan diamonds (Guinea) I: From the lithosphere down to the transition zone. Contrib. Mineral. Petrol. 140, 1–15.

Taylor, W.R., Jaques, A.L., Ridd, M., 1990. Nitrogen-defect aggregation characteristics of some Australasian diamonds: time–temperature constraints on the source regions of pipe and alluvial diamonds. Am. Mineral. 75 (11–12), 1290–1310.

Taylor, W.R., Bulanova, G.P., Milledge, H.J., 1995. Quantitative nitrogen aggregation study of some Yakutian diamonds: constraints on the growth, thermal and deformation history of peridotitic and eclogitic diamonds. Sixth International Kimberlite Conference, Novosibirsk, Russia. pp. 608–610.

Taylor, W.R., Canil, D., Milledge, H.J., 1996. Kinetics of Ib to IaA nitrogen aggregation in diamond. Geochim. Cosmochim. Acta 60 (23), 4725–4733.

Vance, E.R., Harris, J.W., Milledge, H.J., 1973. Possible origins of α-particle damage in diamonds from kimberlite and alluvial sources. Min. Mag. 252, 35–37.

Wilding, M.C., Harte, B., Fallick, A.E., Harris, J.W., 1994. Inclusion chemistry, carbon isotopes and nitrogen distribution in diamonds from the Bultfontein Mine, South Africa. In: Meyer, H.O.A., Leonardos, O.H. (Eds.), Diamonds: Characterization, Genesis and Exploration. CPRM Spec Publ Jan/94, Brasilia, pp. 116–126.

Available online at www.sciencedirect.com

Lithos 71 (2003) 505–527

ELSEVIER

LITHOS

www.elsevier.com/locate/lithos

The electrical structure of the Slave craton

Alan G. Jones[a,*,1], Pamela Lezaeta[b], Ian J. Ferguson[c], Alan D. Chave[b], Rob L. Evans[d], Xavier Garcia[a,b], Jessica Spratt[a,1]

[a] *Geological Survey of Canada, National Resources Canada, 615 Booth Street, Room 218, Ottawa, Ontario, Canada K1A 0E9*
[b] *Deep Submergence Laboratory, Department of Applied Ocean Physics and Engineering, Woods Hole Oceanographic Institution, Woods Hole, MA 02543, USA*
[c] *Department of Geological Sciences, University of Manitoba, Winnipeg, Manitoba, Canada R3T 2N2*
[d] *Department of Geology and Geophysics, Woods Hole Oceanographic Institution, Woods Hole, MA 02543, USA*

Abstract

The Slave craton in northwestern Canada, a relatively small Archean craton (600×400 km), is ideal as a natural laboratory for investigating the formation and evolution of Mesoarchean and Neoarchean sub-continental lithospheric mantle (SCLM). Excellent outcrop and the discovery of economic diamondiferous kimberlite pipes in the centre of the craton during the early 1990s have led to an unparalleled amount of geoscientific information becoming available.

Over the last 5 years deep-probing electromagnetic surveys were conducted on the Slave, using the natural-source magnetotelluric (MT) technique, as part of a variety of programs to study the craton and determine its regional-scale electrical structure. Two of the four types of surveys involved novel MT data acquisition; one through frozen lakes along ice roads during winter, and the second using ocean-bottom MT instrumentation deployed from float planes.

The primary initial objective of the MT surveys was to determine the geometry of the topography of the lithosphere–asthenosphere boundary (LAB) across the Slave craton. However, the MT responses revealed, completely serendipitously, a remarkable anomaly in electrical conductivity in the SCLM of the central Slave craton. This Central Slave Mantle Conductor (CSMC) anomaly is modelled as a localized region of low resistivity (10–15 Ω m) beginning at depths of ~ 80–120 km and striking NE–SW. Where precisely located, it is spatially coincident with the Eocene-aged kimberlite field in the central part of the craton (the so-called "Corridor of Hope"), and also with a geochemically defined ultra-depleted harzburgitic layer interpreted as oceanic or arc-related lithosphere emplaced during early tectonism. The CSMC lies wholly within the NE–SW striking central zone defined by Grütter et al. [Grütter, H.S., Apter, D.B., Kong, J., 1999. Crust–mantle coupling; evidence from mantle-derived xenocrystic garnets. Contributed paper at: The 7th International Kimberlite Conference Proceeding, J.B. Dawson Volume, 1, 307–313] on the basis of garnet geochemistry (G10 vs. G9) populations.

Deep-probing MT data from the lake bottom instruments infer that the conductor has a total depth-integrated conductivity (conductance) of the order of 2000 Siemens, which, given an internal resistivity of 10–15 Ω m, implies a thickness of 20–30 km. Below the CSMC the electrical resistivity of the lithosphere increases by a factor of 3–5 to values of around 50 Ω m. This change occurs at depths consistent with the graphite–diamond transition, which is taken as consistent with a carbon interpretation for the CSMC.

Preliminary three-dimensional MT modelling supports the NE–SW striking geometry for the conductor, and also suggests a NW dip. This geometry is taken as implying that the tectonic processes that emplaced this geophysical–geochemical body are likely related to the subduction of a craton of unknown provenance from the SE (present-day coordinates) during 2630–2620

---

* Corresponding author. Fax: +1-613-943-9285.
E-mail address: ajones@nrcan.gc.ca (A.G. Jones).
[1] Now at: Dublin Institute for Advanced Studies, 5 Merrion Square, Dublin 2, Ireland.

Ma. It suggests that the lithospheric stacking model of Helmstaedt and Schulze [Helmstaedt, H.H., Schulze, D.J., 1989. Southern African kimberlites and their mantle sample: implications for Archean tectonics and lithosphere evolution. In Ross, J. (Ed.), Kimberlites and Related Rocks, Vol. 1: Their Composition, Occurrence, Origin, and Emplacement. Geological Society of Australia Special Publication, vol. 14, 358–368] is likely correct for the formation of the Slave's current SCLM.

Keywords: Slave craton; Magnetotelluric method; Archean tectonics; Canadian Shield; Electromagnetic survey; Geophysics

1. Introduction

The geological core of North America, the Canadian Shield, comprises an amalgam of Archean cratons and cratonic fragments welded together by Paleoproterozoic orogenies (Hoffman, 1988). A component of the Canadian Shield is the Slave craton (Fig. 1), in the northwestern part of the shield, that is approximately 600 km (N–S) × 400 km (E–W) in exposed areal extent and hosts the Acasta gneisses, currently the oldest dated rocks on Earth (4.027 Ga, Stern and Bleeker, 1998). In contrast to many other Archean cratons, excellent exposure has resulted in high quality geological bedrock maps being available. In addition, following Fipke's discovery of diamondiferous kimberlite pipes in the centre of the craton in 1991 (Fipke et al., 1995), the sub-continental lithospheric mantle (SCLM) structure of the craton has been, and is being, extensively studied both geochemically and geophysically.

In a Slave compilation published a decade ago, that predated the extraordinary diamond exploration activities of the 1990s, Padgham and Fyson (1992) contended that the Slave craton possesses several features that make it distinct compared to other Archean cratons, and particularly when compared to the Superior craton. These include high abundances of sedimentary rocks relative to volcanic rocks, high abundances of felsic to mafic rocks, high abundances of sialic basement, and high abundances of potassium-rich granite. Also strikingly is that a terrane classification, successfully applied to the Superior craton, cannot be as readily applied to the Slave craton. Kusky's (1989) attempt to do so has been demonstrated to be invalid through the mapping of a continuous single Mesoarchean (3.2–2.8 Ga) basement complex in the western half of the craton (the Central Slave Basement Complex) by Bleeker et al. (1999a,b).

Concomitant with detailed geological mapping of the surface of the Slave craton (Bleeker and Davis, 1999), which leads primarily to models for the formation of the Slave's crust, geochemical, geophysical and petrological studies have been undertaken over the last 5–10 years to image the Slave's SCLM with a view towards understanding its formation and evolution. The geophysical investigations included reflection profiling (Cook et al., 1999), a major refraction experiment (Viejo et al., 1999), teleseismic studies (Bostock, 1998; Bostock and Cassidy, 1997; Bank et al., 2000; Snyder et al., 2002), and a series of four magnetotelluric (MT) experiments. Geochemical studies of mantle samples have been undertaken by Kopylova et al. (1997), Cookenboo (1999), Griffin et al. (1999a,b), Grütter et al. (1999), MacKenzie and Canil (1999), Carbno and Canil (2002), and Heaman et al. (2002).

There are a number of competing models for the assembly of Archean lithosphere, with the proposed processes ranging from cycles of differentiation and collisional thickening (Jordan, 1988) to collision of island arcs comprising depleted material (Ashwal and Burke, 1989) to buoyant subduction and imbrication by lithospheric-scale stacks (Helmstaedt and Schulze, 1989; Kusky, 1989; Kusky and Polet, 1999) to basal accretion by cooling asthenospheric material (Thompson et al., 1996). Deep probing electromagnetic methods are particularly well suited for constraining lithospheric structure (Jones, 1999) as electrical conductivity is influenced by both thermal state and compositional variations. In particular, high precision MT data can resolve the depth to the lithosphere–asthenosphere boundary (LAB) to better than 10% (see, e.g., Jones, 1999), as electrical conductivity rises by two or more orders of magnitude at the initiation of even very low orders of partial melt (0.1%) due to the efficient interconnectivity of the melt (Nakano and Fujii, 1989; Minarik

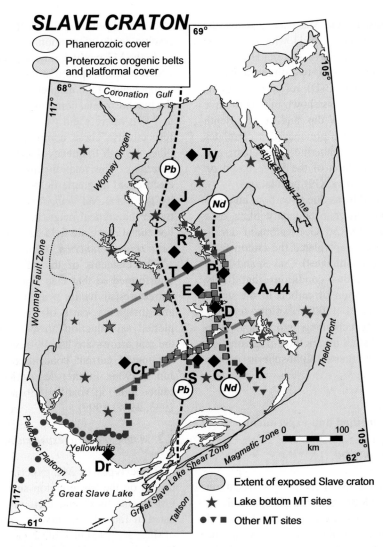

Fig. 1. The Slave craton together with the locations of the MT sites. Dots: 1996 all-weather road sites. Squares: 1998, 1999 and 2000 winter road sites. Stars: 1998–1999 and 1999–2000 lake bottom sites. Significant kimberlite pipes (diamonds): Dr: Drybones; Cr: Cross Lake; S: Snap; C: Camsell; K: Kennady; D: Diavik pipes; E: Ekati mine; P: Point Lake; A-44; A-44 pipe; T: Torrie; R: Ranch; J: Jericho; Ty; Tenacity. Also shown are the north–south Pb and Nd isotope boundaries and the geochemical boundaries of Grütter (two grey dashed lines projecting NE–SW). MT sites in red are those lying above the Central Slave Mantle Conductor.

and Watson, 1995; Drury and Fitz Gerald, 1996; Schilling et al., 1997).

In this paper we discuss electrical resistivity models obtained from deep-probing electromagnetic surveys of the Slave craton using the natural-source magnetotelluric (MT) technique. In total, MT measurements have been made at 138 locations across the craton, predominantly in the southern half. Three previous publications discuss the interpretation of

subsets of these data. Jones and Ferguson (2001) concentrate on the data from west of Yellowknife, and show that the seismically defined base of the crust correlates with a step-like change in electrical resistivity. Jones et al. (2001a) discuss the first results of the craton response from the Slave, focussing on the discovery of a conductive anomaly in the upper mantle beneath the central part of the craton, named the Central Slave Mantle Conductor (CSMC). Finally,

Wu et al. (2002) discuss the MT data crossing the Great Slave Lake shear zone (Fig. 1), which forms the southern boundary of the Slave craton. Herein we present one-dimensional (1-D), two-dimensional (2-D) and three-dimensional (3-D) resistivity models from the craton. The CSMC is shown in 3-D to have a NE–SW strike, and a NW dip, and lie at a depth of ~ 80–100 km. As shown previously, it correlates spatially with the Eocene kimberlite magmatism, and with a geochemical zonation of the Slave based on garnet geochemistry (G10 vs. G9) populations.

We interpret the CSMC as due to carbon, either as graphite or as carbon on grain boundary films, as a relic of Neoarchean (2.8–2.5 Ga) tectonism during which an exotic terrane underplated the craton from the SE (in present-day coordinates). Craven and Jones (2001) discussed the redox conditions that likely resulted in the precipitation of carbon in the upper mantle, and Davis et al. (2003) explore tectonic implications further by compiling information from a variety of sources that all infer a three-part zonation of the Slave's sub-continental lithospheric mantle (SCLM).

2. Magnetotelluric method and experiments

2.1. Magnetotelluric method

The magnetotelluric (MT) method is based on measurement of the electromagnetic (EM) effects of electric currents induced in Earth by natural external sources, such as world-wide lightning activity and the interaction of solar plasma, from solar flares, with the Earth's magnetosphere (Vozoff, 1991). At the frequencies used (typically 20 kHz to 0.0001 Hz, or periods of 0.0005 to 10,000 s), EM propagation is mathematically described by a diffusion equation rather than a wave equation. However, although at these low frequencies the EM fields propagate diffusively, there are significant differences in the technique compared to diffusion by thermal, gravity or static magnetic fields. In particular, there are formal uniqueness solutions to the MT equations for one-dimensional (1-D) (Bailey, 1970) and two-dimensional (2-D) (Weidelt, 2000, pers. comm.) Earths. Unlike potential field methods, the MT method is not inherently non-unique. In particular, as a consequence of frequency dependence

(the *skin depth* effect), MT data have the ability to resolve depth information.

In the MT method, time-varying components of the EM field are measured at the Earth's surface; all three components of the magnetic field (Hx, Hy and Hz) and the two horizontal components of the electric field (Ex and Ey), where x and y usually denote north and east, respectively. These components are related to each other through frequency-dependent, complex *transfer functions*. The relationship between the horizontal electric and magnetic field components are described by the 2×2 MT *impedance tensor*, $\mathbf{Z}(\omega)$, and between the vertical magnetic field component and the horizontal magnetic field components by the *geomagnetic transfer function*, $\mathbf{T}(\omega)$.

The elements of the MT impedance tensor are transformed so that their scaled magnitudes give the correct resistivity for a uniform half space. The real and imaginary parts of the elements of $\mathbf{T}(\omega)$ are plotted as induction arrows, where by convention the real arrows are usually reversed to point towards regions of current concentration (Parkinson, 1962; Jones, 1986). Further description of the MT method can be found in Vozoff (1986, 1991) and Jones (1992, 1993, 1998, 1999).

2.2. Magnetotelluric experiments

Three types of MT experiments have been conducted on the Slave craton since 1996. The types of instruments used, and their observational range, are listed in Table 1, the surveys, with dates and numbers of sites, are listed in Table 2 and the sites are plotted in Fig. 1. In total, MT measurements have been made at 138 locations across the craton, with the majority being in the southern third of the craton.

The land-based surveys involved deployments of conventional MT acquisition. The initial MT survey, in Autumn 1996, comprised broadband acquisition using both V5 and LiMS instruments every 10 km along the only all-weather road on the craton (filled dots in Figs. 1 and 6). This profile is located in the SW corner of the Slave running E–W approximately 100 km either side of the city of Yellowknife (Fig. 1). The 2000 Targeted Geoscience Initiative (TGI) survey consisted of long period systems at 15 locations installed by helicopter and float plane (inverted triangles in Figs. 1 and 6).

Table 1
Details of the MT instrumentation used

Name	Manufacturer	Frequency range (period range)	Deployment time
V5	Phoenix Geophysics	10,000–0.00055 Hz (0.0001–1820 s)	3 nights
V5-2000	Phoenix Geophysics	384–0.00055 Hz (0.0026–1820 s)	2 nights
GMS-06	Metronix	AMT range: 10,000–10 Hz, MT range: 200–0.0005 Hz (0.005–1860 s)	3 nights
LiMS (Long Period MT System)	Geological Survey of Canada	0.05–0.0001 Hz (20–10,000 s)	4 weeks
OBMT (Ocean Bottom MT)	Woods Hole Oceanographic Institution	0.012–0.00004 Hz (85–26,000 s)	1 year

The winter road surveys (squares in Fig. 1) comprised unconventional MT acquisition during wintertime along the ice roads of the Slave craton that are used to supply various exploration and mining camps. The procedure involved separate acquisition of the electric and magnetic fields, with the five electric field sensors (electrodes) being lowered to the bottoms of lakes through holes cut through the ice. Magnetic acquisition directly on the ice of the lakes was severely contaminated by ice movement at 10–100 s periods (McNeice and Jones, 1998), so the magnetometer sensors were installed on the nearest shoreline. Fortuitously, the first winter road survey, in 1998, coincided with the completion of development of a new generation 24-bit MT acquisition system with separate electric and magnetic field recorders, and this survey was the first conducted using these systems by Phoenix Geophysics.

The two lake bottom surveys employed low-power instruments designed for deployment on the continental shelf (Petitt et al., 1994) that were installed and recovered by Twin Otter float planes (stars in Figs. 1 and 6). The electrode chemistry was modified to suit fresh water installation, but no other instrument changes were required for this application. The instruments were deployed in August of each year when the lakes were not ice covered, then retrieved the following July. The digitising rate was set at 2.8 s for the first month, then automatically changed to 28 s for the remaining 10 months. Ten instruments were deployed twice at the 19 locations in Fig. 1, with a year's recording at each location. Some data losses occurred for the electric field systems, but in all cases the magnetic fields were recorded. The orientation on the lake bottom was determined through weekly measurements of a recording compass.

2.3. Data processing

Data processing involved rotating the measured time series into geographic coordinates and estimating the MT response functions relating the horizontal electric field components to the horizontal magnetic field components and the vertical field transfer functions (TF) relating the vertical magnetic field component to the horizontal magnetic field components. The codes used were robust, multi-remote reference codes of Jones (Jones and Jödicke, 1984; method 6 in Jones

Table 2
MT surveys

Date	Style	Major funding agencies	Instrumentation	Number of sites
1996 July–September	Land	LITHOPROBE	V5	60 (13 on craton)
		GSC	LiMS	56 (12 on craton)
1998 March–April	Winter road	LITHOPROBE	V5-2000	12
		GSC	LiMS	11
1999 March–April	Winter road	LITHOPROBE	V5-2000	19
		GSC	LiMS	18
2000 March–April	Winter road	LITHOPROBE	GMS-06 – MT	12
		GSC	LiMS	12
2000 April	Winter road	GSC (EXTECH-III)	GMS-06 AMT + MT	8
2000 July–August	Land (helicopter)	GSC (TGI)	LiMS	15
1998–1999 August–July	Lake bottom	NSF, LITHOPROBE	OBMT	9
1999–2000 August–July	Lake bottom	NSF, LITHOPROBE	OBMT	9

et al., 1989) and Chave and Thomson (2003). Additionally, data segments were selected with low vertical field variations in order to avoid source field effects on the data (Garcia et al., 1997; Jones and Spratt, 2002).

3. Average 1-D craton electrical structure

An estimate of the average 1-D electrical structure of the craton can be obtained by averaging the MT responses, in both azimuthal directions, from all sites on the craton. At each site and frequency, the azimuthal average is obtained from the arithmetic average of the response in the two orthogonal directions, the so-called Berdichevsky, or arithmetic, average (Berdichevsky and Dmitriev, 1976). This estimate is rotationally invariant so is not biased by the orienta-

tion of the deployment. Given that the different MT systems resulted in MT impedance tensor estimates with different period sets, the data from the 138 sites were averaged into period bins that were 1/6 of a decade wide. For both the log(apparent resistivity) and phase data, the averages were obtained robustly by determining the median values within each bin. (A lognormal distribution for apparent resistivities was shown by Bentley, 1973.) Conservative estimates of variance were obtained using jackknife methods (Chave and Thomson, 2003), and the 95% confidence intervals were calculated assuming a Student-t distribution for the variances (e.g., Bendat and Piersol, 1971, p. 112).

The resulting craton-average MT responses are shown in Fig. 2 (solid circles). There is a visible "tear" in the apparent resistivity curve at periods above and below 0.0025 s (400 Hz). This is because

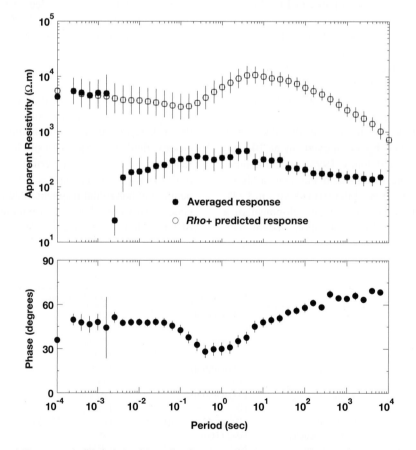

Fig. 2. Averaged Slave MT response (solid circles), with predicted apparent resistivity curve (open circles) using Parker and Booker's (1996) *Rho+* algorithm.

the only high frequency responses come from the land-based sites close to Yellowknife on the Anton complex of the Central Slave Basement Complex, a region with little conducting material in the crust (Jones and Ferguson, 2001). The majority of the Slave stations were located either in lakes (OBMT) or with the electrodes on the lake bottoms (winter road surveys). The accumulation of till on the lake bottoms causes a downward shift of the apparent resistivity curves. This is called the *static shift* effect (Jones, 1988; Sternberg et al., 1988), and is caused by the electrical charges on the boundaries of the till reducing the amount of electric field. In contrast, the phases do not suffer from distortions due to these charges, and accordingly the lateral phase variations are reasonably smooth.

This problem with static shifts associated with three-dimensional geometries means that not only is the actual level in question, but that the *shape* of the apparent resistivity curve could be distorted. This can be tested by predicting the apparent resistivity curve from the phase data, using the *Rho+* algorithm of Parker and Booker (1996). Assuming that the highest frequency apparent resistivity values are at the correct level, then the predicted apparent resistivity data are those shown by the open circles in Fig. 2. The error estimates are 95% confidence intervals derived from the *Rho+* prediction. Using Parker's *D+* analysis, which yields the best possible fitting layered (1-D) model (Parker, 1980; Parker and Whaler, 1981), demonstrates that a normalized RMS (root mean square) misfit of 0.58 is the lowest possible for the open circle data in Fig. 2 (statistically, a value of unity is desired). However, the largest misfit is on the longest period value (10,000 s); deleting these apparent resistivity and phase data results in a minimum possible normalized RMS misfit of 0.34. This suggests that the conservative error estimates for the averaged means of apparent resistivity are, on average, a factor of three too large.

The measured phase data together with the measured high frequency apparent resistivities and the low frequency predicted apparent resistivities can be inverted to estimate the average 1-D structure of the craton. The 1-D layered-Earth algorithm of Fischer et al. (1981) yields a five-layer model with a moderately resistive upper crust, a resistive lower crust and upper mantle to ~ 90 km, then a moderately resistive

mantle underlain by more conducting mantle layers at some hundreds of kilometres. The best-fitting five-layer model, derived using the minimization scheme of Fischer and Le Quang (1981), has a reduced RMS misfit of 0.32. The best-fitting four-layer model has a reduced RMS misfit that is a factor of five larger, and the best-fitting six-layer model has a reduced RMS misfit that is 5% larger. Thus, a five-layer model represents the minimum number of homogeneous layers that can fit the data.

Still in doubt however is the correct level of the apparent resistivity curve that in Fig. 2 was defined by the highest frequency response from the sites on the resistive Anton complex. This level affects both the resistivities and the layer depths of the 1-D model. Given the suggestion of a step change in resistivity at the base of the crust by Jones and Ferguson (2001), the apparent resistivity data can be scaled to result in the interface at 88 km rising to the average Moho depth for the craton. From the teleseismic work of Bank et al. (2000), and the more recent studies by Snyder et al. (2002), the Moho thickness varies across the craton from a maximum of 46 km at the southeast, to a minimum of 37 km in the northwest (see Davis et al., 2003), with an average depth of around 40 km. Scaling the apparent resistivity curve to result in an interface at 40 km requires that the data be shifted by a factor of $(40/90)^2$, or ~ 0.2 (Jones, 1988).

The shifted curve is shown in Fig. 3, together with the 1-D model that fits this curve. The parameters of the model are listed in Table 3. Sensitivity analysis, using singular value decomposition (Edwards et al., 1981; Jones, 1982), shows that the eight model parameters (five layer resistivities, ρ, and three depths, d, with one depth (d_2 = Moho) held constant) are all resolved, in the order ρ_1, d_3, ρ_3, d_1/ρ_5, h_1, h_4, ρ_2 with ρ_4 being least well resolved. The standard deviation ranges of the model parameters are also listed in Table 3. The smoothest model that fits the data, to within 10% of the minimum possible misfit (i.e., 0.37), is also shown in Fig. 3. This smooth model trades off finding the best-fitting model with minimizing the vertical gradient in resistivity (Constable et al., 1987). Converting the responses from apparent resistivities to Schmucker's C response (Schmucker, 1970) shows that the maximum depth of penetration, given by the maximum eddy current flow (Weidelt, 1972), is ap-

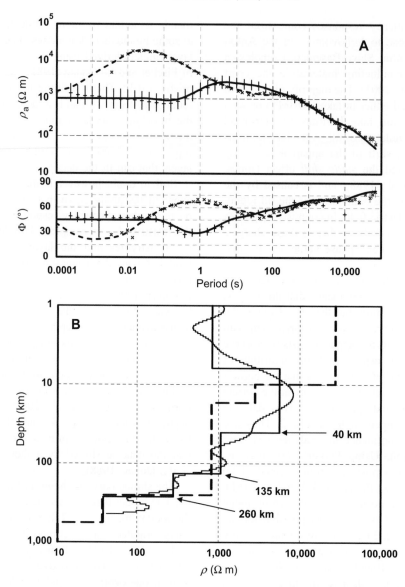

Fig. 3. Modified averaged Slave MT responses with 95% error bounds together with the layered-Earth (solid line) and smooth (light solid line) models that best fit the responses (pluses). Also shown are the MT responses from the central part of the Superior craton (crosses) obtained by Schultz et al. (1993), and their best-fitting layered-Earth model (dashed line).

proximately 300 km. Thus, the data should, on average, penetrate through the sub-continental lithospheric mantle in its entirety.

The inference from the average MT response is that the lithosphere across the Slave is, on average, some 260 km thick. This is consistent with the petrologically defined lithospheric thickness, from analyses of mantle xenoliths, of 260 km for the southern part of

the craton (Kopylova, 2002), but is thicker than the petrologically defined lithospheric thicknesses of 190–200 km for the northern Slave (Jericho, Kopylova et al., 1997) and central Slave (Lac de Gras pipes, Pearson et al., 1999). The majority of the MT sites on the Slave used for this average are located on the southern part of the craton, so there is an inherent bias towards deeper thicknesses for the lithosphere.

Table 3
Parameters of scaled 1-D Slave craton model

Layer	Resistivity (Ω m)	Depth to base (km)
Layer 1: Upper crust	850 (750–950)	6 (4.2–8.6)
Layer 2: Lower crust	5300 (3400–8200)	40 (fixed)
Layer 3: Upper SCLM	1100 (870–1400)	130 (95–185)
Layer 4: Lower SCLM	300 (150–600)	260 (225–310)
Layer 5: Asthenosphere	75 (55–95)	

3.1. Lac de Gras response

Below we discuss the conductivity anomaly discovered in the central part of the Slave craton, which we name the Central Slave Mantle Conductor (CSMC). Due to the presence of this conductor, the incident fields at the longest periods of the winter road data in the middle of the craton do not penetrate through the CSMC with sufficient energy to determine the structure of the mantle below. However, the responses derived from the 1 year of data from the Lac de Gras (Fig. 1) lake bottom site are of high quality to ~ 8000 s, and have a penetration depth of ~ 250–300 km, based on Weidelt's (1972) depth of maximum eddy current flow, and 257 km from Parker's $D+$ model (Parker, 1980; Parker and Whaler, 1981). Fig. 4 shows the averaged apparent resistivity and phase MT curves from the Lac de Gras lake bottom site. Note that the phases are plotted on an expanded scale of 60–90° compared to Figs. 2 and 3. The phase curve displays a subtle minimum at periods of 500–3000 s, which are the periods sensitive to resistivity at depths of the order of 120–200 km. Also shown in Fig. 4 are two smooth models that fit the response to an RMS of 0.90; one model is a continuously smooth model, whereas the other allows a step-change in resistivity at the petrologically defined base of the lithosphere at ~ 200 km (Pearson et al., 1999). Both models show a resistivity minimum within the SCLM of 13 Ω m at a depth of 123 km, consistent with the CSMC. The $D+$ analysis suggests that the data are best fit with a conducting zone centered on 128 km with a depth-integrated conductivity (conductivity–thickness product) of ~ 2000 Siemens. For an internal resistivity of 13 Ω m, the conducting anomaly is 26 km thick, ranging over 115–141 km in extent.

At greater depths resistivity increases to a value of ~ 50 Ω m, then decreases sharply to a poorly defined value but <5 Ω m. Sensitivity analysis of a layered Earth model that fits the data, with interfaces at 40 km (base of crust), 115 km (top of conducting layer), 141 km (base of CSMC), and 200 km (base of lithosphere), shows that the internal resistivity of the CSMC is in the range 10–15 Ω m, and the underlying deep lithosphere resistivity in the range 17–141 Ω m. We take this fourfold increase in resistivity as the transition out of the CSMC conducting anomaly with depth, and interpret it as evidence for a change in electrical properties of carbon when crossing the graphite–diamond stability field.

The base of the lithosphere is suggested by Parker's $D+$ model to be 210 km where a conductance of 13,000 Siemens is seen. This is consistent with the 200 km value for lithospheric thickness reported by Pearson et al. (1999).

3.2. Comparison with the Superior craton

Fig. 3 shows the average response for the Slave craton, after scaling the MT apparent resistivity curve to give an interface at 40 km (see above), with that for the central part of the Superior craton obtained by Schultz et al. (1993). The MT responses obtained by Schultz et al. (1993) are the most precise for any craton, and indeed for mantle-probing depths anywhere in the world, and are the result of 2 years of acquisition using a large electrode array (1 km long lines) in a lake (Carty Lake) in the Kapuskasing region of northern Ontario, Canada. The crust of the Kapuskasing region is geologically highly complex, but very simple in its electrical response (Kurtz et al., 1993), thus providing an excellent window on the mantle below. Due to the high precision of the Carty Lake MT response estimates, the models obtained from them have high resolving power.

Astoundingly, the Slave and Superior phase responses coalesce at periods of 30 s, and the apparent resistivity responses coalesce at a period of 300 s. This suggests that whereas the crustal and upper SCLM electrical structures are clearly different between the two cratons, on average the deeper mantle of the two is the same electrically. The two apparent resistivity curves were scaled in very different ways: the Slave apparent resistivity curve is scaled to result in an interface at the crust–mantle interface, whereas the Superior apparent resistivity curve was scaled to be consistent with observations from nearby geomagnetic

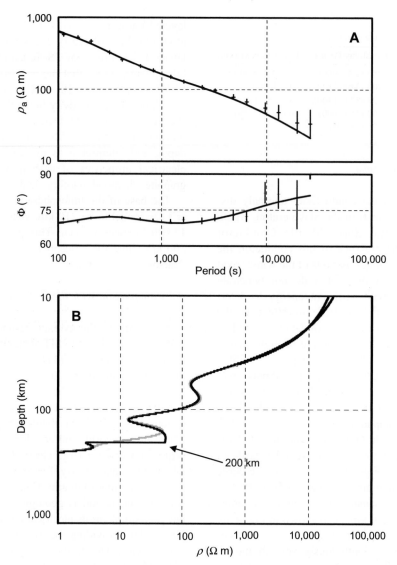

Fig. 4. MT responses from the lake bottom instrument installed in Lac de Gras together with the smooth models that best-fit the responses. One model has no interfaces, whereas the other permits a step-change in resistivity at 200 km depth (base of the lithosphere).

observatories. That these two scalings result in the same apparent resistivity at long periods suggests that the approach used on the Slave data is valid.

The layered-Earth model, with the minimum acceptable number of homogeneous layers, that fits the Superior response is also shown in Fig. 3, and can be compared with the Slave model. Both show a step-change in resistivity at 250–260 km. Given the high precision of the responses from the Superior craton, the thickness of the lithosphere can be far more

precisely determined than for the Slave, and linear sensitivity SVD analysis suggests that it is 253 ± 6 km, with asthenospheric resistivity of 37 ± 7 Ω m. The statistical uncertainty in the base of the lithosphere of ± 6 km is correct under the assumption of a layered 1-D Earth with discrete resistivity changes. If the lithosphere–asthenosphere boundary zone is transitional electrically, then it can occur over more like 50 km (see, e.g., Cavaliere and Jones, 1984). However, laboratory studies suggest that a decrease of >1.5

orders of magnitude in resistivity occurs for a 50 °C change in temperature at the onset of partial melting (Partzsch et al., 2000), and, for a typical continental cratonic geotherm valid for the Slave craton (e.g., Kopylova et al., 1997), a 50 °C change will occur over <20 km.

The resistivity in the deep lithosphere is consistent with a reduced water content compared to tectonically activated or oceanic lithosphere, which may play a role in stabilizing Archean lithosphere against convective erosion (Hirth et al., 2000).

4. Phase maps

MT phase maps are an excellent qualitative indicator of structure, as MT phase is unaffected by electric galvanic distortion that can plague apparent resistivity estimates. Fig. 5 shows phase maps for six periods over the 30 to 10,000 s band, which represent increasing depth penetration within the lithospheric mantle. The phase contoured is the arithmetic (Berdichevsky) average of the two orthogonal phases, and is a rotationally invariant phase. Also shown on the maps are the locations of the geochemically defined mantle domain boundaries of Grütter et al. (1999), and the Pb and Nd isotope boundaries of Thorpe et al. (1992), Davis and Hegner (1992) and Davis et al. (1996).

A coarse guide to the approximate penetration depths of these MT phase maps can be derived from the averaged craton response discussed above. Weidelt's depth of maximum eddy current flow are: 30 s ≈ 65 km; 100 s ≈ 110 km; 300 s ≈ 165 km; 1000 s ≈ 240 km; 3000 s ≈ 310 km; and 10,000 s ≈ 335 km. However, such conversion must be treated with caution, as the highly non-linear mapping from period to depth is a function of the subsurface resistivity and can vary wildly across the map area given the vast range of electrical resistivity. This is particularly true for data from sites located on top of the CSMC; the conductor absorbs and attenuates the EM fields restricting penetration strongly.

The averaged phase data have been smoothed using the nearest neighbour algorithm implemented in the Generic Mapping Tools (GMT) package (Wessel and Smith, 1991). The 30 and 100 s maps have a different colour range than the other maps, from 45°

to 75° rather than from 55° to 85°. There are no 30 s data from the lake bottom sites, and, conversely, the 10,000 s map is almost entirely dictated by the lake bottom responses.

There is a marked increase in phase between the 30 s map and the 100 s map for sites in the centre of the craton lying predominantly within the mantle zone boundaries defined by Grütter et al. (1999). This is caused by the presence of the Central Slave Mantle Conductor (CSMC). High phases are also observed at the eastern boundary of the Slave; these are due to the conducting material within the Thelon orogen (Jones et al., 2001b). Comparing the maps for 300 and 1000 s, a clear NW shift in phases is apparent. This we interpret as evidence for a NW-dip to the CSMC. Note that at these mantle-probing periods, the data do not respect the Pb and Nd isotope boundaries, suggestive of a mantle geometry that is at a high angle to the dominant crustal tectonic boundaries. This dichotomy between the crustally defined structures and those defined for the Slave's sub-continental lithospheric mantle on the basis of petrological, geochemistry and geophysical data is discussed further in Davis et al. (2003).

At the longest periods of 3000 and 10,000 s, there is no clear pattern to the phases, save the suggestion of an east–west difference. At this stage we do not ascribe any significance to this apparent correlation, as it requires further verification from more detailed analyses of the longest period data. In particular, the 10,000 s map may be affected by source field effects (Garcia et al., 1997; Jones and Spratt, 2002).

5. Two-dimensional models

Two-dimensional (2-D) models have been derived for four profiles crossing the Slave province. The locations of the profiles are shown in Fig. 6, together with the locations of the mantle domain boundaries defined by Grütter et al. (1999) and the Pb and Nd isotope boundaries of Thorpe et al. (1992), Davis and Hegner (1992) and Davis et al. (1996). These models were derived independently of each other; further work will occur to ensure the same resistivity structures at intersection points. For all models, strike direction was taken to be perpendicular to the profile direction. This assumption may lead to less resolution

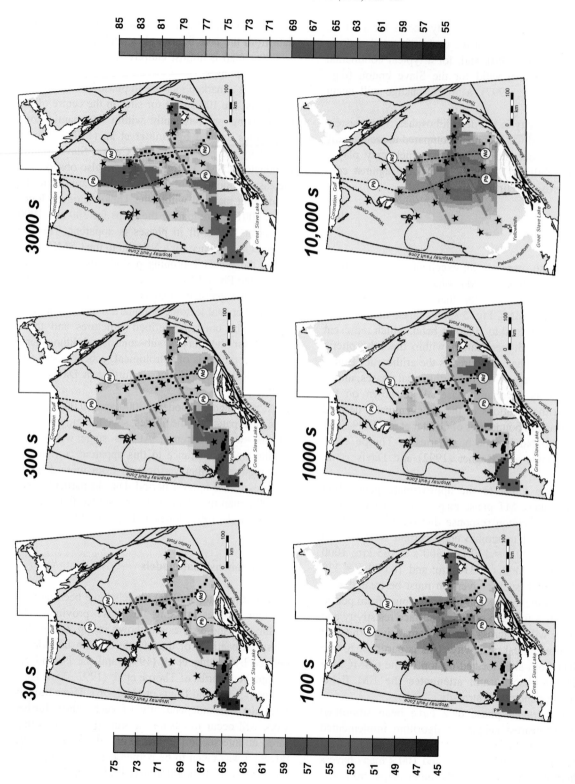

Fig. 5. Contoured averaged phase maps for periods from 30 to 10,000 s. The phases in the shortest frequency maps (30 and 100 s) range from 45° to 75°, whereas the phases in the other four maps range from 55° to 85°.

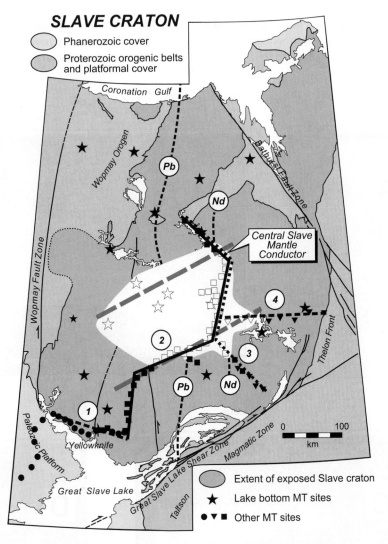

SLAVE CRATON

Fig. 6. The Slave craton together with the locations of the MT sites and the four model profiles. Dots: 1996 all-weather road sites. Squares: 1998 and 1999 winter road sites. Named stars: 1998–1999 and 1999–2000 lake bottom sites. Inverted triangles: 2000 TGI sites. The open MT symbols (squares, inverted triangles and stars) designate sites on top of the CSMC. Also shown are the north–south Pb and Nd isotope boundaries and the geochemical boundaries of Grütter et al. (1999).

than possible in the models, but should not cause erroneous structure as the mantle resistivity structure in the SCLM is slowly varying laterally. Virtually the same results could have been obtained from stitching together 1-D models from all the sites.

Profile 1 is an ~ 150-km-long E–W profile along the all-weather road in the SW part of the craton from Rae in the west to Tibbit Lake at the eastern end of the Ingraham Trail, with the city of Yellowknife at the centre. Profile 2 follows the winter road from Tibbit

Lake at its southern end to the Lupin Mine on the NW end of Contwoyto Lake. It runs north from the eastern end of Profile 1 to the north end of Gordon Lake, where it kinks to the NW along MacKay Lake, then north through the Lac de Gras region to the southern end of Contwoyto Lake, then NW along Contwoyto Lake. Profile 3 starts in the SE part of the Slave at the Kennady exploration property of De Beers and Mountain Province ("K" in Fig. 1), and goes NW until it joins up with Profile 2 at MacKay Lake. Profile 4 runs directly east from the winter road across the northern arm of Aylmer Lake to the eastern boundary of the Slave craton, which is the Thelon orogen.

The models for all four profiles are shown in Fig. 7. The hotter colours represent regions where there are interconnected conducting phases, whereas the colder

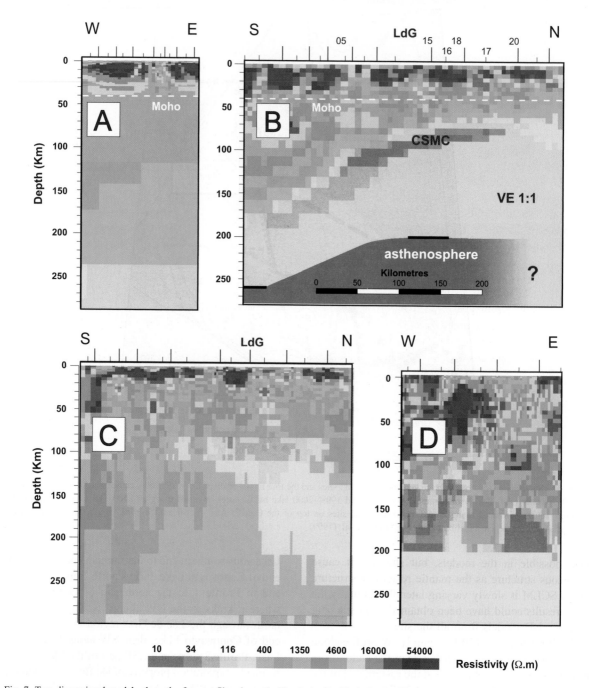

Fig. 7. Two-dimensional models along the four profiles shown in Fig. 6. A = Profile 1; B = Profile 2; C = Profile 3; D = Profile 4. All models plotted with a VE of 1:1 and with the same colour scale.

colours represent regions where conducting material is absent. The models were all obtained using Mackie's RLM2DI code (Rodi and Mackie, 2001), as implemented within the Geotools package, fitting both the strike-parallel (TE-mode) and strike-perpendicular (TM-mode) MT responses simultaneously. The inversion approach is to overparameterize the model, then search for the smoothest model that fits the observations, as pioneered in MT initially in 1-D by Constable et al. (1987), trading off misfit with smoothness using Tikhonov regularization (Tikhonov and Glasko, 1975). The start models for the inversions were either uniform half-spaces, or a layered Earth model based on the 1-D model presented in Fig. 3. The models were all fit to an average misfit of ~ 5° in phase, with a higher misfit on apparent resistivity permitted (25%) to account for static shifts.

Focussing on the SCLM structure, there is a consistent image from profile-to-profile. In the south, there are no significant conducting anomalies in the SCLM, The only crustal anomaly is a region of reduced resistivity to the east of Yellowknife on Profile 1 (Fig. 7A), which may be associated with the mineralisation observed along the Yellowknife fault (Garcia and Jones, 2000). Apart from that anomaly, the crust and SCLM are resistive, and there appears to be a discontinuity at the base of the crust (Jones and Ferguson, 2001). In agreement with the 1-D models, the depth of the base of the lithosphere is of the order of 250 km.

The 2-D model obtained for the data along Profile 2 (Fig. 7b) has been presented and discussed by Jones et al. (2001a), and is included here for completeness. The addition made here is that the area previously described as "region of no penetration" in Fig. 3 of Jones et al. (2001a) has now been sampled by the long period lake bottom data (Fig. 4), and shown to be of higher resistivity than the conductor. Also, the LAB is suggested to be at ~ 200 km. To the north, commencing beneath station 05, there is an upper mantle conducting region that exists to the southern end of Contwoyto Lake (between stations 17 and 20). This Central Slave Mantle Conductor (CSMC) begins at a depth of around 80–100 km, and has an internal resistivity of 30 Ω m or less, with a value of 13 Ω m suggested by the 1-D smooth modelling (Fig. 3).

The model for Profile 3 (Fig. 7C) shows a localized conducting anomaly within the SCLM at a depth of ~ 85 km. It begins to the south at the intersection of the Kennady Lake road and the main winter road, and continues to the southern end of Contwoyto Lake, consistent with the model for Profile 2.

The Profile 4 model (Fig. 7D) also exhibits evidence for the mantle conductor existing as far to the east as the northern arm of Aylmer Lake. There is an apparent conductor deep in the mantle at the eastern end of the line, but this is an artefact caused by the presence of conducting material within the crust of the Thelon orogen to the east, as clearly indicated by the induction arrows (Jones et al., 2001b).

6. Three-dimensional model

A 3-D resistivity model is being developed to explain the MT and TF observations. Herein we present the current version that describes qualitatively the major features predominantly in the TF data. Further work will be undertaken, including 3-D inversion, to fit all available MT and TF data, but we do not anticipate significant changes in the major features and geometries of the final model from the one we present here. The model was constructed by trial-and-error attempting to fit qualitatively the magnetic transfer functions, and some aspects of the MT responses, from the bottom lake sites and some selected winter road sites. Taken into account also were the general background resistivities observed in the 2-D models discussed above. The a priori information regarding the conductive 3-D features has been obtained from the results of a tensor dimensionality analysis scheme, developed by Lezaeta and Haak (2003), which indicates that the long period MT data suggest the existence of an anomalous SW–NE striking 3-D conductor at depth. Sites close to the Thelon Front (site HEA in Fig. 6) are affected by 3-D induction and also by strong current channelling. Both 3-D induction and current channelling effects can be explained with a model of dyke-like conductors of limited lateral extent or anisotropic conductive structures embedded in a more resistive crust (Lezaeta and Haak, 2003).

The model was built using the Geotools model builder, and its forward response was calculated using the code of Mackie and Madden (1993) and Mackie et al. (1993, 1994), with the recent modifications (bi-conjugate gradient stabilized method using an incomplete cholesky decomposition with a new scheme for updating the divergence of the H fields resulting in extremely fast convergence of the solution.) by R.L. Mackie and J.R. Booker (2001, pers. comm.). The model consists of 65 cells in the north direction, 62 cells in the east–west direction, and 24 cells vertically.

A horizontal slice at a depth of 140 km through the 3-D model is shown in Fig. 8, using approximately the same colour scheme as the 2-D models in Fig. 7. Also shown on the figure are the observed (solid arrows) and calculated (open arrows) real induction vectors at a period of 620 s for the lake bottom sites and some selected land and winter road sites. (This period is longer than that for the MT responses as the TF data

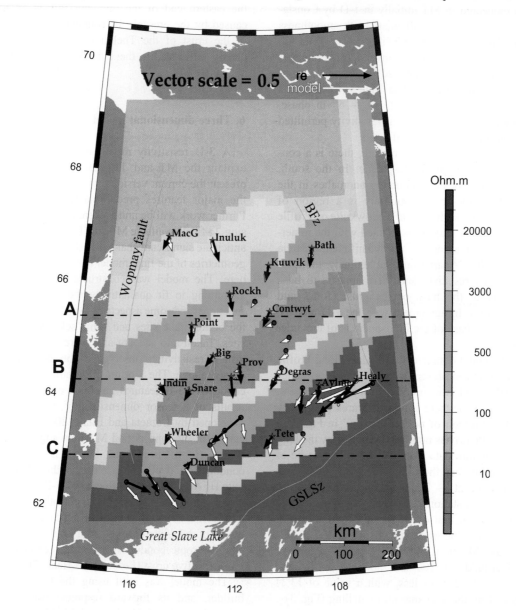

Fig. 8. Horizontal slice of the 3-D resistivity model at 140 km depth. The vectors are the unreversed induction arrows of the measured data (solid arrows) and the calculated data (open arrows).

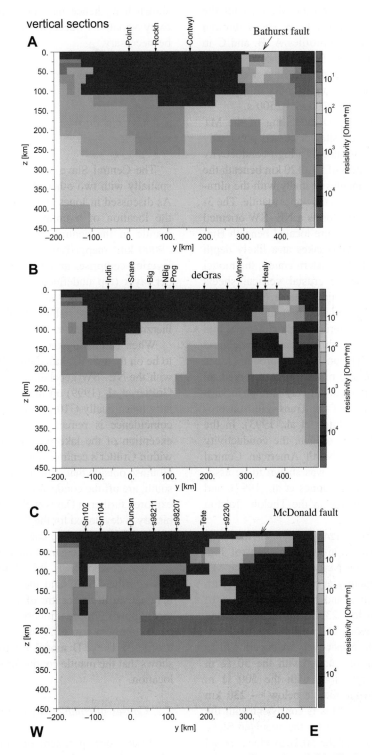

Fig. 9. Vertical slices through the 3-D model along profiles A, B, and C in Fig. 8.

react at longer periods than do the MT data for the same anomaly.) Three vertical E–W slices through the model, corresponding to profiles A, B, and C in Fig. 8, are shown in Fig. 9. Not shown are the comparison plots for other periods that were used during the forward trial-and-error model fitting exercise, which were 120, 320, 1200 and 2500 s.

The interpretations of 2-D modelling of the MT data from the winter road sites discussed previously demonstrate that the existence of a high conductivity zone encountered at depths of 80–120 km beneath the central Slave craton coincides spatially with the ultra-depleted harzburgitic layer in the upper mantle. The 3-D model traces this conductor as a NE–SW oriented mantle structure, with its centre located beneath Big, Providence and Lac de Gras lakes at a likely depth range of 70–250 km. In the eastern craton, the model contains highly conductive crustal vertical sheets beneath the Bathurst fault and a deeper NW dipping structure beneath the McDonald fault, suggesting the presence of conducting material (likely either graphite or sulphides) within a deep dipping fault zone. These we associate with the Thelon–Talston orogen (Jones et al., 2001b), and the size of the effect observed on the easternmost sites is quantitatively similar to sites on the Hearne hinterland of the Trans–Hudson orogen in northern Canada (Jones et al., 1993). In the case of the Trans–Hudson orogen, the conductivity anomaly, the well-known North American Central Plains conductivity anomaly, was shown to be associated with pyrite sulphides (Jones et al., 1997), and the same might also be true in the Thelon–Talston orogen.

The model fits the magnetic transfer functions (see the induction arrows in the plan view of Fig. 8) best, especially at long periods. The fit to the MT phases is partly satisfactory at periods >1000 s (with average RMS between 2 and 5). Two variations of the model structure below 230 km have been tested by setting a homogeneous half space of 50 and 500 Ω m below this depth, respectively. A superior fit to the data is found for the model with the 50 Ω m resistivity block at depth than with the 500 Ω m block, suggesting that the mantle below $\sim$ 230 km depth cannot be resistive. This is consistent with the 1-D analyses presented above for the averaged Slave response. The responses of the 50 Ω m block model have a comparable data fit with that from the model

shown here, hence indicating that the deep conductivity variations traced in the latter may or may not be necessary.

7. Discussion

7.1. Correlation with geochemical zones and Eocene kimberlites

The Central Slave Mantle Conductor corresponds spatially with two other observations from the Slave. As discussed in Jones et al. (2001a), it correlates with the location of a unique, two-layered lithospheric mantle in the central Slave craton extending over >9000 km^2 mapped by Griffin et al. (1999a,b). This layering comprises an ultra-depleted, harzburgitic upper layer (top undefined but shallower than $\sim$ 100 km) separated sharply at 140–150 km depth (approx. graphite–diamond stability field) from a less depleted, lherzolitic lower layer.

When comparing those MT sites that are deemed to be on the CSMC, based on 2-D and 3-D modelling, with the NE–SW geochemical boundaries defined by Grütter et al. (1999) on the basis of garnet geochemistry (specifically G10 garnet populations), the spatial coincidence is remarkable (Fig. 6). With the sole exception of the lake bottom site BIG, all MT sites within Grütter's central zone are on the conductor, and all those outside that zone, either to the north or to the south, are off the conductor. This spatial association is explored further in Davis et al. (2003), where a tectonic history is developed for the Slave craton based, in part, on the geometry of the conductivity anomaly and its association with the garnet distributions.

Also remarkable is the close spatial association between the CSMC and the Eocene-aged kimberlite magmatism in the centre of the craton. The easternmost known Eocene kimberlite is on the western side of the northern arm of Aylmer Lake, in the middle of Profile 3. The 2-D model for that profile (Fig. 7D) shows that the mantle conductor ends precisely at that location.

7.2. Causes of conductivity enhancement

Olivine, orthopyroxene or clinopyroxene-dominated mineralogy would result in electrical resistivities in excess of 100,000 Ω m in the SCLM at likely temperatures of hundreds to a thousand Celcius (Constable and Duba, 1990; Xu et al., 2000). In Jones et al. (2001a), various mechanisms were explored as possible candidates to explain the existence of the Central Slave Mantle Conductor. Two were proposed as favoured; carbon, either as grain-boundary films or as graphite, and diffusion of hydrogen. Others, such as partial melts, sulphides, saline fluids, hydrous mantle minerals, were all rejected. Partial melt we reject because there is no evidence for lithospheric delamination in the central Slave since the Eocene, when the lithosphere was 190 km thick based on petrological analyses of mantle xenoliths (Pearson et al., 1999). With regard to sulphides, for the Kaapvaal craton one could appeal to interconnected sulphides as the cause of enhanced conductivity, due to the observations of interstitial sulphides at the 330 ppm level (Alard et al., 2000); however, there is no evidence for sulphides in mantle xenoliths from the Slave craton (H. Grütter, pers. comm., 2001; B. Doyle, pers. comm., 2001; J. Gurney, pers. comm., 2001). Interconnected saline fluids we reject because of the age of the anomaly: the fluids are gravitationally unstable over geological time scales. Finally, hydrous mantle minerals, such as phlogopite (Boerner et al., 1999), we reject because when observed they are not interconnected, and metasomatism does not appear to be an agent that reduces resistivity (Jones et al., 2002). Also, hydrous crustal minerals do not enhance conductivity in the continental lower crust (Olhoeft, 1981).

The lake bottom data from the Lac de Gras station (Fig. 4) provide invaluable information that suggests the carbon interpretation is correct. First, the internal resistivity of the CSMC appears to be around 10–15 Ω m, which is too low for hydrogen diffusion at such shallow mantle depths (G. Hirth, 2001, pers. comm.). Second, the resistivity increases below the CSMC by a factor of 3–5 to 50 Ω m, consistent with graphite no longer being the dominant conductivity enhancement: the transition from carbon in conducting graphite form above the graphite–diamond stability field to carbon in highly resistive diamond form below the stability field occurs at 130 km (Griffin et al., 1999a). Craven and Jones (2001) demonstrated that the CSMC has a twin in the North Caribou terrane of the western part of the Superior Province, and suggest an environment for the reduction of carbon at upper mantle depths.

Below the graphite–diamond stability field, the resistivity is still orders of magnitude less than an Ol–Cpx–Opx mineralogy would suggest, and possibly hydrogen diffusion (Karato, 1990) may be the valid explanation at these depths.

7.3. Tectonic implications

Griffin et al. (1999a,b) proposed that the unique two-layer petrological lithosphere comprising an upper SCLM harzburgite layer, from about 80–100 to about 140 km, underlain by a lower SCLM lherzolitic layer, was caused by trapped oceanic or arc-related lithosphere associated with the suturing of the western Slave craton to the eastern arc terrane, now dated at ca. 2690 Ma (Davis and Bleeker, 1999). This ultra-depleted lithosphere is thought to have been subsequently underlain by plume lithosphere to explain ultra-deep ferropericlase and Mg-perovskite diamonds (Davies et al., 1999).

We have demonstrated that the Central Slave Mantle Conductor is intimately spatially associated with this ultra-depleted layer, which suggests an ancient provenance for the CSMC. However, the geometry of the CSMC has a NE–SW strike and a NW dip, inconsistent with E–W convergence. Although oblique convergence could have occurred, the spatial orientation of the anomaly with the later post-accretion deformation and plutonism is taken as firm evidence of an association with later tectonism (see Davis et al., 2003). Taken together with other petrological, geochemical and geophysical information from the Slave's SCLM, we conclude that the CSMC is a consequence of subcretion at ca. 2630 Ma by exotic lithosphere from the SE. This is discussed further in Davis et al. (2003).

8. Conclusions

Deep-probing electromagnetic studies, using the magnetotelluric (MT) technique, are capable of contributing significantly to the understanding of the formation and evolution of Archean lithosphere. Herein, we show that such studies on the Slave craton have imaged serendipitously a remarkable anomaly in electrical conductivity located within the upper part of the sub-continental lithospheric mantle in the centre of the craton. This anomaly, named the Central Slave Mantle Conductor (CSMC), correlates with geochemical information obtained from mantle xenoliths and with the known exposures of Eocene-aged kimberlites. Given the internal resistivity of the conductor, 10–15 Ω m, given its limited depth extent to the graphite–diamond stability field, and given the various plausible candidates for explaining enhanced conductivity, we conclude that the anomaly is due to carbon as either graphite or as carbon on grain-boundary films.

We ascribe a Neoarchean age to the CSMC, given the spatial correlation with Griffin et al.'s (1999a,b) harzburgitic ultra-depleted layer and with the occurrence of G10 garnets mapped by Grütter et al. (1999), and suggest that it was emplaced as a consequence of sub-cretion by exotic lithosphere from the SE at ca. 2630–2590 Ma. The spatial association with the Eocene kimberlite magmatism is intriguing, and we speculate that the rheological differences between the different mantle domains may have influenced where the kimberlties were able to erupt.

Features in the northern part of the craton we are less certain about, given the paucity of MT sites. We intend to rectify this with deep-probing MT measurements over the next few years.

Acknowledgements

The electromagnetic experiments on the Slave craton were made possible through the financial support and logistical efforts of many organizations, companies and individuals. Financial support came from LITHOPROBE, Geological Survey of Canada (GSC, under the LITHOPROBE, EXTECH-III and Walmsley Lake Targeted Geoscience Initiative programs), Canadian Federal Department of Indian and Northern Development (DIAND), the U.S. National Science Foundation's Continental Dynamics Program, DeBeers Canada Exploration, Kennecott Exploration and BHP Billiton Diamonds Logistical support was provided by Diavik, BHP Billiton, DeBeers and Winspear, and in Yellowknife by Royal Oak Mines and Miramar Mining. Data acquisition along the all-weather and winter roads in 1996, 1998 and 1999 was undertaken by Phoenix Geophysics, and in 2000 by Geosystem Canada, who are thanked for their support in terms of reduced academic survey rates and close attention to detail which resulted in high quality responses. The staff of the Yellowknife Seismological Observatory is thanked for its support and for providing essential preparation facilities. The lake bottom sites were deployed and retrieved from a Twin Otter float plane with a specially designed winch built for us by Air Tindi. Finally, we wish to thank all our colleagues for enlightening discussions. Particular individuals whom we wish to recognise as contributing to our efforts include Alex Arychuk (Air Tindi), Wouter Bleeker (GSC), Bill Davis (GSC), Buddy Doyle (Kennecott Exploration), Garth Eggenberger (Air Tindi), Leo Fox (Phoenix), Nick Grant (formerly GSC, now University of Victoria), Herman Grütter (formerly DeBeers Canada Exploration, now Mineral Services Canada), Colin Farquharson (UBC), George Jensen (Yellowknife GSC), Leonard Johnson (NSF), Andy Langlois (formerly Yellowknife GSC), Juanjo Ledo (formerly GSC, now University of Barcelona), Grant Lockhart (BHP Billiton Diamonds), Gary McNeice (formerly Phoenix Geophysics, now Geosystem Canada), Brian Roberts (GSC), David Snyder (GSC), and Carolyn Relf (DIAND), and, most especially, Hendrik Falck (C.S. Lord Northern Geoscience Centre).

The data were modelled using 2-D and 3-D codes within the Geotools program, with updates from Randy Mackie. The data were contoured using Paul Wessel's GMT package.

Geological Survey of Canada Contribution No. 2002002. Lithoprobe publication No. 1339. Woods Hole Oceanographic Institution Contribution 11023.

References

Alard, O., Griffin, W.L., Lorand, J.P., Jackson, S.E., O'Reilly, S.Y., 2000. Non-chondritic distribution of the highly siderophile elements in mantle sulphides. Nature 407, 891–894.

Ashwal, L.D., Burke, K., 1989. African lithospheric structure, volcanism, and topography. Earth Planet. Sci. Lett. 96, 8–14.

Bailey, R.C., 1970. Inversion of the geomagnetic induction problem. Proc. R. Soc. Lond., A 315, 185–194.

Bank, C.-G., Bostock, M.G., Ellis, R.M., Cassidy, J.F., 2000. A reconnaissance teleseismic study of the upper mantle and transition zone beneath the Archean Slave craton in NW Canada. Tectonophysics 319, 151–166.

Bendat, J.S., Piersol, A.G., 1971. Random Data: Analysis and Measurement Procedures. Wiley-Interscience, New York.

Bentley, C.R., 1973. Error estimation in two dimensional magneto-telluric analyses. Phys. Earth Planet. Inter. 7, 423–430.

Berdichevsky, M.N., Dmitriev, V.I., 1976. Basic principles of inter-pretation of magnetotelluric sounding curves. In: Adam, A. (Ed.), Geoelectric and Geothermal Studies. KAPG Geophysical Monograph, Akademiai Kiad, pp. 165–221.

Bleeker, W., Davis, W.J., 1999. The 1991–1996 NATMAP slave province project: introduction. Can. J. Earth Sci. 36, 1033–1042.

Bleeker, W., Ketchum, J.W.F., Jackson, V.A., Villeneuve, M.E., 1999a. The Central Slave Basement Complex: Part I. Its struc-tural topology and autochthonous cover. Can. J. Earth Sci. 36, 1083–1109.

Bleeker, W., Ketchum, J.W.F., Davis, W.J., 1999b. The Central Slave Basement Complex: Part II. Age and tectonic significance of high-strain zones along the basement–cover contact. Can. J. Earth Sci. 36, 1111–1130.

Boerner, D.E., Kurtz, R.D., Craven, J.A., Ross, G.M., Jones, F.W., Davis, W.J., 1999. Electrical conductivity in the Precambrian lithosphere of Western Canada. Science 283, 668–670.

Bostock, M.G., 1998. Mantle stratigraphy and evolution of the Slave province. J. Geophys. Res. 103, 21183–21200.

Bostock, M.G., Cassidy, J.F., 1997. Upper mantle stratigraphy beneath the southern Slave Craton. Can. J. Earth Sci. 34, 577–587.

Carbno, G.B., Canil, D., 2002. Mantle structure beneath the SW Slave Craton, Canada; constraints from garnet geochemistry in the Drybones Bay Kimberlite. J. Petrol. 43, 129–142.

Cavaliere, T., Jones, A.G., 1984. On the identification of a transi-tion zone in electrical conductivity between the lithosphere and asthenosphere: a plea for more precise phase data. J. Geophys. 55, 23–30.

Chave, A.D., Thomson, D.J., 2003. A bounded influence regression estimator based on the statistics of the hat matrix. J. Roy. Stat. Soc. Series C, Appl. Stat. 52, 307–322.

Constable, S.C., Duba, A., 1990. Electrical conductivity of olivine, a dunite, and the mantle. J. Geophys. Res. 95, 6967–6978.

Constable, S.C., Parker, R.L., Constable, C.G., 1987. Occam's inversion: a practical algorithm for generating smooth mod-els from electromagnetic sounding data. Geophysics 52, 289–300.

Cook, F.A., van der Velden, A.J., Hall, K.W., Roberts, B.J., 1999. Frozen subduction in Canada's Northwest Territories; lithoprobe deep lithospheric reflection profiling of the western Canadian Shield. Tectonics 18, 1–24.

Cookenboo, H.O., 1999. History and process of emplacement of the Jericho (JD-1) kimberlite pipe, northern Canada. In: Gurney, J.J., Gurney, J.L., Pascoe, M.D., Richardson, S.H. (Eds.), The J.B. Dawson Volume; Proceedings of the VIIth International Kimberlite Conference; Volume 1. Proceedings of the Interna-tional Kimberlite Conference. 7, vol. 1, pp. 125–133.

Craven, J.A., Jones, A.G, 2001. Comparison of Slave and Superior electrical lithospheres. Presented at the Slave-Kaapvaal Work-shop, Merrickville, Ontario, Canada, September 2001.

Davies, R.M., Griffin, W.L., Pearson, N.J., Andrew, A.S., Doyle, B.J., O'Reilly, S.Y., 1999. Diamonds fron the deep: Pipe DO-27, Slave craton, Canada. The 7th International Kimberlite

Conference Proceedings. J.B. Dawson Volume, vol. 1, pp. 148–155.

Davis, W.J., Bleeker, W., 1999. Timing of plutonism, deformation, and metamorphism in the Yellowknife Domain, Slave Province, Canada. Can. J. Earth Sci. 36, 1169–1187.

Davis, W.J., Hegner, E., 1992. Neodymium isotopic evidence for the accretionary development of the Late Archean Slave Prov-ince. Contrib. Mineral. Petrol. V. 111, 493–504.

Davis, W.J., Gariepy, C., van Breeman, O., 1996. Pb isotopic com-position of late Archean granites and the extent of recycling early Archean crust in the Slave Province, northwest Canada. Chem. Geol. 130, 255–269.

Davis, W.J., Jones, A.G., Bleeker, W., Grütter, H., 2003. Litho-spheric development in the Slave Craton: a linked crustal and mantle perspective. Lithos 71, 575–589 (this issue).

Drury, M.R., Fitz Gerald, J.D., 1996. Grain boundary melt films in an experimentally deformed olivine–orthopyroxene rock; impli-cations for melt distribution in upper mantle rocks. Geophys. Res. Lett. 23, 701–704.

Edwards, R.N., Bailey, R.C., Garland, G.D., 1981. Conductivity anomalies: lower crust or asthenosphere? Phys. Earth Planet. Inter. 25, 263–272.

Fipke, C.E., Dummett, H.T., Moore, R.O., Carlson, J.A., Ashley, R.M., Gurney, J.J., Kirkley, M.B., 1995. History of the discov-ery of diamondiferous kimberlites in the Northwest Territories, Canada. Sixth International Kimberlite Conference; Extended Abstracts. Proceedings of the International Kimberlite Confer-ence, vol. 6, pp. 158–160.

Fischer, G., Le Quang, B.V., 1981. Topography and minimization of the standard deviation in one-dimensional magnetotelluric modelling. Geophys. J. R. Astron. Soc. 67, 279–292.

Fischer, G., Schnegg, P.-A., Peguiron, M., Le Quang, B.V., 1981. An analytic one-dimensional magnetotelluric inversion scheme. Geophys. J. R. Astron. Soc. 67, 257–278.

Garcia, X., Jones, A.G., 2000. Regional scale survey of the Yellow-knife fault. Contributed paper at: Yellowknife Geoscience Fo-rum, Yellowknife, NWT, 22–24 November.

Garcia, X., Chave, A.D., Jones, A.G., 1997. Robust processing of magnetotelluric data from the auroral zone. J. Geomagn. Geo-electr. 49, 1451–1468.

Griffin, W.L., Doyle, B.J., Ryan, C.G., Pearson, N.J., O'Reilly, S.Y., Davies, R., Kivi, K., van Achterbergh, E., Natapov, L.M., 1999a. Layered mantle lithosphere in the Lac de Gras area, Slave craton: composition, structure and origin. J. Petrol. 40, 705–727.

Griffin, W.L., Doyle, B.J., Ryan, C.G., Pearson, N.J., O'Reilly, S.Y., Natapov, L., Kivi, K., Kretschmar, R., Ward, J., 1999b. Lithosphere structure and mantle terranes: Slave Craton, Cana-da. The 7th International Kimberlite Conference Proceeding. J.B. Dawson Volume, vol. 1, pp. 299–306.

Grütter, H.S., Apter, D.B., Kong, J., 1999. Crust–mantle coupling; evidence from mantle-derived xenocrystic garnets. Contributed paper at the 7th International Kimberlite Conference Proceeding. J.B. Dawson Volume, vol. 1, pp. 307–313.

Heaman, L.M., Creaser, R.A., Cookenboo, H.O., 2002. Extreme enrichment of high field strength elements in Jericho eclogite xenoliths; a cryptic record of Paleoproterozoic subduction, partial melting, and metasomatism beneath the Slave Craton, Canada. Geology 30, 507–510.

Helmstaedt, H.H., Schulze, D.J., 1989. Southern African kimberlites and their mantle sample: implications for Archean tectonics and lithosphere evolution. In: Ross, J. (Ed.), Kimberlites and Related Rocks, vol. 1: Their Composition, Occurrence, Origin, and Emplacement. Geological Society of Australia Special Publication, vol. 14, pp. 358–368.

Hirth, G., Evans, R.L., Chave, A.D., 2000. Comparison of continental and oceanic mantle electrical conductivity: is the Archean lithosphere dry? Geochem. Geophys. Geosyst. 1, paper number 2000GC000048.

Hoffman, P., 1988. United plates of America, the birth of a craton: early Proterozoic assembly and growth of Proto–Laurentia. Annu. Rev. Earth Planet. Sci. 16, 543–603.

Jones, A.G., 1982. On the electrical crust–mantle structure in Fennoscandia: no Moho and the asthenosphere revealed? Geophys. J. R. Astron. Soc. 68, 371–388.

Jones, A.G., 1986. Parkinson's pointers' potential perfidy! Geophys. J. R. Astron. Soc. 87, 1215–1224.

Jones, A.G., 1988. Static shift of magnetotelluric data and its removal in a sedimentary basin environment. Geophysics 53, 967–978.

Jones, A.G., 1992. Electrical conductivity of the continental lower crust. In: Fountain, D.M., Arculus, R.J., Kay, R.W. (Eds.), Continental Lower Crust. Elsevier, Amsterdam, pp. 81–143. Chap. 3.

Jones, A.G., 1993. Electromagnetic images of modern and ancient subduction zones. In: Green, A.G., Kroner, A., Gotze, H.-J., Pavlenkova, N. (Eds.), Plate Tectonic Signatures in the Continental Lithosphere. Tectonophysics, vol. 219, pp. 29–45.

Jones, A.G., 1998. Waves of the future: superior inferences from collocated seismic and electromagnetic experiments. Tectonophysics 286, 273–298.

Jones, A.G., 1999. Imaging the continental upper mantle using electromagnetic methods. Lithos 48, 57–80.

Jones, A.G., Ferguson, I.J., 2001. The electric Moho. Nature 409, 331–333.

Jones, A.G., Jödicke, H., 1984. Magnetotelluric transfer function estimation improvement by a coherence-based rejection technique. Contributed paper at 54th Society of Exploration Geophysics Annual General Meeting. Atlanta, Georgia, U.S.A., December 2–6, pp. 51–55. Abstract volume.

Jones, A.G., Spratt, J., 2002. A simple method for deriving the uniform field MT responses in auroral zones. Earth Planets Space 54, 443–450.

Jones, A.G., Chave, A.D., Egbert, G., Auld, D., Bahr, K., 1989. A comparison of techniques for magnetotelluric response function estimation. J. Geophys. Res. 94, 14201–14213.

Jones, A.G., Craven, J.A., McNeice, G.A., Ferguson, I.J., Boyce, T., Farquharson, C., Ellis, R.G., 1993. The North American Central Plains conductivity anomaly within the Trans–Hudson orogen in northern Saskatchewan. Geology 21, 1027–1030.

Jones, A.G., Katsube, J., Schwann, P., 1997. The longest conductivity anomaly in the world explained: sulphides in fold hinges causing very high electrical anisotropy. J. Geomagn. Geoelectr. 49, 1619–1629.

Jones, A.G., Ferguson, I.J., Chave, A.D., Evans, R.L., McNeice,

G.W., 2001a. The electric lithosphere of the Slave craton. Geology 29, 423–426.

Jones, A.G., Snyder, D., Spratt, J., 2001b. Magnetotelluric and teleseismic experiments as part of the Walmsley Lake project: experimental designs and preliminary results. Geol. Surv. Can. Curr. Res. C6, 10-1–10-4.

Jones, A.G., Snyder, D., Hanmer, S., Asudeh, I., White, D., Eaton, D., Clarke, G., 2002. Magnetotelluric and teleseismic study across the Snowbird Tectonic Zone, Canadian Shield: a Neoarchean mantle suture? Geophys. Res. Lett. 29 (10), 10-1–10-4 doi: 10.1029/2002GL015359.

Jordan, T.H., 1988. Structure and formation of the continental tectosphere. J. Petrol., Special Lithosphere Issue, 11–37.

Karato, S., 1990. The role of hydrogen in the electrical conductivity of the upper mantle. Nature 347, 272–273.

Kopylova, M.G., 2002. The deep structures of the Slave Craton. Presented at "Diamond Short Course", held in Vancouver, Canada, B.C., 21 February.

Kopylova, M.G., Russel, J.K., Cookenboo, H., 1997. Petrology of peridotite and pyroxenite xenoliths from Jericho Kimberlite; implications for the thermal state of the mantle beneath the Slave Craton, northern Canada. J. Petrol. 40, 79–104.

Kurtz, R.D., Craven, J.A., Niblett, E.R., Stevens, R.A., 1993. The conductivity of the crust and mantle beneath the Kapuskasing Uplift: electrical anisotropy in the upper mantle. Geophys. J. Int. 113, 483–498.

Kusky, T.M., 1989. Accretion of the Archean Slave province. Geology 17, 63–67.

Kusky, T.M., Polet, A., 1999. Growth of granite–greenstone terranes at convergent margins, and stabilization of Archean cratons. Tectonophysics 305, 43–73.

Lezaeta, P., Haak, V., 2003. Beyond MT decomposition: Induction, current channeling and magnetotelluric phases over 90°. J. Geophys. Res. 108 (B6), 2305. doi:10.1029/2001JB000990

MacKenzie, J.M., Canil, D., 1999. Composition and thermal evolution of cratonic mantle beneath the central Archean Slave Province, NWT, Canada. Contrib. Mineral. Petrol. 134, 313–324.

Mackie, R.L., Madden, T.R., 1993. Conjugate direction relaxation solutions for 3-D magnetotelluric modeling. Geophysics 58, 1052–1057.

Mackie, R.L., Madden, T.R., Wannamaker, P.E., 1993. Three-dimensional magnetotelluric modeling using difference equations; theory and comparisons to integral equation solutions. Geophysics 58, 215–226.

Mackie, R.L., Smith, J.T., Madden, T.R., 1994. Three-dimensional electromagnetic modeling using finite difference equations: the magnetotelluric example. Radio Sci. 29, 923–935.

McNeice, G.W., Jones, A.G., 1998. Magnetotellurics in the frozen north: measurements on lake ice. Contributed paper at 14th EM Induction Workshop, Sinaia, Romania, August 16–23.

Minarik, W.G., Watson, E.B., 1995. Interconnectivity of carbonate melt at low melt fraction. Earth Planet. Sci. Lett. 133, 423–437.

Nakano, T., Fujii, N., 1989. The multiphase grain control percolation; its implication for a partially molten rock. J. Geophys. Res. 94, 15653–15661.

Olhoeft, G.R., 1981. Electrical properties of granite with implications for the lower crust. J. Geophys. Res. 86, 931–936.

Padgham, W.A., Fyson, W.K., 1992. The Slave Province: a distinct Archean craton. Can. J. Earth Sci. 29, 2072–2086.

Parker, R.L., 1980. The inverse problem of electromagnetic induction: existence and construction of solutions based on incomplete data. J. Geophys. Res. 85, 4421–4425.

Parker, R.L., Booker, J.R., 1996. Optimal one-dimensional inversion and bounding of magnetotelluric apparent resistivity and phase measurements. Phys. Earth Planet. Inter. 98, 269–282.

Parker, R.L., Whaler, K.A., 1981. Numerical methods for establishing solutions to the inverse problem of electromagnetic induction. J. Geophys. Res. 86, 9574–9584.

Parkinson, W.D., 1962. The influence of continents and oceans on geomagnetic variations. Geophys. J. R. Astron. Soc. 6, 441–449.

Partzsch, G.M., Schilling, F.R., Arndt, J., 2000. The influence of partial melting on the electrical behavior of crustal rocks: laboratory examinations, model calculations and geological interpretations. Tectonophysics 317, 189–203.

Pearson, N.J., Griffin, W.L., Doyle, B.J., O'Reilly, S.Y., Van Achterbergh, E., Kivi, K., 1999. Xenoliths from kimberlite pipes of the Lac de Gras area, Slave craton, Canada. In: Gurney, J.J., et al. (Eds.), Proceedings of the 7th international Kimberlite conference. P.H. Nixon, vol. 2. Red Roof Design, Cape Town, pp. 644–658.

Petitt Jr., R.A., Chave, A.D., Filloux, J.H., Moeller, H.H., 1994. Electromagnetic field instrument for the continental shelf. Sea Technol. 35, 10–13.

Rodi, W., Mackie, R.L., 2001. Nonlinear conjugate gradients algorithm for 2-D magnetotelluric inversion, in press. Geophysics 66, 174–187.

Schilling, F.R., Partzsch, G.M., Brasse, H., Schwarz, G., 1997. Partial melting below the magmatic arc in the central Andes deduced from geoelectromagnetic field experiments and laboratory data. Phys. Earth Planet. Inter. 103, 17–31.

Schmucker, U., 1970. Anomalies of geomagnetic variations in the southwestern United States. Bull. Scripps Inst. Oceanogr., vol. 13. Univ. Calif. Press, Berkeley.

Schultz, A., Kurtz, R.D., Chave, A.D., Jones, A.G., 1993. Conductivity discontinuities in the upper mantle beneath a stable craton. Geophys. Res. Lett. 20, 2941–2944.

Snyder, D., Asudeh, I., Darbyshire, F., Drysdale, J., 2002. Field-based feasibility study of teleseismic surveys at high northern latitudes: Northwest Territories and Nunavut. Geological Survey of Canada Current Research, 2002-C03. 10 pp.

Stern, R.A., Bleeker, W., 1998. Age of the world's oldest rocks refined using Canada's SHRIMP: the Acasta Gneiss Complex, Northwest Territories, Canada. Geosci. Can. 25, 27–31.

Sternberg, B.K., Washburne, J.C., Pellerin, L., 1988. Correction for the static shift in magnetotellurics using transient electromagnetic soundings. Geophysics 53, 1459–1468.

Thompson, P.H., Judge, A.S., Lewis, T.J., 1996. Thermal evolution of the lithosphere in the central Slave Province: implications for diamond genesis. In: LeCheminant, A.N., Richardson, D.G., DiLabio, R.N.W., Richardson, K.A. (Eds.), Searching for Diamonds in Canada. Geol. Surv. Canada, pp. 151–160. Open File 3228.

Thorpe, R.I., Cumming, G.L., Mortensen, J.K., 1992. A significant Pb isotope boundary in the Slave Province and its probable relation to ancient basement in the western Slave Province. Project Summaries, Canada-Northwest Territories mineral Development Subsidiary Agreement. Geol. Surv. Canada, pp. 279–284. Open-File Report 2484.

Tikhonov, A.N., Glasko, V.B., 1975. Application of the regularization method to geophysical interpretation problems. Phys. Solid Earth 11, 25–32.

Viejo, G.F., Clowes, R.M., Amor, J.R., 1999. Imaging the lithospheric mantle in northwestern Canada with seismic wide-angle reflections. Geophys. Res. Lett. 26, 2809–2812.

Vozoff, K. (Ed.), 1986. Magnetotelluric Methods. Soc. Expl. Geophys. Reprint Ser., vol. 5. Tulsa, OK, ISBN 0-931830-36-2.

Vozoff, K., 1991. The magnetotelluric method. Electromagnetic Methods in Applied Geophysics—Applications. Society of Exploration Geophysicists, Tulsa, OK, pp. 641–712. Chap. 8.

Weidelt, P., 1972. The inverse problem of geomagnetic induction. Z. Geophys. 38, 257–289.

Wessel, P., Smith, W.H.F., 1991. Free software help map and display data. EOS 72, 441.

Wu, X., Ferguson, I.J., Jones, A.G., 2002. Magnetotelluric response and geoelectric structure of the Great Slave Lake shear zone. Earth Planet. Sci. Lett. 196, 35–50.

Xu, Y., Shankland, T.J., Poe, B.T., 2000. Laboratory-based electrical conductivity of the Earth's mantle. J. Geophys. Res. 105, 27865–27875.

Available online at www.sciencedirect.com

Lithos 71 (2003) 529–539

ELSEVIER

LITHOS

www.elsevier.com/locate/lithos

Two anisotropic layers in the Slave craton

D.B. Snyder[a,*], M.G. Bostock[b], G.D. Lockhart[c]

[a] *Geological Survey of Canada, 615 Booth Street, Ottawa, ON, Canada K1A 0E9*
[b] *Department of Earth and Ocean Sciences, University of British Columbia, Vancouver, BC, Canada V6T 1Z1*
[c] *BHP-Billiton, BHP Diamonds, Inc., Kelowna, BC, Canada V1X 4L1*

Abstract

Four years of recording global earthquakes using a broadband seismometer located at the Ekati diamond mine revealed variations with earthquake azimuth in the arrival of SKS phases. These variations can be modeled assuming two distinct layers of anisotropy in the lithosphere. The lower layer probably lies in the mantle, and the anisotropy aligns with both North American plate motion and the strike of mantle structures identified by previous conductivity and geochemical analyses, at ~N50°E. The upper layer is hypothesized to result from regional structures in the uppermost mantle and the crust; these trends are distinct from the mantle trends.
© 2003 Elsevier B.V. All rights reserved.

Keywords: Seismic anisotropy; Mantle structure; Teleseismic studies; Slave craton; Lithosphere

The influence of anisotropy on seismic waves from distant earthquakes that pass through continental lithosphere is used increasingly to study deep and large-scale lithospheric structure. The anisotropy can cause variation with azimuth in the phase and polarization of SKS-waves, waves originally generated from upgoing P-waves at the core–mantle boundary. Typically, the anisotropy is attributed to strain-induced crystal alignment at mineral scales, fluid-filled cracks, or to alternating rock layers with different velocities occurring over scales of kilometers or tens of kilometers (e.g., Backus, 1962; Vinnik et al., 1992; Silver, 1996). Because of the vertical wavepaths involved, it is not always possible to determine at what depths the anisotropy occurs, but recent analytical and numerical studies provide important guidelines and clues (Silver

and Savage, 1994; Rümpker and Silver, 1998; Rümpker et al., 1999; Saltzer et al., 2000). The key to determine layered structure is an observational data set with seismic waves arriving from earthquakes at a large, well-distributed range of azimuths. Two years is usually a minimal observation window with which to fulfill this requirement, and most studies do not have the luxury of such long-term deployments.

Teleseismic studies of the Slave craton are in their infancy (e.g., compare Bostock and Cassidy, 1997; Bank et al., 2000 with Silver et al., 2001), but here we report some early results of a more extensive study planned for the next few years. The Slave hosts most of the rapidly growing Canadian diamond mining industry and an extensive geological, geochronological and geochemical knowledge base from several decades of multidisciplinary research (e.g., Bleeker et al., 1999). This combination of recent industry- and academic-based scientific interest in the craton provides a strong foundation on which to test currently evolving ideas

---

* Corresponding author. Tel.: +1-613-992-9240; fax: +1-613-943-9285.
E-mail address: dsnyder@nrcan.gc.ca (D.B. Snyder).

0024-4937/$ - see front matter © 2003 Elsevier B.V. All rights reserved.
doi:10.1016/j.lithos.2003.09.001

about anisotropy in continental lithosphere (e.g., Levin et al., 1999; Debayle and Kennett, 2000).

1. Setting

The Slave craton is unusual in that crustal studies suggest roughly north–south trends in large-scale structure, whereas several independent mantle studies reveal ENE–WSW trends (Fig. 1) (Davis and Hegner, 1992; Bleeker et al., 1999; Grütter et al., 1999; Griffin et al., 1999; Jones et al., 2001). The relevant crustal observations include isotopic studies of large intrusive bodies that show Nd and Pb isotope trends with roughly north–south contours (Fig. 1) (Davis and Hegner, 1992; Thorpe et al., 1992). Regional mapping

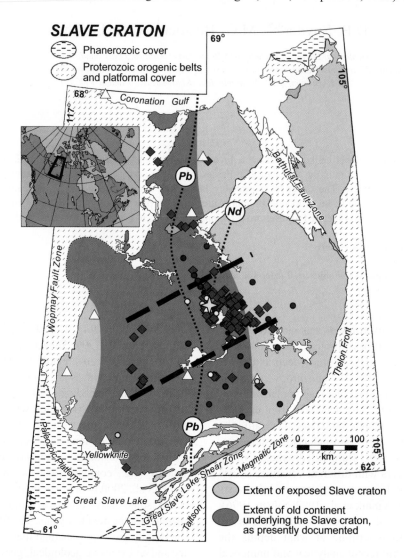

Fig. 1. Map of the Slave craton. Black diamonds locate kimberlites, white triangles are the 13 temporary teleseismic stations operated by UBC; small circles are existing (white) and proposed teleseismic sites operated by the Geological Survey of Canada (GSC). The white star near the center locates the EKAT teleseismic station. The dashed lines labelled Pb and Nd are isotopic trend interpreted to map the eastern border of the Central Slave Basement Complex (Bleeker et al., 1999 and references therein). Thicker dashed lines are boundaries between mantle with distinct geochemical characteristics as determined using garnets from kimberlites (Grütter et al., 1999). Note that these mantle trends are distinct from the crustal ones.

and closely integrated age dating of basement rocks have shown that a composite, 4.04–2.83-Ga gneiss and granitoid complex overlain by a cover sequence of quartzite and banded iron formation (Central Slave Basement Complex of Bleeker et al., 1999) lies to the west of these isotopic lines, whereas isotopically juvenile basement lies to the east (Davis et al., 1996). In addition, about a fifth of the craton is now mapped in sufficient detail to show the trends of large-scale F1 folds (Bleeker et al., 1999) that have sufficiently large wavelengths to possibly influence teleseismic waves. Significantly, these fold trends vary greatly across the exposed craton and therefore provide a good correlation test for any possible anisotropy hosted by the upper crust.

Mantle trends are different and generally strike ENE–WSW (Fig. 2). This trend was first identified using population statistics of individual garnet crystals recovered from surface till samples or xenoliths from kimberlites (Grütter et al., 1999). Geochemical analysis of these crystals revealed a zonation, with so-called G10 garnets associated with a band centered roughly on the Lac de Gras area. Other geochemical observations identify an ultra-depleted, olivine-rich layer at 70–150 km depth separated from a deeper less depleted layer by a sharp boundary (Griffin et al., 1999). Magnetotelluric observations indicate that the upper olivine-rich layer forms a prominent conductor centered, and perhaps limited to, the Lac de Gras region (Jones et al., 2001). Structural and geochemical analysis of xenoliths from kimberlites within and on either side of the G10 garnet band indicate different depths for the base of the lithosphere at the time of kimberlite eruption; estimates range from 160–190 to 230 km (Kopylova and Russell, 2000).

2. The SKS data set

Seismologists based at the University of British Columbia (UBC) established 13 temporary teleseismic stations within the Slave craton between November 1996 and May 1998; their analysis included SKS anisotropy, receiver functions and tomography (Bank et al., 2000). Logistical considerations in this cold, remote region limited the observation windows and resulted in between 3 and 24 earthquakes that were suitable for SKS splitting analysis at any individual station (e.g., Silver, 1996). The regional-scale results possible with these restricted data indicated that fast anisotropy directions generally coincided with North American plate motion direction and a consistent, smoothly varying Moho depth of 37–46 km that is also consistent with broad variations in the observed gravity field. Tomography resolved higher velocities (< 1% slowness perturbation) beneath the oldest core of the craton, the Central Slave Basement Complex (Bleeker et al., 1999), and indications of lower velocities directly beneath the Lac de Gras kimberlite field.

During the past 4 years, the Geological Survey of Canada (GSC) operated five seismic stations on the Slave craton at the Ekati, Snap Lake and Kennady Lake mine/exploration sites (Fig. 2). Each station consists of a three-component broadband sensor (Guralp CMG-40T) and Orion data logger. Early results from anisotropy and receiver function analysis are consistent with the previous results for average site anisotropy and Moho depth determinations (Fig. 2).

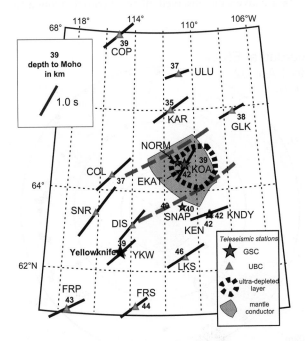

Fig. 2. Map of Moho depth determinations and average SKS anisotropy vectors (Bank et al., 2000). Also shown are the approximate outlines of a strong uppermost mantle conductor (Jones et al., 2001) and ultra-depleted harzburgite layer (Griffin et al., 1999). Other symbols as in Fig. 1; GSC operated stations appear as stars and have associated Moho depth determinations indicated.

One station, EKAT, has operated continuously since March 1999, and this more extensive data set provides a sufficient distribution of earthquake sources (Fig. 3) with which to study variations with azimuth of the incoming seismic waves (Figs. 4 and 5). The 4 years of data studied to date provided forty earthquakes with reliable SKS splitting measurements in addition to the eight reported by Bank et al. (2000) from a site less than 1000 m distant (Table 1). Four of the new events were studied at various frequency bandwidths and a window of 0.08–2.0 Hz was determined as the most stable (e.g., Rümpker et al., 1999). Useful earthquakes were located at distances of 79–113° and arrived from back-azimuths other than a 0–60° gap (Figs. 3 and 5). The other new or reoccupied sites have limited operational windows to date and the two to nine observations available at present are used as average values (Table 2). The anisotropy parameters for nine earthquakes recorded at one station, NORMS, located 6 km northwest of EKAT, are also tabulated (Table 1) and plotted (Fig. 5) along with the EKAT data to demonstrate the consistency of the results from indi-

vidual earthquakes and in general. Station EKTN replaced EKAT at the same location, with a 6-month overlap period.

3. Analysis and interpretation

New estimates of shear wave splitting were made using SKS and SKKS phases from forty teleseismic events recorded at station EKAT. Mode conversion at the core–mantle boundary ideally removes source side anisotropy effects and produces a radially polarized S-wave at the core–mantle boundary. We use the methodology of Silver and Chan (1991), which models the lithosphere as a homogeneous and weakly anisotropic medium. In this medium, near-vertical S-wave propagation is characterized by ϕ, the azimuth of the shear wave polarization with maximum velocity, and dt, the accumulated arrival time difference between fast and slow polarizations (Backus, 1965). Analysis consists of finding values of ϕ and dt which best convert the observed elliptical particle motion to

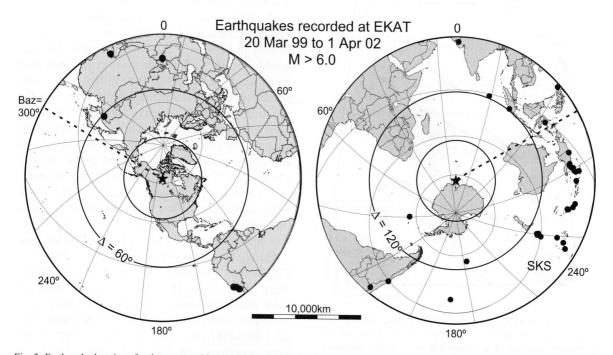

Fig. 3. Earthquake locations for the events with magnitudes greater than 6.0 recorded at EKAT (star at center of projection) and used in this study. Note the gap in events at distances between 80° and 115° for azimuths 0–60° and 180–240°; few earthquakes occur in these regions of the globe.

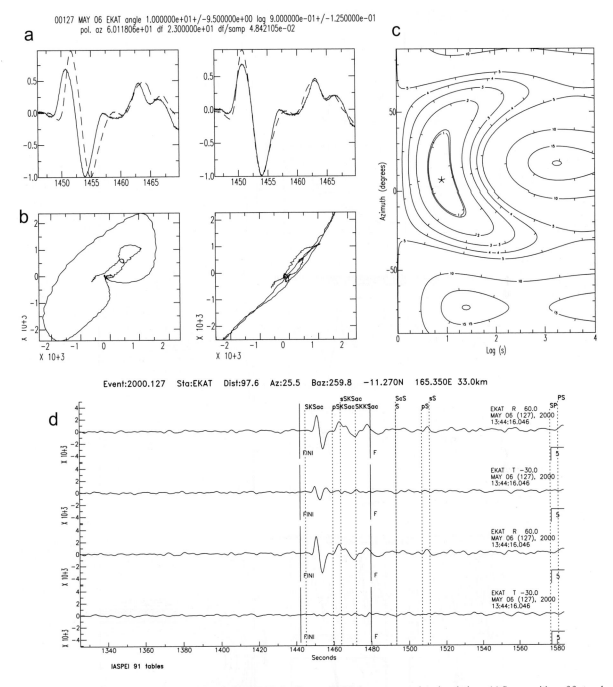

Fig. 4. Examples of SKS splitting analysis using station EKAT. See Savage (1999) for more complete descriptions. (a) Superposition of fast and slow (dashed) components uncorrected (left) and corrected (right); (b) corresponding particle motion diagrams. Note that the elliptical particle motion becomes linear when corrected. (c) Contour plot of the energy on the corrected transverse component showing minimum value (star) and 95% confidence region (double contour) and multiples of that contour interval. (d) Original, filtered radial (R) and transverse (T) component seismograms above the corrected counterparts. Note that the corrected transverse component has been minimized. (e–g) Plots as in (a–c) for earthquakes in Tibet, Papua New Guinea and Indonesia, respectively. Note the relatively large uncertainty in lag time in (g), typical of back-azimuths near the fast (or slow) polarization direction, called a node.

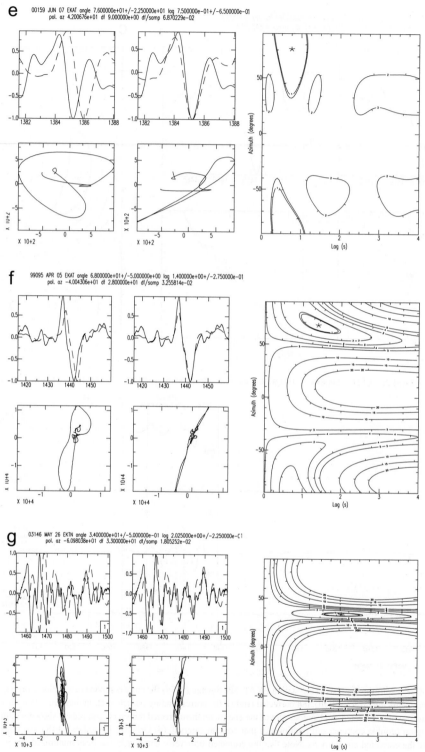

Fig. 4 (*continued*).

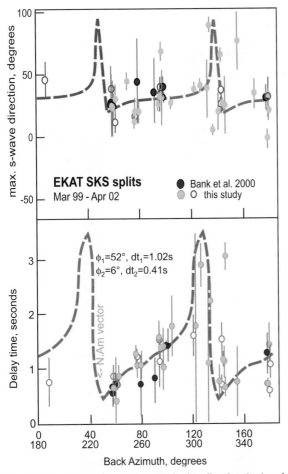

Fig. 5. Plots of the apparent fast polarization direction (top) and delay or splitting time (bottom) as a function of back-azimuth for station EKAT assuming a frequency of 0.08 Hz. Splitting parameters repeat every 180° so that the observations from both hemispheres are superimposed. The observations have standard uncertainties (e.g., Silver, 1996) shown by vertical bars. Grey dots are new measurements using EKAT, the open circles are from the nearby station NORM and are shown to demonstrate consistency of the results. The wide, dashed line shows the predicted variation of these parameters for a two-layer model (Silver and Savage, 1994) with the parameters indicated in the lower graph. The predictive model repeats every 90° so that the observational gap from 0° to 60° is not debilitating.

a linear particle motion. These assumptions require ϕ and dt to show no variation with back-azimuth, the direction to the earthquake from the station.

The estimated splitting parameters for station EKAT show a statistically meaningful and systematic variation with back-azimuth (Fig. 5). Four groups of events provide robust and statistically distinct values

for ϕ and dt at back-azimuths of 239°, 257°, 277°, and 144° (Table 3). The first three groups define a clear increasing trend in dt over back-azimuths of 225–290°. Minimums in both ϕ and dt occur at 240° and 145°. Splitting parameter determinations are unstable at 120–134° (300–314°), typical of nulls associated with alignment of back-azimuth and the fast direction of the anisotropy (e.g. Savage, 1999). All of the above observations indicate that a single overall value for either splitting parameter is not appropriate at EKAT so that the assumption of a homogeneous single layer of anisotropy is too simplistic.

Azimuthal variation of delay or splitting times between the fast and slow SKS-wave arrivals is characteristic of dipping or multiple layers of anisotropy (Fig. 5) (Silver and Savage, 1994). The variation of the apparent fast direction, while less diagnostic, is also indicative of multiple layers. The analysis and interpretation of SKS splitting is a rapidly evolving subject, and several recent papers have provided important analytical and numerical solutions for a few idealized anisotropy distributions within the mantle lithosphere (Rümpker and Silver, 1998; Rümpker et al., 1999; Saltzer et al., 2000 and references therein). The synthetic azimuthal variation of delay times and apparent fast polarization directions suggests that the EKAT site is underlain by weakly heterogeneous anisotropy that is either slowly varying with depth (Rümpker and Silver, 1998; Saltzer et al., 2000) or largely confined to two layers (Silver and Savage, 1994). Patterns characteristic of dipping layers are not observed. Because of our limited understanding of the Slave mantle structure, here the simpler of these two choices will be assumed. The numerical models show that the fast polarization direction of the thicker anisotropic layer will coincide with the azimuth where the minimum observed delay time occurs (Rümpker et al., 1999), here about 47–52° (Fig. 5). A grid search of the four parameters associated with two anisotropy layers indicates that most of the anisotropy resides in a lower layer with a fast direction of $\phi \approx 51 \pm 1°$ and a delay time of d$t \approx 0.96 \pm 0.05$ s. The preferred splitting parameters for the upper layer are $\phi = 10°$ and d$t = 0.39$ s (Fig. 5). These four values have an estimated χ^2 (chi squared) misfit of 37 to the EKAT observations. Thus, over two-thirds of the total anisotropy observed at EKAT resides in the assumed lower of the two layers.

Table 1
Earthquakes used for anisotropy studies at the EKAT (E), EKTN, and NORM (N) sites

Event	Δ	BAZ	φ	±	dt	±	Station	Earthquake location
14 October 1996 23:26	98.1	270.5	36	26	0.80	0.40		Bank: Solomon Is.
21 March 1997 12:07	110.0	238.9	24	11.5	0.70	0.23		Bank: Kermadec Is.
5 April 1997 12:23	101.0	278.46	31	2.5	1.40	0.08		Bank: Papua N. Guinea
21 April 1997 12:02	98.3	258.51	44	34	0.70	0.53		Bank: Santa Cruz Is.
3 May 1997 16:46	110.2	237.80	27	28.5	0.55	0.28		Bank: Kermadec Is.
25 May 1997 23:22	110.8	238.67	26	5	0.65	0.08		Bank: Kermadec Is.
10 June 1997 21:53	100.5	177.85	31	10	1.25	0.38		Bank: E. Pacific Rise
12 June 1997 12:07	100.6	278.54	40	8	1.40	0.30		Bank: Papua N. Guinea
3 April 1999 6:17	86.1	143.77	40	5	1.10	0.30		S. Peru (e only)
5 April 1999 11:08	99.2	276.48	68	8	1.40	0.29		Papua N. Guinea
15 November 1999 5:43	113.0	326.45	5	3	3.00	0.25		Sumatra
19 November 1999 13:56	100.25	276.91	31	22	1.00	0.40		Papua N. Guinea
11 December 1999 18:03	91.15	312.14	38	23	1.10	1.50		Philippines
8 January 2000 16:47	94.51	239.36	23	23	0.40	0.80		Fiji
6 February 2000 11:33	98.89	275.48	32	5	1.35	0.23		Papua N. Guinea
4 May 2000 4:21	105.6	302.64	38	4	1.75	2.90		Sulawesi
6 May 2000 13:44	97.56	259.8	20	9	1.05	0.25		Papua N. Guinea
7 June 2000 21:46	86.2	335.5	76	23	0.75	0.65		Tibet
17 July 2000 22:53	79.35	358.7	33	5	0.70	0.10		Afghanistan
4 October 2000 16:58	100.6	256.6	12	5	1.25	0.18		N. Hebrides
7 November 2000 00:18	134.8	126.8	41	5	2.85	0.40	E	Scotia Arc
16 November 2000 04:54	98	274	34	14	1.10	0.45	N	Papua N. Guinea
17 November 2000 21:01	98.2	274.6	26	5	1.55	0.30	E	Papua N. Guinea
			40	4	1.5	0.18	N	
18 December 2000 01:19	100.2	241.9	30	12	0.85	0.20	E	Fiji
9 January 2001 16:49	100.1	256.6	20	11	1.05	0.23	E	N. Hebrides
			16	5	1.2	0.15	N	
26 January 2001 3:16	92.2	359.2	31	16	1.05	0.35	N	Pakistan
13 February 2001 19:28	116	323	37	7	0.70	0.18	N	Sumatra
25 February 2001 21:59	79.2	358.8	−2	8	0.70	0.10	E	Afghanistan
28 April 2001 04:48	97	241	12	9	0.70	0.10	N	Kermadec
3 June 2001 02:41	108	238	26	17	0.65	0.30	E	Tonga
			38	23	0.6	1.70	N	
5 July 2001 13:53	85	144	27	4	1.15	0.13	E	Peru
6 August 2001 03:52	120	188	46	14	0.75	0.43	N	S. Pacific
10 January 2002 11:15	100	284	27	5	1.75	0.55	E	Papua N. Guinea
28 March 2002 04:56	92	141	20	15	0.75	0.30	E	Chile
1 April 2002 19:59	99	146	25	23	0.65	0.18	E	Chile
18 April 2002 16:08	97	144	21	16	0.6	0.23	EKTN	Chile
			11	19	0.75	0.3	E	
17 June 2002 21:26	98	258	7	10	1.3	0.23	E	Santa Cruz Is.
18 June 2002 13:56	100	146	16	9	0.75	0.15	EKTN	Chile
			10	12	0.75	0.2	E	
13 September 2002 22:28	100	336	32	8	0.73	0.14	EKTN	India
8 September 2002 18:44	100	284	24	7	2.08	0.64	EKTN	Papua N. Guinea
11 March 2003 22:32	97	274	27	4	1.93	0.3	EKTN	Papua N. Guinea
27 April 2003 16:03	104	252	20	9	1.08	0.21	EKTN	Vanuatu
13 May 2003 21:21	102	256	15	6	1.1	0.15	EKTN	Vanuatu
26 May 2003 19:23	102	300	34	1	2.03	0.23	EKTN	Halmahera
12 June 2003 8:59	97	271	28	4	1.43	0.18	EKTN	Papua N. Guinea
28 June 2003 15:29	99	281	32	2	1.65	0.14	EKTN	Papua N. Guinea
25 July 2003 9:37	95	278	31	3	1.53	0.16	EKTN	Papua N. Guinea
27 July 2003 11:41	92	137	5	11	0.8	0.3	EKTN	Bolivia

Table 2
Splitting parameters for temporary stations on the Slave craton used in this study

Station	Latitude (°N)	longitude (°E)	ϕ	dt	N
EKAT	64.6985	− 110.6097	33	1.2	32
BHPN	64.7328	− 110.6703	20	1.1	7
NORM	64.7797	− 110.7865	31	1.0	9
SNAP	63.5951	− 110.8684	22	0.9	3
KNDY	63.4380	− 109.1916	36	0.5	2
PRLD	62.5772	− 114.048	5	3.0	4
COP	67.827	115.09	70	0.4	10
ULU	66.895	111.00	80	0.8	4
KAR	66.029	111.47	66	0.7	3
GLK	65.920	107.46	71	0.8	3
KOA	64.692	110.63	43	1.0	8
COL	64.411	115.10	50	1.1	9
SNR	63.508	116.01	40	1.5	3
KEN	63.436	109.21	65	0.8	3
DIS	63.188	113.89	41	0.8	3
YKW	62.493	114.51	56	0.8	24
LKS	62.406	110.74	59	1.2	4
FRS	61.175	113.68	59	1.2	5
FRP	61.050	117.44	64	1.1	5

ϕ is the fast anisotropy direction (in degrees); dt is the delay or splitting time (in degrees); N is the number of observations used.

With limited data sets, azimuthal dependence cannot be investigated, and anisotropy parameters for a single station are typically averaged in order to produce representative values for the relatively large region of the lithosphere being studied. Similarly, an average can also be obtained by combining all the sites on the Slave craton studied to date. The average fast polarization direction for the whole Slave craton is $51 \pm 20°$, the average delay time is 1.0 ± 0.3 s. These values are the same, within uncertainties, as the splitting parameters for the lower anisotropic layer described above. Bank et al. (2000) noted that most of their station averages grouped around $\phi = 50°$ and that this coincided with the direction of absolute plate motion of North America ($227° \Rightarrow \phi \approx 47°$).

If this anisotropy resides in the ultramafic mantle beneath the Slave craton and derives from strain related to plate motion, then it is consistent with most published interpretations of SKS anisotropy from continental cratonic regions (e.g. Vinnik et al., 1992;

Fouch et al., 2000). The observed 1.0-s delay time suggests roughly 100 km of mantle with 4% anisotropy that is consistently aligned (Silver, 1996). Recent surface wave studies indicate that, to 200–250 km depth beneath Australia, anisotropy correlates with plate motion and is partitioned between flow-related alignment of minerals within the asthenosphere and strain-induced recrystallization and alignment of minerals as shallow as 150 km within the lithosphere (Debayle and Kennett, 2000). In the Slave province, this NE–SW (50°) trend is also that of the Great Slave shear zone located along the southern margin of the craton (Eaton et al., 2001). This shear zone is interpreted to represent hundreds of kilometers of displacement and consistent strain indicators are observed many hundreds of kilometers to the north of its trace within the craton. This general trend also coincides with that of the G10 garnet zones and the strike of the regional mantle conductor and ultra-depleted harzburgite layer discussed above.

Given the observed conductor (Jones et al., 2003) and depleted layer (Griffin et al., 1999) recognized in the mantle beneath EKAT via other observations, at present, it appears most reasonable to hypothesize that the upper anisotropic layer beneath EKAT lies above the 90-km depth marking the top of the conductor and anomalous mantle layer, up to and probably into the crust. Regional compilations of the surface geology are available for much of the Slave craton. If the 'residual' values are further restricted and hypothesized to derive from anisotropy in the upper crust, then the fast polarization directions may show trends that match either fracture trends or the axes of large-scale, first-order F1 fold structures mapped near some of the seismic stations (Bleeker et al., 1999). The recognized fractures are generally aligned with the apparent fast splitting direction of the lower layer of anisotropy, not that of the upper layer. Some residuals do match fold axis trends; others differ significantly. Further surface mapping is required to define these structural trends over larger areas; further seismic observations are required to formally determine the splitting parameters of a possible upper layer at more

Note to Table 1:
Event is the date and time (hour:minute) of the earthquake considered. Δ is the distance in degrees between EKAT and the earthquake; BAZ is the back-azimuth, the direction pointing back to the earthquake from EKAT. ϕ is the fast apparent anisotropy direction (in degrees); dt is the delay or splitting time (in seconds). Bank indicates a result taken from Bank et al. (2000).

Table 3
Splitting parameters for groups of earthquakes at similar back-azimuths

Back-azimuth	Number	Average dt	S.D.	Average ϕ	S.D.
144	9	0.81	0.18	20	10
239	7	0.64	0.13	24	5
257	7	1.08	0.18	20	12
277	12	1.46	0.24	34	11

S.D. is the standard deviation of this population.

seismic stations. Until then, this preliminary correlation between F1 fold trends and crustal anisotropy must be considered highly speculative.

The existence of two layers of anisotropy beneath the Ekati diamond mine does appear robust and the lower layer fast direction does match several independently derived trends within the mantle. To our knowledge, this combination of observations has not been reported from other cratonic regions worldwide and suggests a promising new line of research for mantle exploration.

Acknowledgements

Without the able assistance and logistical support of the exploration staff at BHP-Billiton's Ekati property, at DeBeers Canada's Kennady Lake and Snap Lake (formerly Winspear Resources) properties and at Natural Resources Canada's Yellowknife Seismological Observatory, these data would not have been acquired. Isa Asudeh (NRCan, Geological Survey of Canada) provided invaluable help with the acquisition and processing of data. C.-G. Bank (University of British Columbia) kindly provided data from earlier surveys. Geological Survey of Canada contribution number 2002121.

References

Backus, G.E., 1962. Long-wave elastic anisotropy produced by horizontal layering. Journal Geophysical Research 67, 4427–4440.
Backus, G.E., 1965. Possible forms of seismic anisotropy of the uppermost mantle under oceans. Journal Geophysical Research 70, 3429–3439.
Bank, C.-G., Bostock, M.G., Ellis, R.M., Cassidy, J.F., 2000. A reconnaissance teleseismic study of the upper mantle and tran-

sition zone beneath the Archean Slave craton in NW Canada. Tectonophysics 319, 151–166.
Bleeker, W., Ketchum, J.W.F., Jackson, V.A., Villeneuve, M., 1999. The Central Slave Basement Complex: part I. Its structural topology and autochthonous cover. Canadian Journal of Earth Science 36, 1083–1109.
Bostock, M.G., Cassidy, J.F., 1997. Upper mantle stratigraphy beneath the southern Slave Craton. Canadian Journal of Earth Science 34, 577–587.
Davis, W.J., Hegner, E., 1992. Neodymium isotopic evidence for the accretionary development of the Late Archean Slave Province. Contributions in Mineralogy and Petrology 111, 493–504.
Davis, W.J., Gariepy, C., van Breeman, O., 1996. Pb isotopic composition of Late Archean granites and the extent of recycling early Archean crust in the Slave Province, northwest Canada. Chemical Geology 130, 255–269.
Debayle, E., Kennett, B.L.N., 2000. Anisotropy in the Australian upper mantle from Love and Rayleigh waveform inversion. Earth and Planetary Science Letters 184, 339–351.
Eaton, D., Fergusin, I., Jones, A.G., Hope, J., Wu, X., 2001. A geophysical shear-sense indicator and the role of mantle lithosphere in transcurrent faulting. Programme and Abstracts for Pan-Lithoprobe Workshop III, Banff, Alberta, Canada. Lithoprobe Report, vol. 81, pp. 12–15.
Fouch, M.J., Fischer, K.M., Parmentier, E.M., Wysession, M.E., Clarke, T.J., 2000. Shear wave splitting, continental keels and patterns of mantle flow. Journal Geophysical Research 105, 6255–6276.
Griffin, W.L., Doyle, B.J., Ryan, C.G., Pearson, N.J., O'Reilly, S.Y., Davies, R.M., Kivi, K., van Achterbergh, E., Natapov, L.M., 1999. Layered mantle lithosphere in the Lac de Gras area, Slave Craton: composition, structure, and origin. Journal of Petrology 40, 705–727.
Grütter, H.S., Apter, D.B., Kong, J., 1999. Crust–mantle coupling: evidence from mantle-derived xenocrystic garnets. In: Gurney, J.J., Richardson, S.R. (Eds.), Proc. 7th Kimberlite Conference. Red Roof Designs, Cape Town, pp. 307–312.
Jones, A.G., Ferguson, I.J., Chave, A.D., Evans, R.L., McNeice, G.W., 2001. Electric lithosphere of the Slave craton. Geology 29, 423–426.
Jones, A.G., Lezaeta, P., Ferguson, I.J., Chave, A.D., Evans, R.L., Garcia, X., Spratt, J., 2003. The electrical structure of the slave craton. Lithos 71, 505–527 (this issue).
Kopylova, M.G., Russell, J.K., 2000. Chemical stratification of cratonic lithosphere: constraints from the northern Slave craton, Canada. Earth and Planetary Science Letters 181, 71–87.
Levin, V., Menke, W., Park, J., 1999. Shear wave splitting in the Appalachians and the Urals: a case for multilayered anisotropy. Journal Geophysical Research 104, 17975–17999.
Rümpker, G., Silver, P.G., 1998. Apparent shear-wave splitting parameters in the presence of vertically varying anisotropy. Geophysical Journal International 135, 790–800.
Rümpker, G., Tommasi, A., Kendall, J.-M., 1999. Numerical simulations of depth-dependent anisotropy and frequency-dependent wave propagation effects. Journal of Geophysical Research 104, 23141–23153.

Saltzer, R.L., Gaherty, J.B., Jordan, T.H., 2000. How are vertical shear wave splitting measurements affected by variations in the orientation of azimuthal anisotropy with depth? Geophysical Journal International 141, 374–390.

Savage, M.K., 1999. Seismic anisotropy and mantle deformation: what have we learned from shear wave splitting. Reviews of Geophysics 37, 65–106.

Silver, P.G., 1996. Seismic anisotropy beneath the continents: probing the depths of geology. Annual Review of Earth and Planetary Sciences 24, 385–432.

Silver, P.G., Chan, W.W., 1991. Shear wave splitting and subcontinental mantle deformation. Journal Geophysical Research 96, 16429–16454.

Silver, P.G., Savage, M.K., 1994. The interpretation of shear wave splitting parameters in the presence of two anisotropic layers. Geophysical Journal International 119, 949–963.

Silver, P.G., Gao, S.S., Liu, K.H., the Kaapvaal Seismic Group, 2001. Mantle deformation beneath southern Africa. Geophysical Research Letters 28, 2493–2496.

Thorpe, R.I., Cumming, G.L., Mortensen, J.K., 1992. A significant Pb isotope boundary in the Slave Province and its probable relation to ancient basement in the western Slave Province. Project Summaries, Canada–Northwest Territories Mineral Development Subsidiary Agreement. Geological Survey Canada Open-File Report, vol. 2484, pp. 279–284.

Vinnik, L.P., Makayeva, L.I., Milev, A., Usenko, A.Y., 1992. Global patterns of azimuthal anisotropy and deformations in the continental mantle. Geophysical Journal International 111, 433–447.

Available online at www.sciencedirect.com

Lithos 71 (2003) 541–573

ELSEVIER

LITHOS

www.elsevier.com/locate/lithos

Petrology and U–Pb geochronology of lower crustal xenoliths and the development of a craton, Slave Province, Canada

W.J. Davis[a,*], D. Canil[b], J.M. MacKenzie[c], G.B. Carbno[d]

[a] Geological Survey of Canada, 601 Booth Street, Ottawa, On, Canada K1A 0E8
[b] School of Earth and Ocean Sciences, University of Victoria, 3800 Finnerty Road, Victoria, BC, Canada V8W 3P6
[c] Redlen Crystals Ltd., 1290 Broad Street, Victoria, BC, Canada V8W 2A5
[d] Ericsson Research, 8400 Decarie Boulevard, Mount Royal, QC, Canada H4P 2N2

Abstract

Lower crustal xenoliths recovered from Eocene to Cambrian kimberlites in the central and southern Slave craton are dominated by mafic granulites (garnet, clinopyroxene, plagioclase ± orthopyroxene), with subordinate metatonalite and peraluminous felsic granulites. Geothermobarometry indicates metamorphic conditions of 650–800 °C at pressures of 0.9–1.1 GPa. The metamorphic conditions are consistent with temperatures expected for the lower crust of high-temperature low-pressure (HT-LP) metamorphic belts characteristic of Neoarchean metamorphism in the Slave craton. U–Pb geochronology of zircon, rutile and titanite demonstrate a complex history in the lower crust. Mesoarchean protoliths occur beneath the central Slave supporting models of an east-dipping boundary between Mesoarchean crust in the western and Neoarchean crust in the eastern Slave. At least, two episodes of igneous and metamorphic zircon growth occurred in the interval 2.64–2.58 Ga that correlate with the age of plutonism and metamorphism in the upper crust, indicating magmatic addition to the lower crust and metamorphic reworking during this period. In addition, discrete periods of younger zircon growth at ca. 2.56–2.55 and 2.51 Ga occurred 20–70 my after the cessation of ca. 2.60–2.58 Ga regional HT-LP metamorphism and granitic magmatism in the upper crust. This pattern of younger metamorphic events in the deep crust is characteristic of the Slave as well as other Archean cratons (e.g., Superior). The high temperature of the lower crust immediately following amalgamation of the craton, coupled with evidence for continued metamorphic zircon growth for >70 my after 'stabilization' of the upper crust, is difficult to reconcile with a thick (200 km), cool lithospheric mantle root beneath the craton prior to this event. We suggest that thick tectosphere developed synchronously or after these events, most likely by imbrication of mantle beneath the craton at or after ca. 2.6 Ga. The minimum age for establishing a cratonic like geotherm is given by lower crustal rutile ages of ca. 1.8 Ga in the southern Slave. Transient heating and possible magmatic additions to the lower crust continued through the Proterozoic, with possible additional growth of the tectosphere.
© 2003 Elsevier B.V. All rights reserved.

Keywords: Archean; Slave Province; Lower crustal xenoliths; Absolute age U/Pb-zircon; *P–T* conditions; Thermal history

1. Introduction

* Corresponding author. Tel.: +1-613-943-8780; fax: +1-613-995-7997.
E-mail address: bidavis@NRCan.gc.ca (W.J. Davis).

Deep mantle roots are recognized beneath many Archean cratons and likely played an essential role in the origin and stability of Earth's first continents

0024-4937/$ - see front matter © 2003 Elsevier B.V. All rights reserved.
doi:10.1016/S0024-4937(03)00130-0

(Jordan, 1975). Most constraints on the development of Archean cratons and their mantle roots are derived from kimberlite-hosted mantle xenoliths that sample the deeper lithosphere, and the geology of surface rocks that record orogenic processes in the upper crust. The lower crust is an important interface linking the top and bottom of the lithosphere but, with few exceptions, is poorly represented in discussions of the geological evolution of Archean cratons (e.g., Rudnick, 1992; Moser et al., 1996; Davis, 1997; Schmitz and Bowring, 2000; Moser et al., 2001). This is mainly due to the relative paucity of information on lower crust beneath cratons, in comparison to a much larger database on mantle xenoliths and surface geology.

Isotopic studies of mantle rocks hosted in kimberlites from several Archean cratons show that the original formation of cratonic mantle lithosphere dominantly occurred in the Archean (Pearson, 1999) in tectonic environments that are debated (Herzberg, 1993; Kelemen et al., 1998; Walter, 1999; Canil, 2002). Current models predict that construction of a thick mantle root beneath Archean cratons by either in situ subcretion from below, or shallow subduction of buoyant oceanic plates (Helmstaedt and Schulze, 1989; Abbott, 1991; DeWit et al., 1992), would cease during the last pulse of orogenic activity recorded in the surface geology. However, lower crustal granulites from at least three Archean cratons (Superior, Slave and Kaapvaal) record metamorphic ages and/or deformational structures that post-date the formation and deformation ages recorded in surface rocks in greenstone belts by tens to hundreds of million years (Krogh, 1993; Davis, 1997; Moser et al., 1996; Moser and Heaman, 1997; Schmitz and Bowring, 2000; Moser et al., 2001). How lower crustal rocks reach granulite grade in terranes that are thought to be underlain by stable, cool and insulating mantle roots raises the question as to when the cool, insulating mantle root actually forms beneath a craton (Moser et al., 2001).

This paper describes the petrology and geochronology of the lower crust as represented by xenoliths hosted in several kimberlite pipes from the southern part of the Slave craton. The samples preserve a complex record of lower crustal development from ca. 3.2 Ga through the Neoarchean and Paleoproterozoic that bears on the timing of mantle root forma-

tion beneath the Slave Province, Canada. The Slave craton is an excellent location to address this problem because it has recently been the subject of several geological, geochronological and geophysical investigations of both mantle (Griffin et al., 1999; Kopylova and Russell, 2000; Aulbach et al., 2001; Irvine et al., 2001; Carbno and Canil, in press) and crust (Bleeker and Davis, 1999 and references therein).

2. The Slave Province

The Archean Slave Province is a small, well-exposed craton in northwestern Canada with a rock record spanning over 2.0 Ga (Fig. 1; Isachsen and Bowring, 1994; Bleeker and Davis, 1999). The exposed crust is dominated by metasedimentary and lesser metavolcanic rocks of the Yellowknife Supergroup dated at 2.71–2.61 Ga intruded by extensive metaluminous and peraluminous granitoid plutons at 2.64–2.58 Ga (Padgham and Fyson, 1992; van Breemen et al., 1992). Regional deformation and associated high-temperature low-pressure (HT-LP) metamorphism accompanied plutonism at ca. 2.6–2.58 Ga (Fyson and Helmstaedt, 1988; Thompson, 1989a,b; Davis and Bleeker, 1999). The western and eastern parts of the craton are distinct, with a Mesoarchean (>2.9 Ga) basement and cover sequence, termed the Central Slave Basement Complex, recognized only in the western half of the craton (Bleeker et al., 1999), and generally more isotopically juvenile terranes in the east. These two parts of the craton are divided by distinct isotopic boundaries, referred to as the Pb and Nd isotopic lines (Fig. 1), which show changes in the isotopic composition of Pb and Nd in granitoids (Davis and Hegner, 1992; Davis et al., 1996; Yamashita et al., 1999), and Pb in sulfides (Thorpe et al., 1992).

Eocene to Cambrian-aged kimberlites occur throughout the Slave Province (Fig. 1) and host mantle xenoliths and xenocrysts that show evidence for a ~ 200-km-thick lithospheric root, with considerable lateral and vertical variability in its chemical characteristics (Griffin et al., 1999; MacKenzie and Canil, 1999; Kopylova and Russell, 2000; Carbno and Canil, in press). Mantle garnet concentrates in surface till samples gathered for diamond exploration purposes (Grütter et al., 1999) are consistent with three

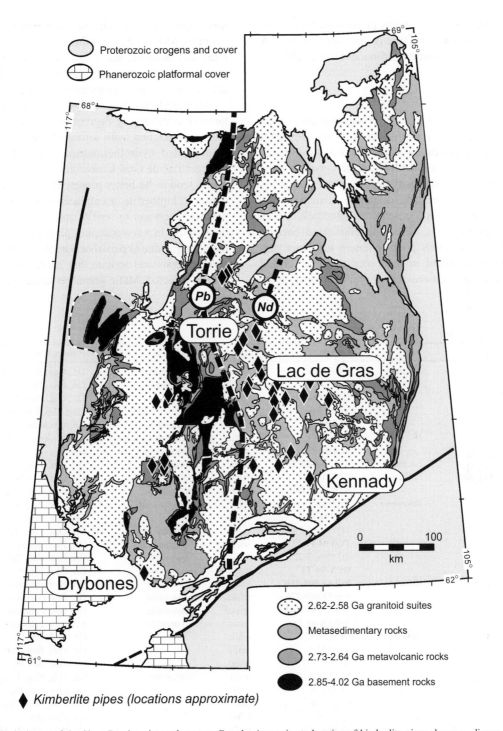

♦ *Kimberlite pipes (locations approximate)*

Fig. 1. Geological map of the Slave Province in northwestern Canada. Approximate location of kimberlite pipes shown as diamonds. Also shown is the Pb isotopic boundary (Thorpe et al., 1992) dividing the eastern and western parts of the craton, and the Nd isotopic boundary in granites (Davis and Hegner, 1992). The samples from this study are all from drill cores of kimberlites at Drybones, Kennady Lake, Torrie and the Grizzly, Fox and Koala pipes, which are all in the Lac de Gras area.

broadly NE trending domains in the underlying mantle, but this domain structure is interrupted in the SW region near Yellowknife (Carbno and Canil, 2002).

2.1. Samples

The lower crustal rocks were sampled as 2–6 cm diameter xenoliths in drill cores from six kimberlite pipes with emplacement ages from Eocene to Cambrian (Fig. 1, Table 1). Most of the xenoliths were identified with easily recognizable garnet and/or pyroxene. Mafic granulites dominate felsic types, a global trend in granulite xenoliths (Rudnick, 1992), but a sampling bias cannot be eliminated because some felsic granulites poor or absent in garnet may have been identified in drill core as upper crustal granitoids. In addition, felsic xenoliths tend to be

much more strongly altered compared to the mafic varieties. Most of the granulites are equigranular (1–3 mm) and massive, but a few samples have varying proportions of pyroxenes, plagioclase and garnet that define millimeter- to centimeter-sized bands or layers.

All xenoliths show well-developed granoblastic textures with varying degrees of secondary alteration due to interaction with kimberlite melt or emplacement-related hydrothermalism. Xenoliths from the younger Lac de Gras kimberlites (Fox, Grizzly, Koala pipes) tend to be better preserved with minimal infiltration of kimberlite or emplacement-related low temperature alteration to sericite and chlorite along grain boundaries. In a few samples, plagioclase is chalky in hand sample due to partial or complete replacement by clay minerals and sericite, but garnet and pyroxenes are not affected. Mafic granulites contain mainly Grt,

Table 1
Granulite sample data

Kimberlite pipe	Age	Sample	Type	Rock type	Primary mineralogy[a]
Lac de Gras-Grizzly	Eocene	**GR-95-36**[b]	felsic	metatonalite	Grt, Opx, Ilm, Mt (Plag)
		GR9557	mafic	metagabbro	Grt, Cpx, Plag
		GR9500	mafic	metagabbro	Grt, Cpx, Plag, Ilm, Ap, Rt, Ttn
		X9501BB	felsic		Plag, Ksp, Cpx, Opx, Ttn, Qz
		92-2-942[b]	mafic	metagabbro	Grt, Cpx, Plag
		92-2-732[b]	mafic	altered metagabbro	Grt, Hb, Cpx?, Plag?
		92-2-865[b]	mafic	altered	Grt?
Lac de Gras-Fox	Eocene	**FUC-4-6-42.26**[b]	mafic	metagabbro	Grt, Cpx, Plag
Lac de Gras-Koala	Eocene	KDC-13-131	felsic	metagranitoid	Grt, Opx, Cpx, Plag, Biot
		KDC-13-141[b]	mafic	metagabbro	Grt, Opx, Cpx, Plag, Ilm
Torrie	Paleocene	TQ9511	felsic	metagranitoid	Grt, Cpx, Plag, Ilm
		TQY94-17-9	mafic		Grt, Cpx, Plag
		TQY94-17-22	felsic	metagranitoid	Grt, Opx, Cpx, Plag, Kspar, Ilm
		TQY94-17-22E[b]	felsic	metagranitoid	Grt, Herc, Kspar, Sill, Biot
		TQY94-17-22J	felsic	metagranitoid	Grt, Cpx, Plag, Ilm
		DRA-94-T1[b]	mafic	metagabbro	Grt, Cpx, Plag
Kennady Lake	~550 Ma	**96BAK009-416**[b]	felsic	metatonalite	Grt, Biot inclusions
		96BAK009-226	felsic	metatonalite	Grt, Biot inclusions
		96BAK004-263[b]	felsic	metatonalite	Grt, Biot inclusions
		96BAK-41-166[b]	felsic	metatonalite	Grt, Biot inclusions
Drybones Bay	~450 Ma	96.21.326A	mafic		Grt, Plag, Ilm
		96.17.318	mafic		Grt, Plag, Ilm, (Opx)
		96.21.390.7	felsic		Grt, Plag, Ilm, Rt
		95.17.298	mafic	metagabbro	Grt, Cpx, Plag, Ilm, Rt,
		96.20.274	mafic		Grt, Plag, Ilm, Rt, (Opx)
		96.15.244.7	mafic		Grt, Plag, Ilm, (Opx)
		96.15.455.7	felsic		Grt, Plag, Ilm
		96.16.438.5	mafic		Grt, Plag, Ilm, Rt, Mt, (Cpx)
		96.21.326.5	mafic		Grt, Plag, Ilm, (Opx)

[a] Pseudomorphed primary mineral given in brackets.
[b] Samples with geochronology data in bold font.

Cpx and Plag, with subordinate Qz ± Opx ± Ilm ± Rt ± Tit. All felsic samples are metagranitoids with Grt ± Plag ± Kspar ± Qz ± Opx ± Cpx ± Ilm ± Rt. In the latter samples, K-feldspar shows coarse perthitic textures and one peraluminous sample (TQY94-17-22E) contains blocky sillimanite grown on hercynite-rich spinel.

Granulite xenoliths from the older kimberlites at Kennady Lake and Drybones Bay are more severely altered (Table 1). Only primary garnet and plagioclase remain in the Drybones Bay samples, with orthopyroxene and clinopyroxene completely pseudomorphed by chlorite/serpentine and chlorite/tremolite mixtures, respectively. Primary clinopyroxene is preserved in the cores of a few grains in one Drybones Bay sample (95.17.298). In the Kennady Lake samples, all primary phases except garnet (and its inclusions of biotite and quartz) are altered to clay minerals, carbonates and serpentine/chlorite mixtures. Nonetheless, primary granoblastic textures are preserved by the primary mineral pseudomorphs.

Mineral chemical data are given in Appendix A. The primary minerals in the xenoliths are chemically homogeneous and minor zoning is recognized only in plagioclase, as shown by an increase in Or content towards margins. This zoning is attributed to interaction with host kimberlite liquid. In many samples, magnetite is broken down to ilmenite–ulvospinel by oxy-exsolution as observed in many granulites (Frost and Chacko, 1989).

3. Thermobarometry

The homogeneous compositions and textures for most minerals in the granulite xenoliths suggest they were well-equilibrated before entrainment in their host kimberlite. A few samples have mineral assemblages amenable to TWQ multi-equilibrium thermobarometry (Berman, 1991), which uses a common intersection of the calculated *P–T* arrays for reactions among components of all minerals to represent the likely *P–T* condition recorded by the rock. The results of TWQ calculations for four Lac de Gras samples are shown in Fig. 2 with the limiting reactions used for the *P–T* intersections in each sample given in Table 2. These samples show equilibration conditions between 0.9 and 1.1 GPa and 650 and 800 °C.

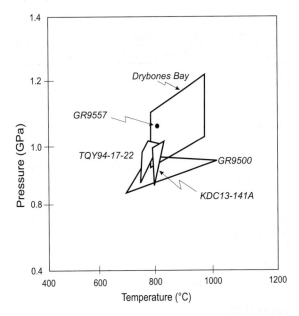

Fig. 2. Estimated pressure–temperature conditions using TWQ thermobarometry (Berman, 1991) for four granulite xenoliths from the Lac de Gras area, and estimated conditions for Drybones Bay samples in which primary orthopyroxene is not preserved (Table 2). Mineral data in Appendix A.

In most of the granulite xenoliths, the rock contained too few phases, or phases that were not demonstrably in equilibrium, to converge on a unique *P–T* location using TWQ. In these samples, exchange geothermometers were used to calculate the temperatures of equilibration at an assumed *P* of 1.0 GPa, which is near that calculated using TWQ for four samples (Table 2) and presented as a histogram in Fig. 3. The Fe–Mg exchange temperatures between garnet and pyroxenes for several xenoliths are between 650 and 800 °C. The results shown in Fig. 3 differ by less than ± 50 °C using different versions of Fe–Mg geothermometers. Almost all granulites show evidence for re-setting of Fe–Mg temperatures due to slow cooling (Frost and Chacko, 1989), and these conditions should be considered minimum temperatures for metamorphism. Two-pyroxene solvus temperatures are similar to Fe–Mg exchange temperatures, even though exchange of Ca between two pyroxenes is much slower than Fe–Mg exchange (Smith and Barron, 1991). No attempt was made to re-correct Fe–Mg temperatures for effects of retrograde exchange. That temperatures greater than 800

Table 2
Thermobarometry for granulite samples

Sample	T (°C) cpx-grt[a]	T (°C) opx-cpx[b]	T (°C) opx-grt[c]	T (°C) grt-biot[d]	TWQ reactions[e]
GR-95-36[f]			770		
GR9557	734				2Gr + Py + 3Qz↔3Di + 3An
					Alm + 2Gr + 3Qz↔3Hed + 3An
					Alm + 3Di↔Py + 3Hed
GR9500	723				2Rt + Gr + Alm↔2An + Hed + 2Ilm
					2Rt + Py + 2Hed + Gr↔2An + 3Di + 2Ilm
					4Rt + Hed + Gr↔An + Ilm + 3Ttn
KDC-13-131	754	703	698	714	
KDC-13-141A	749	741	731		Qz + Py + Di↔An + Fsl + 3En
					Alm + 3Di↔Py + 3Fsl
					Di + Fsl↔En + Hed
TQ9511	695				
TQY94-17-9	796				
TQY94-17-22	714	677	656		Alm + Gr + Py + 3Ttn↔3Ilm + 3Di + 3An
					Gr + 3Hed + 2Py + 3ttn↔3Ilm + 6Di + 3An
					Alm + Ttn↔Ilm + 2Fsl + An
					Alm + 3En↔Py + 3Fsl
					Di + Fsl↔En + Hed
TQY94-17-22E[f]				566	
TQY94-17-22J	702				
X9501BB		654			
KDC13-141B	771				
96BAK041-229				500	
96BAK009-416[f]				582	
95.17.298	760				

All thermometers calculated at assumed P of 1.0 GPa.

[a] Ellis and Green (1979).

[b] Brey and Kohler (1990).

[c] Harley (1984).

[d] Perchuk and Lavrent'eva (1983).

[e] Limiting reactions used for P–T polygon plotted in Fig. 3 (Berman, 1991).

[f] Samples with geochronology data in bold font.

°C have been attained in the lower crust is evident in at least one sample (TQY94-17-22E) by overgrowth of blocky sillimanite on hercynite that is included or surrounded by larger almandine-rich garnets. This texture likely records cooling of the rock through the equilibrium:

2Sill + Alm = 3Hc + 5Qz

which occurs in the end-member FeO–Al_2O_3–SiO_2 system at ~1000 °C and 0.8 GPa.

In all the Kennady Lake samples and in one Torrie sample, biotite included in garnets record temperatures of less than 600 °C using the garnet–biotite Fe–Mg exchange thermometer of Perchuk and Lavrent'eva (1983). This effect is likely due to a lower closure temperature for this system combined with the

high mass of garnet to biotite in all samples that would accentuate retrograde Fe–Mg exchange. A higher T garnet–biotite pair is preserved in only one granulite sample (KDC-13-131).

Both TWQ and exchange thermobarometry could not be applied to all but one of the Drybones Bay samples because they do not contain primary pyroxenes. To make some estimation of the P–T conditions for these rocks, the Fe and Mg contents of orthopyroxenes that would be in equilibrium with the preserved, primary garnets at temperatures of 650–1000 °C were calculated using experimentally measured K_d for Fe–Mg exchange between garnet and orthopyroxene (Harley, 1984). This temperature range was selected because it represents wide limits on Fe–Mg temperatures recorded by the other granulite xenoliths

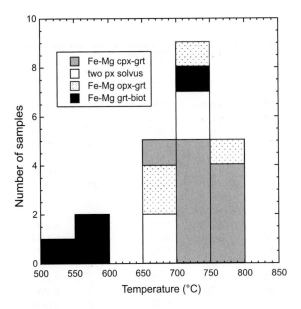

Fig. 3. Histogram showing Fe–Mg exchange temperatures calculated at 1.0 GPa using garnet–biotite (Perchuk and Lavrent'eva, 1983), garnet–orthopyroxene (Harley, 1984), garnet–clinopyroxene (Ellis and Green, 1979) and two-pyroxene solvus (Brey and Köhler, 1990) thermometry for granulite xenoliths from this study (Table 2).

in which all primary phases are preserved (Fig. 3). These fictive Fe–Mg orthopyroxene compositions were then used in the reaction:

$$3Qz + 2Py + Gr = 3An + 6Fsl$$

to estimate P over the T range 650–1000 °C. The pressures so derived are maxima because Qz is not present in the samples. The Ca and Al contents of the 'fictive' orthopyroxenes could not be calculated by this method, but these are minor components in crustal orthopyroxenes and not likely to greatly affect the P estimates.

Furthermore, the technique could only be applied to samples that originally contained orthopyroxene (now occurring as serpentine pseudomorphs) because the larger Ca and Al contents of the primary clinopyroxenes (now tremolite/chlorite pseudomorphs) could not be calculated with any confidence by this approach. The combined P–T results for all Drybones Bay granulite samples are similar to the results for granulite xenolith samples from the Lac de Gras area in which all primary phases are preserved (Fig. 2).

4. Geochronology

Eleven xenolith samples were studied for U–Pb geochronology using isotope dilution thermal ionization mass spectrometry (ID TIMS) and three of these were further studied using the ion microprobe (SHRIMP). Sample weights ranged from 10 to 140 g (Table 3). Analytical methods are given in Appendix B and the data presented in Tables 3 and 4.

4.1. Intermediate to felsic xenoliths

Sample GR-95-36 is a tonalite gneiss from the Grizzly pipe with mm-thick mafic bands defined by orthopyroxene, garnet and accessory ilmenite and magnetite (Table 1). Zircons typically occur as subhedral, round to prismatic grains. Internal structures revealed by CL imaging demonstrate a range of zoning types including homogenous, unzoned cores, finely to broadly oscillatory-zoned grains and polygonal sector zoned grains (Fig. 4a–c). Some grains are mantled by overgrowths of variable thickness (Fig. 4a–c). Overgrowths can also be seen under the optical microscope and when possible these were physically separated for TIMS analyses. Individual overgrowth

Notes to Table 3:

Pb_c = total common Pb in analysis corrected for spike and fractionation.

Concentration uncertainty varies with sample weight: >10% for sample weights < 10 μg, <10% for sample weights above 10 μg.

Ratios corrected for spike, fractionation, blank and initial common Pb, except $^{206}Pb/^{204}Pb$ ratio corrected for spike and fractionation only. Errors are 1σ in %.

$^{206}Pb/^{238}U$ age and $^{207}Pb/^{206}Pb$ age with 2σ absolute error in Ma.

[a] Abbreviations: ovg = overgrowth, fg = fragment, eu = euhedral, an = anhedral, ro = round, sub = subhedral, eq = equant, pr = prismatic, clr = clear, bl = black, br = brown, pbr = pale brown, dk = dark, op = opaque, irr = irregular shape.

[b] Refers to number of grains in fraction.

* Radiogenic Pb.

Table 3
U–Pb analytical data, isotope dilution analyses

Fraction	Description[a]	# grains[b]	Weight (µg)	U (ppm)	Pb (ppm)*	Th/U	$^{206}Pb/^{204}Pb$	Pb_c (pg)	$^{208}Pb/^{206}Pb$
GR-95-36 (sample weight = 57 g) GSC lab number 4325									
F1-A1	clr, co, ovg, fg	2	2	821	409	0.065	1156	32	0.0181
F1-A2	clr, co, ovg, fg	1	2	501	257	0.152	508	46	0.0423
Z1-C1	clr, core	1	4	644	323	0.055	17797	4	0.0153
Z1-C2	clr, core	1	3	754	378	0.058	5717	11	0.0163
Z2-A	clr, co, ovg	1	4	82	56	1.655	1919	5	0.4616
Z2-C	clr, core	1	2	347	178	0.103	2905	5	0.0287
Z7A	clr, eu	1	4	170	98	0.691	674	27	0.1927
96-BAK-41-166 (138 g) GSC lab number 5166									
Z2A	ovg	1	4	155	91	0.769	5112	3	0.2142
Z2B	ovg	1	2	120	71	0.805	2037	4	0.2244
Z2C	ovg	1	4	82	45	0.477	1254	8	0.1331
Z3A	eu, eq	1	7	198	112	0.605	6593	6	0.1686
Z3B	eu, eq	1	5	218	122	0.548	21999	2	0.1527
Z3C	eu, eq	1	4	170	98	0.711	5869	3	0.1981
Z1B	eu, eq	1	2	163	85	0.157	5798	2	0.0438
R1	bl, op, an, 250	>20	463	14	4	0.018	2332	52	0.0053
R2	bl, op, an, 175	>20	468	14	4	0.022	2345	48	0.0066
R3	bl, op, an, 225	>20	282	12	4	0.002	7343	9	0.0005
R4	br, clr, eu, 100	>20	90	7	2	0.064	215	52	0.0187
96-BAK-009-416 (93.9 g) GSC lab number 5164									
Z1A	clr, eu, ro	1	2	45	27	0.640	1549	2	0.1777
Z1B	clr, eu, ro	2	2	161	86	0.344	3911	3	0.0959
Z2A	clr, eu, pr	1	3	120	57	0.197	926	9	0.0554
Z2C	clr, eu, pr	1	3	235	135	0.620	4579	4	0.1727
TQY-94-17-22E (18.8 g) GSC lab number 5167									
Z1A	clr, eu, eq	1	5	407	242	0.032	13817	5	0.0088
Z1B	clr, eu, eq	1	4	393	240	0.015	13601	4	0.004
Z1C	clr, eu, eq	1	6	117	74	0.415	9766	2	0.1139
Z1D	clr, eu, eq	1	2	278	160	0.359	1640	10	0.0994
Z1E	clr, eu, eq	1	2	429	252	0.498	6892	3	0.138
Z1F	eu, ro, clr,	3	2	268	162	0.698	4728	4	0.1937
Z3A	clr, ovg	1	2	163	84	0.158	2369	5	0.0439
Z3B	br, ovg	1	2	491	310	0.290	13676	2	0.0792
96-BAK-004-263 (130.3 g) GSC lab number 5165									
T1	br, dk, sub	11	17	91	54	1.070	2370	18	0.2995
T2	br, sub	16	13	179	106	0.988	5793	11	0.2766
KDC-13-141 (30.4 g) GSC lab number 5168									
Z1A	clr, eq, eu	1	3	163	96	0.821	1735	8	0.2288
Z2A	clr, eq, eu	1	6	250	139	0.537	12450	4	0.1497
Z2B	clr, eq, eu	1	3	163	89	0.493	6345	3	0.1374
Z2C	clr, eq, eu	1	5	101	58	0.695	8249	2	0.1941
R1	bl, op, sub, 250		367	46	9	0.206	2594	75	0.0622
R2	bl, op, sub, 150	44	259	44	7	0.005	1755	70	0.0014
R3	bl, op, an, 250	33	156	50	9	0.001	3807	24	0.0002

^{206}Pb/^{238}U	1σ error	^{207}Pb/^{235}U	1σ error	^{207}Pb/^{206}Pb	1σ error	Apparent age (Ma)			
						^{206}Pb/^{238}U	^{207}Pb/^{206}Pb	2σ	% disc.
0.4857	0.0008	11.722	0.019	0.17506	0.00012	2551.9	2606.6	2.5	2.5%
0.4903	0.0013	11.634	0.028	0.17211	0.00024	2571.9	2578.3	4.7	0.3%
0.4899	0.0004	11.723	0.012	0.17353	0.00005	2570.5	2592.0	1.0	1.0%
0.4898	0.0004	11.698	0.012	0.17321	0.00007	2570.0	2588.9	1.2	0.9%
0.4880	0.0007	11.624	0.027	0.17274	0.00028	2562.2	2584.4	5.4	1.0%
0.4964	0.0006	12.034	0.017	0.17583	0.00011	2598.2	2613.9	2.1	0.7%
0.4883	0.0011	11.698	0.025	0.17375	0.00021	2563.2	2594.1	4.0	1.4%
0.4918	0.0005	11.711	0.015	0.17271	0.00010	2578.4	2584.1	1.9	0.3%
0.4921	0.0005	11.746	0.014	0.17311	0.00009	2579.8	2588.0	1.5	0.4%
0.4877	0.0006	11.615	0.016	0.17274	0.00009	2560.5	2584.4	1.8	1.1%
0.4902	0.0005	11.739	0.014	0.17370	0.00007	2571.3	2593.6	1.4	1.0%
0.4906	0.0004	11.728	0.013	0.17336	0.00007	2573.4	2590.3	1.4	0.8%
0.4895	0.0005	11.695	0.014	0.17328	0.00007	2568.6	2589.6	1.4	1.0%
0.4978	0.0005	12.063	0.014	0.17573	0.00009	2604.5	2613.0	1.7	0.4%
0.3172	0.0003	5.185	0.007	0.11856	0.00007	1775.9	1934.7	2.2	9.4%
0.2936	0.0003	4.465	0.006	0.11032	0.00007	1659.3	1804.7	2.2	9.1%
0.3181	0.0003	5.009	0.006	0.11422	0.00006	1780.4	1867.6	1.7	5.3%
0.2922	0.0005	4.445	0.027	0.11034	0.00054	1652.4	1805.0	17.9	9.6%
0.5063	0.0009	12.470	0.022	0.17864	0.00011	2640.7	2640.2	2.0	− 0.0%
0.4854	0.0008	11.736	0.020	0.17538	0.00011	2550.6	2609.6	1.8	2.7%
0.4466	0.0007	10.921	0.019	0.17735	0.00011	2380.1	2628.3	2.1	11.3%
0.4952	0.0008	12.142	0.021	0.17782	0.00011	2593.2	2632.7	1.8	1.8%
0.5651	0.0006	16.646	0.020	0.21365	0.00009	2887.6	2933.4	1.3	1.9%
0.5814	0.0005	17.486	0.019	0.21811	0.00009	2954.7	2966.8	1.3	0.5%
0.5559	0.0005	15.834	0.019	0.20659	0.00008	2849.6	2879.0	1.4	1.3%
0.5190	0.0005	13.637	0.018	0.19058	0.00011	2694.9	2747.1	1.8	2.3%
0.5143	0.0005	13.553	0.016	0.19113	0.00008	2675.0	2751.9	1.4	3.4%
0.5076	0.0005	12.976	0.014	0.18541	0.00006	2646.4	2701.8	1.1	2.5%
0.4937	0.0005	12.166	0.016	0.17873	0.00009	2586.5	2641.2	1.7	2.5%
0.5689	0.0006	16.589	0.020	0.21148	0.00008	2903.4	2916.9	1.3	0.6%
0.4730	0.0004	11.050	0.012	0.16944	0.00007	2496.6	2552.1	1.3	2.6%
0.4724	0.0004	11.020	0.011	0.16920	0.00005	2493.9	2549.7	1.1	2.6%
0.4917	0.0005	11.669	0.013	0.17211	0.00009	2578.1	2578.3	1.6	0.0%
0.4886	0.0004	11.545	0.013	0.17139	0.00007	2564.4	2571.3	1.4	0.3%
0.4875	0.0005	11.492	0.014	0.17098	0.00009	2559.9	2567.2	1.5	0.3%
0.4844	0.0005	11.307	0.014	0.16932	0.00007	2546.2	2550.9	1.5	0.2%
0.1919	0.0002	2.281	0.003	0.08623	0.00005	1131.4	1343.4	2.5	17.2%
0.1786	0.0002	2.036	0.003	0.08267	0.00007	1059.5	1261.5	3.2	17.4%
0.1937	0.0002	2.320	0.003	0.08686	0.00004	1141.5	1357.5	2.1	17.4%

(continued on next page)

Table 3 (*continued*)

Fraction	Description[a]	# grains[b]	Weight (µg)	U (ppm)	Pb (ppm)*	Th/U	$^{206}Pb/^{204}Pb$	Pb_c (pg)	$^{208}Pb/^{206}Pb$
KDC-13-141 (30.4 g) GSC lab number 5168									
R4	bl, op, an, 125	32	144	45	8	0.017	7279	11	0.0052
R5	bl, op, eu, 100	45	111	43	7	0.002	3523	14	0.0007
R6	br, sub, op, 150	29	45	15	2	0.243	43	234	0.0742
R7	br, eu, clr, 75	50	21	99	17	0.056	141	182	0.0169
92-942 (100 g) GSC lab number 4324									
Z4B	clr, sub, eq	1	1	329	200	0.878	224	49	0.2445
Z4C	clr, sub, eq	1	2	87	55	1.154	223	19	0.3212
Z6B	clr, an ovg, fg	1	3	91	49	0.514	836	11	0.1436
Z6C	clr, an ovg, fg	1	5	83	46	0.575	280	42	0.1607
92-2-865 (12.7 g) GSC lab number 7444									
ZO-1	clr, ovg, fg	1	2	1112	607	0.083	5201	10	0.0229
Z1	clr, sub	1	2	782	466	0.146	4334	9	0.0401
Z3A	clr, sub	1	2	312	177	0.204	1525	12	0.0564
Z3B	clr, sub	1	2	324	191	0.275	736	30	0.0758
FUC 4-6 42.26 (22 g) GSC lab number 4092									
Z1	clr, ovg, fg	1	1	214	113	0.463	257	26	0.1295
Z2B	eu, clr, pr	1	2	60	33	0.479	179	22	0.1336
Z3A	clr, ovg, fg	1	2	475	266	0.644	946	30	0.1799
Z3C	clr, ovg, fg	1	1	388	207	0.470	321	38	0.1314
Z3C	clr, ovg, fg	1	1	388	209	0.495	322	38	0.1383
Z3D	clr, ovg, fg	1	1	282	147	0.377	464	19	0.1054
R1	br, eu	30	61	26	3	0.004	1635	9	0.0012
R2	br, eu	40	125	27	4	0.006	1284	24	0.0019
T1	pbr, fg	20	498	20	14	4.838	414	456	1.4147
T2	pbr, fg	18	240	21	15	4.993	385	252	1.4547
92-2-732 (11.9 g) GSC lab number 4326									
Z1	pbr, ovg, fg	1	2	274	143	0.199	1096	14	0.0553
Z3A	eu, pbr, eq	1	2	268	140	0.212	4486	4	0.0591
DRA-94-T1 (10 g) GSC lab number 3863									
Z1	eu, pr, clr	1	2	130	82	0.987	1312	5	0.2745
Z2	eu, clr, eq	1	1	466	277	0.678	2696	5	0.1885
Z3	an core	1	5	81	52	1.133	2269	5	0.3149
Z4	ovg, fg	2	2	79	48	1.013	774	5	0.2827
Z7	ovg, fg	1	1	450	271	0.982	4762	3	0.274
Z8	ovg, fg	1	3	113	87	2.370	2036	5	0.6608
Z9	ovg, fg	1	3	438	250	0.859	4472	7	0.2407
Z12	ovg, fg	1	1	344	208	0.898	1501	7	0.2501
Z14A	pbr, irr, fg	1	6	202	125	1.110	7738	5	0.3099
Z14C	pbr, irr, fg	1	4	546	321	0.865	16792	4	0.2415
Z13B	eu, clr, eq	1	2	343	203	0.905	4347	5	0.2526

fragments yield ages of 2580–2590 Ma (Table 3, Fig. 6a). Analyses of single euhedral grains, as well as strongly abraded cores yield slightly older, discordant ages of 2590–2615 Ma. The range in ages of overgrowth fragments may indicate that complete separation from older core materials was not achieved. SHRIMP analyses targeted on individual growth zones reveal a wider range of ages up to 3.2 Ga (Fig. 6b). The oldest grains have oscillatory zoning, suggesting an igneous origin, and have ages of 3.19,

$^{206}Pb/^{238}U$	1σ error	$^{207}Pb/^{235}U$	1σ error	$^{207}Pb/^{206}Pb$	1σ error	Apparent age (Ma)			
						$^{206}Pb/^{238}U$	$^{207}Pb/^{206}Pb$	2σ	% disc.
0.2011	0.0002	2.447	0.003	0.08827	0.00004	1181.2	1388.4	1.7	16.3%
0.1720	0.0002	1.882	0.002	0.07937	0.00005	1022.9	1181.3	2.2	14.5%
0.1567	0.0020	1.612	0.092	0.07461	0.00358	938.5	1058.0	8.3	12.1%
0.1797	0.0005	1.991	0.022	0.08033	0.00077	1065.5	1205.1	37.8	12.6%
0.4960	0.0031	12.180	0.069	0.17809	0.00057	2596.7	2635.1	11.0	1.8%
0.4953	0.0033	12.137	0.078	0.17772	0.00068	2593.6	2631.8	12.5	1.8%
0.4837	0.0011	11.301	0.025	0.16946	0.00017	2543.3	2552.3	3.5	0.4%
0.4812	0.0024	11.335	0.051	0.17086	0.00046	2535.3	2566.1	9.0	1.6%
0.5197	0.0008	14.258	0.024	0.19897	0.00012	2698.0	2817.8	2.0	5.2%
0.5551	0.0009	15.905	0.027	0.20779	0.00012	2846.5	2888.4	1.9	1.8%
0.5286	0.0008	14.044	0.024	0.19272	0.00012	2735.3	2765.5	2.0	1.3%
0.5382	0.0014	14.684	0.037	0.19789	0.00022	2775.8	2808.8	3.6	1.4%
0.4727	0.0025	10.807	0.056	0.16580	0.00050	2495.6	2515.7	10.0	1.0%
0.4905	0.0039	11.487	0.088	0.16986	0.00071	2572.8	2556.2	14.0	−0.8%
0.4832	0.0008	11.236	0.019	0.16864	0.00013	2541.4	2544.2	2.7	0.1%
0.4782	0.0020	11.000	0.045	0.16683	0.00038	2519.4	2526.0	7.8	0.3%
0.4789	0.0020	11.012	0.046	0.16677	0.00040	2522.6	2525.5	8.0	0.1%
0.4765	0.0014	10.862	0.033	0.16535	0.00026	2511.9	2511.1	5.4	−0.0%
0.1472	0.0003	1.541	0.008	0.07596	0.00033	885.0	1094.0	17.7	20.4%
0.1510	0.0002	1.594	0.004	0.07654	0.00014	906.6	1109.2	7.1	19.6%
0.3078	0.0003	4.648	0.015	0.10953	0.00028	1729.8	1791.6	9.3	3.9%
0.3208	0.0004	5.150	0.017	0.11642	0.00030	1793.7	1901.9	9.4	6.5%
0.4937	0.0008	11.834	0.020	0.17386	0.00012	2586.6	2595.1	2.4	0.4%
0.4932	0.0008	11.774	0.020	0.17313	0.00010	2584.6	2588.1	2.0	0.2%
0.5000	0.0010	12.233	0.031	0.17742	0.00027	2614.0	2628.9	5.1	0.7%
0.5034	0.0008	12.650	0.022	0.18223	0.00015	2628.6	2673.3	2.7	2.0%
0.5022	0.0008	12.360	0.021	0.17851	0.00011	2623.1	2639.1	1.8	0.7%
0.4852	0.0012	11.321	0.028	0.16921	0.00019	2550.1	2549.8	3.8	−0.0%
0.4857	0.0008	11.351	0.019	0.16950	0.00010	2552.1	2552.7	1.8	0.0%
0.4889	0.0008	11.516	0.043	0.17083	0.00050	2566.0	2565.8	9.8	−0.0%
0.4712	0.0007	10.757	0.023	0.16558	0.00022	2488.7	2513.5	4.3	1.2%
0.4945	0.0022	11.581	0.052	0.16987	0.00010	2590.0	2556.4	1.9	−1.6%
0.4851	0.0008	11.389	0.019	0.17028	0.00010	2549.4	2560.4	1.8	0.5%
0.4833	0.0008	11.276	0.019	0.16923	0.00010	2541.4	2550.0	1.8	0.4%
0.4843	0.0008	11.348	0.019	0.16995	0.00010	2545.8	2557.2	1.8	0.5%

2.98 and 2.74 Ga (Figs. 4c and 6b, Table 4). These grains are interpreted to be inherited and their ages accord with well-documented ages in the Central Slave Basement complex and cover sequence. Two of these inherited grains have texturally distinct over-growths with ages of 2.61–2.62 Ga (Fig. 4c, Table 4, grains 6 and 10)). The majority of the grains have $^{207}Pb/^{206}Pb$ ages between 2.56 and 2.65 Ga, including both homogenous and polygonally zoned core and rim material with two modes at 2564 ± 21 ($n=3$) and

Table 4
SHRIMP analytical data

Labels[a]	Description	U (ppm)	Th (ppm)	Th/U	Pb (ppm)*	^{204}Pb (ppb)	^{204}Pb/^{206}Pb	± 204/206	f206[b]	^{208}Pb/^{206}Pb
92-2-865										
1.1	core	663	436	0.657	472	10	3.11E − 05	1.58E − 05	5.40E − 04	0.1807
1.3	core	448	265	0.591	297	7	3.07E − 05	1.65E − 05	5.30E − 04	0.1613
1.2	rim	1081	11	0.010	574	4	8.55E − 06	5.76E − 06	1.50E − 04	0.0026
2.1	core	1971	20	0.010	1019	6	6.85E − 06	9.50E − 06	1.20E − 04	0.0029
2.2	core	945	92	0.098	498	5	1.10E − 05	1.77E − 05	1.90E − 04	0.0268
2.3	rim	772	95	0.123	394	11	3.49E − 05	1.58E − 05	6.00E − 04	0.0324
GR-95-36										
1.1	core	122	78	0.642	71	6	1.25E − 04	9.63E − 05	2.16E − 03	0.1854
1.2	dark rim	206	127	0.615	112	2	2.98E − 05	2.89E − 05	5.20E − 04	0.1696
2.1	core	52	61	1.161	33	9	3.90E − 04	2.05E − 04	6.76E − 03	0.3249
3.1	bright rim	42	55	1.313	26	3	2.02E − 04	1.91E − 04	3.51E − 03	0.3599
3.2	core	153	90	0.586	85	5	7.47E − 05	3.98E − 05	1.30E − 03	0.1569
5.1	core	222	209	0.941	142	12	1.23E − 04	7.37E − 05	2.13E − 03	0.2587
5.2	core	237	236	0.999	148	5	5.13E − 05	3.42E − 05	8.90E − 04	0.2711
6.1	rim	2752	63	0.023	1395	36	3.07E − 05	9.37E − 06	5.30E − 04	0.0063
6.2	core	1225	70	0.057	831	24	3.67E − 05	2.55E − 05	6.40E − 04	0.0135
6.3	core	289	35	0.122	197	29	1.93E − 04	6.42E − 05	3.34E − 03	0.0333
7.1	core	749	213	0.285	506	12	3.22E − 05	2.97E − 05	5.60E − 04	0.0805
7.2	core	956	69	0.072	672	28	5.22E − 05	8.64E − 06	9.10E − 04	0.0180
8.1	core	658	68	0.103	331	25	8.96E − 05	2.91E − 05	1.55E − 03	0.0259
9.1	core	186	160	0.858	108	12	1.61E − 04	7.65E − 05	2.80E − 03	0.2277
9.2	core	171	102	0.601	93	15	2.08E − 04	8.94E − 05	3.60E − 03	0.1587
10.1	core	1072	90	0.084	577	24	5.04E − 05	1.41E − 05	8.70E − 04	0.0223
10.2	bright rim	96	47	0.488	54	27	6.37E − 04	1.52E − 04	1.10E − 02	0.1015
DRA-94-T1										
1.1	inner zone	186	166	0.895	111	3	3.98E − 05	3.94E − 05	0.00069	0.2464
1.2	inner zone	237	200	0.845	138	7	7.34E − 05	5.9E − 05	0.00127	0.2284
2.1	core	311	290	0.931	192	5	3.86E − 05	3.86E − 05	0.00067	0.2535
3.1	core	286	184	0.642	160	6	4.83E − 05	3.98E − 05	0.00084	0.1753
3.2	core	383	305	0.795	203	8	5.27E − 05	2.58E − 05	0.00091	0.2147
4.1	core	256	221	0.865	143	3	3.29E − 05	4.93E − 05	0.00057	0.2341

[a] Labels identify grain number on mount. Number after decimal indicates spot number.

[b] f206 indicates mole fraction of ^{206}Pb that is due to common Pb. Data corrected for common Pb following methods outlined in Stern (1997).

[c] Uncertainties (±) are given at 1σ absolute and are calculated by numerical propagation of all known error (Stern, 1997).

[d] Pb/U ratios corrected for measurement bias relative to Kipawa zircon standard (993 Ma). Uncertainty in calibration of standard is ± 1.0%.

[e] Conc. (%) = 100 × (^{206}Pb/^{238}U age)/(^{207}Pb/^{206}Pb age).

* Radiogenic Pb.

2623 ± 13 Ma ($n = 8$). The older age is interpreted to be the igneous crystallization age of the tonalite and the younger age records later metamorphic growth at 2564 ± 21 Ma. The difference in age between the overgrowths analyzed by TIMS and those analyzed by SHRIMP may reflect two different ages of over-growths or mixing of the two age populations within the analyzed overgrowth fragments.

Sample 96-BAK-41-166, a garnet–biotite metato-nalite from the Kennady Lake pipe, contains domi-nantly pale brown, equant zircon some of which have pale brown zircon overgrowths. Individual

$\pm^c$	$^{206}Pb/^{238}U^d$	$\pm^c$	$^{207}Pb/^{235}U$	$\pm^c$	$^{207}Pb/^{206}Pb$	$\pm^c$	Apparent ages				Conc. $(\%)^e$
							$^{206}Pb/^{238}U$	$\pm^c$	$^{207}Pb/^{206}Pb$	$\pm^c$	
0.0017	0.5885	0.0271	18.28	0.89	0.22525	0.00254	2983	111	3019	18	98.8
0.0025	0.5559	0.0113	17.11	0.50	0.22319	0.00422	2849	47	3004	31	94.9
0.0003	0.5161	0.0070	13.81	0.22	0.19401	0.00143	2683	30	2776	12	96.6
0.0005	0.5095	0.0079	12.49	0.26	0.17770	0.00214	2655	34	2632	20	100.9
0.0012	0.5092	0.0081	12.38	0.24	0.17629	0.00166	2653	35	2618	16	101.3
0.0010	0.4930	0.0085	11.66	0.23	0.17148	0.00115	2584	37	2572	11	100.4
0.0063	0.4931	0.0170	12.25	0.49	0.18017	0.00301	2584	74	2654	28	97.3
0.0031	0.4693	0.0094	11.04	0.25	0.17058	0.00152	2481	42	2563	15	96.8
0.0123	0.4920	0.0197	11.27	0.58	0.16606	0.00464	2580	86	2518	48	102.4
0.0145	0.4722	0.0282	11.28	0.84	0.17319	0.00649	2493	124	2589	64	96.3
0.0034	0.4837	0.0129	11.36	0.35	0.17036	0.00205	2543	56	2561	20	99.3
0.0049	0.5163	0.0138	12.65	0.40	0.17776	0.00246	2683	59	2632	23	101.9
0.0032	0.5009	0.0133	12.19	0.36	0.17651	0.00179	2618	57	2620	17	99.9
0.0004	0.4984	0.0118	12.11	0.31	0.17625	0.00110	2607	51	2618	10	99.6
0.0012	0.6237	0.0127	21.59	0.50	0.25105	0.00210	3125	51	3191	13	97.9
0.0029	0.6160	0.0172	21.40	0.66	0.25192	0.00238	3094	69	3197	15	96.8
0.0022	0.6037	0.0116	18.26	0.41	0.21934	0.00199	3045	47	2976	15	102.3
0.0006	0.6519	0.0109	21.20	0.37	0.23582	0.00087	3236	43	3092	6	104.6
0.0014	0.4862	0.0075	11.88	0.21	0.17721	0.00136	2554	32	2627	13	97.2
0.0050	0.4779	0.0131	11.66	0.40	0.17702	0.00316	2518	57	2625	30	95.9
0.0048	0.4787	0.0151	11.28	0.41	0.17097	0.00234	2522	66	2567	23	98.2
0.0007	0.5160	0.0110	13.53	0.31	0.19012	0.00119	2682	47	2743	10	97.8
0.0066	0.5055	0.0228	12.20	0.62	0.17509	0.00329	2638	98	2607	32	101.2
0.0044	0.4895	0.0108	11.49	0.30	0.17018	0.00198	2569	47	2559	20	100.4
0.0038	0.4848	0.0106	11.39	0.29	0.17039	0.00173	2548	46	2562	17	99.5
0.0033	0.5022	0.0132	11.91	0.35	0.17204	0.00183	2623	57	2578	18	101.8
0.0025	0.4835	0.0087	11.14	0.23	0.16703	0.00135	2543	38	2528	14	100.6
0.0025	0.4434	0.0095	10.53	0.25	0.17223	0.00119	2366	43	2579	12	91.7
0.0041	0.4613	0.0093	10.79	0.27	0.16964	0.00215	2445	41	2554	21	95.7

grains show complex internal structures with oscillatory zoned cores overgrown by patchy sector zoned zircon and homogeneous overgrowths (Fig. 4d,e). Optically visible overgrowths were physically separated from cores and both types analyzed separately. Single euhedral grains with no optically visible overgrowths were strongly abraded to remove the outer part of the grain prior to analysis. The euhedral grains are slightly older than the overgrowths with $^{207}Pb/^{206}Pb$ ages of 2590–2615 Ma (Fig. 6c). Overgrowth fragments have younger ages between 2584 and 2588 Ma demonstrating metamorphic zircon growth at 2.58 on older ca. 2.62 Ga zircons (Fig. 6c, Table 3).

Rutile in this sample consists dominantly of large black opaque anhedral grains. $^{207}Pb/^{206}Pb$ ages are discordant and range from 1805 to 1935 Ma (Fig. 7). The two oldest fractions consist of larger grains and were abraded for a short period of time (15 min), whereas the two younger fractions are smaller and were not abraded. The data indicate closure of Pb diffusion in these rutile occurred by at least 1.8 Ga.

Zircon in sample 96-BAK-009-416, a garnet biotite metatonalite also from Kennedy Lake, exhibits two

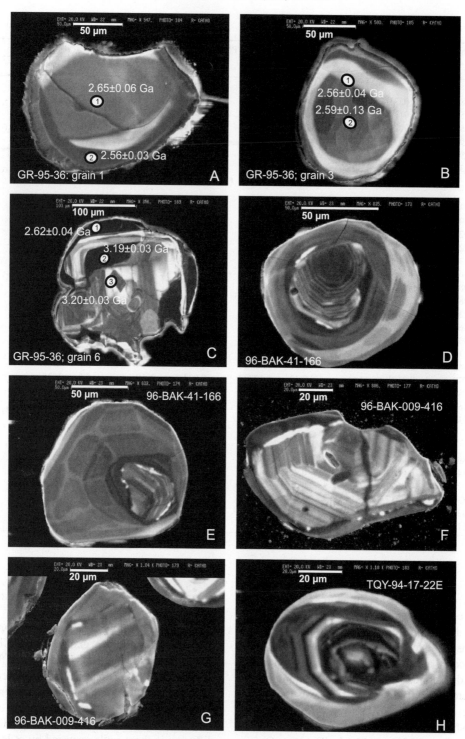

Fig. 4. Cathodoluminesence images of zircon from selected xenoliths. Zircons were mounted in epoxy and polished to expose approximate midsections. White circles identify position and true size of SHRIMP analytical spots with measured age and 95% confidence error as indicated. Scale is indicated by white bar.

distinct morphologies; small (<100 μm), round, equant grains and prismatic, subround grains with aspect ratio of 3–4:1. The CL images of the grains are characterized by oscillatory-zoned cores that are resorbed with homogeneous dark rims (Fig. 4f,g). An analyses of a single round, equant zircon yielded a concordant age of 2640 ± 2 Ma and a collection of two grains yielded a discordant (2.7%) age of 2610 Ma (Fig. 6d, Table 3). Two analyses of single prismatic grains yield ages of 2633 (1.8% disc.) and 2628 Ma (11% disc.). The spread in the data indicate complex U–Pb systematics and a precise age cannot be determined with the available data. The protolith may be as old as the concordant point at 2640 ± 2 Ma with the dispersion to younger ages in the other grains indicating early Pb loss, recrystallization or younger metamorphic growth as observed in CL imaging. The younger outer part of the grains seen in Fig. 4f and g were not directly analyzed. The ca. 2.63–2.64 Ga ages for the core material are similar to the age of early diorite to tonalite magmatism in the south central Slave (van Breemen et al., 1992; Davis and Bleeker, 1999) and the sample may represent, either directly or indirectly, magmatic material emplaced within the lower crust of the Slave Province at that time.

Sample TQY-94-17-22E is a peraluminous metagranitoid from the Torrie pipe containing hercynite overgrown by sillimanite. Zircon dominantly occurs as small (<100 μm) round, euhedral, colourless to pale brown grains, some of which have colourless overgrowths. CL images reveal oscillatory growth zoning with some grains containing thick homogenous overgrowths that truncate the inner zones (Fig. 4h). Analyses of abraded single euhedral grains yield a range of discordant ages from 2747 to 2967 Ma (Fig. 6e, Table 3). An analysis of a colourless thin fragment, interpreted to represent the younger rim material, yielded a significantly younger but discordant age of 2641 Ma. The analyses plot along or beneath a reference mixing line between ca. 2.97 and 2.6 Ga (Fig. 6e), consistent with two-component mixing and Pb loss (both Neoarchean and recent). However, the age of the sample cannot be determined with the available data. One possibility is that most of the zircon in the rock is inherited, a feature common to S-type granites (e.g., Williams, 1992) and the maximum age for the sample is given by the age of

the overgrowth material at ca. 2.64 Ga. If this is the case, then the metagranitoid may represent the melt product of sedimentary rocks dominated by Mesoarchean protoliths and the overgrowths represent igneous zircon growth at or after 2.64 Ga. Alternatively, the sample may be Mesoarchean in age (>2967 Ma) with zircon undergoing extensive recrystallization or new zircon growth and mixing during post-2.64 Ga metamorphism.

Metatonalite sample 96-BAK-004-263 from the Kennady Lake pipe did not yield zircon but contained a small amount of brown blocky titanite. Two analyses yield overlapping discordant ages of 2.55 Ga (Fig. 6f, Table 3). Since the titanite are 2.6% discordant this is a minimum age for closure of the U–Pb system in these titanite, indicating that this rock cooled below 600 and 700 °C (e.g., Cherniak, 1992; Scott and St-Onge, 1995) by at least 2.55 Ga.

4.2. Mafic granulite xenoliths

Sample KDC-13-141 is a mafic granulite from the Koala pipe at Lac de Gras that yielded many zircons with equant, round morphologies typical of metamorphic zircon in mafic rocks. The internal structures of the grains are complex. Some have small cores with oscillatory-zoning truncated by homogenous overgrowths (Fig. 5a,b). Others have sector zoned cores or irregular wispy zonings. Four analyses, each consisting of a single abraded zircon crystal resulted in a range of concordant ages from 2578 ± 2 to 2551 ± 2 Ma (Fig. 8a, Table 3). The spread in ages along the concordia curve is consistent with two or more episodes of Neoarchean zircon growth in the sample, which is also indicated by the internal growth structure. The minimum age for the older component, possibly represented by the oscillatory-zoned cores, is 2578 Ma and is interpreted to indicate igneous crystallization in the lower crust. As this represents a minimum age, the true crystallization age could be considerably older, perhaps as old as early diorite–tonalite magmatism at 2.64–2.61 Ga. The younger component represents new growth or recrystallization at or after 2550 Ma.

Ages of rutile from this sample are discordant with a range of $^{207}Pb/^{206}Pb$ ages from 1058 to 1388 Ma (Fig. 7, Table 3). Rutile ranges from brown, translucent, euhedral grains to larger, black opaque

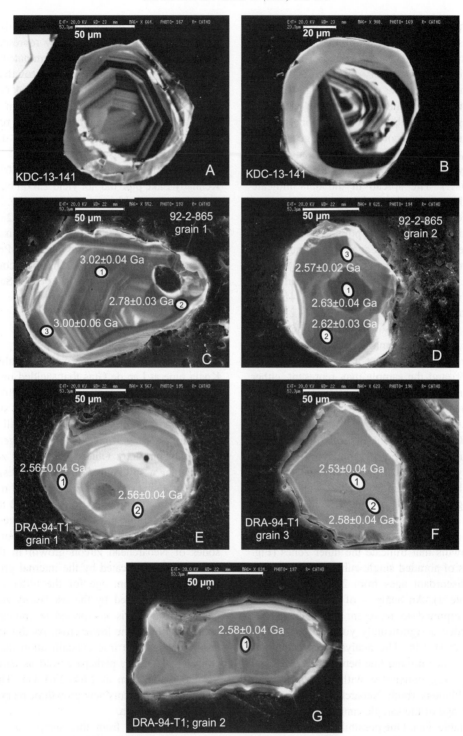

Fig. 5. Cathodoluminesence images of zircon from selected xenoliths. Zircons were mounted in epoxy and polished to expose approximate midsections. White circles identify position and true size of SHRIMP analytical spots with measured age and 95% confidence error as indicated. Scale is indicated by white bar.

grains. The larger black grains (200–300 µm) have lower common Pb contents and yield relatively more precise, and older ages compared to the brown grains. The data do not regress to a single line indicating the effects of complex Pb loss.

Sample 92-2-942 is a mafic garnet-cpx granulite from the Grizzly pipe at Lac de Gras. Zircons consist of euhedral, prismatic to equant grains with poorly developed terminations. Many of the zircons have colourless overgrowths that were easily observed under binocular microscope. Four analyses were undertaken, two of individual overgrowth fragments and two single prismatic grains. The prismatic grains yield discordant, relatively imprecise, overlapping ages of 2635 ± 11 and 2632 ± 13 Ma (Fig. 8b, Table 3). Overgrowth fragments have distinctly younger ages of 2552 ± 4 ($< 0.4\%$ discordant) and 2566 ± 9 Ma (Fig. 8b, Table 3), the younger of which is considered the maximum age for the overgrowth. The age of the xenolith may be given by the ca. 2.63 Ga prismatic grains, with metamorphic zircon growth occurring at or after ca. 2.55 Ga. Alternatively, the ca. 2.63 Ga zircons may be inherited in which case the mafic granulite may represent new addition to the crust at or after 2.55 Ga.

Sample 92-2-865 is a highly altered mafic xenolith of uncertain protolith consisting of 5% garnet in a matrix of chlorite and clay minerals pseudomorphing the original mineralogy. Biotite is present but appears to be secondary in origin. Zircons consist of subhedral round equant to prismatic crystals. Some of the zircons have thin irregular rims of baddeleyite, possibly developed during alteration by interaction with the kimberlite. In CL imaging, the grains exhibit oscillatory to broad growth zoning. In some grains, this zoning is resorbed and one or more overgrowths are present (Fig. 5c,d). TIMS data on pale brown, anhedral to rounded single grains yielded a range of discordant $^{207}Pb/^{206}Pb$ ages from 2766 to 2888 Ma (Fig. 8c, Table 3). Given the textural evidence for multiple episodes of zircon growth, the data are interpreted to indicate mixing of zircon of different ages. This is confirmed by two SHRIMP analyses of an oscillatory-zoned core that yielded a weighted average age of 3015 ± 30 Ma (Figs. 5c and 8c, Table 4, grain 1). The oscillatory-zoning in this grain is resorbed and overgrown by a homogeneous zone with an age of 2776 ± 24 Ga

(Figs. 5c and 8c). Two analyses of the core zone of a second grain with poorly developed zoning and a bright overgrowth yielded a much younger weighted mean age of 2623 ± 25 Ga (Figs. 5d and 8c, Table 4, grain 2). The youngest zone gives an age of ca. 2.57 Ga. The age of the xenolith has not be determined; it may be as old as 3015 ± 30 Ma as indicated by the SHRIMP age of the oscillatory-zoned grain, and experienced at least three episodes of subsequent zircon resorption and growth at 2.78, 2.62 and 2.57 Ga.

Sample FUC-4-6-42.26 is from the Fox kimberlite at Lac des Gras. The sample is a relatively fresh rock consisting of clinopyroxene, plagioclase and garnet with accessory rutile, titanite and ilmenite. Zircon forms euhedral prismatic grains with surrounded terminations. A population of clear, colourless, thin zircon fragments interpreted to be walls of zircon overgrowths was also found. A single analysis of a prismatic zircon yielded an imprecise age of 2556 ± 14 Ma (Fig. 8d, Table 3). The pale overgrowth fragments yielded a range of younger ages from 2544 to 2508 Ma (Fig. 8d). The data are interpreted to indicate at least two periods of zircon growth in this sample, one at >ca. 2.56 Ga and again at 2.51 Ga. Intermediate ages indicate mixing of the different age populations.

Small euhedral titanite grains yield significantly younger, discordant ages of 1902 and 1792 Ma (Fig. 7b, Table 3). The range of ages indicates differential amounts of Pb loss or recrystallization between the two multi-grain fractions. A two-point reference line through the two fractions intersects concordia at ca. 1.7 and 2.6 Ga, consistent with partial resetting of Neoarchean titanite. This sample either cooled below titanite closure temperature (ca. $>600–700$ °C; Cherniak, 1992; Scott and St-Onge, 1995) at ca. 1.7–1.9 Ga was heated above the closure temperature for Pb diffusion in titanite, or crystallized titanite in the Proterozoic. The titanite in this sample is significantly younger than titanite from the Kennady Lake tonalite (96-BAK-004-263) sample (Fig. 6f) suggesting that transient heating in the Proterozoic may have been more pronounced in the central Slave.

Rutile in the sample gives much younger discordant ages of ca. 1100 Ma (Fig. 7c, Table 3). These ages have been interpreted to indicate transient heat-

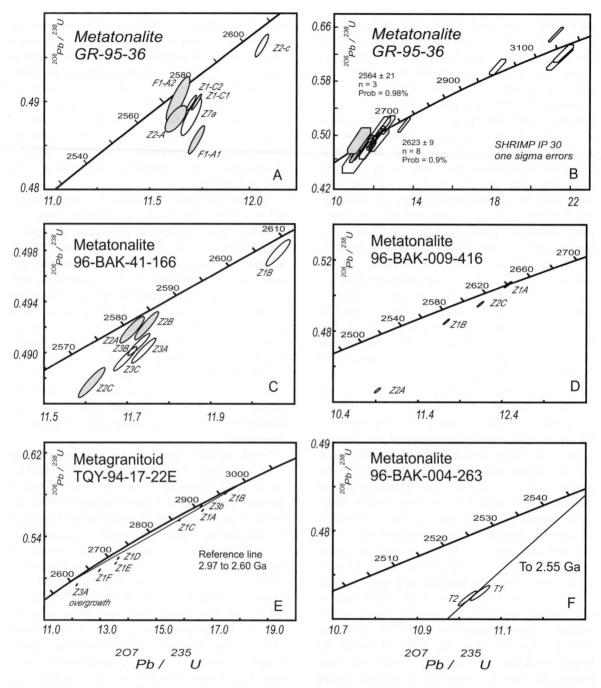

Fig. 6. U–Pb concordia plots for felsic granulite xenoliths. Data are results of ID TIMS analysis except (B) which are results of SHRIMP analyses. Analyses of overgrowth fragments are shaded, single grain and cores are open ellipses. Labels cross-referenced to Table 3.

ing of the crust during the Mackenzie igneous event (Davis, 1997). Alternatively, the difference between rutile and titanite ages could approximate slow cool-

ing of the lower crust from temperatures of ca. 600 °C at ca. 1.8 Ga to temperatures below 450 °C by 1100 Ma, a cooling rate of approximately 0.3 °C/my.

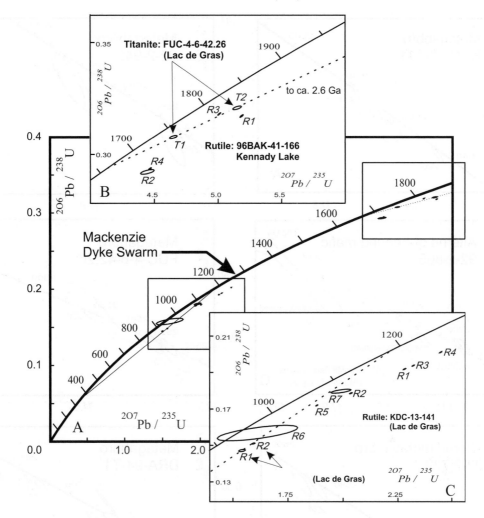

Fig. 7. (A) U–Pb concordia plot of titanite and rutile data. Discordia lines and individual analyses identified in B and C. Age of Mackenzie plume event indicated by arrow. (B) Expanded view of titanite (FUC-4-6-42.26) and rutile (96BAK-41-166) data. Dashed line through two titanite analyses from sample FUC-4-6-42.26 intersects concordia at ~2.6 and 1.67 Ga. Titanite data indicated by T; rutile data by R. (C) Expanded view of rutile data for samples FUC-4-6-42.26 and KDC-13-141. Dashed line is reference discordia for rutile data from Lac des Gras area between 0.35 and 1.19 Ga (Davis, 1997, unpublished data).

Sample 92-2-732 is a highly altered garnet and amphibole-bearing mafic xenolith from the Grizzly pipe. Zircons form equant to round grains with generally homogeneous bright interiors in CL, some of which have darker homogenous overgrowths (not shown). Analyses of a strongly abraded euhedral grain and a fragment of overgrowth material yielded ages of 2588 and 2595 Ma, respectively (Fig. 8e, Table 3).

Sample DRA-94-T1 is a mafic granulite consisting of garnet, clinopyroxene and plagioclase from the Torrie pipe. Zircons form euhedral, equant to prismatic grains some with thin overgrowths visible under binocular microscope. The grains have relatively dark homogenous internal zones in CL (Fig. 5e,f,g). Two single grain analyses of the prismatic zircon yielded slightly discordant ages of 2629 and 2673 Ma (Fig. 8f, Table 3). A piece of core material from which an overgrowth had been broken off yielded an age of 2639 Ma. Seven analyses of overgrowth fragments, either broken off of core material or identified as

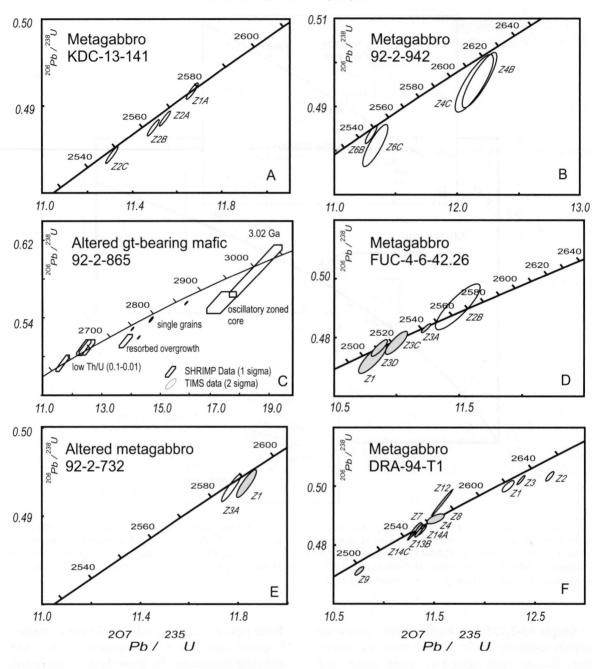

Fig. 8. U–Pb concordia plots for mafic granulite xenoliths. Data are results of ID TIMS analysis except (C) which includes both types of data. SHRIMP data are plotted with polygonal error envelopes, TIMS data as smooth ellipses. Analyses of overgrowth fragments are shaded; single grain and cores are represented by open ellipses. Labels cross-referenced to Table 3.

fragments with negative crystal faces under binocular microscope yielded a cluster of ages at around 2.55–2.56 Ga. This age of zircon growth is supported by a

limited number of targeted spot analyses using the SHRIMP. A total of six analyses (Table 4), mostly on homogenous grain interiors, yield a weighted mean age of 2561 ± 23 Ma (MSWD = 1.9) within error of the age for the overgrowth fragments determined by TIMS.

5. Discussion

The temperatures in the lower crust recorded by the granulite xenoliths lie far above those predicted for a cratonic geotherm constructed from P–T arrays for kimberlite-hosted mantle xenoliths from the Slave Province (Fig. 9). The latter geotherm would predict temperatures of 400–500 °C at Moho depths of 35–40 km for the Slave Province (Bostock, 1998; Cassidy, 1995), about 300–400 °C below those recorded by mineral equilibria in the granulite xenoliths. These low temperatures could never be recorded by xenoliths based on considerations of diffusion kinetics (Harte et al., 1981), and it is clear that the xenoliths record fossil or frozen–in conditions, rather than ambient conditions in the craton at the time of kimberlite emplacement. The metamorphic conditions recorded by the xenoliths are consistent with proposed geothermal field gradients at 2.6–2.58 Ga determined from metamorphic studies of surface rocks in the craton (e.g., Thompson, 1989b; Thompson et al., 1996).

The high closure temperature for Pb diffusion in zircon combined with the preservation of distinct textural zoning and sharp overgrowths demonstrate that zircon is not systematically reset under the granulite conditions documented in the lower crust (Hanchar and Rudnick, 1995; Vavra et al., 1996). Multiple ages of zircon within a single xenolith record a complex, polyphase history. The exact cause and significance of zircon growth or resorption in metamorphic rocks is a function of poorly understood phase equilibria involving major and accessory minerals and fluid or melt phases. For this reason, directly correlating episodes of zircon growth or resorption to specific metamorphic reactions or conditions remains largely conjectural. In mafic granulites zircon has been interpreted to form during prograde mineral reactions involving the breakdown of amphibole to pyroxene (e.g., Schmitz and Bowring, 2000), but also to exsolution of Zr from clinopyroxene and or garnet (Fraser et al., 1997;

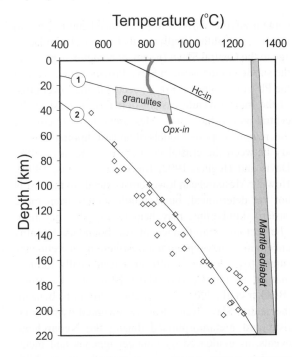

Fig. 9. Temperature–depth diagram showing P–T conditions for lower crustal granulite xenoliths from this study (box), compared to fluid-absent melting conditions for pelites, defined by the "Opx-in" curve (Vielzeuf and Montel, 1994), and to the "Hc-in" reaction: 2 Sill + Alm = 3 Hc + 5 Qz, calculated using the Berman (1988) database. A cratonic geotherm (2) with radiogenic heat production of 0.6 μW in the crust, and surface heat flow of 41 mW m^{-2} fitted through the P–T array (diamonds) for kimberlite-hosted mantle xenoliths (Boyd and Canil, 1997; Kopylova et al., 1998; MacKenzie and Canil, 1999) using the approach described in MacKenzie and Canil (1999) shows the base of the lithosphere (defined by the 1300 °C adiabat) at ~220 km at the time of kimberlite eruption. A geotherm consistent with P–T conditions of the Archean granulites is fitted with two times the radiogenic heat production in the crust (1.2 μW) and surface heat flow of 90 mW m^{-2} and is consistent with a lithosphere thickness of only ~60 km at the time of granulite grade metamorphism.

Schmitz and Bowring, 2000). In more felsic lithologies, growth of metamorphic zircon may be favoured by the presence of a melt phase (e.g., Schaltegger et al., 1999; Vavra et al., 1999; Roberts and Finger, 1997).

5.1. Mesoarchean protoliths in the lower crust of the Central Slave craton

Three of the xenolith samples from the Torrie and Grizzly pipes (GR-95-36; 92-2-865; TQY-94-17-22E) contain zircons derived from Mesoarchean (>2.9 Ga) igneous protoliths, implying that material similar in age to the Central Slave Basement complex exposed at the surface in the western Slave Province (i.e., west of the Pb isotope line on Fig. 1) occurs in the lower crust of the more juvenile eastern domains, at least in the central part of the craton. Mesoarchean lower crust in this region supports models of an east dipping boundary between the crustal domains (e.g., Kusky, 1989; Davis and Hegner, 1992; Bleeker and Davis, 1999). How far Mesoarchean lower crust extends to the east is not yet determined, but xenolith studies from more easterly kimberlites may help resolve this question. The tectonic significance of the boundary remains uncertain, it may represent a west-verging accretionary suture (e.g., Kusky, 1989), or a tectonically thinned margin to the Central Slave Basement Complex (Bleeker et al., 2001). Mesoarchean protoliths beneath the central Slave were strongly reworked by magmatism and metamorphosed during the Neoarchean events, as evidenced by younger igneous and metamorphic zircon growth or overgrowths at ca. 2.78, 2.64–2.62 and 2.56 Ga.

5.2. Neoarchean history of the lower crust

Neoarchean zircon growth in the granulite xenoliths cluster into at least four intervals: 2.64–2.61, 2.59–2.58, 2.56–2.55 and ca. 2.51 Ga (Fig. 10). The intervals 2.64–2.61 and 2.59–2.58 Ga are significant and can be directly correlated with orogenic events in the surface geology (Fig. 10). The 2.64–2.61 Ga ages represented by igneous grains and core material in metatonalite and some mafic granulite xenoliths correspond to ages of diorite and tonalite plutons exposed throughout the craton (van Breemen et al., 1992), whereas the younger interval of 2.59–2.58 Ga, represented largely by zircon overgrowths in both felsic and mafic xenoliths, correspond closely to crystallization ages of K-feldspar megacrystic biotite granite and peraluminous, two-mica granitoids and their associated regional greenschist to granulite grade HT-LP metamorphism (van Breemen et al., 1992; Davis and Bleeker, 1999; Pehrsson et al., 2000). The frequency of ca. 2.64–2.61 Ga ages in igneous zircon populations in the lower crustal granulites argues for a component of magmatic addition to the lower crust at that time. Extensive melting and intrusion of granitic plutons

into the mid crust at 2.60–2.58 Ga may be represented by some zircon overgrowths on the early formed zircon, as well as new zircon growth in some xenoliths. This metamorphic zircon growth may have been facilitated by a melt phase. Curiously, very little evidence of ca. 2.71–2.68 Ga zircon, the dominant age of Neoarchean crust in surface rocks, has been documented in the lower crustal xenoliths examined so far.

Younger ages for zircon overgrowths at 2.56–2.55 and ca. 2.51 Ga in several of the granulite xenoliths are enigmatic, as they do not correspond to any significant tectonic events dated in the surface geology of the Slave Province. The ca. 2.56–2.55 Ga age is particularly common for zircon overgrowths in the xenoliths, and its preservation may be as, or more significant than the earlier 2.60–2.58 Ga metamorphism, generally considered as the peak metamorphic period in the Slave craton (e.g., Thompson, 1989a,b; Davis and Bleeker, 1999; Pehrsson et al., 2000). The manifestation of these events as texturally discrete zircon overgrowths indicates punctuated, as opposed to continuous zircon growth.

It remains uncertain as to what caused crystallization of the zircon overgrowths, and whether they record peak granulite metamorphic conditions (prograde), or whether they formed during a later event in the lower crust, perhaps at somewhat lower metamorphic grade. If they are associated with peak granulite assemblages then this implies that granulite conditions persisted or developed in the lower crust well after cessation of metamorphism in the mid-upper crust.

Alternatively, the zircon growth events may have occurred at post-peak conditions. Textural evidence for cooling or decompression reactions, for example, exsolution of Zr from clinopyroxene and/or garnet during cooling of the lower crust (Fraser et al., 1997) are not documented, and the punctuated nature of the zircon growth argues for a discrete rather than a protracted process. Zircon may have recrystallized due to fluid or melt-mediated reactions at post-peak conditions (Roberts and Finger, 1997; Schmitz and Bowring, 2000).

The pattern of younger metamorphic events in the deep crust (20–40 km) relative to the shallow crust (6–15 km) is characteristic of other Archean cratons (e.g., Superior and Kaapvaal). For example, Krogh (1993) and Moser et al. (1996) describe repeated periods of granulite metamorphism and extension within the lower crust beneath the Archean Superior province for

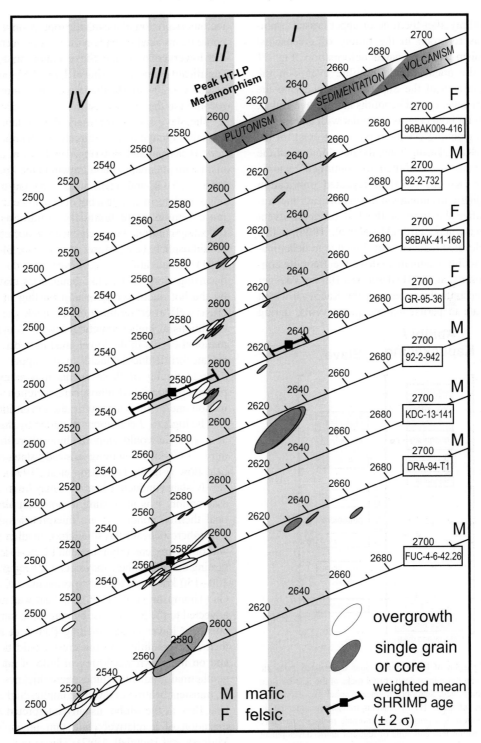

Fig. 10. Summary of Neoarchean ID-TIMS ages for granulite xenoliths shown on stacked concordia diagram. Ages of principal tectonic events recorded in the upper crust are schematically indicated at the top of the figures. Four dominant periods of zircon growth in the lower crustal xenoliths are indicated by shaded zones. Average SHRIMP ages are indicated by filled box and error bar.

close to 100 my after the time of upper crustal 'stabilization', as recorded in the timing of greenschist metamorphism and associated deformation. Fig. 11 compares the documented granulite events in the Kapuskasing area of the Superior Province with the results from lower crustal xenoliths in the Slave craton. The pattern for both cratons is similar with at least two punctuated granulite events in the lower crust. In the case of the Kapuskasing zone, the development of the younger events is attributed to continued tectonic imbrication (Krogh, 1993) or repeated imbrication and lithospheric delamination events beneath the craton well after the time of the last orogenic event recorded in the upper crust (Moser et al., 1996). In this scenario, elevated reduced heat flow due to delamination and possibly magmatism drive the granulite conditions. Moser et al. (2001) describe a 100 my younger granulite metamorphism beneath the Kaapvaal which they attribute to partial delamination events, during

tectonic accretion of the mantle root. A similar scenario could be invoked to explain the ages of overgrowths in the lower crust of the Slave craton with the most significant events occurring at 2.56–2.55 Ga and again at 2.51 Ga, 20–70 Ma after the last metamorphic event recorded in the upper crust.

The Slave craton, as preserved, is certain to be only a small fragment of an originally much larger orogenic zone (Isachsen and Bowring, 1994). Continued tectonism with attendant zircon growth in the lower crust at 2.56–2.55 Ga and again at ca. 2.51 Ga may reflect ongoing deformation in a hot ductile lower crust due to more remote plate boundary conditions, perhaps reflecting continued convergence and accretion within the lithosphere (e.g., Moser et al., 1996; Moser et al., 2001). Reactivation of the lower crust due to events occurring along distal plate boundaries may be aided by the hot ductile conditions persisting in the lower crust well after termination of peak metamorphic conditions at higher structural levels. Younger deformational events in the lower crust may be manifested as late brittle/ductile faults in the upper crust. The discrete timing of these events may correspond to periods of continued lithosphere accretion and deformation and/or fluid activity in the lower crust.

The high $P–T$ conditions recorded by the xenoliths at ~ 2.60 Ga would support a lithosphere thickness of only ~ 60 km, assuming a steady-state conductive heat flow (Fig. 9; Thompson et al., 1996). Although steady-state conductive heat flow may be an unrealistic assumption, it is interesting to note that almost identical thermal conditions are interpreted today in the lithosphere beneath the southern Canadian Cordillera and other orogens (Hyndman and Lewis, 1999), suggesting that lithosphere may not have been greater than 100–150 km thick following orogenic activity at 2.6 Ga (Thompson et al., 1996). As heat was most likely advected to the lower crust by magmatic intrusion and attendant lower crustal melting (Davis et al., 1994), then $P–T$ conditions in the crust would be transient and on their own would reveal little of late Archean geothermal gradient or lithosphere thickness. However, transient heating of the crust implies higher reduced heat flow at the Moho, usually interpreted as a manifestation of a relatively hot, thin lithosphere (e.g., Midgley and Blundell, 1997). This is most consistent with development of thick lithosphere to present day values of ~ 200 km sometime after 2.6 Ga.

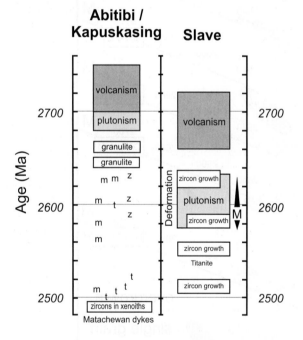

Fig. 11. Comparison of lower crustal ages determined from the Slave craton with those from deep crustal rocks of the Kapuskasing area of the Abitibi belt, Superior Province. Capital M with wedges indicates time of peak HT-LP metamorphism documented in exposed medium to low pressure metamorphic rocks in the Slave craton. Deformation indicates time of regional fabric development in exposed crustal rocks. Abbreviations: m = monazite age, t = titanite age, z = zircon age. Abitibi data sources: Krogh (1993), Moser et al. (1996) and Moser and Heaman (1997).

Other constraints on the age and development of the mantle root beneath the Slave Province could potentially derive from direct age estimates of samples comprising the root. The use of Re–Os isotope systematics permits minimum estimates for the age of melt depletion of mantle samples. For example, peridotite xenoliths from the Jericho kimberlite in the north central Slave indicate a range of Archean ages with a significant number of post-Archean Re depletion ages (50%) in the deep lithosphere (Irvine et al., 1999; 2001). This may reflect Proterozoic additions to the mantle. Conversely, Aulbach et al. (2001) argue for thick Mesoarchean lithosphere to have underlain the central part of the craton throughout the Neoarchean orogenesis, based on their Re–Os data from the Lac des Gras area. We feel that a thick Mesoarchean lithosphere is difficult to reconcile with the lower crustal metamorphic history in this area and older mantle components within the lithosphere may have been subcreted beneath the craton after 2.6 Ga (Davis et al., 2003). Additional and more widely distributed age dating of mantle xenoliths is required to evaluate the extent to which the tectosphere beneath the Slave Province formed during pre-2.6 Ga times, and its role in Neoarchean stabilization of the craton.

Rutile ages for a Kennady Lake xenolith (96-BAK-41-166) demonstrate that the lower crust of the southeastern Slave craton cooled through the blocking temperature for Pb diffusion in rutile by at least ca. 1.8 Ga. Recent experimental work suggests that rutile blocking temperature may be as high as 600 °C (Cherniak, 2000), although the empirically derived value is considered to be significantly lower at ca. 400–500 °C (Mezger et al., 1989). Temperatures below rutile closure are expected in the lower crust along a steady-state cratonic geotherm, as observed in the North American craton today by geothermic and seismological investigations (Hyndman and Lewis, 1999) and predicted by extension of a Slave Province mantle xenolith paleogeotherm to crustal depths (Fig. 9). The ca. 1.8 Ga rutile ages thus represent a minimum age for establishment of a 'craton-like' crustal geotherm. Titanite ages of ca. 2.55 Ga from a xenolith from the Kennady Lake pipe indicate that initial cooling from minimum lower crustal temperatures of 650–800 °C to temperatures below closure for Pb diffusion in titanite

(600–700 °C; Cherniak, 1992; Scott and St-Onge, 1995) occurred relatively rapidly, perhaps within 30 my of widespread lower crustal melting. Initial cooling may have initiated during crustal extension at or after 2585 Ma (e.g., James and Mortensen, 1992; Pehrsson et al., 2000). Ar–Ar dating of metamorphic minerals at mid-crustal levels (9–15 km) in the Yellowknife domain, near the Drybones kimberlite, indicates an initial period of rapid cooling (30–10 °C/Ma) between 2600 and 2500 Ma, followed by slower cooling (3–1 °C) between 2550 and 2400 Ma over the temperature interval of 350–250 °C (Bethune et al., 1999). The rutile data from lower crustal xenoliths at Kennady Lake would be consistent with slow cooling of the lower crust from 2.55 Ga to cratonic geotherms similar to those of present day values by at least ca. 1.8 Ga (e.g., Thompson et al., 1996). This would approximate a minimum integrated cooling rate of 0.5 °C/my over this interval. The ca. 1.8 Ga time of initial establishment of a 'cratonic-like' geotherm with lower crustal temperatures of 400–500 °C must be considered a minimum age as the lower crust may have been periodically disturbed by transient heating during mafic dyke swarm events at 2.2 and 2.0 Ga (e.g., Davis, 1997).

Younger rutile ages in one sample from Kennady Lake (KDC-13-141) and samples from the central Slave Province suggest Pb loss due to a later, and separate transient heating of lower crust (Davis, 1997), possibly during intrusion of the Mackenzie diabase dyke swarm, which was emplaced at 1.2 Ga from a plume centre located northwest of the Slave Province (LeCheminant and Heaman, 1989; Ernst and Baragar, 1992). Rutile and titanite from samples at Kennady Lake are consistently older than those from the Lac de Gras pipes suggesting that the regional effect of the ca. 1.2 Ga thermal resetting was more intense in the central part of the craton closer to the Mackenzie plume centre (Davis, 1997).

6. Conclusions

The petrologic and geochronologic data described above support a scenario in which lower crust beneath the central Slave craton initially formed by at least 2.9 Ga with components as old 3.2 Ga as suggested by inherited oscillatory-zoned zircons from sample GR-95-36. The lower crust was extensively intruded by magmas and metamorphosed in a series of punctuated events beginning at 2.78 Ga and most significantly between 2.64 and 2.55 Ga as recorded by both igneous and metamorphic zircon growth in lower crustal xenolith samples. An origin for some of the mafic granulites as lower crustal equivalents of the ca. 2.64–2.61 Ga intrusive rocks, suggests transient heating and metamorphism of the lower crust. At least, some of the mafic and tonalitic granulites may represent the direct crystallization products of magmas intruded within the lower crust during this time period, perhaps as a result of subduction beneath the craton (Davis et al., 2003). Interaction of these magmas with Mesoarchean crust resulted in inheritance of older zircons within the younger magmatic rocks. The onset of widespread crustal melting as indicated by the extensive mid-crustal granite plutonism followed shortly after at 2.60–2.58 Ga, and is recorded in metamorphic zircon growth in many of the lower crustal xenoliths. The cause of this metamorphism is uncertain, with continued thickening of the crust and/or increased reduced heat flow due to lithospheric thinning or delamination suggested as possibilities (Thompson, 1989a; Kusky, 1993; Davis et al., 1994). Lithospheric thickness during this interval may not have exceeded 100 km (e.g., Davis et al., 1994; Thompson et al., 1996).

Continued shortening and eventually tectonic subcretion of lithosphere (Davis et al., 2002) may have established a cratonic root by ca. 2.60–2.55 Ga. Evidence for continued metamorphic zircon growth in the lower crust at ca. 2.56–2.55 and 2.51 Ga may be related to continued tectonic activity in the deep crust associated with construction of the lithopheric root, due to boundary conditions well outside the present boundaries of the craton. Thick lithosphere to at least 200 km depth with a stable cratonic geotherm was established by at least 1.8 Ga, in the southeastern Slave Province as recorded by the cooling of rutile through their blocking temperature at conditions expected along a cratonic geotherm. Whether this occurred at a similar time in the central Slave cannot be demonstrated as rutile from that area are considerably younger and are interpreted to reflect higher transient heating in the central and northern Slave craton due to the effects of the Mackenzie plume event.

Acknowledgements

We thank U. Kretschmar and D. Smith, J. Carlson (BHP Resources), E. Schiller (Tanqueray Resources), M. Kirkley and H. Cookenboo (formerly at Canamera) for access to drill core of the Drybones Bay, Lac de Gras, Torrie and Kennady Lake (5034) pipes, respectively. E. Essene and T. Chacko are thanked for their advice on thermobarometry. Richard Stern is thanked for his advice and help in acquiring the SHRIMP ion probe data at the G.S.C. Comments on an earlier version of the manuscript by Vicki McNicoll helped to improve the clarity of the presentation. Julie Peressini helped in final manuscript preparation. Roberta Rudnick and Mark Schmitz are thanked for constructive reviews. This research was funded by Lithoprobe grants to DC and by the Geological Survey of Canada contribution number 2001217.

Appendix A. Mineral analyses for granulite xenoliths

Sample	Na$_2$O	MgO	Al$_2$O$_3$	SiO$_2$	K$_2$O	CaO	TiO$_2$	Cr$_2$O$_3$	MnO	FeO	NiO	P$_2$O$_5$	Total
Grizzly													
GR9536													
GR9536 Grt	0.02	6.72	21.74	37.73	0.00	4.04	0.03	0.06	1.25	27.80	0.02	0.00	99.41
GR9536 Opx	0.01	18.26	2.65	49.49	0.00	0.35	0.13	0.09	0.42	27.66	0.03	0.01	99.10
GR9536 Ilm host	0.02	1.05	0.05	0.08	0.00	0.03	43.36	0.19	0.31	51.62	0.00	0.02	96.73
GR9536 Ulv exsol	0.01	1.03	0.11	0.03	0.00	0.02	29.63	0.51	0.36	63.98	0.00	0.03	95.73
GR9536 Mt	0.00	0.17	0.38	0.05	0.01	0.22	0.09	2.82	0.04	89.22	0.00	0.03	93.02

Appendix A (*continued*)

Sample	Na$_2$O	MgO	Al$_2$O$_3$	SiO$_2$	K$_2$O	CaO	TiO$_2$	Cr$_2$O$_3$	MnO	FeO	NiO	P$_2$O$_5$	Total
Grizzly													
GR9557													
GR9557 Grt	0.01	6.41	21.67	37.81	0.01	7.11	0.08	0.14	0.84	25.96	0.01	0.04	100.09
GR9557 Cpx	0.61	13.11	2.87	50.79	0.00	22.43	0.53	0.08	0.11	8.71	0.03	0.03	99.30
GR9557 Plag	5.20	0.01	28.00	54.26	0.25	10.73	0.02	0.03	0.01	0.01	0.02	0.04	98.58
GR9500													
GR9500 Grt	0.01	6.50	21.59	37.94	0.01	7.44	0.09	0.10	0.80	25.75	0.01	0.04	100.26
GR9500 Cpx	0.62	13.25	2.83	51.28	0.00	22.61	0.42	0.09	0.10	8.10	0.04	0.02	99.35
GR9500 Plag	5.34	0.01	28.03	54.59	0.20	10.55	0.02	0.03	0.00	0.03	0.01	0.02	98.84
GR9500 Ilm	0.00	2.39	0.08	0.01	0.00	0.03	53.49	0.04	0.38	44.22	0.01	0.00	100.66
GR9500 Ttn	0.01	0.02	2.06	30.24	0.01	28.77	37.33	0.04	0.03	0.40	0.00	0.00	98.92
GR9500 Rut	0.00	0.02	0.12	0.02	0.00	0.20	97.81	0.26	0.01	0.70	0.01	0.01	99.18
X9501BB													
X9501BB Cpx	0.41	11.81	1.45	51.17	0.00	22.23	0.21	0.00	0.28	12.06	0.01	0.00	99.63
X9501BB Opx	0.01	16.24	0.83	50.05	0.01	0.89	0.12	0.01	0.66	31.08	0.05	0.01	99.96
X9501BB Plag	7.04	0.02	25.76	58.84	0.32	7.50	0.01	0.00	0.00	0.05	0.00	0.03	99.57
X9501BB Kspar	0.73	0.02	18.64	63.86	15.90	0.04	0.01	0.00	0.00	0.04	0.02	0.01	99.26
Koala													
KDC-13-131													
KDC-13-131 Grt	0.03	9.26	22.32	38.92	0.00	6.19	0.06	0.05	0.43	22.52	0.00	0.00	99.78
KDC-13-131 Opx	0.06	25.18	1.78	53.16	0.00	1.09	0.03	0.04	0.07	18.62	0.01	0.00	100.03
KDC-13-131 Cpx	1.36	13.81	3.45	52.66	0.01	21.47	0.22	0.09	0.03	6.39	0.01	0.00	99.49
KDC-13-131 Plag	7.71	0.00	24.23	61.10	0.47	5.83	0.00	0.00	0.00	0.04	0.00	0.00	99.39
KDC-13-131 Biot	0.08	18.67	13.82	39.64	9.91	0.02	4.03	0.10	0.01	8.66	0.05	0.00	94.98
KDC-13-141A													
KDC-13-141A Grt	0.02	7.85	21.97	38.56	0.00	6.06	0.10	0.09	0.64	24.30	0.00	0.00	99.59
KDC-13-141A Opx	0.03	22.68	1.26	52.29	0.02	0.36	0.04	0.01	0.19	22.92	0.04	0.00	99.85
KDC-13-141A Cpx	1.22	13.32	3.56	51.83	0.01	20.93	0.27	0.04	0.08	7.74	0.03	0.00	99.02
KDC-13-141A Plag	7.38	0.01	25.15	60.21	0.26	6.76	0.00	0.00	0.00	0.02	0.00	0.00	99.83
KDC-13-141A Ilm	0.00	2.60	0.20	0.02	0.00	0.01	51.07	0.07	0.19	42.75	0.12	0.00	97.06
Torrie													
TQ9511													
TQ9511 Grt	0.01	5.61	21.51	37.75	0.00	7.03	0.04	0.04	0.79	27.62	0.00	0.04	100.43
TQ9511 Cpx	0.74	13.15	1.85	52.12	0.01	22.23	0.14	0.03	0.09	9.23	0.04	0.05	99.69
TQ9511 Plag	6.99	0.03	25.99	58.43	0.14	7.70	0.00		0.01	0.03	0.01	0.02	99.35
TQ9511 Kspar	0.97	0.09	17.88	63.72	15.18	0.11	0.03	0.01	0.03	0.75	0.00	0.01	98.78
TQ9411 Ilm	0.01	1.81	0.08	0.02	0.00	0.02	50.27	0.04	0.23	46.92	0.04	0.01	99.45
TQY94-17-9													
TQY94-17-9 Grt	0.01	6.28	21.39	37.73	0.00	6.52	0.08	0.02	0.85	25.92	0.00	0.00	98.81
TQY94-17-9 Cpx	0.82	11.57	4.85	49.93	0.00	20.75	0.61	0.03	0.16	10.09	0.01	0.00	98.80
plag not analyzed													
TQY94-17-22													
TQY94-17-22 Grt	0.02	5.57	21.51	37.76	0.00	6.74	0.07	0.04	0.76	26.45	0.00	0.00	98.92
TQY94-17-22 Opx	0.01	20.33	1.19	51.62	0.00	0.44	0.09	0.04	0.19	26.17	0.04	0.00	100.13
TQY94-17-22 Cpx	0.71	12.71	2.66	51.57	0.01	21.96	0.27	0.08	0.09	9.30	0.01	0.00	99.36
TQY94-17-22 Plag	5.52	0.02	27.08	55.99	0.32	9.69	0.00	0.00	0.00	0.02	0.00	0.00	98.64

(continued on next page)

Appendix A (*continued*)

Sample	Na$_2$O	MgO	Al$_2$O$_3$	SiO$_2$	K$_2$O	CaO	TiO$_2$	Cr$_2$O$_3$	MnO	FeO	NiO	P$_2$O$_5$	Total
Torrie													
TQY94-17-22													
TQY94-17-22 Kspar	1.49	0.15	17.01	65.39	13.79	0.11	0.00	0.00	0.00	0.67	0.00	0.00	98.61
TQY94-17-22 Ilm	0.00	1.74	0.00	0.02	0.00	0.00	47.37	0.11	0.20	47.48	0.02	0.00	96.95
TQY94-17-22E													
TQY94-17-22E Grt	0.01	13.12	22.81	39.37	0.00	1.18	0.05	0.11	0.39	22.29	0.00	0.00	99.34
TQY94-17-22E Herc	0.00	11.92	57.40	0.01	0.00	0.01	0.00	5.37	0.01	19.70	0.20	0.00	97.70
TQY94-17-22E Kspar	3.08	0.79	18.33	64.37	11.27	0.31	0.00	0.00	0.00	0.39	0.00	0.00	98.64
TQY94-17-22E Sill	0.01	0.01	62.17	36.25	0.00	0.01	0.02	0.24	0.01	0.21	0.00	0.00	99.50
TQY94-17-22E Biot	0.17	22.91	13.40	40.31	10.40	0.05	2.23	0.18	0.01	4.71	0.09	0.00	102.31
TQY94-17-22J													
TQY94-17-22J Grt	0.01	5.39	21.78	37.68	0.00	6.88	0.06	0.00	0.74	27.41	0.00	0.00	99.96
TQY94-17-22J Cpx	0.63	12.86	1.83	51.97	0.01	21.75	0.19	0.04	0.10	9.61	0.02	0.00	99.01
TQY94-17-22J Ilm	0.00	1.60	0.00	0.01	0.00	0.04	50.86	0.01	0.21	44.32	0.03	0.00	97.15
plag not analyzed													
Kennady Lake													
96BAK002-226													
96BAK002-229 Grt	0.02	12.02	22.77	39.25	0.00	1.77	0.02	0.09	0.30	23.12	0.00	0.00	99.36
96BAK002-229 Biot	0.33	22.39	13.71	39.74	10.08	0.04	3.85	0.19	0.03	4.09	0.02	0.00	94.48
96BAK009-416													
96BAK009-416 Grt	0.02	10.62	22.55	38.78	0.00	1.17	0.03	0.08	0.27	25.92	0.00	0.00	99.43
96BAK009-416 Biot	0.20	19.21	14.71	38.79	9.82	0.04	5.50	0.11	0.02	6.84	0.07	0.00	95.31
Drybones Bay													
96.21.326A													
96.21.326A Grt	0.01	9.43	21.90	38.82		6.19	0.16	0.05	0.76	22.26	0.01		99.59
96.21.326A Plag	6.74		25.83	58.64	0.30	7.91				0.06			99.49
96.21.326A Ilm		6.03					43.97		0.22	46.55			96.78
96.17.318													
96.17.318 Grt	0.01	9.61	21.96	38.79		6.14	0.22	0.07	0.72	21.80	0.02		99.34
96.17.318 Plag	6.89		25.99	58.55	0.32	7.88				0.06			99.69
96.17.318 Serp (Opx)	0.18	30.11	7.00	42.53	0.14	0.29	0.29	0.07	0.02	4.23	0.06		84.93
96.17.318 Ilm		10.35					50.26		0.39	34.71			95.72
96.21.390.7													
96.21.390.7 Grt	0.01	10.03	22.04	38.90		6.11	0.20	0.05	0.40	21.37	0.01		99.12
96.21.390.7 Plag	6.14		26.72	57.12	0.45	8.94				0.03			99.39
96.21.390.7 Ilm		2.26					45.13		0.11	49.74			97.24
96.21.390.7 Rut		0.01					99.22	0.04	0.00	0.65			99.92
95.17.298													
95.17.298 Grt	0.01	8.15	22.13	38.76		5.91	0.15	0.05	0.46	23.52			99.13
95.17.298 Cpx	1.44	13.33	3.99	51.83	0.00	20.98	0.49	0.04	0.06	7.54			99.70
95.17.298 Plag	6.24	0.00	26.31	59.58	0.47	7.88				0.07			100.54
95.17.298 Ilm		2.30	0.00	0.02		0.03	44.02	0.08	0.15	50.06	0.02		96.67
95.17.298 Rut		0.00	0.00	0.06		0.10	97.86	0.02	0.02	1.09	0.02		99.16

Appendix A (*continued*)

Sample	Na$_2$O	MgO	Al$_2$O$_3$	SiO$_2$	K$_2$O	CaO	TiO$_2$	Cr$_2$O$_3$	MnO	FeO	NiO	P$_2$O$_5$	Total
Drybones Bay													
96.20.274													
96.20.274 Grt	0.01	9.18	21.96	38.69		6.28	0.23	0.06	0.50	22.74	0.01		99.67
96.20.274 Serp (Opx)	0.18	27.16	7.16	45.22	0.35	0.30	0.08	0.05	0.02	6.53	0.10		87.14
96.20.274 Plag	5.74		27.68	56.09	0.33	9.76				0.04			99.65
96.20.274 Ilm		2.28					45.11		0.24	49.57			97.20
96.20.274 Rut		0.01					99.03	0.09	0.01	0.65			99.79
96.15.244.7													
96.15.244.7 Grt	0.02	10.46	22.35	38.76		5.65	0.24	0.06	0.43	21.39	0.02		99.38
96.15.244.7 Serp (Opx)	0.18	29.41	8.39	42.14	0.14	0.35	0.46	0.05	0.02	4.81	0.16		86.11
96.15.244.7 Plag	6.56		26.31	58.08	0.24	8.32				0.03			99.54
96.15.244.7 Ilm		2.24					45.52		0.13	49.26			97.16
96.15.455.7													
96.15.455.7 Grt	0.01	10.40	22.31	38.92		6.67	0.20	0.04	0.42	20.58	0.00		99.55
96.15.455.7 Plag	6.03		26.70	57.93	0.42	8.67				0.03			99.78
96.15.455.7 Ilm		5.05					41.69		0.29	49.07			96.10
96.16.438.5													
96.16.438.5 Grt	0.01	9.27	22.37	39.02		6.43	0.19	0.07	0.45	21.39			99.22
96.16.438.5 Trem (Cpx)	2.78	0.06	27.92	45.75	0.21	17.46	0.07	0.01	0.00	0.10			94.35
96.16.438.5 Plag	5.70	0.01	26.62	58.85	0.72	8.43				0.14			100.47
96.16.438.5 Ilm		2.61	0.00	0.01		0.01	43.68	0.08	0.06	50.13	0.16		96.73
96.16.438.5 Rut		0.00	0.00	0.00		0.01	99.40	0.01	0.00	0.36	0.02		99.81
96.16.438.5 Ulv		0.46	0.25	0.02		0.01	19.20	0.44	0.03	71.57	0.09		92.06
96.21.326.5													
96.21.326B Grt	0.01	10.11	22.31	39.07		5.91	0.17	0.02	0.75	21.26	0.00		99.61
96.21.326B Serp (Opx)	0.06	28.21	1.00	42.23	0.09	0.24	0.02	0.01	0.12	8.03	0.10		80.11
96.21.326B Plag	6.92		25.65	58.97	0.29	7.61				0.05			99.49
96.21.326B Ilm		6.03					43.85		0.26	46.68			96.81

Appendix B. Analytical techniques

Xenolith samples were extracted from kimberlite with a fine diamond saw and all surfaces were cleaned by grinding off any adhering kimberlite, and cleaning in ultrasonic bath in water. In the cases where the kimberlite was friable, xenoliths were broken directly out of the kimberlite. Following cleaning, samples were either hand crushed in a mortar and pestle, or in a hydraulic press; the fines were eluted in water and the heavy minerals concentrated with methyl iodine (s.g. 3.2–3.3).

B.1. Thermal ionization mass spectrometry methods

Thermal ionization mass spectrometry was performed at the Geological Survey of Canada. All zircon fractions and selected titanite and rutile fractions were air abraded (Krogh, 1982). Grains with overgrowths were aggressively abraded to try and remove the outer zircon. Overgrowths when recognized, and large enough, were separated from cores by fracturing with tweezers under a binocular microscope in alcohol. In many cases, the fracture followed the discontinuity surface between the two zircon zones, although it could not be determined with certitude whether the separation was perfect. Overgrowths were also found as thin fragments that were presumably separated from core material during the crushing process. These fragments were identified as overgrowths based on negative crystal faces and cup-like appearance.

Analytical methods for U–Pb analyses of zircon are summarized in Parrish et al. (1987), and for titanite in Davis et al., (1997). Analytical errors are determined based on error propagation methods of Roddick (1987), assuming a mass fractionation value of 0.09% ± 0.036% based on replicate analyses of NBS 981 and 983 standards. Samples were corrected for laboratory blank and initial common Pb (Cumming and Richards, 1975). Regressions based on modified linear regression (York, 1969). All age errors are reported and plotted at 95% confidence interval.

B.2. Ion probe methods

U–Pb ion probe data were acquired using the Geological Survey of Canada Sensitive High Resolution Ion Microprobe (SHRIMP II). Analytical methods are provided in greater detail in Stern (1997) and Stern (1998). Zircons were mounted and polished along with grains of Kipawa zircon standard in a 2.5-cm epoxy puck. Back scatter electron and cathodoluminescence images were made using a scanning electron microscope to fully characterize the internal structures of the grains and aid in beam positioning. Data was collected using a secondary electron multiplier operating in pulse counting mode. No mass fractionation correction was applied, and Pb isotopic data are corrected for the presence of small amounts of surface common Pb based on measured ^{204}Pb abundances. Bias in the measured Pb/U values was corrected relative to the Kipawa zircon standard ($^{206}Pb/^{238}U$ age = 993 Ma) using a linear calibration curve determined between $^{206}Pb^+/^{238}U^+$ and $^{254}U^+/^{238}U^+$. An uncertainty of ± 1.0% (1σ) in the calibration curve was propagated along with counting statistic errors to determine errors of unknowns. Final U–Pb ages are reported and plotted with 1σ errors in Table 4 and Figs. 6b and 8c. Ages reported in text are given at 2σ.

References

Abbott, D., 1991. The case for accretion of the tectosphere by buoyant subduction. Geophys. Res. Lett. 18, 585–588.

Aulbach, S., Griffin, W.L., Pearson, N.J., O'Reilly, S.Y., Doyle, B.J., Kivi, K., 2001. Re–Os isotope evidence for Meso-Archean mantle beneath 2.7 Ga Contwoyto Terrane, Slave craton, Canada: implications for the tectonic history of the Slave Craton. Abstract. In: Carlson, R., Jones, A.G. (Eds.), The Slave Kaapvaal Workshop, Merrickville, Ontario September 5–9.

Bethune, K.M., Villeneuve, M.E., Bleeker, W., 1999. Laser $^{40}Ar/^{39}Ar$ thermochronology of Archean rocks in the Yellowknife Domain, southwestern Slave Province: insights into the cooling history of an Archean granite–greenstone terrrane. Can. J. Earth Sci. 36, 1189–1206.

Berman, R.G., 1988. Internally consistent thermodynamic data for minerals in the system $Na_2O-K_2O-CaO-MgO-FeO-Fe_2O_3-Al_2O_3-SiO_2-TiO_2-H_2O-CO_2$. J. Petrol. 29, 445–522.

Berman, R.G., 1991. Thermobarometry using multi-equilibrium calculations: a new technique, with petrological applications. Can. Mineral. 29, 833–855.

Bleeker, W., Davis, W.J., 1999. The 1991–1996 NATMAP Slave Province Project: introduction. Can. J. Earth Sci. 36, 1033–1042.

Bleeker, W., Ketchum, J.W.F., Jackson, V.A., Villeneuve, M.E., 1999. The Central Slave Basement Complex: Part I. Its structural topology and autochthonous cover. Can. J. Earth Sci. 36, 1083–1109.

Bleeker, W., Davis, W.J., Ketchum, J.W., Sircombe, K., Stern, R.A., 2001. Tectonic evolution of the Slave craton. In: Cassidy, K.F., Dunphy, J.M., Van Kranendonk, M.J. (Eds.), 4th International Archaean Symposium 2001, Extended Abstracts, AGSO-Geoscience Australia, Record 2001/37, pp. 288–290.

Bostock, M.G., 1998. Mantle stratigraphy and evolution of the Slave Province. J. Geophys. Res. 103, 21183–21200.

Boyd, F.R., Canil, D., 1997. Peridotite xenoliths from the Slave craton, NWT. LPI Contrib. 921, 34.

Brey, G., Köhler, T., 1990. Geothermobarometry in four-phase lherzolites: II. New thermobarometers, and practical assessment of existing thermobarometers. J. Petrol. 31, 1353–1378.

Canil, D., 2002. Vanadium in peridotites, mantle redox and tectonic environments; Archean to present. Earth Planet. Sci. Lett. 195, 75–90.

Carbno, G.B., Canil, D., 2002. Mantle structure beneath the southwest Slave craton, Canada: constraints from garnet geochemistry in the Drybones Bay kimberlite. J. Petrol. 43, 129–142.

Cassidy, J.F., 1995. A comparison of the receiver structure beneath stations of the Canadian National Seismograph Network. Can. J. Earth Sci. 32, 938–951.

Cherniak, D.J., 1992. Lead diffusion in titanite and preliminary results on the effects of radiation damage on Pb transport. Chem. Geol. 110, 177–194.

Cherniak, D.J., 2000. Pb diffusion in rutile. Contrib. Mineral. Petrol. 139, 198–207.

Cumming, G.L., Richards, J.R., 1975. Ore lead in a continuously changing earth. Earth Planet. Sci. Lett. 28, 155–171.

Davis, W.J., 1997. U–Pb zircon and rutile ages from granulite xenoliths in the Slave Province: evidence for mafic magmatism in the lower crust coincident with Proterozoic dike swarms. Geology 25, 343–346.

Davis, W.J., Bleeker, W., 1999. Timing of plutonism, deformation, and metamorphism in the Yellowknife Domain, Slave Province, Canada. Can. J. Earth Sci. 36, 1169–1187.

Davis, W.J., Hegner, E., 1992. Neodymium isotopic evidence for the tectonic assembly of late Archean crust in the Slave Province, northwest Canada. Contrib. Mineral. Petrol. 111, 493–504.

Davis, W.J., Fryer, B.J., King, J.E., 1994. Geochemistry and evolution of late Archean plutonism and its significance to the tectonic development of the Slave craton. Precambrian Res. 67, 207–241.

Davis, W.J., Gariepy, C., van Breemen, O., 1996. Pb isotopic composition of late Archaean granites and the extent of recycling early Archaean crust in the Slave Province, northwest Canada. Chem. Geol. 130, 255–269.

Davis, W.J., McNicoll, V., Bellerive, D., Scott, D.J., 1997. Modified chemical procedures for the extraction and purification of uranium from titanite, allanite and rutile in the Geochronology Laboratory, Geological Survey of Canada. Radiogenic Age and Isotope Studies: Report 10. Geological Survey of Canada, Paper 1997-F, pp. 33–36.

Davis, W.J., Jones, A.G., Bleeker, W., Grütter, H., 2003. Lithospheric development in the Slave Craton: a linked crustal and mantle perspective. Lithos 71, 575–589, this issue.

DeWit, M.J., Roering, C., Hart, R.J., Armstrong, R.A., de Ronde, C.E.J., Green, R.W.E., Tredoux, M., Peberdy, E., Hart, R.A., 1992. Formation of an Archean continent. Nature 357, 553–562.

Ellis, D.J., Green, D.H., 1979. An experimental study of the effect of Ca upon the garnet–clinopyroxene exchange equilibria. Contrib. Mineral. Petrol. 71, 13–22.

Ernst, R.E., Baragar, W.R.A., 1992. Evidence from magnetic fabric for the flow pattern of magma in the Mackenzie giant radiating dike swarm. Nature 356, 511–513.

Fraser, G., Ellis, D., Eggins, S., 1997. Zirconium abundance in granulite-facies minerals, with implications for zircon geochronology in high-grade rocks. Geology 25, 607–610.

Frost, B.R., Chacko, T., 1989. The granulite uncertainty principle: limitations on thermobarometry of granulites. J. Geol. 97, 435–450.

Fyson, W.K., Helmstaedt, H., 1988. Structural patterns and tectonic evolution of supracrustal domains in the Archean Slave Province, Canada. Can. J. Earth Sci. 25, 301–315.

Griffin, W.L., Doyle, B.J., Ryan, C.G., Pearson, N.J., O'Reilly, S., Davies, R., Kivi, K., van Achterbergh, E., Natapov, L., 1999. Layered mantle lithosphere in the Lac de Gras area, Slave craton: composition, structure and origin. J. Petrol. 40, 705–727.

Grütter, H.S., Apter, D.B., Kong, J., 1999. Crust–mantle coupling: evidence from mantle-derived xenocrystic garnets. Proceedings of the 7th International Kimberlite Conference. Red Roof Design, Cape Town., pp. 307–313.

Hanchar, J.L., Rudnick, R.L., 1995. Revealing hidden structures: the application of cathodoluminescence and back-scattered electron imaging to dating zircons from lower crustal xenoliths. Lithos 36, 289–303.

Harley, S.L., 1984. An experimental study of the partitioning of Fe and Mg between garnet and orthopyroxene. Contrib. Mineral. Petrol. 86, 359–373.

Harte, B., Jackson, P.M., Macintyre, R.M., 1981. Age of mineral equilibria in granulite facies nodules from kimberlites. Nature 291, 159–160.

Helmstaedt, H., Schulze, D.J., 1989. Southern African kimberlites and their mantle sample: implications for the Archaean tectonics and lithosphere evolution. Geol. Soc. Aust. Spec. Publ. 14, 358–368.

Herzberg, C., 1993. Lithosphere peridotites of the Kaapvaal craton. Earth Planet. Sci. Lett. 120, 13–29.

Hyndman, R.D., Lewis, T.J., 1999. Geophysical consequences of the Cordillera–Craton thermal transition in southwestern Canada. Tectonophysics 306, 397–422.

Irvine, G.J., Kopylova, R.W., Carlson, R.W., Pearson, D.G., Shirey, S.B., 1999. Age of the lithospheric mantle beneath and around the Slave craton: a rhenium–osmium-isotopic study of peridotite xenoliths from the Jericho and Somerset Island kimberlites. 9th Annual V. M. Goldschmidt Conference, LPI Contribution No. 971. Lunar and Planetary Institute, Houston, pp. 134–135.

Irvine, G.J., Pearson, D.G., Kopylova, R.W., Carlson, R.W., Kjarsgaard, B.A., Dreibus, G., 2001. The age of two cratons: a PGE and Os-isotopic study of peridotite xenoliths from the Jericho kimberlite (Slave craton) and the Somerset Island kimberlite field (Churchill Province). In: Carlson, R., Jones, A.G. (Eds.), The Slave Kaapvaal Workshop, Merrickville, Ontario September 5–9.

Isachsen, C., Bowring, S.A., 1994. Evolution of the Slave craton. Geology 22, 917–920.

Jordan, T.H., 1975. The continental tectosphere. Rev. Geophys. Space Phys. 13, 1–12.

James, D.T., Mortensen, J.K., 1992. An Archean metamorphic core complex in the southern Slave Province; basement-cover structural relations between the Sleepy Dragon Complex and the Yellowknife Supergroup. Can. J. Earth Sci. 29, 2133–2145.

Kelemen, P.B., Hart, S.R., Bernstein, S., 1998. Silica enrichment in the continental upper mantle via melt/rock reaction. Earth Planet. Sci. Lett. 164, 387–406.

Kopylova, M., Russell, J.K., 2000. Chemical stratification of cratonic lithosphere: constraints from the northern Slave craton, Canada. Earth Planet. Sci. Lett. 181, 1–87.

Kopylova, M.G., Russell, J.K., Cookenboo, H., 1998. Petrology of peridotite and pyroxenite xenoliths from the Jericho kimberlite: implications for the thermal state of the mantle beneath the Slave craton. J. Petrol. 40, 79–104.

Krogh, T.E., 1982. Improved accuracy of U–Pb zircon ages by the creation of more concordant systems using an air abrasion technique. Geochim. Cosmochim. Acta 46, 637–649.

Krogh, T.E., 1993. High precision U–Pb ages for granulite metamorphism and deformation in the Archean Kapuskasing structural zone, Ontario: implications for structure and development of the lower crust. Earth Planet. Sci. Lett. 119, 1–18.

Kusky, T.M., 1989. Accretion of the Archean Slave Province. Geology 17, 63–67.

Kusky, T.M., 1993. Collapse of Archean orogens and the generation of late- to postkinematic granitoids. Geology 21, 925–928.

LeCheminant, A.N., Heaman, L.M., 1989. Mackenzie igneous events, Canada: middle Proterozoic hotspot magmatism associated with ocean opening. Earth Planet. Sci. Lett. 96, 38–48.

Midgley, J.P., Blundell, D.J., 1997. Deep seismic structure and thermo-mechanical modelling of continental collision zones. Tectonophysics 273, 155–167.

Mezger, K., Hanson, G.N., Bohlen, S.R., 1989. High-precision U–Pb ages of metamorphic rutile: application to the cooling history of high-grade terranes. Earth Planet. Sci. Lett. 96, 106–118.

MacKenzie, J.M., Canil, D., 1999. Composition and thermal evolution of cratonic mantle beneath the central Archean Slave Province, NWT, Canada. Contrib. Mineral. Petrol. 134, 313–324.

Moser, D.E., Heaman, L.M., 1997. Proterozoic zircon growth in Archean lower crustal xenoliths, southern Superior craton—a consequence of Matachewan ocean opening. Contrib. Mineral. Petrol. 128, 164–175.

Moser, D.E., Heaman, L.M., Krogh, T.E., Hanes, J.A., 1996. Intracrustal extension of an Archean orogen revealed using single-grain U–Pb zircon geochronology. Tectonics 15, 1093–1109.

Moser, D.E., Flowers, R.M., Harte, R.J., 2001. Birth of the Kaapvaal tectosphere 3.08 billion years ago. Science 291, 465–468.

Padgham, W.A., Fyson, W.K., 1992. The Slave Province: a distinct Archean craton. Can. J. Earth Sci. 29, 2072–2086.

Parrish, R.R., Roddick, J.C., Loveridge, W.D., Sullivan, R.W., 1987. Uranium–lead analytical techniques at the geochronology laboratory, Geological Survey of Canada. Radiogenic Age and Isotopic Studies: Report 1, Geological Survey of Canada, Paper 87-2, pp. 3–7.

Pearson, D.G., 1999. The age of continental roots. Lithos 48, 171–194.

Pehrsson, S.J., Chacko, T., Pilkington, M., Villeneuve, M.E., Bethune, K., 2000. Anton Terrane revisited; late Archean exhumation of a moderate-pressure granulite terrane in the western Slave Province. Geology 28, 1075–1078.

Perchuk, L.L., Lavrent'eva, I.V., 1983. Experimental investigation of exchange equilibria in the system cordierite–garnet–biotite. In: Saxena, S. (Ed.), Kinetics and Equilibrium in Mineral Reactions. Advances in Physical Geochemistry. Springer-Verlag, Berlin, pp. 199–239.

Roberts, M.P., Finger, F., 1997. Do U–Pb zircon ages from granulites reflect peak metamorphic conditions? Geology 25, 319–322.

Roddick, J.C., 1987. Generalized numerical error analysis with applications to geochronology and thermodynamics. Geochim. Cosmochim. Acta 51, 2129–2135.

Rudnick, R.L., 1992. Xenoliths—samples of the lower continental crust. In: Fountain, D.M., Arculus, R., Kay, R.W. (Eds.), The Continental Lower Crust. Developments in Geotectonics, vol. 23. Elsevier, Amsterdam, pp. 269–316.

Schaltegger, U., Fanning, C.M., Gunther, D., Maurin, J.C., Schulmann, K., Gebauer, D., 1999. Growth, annealing and recrystallization of zircon and preservation of monazite in high-grade metamorphism: conventional and in-situ U–Pb isotope, cathodoluminescence and microchemical evidence. Contrib. Mineral. Petrol. 134, 186–201.

Schmitz, M.D., Bowring, S.A., 2000. The significance of U–Pb zircon dates in lower crustal xenoliths from the southwestern margin of the Kaapvaal craton, southern Africa. Chem. Geol. 172, 59–76.

Scott, D.J., St-Onge, M.R., 1995. Constraints on Pb closure temperature in titanite based on rocks from the Ungava Orogen, Canada; implications for U–Pb geochronology and *P–T–t* path determinations. Geology 23, 1123–1126.

Smith, D., Barron, B.R., 1991. Pyroxene–garnet equilibration during cooling in the mantle. Am. Mineral. 76, 1950–1963.

Stern, R.A., 1997. The GSC Sensitive High Resolution Ion Microprobe (SHRIMP): analytical techniques of zircon U–Th–Pb age determinations and performance evaluation. Radiogenic Age and Isotope Studies: Report 10. Geological Survey of Canada, Paper 1997-F, pp. 1–31.

Stern, R.A., 1998. High-resolution SIMS determination of radiogenic tracer-isotope ratios in minerals. In: Cabri, L.J., Vaughan, D.J. (Eds.), Modern Approaches to Ore and Environmental Mineralogy: Short Course Series, Mineralogical Association of Canada, vol. 27, pp. 241–268.

Thompson, P.H., 1989a. Moderate overthickening of thinned sialic crust and the origin of granitic magmatism and regional metamorphism in low-*P* high-*T* terranes. Geology 17, 520–523.

Thompson, P.H., 1989b. An empirical model for metamorphic evolution of the Archaean Slave Province and adjacent Thelon tectonic zone, north-western Canadian Shield. In: Daly, J.S., Cliff, R.A., Yardley, B.W.D. (Eds.), Evolution of metamorphic belts. Geol. Soc., London, Spec. Publ., pp. 245–263.

Thompson, P.H., Judge, A.S., Lewis, T.J., 1996. Thermal evolution of the lithosphere in the central Slave Province: Implications for diamond genesis. In: LeCheminant, A.N., Richardson, D.G., DiLabio, R.N.W., Richardson, K.A. (Eds.), Searching for Diamonds in Canada. Geological Survey of Canada, Open File 3228, pp. 151–160.

Thorpe, R.I., Cumming, G.L., Mortensen, J.K., A significant Pb isotope boundary in the Slave Province and its probable relation to ancient basement in the western Slave Province. Geol. Surv. Canada Open File 2484, pp. 179–184.

Vavra, G., Gebauer, D., Schmid, R., Compston, W., 1996. Multiple zircon growth and recrystallization during polyphase late Carboniferous to Triassic metamorphism in granulites of the Ivrea Zone (southern Alps): an ion microprobe (SHRIMP) study. Contrib. Mineral. Petrol. 122, 337–358.

Vavra, G., Schmid, R., Gebauer, D., 1999. Internal morphology, habit and U–Th–Pb microanalysis of amphibolite-to-granulite facies zircons: geochronology of the Ivrea Zone (Southern Alps). Contrib. Mineral. Petrol. 134, 380–404.

van Breemen, O., Davis, W.J., King, J.E., 1992. Temporal distribution of granitoid plutonic rocks in the Archean Slave Province, northwest Canadian Shield. Can. J. Earth Sci. 22, 2186–2199.

Vielzeuf, D., Montel, J.M., 1994. Partial melting of metagreywackes: Part I. Fluid-absent experiments and phase relationships. Contrib. Mineral. Petrol. 117, 375–393.

Walter, M.J., 1999. Melting residues of fertile peridotite and the origin of cratonic lithosphere. In: Fei, Y., Bertka, C., Mysen, B.O. (Eds.), Mantle Petrology: Field Observations and High Pressure Experimentation. The Geochemical Society Special Publication, vol. 6, pp. 225–240.

Williams, I.S., 1992. Some observations on the use of zircon U–Pb geochronology in the study of granitic rocks. Trans. R. Soc. Edinb. 83, 447–458.

Yamashita, K., Creaser, R.A., Stemler, J.U., Zimaro, T.W., 1999. Geochemical and Nd–Pb isotopic systematics of late Archean granitoids, southwestern Slave Province, Canada: constraints for granitoid origin and crustal isotopic structure. Can. J. Earth Sci. 36, 1131–1147.

York, D., 1969. Least squares fitting of a straight line with correlated errors. Earth Planet. Sci. Lett. 5, 320–324.

Available online at www.sciencedirect.com

Lithos 71 (2003) 575–589

LITHOS

www.elsevier.com/locate/lithos

ELSEVIER

Lithosphere development in the Slave craton: a linked crustal and mantle perspective

W.J. Davis[a,*], A.G. Jones[a], W. Bleeker[a], H. Grütter[b]

[a]*Geological Survey of Canada, 601 Booth St., Ottawa, Ontario, Canada K1A 0E8*
[b]*Mineral Services Canada, #1300-409 Granville Street, Vancouver, British Columbia, Canada V6C 1T2*

Abstract

The late tectonic evolution of the Slave craton involves extensive magmatism, deformation, and high temperature-low pressure (HT-LP) metamorphism. We argue that the nature of these tectonic events is difficult to reconcile with early, pre-2.7 Ga development and preservation of a thick tectosphere, and suggest that crust–mantle coupling and stabilization occurred only late in the orogenic development of the craton. The extent and repetitiveness of the tectonic reworking documented within the Mesoarchean basement complex of the western Slave, together with the development of large-volume, extensional mafic magmatism at 2.7 Ga within the basement complex argue against preservation of a widespread, thick, cool Mesoarchean tectosphere beneath the western Slave craton prior to Neoarchean tectonism. Broad-scale geological and geophysical features of the Slave craton, including orientation of an early F1 fold belt, distribution of ca. 2.63–2.62 Ga plutonic rocks, and the distribution of geochemical, petrological and geophysical domains within the mantle lithosphere collectively highlight the importance of an NE–SW structural grain to the craton. These trends are oblique to the earlier, ca. 2.7 Ga north–south trending boundary between Mesoarchean and Neoarchean crustal domains, and are interpreted to represent a younger structural feature imposed during northwest or southeast-vergent tectonism at ca. 2.64–2.61 Ga. Extensive plutonism, in part mantle-derived, crustal melting and associated HT-LP metamorphism argue for widespread mantle heat input to the crust, a feature most consistent with thin (<100 km) lithosphere at that time. We propose that the mantle lithosphere developed by tectonic imbrication of one or more slabs subducted beneath the craton at the time of development of the D1 structural grain, producing the early 2.63–2.62 Ga arc-like plutonic rocks. Subsequent collision (external to the present craton boundaries) possibly accompanied by partial delamination of some of the underthrust lithosphere, produced widespread deformation (D2) and granite plutonism throughout the province at 2.6–2.58 Ga. An implication of this model is that diamond formation in the Slave should be Neoarchean in age.
© 2003 Elsevier B.V. All rights reserved.

Keywords: Archean; Slave province

1. Introduction

One of the defining features of Archean cratons is the presence of a thick (>150 km) lithospheric mantle keel, termed tectosphere by Jordan (1988), characterized by high P-wave velocities, low geothermal gradients and chemically depleted compositions. Debate

* Corresponding author. Tel.: +1-613-943-8780; fax: +1-613-995-7997.
E-mail address: bidavis@nrcan.gc.ca (W.J. Davis).

0024-4937/$ - see front matter © 2003 Elsevier B.V. All rights reserved.
doi:10.1016/S0024-4937(03)00131-2

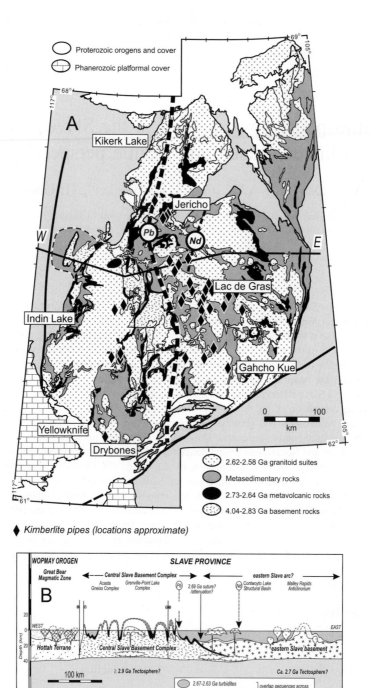

Fig. 1. (A) Geological map of the Slave craton showing distribution of Mesoarchean basement and isotopic boundaries defined by Pb in VMS deposits (Thorpe et al., 1992) and Nd in granites (Davis and Hegner, 1992). (B) E–W cross section of the central Slave craton, illustrating east-dipping boundary between Mesoarchean crustal block in west and Neoarchean crustal domain in east (Bleeker and Davis, 1999).

continues about the genesis of these keels, and models include repeated cycles of differentiation and collisional thickening (Jordan, 1975, 1988), collision of island arcs comprising depleted material (Ashwal and Burke, 1989), buoyant subduction and imbrication by lithospheric-scale stacks (Helmstaedt and Schulze, 1989), and basal accretion by cooling asthenospheric material (Thompson et al., 1996).

Equally important, however, and arguably less well understood, is the genetic relationship between these thick, depleted lithospheric Archean keels and their overlying crustal sections. Re–Os isotopic studies of xenolith samples from different Archean cratons indicate that significant portions of the tectosphere were initially depleted synchronously with, or within a short period following, formation of the overlying crustal section (Pearson, 1999). The broad similarity in timing of crust formation and mantle depletion is interpreted to indicate either (1) a temporal and genetic link and significant coupling between crust and subcontinental lithospheric mantle formation, or (2) that the Archean crust is preserved as a consequence of protection by deep lithospheric keels, which may be coupled to the crust somewhat later than the age of crust formation (e.g., Pearson, 1999; Moser et al., 2001). As it is often challenging to resolve lithospheric age differences at time scales of less than 200 my using Re–Os model age or isochron methods, establishing the direct temporal relationship between the crust and mantle at time scales appropriate to the cycle of orogenic processes is problematical. Therefore, it remains difficult to differentiate between these two competing possibilities.

Over the past decade the Slave craton, in northwestern Laurentia, has emerged as a major diamond producing province (Fipke et al., 1995; Rylatt and Popplewell, 1999). The extensive and well-documented geological record of the Slave craton (Fig. 1; Padgham, 1992; Isachsen and Bowring, 1994; Bleeker and Davis, 1999, and references therein) provides an important new crustal, as well as emerging mantle perspective (Grütter et al., 1999; Griffin et al., 1999; Bank et al., 2000; Kopylova and Russell, 2000; Carbno and Canil, 2002) on the development of diamond-bearing tectosphere. The late tectonic evolution of Archean cratons, such as the Slave, is complex and involves extensive rifting, magmatism, compressional deformation, and metamorphism that in many cases significantly post-dates the timing of initial crust formation by 10 to >100 my. The Slave's Neoarchean orogenesis is characterized by high temperature-low pressure metamorphic conditions (HT-LP) and the intrusion of voluminous granitoid plutons within a short time interval (Fyson and Helmstaedt, 1988; Thompson, 1989; van Breemen et al., 1992). In modern tectonic settings, the association of HT-LP metamorphism with compressional regimes is generally thought to require additions of mantle-derived heat to the crust, either directly through intrusion of mantle melts, or by delamination or lithospheric thinning processes (e.g., Midgley and Blundell, 1997). This implies at least partial removal of pre-existing mantle lithosphere, with the total replacement of the mantle section in extreme cases.

Such a tectonic style is difficult to reconcile with the notion of a relatively cool, thick mantle tectosphere coupled to the crust beneath the Slave craton throughout its Neoarchean evolution. Thus, the crustal perspective on tectosphere development and stabilization presents a fundamental paradox: Can extensive plutonism, including mantle-derived magmatism, and HT-LP metamorphism characteristic of the Slave craton and many other Neoarchean terrains develop above previously stabilized, thick tectosphere? This question is particularly relevant to understanding the development of the Slave craton, as initial Re–Os studies of xenoliths from kimberlites suggest that at least parts of the Slave mantle lithosphere may be Mesoarchean in age down to a considerable thickness and remained coupled with the overlying crust throughout the extensive tectonic reworking in the Neoarchean (Aulbach et al., 2001).

In this paper we discuss critical petrological, geophysical and geochemical observations and first-order geological observations that are relevant to this debate. We conclude that these observations can be best explained if thick tectosphere developed only relatively late during collisional orogenesis, most likely by tectonic imbrication (e.g., Helmstaedt and Schulze, 1989).

2. Geological background

The Slave is a small craton, ~ 700 × 500 km in exposed areal extent, bounded by Paleoproterozoic belts to the south, east and west and covered by younger rocks to the north (Padgham, 1992; Isachsen and Bowring, 1994; Bleeker and Davis, 1999). The craton is characterized throughout its western part by a Mesoarchean basement (4.0–2.9 Ga), referred to as the Central Slave Basement Complex (Bleeker et al., 1999b), with isotopically juvenile (<2.85 Ga?) but undefined basement in the east (Fig. 1A; Thorpe et al., 1992; Davis and Hegner, 1992; Davis et al., 1996). Isotopic data from granites and lower crustal xenoliths suggest that the Mesoarchaen basement dips to the east and underlies the central part of the craton at depth, although its eastern extent remains undefined (Davis et al., 1996, 2003; Davis and Hegner, 1992; Fig. 1B).

This east–west asymmetry has received considerable attention in tectonic models for the Slave's cratonic development. In part, it forms the basis for arc-continent collisional models of Kusky (1989) and Davis and Hegner (1992). The detailed structural and stratigraphic data to support these generalized models are lean, with the dominant structures being considerably younger and affecting equally the eastern and western parts of the craton (e.g., Fyson and Helmstaedt, 1988; Padgam, 1992; Padgham and Fyson, 1992; Isachsen and Bowring, 1994; Bleeker et al., 1999a; Bleeker, 2001). The origin of the asymmetry in crustal age domains remains uncertain. A collisional suture remains a possibility but such a structure must be early and predate 2.69 Ga (Bleeker et al., 1999a). Alternatively, the eastern Slave may represent highly attenuated and modified Mesoarchean lithosphere that developed during rifting at ca. 2.85–2.70 Ga (Bleeker, 2003). If one assumes that some thickness of mantle lithosphere was coupled to the isotopically distinct crustal domains, then mantle lithosphere under the western Slave could be significantly older, perhaps by up to 400 my, than that underlying the eastern Slave, regardless of the exact relationship between the domains (Grütter et al., 2000).

The composite basement preserves a complex polymetamorphic and magmatic history with at least 10 distinct magmatic and/or metamorphic "events" between 4.0 and 2.85 Ga (Isachsen and Bowring, 1994; Bowring and Williams, 1999; Bleeker and Davis, 1999; Ketchum and Bleeker, 2001). The extent and repetitiveness of this tectonic reworking on a ca. 100 Ma interval is uncharacteristic of the stability

generally attributed to cratons underlain and protected by thick lithosphere. Development of a thin cover sequence consisting of fuchsitic quartzite and banded iron formation on the basement at 2850–2800 Ma marks the first indication of widespread, but transient stability within the basement (Bleeker et al., 1999b; Sircombe et al., 2001).

Thick, tholeiitic submarine volcanic sequences were extruded over the quartzites and Central Slave Basement Complex between 2.73 and 2.70 Ga, with no correlative volcanic sequences as yet documented in the eastern Slave (Padgham, 1992; van Breemen et al., 1992; Isachsen and Bowring, 1994; Bleeker et al., 2001). Mafic magmatic rocks cover an area of at least 100,000 km^2 with a typical thickness of 1–6 km, approaching proportions comparable to modern large igneous provinces (LIPs; Eldholm and Coffin, 2000). Such voluminous magmatism suggests it may be associated with large-scale mantle plume or mantle overturn events (Bleeker et al., 2001). Granitoids of similar age occur within the basement as a result of localized crustal melting.

Widespread calc-alkaline volcanism followed between 2.70 and 2.66 Ga in both the eastern and western Slave (van Breemen et al., 1992), and was terminated by deposition of thick turbidite sequences over the entire exposed craton at 2.66–2.63 Ga (Bleeker and Villeneuve, 1995; Pehrsson and Villeneuve, 1999) The post-2.69 Ga volcanic rocks represent the first sequence that can be correlated across the entire exposed craton, and provide the earliest evidence of linkage between the eastern and western Slave domains (Bleeker, 2001).

The dominant tectono-metamorphic structures recorded in exposed crustal rocks developed between 2.64 and 2.58 Ga, 20–80 my after deposition of the principal volcanic sequences, and at least several 100 my after development of the Mesoarchean Central Slave Basement Complex. Post-2.64 Ga structures are dominated by at least three regional folding events at shallow to mid-crustal levels (D1, D2, D3), accompanied by a systematic temporal variation in the composition of associated plutonic rocks (Relf, 1992; van Breemen et al., 1992; Davis and Bleeker, 1999; Pehrsson et al., 2000). The deformation events record large horizontal shortening and show little or no apparent spatial correlation with the location of known or inferred Mesoarchean basement. Pehrsson

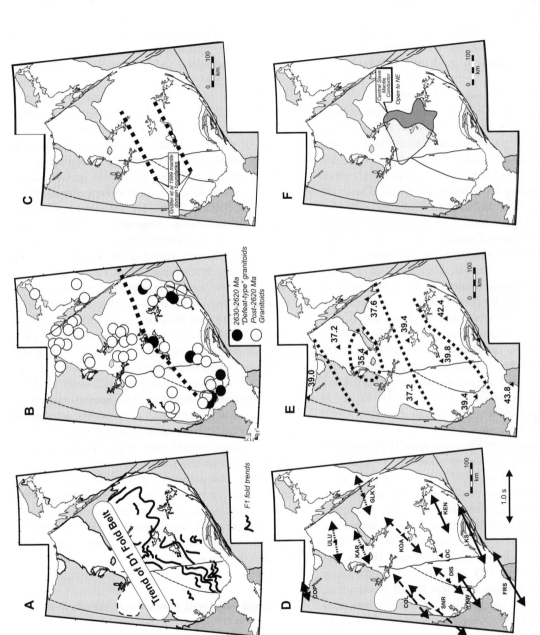

Fig. 2. Location and orientation of a number of geological, geochemical and geophysical characteristics of the Slave craton. (A) Inferred trend lines of the F1 fold belt (Bleeker et al., 1999b, 2001). (B) Distribution of dated plutons within the craton (open circles) with dated plutons between 2620 and 2635 Ma represented by filled circles documented only in the south and southeastern part of the craton (Davis and Bleeker, 1999). (C) Trends of geochemical mantle domains based on garnet chemistry (Grütter et al., 1999). (D) Summary of teleseismic anisotropy data (from Bank et al., 2000). (E) Crustal thickness estimates from seismic data (Bank et al., 2000). (F) Location and extent of mantle conductor in central Slave (Jones et al., 2001).

et al. (2000) suggest that widespread, medium-pressure granulite-facies rocks in the western Slave may be preferentially exposed owing to the presence of basement rocks in that area.

Folding cannot be related to events internal to the craton, such as previously inferred in arc/microcontinent collision models (e.g., Kusky, 1989), and is interpreted to reflect tectonic forces that originated outside the preserved area of the craton.

The orientation of D1 fold structures in the central and southern Slave province define an approximately NE–SW trending fold belt, after taking into account the effects of later D2 folding (Fig. 2A, Bleeker et al., 1999b). Padgham (1985, 1992) previously highlighted NE–SW trending zones within the craton. The orientation of the inferred fold belt is at relatively high angle to the inferred N–S trending boundary between contrasting basement domains (Bleeker et al., 1999b). The timing of D1 shortening is constrained in the Yellowknife area to pre-date intrusion of ca. 2.63 Ga diorite to granodiorite plutons of the Defeat plutonic Suite (Davis and Bleeker, 1999). In the north and central Slave, a minimum age for this event is only loosely bracketed to be older than ca. 2.615–2.608 Ga (e.g., Relf, 1992; van Breemen et al., 1992). The D1 event is established to be diachronous as sedimentary rocks in the Indin Lake area in the westernmost Slave craton were deposited after initiation of D1 folding in the Yellowknife area (Davis and Bleeker, 1999; Pehrsson and Villeneuve, 1999).

In the southeastern Slave, the post-D1 plutonism is characterized by diorite–granodiorite compositions (van Breemen et al., 1992; Davis and Bleeker, 1999). This plutonism is regionally diachronous, with >2.62 Ga plutonic rocks occurring in the south and southeastern parts of the craton, roughly paralleling the trend of the D1 fold belt, and younger, 2.62–2.60 Ga plutons to the north and northwest (Davis and Bleeker, 1999, Fig. 2B). Although the tectonic cause of this event remains uncertain the most primitive, gabbro to diorite compositions require a subduction-enriched mantle component, and thus a melting event in the mantle beneath the Slave craton at ca. 2.630–2.605 Ga (Davis et al., 1994; Yamashita et al., 1999). Geochemical signatures of these plutons are consistent with a 'subduction-modified' mantle source (Davis et al., 1994; Yamashita et al., 1999). Griffin et al. (1999) proposed a plume model to drive this

event; however, the temporal and spatial relationships between regional deformation and plutonism are consistent with a subduction/collisional origin. The early 2.63–2.62 Ga plutons have compositional characteristics of arc-related plutons (Yamashita et al., 1999) and these are followed by intrusion of ca. 2.61 Ga diorites in the central and northern Slave with LREE-enriched high-Mg andesite compositions commonly found in arc or post-collisional settings, and interpreted to be related to lithospheric delamination (Davis et al., 1994; Sajona et al., 2000).

Major regional shortening continued through the interval 2610–2585 Ma and was accompanied by voluminous two-mica and K-feldspar granite plutonism throughout the craton (van Breemen et al., 1992; Davis and Bleeker, 1999). The D2 structures indicate east–west shortening, suggesting a change in the orientation of the principal shortening direction or an oblique geometry (Bleeker and Beaumont-Smith, 1995). Although spanning 20 my, the granite plutonism shows no resolvable regional diachroneity, regardless of the timing of the earlier ca. 2605–2630 Ma plutonism (van Breemen et al., 1992; Davis and Bleeker, 1999). Furthermore, the distribution of these younger granites shows no relationship to the distribution of basement domains, although the two-mica granites are certainly associated with areas of thickened sedimentary sequences. This intense craton-wide "granite bloom" argues for a widespread thermal disturbance, the exact cause of which remains speculative. Various models have been suggested for this event, including lithospheric delamination (Davis et al., 1994), post-collisional extension (Kusky, 1993), interaction with a mantle plume (Griffin et al., 1999) and crustal thickening of thinned, warm lithosphere (Thompson, 1989). These models predict a relatively thin (< diamond stability window) mantle lithosphere beneath the craton at 2.6 Ga.

3. Geophysical and geochemical mantle domains

As discussed above, prior deliberation of the Slave's tectonic history has been dominated by the obvious east–west disparity in exposed bedrock geology. However, we contend that this geometry is only a feature of the Slave's crust, and that its subcontinental lithospheric mantle exhibits a NW–SE mantle zonation comprising three regions with distinctive geochemical and geophysical characteristics.

3.1. Geochemical boundaries

The abundance, distribution and "stratigraphy" of lithologies within subcontinental mantle lithosphere can be constrained in space and time by detailed geochemical investigation of mantle-derived xenoliths and xenocrysts (e.g., O'Reilly and Griffin, 1996). Mantle lithologies are commonly defined with reference to garnet compositions because garnet shows extensive solid solution and is a stable mineral in a large variety of lithospheric bulk compositions at pressures exceeding 1.6–2.0 GPa (Boyd, 1970; Sobolev, 1977). Griffin et al. (1999) utilized minor and trace element compositions of Cr-pyrope garnet to identify and describe a unique ultradepleted layer (henceforth UDL) dominated by clinopyroxene-free, garnet harzburgite that underlies the shallow mantle lithosphere in the central Slave craton. This UDL occurs at mantle temperatures less than ~ 950 °C and is replaced by moderately depleted lherzolite-dominated lithologies at temperatures of ~ 950 to ~ 1200 °C. Xenolith thermobarometry constrains the base of the UDL at ~ 140 km depth and shows that the moderately depleted central Slave lithosphere extends to a depth of ~ 200 km (Pearson et al., 1999). The UDL contains Cr-pyrope garnets with distinctively low Cr_2O_3 subcalcic major element compositions (the G10-1 population of Grütter and Anckar, 2001) that are known to occur with regularity in kimberlites and till samples within a ~ 140 km wide and ~ 220 km long east–northeast trending zone in the central Slave craton (Fig. 2C). Similar low Cr_2O_3 subcalcic garnet compositions are extremely rare in kimberlite or till samples outside this zone (Grütter et al., 1999), indicating that the UDL occurs as a distinct east–northeast trending unit at shallow depth within the central Slave craton and that the stratigraphic relations and mutual proportions of garnet-bearing mantle lithologies below the crust of the northern and southern Slave craton differ from that in the central Slave craton (see also Kopylova and Caro, 2001). Carbno and Canil (2002) suggest that the ultradepleted layer may extend to the southeastern Slave (Drybones area) but the deeper lithosphere is of different composition than in the east. This may reflect modification during Paleoproterozoic craton margin events (Carbno and Canil, 2002).

Mantle xenoliths from the diamondiferous Jericho kimberlite in the northern Slave craton show that garnet-bearing mantle lithosphere occurs within a depth range of ~ 80 to ~ 200 km and that eclogitic and pyroxenitic lithologies are comparatively common within a lherzolite-dominated lithospheric section (Kopylova et al., 1998). A relatively limited number of garnet xenocryst populations have been described from the northern Slave craton, but those that are available suggest the lithospheric section may contain an above-average proportion of low-Cr_2O_3 eclogite and that G10-bearing garnet harzburgite is very rare (e.g., Fig. 2(F) of Grütter et al., 1999). G10 garnets are also not described as a prominent xenocryst component in several recently discovered diamondiferous kimberlites within the Coronation district in the far northwestern Slave craton (data in Armstrong, 2002, but also based on an informal survey of press releases of various diamond exploration companies).

The southern Slave craton contains a number of ~ 530 Ma old diamondiferous kimberlites that have sampled garnet-facies mantle to extreme depths of ~ 250 km (Kopylova and Caro, 2001; McLean et al., 2001). Garnet xenocryst assemblages described from the Snap Lake (McLean et al., 2001), CL-25 (Pokhilenko et al., 1997), MZ dyke (Mountain Province Diamonds, 2001) and Gahcho Kue kimberlites (Grütter et al., 2000) document a lherzolite-dominated lithospheric section with subordinate eclogite and occasional G10 garnets with moderate-Cr_2O_3 which are different in composition to G10 garnets in the UDL. A compositionally distinct high-Cr_2O_3, moderate-CaO subcalcic garnet xenocryst population occurs with low frequency within these kimberlites (Grütter et al., 1999). Essentially identical garnet compositions are now also recognized as a low-abundance component derived from extreme lithospheric depths below the central Slave craton (the G10-3 population of Grütter and Anckar, 2001). These compositional and depth attributes indicate that the known lithospheric section of the southern Slave craton (east of longitude 111° W) is dissimilar to that of the central Slave craton at typical UDL depths, but that a mutually common high-Cr_2O_3 garnet harzburgite component exists at extreme depth. Hence, a combination of three different lithospheric sections is required to describe the geochemical features of the northern, central and southern Slave mantle. A schematic cross section of the geochemical architecture (Fig. 3) requires a three-fold division at UDL depths, but shows a similar G10-3 component between the central and southern Slave mantles at extreme depths within the lithospheric keel.

3.2. Teleseismic SKS splitting observations

Determination of shear wave splitting (SKS) directions for stations on the Slave craton (Table 1) by Bank et al. (2000) were interpreted to show relatively uniform characteristics (Fig. 2D) similar in orientation to the North American plate vector motion. In particular, the northern two stations, COP and ULU, were considered to exhibit no evidence of deviation from other values on the Slave craton, and this was taken as lack of evidence for any MacKenzie plume modification of the underlying lithospheric mantle as suggested by Ernst and Baragar (1992).

Using statistics appropriate for directional data (Mardia, 1972), and taking the 90° ambiguity into account, the weighted average of the SKS directions for the northern two stations (14 data) is 074° with a standard error of 1.25°, and for all 13 Slave stations (84 data) is 055° ± 3.52°. The t-value to test whether the difference of these means is significant is 19.91, which indicates that the null hypothesis that these means are the same can be rejected. Similarly, the time delays show a statistically significant difference, with the two northernmost stations giving a weighted

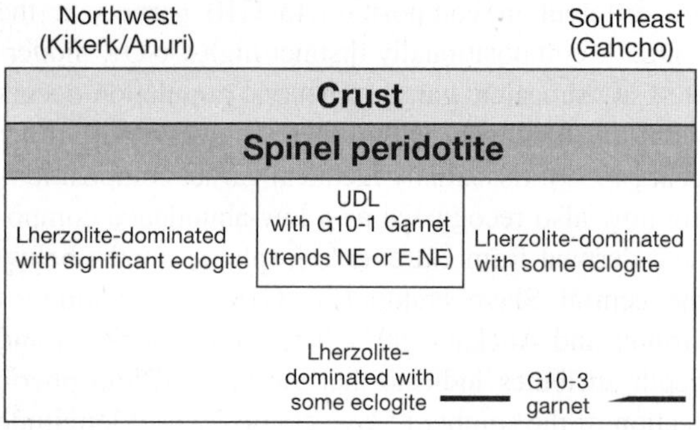

Table 1
SKS directions and time delays for Slave sites (taken from Bank et al., 2000) and statistical analyses

Site	No.	Phi	sd	Av	sd	dt	sd	Av	sd
COP	10	70	7	71	1.2	0.4	0.2	0.68	0.17
ULU	4	80	10			0.8	0.1		
KAR	3	66	9			0.7	0.4		
GLK	3	71	9			0.8	0.5		
KOA	8	43	9	42	0.8	1.0	0.2	1.09	0.19
COL	9	50	9			1.1	0.3		
SNR	3	40	2			1.5	0.6		
DIS	3	41	11			0.8	1.3		
KEN	3	65	10	62	0.9	0.8	0.5	1.01	0.19
YKW	24	56	10			0.8	0.3		
LKS	4	65	9			1.2	0.4		
FRS	5	59	12			1.2	0.3		
FPR	5	64	7			1.1	0.2		

average of 0.67 ± 0.2 s compared to the total Slave average of 0.90 ± 0.26 s, giving a t-value of 3.18 which rejects the null hypothesis at below the 0.5% level.

Closer inspection of the SKS azimuths (Table 1) shows a statistically significant three-part subdivision of the Slave SKS results into northern sites (COP, ULU, KAR, GLK), central sites (KOA, COL, SNR, DIS) and southern sites (KEN, YKW, LKS, FRS, FPR). The weighted azimuthal averages, and their estimated standard errors, are listed in Table 1. The time delays also show a similar subdivision, with the northern sites statistically different from the central and southern sites. The t-value for the northern and central groups is 7.41, which for 43 degrees of freedom is larger than the 0.1% t-distribution value of 3.55 and implies that the null hypothesis can be rejected with high confidence.

3.3. Crustal thickness

Crustal thickness was estimated by Bank et al. (2000) using receiver functions, and the estimated Moho depths are shown in Fig. 2E. There is a distinct NE–SW striking variation of crustal thickness through the Slave craton. The northwestern part of the exposed craton has crustal thickness of 37.3 ± 0.2 km (ignoring the anomalously low value for station KAR). The central Slave craton has crustal thicknesses of 39.5 ± 0.2 km, and the SE part of the craton has a crustal thickness in excess of 42 km. The

Fig. 3. Inferred geochemical architecture of the Slave craton lithosphere summarized in NW–SE schematic cross section through the central Slave province based on garnet and xenolith data referenced in text.

thickest part of the craton occurs in the area of the early 2.63–2.62 Ga plutonic belt.

3.4. Electromagnetic anomaly

The mapped location of the central Slave mantle conductor (Jones et al., 2001, 2003) is shown in Fig. 2F, together with its inferred extension to the west to account for the high magnetotelluric phases observed there (Jones et al., 2003). The central Slave mantle conductor lies almost wholly within the NE-trending geochemical boundaries identified by Grütter et al. (1999) as shown in Fig. 2C. Although the cause of the observed enhanced electrical conductivity is unknown, the spatial association of the anomaly with Griffin et al.'s (1999) ultradepleted harzburgitic layer and with Grütter's mantle domain boundaries suggests an ancient origin, not one associated with the Eocene kimberlite emplacement event. Based on existing knowledge, Jones et al. (2003) interpret the central Slave mantle conductor as due to carbon in graphite form above the diamond stability field.

4. Discussion

The Slave craton has a well-documented crustal history from 4.0 to 2.6 Ga (Padgham, 1992; Isachsen and Bowring, 1994; Bleeker and Davis, 1999) but it is uncertain how persistent lithospheric mantle was during this interval. Did relatively thick mantle lithosphere stabilize at the same time as the crustal sections during the Mesoarchean, or was early formed mantle lithosphere modified and/or destroyed during the subsequent tectonic events? Based on the distribution of crustal age domains, the former hypothesis would predict older, Mesoarchean mantle depletion ages in the west beneath the Mesoarchean terrain and younger lithosphere in the east beneath the eastern domains (Fig. 1B; Grütter et al., 2000), with the structure within the mantle in part controlled by the distribution of Mesoarchean lithosphere. At present, the extent of Re–Os model age mapping of the lithosphere is insufficient to fully evaluate this possibility. Dominantly Mesoarchean depletion ages are determined beneath the central Slave area (Aulbach et al., 2001). Data from Jericho in the north-central Slave indicate dominantly Neoarchean or younger ages,

with few samples having depletion ages >3.0 Ga (Irvine et al., 1999, 2001). This, in combination with the petrological differences described above argues for a lithospheric break or transition between these sites. The orientation of this boundary is not constrained, although it may correspond to NE–SW compositional boundaries shown in Figs. 2C and 3.

Similarly, studies of xenolith and xenocryst suites, along with geophysical imaging document important regional variations in the composition and structure of the Slave lithospheric mantle (Grütter et al., 1999; Griffin et al., 1999; Kopylova and Russell, 2000; Jones et al., 2001, 2003; Kopylova and Caro, 2001; Carbno and Canil, 2002). As described above, and originally proposed by Grütter et al. (1999), the Slave lithosphere can be divided into three approximately E–NE oriented zones, each defined by distinct garnet chemistry (Figs. 2C and 3). Importantly, the orientation of these zones is subparallel to the D1 structural grain of the craton (Fig. 2A, Bleeker et al., 1999a,b), and at high angle to north–south isotopic boundaries mapped in the crust (Fig. 1). Since the present distribution of the mantle domains appears to transect the east–west crustal age asymmetry, it is inferred to be a younger feature that probably developed after ca. 2.7 Ga. This would imply that at least the garnet-facies mantle beneath the craton was established late in its evolution, after the time of initial crust formation.

Absence of a pre-2.7 Ga, thick, buoyant lithosphere would be consistent with the repeated episodes of magmatism and metamorphism within the Central Slave Basement Complex throughout the 3.6–2.85 Ga interval (Isachsen and Bowring, 1994; Bleeker and Davis, 1999; Ketchum and Bleeker, 2001). As noted above, the Central Slave Basement complex does not exhibit the tectonic stability generally associated with continental areas underlain by thick tectosphere (Ketchum and Bleeker, 2001). At least two periods of extensional volcanism developed on the Mesoarchean crust; at 2.85 Ga, and perhaps more significantly at 2.73–2.70 Ga. Interpretation of the ca. 2.73–2.70 Ga tholeiitic volcanism in terms of LIP-scale basaltic volcanism (Bleeker et al., 2001) suggests that pre-existing lithosphere may have been substantially modified and/or thinned by the impinging of upwelling asthenosphere (plume?) during extensional magmatism. If the tholeiites were sourced beneath the Mesoarchean crustal block, then segregation at relatively shallow pressures within spinel facies is implied, consistent with a lithospheric thickness of less than 100 km at 2.7 Ga (e.g., White and McKenzie, 1995). Yamashita et al. (1999) suggested that the Mesoarchean basement terrains in the west-central Slave represent highly dismembered crustal segments with intervening dominantly juvenile ca. 2.70 Ga marginal basins. Their model equally suggests a high degree of lithospheric attenuation at 2.7 Ga, and in such a scenario, preservation of ancient mantle lithosphere is likely to be fragmentary, and relegated to the shallowest, spinel peridotite lithosphere.

The absence of thick lithosphere at ca. 2.7 Ga is consistent with the subsequent metamorphic and magmatic history of the craton. The metamorphic conditions attained at 2.6 Ga are characteristic of HT-LP metamorphic belts, with lower crustal temperatures of >700 °C at 0.9–1.1 GPa (Davis et al., 2003). Based on a conductive model with crustal heat production and metamorphic thermal conditions, Thompson et al. (1996) argued that a thermally stabilized lithosphere beneath the Slave could be no thicker than 100 km at 2.6 Ga, and suggested that the lithosphere grew by accretion of asthenosphere at its base between 2.6 and 1.8 Ga. Their model did not attempt to account for any chemical variation or lateral structure within the lithosphere, as is now indicated by geophysical and geochemical data sets.

Thermal models of shortening and thickening of continental lithosphere indicate that development of HT-LP metamorphism and widespread crustal melting are most sensitive to three parameters: (1) the total radiogenic heat production and its distribution in the crust; (2) the thermal structure of the crust prior to thickening; and (3) the reduced heat flow at the base of the crust (e.g., Midgley and Blundell, 1997). Lithospheric thickness and its control on reduced heat flow to the crust is arguably the most significant parameter in these thermal models and may be essential to generate high temperature conditions in modern orogens (e.g., Midgley and Blundell, 1997). Geologically, this may be the result of lithospheric thinning or delamination events, bringing hot asthenospheric material to shallow depths (e.g., Bird, 1979; Houseman et al., 1981; Nelson, 1992). These models argue against the presence of thick cool lithosphere beneath HT-LP metamorphic belts.

HT-LP metamorphism is by no means unique to a specific time period in Earth's history, but it is particularly common in the Archean (Sandiford, 1989). An important consideration is that Archean crust will have at least twice the heat production (e.g., Pollack, 1997) owing to the greater proportion of radiogenic heat-producing elements in the past, favouring higher metamorphic temperatures and crustal melting during shortening (e.g., McLaren et al., 1999). Certainly, the Slave crust is characterized by generally high heat production, particularly the late granites, although most units, such as the basement and volcanic rocks, are not anomalously rich in heat-producing elements (Thompson et al., 1996; Kopylova et al., 1999). Can greater heat production within the crust permit HT-LP metamorphic belts to develop above areas of thick, cool lithosphere? Although this possibility cannot be eliminated by thermal arguments alone (e.g., McLaren et al., 1999), it is not favoured for the Slave craton because it fails to account for the occurrence of the mantle-derived magmatism between 2.630 and 2.605 Ga (Davis et al., 1994; Yamashita et al., 1999). In many parts of the craton these plutons were intruded prior to, or early during D2 regional shortening and peak metamorphism, and thus argue for a role for transient heating within the crust (King et al., 1992). Certainly, greater crustal heat production would contribute to the observed steep metamorphic field gradients but the sequence of early mantle-derived magmatism followed by dominantly crustal melts argues for a significant mantle component to the heat budget.

4.1. Development of the Slave mantle lithosphere by subcretion

If the HT-LP metamorphism and magmatism at 2.6 Ga reflect a thinner lithosphere and transient heating, then thick, cool lithosphere must have developed sometime after ca. 2.6 Ga (e.g., Isachsen and Bowring, 1994; Thompson et al., 1996). Although admittedly speculative, our preference is for a model in which the mantle lithosphere developed by subcretion during NW, or possibly SE-vergent subduction beneath the Slave craton during D1 shortening and the early 2.63–2.61 Ga plutonism (Fig. 4). This would impart a NE–SW structural grain in the lithosphere during development of the D1 fold belt and early

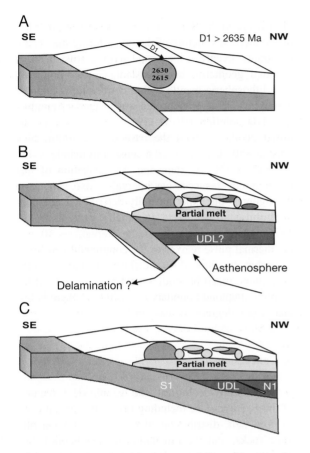

Fig. 4. Cartoon model for development of Slave lithosphere by tectonic imbrication of lithosphere during deformation and plutonism at ca. 2.6 Ga. (A) Subduction beneath the craton at ca. 2.64–2.61 Ga generates early SW–NE trending F1 fold belt and subparallel plutonic belt in SE Slave. Development of mantle domains may have been initially established at this time. (B) Subsequent collision (external to the present craton boundaries) possibly accompanied by partial delamination of some of the underthrust lithosphere produced widespread deformation (D2) and granite plutonism throughout the province at 2.6–2.58 Ga. C) termination of orogenesis and subcretion of deeper mantle lithosphere.

2.63–2.61 Ga plutonism, and would imply decoupling of Mesoarchean–Neoarchean crustal boundaries from the underlying deeper (garnet-bearing) subcontinental lithospheric mantle (Grütter et al., 1999). It is important to note that the Slave craton, as exposed, represents only a small fragment of a presumably much larger craton (Bleeker, 2003; Isachsen and Bowring, 1994), and the location of the preserved craton within a framework of possible Neoarchean

plate boundaries is unconstrained. The subcreted mantle component may include oceanic lithosphere or arc-wedge material. The ultradepleted component characteristic of the central part of the Slave lithosphere may represent the latter material, as suggested by Griffin et al. (1999), structurally separated from adjacent zones. Subsequent collision (external to the present craton boundaries) possibly accompanied by partial delamination of some of the underthrust lithosphere, produced widespread deformation (D2) and granite plutonism throughout the province at 2.6–2.58 Ga, with continued metamorphism (extension?) in the lower crust to at least 2.56 Ga (Davis et al., 2003).

A prediction of this model is that the Slave mantle lithosphere was dominantly stabilized in the latest Archaean or younger times. It is however at odds with the documentation of extensive regions of Mesoarchean lithosphere beneath the central Slave to depths of 150–200 km (Aulbach et al., 2001). Certainly, accreted oceanic lithosphere is expected to be somewhat older than the time of its emplacement, perhaps by up to 150 my in modern systems. Significantly older components (i.e., >2750 Ma) could represent older parts of the oceanic lithosphere that were decoupled from their crust and imbricated beneath the craton during collisional events, or perhaps remnants of ancient Slave lithosphere caught up in the subcreted collage. Greater buoyancy of ultradepleted oceanic lithosphere in the Archean may permit longer cycles for recycling of oceanic lithosphere.

Further modification and addition to the mantle is thought to have occurred through imbrication accompanying Proterozoic accretion to the western craton margin (Cook et al., 1999; Bostock, 1997; Carbno and Canil, 2002), which may have disturbed a primary lithospheric architecture of Neoarchean age.

4.2. On the occurrence of Archean diamonds

If a thin lithosphere and elevated reduced heat flow is required to account for the metamorphic and magmatic history of the craton at ca. 2.6 Ga, what does this imply for the age of diamonds? Thompson et al. (1996) argued, on the basis of paleogeotherms and crustal heat production, that diamonds could not be stabilized within the Slave lithosphere until >500 my after the last tectonothermal event to have affected the craton. Although not yet proven to be present in the

Slave, diamonds of Archean age have been identified in other cratons, implying that thermal conditions appropriate for diamond stability were established relatively early, perhaps within 100–200 my of the last major tectono-metamorphic event recorded in the crust (e.g., Richardson et al., 2001). A logical implication of the model presented above is that diamond growth in eclogite and/or peridotite occurred contemporaneously with the subcretion event, or at younger times (e.g., Kesson and Ringwood, 1989a,b). Subcretion of relatively cool mantle will serve to cool the lithospheric section permitting the preservation or growth of diamond. A prediction of the model is that diamonds beneath the Slave craton formed at or after 2.6 Ga, within slightly older mantle lithosphere.

One question that can be posed is whether older diamonds in subcreted lithosphere can survive the thermal pulse from the overlying hot crust. This may be specifically relevant to the case of subcretion or other addition of a significantly older, cold buoyant

lithosphere to the Slave after ca. 2.6. Using estimates of the thermal structure of the crust during the ca. 2.6 Ga granite event we have modeled the crust, with an elevated geotherm, being instantaneously underlain by a lithospheric mantle with a conventional cratonic geotherm. The approach used was a standard conductive 1-D solution (Wang, 1999). Fig. 5 shows the initial geotherm, with the base of the 50-km-thick crust at 850 °C juxtaposed against cold mantle at 450 °C, i.e., a 350 °C step, and the relaxation of that geotherm over successive intervals. Also shown on the figure is the experimentally determined graphite-diamond stability field (Kennedy and Kennedy, 1976). Over a relatively short interval, ~ 10 my, the thermal pulse relaxes to the continental geotherm. Note that its effects do not diffuse into the subcreted lithosphere much beyond ~ 75 km depth, and at the graphite-diamond boundary (~ 140 km), there is less than a few degrees increase in temperature.

5. Conclusions

Broad-scale geological and geophysical features of the Slave craton, including orientation of an early F1 fold belt, distribution of ca. 2.62–2.63 Ga plutonic rocks, and the orientation of geochemical and geophysical domains within the mantle lithosphere collectively highlight the importance of a NE–SW structural grain to the craton. This structural grain is oblique to the north–south crustal age domain boundaries directly mapped by exposures of Mesoarchean crust and indirectly by the isotopic composition of VMS deposits and late granites. We interpret this to indicate that the subcontinental lithospheric mantle architecture post-dates events that lead to the crustal age asymmetry (suture?) as well as the extensive plume or rift-related LIP-type volcanism at 2.7 Ga. The lithosphere developed beneath the craton late in the orogenic cycle, most likely as a result of tectonic imbrication of buoyant lithosphere. An implication of the model is that diamond formation occurred at the earliest in the latest Archean, within only slightly older lithosphere. Improving the resolution of mantle domains and reconciling their age and structural geometry with crustal structures is essential to develop more refined models of tectosphere formation.

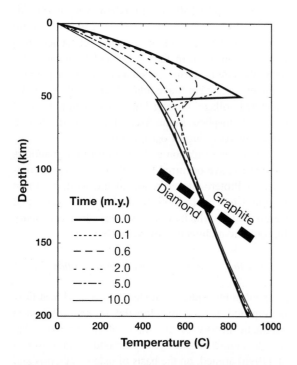

Fig. 5. Conductive thermal relaxation modelling of cold lithosphere, with a cratonic geotherm, subcreted beneath hot crust. The geotherms at time intervals of 0.1, 0.6, 2, 5 and 10 my after subcretion are shown. Also shown is the boundary between stability fields of graphite and diamond (Kennedy and Kennedy, 1976).

Acknowledgements

We wish to acknowledge the logistical and financial assistance of all the many organizations and companies that made our work possible in the Slave craton. These include LITHOPROBE, the Geological Survey of Canada (GSC contribution number 2001219, under the NATMAP, LITHOPROBE, EXTECH-III and Targeted Geoscience Initiative programs), the Department of Indian and Northern Development (DIAND), the U.S. National Science Foundation's Continental Dynamics Program, DeBeers Canada Exploration, Kennecott Exploration and BHP Billiton Diamonds, Diavik, DeBeers, Winspear, Royal Oak Mines and Miramar Mining. Secondly, we wish to acknowledge our many colleagues, within government, academia and industry, for their insightful discussions that prompted us to carefully construe our arguments. Kelin Wang is thanked for providing us with his thermal modelling code, and John Percival for comments on an earlier version of this manuscript. Sam Bowring, an anonymous journal reviewer, and volume editor Ric Carlson, provided useful comments.

References

Armstrong, J.P., 2002. Diamond exploration data in the North Slave Craton, Nunavut. DIAND NU Open Report 2002-01 (CD-ROM).

Ashwal, L.D., Burke, K., 1989. African lithospheric structure, volcanism, and topography. Earth and Planetary Science Letters 96, 8–14.

Aulbach, S., Griffin, W.L., Pearson, N.J., O'Reilly, S.Y., Doyle, B.J., Kivi, K., 2001. Re–Os isotope evidence for Meso-archean mantle beneath 2.7 Ga Contwoyto Terrane, Slave craton, Canada: implications for the tectonic history of the Slave Craton. Abstract. In: Carlson, R., Jones, A.G. (convenors), The Slave Kaapvaal Workshop, Merrickville, Ontario, September 5–9, 2001.

Bank, C.G., Bostock, M.G., Ellis, R.M., Cassidy, J.F., 2000. A reconnaissance teleseismic study of the upper mantle and transition zone beneath the Archean Slave Craton in NW Canada. Tectonophysics 319, 151–166.

Bird, P., 1979. Continental delamination and the Colorado Plateau. Journal of Geophysical Research 84 (B13), 7561–7571.

Bleeker, W., 2001. The ca. 2680 Ma Raquette Lake Formation and correlative units across the Slave Province, Northwest Territories: evidence for a craton-scale overlap sequence. Current Research - Geological Survey of Canada 2001-C7 (26 pp.).

Bleeker, W., 2003. The late Archean record: a puzzle in ca. 35 pieces. Lithos 71, 99–134 (this issue).

Bleeker, W., Beaumont-Smith, C., 1995. Thematic structural studies in the Slave Province, Northwest Territories; the Sleepy Dragon Complex. Current Research - Geological Survey of Canada 1995-C, 87–96.

Bleeker, W., Davis, W.J., 1999. The 1991–1996 NATMAP Slave Province Project: introduction. Canadian Journal of Earth Sciences 36, 1033–1042.

Bleeker, W., Villeneuve, M., 1995. Structural studies along the Slave portion of the SNORCLE Transect. In: Cook, F., Erdmer, P. (compilers), Slave-NORthern Cordillera Lithospheric Evolution (SNORCLE). Report of 1995 Transect Meeting, April 8–9, University of Calgary, 8–14 pp.

Bleeker, W., Ketchum, J.W.F., Davis, W.J., 1999a. The Central Slave Basement Complex: Part II. Age and tectonic significance of high-strain zones along the basement-cover contact. Canadian Journal of Earth Sciences 36, 1111–1130.

Bleeker, W., Ketchum, J.W.F., Jackson, V.A., Villeneuve, M.E., 1999b. The Central Slave Basement Complex: Part I. Its structural topology and autochthonous cover. Canadian Journal of Earth Sciences 36, 1083–1109.

Bleeker, W., Davis, W.J., Ketchum, J.W., Sircombe, K., Stern, R.A., 2001. Tectonic evolution of the Slave craton. In: Cassidy, K.F., et al. (Eds.), 4th International Archaean Symposium 2001, Extended Abstracts. AGSO-Geoscience Australia, Record 2001/37, pp. 288–290.

Bostock, M.E., 1997. Anisotropic upper-mantle stratigraphy and architecture of the Slave Craton. Nature 390, 392–395.

Bowring, S.A., Williams, I.S., 1999. Priscoan (4.00–4.03 Ga) orthogneisses from northwestern Canada. Contributions to Mineralogy and Petrology 134, 3–16.

Boyd, F.R., 1970. Garnet peridotites and the system $CaSiO3$–$MgSiO3$–$Al2O3$. Special Paper - Mineralogical Society of America 3, 63–75.

Carbno, G.B., Canil, D., 2002. Mantle structure beneath the southwest Slave craton, Canada: constraints from garnet geochemistry in the Drybones Bay kimberlite. J. Petrol. 43, 129–142.

Cook, F.A., van, d.V.A.J., Hall, K.W., Roberts, B.J., 1999. Frozen subduction in Canada's Northwest Territories; lithoprobe deep lithospheric reflection profiling of the western Canadian Shield. Tectonics 18, 1–24.

Davis, W.J., Bleeker, W., 1999. Timing of plutonism, deformation, and metamorphism in the Yellowknife domain, Slave province, Canada. Canadian Journal of Earth Sciences 36, 1169–1187.

Davis, W.J., Hegner, E., 1992. Neodymium isotopic evidence for the accretionary development of the Late Archean Slave Province. Contributions to Mineralogy and Petrology 111, 493–503.

Davis, W.J., Fryer, B.J., King, J.E., 1994. Geochemistry and evolution of late Archean plutonism and its significance to the tectonic development of the Slave craton. Precambrian Research 67, 207–241.

Davis, W.J., Gariepy, C., van Breemen, O., 1996. Pb isotopic composition of late Archaean granites and the extent of recycling early Archaean crust in the Slave Province, northwest Canada. Chemical Geology 130, 255–269.

Davis, W.J., Canil, D., Mackenzie, J.M., Carbno, G.B., 2003. Petrology and U–Pb geochronology of lower crustal xenoliths and the development of a craton, Slave Province, Canada. Lithos 71, 541–573 (this issue).

Eldholm, O., Coffin, M.F., 2000. Large igneous provinces and plate tectonics. In: Richards, M.A., Gordon, R.G., van der Hilst, R.D. (Eds.), The History and Dynamics of Global Plate MotionsGeophysical Monograph, vol. 121. American Geophysical Union, Washington, DC, pp. 309–326.

Ernst, R.E., Baragar, W.R.A., 1992. Evidence from magnetic fabric for the flow pattern of magma in the Mackenzie giant radiating dyke swarm. Nature 356, 511–513.

Fipke, C.E., Dummett, H.T., Moore, R.O., Carlson, J.A., Ashley, R.M., Gurney, J.J., Kirkley, M.B., 1995. History of the discovery of diamondiferous kimberlites in the Northwest Territories, Canada. Sixth International Kimberlite Conference, Extended Abstracts. Proceedings of the 6th International Kimberlite, pp. 158–160.

Fyson, W.K., Helmstaedt, H., 1988. Structural patterns and tectonic evolution of supracrustal domains in the Archean Slave Province, Canada. Canadian Journal of Earth Sciences 25, 301–315.

Griffin, W.L., Doyle, B.J., Ryan, C.G., Pearson, N.J., O'Reilly, S.Y., Davies, R., Kivi, K., van Achterbergh, E., Natapov, L.M., 1999. Layered mantle lithosphere in the Lac de Gras area, Slave craton: composition, structure and origin. Journal of Petrology 40, 705–727.

Grütter, H.S., Anckar, E., 2001. Cr–Ca characteristics of mantle pyropes from the central Slave craton—a comparison with the type areas on the central Kaapvaal craton, with implications for carbon in peridotite. Presented at the Slave-Kaapvaal Workshop, Merrickville, Ontario, Canada, September 2001.

Grütter, H.S., Apter, D.B., Kong, J., 1999. Crust–mantle coupling; evidence from mantle-derived xenocrystic garnets. In: Gurney, J.J., et al. (Eds.), Proceedings of the 7th international Kimberlite Conference. J.B. Dawson Volume 1 Red Roof Design, Cape Town, pp. 307–313.

Grütter, H.S., Davis, W.J., Jones, A., 2000. Chemical and physical images of the Slave craton lithosphere. Pan-Lithoprobe 2 Moho Workshop Abstract, Banff, Alberta.

Helmstaedt, H.H., Schulze, D.J., 1989. Southern African kimberlites and their mantle sample: implications for Archean tectonics and lithosphere evolution. In: Ross, J. (Ed.), Kimberlites and Related Rocks: Volume 1. Their Composition, Occurrence, Origin, and Emplacement. Geological Society of Australia Special Publication, vol. 14, pp. 358–368.

Houseman, G.A., McKenzie, D.P., Molnar, P., 1981. Convective instability of a thickened boundary layer and its relevance for the thermal evolution of continental convergent belts. Journal of Geophysical Research 86, 6115–6132.

Irvine, G.J., Kopylova, R.W., Carlson, R.W., Pearson, D.G., Shirey, S.B., 1999. Age of the lithospheric mantle beneath and around the Slave craton: a rhenium–osmium-isotopic study of Peridotite Xenoliths from the Jericho and Somerset Island kimberlites. Ninth Annual V.M. Goldschmidt Conference LPI Contribution No. 971. Lunar and Planetary Institute, Houston, pp. 134–135.

Irvine, G.J., Pearson, D.G., Carlson, R.W., 2001. Lithospheric mantle evolution of the Kaapvaal craton: a Re–Os isotope study of peridotite xenoiths from Lesotho kimberlites. Geophysical Research Letters 28, 2505–2508.

Isachsen, C., Bowring, S.A., 1994. Evolution of the Slave craton. Geology 22, 917–920.

Jones, A.G., Ferguson, I.J., Evans, R., Chave, A.D., McNeice, G.W., 2001. The electric lithosphere of the Slave craton. Geology 29, 423–426.

Jones, A.G., Lezaeta, P., Ferguson, I.J., Chave, A.D., Evans, R.L., Garcia, X., Spratt, J., 2003. The electrical structure of the Slave craton. Lithos 71, 505–527 (this issue).

Jordan, T.H., 1975. The continental tectosphere. Reviews of Geophysics and Space Physics 13, 1–12.

Jordan, T.H., 1988. Structure and formation of the continental tectosphere, in Oceanic and continental lithosphere; similarities and differences. In: Menzies, M.A., Cox, K.G. (Eds.), Journal of Petrology, Special Lithosphere Issue, pp. 11–37.

Kennedy, C.S., Kennedy, G.C., 1976. The equilibrium boundary between graphite and diamond. Journal of Geophysical Research 81, 2467–2470.

Kesson, S.E., Ringwood, A.W., 1989a. Slab–mantle interactions: 1. Sheared and refertilised garnet peridotite xenoliths; samples of Wadati–Benioff zones? Chemical Geology 78, 83–96.

Kesson, S.E., Ringwood, A.W., 1989b. Slab–mantle interactions: 2. The formation of diamonds. Chemical Geology 78, 97–118.

Ketchum, J.W.F., Bleeker, W., 2001. Genesis of the 4.03–2.85 Ga Salve protocraton, northwestern Canada. In: Cassidy, K.F., et al. (Eds.), 4th International Archaean Symposium 2001, Extended Abstracts. AGSO-Geoscience Australia, Record 2001/37, pp. 514–515.

King, J.E., Davis, W.J., Relf, C., 1992. Late Archean tectono-magmatic evolution of the central Slave Province, Northwest Territories. Canadian Journal of Earth Sciences 29, 2156–2170.

Kopylova, M.G., Caro, G., 2001. Lithospheric mapping of the Slave craton: contrasting North and South. Extended Abstract, Slave-Kaapvaal Workshop, 5–9 September, Merrickville, Ontario.

Kopylova, M.G., Russell, J.K., 2000. Chemical stratification of cratonic lithosphere; constraints from the northern Slave Craton, Canada. Earth and Planetary Science Letters 181, 71–87.

Kopylova, M.G., Russel, J.K., Cookenboo, H., 1998. Upper-mantle stratigraphy of the Slave craton, Canada: insights into a new kimberlite province. Geology 26, 315–318.

Kopylova, M.G., Russell, J.K., Cookenboo, H., 1999. Petrology of peridotite and pyroxenite xenoliths from Jericho Kimberlite; implications for the thermal state of the mantle beneath the Slave Craton, northern Canada. Journal of Petrology 40, 79–104.

Kusky, T.M., 1989. Accretion of the Archean Slave Province. Geology 17, 63–67.

Kusky, T.M., 1993. Collapse of Archean orogens and the generation of late- to postkinematic granitoids. Geology 21, 925–928.

Mardia, K.V., 1972. Statistics of Directional Data. Academic Press, London. ISBN 0-12-471150-2.

McLaren, S., Sandiford, M., Hand, M., 1999. High radiogenic heat-producing granites and metamorphism; an example. Geology 27, 679–682.

McLean, R.C., Pokhilenko, N.P., Hall, A.E., Luth, R., 2001. Pyropes and chromites from kimberlites of the Snap Lake area, southeast Slave craton: garnetization reaction of depleted peridotites at extremely deep levels of the lithospheric mantle. Extended Abstract, Slave-Kaapvaal Workshop, 5–9 September, Merrickville, Ontario.

Midgley, J.P., Blundell, D.J., 1997. Deep seismic structure and thermo-mechanical modelling of continental collision zones. Tectonophysics 273, 155–167.

Moser, D.E., Harte, R.J., Flowers, R., 2001. Birth of the Kaapvaal Tectosphere 3.08 Billion Years Ago. Science 291, 465–468.

Mountain Province Diamonds, 2001. De Beers reports on MZ Lake kimberlites: three sills diamondiferous. Public news release dated September 10, 2001.

Nelson, K.D., 1992. Are crustal thickness variations in old mountain belts like the Appalachians a consequence of lithospheric delamination? Geology 20, 498–502.

O'Reilly, S.Y., Griffin, W.L., 1996. 4-D lithosphere mapping; methodology and examples. Tectonophysics 262, 3–18.

Padgham, W.A., 1985. Observations and speculations on supracrustal successions in the Slave Structural Province. In: Ayres, L.D., Thurston, P.C., Card, K.D., Weber, W. (Eds.), Evolution of Archean Sequences. Geological Association of Canada, Special Paper, vol. 28, pp. 133–151.

Padgham, W.A., 1992. The Slave structural Province, North America: a discussion of tectonic models. In: Glover, J.E., Ho, S.E. (Eds.), Proceedings Volume for the Third International Archaean Symposium. Geology Dept. and University Extension, The University of Western Australia, Publication, vol. 22, pp. 381–394.

Padgham, W.A., Fyson, W.K., 1992. The Slave province: a distinct Archean craton. Canadian Journal of Earth Sciences 29, 2072–2086.

Pearson, D.G., 1999. The age of continental roots. Lithos 48, 171–194.

Pearson, N.J., Griffin, W.L., Doyle, B.J., O'Reilly, S.Y., Van Achterbergh, E., Kivi, K., 1999. Xenoliths from kimberlite pipes of the Lac de Gras area, Slave craton, Canada. In: Gurney, J.J., et al. (Eds.), Proceedings of the 7th International Kimberlite Conference. P.H. Nixon Volume 2 Red Roof Design, Cape Town, pp. 644–658.

Pehrsson, S.J., Villeneuve, M.E., 1999. Deposition and imbrication of a 2670–2629 Ma supracrustal sequence in the Indin Lake area, southwestern Slave Province, Canada. Canadian Journal of Earth Sciences, 1149–1168.

Pehrsson, S.J., Chacko, T., Pilkington, M., Villeneuve, M.E., Bethune, K., 2000. Anton Terrane revisited; late Archean exhumation of a moderate-pressure granulite terrane in the western Slave Province. Geology 28, 1075–1078.

Pokhilenko, N.P., McDonald, J.A., Melnik, U., McCorquodale, J., Reimers, L.F., Sobolev, N.V., 1997. Indicator minerals from the CL-25 kimberlite pipe, Slave craton, Northwest Territories, Canada. Russian Geology and Geophysics 38, 550–558.

Pollack, H.N., 1997. Thermal characteristics of the Archaean Greenstone belts. In: De Wit, M.J., Ashwal, L.D. (Eds.), Greenstone Belts. Oxford Monographs on Geology and Geophysics, vol. 35, pp. 223–232.

Relf, C., 1992. Two distinct shortening events during late Archean orogeny in the west-central Slave Province, Northwest

Territories, Canada. Canadian Journal of Earth Sciences 29, 2104–2117.

Richardson, S.H., Shirey, S.B., Harris, J.W., Carlson, R.W., 2001. Archean subduction recorded by Re–Os isotopes in eclogitic sulfide inclusions in Kimberley diamonds. Earth and Planetary Science Letters 191, 257–266.

Rylatt, M.G., Popplewell, G.M., 1999. Ekati diamond mine—background and development. Mining Engineering 51, 37–43.

Sajona, F.G., Maury, R.C., Pubellier, M., Leterrier, J., Bellon, H., Cotton, J., 2000. Magmatic source enrichment by slab-derived melts in a young post-collision setting, central Mindanao (Philippines). Lithos 54, 173–206.

Sandiford, M., 1989. Secular trends in the thermal evolution of metamorphic terrains. Earth and Planetary Science Letters 95, 85–96.

Sircombe, K.N., Bleeker, W., Stern, R.A., 2001. Detrital zircon geochronology and grain-size analysis of a approximately 2800 Ma Mesoarchean proto-cratonic cover succession, Slave Province, Canada. Earth and Planetary Science Letters 189, 207–220.

Sobolev, N.V., 1977. Deep Seated Inclusions in Kimberlites and the Problem of the Composition of the Upper Mantle (English Translation by D.A. Brown). American Geophysical Union, Washington, DC, pp. 1–279.

Thompson, P.H., 1989. An empirical model for metamorphic evolution of the Archaean Slave Province and adjacent Thelon tectonic zone, north-western Canadian Shield. In: Daly, J.S., Cliff, R.A., Yardley, B.W.D. (Eds.), Evolution of Metamorphic Belts. Geological Society Special Publications Geological Society of London, London, UK, pp. 245–263.

Thompson, P.H., Judge, A.S., Lewis, T.J., 1996. Thermal evolution of the lithosphere in the central Slave Province: implications for diamond genesis. In: LeCheminant, A.N., et al. (Eds.), Searching for Diamonds in Canada. Geological Survey of Canada, Open File 3228, pp. 151–160.

Thorpe, R.I., Cumming, G.L., Mortensen, J.K., 1992. A significant Pb isotope boundary in the Slave Province and its probable relation to ancient basement in the western Slave Province. Project Summaries, Canada Northwest Territories Mineral Development Subsidiary Agreement. Geological Survey of Canada Open File Report 2484, pp. 279–284.

Wang, K., 1999. Personal communication.

White, R.S., McKenzie, D., 1995. Mantle plumes and flood basalts. Journal of Geophysical Research 100B, 17543–17585.

van Breemen, O., Davis, W.J., King, J.E., 1992. Temporal distribution of granitoid plutonic rocks in the Archean Slave Province, northwest Canadian Shield. Canadian Journal of Earth Sciences 22, 2186–2199.

Yamashita, K., Creaser, R.A., Stemier, J.U., Zimaro, T.W., 1999. Geochemical and Nd–Pb isotopic systematics of late Archean granitoids, southwestern Slave Province, Canada; constraints for granitoid origin and crustal isotopic structure. Canadian Journal of Earth Sciences 36, 1131–1147.

Author index

doi:10.1016/S0024-4937(03)00207-X